The Physics and Chemistry of SiO_2 and the Si-SiO_2 Interface

The Physics and Chemistry of SiO_2 and the $Si\text{-}SiO_2$ Interface

Edited by

C. Robert Helms

Stanford University
Stanford, California

and

Bruce E. Deal

Fairchild Research Center
National Semiconductor
Santa Clara, California

Springer Science+Business Media, LLC

Library of Congress Cataloging in Publication Data

Symposium on the Physics and Chemistry of SiO_2 and the Si-SiO_2 Interface (1988: Atlanta, Ga.)

The physics and chemistry of SiO_2 and the Si-SiO_2 interface / edited by C. Robert Helms and Bruce E. Deal.

p. cm.

"Proceedings of the Symposium on the Physics and Chemistry of SiO_2 and the Si-SiO_2 Interface, held May 15-20, 1988, at the 173rd Meeting of The Electrochemical Society, in Atlanta, Georgia"—T.p. verso.

Includes bibliographies and index.

ISBN 978-1-4899-0776-9 ISBN 978-1-4899-0774-5 (eBook)

DOI 10.1007/978-1-4899-0774-5

1. Silica—Congresses. 2. Surfaces (Physics)—Congresses. 3. Surface chemistry—Congresses. I. Helms, C. Robert. II. Deal, Bruce E. III. The Electrochemical Society, Meeting (173: 1988: Atlanta, Ga.) IV. Title.

QC585.75.S55S94 1988 88-28168

546'.6832—dc19 CIP

This limited facsimile edition has been issued for the purpose of keeping this title available to the scientific community.

10 9 8 7 6 5 4 3

Proceedings of the symposium on The Physics and Chemistry of SiO_2 and the Si-SiO_2 Interface, held May 15-20, 1988, at the 173rd meeting of The Electrochemical Society, in Atlanta, Georgia

Originally published by Plenum Press, New York in 1988

Proceedings of the Symposium held at the 173rd Meeting of The Electrochemical Society in Atlanta, Georgia on May 15-20, 1988

CONFERENCE CHAIRMAN

C. Robert Helms
Stanford University

CONFERENCE CO-CHAIRMEN

Bruce E. Deal	Edward H. Poindexter	Eugene A. Irene
National Semiconductor	U.S. Army ET & DL	University of North Carolina

CONFERENCE ORGANIZING COMMITTEE

K. L. Brower, Sandia
R. A. Devine, CNET
D. J. Di Maria, IBM
R. H. Doremus, RPI
P. V. Dressendorfer, Sandia
A. H. Edwards, Army ET & DL
F. J. Feigl, Lehigh
L. C. Feldman, AT & T Bell Labs
D. L. Griscom, NRL
F. J. Grunthaner, JPL
P. O. Hahn, Wacker
F. D'Heurle, IBM
G. Hollinger, Univ. of Lyon
W. Holton, SRC
N. M. Johnson, Xerox
R. DeKeersmaecker, IMEC
J. Maserjian, JPL
E. H. Nicollian, Univ. of N. C.
Y. Nishi, HP
S. I. Raider, IBM, ECS
S. Rigo, Univ. of Paris
J. Ruzyllo, Pennsylvania State
M. Schulz, Univ. of Erlangen
M. Stroscio, ARO
T. Sugani, Univ. of Tokyo
J. B. Wagner, Arizona State, ECS
G. Wright, ONR
D. R. Young, Lehigh

PREFACE

The properties of SiO_2 and the Si-SiO_2 interface provide the key foundation onto which the majority of semiconductor device technology has been built. Their study has consumed countless hours of many hundreds of investigators over the years, not only in the field of semiconductor devices but also in ceramics, materials science, metallurgy, geology, and mineralogy, to name a few. These groups seldom have contact with each other even though they often investigate quite similar aspects of the SiO_2 system. Desiring to facilitate an interaction between these groups we set out to organize a symposium on the Physics and Chemistry of SiO_2 and the Si-SiO_2 Interface under the auspices of The Electrochemical Society, which represents a number of the appropriate groups. This symposium was held at the 173rd Meeting of The Electrochemical Society in Atlanta, Georgia, May 15-20, 1988.

These dates nearly coincided with the ten year anniversary of the "International Topical Conference on the Physics of SiO_2 and its Interfaces" held at IBM in 1978. We have modeled the present symposium after the 1978 conference as well as its follow on at North Carolina State in 1980. Of course, much progress has been made in that ten years and the symposium has given us the opportunity to take a multidisciplinary look at that progress. One of the major accomplishments of the IBM conference was the preparation of a conference proceedings which still serves as an exceedingly valuable reference, even after ten years. It is our hope and expectation that the present volume will be just as useful for the coming ten years.

In addition to ourselves, Ed Poindexter and Gene Irene served as symposia co-chairmen. With help from the organizing committee listed on a previous page, we set out to construct a program for the symposium leading to papers for this book that bridged the appropriate disciplines, provided sufficient review and tutorial material and at the same time presented the latest advancements in our understanding of the Si-SiO_2 system. Of the sixty articles published here, twenty seven were invited papers providing both for review and discussion of new hot topics.

The symposium was divided into seven sessions leading to six chapters in this book. Preceding the papers themselves each chapter is introduced and summarized by a member of the organizing committee who we felt would provide a critical assessment of the body of work. The first chapter deals with the processes used to grow SiO_2 on silicon and mechanisms responsible for those growth processes. This chapter, introduced and summarized by J. Blanc, contains what to some might seem a surprising cross-section of divergent views of the topic of Si oxidation.

In the second chapter, introduced by Dave Griscom, we discuss the general topic of the thermal and structural properties of SiO_2 with particular emphasis on the role of hydrogen and the effect of processing on oxide chemistry and bonding.

As important as the chemistry and bonding of SiO_2 are to silicon device structures, a key factor that makes the Si-SiO_2 system so well behaved is the unique properties of the interface itself. In the third chapter, progress in our understanding of the structure, chemistry, bonding and morphology of the Si-SiO_2 interface is reviewed. Key findings are presented relating to the role of SiO_x bonding configurations and the importance of crystalline SiO_2 near the interface. In addition, recent theoretical progress in our understanding of the electronic structure and bonding of the Si-SiO_2 interface is presented.

Although the level of chemical perfection of the Si-SiO_2 interface that can be achieved is hard to comprehend, it still contains defects which can control defect properties. The nature of these defects is the subject of the fourth chapter introduced by Ed Poindexter and Max Schulz. Recent breakthroughs in our understanding of the relationship between SiO_2 defects and those which control MOS device properties are discussed along with recent reports of measurements of the trapping and detrapping of single e^- defects.

A major new focus in the last few years has been the effect of the silicon surface condition prior to oxidation as well as the effect of the oxidation process itself on other important phenomena especially relating to defects. These topics are discussed in Chapter 5, introduced by Gene Irene. Finally in Chapter 6, introduced by Paul Dresssendorfer, we discuss the current state of the art in understanding carrier transport and tunneling in SiO_2.

The success of the symposium and the excellent quality of the papers in this book was due to the efforts of our co-chairman, the organizing committee, and of course all the authors, who we would like to take this opportunity to thank. In addition we would also like to acknowledge The Electrochemical Society for providing us with an administrative framework and the financial support which made our job so much easier.

C. Robert Helms
Bruce E. Deal

Palo Alto, California
June, 1988

CONTENTS

I. GROWTH MECHANISMS OF SIO_2 FILMS

II. THERMAL AND STRUCTURAL PROPERTIES OF SiO_2

III. THE ATOMIC AND ELECTRONIC STRUCTURE OF THE Si-SiO_2 INTERFACE

IV. DEFECTS, IMPURITIES, AND DAMAGE MECHANISMS IN Si-SiO_2 SYSTEMS

V. EFFECTS OF PREOXIDATION SUBSTRATE QUALITY, SURFACE TREATMENTS, AND OXIDATION INDUCED POINT DEFECT GENERATION

VI. ELECTRON AND HOLE TRANSPORT AND TUNNELING IN SiO_2

I. GROWTH MECHANISMS OF SIO_2 FILMS

INTRODUCTION

J. Blanc

12 Willow Street
Princeton, NJ 08542

"The linear-parabolic model presented in the early 1960's has been demonstrated to characterize the kinetics of silicon thermal oxidation to a reasonable degree...However... several questions have not been satisfactorily answered."

B. E. Deal

The important properties of the Si-SiO_2 system that dominate the behavior of silicon devices are always extremely sensitive to the method used to grow or deposit the SiO_2. In this chapter we review the field of thermal SiO_2 growth and discuss new growth methods. Aside from Bruce Deal's usual magisterial review of the state-of-the-art from genesis to the present at the beginning of this chapter, the papers presented in this chapter fall into three intertwined categories which in turn are centrally interconnected to much of the remainder of this volume:

(1) Physical and Chemical Models for the thermal oxidation of silicon and silicides, largely by dry oxygen (Doremus, Helms and de Larios, Ling, Dupas and De Meyer and d'Heurle);

(2) Recent Physical and Chemical Experimental Studies Relevant to thermal oxidation by Rigo and Irene;

(3) Attempts to Produce Silica Films on Silicon by means other than conventional thermal oxidation (Sakon and Wagner, Long, Xu and Zheng, Hoff and Ruzyllo, Tsu, Kim and Lucovsky, and Sayyah.

In terms of modeling, the current state of confusion is perhaps best seen by comparing the views of Nicollian and Reisman[1], on the one hand, to those of Doremus on the other. Without going into detail, which the reader can learn from the respective articles, the former insist that dry oxidation can best be understood on the assumption that the process is entirely controlled by the interfacial reaction while the latter argues with equal vigor that the kinetics are completely dominated by the transport of oxygen molecules to the interface. As the kinetics are neither linear nor parabolic (this is perhaps the only point on which all disputants are agreed), both must invoke other physical assumptions, whose validity are perhaps not terribly evident to an unbiased observer.

On still another hand, Helms and de Larios seek to explain the observed pressure dependence of oxidation for small oxide thickness. They supplement the Han-Helms model of parallel "Deal-Grove" reactions with notions from earlier workers that at least one of their oxidants is a mono-atomic species of oxygen, possibly electrically charged. The Nicollian-Reisman model appears to be silent on *a priori* predictions of the pressure dependence of the rates; the Doremus model clearly predicts the rates to be directly proportional to oxygen pressure. Doremus[2] believes that all prior reports of sub-linear pressure dependences are based on incorrect experimental inferences.

Raider focuses on an entirely different aspect of the question. He seeks to identify the actual oxidant molecular species by indirect means and concludes that transport of molecular SiO is central to the oxidation of Si by oxygen, as well as to the decomposition of SiO_2 at low oxygen pressures.

The large uncertainties which arise in quantities derived from experimentally observed thickness vs time are analyzed in great detail in the first of the Ling-Dupas-DeMeyer papers. In the second of their two papers the same authors present a detailed and interesting model for the oxidation in the presence of chlorine.

d'Heurle presents a review of the elegant experiments and analyses that he and his colleagues have performed over the years on the oxidation of silicides. From the point of view of understanding the oxidation of silicon itself, perhaps the most suggestive conclusion is the finding that oxidation of "large silicides is invariably more rapid. In his oral presentation, d'Heurle argued that workers in the field of silicon oxidation should seriously consider the silicides as independently controlled experimental variables, bringing much additional information to the analysis of silicon oxidation; another limiting point of view is that oxidation of silicon is itself sufficiently confusing that there is no profit in yet another complete family of complications.

What is one to make of this array of confusing and contradictory models and opinions? From one point of view, the original Deal and Grove model (in spite of its quantitative failings for dry oxidation) still succeeds in providing qualitative predictions while using a readily understandable physical argument; perhaps we should ask no more of a model. On the other hand, there are those of us who will remain unhappy with less than quantitative descriptions and predictions. The only generally agreed point for all members of the latter school seems to be that experiments other than accurate measurements of the time dependence of the oxide thickness have to be brought to the fore in an essential way if we hope to achieve a physically meaningful model; sadly, there appears to be no similarly general agreement as to what experiments are most crucial or feasible.

Rigo reviews the beautiful work he and his collaborators have done for the past decade using different oxygen isotopes in sequential oxidations. For wet oxidation the growth occurs by interstitial diffusion of water molecules reacting with the silica network. This is, of course, entirely consistent with the conventional wisdom, but it is certainly reassuring that there now is direct experimental confirmation of the process; to me, it still appears almost miraculous that such a complex set of events is so accurately described by the Deal-Grove model. Inevitably, the oxidation by dry oxygen is far more complex: interstitial transport of oxygen molecules for thick oxide films, with step by step motion of network oxygen increasing in importance as the thickness decreases. Rigo discusses various possible microscopic mechanisms for these results, without coming to a definite conclusion.

Rather than concentrating on a single technique, Irene and co-workers have utilized a wide variety of experimental methods including effects of cleaning procedures (to which a separate session of the symposium is devoted), contact angle measurements, stress measurements and the usual time dependence of oxide thickness. Irene views his results in terms of three distinct regimes: the very initial oxidation (formation of the "native" oxide); the initial regime (up to roughly 30 nm); and the final regime where he believes the Deal-Grove model to correctly describe the situation. For thinner oxides, a number of distinct effects are operative (surface impurities, thermionic oxidation, interfacial impurities, substrate orientation, stress, etc.) although it is not evident at this time that such effects can be ordered in terms of relative importance for the overall process.

In an effort to produce "a thin and uniform gate oxide", Sakon and Wagner have studied the oxidation of silicon in dry carbon dioxide. They report that the oxidation is independent of the pressure of both carbon monoxide and oxygen, depending only on carbon dioxide pressure. In their oral presentation at the symposium, however, they concluded that the amount of carbon present in the grown films was not quantitatively compatible with their model. These results, as well as those below, deserve careful scrutiny.

Long, Xu and Zheng report that the addition of HF to an oxygen stream increases the oxidation rate compared to pure oxygen by a factor of 40 or more compared to pure oxygen; these results cannot be ascribed solely to water formation in the gas as this would enhance the rate by only 30%. The observed kinetics are said to be describable by Deal-Grove kinetics, but a detailed kinetic model is not presented. After sufficiently prolonged annealing, MOS capacitors exhibited breakdown fields greater than 9.5 MC/cm. The authors ascribe their results immediately after oxidation to the presence of both Si-F and Si-OH bonds, and the results after annealing to elimination of HF from these two types of bonding configuration.

Hoff and Ruzyllo have used an afterflow (oxygen excited in a microwave discharge) to produce O atoms; this athermal atomic oxidation process was further enhanced by addition of small additions of NF_3, which can be used to grow the oxide in reasonable times at temperatures down to about 700°C; at temperatures below approximately 525°C, etching dominates growth and the process becomes impractical. Some of the electrical properties are promising, but fixed charge densities of order $10^{11}cm^{-2}$, were obtained.

The last two papers of this section abandon thermal oxidation altogether. Tsu, Kim and Lucovsky describe an elegant process which allows them to selectively produce a plasma rich in highly excited oxygen, nitrogen or ammonia, while using ground-state silane as the silicon source for chemical deposition of oxide, nitride or oxynitride of known and controllable stoichiometry. Detailed studies of the process chemistry in the formation of such films are beginning, and unanswered questions remain. "Device-quality" films of silica have been successfully deposited on silicon, with breakdown fields in excess of 8 MV/cm and defect densities of less than $3x10^{10}/cm^2$.

Finally, Sayyah describes the anodics oxidation of Si at room temperature; as grown, these silica films are clearly non-stoichiometric and/or porous as judged by infra-red absorption; a forming gas anneal at 450°C seems to bring the films about halfway to the structural properties of a thermally grown oxide. In spite of this, 450°C anneals in the presence of A1 yields defect densities of the order of $5x10^{10}/cm^2$ or less and mean breakdown fields of the order of 10 MV/cm. For reasons which are quite unknown at present, films of thickness larger than 100 nm have seriously degraded electrical characteristics.

It seems certain that because of the technological necessity to process at ever lower temperatures and shorter times, industrially acceptable processes other than thermal oxidations will become inevitable at some time in the future. None of the processes described above yet reaches such goals. I would guess that these alternative processes will not receive the close attention, analysis and criticism that thermal oxides are subject to until they come closer to meeting manufacturing needs in a relatively general fashion.

REFERENCES

[1] E. H. Nicollian and A. Reisman, paper presented at symposium and published in J. Electronic Materials (July 1988).

[2] R. H. Doremus, private communication.

HISTORICAL PERSPECTIVES OF SILICON OXIDATION

Bruce E. Deal

Fairchild Research Center
National Semiconductor Corporation
Santa Clara, CA 95052

ABSTRACT

Thermal silicon dioxide has been a critical part of silicon semiconductor technology since the late 1950's. In 1965, oxidation kinetics were characterized by the linear-parabolic expression for both dry oxygen and steam over a wide range of process variables. Since then, numerous attempts have been made to model the dry O_2 oxidation process in the 0-30 nm thickness range where data have indicated a departure from the linear-parabolic relationship. These and other historical trends are reviewed, with emphasis being placed on the factors and process variables which affect oxidation kinetics and associated oxide properties. Future device directions and corresponding requirements for thermal oxides are also reviewed.

INTRODUCTION

Thermally formed silicon dioxide has been used in the fabrication of silicon devices since the late 1950's, when scientists at Bell Laboratories first used SiO_2 films to selectively mask against dopants in diffused junction silicon transistors (1). Later on in the process, however, the oxide was removed leaving an active, exposed junction at the surface of these so-called MESA structures. During this same time period, other groups at Bell Laboratories characterized the various properties of thermally oxidized silicon (2-4), thus providing much of the foundation for subsequent developments. In 1959, Hoerni and co-workers at Fairchild invented the PLANAR process (5), in which the masking thermal oxide was left on the surface - thus passivating the junctions. This process is used even today to fabricate essentially all silicon devices and integrated circuits. At about the same time that the PLANAR process was developed, Kahng and Atalla reported the fabrication of the first MOS (metal-oxide-semiconductor) transistor, in which a thin thermal oxide was used as the gate dielectric (6). Since that time, many other applications for thermal oxides in silicon device fabrication have evolved, such as thin tunneling dielectrics, device isolation, interconnect (poly-Si) insulation, capacitors, and gettering and planarizing using phosphorus-doped oxides. It is possible also to thermally oxidize other silicon-containing materials, including metal silicides, silicon nitride, and silicon carbide, which can also be utilized in device structures.

OXIDATION KINETICS - THE LINEAR-PARABOLIC MODEL

Early attempts to characterize silicon oxidation kinetics for steam and oxygen ambients resulted in a simple parabolic expression [$x_o^2 = Bt$] (2,7). This is not surprising since most silicon oxides were produced at temperatures above 1000°C and were greater than 0.5 um in thickness. However, as the ranges of oxide thickness and oxidation temperature were expanded, other more complex expressions were proposed to satisfy the data observed. The most satisfactory and accepted model was the linear-parabolic relationship, first proposed in 1963 (8) and characterized in detail in 1965 (9) using the following expression:

$$x_o^2 + Ax_o = B\ (t + \tau) \qquad (1)$$

where x_o = oxide thickness at time t, and A, B, and τ are constants.

Equation (1) is based on the assumption that the thermal oxidation of silicon proceeds by (a) the absorption of an oxidizing species, either O_2 or H_2O, into the outer surface of the oxide being formed; (b) the diffusion of that species through the oxide; and (c) the reaction of the oxidizing species with the silicon at the Si-SiO_2 interface to form SiO_2. It also assumes that the flux for each reaction (a), (b), and (c) are equal at steady state and that the oxidation process reaches steady state very rapidly. A schematic representation of the oxidation process is shown in Fig. 1, where the three reactions or fluxes described above are indicated.

Two limiting cases of the oxidation process as represented by Eq. (1) result in either a parabolic expression for thicker oxides ($x_o^2 = Bt$), where B is the parabolic rate constant, (diffusion controlled), or a linear expression for thinner oxides, [$x_o = B/A(t + \tau)$], where B/A is the linear rate constant (interface reaction controlled). Based on the derivation of Eq. (1), the constants A, B, and B/A are defined as follows:

$$A = 2\ D\ (1/k + 1/h) \qquad (2)$$

$$B = 2\ DC^*/N_1 \qquad (3)$$

$$B/A = \frac{C^*}{N}\left(\frac{kh}{k+h}\right) \qquad (4)$$

where D = oxidizing species diffusion coefficient
C^* = oxidant equilibrium concentration in oxide
N_1 = number of oxidant molecules in a unit volume of oxide
k = rate constant at Si-SiO_2 interface
h = rate constant at outer oxide surface.

While the linear-parabolic expression was found to characterize silicon oxidation in steam over a wide range of temperature and oxide thickness, it was found that dry O_2 oxidation in the range 0-30 nm is faster than predicted. This is shown in Fig. 2. Since oxide thicknesses employed for devices in the 1960's were of the order of 100 nm or more, this deviation did not cause a problem, and Eq. (1) used the constant τ to account for this region. Its value is expressed as

$$\tau = (x_i^2 + Ax_i)/B \qquad (5)$$

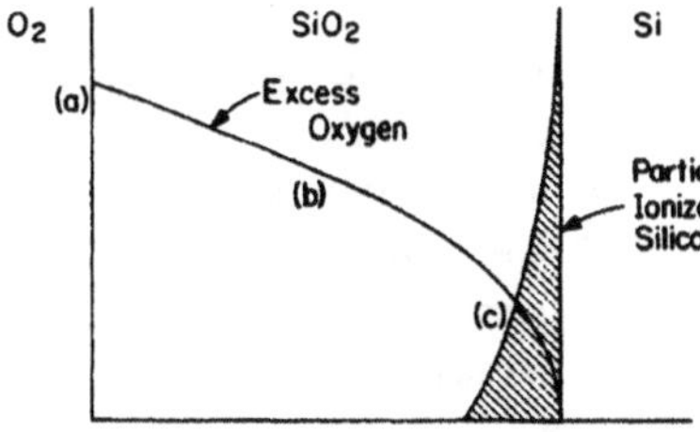

Fig. 1 Schematic cross-section of silicon thermal oxidation process.

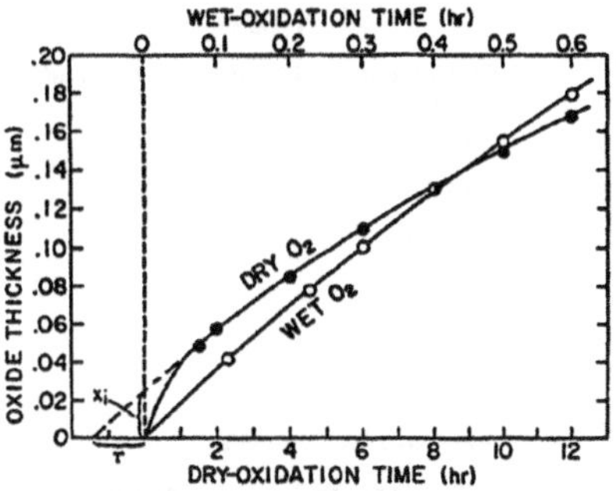

Fig. 2 Comparison of silicon thermal oxidation in dry oxygen and steam at 920 C, showing deviation of dry oxygen process from linear-parabolic model in initial 30 nm regime.

where x_i is the effective thickness at the start of an oxidation, either real or implied by the increased initial oxidation rate.

The linear-parabolic model permitted scientists and engineers to obtain oxidation data over a reasonable range of temperature and oxide thickness and to incorporate these data into process modeling programs. This was accomplished by determining values of B and B/A for various process conditions and preparing Arrhenius plots. Activation energies calculated from these plots helped to shed more light on the basic mechanisms involved in the thermal oxidation process. Also, the validity of Eq. (1) was strengthened by comparing results with those obtained previously by investigations of gas permeation through glasses (10, 11).

Another positive contribution of the linear-parabolic oxidation model was its relationship to and dependence on various process variables. Some of these are listed in Table I. Thus, it was possible to explain different effects caused by variations in silicon orientation, surface doping, and device geometries, and to model, predict, and control the oxidation process to achieve desired results.

In certain cases, the oxidation process also was found to affect and control certain properties of the oxidized silicon surface. These are listed in Table II and include surface impurity profiles (impurity redistribution) and oxide charges. In both cases, a direct relationship apparently exists between them and the oxidation process. The location and origin of oxide charges is depicted in Fig. 3, where their relationship to the oxidation process can be observed for two or more of the charges (Q_f, Q_{it}, Q_o).

QUESTIONS REGARDING THE LINEAR-PARABOLIC MODEL

While the linear-parabolic oxidation expression, Eq. (1), was found to accurately characterize silicon thermal oxidation over a wide range of process conditions, deviations were noted under certain conditions as these ranges were extended or new variables investigated. While many of these effects were noted in the early 1960's, many of the questions regarding details and mechanisms remain

TABLE I

PROCESS VARIABLES WHICH CAN AFFECT SILICON OXIDATION KINETICS

- OXIDATION TEMPERATURE
- OXIDANT TYPE
- OXIDATION PRESSURE
- SILICON ORIENTATION
- DOPANT CONCENTRATION
- STRUCTURAL VARIATIONS
- AMBIENT ADDITIONS (Cl, F, H_2O, etc)
- OPTICAL, ELECTRICAL, RADIATION EXCITATION
- PRE-OXIDATION CLEANING TREATMENTS

TABLE II

SILICON SURFACE PROPERTIES AFFECTED BY THE THERMAL OXIDATION PROCESS

- DOPANT PROFILE (REDISTRIBUTION)
- OXIDE CHARGES
- CARRIER MOBILITIES
- SURFACE TOPOGRAPHY
- SILICON DEFECTS, STACKING FAULTS
- JUNCTION DEPTHS (OED)

unanswered today. Some of these questions are tabulated in Table III. The first of these listed has to do with the apparent deviation of oxidation thickness-time data in the initial 30 nm for dry O_2 oxidation. The inability to model dry O_2 kinetics for thin oxides was not a problem in the early years and the faster oxidation could be represented by the constant τ in Eq. (1). However, as I.C. component densities increased, and feature sizes and oxide thicknesses scaled downwards (see Fig. 4), the need for a proper understanding of mechanisms involved in producing thin oxides became more critical. This subject is discussed further in the next section.

Another process variable affecting oxidation kinetics, and first reported in 1961 by Ligenza at BTL (12), is silicon crystal orientation. While Ligenza's results were for higher steam pressures, later data (13) indicated that in general the linear rate constant B/A depends on the silicon orientation as follows: (111)>(110)>(100). Similar dependencies were found for oxide charge densities (14). More recently, it has been found that under certain conditions (small oxide thicknesses or low temperatures), departures from the above orientation order occur (15-16). Such a departure or "cross-over" is shown in Fig. 5, where (110) silicon oxidizes faster than (111) below 20 nm at 900°C in dry O_2. Attempts to explain these discrepancies and even the orientation effect itself in terms of surface atom densities or other mechanisms have not yet been successful.

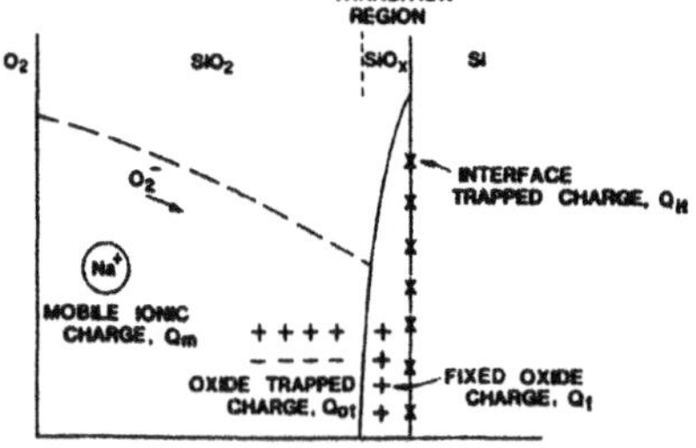

Fig. 3 Designation and location of four types of charges associated with the thermally oxidized silicon system.

TABLE III

UNANSWERED QUESTIONS CONCERNING LINEAR-PARABOLIC OXIDATION RELATIONSHIP

- INITIAL REGIME (0-30 nm) FOR DRY O_2 OXIDATION
- ORIENTATION EFFECT
- PRESSURE DEPENDENCE FOR DRY O_2 LINEAR RATE CONSTANT
- BREAK IN ARRHENIUS PLOT FOR PARABOLIC RATE CONSTANT AT 950°C
- MECHANISM FOR OXIDANT DIFFUSION THRU OXIDE
- PRE-OXIDATION CLEAN EFFECT
- DEPENDENCE OF OXIDE CHARGES (Q_f, Q_{it}) ON OXIDATION PROCESS
- TIME DEPENDENCE OF RATE CONSTANTS

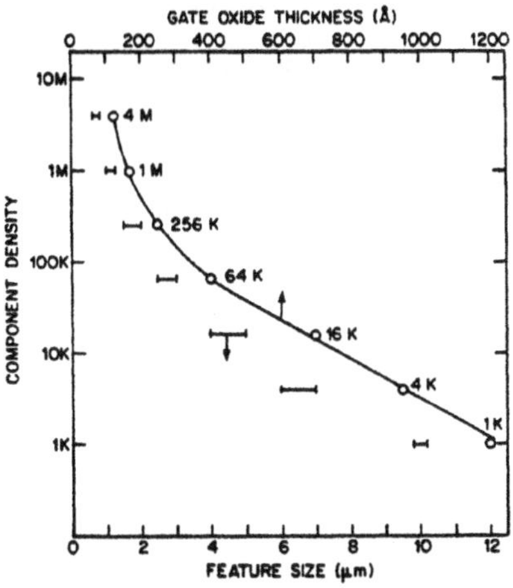

Fig. 4 Scaling of gate oxides in MOS D-RAMs.

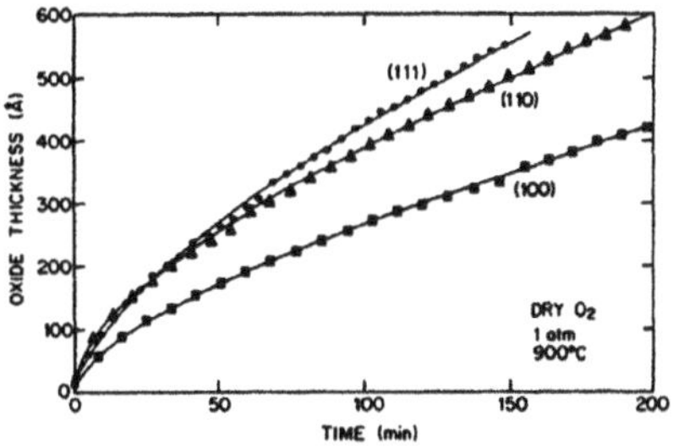

Fig. 5 Silicon orientation effect on thermal oxidation in initial thickness regime (15).

Another process variable modeled in linear-parabolic expression is oxidant pressure. The pressure should be directly proportional to B (and thus B/A) for both dry O_2 and steam. Initially this was found to be the case (9), but as more detailed investigations were carried out, it was found that the linear rate constant B/A for dry O_2 is proportional to $P^{0.7}$ (Fig. 6). An explanation for this deviation from first order dependence has not yet been presented.

Still another departure of oxidation kinetics from the linear-parabolic model is the parabolic rate constant dependence on temperature. At about 900°C, a break in the Arrhenius plot is normally observed, which has been attributed to the viscosity change for glasses (18).

Other questions involving oxidant diffusing species, pre-oxidation cleaning effects on kinetics, relationship of oxide charges to oxidation kinetics, possible dependence of rate constants on oxide thickness, and surface doping effects, all are still not answered after more than twenty-five years.

PROPOSED MODELS FOR THIN DRY O_2 OXIDATION

As indicated above, the general linear-parabolic expression, Eq. (1), does not accurately model dry O_2 oxidation process in the 0-30 nm range, and a number of individuals or groups have proposed appropriate kinetic models. Some of these have included physical mechanisms - others have been straight mathematical expressions. A number of these models are tabulated in Table IV. A comparison and discussion concerning the various proposed models has been reported (19-20) and they will not be discussed in detail here. However, it is possible to classify the many models listed in Table IV into six general areas. These are: (i) space charge/electrical effects, (ii) oxide structural effects, (iii) stress effects, (iv) silicon surface and substrate effects, (v) parallel oxidation processes, and (vi) other mathematical expressions. As can be noted, the first four of these represent physical mechanisms or phenomena, the last one is purely mathematical, while (v) is more mathematical but undoubtedly reflects two simultaneous processes or reactions occurring in the initial stages of oxidation. Considerable overlap among a number of these classes obviously exists.

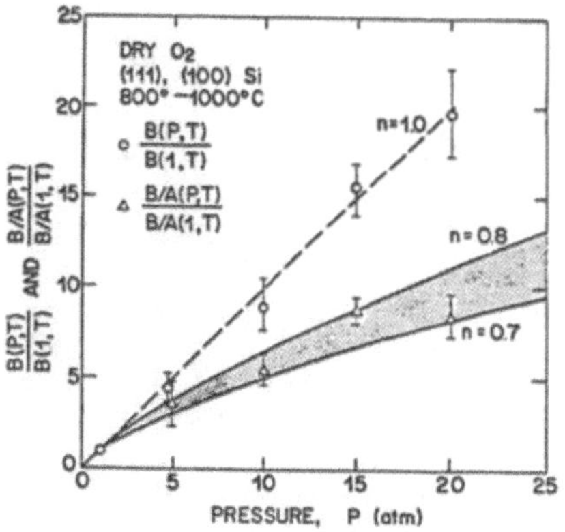

Fig. 6 Pressure dependence of linear and parabolic rate constants for dry O_2 oxidation (17).

TABLE IV

PROPOSED MECHANISMS FOR INITIAL SILICON OXIDATION IN DRY O_2

ELECTRICAL	a. Space Charge/Electrochemical Cabrera and Mott (21) Deal and Grove (9) Jorgensen (22) b. O_2 Dissociation/Oxygen Vacancy Blanc (23) Hu (24) Irene (25) c. Positive Oxide Charge Hamasaki (26) Schafer and Lyon (27)
STRUCTURAL	d. Microchannels/Pores Revesz and Evans (28) Irene (29) e. Blocking Layer Tiller (30) f. Oxygen Solubility Derbenwick and Anderson (31)
STRESS	g. Stress EerNisse (32) Irene (33) Doremus (34) Fargeix and Ghibaudo (35)
SILICON SURFACE/ SUBSTRATE	h. Si Substrate Massoud and Plummer (36) Murali and Murarka (37) i. Si Surface Preparation Grunthaner (38) Schwettmann (39) Morita (40)
PARALLEL OXIDATION	j. Parallel Oxidation Hopper et al (41) van der Meulen (42) Han and Helms (43) Rigo et al (20b)
MATHEMATICAL	k. Power Law Reisman and Nicollian (44)

Space Charge/Electrical Effects - In examining the many proposed models listed in Table IV, mechanisms (a), (b), and (c) fall into the space charge/electrical effects classification. A number of observed effects initially made this type of mechanism attractive for explaining the initial deviation from the linear-parabolic expression. Later, it was less popular but seems to be finding some favor again.

Structural Effects - The oxide structural effects category includes models (d), (e), and (f) in Table IV. While these mechanisms are feasible explanations for non- or modified linear-parabolic oxidation the difficulty in observing and identifying them and distinguishing them from other defects and impurities makes it difficult to support this concept.

Stress Effects - Many investigators (32-35) have provided evidence that stress effects in thermally oxidized silicon can play an important roll in the kinetics of silicon thermal oxidation - especially in the initial stages. This phenomenon is also undoubtedly related to the break observed in the activation energy plot of the parabolic rate constant at about 950°C. It is not clear, however, why differences are not noted between initial oxidation kinetics above and below this temperature. If stress and viscous flow is relieved at the higher temperatures, then initial deviation from the linear-parabolic kinetics should be much less. At this point, more information is evidently required to determine contributions of stress.

Silicon Surface and Substrate Effects - Some very interesting observations have been reported regarding effects of pre-oxidation cleaning treatments on dry O_2 oxidation kinetics (38-40, 45). Likewise, other possible contributors to kinetic deviation due to dissolved oxygen or silicon structure have been proposed (36-37). While these models are somewhat speculative, there is little question that the nature of the silicon itself is extremely important and can have considerable effect on subsequent oxidation processes. An example of oxidation data in the thin dry O_2 thickness range and the modified linear-parabolic expressions proposed in Ref. (19, 36) is shown in Fig. 7.

Parallel Oxidation Processes - An oxidation process that initially incorporates two separate but parallel reactions has been proposed to account for the so-called "rapid" oxidation observed for the 0-30 nm range of silicon oxidation in dry O_2 (41-43). This model uses a modified linear-parabolic growth law and takes the form:

$$\frac{dx_o}{dt} = \frac{B_1}{2x_o+A_1} + \frac{B_2}{2x_o+A_2} \tag{6}$$

where B_1 and B_2, and A_1 and A_2, are the respective values of the constants in Eq. (1) for processes 1 and 2. Initially, for very thin oxides, one of the two parallel processes controls the oxidation process. It may be based on a parabolic mechanism. Fairly rapidly, the second parallel process "takes over" and the kinetics are in the linear regime of the linear parabolic expression. As oxide thickness x_o goes to infinity, Eq. (5) approaches a parabolic growth law, as does the conventional linear-parabolic expression [Eq. (1)]. This expression has been used to plot literature data in Fig. 8 by Han and Helms (43) and the good agreement between experiment and theory is indicated.

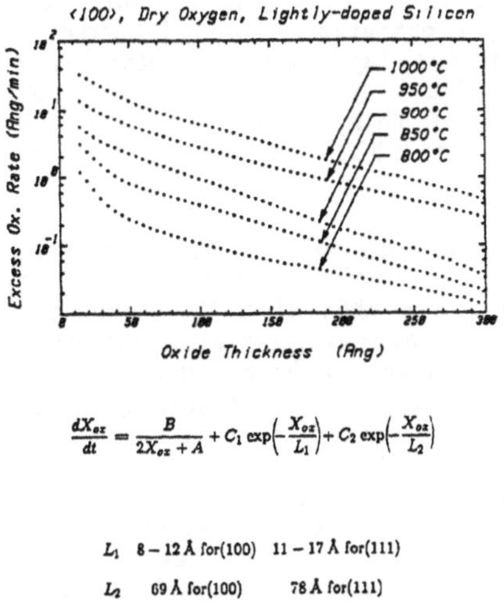

Fig. 7 Model for silicon thermal oxidation in dry O_2 in initial thickness regime [after Massoud et al (36)].

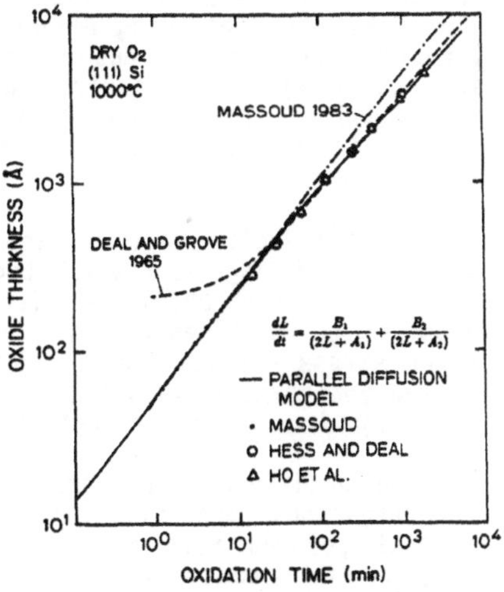

Fig. 8 Parallel oxidation model for silicon thermal oxidation in dry O_2 [after Han and Helms (43)].

It has been observed that such a parallel growth law satisfies dry O_2 oxidation data over a wide range of thickness and temperature. Han and Helms have proposed that the second parallel reaction contributing to the oxidation kinetics in the thin region may be a diffusion reaction of a species such as atomic oxygen or an oxygen vacancy. It is also possible that the parallel model may be related to several of the other proposed mechanisms listed in Table IV. More information must be obtained before this can be definitely determined.

Other Mathematical Expressions - It has been tempting to various authors to derive his or her own oxidation kinetics expression to account for the observed deviation of dry O_2 oxidation data in the 0-30 nm range from the general linear-parabolic model. This has been done, for instance, by assigning a power law to the data obtained by various investigators (44). Such an expression may well fit the data, but provides little guidance or insight into mechanisms involved or effects of process variations. An interesting discussion on the subject of modeling in general of the thin regime of silicon oxidation has been presented by Blanc (46), and such "philosophy" is indeed worthwhile reading for the many persons involved in the study of silicon oxidation kinetics.

It should also be noted that recently Blanc has proposed yet another model for the thin oxidation kinetics based on "percolation theory" (47). There seem to be no end of possible models or mechanisms to explain silicon oxidation kinetics.

OTHER CONSIDERATIONS

In reviewing the factors which might influence the mechanism of silicon thermal dry O_2 oxidation in the initial stages, some other considerations should be noted. First, as mentioned earlier, it has been observed for several years that particular chemical cleaning treatments can have a considerable effect on oxide growth rate (38-39). Recent investigations have confirmed the importance of pre-oxidation cleaning treatments (45, 48). It has furthermore been shown that the retarding effect of pre-oxidation cleans in ammonium peroxide solutions is due to a modified layer on the outer oxide surface (49). Obviously, such effects must be considered and understood if proper characterization of thin oxidation kinetics is to be achieved, and agreement is obtained among various investigators.

Another factor which plays an important roll in thin oxidation kinetics modeling, and which becomes more critical as device dimensions are scaled below one micrometer, is that of multi-dimensional processing. Combined with the trend to lower processing temperatures where stress effects are much more important, a better understanding of two and three dimensional oxidation mechanisms is crucial. Recent investigations of such effects have been reported (50-51).

One additional consideration which should be noted is that of investigations concerning oxidation mechanisms on an "atomic basis." Several persons have been looking at the interface reaction with respect to atomic species such as vacancies and interstitials (52-53). These species are formed due to the changes in atomic and molecular volumes as silicon is converted to silicon dioxide. They affect other processes such as dopant diffusion and impurity gettering, and they play an important role in the oxidation process itself. Also, these studies help to explain interface stress effects which quite likely determine and control the oxidation kinetics.

CONCLUSIONS AND FUTURE DIRECTIONS

The linear-parabolic model presented in the early 1960's, has been demonstrated to characterize the kinetics of silicon thermal oxidation to a reasonable degree. This model, which is based on physical reactions occurring during oxidation, has permitted scientists and engineers to accurately predict the effects of process variables on the oxidation process and to characterize oxidation kinetics in general. These variables include ambient composition and pressure, silicon orientation, dopant concentrations, and several others. However, as process conditions have been expanded, several questions regarding certain aspects of the model have not been satisfactorily answered.

Perhaps the most important question involves the deviation of experimental data from the original linear-parabolic model for dry O_2 oxidation in the 0-30 nm thickness range. A number of models primarily based on extensions of the linear-parabolic expression have been proposed, but to date no single one appears to satisfy all conditions and provide an good indication of the physical mechanisms involved. Results thus far indicate that some sort of parallel oxidation model in the initial stages of dry O_2 oxidation may best satisfy the experimental data.

One of the difficulties in properly characterizing the initial stages of silicon oxidation has been due to the observed effects of the pre-oxidation cleaning treatments. Several investigations now underway, especially those in conjunction with molecular beam epitaxy processing, will undoubtedly help to provide the understanding necessary to produce more controlled silicon surfaces. This should lead the way to resolving the thin oxidation kinetics question. Other issues have arisen due to lower temperature processing required by scaling of device densities and feature sizes. These include increased stress and orientation effects on the oxidation process, pressure dependence of rate constants, and structural variations related to complex multidimensional geometries. Oxide integrity is also becoming a critical issues. All of these areas are being investigated, and results of these studies should shed more light on such questions.

Even with all the above problems, currently unanswered questions, and more demanding device requirements, it can be assumed that the proper knowledge and understanding of the oxidation process will be obtained, if developments over the past thirty years are considered. During this period device and process scientists have continued to meet the challenges presented to them and will undoubtedly do so in the future.

REFERENCES

1. C.J. Frosch and L. Derick, J. Electrochem. Soc., 104, 547 (1957); 105, 695 (1958).

2. M.M. Atalla, E. Tannenbaum, and J. Scheibner, Bell Sys. Tech. J., 38, 749 (1959).

3. J. R. Ligenza and W.G. Spitzer, J. Phys. Chem. Solids, 14, 131 (1960).

4. J.R. Ligenza, J. Electrochem. Soc., 109, 73 (1962).

5. J.A. Hoerni, IRE Electron Devices Meeting, Washington, DC, 1960; U.S. Patents 3,025,589 (1962) and 3,064,167 (1962).

6. D. Kahng and M.M. Atalla, IRE Solid State Research Conference, Carnegie Institute of Technology, Pittsburgh, PA, 1960; D. Kahng, U.S. Patent 3,102,230 (1963).

7. J.T. Law, J. Phys. Chem. 62, 1200 (1957).

8. B.E. Deal, J. Electrochem. Soc., 110, 527 (1963).

9. B.E. Deal and A.S. Grove, J. Appl. Phys., 36, 3770 (1965).

10. F.J. Norton, Nature, 171, 701 (1961).

11. A.J. Moulson and J.P. Roberts, Trans. Faraday Seoc., 57, 1208 (1961).

12. J.R. Ligenza, J. Phys. Chem., 65, 2011 (1961).

13. W.A. Pliskin, IBM, J. Rsch., 10, 198 (1966).

14. B.E. Deal, M. Sklar, A.S. Grove, and E.H. Snow, J. Electrochem. Soc., 114, 266 (1967).

15. E.A. Irene, H.Z. Massoud, and E. Tierney, J. Electrochem. Soc., 133, 1253 (1986).

16. E.A. Lewis and E.A. Irene, J. Electrochem. Soc., 134, 2332 (1987).

17. L.N. Lie, R.R. Razouk, and B.E. Deal, J. Electrochem. Soc., 129, 2828 (1982).

18. E.P. EerNise, App. Phys. Lett., 30, 290 (1977).

19. H.Z. Massoud, J.D. Plummer, and E.A. Irene, J. Electrochem. Soc., 132, 2693 (1985).

20. (a) S.P. Murarka, in "ULSI Science and Technology - 1987," S. Broydo and C.M. Osburn, Eds., pp. 87-100, Vol. 87-11, The Electrochemical Society, Pennington, NJ (1987); (b) S. Rigo in "Instabilities in Silicon Devices, Vol. 1," G. Barbottin and A. Vapaille, Eds., pp. 5-100, Elsevier Science Publishers, Amsterdam (1986).

21. N. Cabrera and N.F. Mott, Rept. Progr. Phys., 12, 163 (1948).

22. P.J. Jorgensen, J. Chem. Phys., 874 (1962).

23. J. Blanc, Appl. Phys. Lett., 33, 424 (1978).

24. S.M. Hu, Appl. Phys. Lett., 42, 872 (1983).

25. E.A. Irene and Y.J. van der Meulen, J. Electrochem. Soc., 123, 1380 (1976).

26. M. Hamasaki, Solid State Electronics, 25, 479 (1982).

27. S.A. Schafer and S.A. Lyon, Appl. Phys. Lett., 47, 154 (1985).

28. A.G. Revesz and R.J. Evans, J. Phys. Chem. Solids, 30, 551 (1969).

29. E.A. Irene, J. Electrochem. Soc., 125, 1708 (1978).

30. W.A. Tiller, J. Electrochem. Soc., 130, 501 (1983).

31. G.F. Derbenwick and R.E. Anderson, Private Communication.

32. E.P. EerNesse, Appl. Phys. Lett., 35, 8 (1979).

33. E.A. Irene, J. Appl. Phys., 54, 5416 (1983).

34. R.H. Doremus, Thin Solid Films, 122, 191 (1984).

35. A. Fargeix, G. Ghibaudo, and G. Kamarinos, J. Appl. Phys., 54, 2878 (1983).

36. H.Z. Massoud, J.D. Plummer, and E.A. Irene, J. Electrochem. Soc., 132, 1745, 2685 (1985).

37. V. Murali and S.P. Murarka, J. Appl. Phys., 60, 2106 (1986).

38. F.J. Grunthaner and J. Maserjian, IEEE Trans. Nucl. Sci., NS-24, 2108 (1977).

39. F.N. Schwettmann, K.L. Chiang, and W.A. Brown, Paper No. 276 in Spring Meeting of The Electrochemical Society, p. 688 Extended Abstracts 78-1, Seattle, WA (1978).

40. M. Morita, T. Kubo, T. Ishihara, and M. Hirose, Appl. Phys. Lett., 45, 1312 (1985); 47, 253 (1985).

41. M.A. Hopper, R.A. Clarke, and L. Young, J. Electrochem. Soc., 122, 1216 (1975).

42. Y.J. van der Meulen, J. Electrochem. Soc., 119, 530 (1972).

43. C.J. Han and C.R. Helms, J. Electrochem. Soc., 134, 1297 (1987).

44. A. Reisman, E.H. Nicollian, C.K. Williams, and C.J. Merz, J. Electronic Mat., 16, 45 (1987).

45. G. Gould and E.A. Irene, J. Electrochem. Soc., 134, 1031 (1987).

46. J. Blanc, Phil. Mag., 55, 685 (1987).

47. J. Blanc, private communication.

48. J. Ruzyllo, G.T. Duranko, and A.M. Hoff, J. Electrochem. Soc., 134, 2052 (1987).

49. D.B. Kao, B.E. Deal, J.M. deLarios, and C.R. Helms, thi Proceedings, p. (1988).

50. R.B. Marcus and T.T. Sheng, J. Electrochem. Soc., 129, 1278 (1982).

51. D.B. Kao, J.P. McVittie, W.D. Nix, and K.C. Saraswat, IEEE Trans. Electron Devices, ED-34, 1008 (1987); ED-35, 25 (1988).

52. J.D. Plummer, Solid State Tech., 29(3), 61 (1986).

53. R.B. Fair, App. Solid State Science, Suppl. 2B, Academic Press, p. 1 (1981).

OXIDATION OF SILICON: TESTS OF MECHANISMS

Robert H. Doremus

Materials Engineering Department
Rensselaer Polytechnic Institute
Troy, N.Y.

"He said, that new systems of nature were but new fashions, which would vary in every age; and even those who pretend to demonstrate them from mathematical principles, would fluorish but a short period of time, and be out of vogue when that was determined." -Swift

ABSTRACT

Oxidation thicknesses for silicon in oxygen above 1100°C and for certain silicides are proportional to square root time (parabolic kinetics). The absolute value and temperature dependence of these results match closely with measurements of the permeation of molecular oxygen through bulk amorphous silica. Deviations from parabolic kinetics are accompanied by a decrease in the parabolic growth coefficient. No evidence for surface reactions influencing kinetics is found. A model involving strain in the oxide layer can account for the deviations from parabolic kinetics and the decrease in oxygen permeation through the oxide.

INTRODUCTION

Many mechanisms for the oxidation of silicon and silicides have been proposed. How can we choose between them? In this discussion I will try to start with a somewhat unconventional approach. I have collected all oxidation data for silicon and silicides in oxygen that follow simple parabolic kinetics at 850°C and above. These data agree closely with the permeation rates through bulk vitreous silica measured by Norton (1).

Next empirical fits to nonparabolic thickness-time measurements are considered, and it is shown that lower parabolic coefficients always accompany deviations from parabolic behavior. Then the importance of chemical reactions at the oxide-silicon interface is weighed, and no direct evidence that these reactions influence the rates of oxidation is found. If this conclusion is correct, only transport in the oxide film controls oxidation rates; the influence of ions and stress in the film on this transport is then discussed.

The present treatment is limited to oxidation with oxygen because much more data are available for this oxident. Nevertheless similar conclusions can be drawn for oxidation of silicon and silicides by water (2,3).

Parabolic Kinetics

Parabolic oxidation parameters B are plotted in Fig. 1 for thickness-time data that fit a simple parabolic equation:

$$L^2 = Bt \tag{1}$$

in which L is film thickness after oxidation for a time t at a constant temperature. A factor of $-L_o^2$ can be added to the left hand side of Eq. 1 for experiments in which there is an oxide film of thickness L_o already present; for none of the data in Fig. 1 was this necessary.

A simple model for Eq. 1 is the diffusion of a substance in dilute concentration through the oxide film. Experiments show that oxidant diffusion to the oxide-silicon interface controls oxidation of silicon and silicides. If c_L is the concentration of oxidant in the oxide at the oxide-gas surface, and D is the diffusion coefficient of oxidant in the film, then (2)

$$B = 2\ c_L D/\rho \tag{2}$$

in which ρ is the oxygen concentration in the oxide. Eq. 2 assumes that the concentration of oxidant at the oxide-silicon interface is small; see the section on surface reactions.

Gases dissolve and diffuse molecularly in amorphous silica (8). Not only inert gases such as helium, neon, argon, krypton, and xenon permeate this way, but also reactive gases such as hydrogen, water, oxygen and nitrogen. The square root of the activation energy for diffusion of all these gases is proportional to their molecular size (8). Norton (1) measured the permeation of oxygen through bulk amorphous silica, and his data are compared with permeations calculated from parabolic oxidation results in Fig. 1. There is remarkable agreement in absolute B values and activation energies between oxidation and permeation data. Furthermore both permeation and oxidation rates are proportional to oxygen pressure. I conclude that the silicon dioxide formed during oxidation of silicon and silicides is very similar to bulk vitreous silica; this conclusion is also confirmed by the identical refractive index of bulk and oxidation silicas. I also conclude that for the data in Fig. 1 the rate of oxidation is controlled solely by the diffusion of molecular oxygen through the silicon dioxide layer.

Non-parabolic Oxidation

If all oxidation data for silicon and silicides fit the line in Fig. 1, there would be no symposium held on this subject. Fortunately for all of us, the oxidation behavior is not that simple. For oxidation of silicon below about 1100°C the thickness-time relations are not parabolic, and parabolic coefficients become successively lower than expected from an extrapolation of Norton's data to lower temperatures. Even the simple addition of a linear term, or two more parameters A and τ in the form (4):

$$L^2 + AL = B(t + \tau) \tag{3}$$

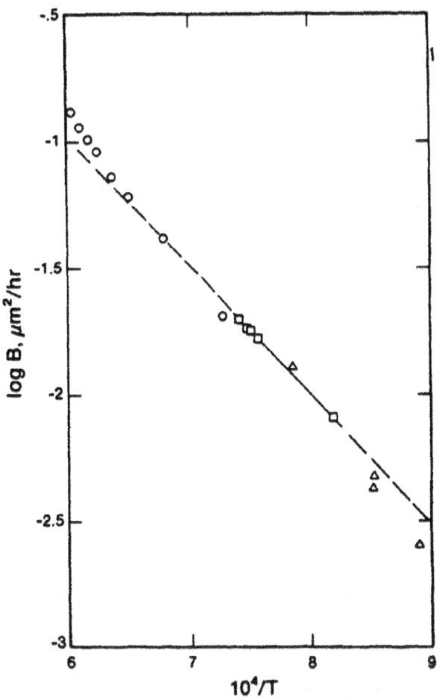

Fig. 1. Parabolic growth coefficients B from Eq. 1 for oxidation to one atm. of oxygen as a function of reciprocal temperature. ○ oxidation of silicon (4,5); △ oxidation of silicides (6,7); calculation from permeation data of Norton (1).

do not fit the data, as shown in Fig. 2.

Two groups of investigators (9,10) independently found that a plot of log L versus log t gives a straight line for all available data for oxidation of silicon and silicides in oxygen. Examples are shown in Fig. 3. Thus a power law gives a two-parameter fit to the data. This relation can be written as

$$L^{(2-m)} = (B + b)t \tag{4}$$

with parameters m and b to emphasize deviations from Eq. 1. Several investigators have proposed equations with logarithmic terms (11); three groups independently proposed the following equation (9,12,13):

$$L^2 + AL - AC\ \ell n(1 + L/C) = Bt \tag{5}$$

with parameters A, B and C; this equation also fits all experimental data on oxidation of silicon and silicides. It also gives linear log L vs. log t plots over wide ranges of conditions. The problem is therefore to understand the origin of the parameters m and b or A and C, and changes from expected values of B.

Interfacial Reactions

In some phase transformations such as precipitation from liquid or solid solution there is evidence that both matrix diffusion and interfacial reactions influence transformation rates (14). Thus it has been natural to explain deviations from parabolic oxidation kinetics as resulting from an interfacial reaction. A first order chemical reaction at the interface introduces a linear term such as the AL term

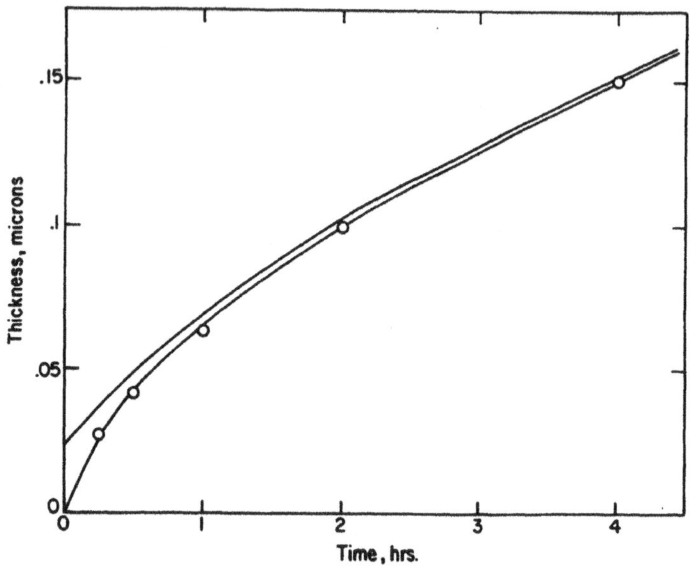

Fig. 2. Thickness of a silica layer on silicon with time of oxidation in one atm. of oxygen at 1000°C. O data of Deal and Grove (4); upper line from Eq. 3 with parameters from Ref. (4); lower line from Eq. 5 with parameters from Ref. (9).

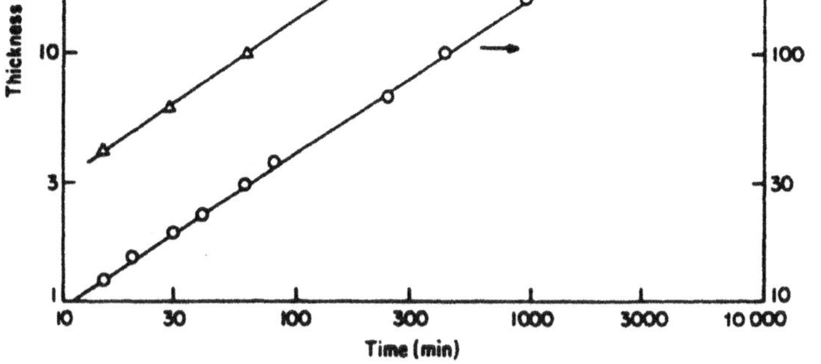

Fig. 3. Log thickness of silica layers on silicon as a function of log time during oxidation in one atm. oxygen. O at 900°, △ at 800°, data from Refs. 4, 17 and 18.

in Eq. 5. Nevertheless there is no direct evidence that such a reaction is important in any oxidation kinetics. Bare metals and silicon react very rapidly with oxidants at temperatures above 500°C, and even at room temperature unoxidized silicon is quickly covered with a very thin protective oxide layer. According to Birks and Meier (19) ..... "there are no known cases where the reaction at the metal-scale interface has been found to play a rate determining role." Wolters and Zegers van Duynhoven also conclude that the interfacial reaction is rapid and transport in the oxide controls (15).

If an interfacial reaction is involved, the oxidant concentration c_i at the interface is given by the relation

$$c_i/c_L = A/(A + 2L) \qquad (6)$$

in which A is the linear parameter from Eq. 5. A value of A of about .07μm was calculated from measurements of silicon oxidation in one atm. of oxygen at 900°C and 800°C (9). Thus for films thinner than about .1μm the interfacial oxidant concentration is appreciably higher than zero. In very thin films tens of nanometers thick the interfacial concentration of oxidant in the oxide should be close to c_L, the concentration at the gas-oxide surface, so the concentration of dissolved oxygen or water is nearly constant through the oxide. Direct measurements of dissolved oxygen or water profiles in the oxide are difficult, but these profiles can be deduced indirectly from profile measurements of oxygen isotopes in the oxide. For example, Costello and Tressler found a decrease in O^{18} reacted with a .1μm oxide film at 1000°C from the gas-oxide surface into the oxide; at about half the film thickness the reacted O^{18} decreased to less than background (16). This profile is inconsistent with an interfacial oxygen concentration higher than zero. Effects of residual water must be included for a full explanation of these profiles (15).

Another fact is that whenever the kinetics deviate from being parabolic, the resulting parabolic coefficient B is smaller than expected from the data in Fig. 1. This decrease in B is not expected from a simple model of an interface reaction; this reaction should have no influence on the diffusion of oxidant in the oxide. It is much more likely that the deviation from parabolic kinetics and the decrease in B both result from a common factor that changes transport in the oxide.

From these considerations I conclude that it is unlikely that the interfacial reaction of oxidant with the silicon is slow enough to influence oxidation kinetics. The decreases of B and deviations from parabolic kinetics thus result from reduction of oxidant transport rates in the oxide.

Changes in Transport

It is sometimes stated that the oxidation rate for the thinnest films is greater than expected. This idea results from a fit of oxidation data in oxygen to Eq. 3, as shown in Fig. 2; it appears from such a plot that deviations at shorter times involve a faster oxidation rate. However, even this early time rate is slower than the oxidation rate calculated from Norton's permeation data. Thus even in the thinnest films below 1100°C oxidation in oxygen is slower than expected from a comparison with vitreous silica results or the data shown in Fig. 1.

Some workers have concluded from the deviations at short times in Fig. 2 that thin silica films on silicon are anomalous or different structurally from thickner films. However other plots, such as in Fig. 3, or fits to Eqs. 4 or 5, give the same functional dependence of L on t over the entire measured thickness range from a few nm to microns. These fits suggest that the transport mechanism is the same, or at least changes continuously, over the measured thickness range; there is nothing anomalous about any thickness range.

The structure of bulk amorphous silica shows no changes at temperatures from 500°C to 1000°C. Some oxide films on silicon show small changes in refractive index and optical absorption from expected bulk values, but these changes can be readily explained in terms of strain in the oxide (see next section). Lattice defects of various kinds are found in the oxide layers on silicon, but there are probably too few of these defects to influence molecular transport in the layers. Therefore it is unlikely that structural changes in the oxide can explain large changes in the permeation of oxygen and water molecules.

Transport of lattice oxygen or ionic species derived from water has been suggested as a contribution to oxidant transport in silica on silicon (12,15). Electrolysis measurements show that ionic transport in silica is controlled by monovalent cations, especially sodium and lithium; transport by lattice oxygen is negligibly small (20). There are similar results on crystalline quartz (21). Isotope exchange experiments with O^{18} in quartz determine a very low maximum diffusion coefficient of lattice oxygen in silicas (22). Experiments on electrical transport in silicate glasses show the following sequence of mobilities: monovalent cations $\gg$ divalent cations $\gg$ electrons (23). There is no evidence for transport of lattice oxygen.

Hydrogen-bearing ions have a diffusion coefficient of about 10^{-10} cm^2/s in amorphous silica at 1000°C (24,25). Their concentration is unlikely to be more than 5 ppm; it is probably much less than this. The product of this concentration and diffusion coefficient is about two orders of magnitude less than required for the measured B factors. All these considerations show that it is unlikely that ionic transport contributes to oxidation of silicon.

Strain

Strain in silica films on silicon reduces the rate of molecular transport in them (26), and also reduces the diffusion coefficients of molecules in bulk amorphous silica (27). The refractive indeces of films grown above about 1100°C are close to that of amorphous silica, whereas at lower temperatures of film growth the refractive indeces of films are greater than for bulk amorphous silica (28). The changes in refractive index become greater at lower growth temperatures, corresponding to greater decreases in B values compared to those expected from Norton's data (28). All these results strongly suggest that strain plays an important role in the deviations from parabolic kinetics.

A prediction from these considerations is that the refractive indeces of films for the data in Fig. 1, which have permeations close to those measured for bulk silica, should be close to that of bulk silica. Increasing deviations from B values expected from these data should be accompanied by increasing deviations in the refractive indeces of the films involved. This prediction is perhaps the most direct way to test the importance of strain in the oxidation of silicon and silicides.

A model for the influence of strain on the oxidation kinetics leads to a linear-parabolic equation (Eq. 5 with C = 0) for oxidation of silicon in water (3), and Eq. 5 for the oxidation of silicon in oxygen (9). The temperature dependencies of B and B/A values calculated from experimental data can be rationalized with the models. Thus strain effects can explain deviations from parabolic kinetics and in B values for the oxidation of silicon and silicides.

Acknowledgements: This work was supported by NSF-MRG Contract No. DMR-8510617.

References

1. F. J. Norton, Nature 191, 701 (1961).
2. R. H. Doremus, J. Phys. Chem. 80, 1773 (1976).
3. R. H. Doremus, Thin Solid Films 122, 191 (1984).
4. B. E. Deal and A. S. Grove, J. Appl. Phys. 36, 3770 (1965).
5. G. H. Schiroky, J. Matls. Sci. 22, 3595 (1987).
6. R. R. Razouk, M. E. Thomas, and S. L. Pressacco, J. Appl. Phys. 53, 5342 (1982).
7. P. J. Jorgensen, J. Appl. Phys. 37, 874 (1962).
8. R. H. Doremus, "Glass Science", Wiley, New York, 1973, Chapt. 8.
9. R. H. Doremus and A. Szewczyk, J. Matls. Sci. 22, 2887 (1987).
10. A. Reisman, E. H. Nicollian, C. K. Williams, and C. J. Merz, J. Electronic Matls. 16, 45 (1987).
11. K. Hirabayashi and J. Iwamura, J. Elect. Soc. 120, 1595 (1973).
12. C. J. Han and C. R. Helms, J. Appl. Phys. 59, 1767 (1986).
13. J. Blanc, Phil. Mag. 55B, (1987); J. Electrochem. Soc. 133, 1981 (1986).
14. R. H. Doremus, "Rates of Phase Transformations", Academic Press, Orlando, FL, 1985.
15. D. R. Wolters and A. T. A. Zegers van Duynhoven in "Silicon Nitride and Silicon Oxide Thin Insulating Films", Electrochem. Soc., Pennington, N.J., 1987, p. 277.
16. J. A. Costello and R. E. Tressler, J. Electrochem. Soc. 131, 1944 (1984).
17. B. E. Deal, D. W. Hess, J. D. Plummer, and C. P. Ho, J. Electrochem. Soc. 125, 339 (1978).
18. H. Z. Massoud, Ph.D. Thesis, Stanford Univ. (1983).
19. N. Birks and G. H. Meier, "Introduction to High Temperature Oxidation of Metals", Edward Arnold, London (1983), p. 55.
20. R. H. Doremus, Phys. Chem. Glasses 10, 28 (1969).
21. A. Kats, Philips Res. Repts. 17, 133, 201 (1962).
22. R. Haul and G. Dumbgen, Z. Electrochem. 66, 636 (1962).
23. W. A. Lanford, to be published in J. Noncryst. Solids.
24. G. Hetherington, K. H. Jack, and M. W. Ramsay, Phys. Chem. Glasses 6, 6 (1965).
25. G. H. Frischat, J. Am. Ceram. Soc. 51, 528 (1970).
26. C. K. Huang, R. J. Jaccodine, and S. R. Butler in "Silicon Nitride and Silicon Dioxide Thin Insulating Films", V. J. Kapoor and K. T. Hankins, eds., Electrochem. Soc., Pennington, N.J. 1987, p. 343.
27. M. Nogami and M. Tomozawa, J. Am. Ceram. Soc. 67, 151 (1984).
28. E. A. Taft, J. Electrochem. Soc. 125, 968 (1978); 127, 993 (1980); 132, 2486 (1985).

SILICON OXIDATION MODELS BASED ON PARALLEL MECHANISMS

C. R. Helms and J. de Larios

Stanford Electronics Laboratories
Stanford University, Stanford, CA 94305

ABSTRACT

It has been suggested by numerous workers over the years that multiple mechanisms acting in parallel might be responsible for the growth of thermal oxides on silicon. The most obvious example of such an effect is oxidation in "wet" O_2 where both H_2O and O_2 act in parallel to oxidize the silicon. In this paper these models will be reviewed in light of more recent data, and suggest a new model based on a combination of these mechanisms.

INTRODUCTION

The search for the important species and mechanisms responsible for silicon oxidation has consumed many investigators for over 20 years. The most popular models are those based on the idea that the rate of a given process will be limited by a combination of diffusion of the oxidizing species through the previously grown oxide or the reaction of that species at the Si-SiO_2 interface.[1] For silicon oxidation in H_2O, such a formulation works exceedingly well in describing the kinetics of the process with the extracted rate constants exhibiting appropriate linear Arrhenius behavior and linear H_2O pressure dependence over a wide range of temperatures, pressures, times, and thicknesses.

For oxidation in dry O_2, it has been well established that such a formulation does not provide an adequate description of the process. Numerous models have been proposed to explain the deviation observed from linear parabolic kinetics. These models have taken various forms and have invoked numerous mechanisms including oxide stress, space charge effects on transport of ionized species, and more complicated chemical mechanisms relating to $O_2 \Leftrightarrow 2O$ reactions and other parallel acting species.

The purpose of this paper is to present a discussion of what we will call the chemically based models. These models assume that the dominant mechanisms relate to the diffusion and reaction of species driven by free energy changes associated with concentration gradients or chemical enthalpy and entropy differences. All other electrochemical fields such as those due to stress or space charge are considered to be of secondary importance. The processes to be considered all involve multiple mechanisms acting in parallel and are thus termed parallel oxidation models.

The idea of oxidation of silicon occurring via two parallel mechanisms goes back to the work of Jorgensen[2] analyzed by Rayleigh[3] who suggested the importance of the diffusion of both ionic as well as molecular species. In an attempt to explain the pressure dependence of the kinetics, Ghez and van der Meulen[4] (also see Blanc[5]) proposed a parallel mechanism involving one diffusing species (presumably O_2) with parallel interfacial reactions with Si, one involving O_2 directly and one involving O atoms. Subsequently, Hopper et al[6] (also see Irene[7]) suggested the diffusion of two oxidant species in parallel. They focused on two ranges of oxidation, one below 200Å and one above 300Å suggesting that one process dominated for the thinner range and one for the thicker. Although not required, they limited their consideration to species (such as O_2 and O_2^-) which diffused at different rates but had equal interface reaction rate constants. This type of formulation has also ben proposed by Irene[8] as well as Revesz[9] and co-workers. In the latter work, a parallel flux term which is independent of thickness is proposed (equivalent to diffusion coefficient proportional to thickness) related to the possible existence of structural channels through the SiO_2. This model leads to linear-parabolic-linear kinetics. A parallel diffusion model based on O_2^{18} tracer experiments was suggested by Han et al[10]. In this model, the interface reaction rate constants of the oxidant species (as well as their permeation constants) were found to be quite different. A determination of the rate constants led to parabolic-linear-parabolic kinetics.

Each of these models clearly can explain a large part of the available data, but none of them by themselves appears completely consistent with all available data. It will be shown, however that if the $O_2 \Leftrightarrow 2O$ interface mechanism proposed by Ghez and van der Meulen[4] is combined with the idea of two diffusing species, a model consistent with the available data is obtained.

Thickness/Time Dependence

The major driving force for developing what seems to be overly complex oxidation models is the observation that the simple linear parabolic model cannot account for the observed kinetics in O_2 and also does not provide a mechanism for the observed pressure dependence. However, although it has been recently argued to the contrary [11], it is clear to these authors that the dominant rate limiting process in the limit of thick oxides is diffusion of molecular oxygen. This has been shown very elegantly by Doremus [12] and will not be discussed here.

It is instructive at this point to deal with the question: "How thick is thick?" We will answer that question in the following way: Consider the thickness of the oxide as a function of time and parameters a, b, c, ..., so that $x = x(t, a, b, c, ...)$ where $x \equiv f(a)t^{1/2}$ for $t \rightarrow \infty$.

We will consider the oxide to be "thick" if a given percentage change in the other parameters (b, c, ...) yields a factor of ten smaller percentage change in x. A similar analysis can be applied to the thin limit (if one exists).

These ranges are shown schematically for steam and dry oxidations in figure 1 calculated from the linear parabolic model. The 10% limits as defined above are indicated along with a 50% line which corresponds to the point for which the linear and parabolic terms are equally important (corresponds to x=A). The first thing we note is that the oxides must be quite thick(thin) for nearly parabolic (linear) behavior to be approached. For most of the thickness range of interest the rate even for steam oxidation is neither linear nor parabolic. For the dry O_2 case the thin limit is never reached as the effect of the so-called initial fast process is clearly in evidence for $x \sim< 300$ Å.

One of the reasons it has been useful to revisit some of the older models is that new data over large thickness ranges is becoming available. This is illustrated in figure 2 whereby taking data from Hess and Deal[13], as well as Massoud, Plummer and Irene[14], we can model results from a few tens of angstroms up to nearly a micron.

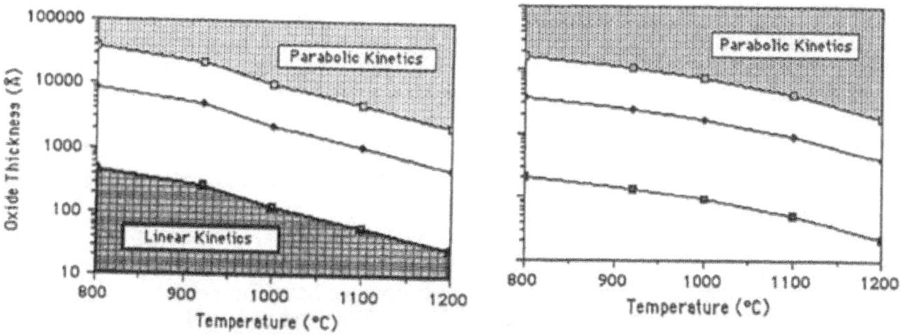

Fig. 1: Illustration of the regions where silicon oxidation can be considered nearly parabolic or linear for both steam (left panel) and dry (right panel) oxidations (based on the linear parabolic model)

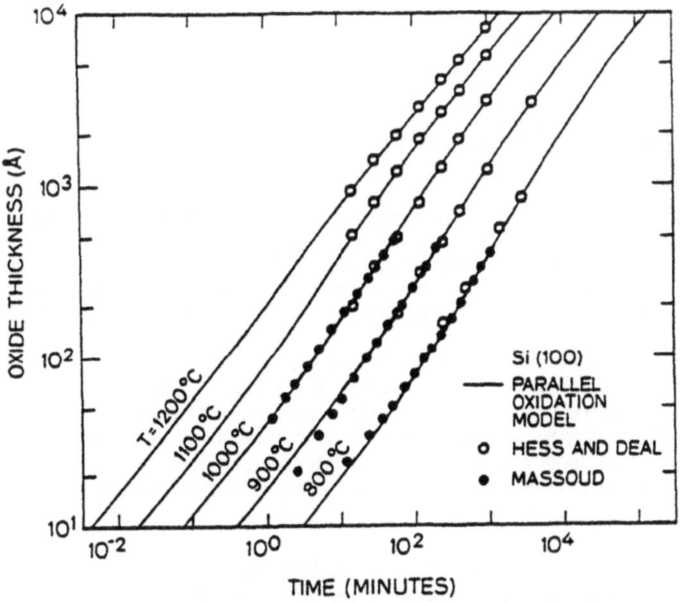

Fig. 2: Data from Massoud et al[14] (solid dots) and Hess & Deal[13] (open circles) for the oxidation of Si(100) in dry oxygen along with a fit to the parallel oxidation model[10] to be discussed later.

Pressure Dependence

One of the major motivations for the original development of the O_2/O model mentioned above [4,5] was to develop a framework which would allow a pressure dependence in the appropriate rate constants that is less than first power in O_2 pressure. Indeed that model allows for pressure dependence between 1/2 and first power in O_2 pressure. Prior to investigating these effects in more detail, a critical question is: "how should the pressure dependence be analyzed?" One standard approach is to collect thickness versus time data, choose ones favorite model, and then extract the appropriate rate constants for various pressures. The pressure dependence of the rate constants can then be determined. For example, using the linear parabolic model Lie, Razouk, and Deal[15] have extracted the pressure dependence of the linear and parabolic rate constants for this model and found them to behave approximately as $p^{0.7}$ and $p^{1.0}$ respectively. It has even been argued that the less than linear pressure dependence is an artifact of the data analysis methods and the true pressure dependence of the correct rate constants is always first power in O_2 pressure[16]. In the analysis to follow we have chosen to investigate the pressure dependence of the oxidation rate (dx/dt) which indeed is not a measured quantity but which we will determine from the available data using a simple interpolation scheme which we will now outline. In this way the actual pressure dependence of the rate is analyzed which will allow us to assess any deviations observed compared to simple first power pressure dependence.

The data is analyzed using a power series approximation to the oxidation rate which has been shown to yield excellent fits to the data over a large range of x and t. This relationship for the rate R is given by:

$$R = \frac{ax + b}{x\,(x + c)}$$

and $$t = \frac{ax^2}{2} + \frac{c-b}{a}x - \frac{(c-b)b}{a^2}\ln\left\{1 + \frac{ax}{b}\right\}$$

a, b, and c are determined for each pressure and then the rate itself recalculated from these values. Of course one could argue that this procedure is artificial and could lead to large correlated errors for values obtained for a, b, and c. However the value of the rate obtained will be much less sensitive to these errors and we believe so long as no extrapolation outside the available (x, t) data set is attempted, the rate obtained should be reliable.

The next step is to extract a quantitative measure of the pressure dependence. As we will see, the pressure dependence is clearly less than first power in O_2 pressure except for "thick" oxides. This can be seen in figure 3a in which the rate at 900°C for (100) Si is plotted versus pressure for various thicknesses where data is available on a log-log plot. Although we do not require that these plots be linear (and indeed do not necessarily associate much significance with this fact) their linearity allows for determination of the exponent n in a relationship $R \propto p^n$.

These values of n are plotted versus thickness in figure 3b. We observed that in the limit of large thickness the exponent approaches unity as expected but for thinner oxides it decreases to a minimum value of about 0.7. From this analysis procedure we feel confident that indeed the pressure dependence of the rate is sublinear and would therefore require any proposed oxidation model to provide for such pressure dependence as shown in the figure.

To our knowledge the only model capable of predicting the pressure dependence observed is due originally to Ghez and van der Meulen simplified by Blanc[4,5] and we will base a large part of our discussion on related models. As discussed by Blanc, his

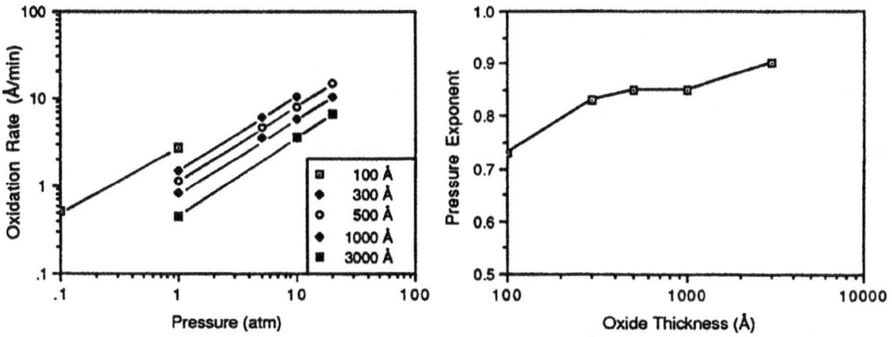

Fig. 3: A plot of the oxidation rate of Si (100) in dry O_2 as a function of pressure extracted from data of references 14, 15 is shown in part a (left panel) on a log-log plot. The slope of the lines (n for R α p^n) is plotted in part b (right panel) as a function of oxide thickness

simplified form (with 3 parameters) does not predict the proper x vs. t dependence and therefore might be dismissed without further discussion. As we will see, however, retaining key features of the model in a somewhat different formalism do provide excellent fits to x vs. t and the original model of Ghez and van der Meulen with 5 parameters can undoubtedly be forced to fit the available x vs. t data (although no attempt has been made to do this).

Review of Previous Models

Parallel Diffusion/Single Interface Model

This model which was developed from the idea that oxidation occurred via diffusive transport of the oxidant, presumably O_2 by two parallel paths, one associated with normal diffusion, one associated with transport through structural micro channels thought to extend considerable distances through the oxide. The most recent treatment of these effects[9] has led to the following relationship for the oxidation rate:

$$R = \frac{K_s(K_o x + K_T)}{(K_o + K_s)\,x + K_T}$$

This model has the remarkable feature that it predicts linear kinetics in the limit of thick oxides, related to a constant flux of oxidant through the proposed structural microchannels. Over the complete range of thickness the behavior might be described as linear-parabolic-linear being linear for thin oxides with R α K_s, tending toward parabolic for intermediate thicknesses, and then reverting to linear for thick oxides with $R\,\alpha\,\frac{K_s K_o}{K_s + K_o}$

Although our current view requires diffusive transport to be rate limiting in the limit of thick oxides which is inconsistent with the above analysis, Revez et al[9] never the less obtained excellent fits of their model to available data even with the linear-parabolic-linear model. Subsequent analysis however for thicker oxides than originally considered by Revesz does not appear consistent with their linear thick limit kinetics.[17]

Single Diffusion/Parallel Interface Model

Prior to the present work, the only attempts at oxidation modeling which implicitly allow for a variable power pressure dependence are based on an interface process involving molecular oxygen/oxygen atom reactions. As originally formulated by Ghez and van der Meulen[4] it could be described by the following equations for the rate of change of interface O_2 and O concentrations:

$$\frac{d[O_2]}{dt} = \frac{D}{X}\left\{[O_2]_o - [O_2]\right\} - k_3[O_2] + k_{-3}[O]^2 - k_1[O_2]$$

$$\frac{d[O]}{dt} = 2k_3[O_2] - 2k_{-3}[O]^2 + k_1[O_2] - k_2[O]$$

$$\frac{dx}{dt} = k_1[O_2] + k_2[O]$$

As we see, this model is mathematically quite complex, containing 6 parameters (reducible to 5 in the x vs. t solution). In the limit of thick oxides, however, these equations reduce to the parabolic limit with the rate equal to $\frac{dx}{dt} = \frac{D[O_2]_o}{X}$ as in the original linear parabolic model

Subsequent to the original formulation, Blanc[5] suggested that the oxidation process could proceed exclusively by a reaction of silicon with oxygen atoms. This amounts to setting $k_1 = O$ in the above equations. As an additional simplification, he also assumed molecular oxygen/atomic oxygen equilibrium at the interface. With these simplifications the above equations can be easily solved to give

$$\frac{dx}{dt} = \frac{k_2^2 k_{-3} x}{2Dk_3}\left\{\left(1 + \frac{4D^2 k_3[O_2]_o}{k_2^2 k_{-3} x^2}\right)^{\frac{1}{2}} - 1\right\}$$

This model also approaches the parabolic limit for thick oxides:

$$\frac{dx}{dt} = \frac{D[O_2]_o}{x}$$

as well as the linear limit for thin oxides

$$\frac{dx}{dt} = \frac{k_2 k_{-3}^{1/2}}{k_3^{1/2}}[O_2]_o^{1/2}$$

In addition, as we see from the above equations, the rate for thick oxides is proportional to O_2 pressure to the one half power for thinner oxides. A major difficulty with this formulation (which has only 2 parameters) is that indeed it does not fit the available kinetics data over a large x vs. t range. If we assume that the thick limit is correct, predicting the rate is determined by O_2 permeation, then the rate predicted is slower than that actually observed for thinner oxides. This is the same shortcoming of the original linear parabolic model! However the original Ghez and van der Meulen formulation, which gives the same parabolic rate in the thick limit, contains the additional O_2 reaction mechanism which will increase the total rate for thinner oxides. Although we have not attempted a full blown fit using the complete 5 parameter equations, qualitatively it will provide an increased rate over the limit determined by Blanc leading to better fits.

Parallel Diffusion/Parallel Interface Models

This type of model dates back to Rayleigh[3] and involves two independent oxidant species with independent interface mechanisms. For such a case the oxidation rate can be written as

$$R = \frac{B_1}{2x + A_1} + \frac{B_2}{2x + A_2} = \frac{B_2}{2x + A_2}\left(1 + \frac{\beta}{2x + A_1}\right)$$

Fits of this relationship to kinetic data lead to the conclusion that one of the A's is typically 5Å and can therefore be neglected for most ranges of interest giving

$$R = \frac{B_1}{2x} + \frac{B_2}{2x + A_2} = \frac{B_2}{2x + A_2}\left(1 + \frac{\beta}{2x}\right)$$

This relationship gives the expected parabolic dependence for thick oxides with $R = \frac{B_1 + B_2}{2x}$ as well as parabolic dependence for thin oxides with $R = \frac{B_1}{2x.}$ and could be termed parabolic-linear-parabolic (although for the values of the parameters obtained the rate is never actually linear). Determination of a reliable set of parameters for this model requires fits over a substantial thickness range which was recently performed by Han and Helms[10]. Their fits for Si(100) are shown in figure 2.

If we conclude that one of the mechanisms (associated with 2) involves the diffusion and interface reaction of molecular oxygen, what is the other associated with? Evidence comes from O_2^{16}/O_2^{18} sequential oxidation studies performed independently by Rochet, Agius & Rigo[19] and Han and Helms[20] who found that the initial rapid oxidation occurred at the outer SiO_2 surface.

The nature of the species that diffuses from the interface back to the surface cannot be determined from the available data. It is most likely a defect such as an oxygen vacancy, silicon interstitial, or other $SiO_x(x<2)$ type center, that reacts rapidly when reaching the outer surface and encountering adsorbed molecular oxygen.

Modified Model

Of the models discussed, two key points can be made. First, the single diffusion/parallel interface model involving molecular and atomic oxygen at the interface can provide fits to kinetic data and also exhibit the proper pressure dependence. However, this model fails to provide a mechanism for the initial fast process associated with the SiO_2 surface. Second, the parallel diffusion/parallel interface model involving one process associated with O_2 diffusion and reaction and another process associated with diffusion of a species (oxygen vacancies, Si interstitials, etc.) from the interface to the surface, provides fits to kinetic data and quantitatively explains surface oxidation observed in O_2^{16}/O_2^{18} tracer experiments. However, this model fails to provide for less than first power O_2 pressure dependence. This has led us to combine features of both models giving a model where molecular oxygen diffuses through previously grown oxide, dissociates into atomic oxygen, reacts with the silicon, thus creating oxygen deficient defects which diffuse to the surface and reacts with adsorbed O_2. The rate equation for such a process (in its simplest form) is given by

$$\frac{dx}{dt} = \left\{\left[A^2\left(1 + \frac{x}{L}\right)^2 + B[O_2]_o\right]^{\frac{1}{2}} - A\left(1 + \frac{x}{L}\right)\right\}\left\{1 + \frac{\beta}{2x}\right\}$$

$$\text{where } A = \frac{k_2^2}{4\,k_{-3}}, \quad B = \frac{k_3 k_2^2}{k_{-3}} \qquad L = \frac{D}{k_3},$$

and β is proportional to the diffusion coefficient of the species diffusing toward the surface. For thin oxides the rate is given approximately as

$$\frac{dx}{dt} \cong \frac{\beta}{2x}\left\{\left[A^2 + B[O_2]_o\right]^{\frac{1}{2}} - A\right\}$$

which is parabolic and has a pressure dependence between half and first power in O_2 pressure, depending on the magnitudes of A & B. For thick oxides the relationship reduces to

$$\frac{dx}{dt} \cong \frac{B[O_2]_o L}{2x} = \frac{D[O_2]_o}{x}$$

which is also parabolic and has first power O_2 pressure dependence.

A fit of this equation to kinetics data for Si(100) @ 900°C is shown in figure 4. In 4a, β has been set equal to zero showing the previously mentioned difficulty of this model fitting thin data. In 4b, β has been chosen for a best fit to the data. To assess the applicability of this model at other pressures, thicknesses as a function of time are then calculated at the other pressures where extensive data exists as shown in figure 5.

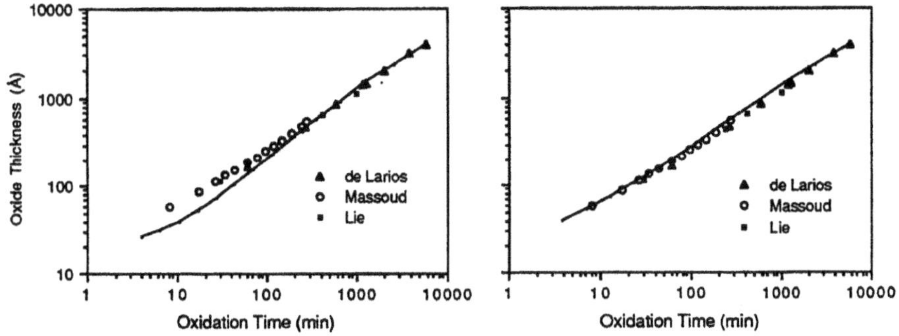

Fig. 4: A plot of oxide thickness vs. time for Si (100) in dry O_2 is shown in part 4a where the parameter β that is related to the thin oxide growth is set equal to zero. In part b, β = 100Å provides a best fit. The model is compared to the data of de Larios[21], Massoud[14], and Lie[15].

It should be stressed that additional fitting over and above what was required at one atmosphere was not attempted. The fits are indeed quite good and it is likely the poorer fit at 0.01 atmosphere for thinner oxides could be corrected (if A_1 in the above equations was not set equal to zero.

Effect of Oxide Stress

Up to this point we have neglected the effect of oxide stress on the oxidation kinetics and indeed it has not been necessary to include stress effects to model the available

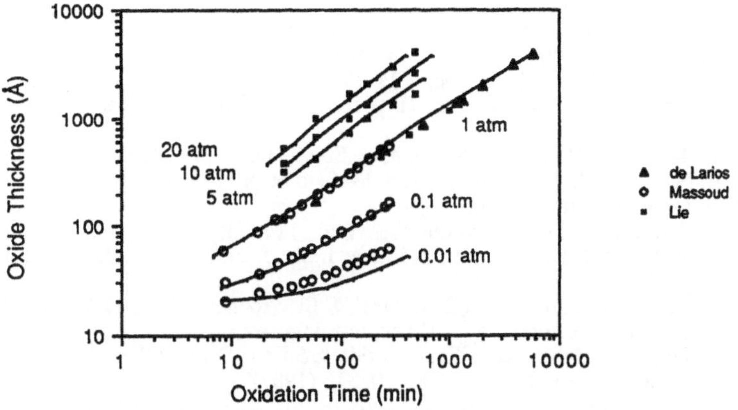

Fig. 5: The comparison of the model with the data as shown in fig. 4b is extended to pressures from 0.01 to 20 atm. In the model A, B, L, and β are held constant and the pressure is varied. The data is that of de Larios[21], Massoud[14], and Lie[15].

data (at least at a constant temperature). As a function of temperature, the oxide grown indeed has different properties as discussed by Rendall and Ngai[22] and Landsberger and Tiller[23] in this volume as well as Taft[24]. This shows up in the kinetic analysis as nonlinear Arrhenius plots for the rate constants[25]. We feel however that although stress effects are certainly very important as a function of temperature, at a constant temperature they play a minimal role in determining the oxidation kinetics.

Acknowledgement

This work has been supported in part by gift funds from National Semiconductor and Semiconductor Research Corporation Contract 87-SJ-101.

References

1. B. E. Deal and A. S. Grove, J. Appl. Phys., 36, 3770 (1965).
2. P. J. Jorgensen, J. Chem. Phys., 37, 874 (1962).
3. D. O. Rayleigh, J. Electrochem. Soc., 113, 782 (1966).
4. R. Ghez and Y. J. van der Meulen, J. Electrochem. Soc., 119, 1100 (1972).
5. J. Blanc, Appl. Phys. Lett., 33, 424 (1978).
6. M. A. Hopper, R.A. Clarke, L. Young, J. Electrochem. Soc., 122, 1216 (1975).
7. E. A. Irene, Appl. Phys. Lett., 40, 74 (1982).
8. E. A. Irene, J. Electrochem. Soc., 125, 1708 (1978).
9. A. G. Revesz, B. J. Mrstik, H. L. Huges, H. L. McCarthy, J. Electrochem Soc., 133, 586 (1986).
10. C. J. Han, C. R. Helms, J. Electrochem. Soc., 134, 1299 (1987).
11. E. H. Nicollian, A. Reisman, J. Elect. Mat., to be published.
12. R. H. Doremus, this volume.
13. D. W. Hess, B. E. Deal, J. Electrochem Soc., 124, 735 (1977).
14. H. Z. Massoud, J. D. Plummer, and E. A. Irene, J. Electrochem. Soc., 132, 2693 (1985).
15. L. N. Lie, R. R. Razouk, and B. E. Deal, J. Electrochem. Soc., 129, 2828 (1982).
16. R. H. Doremus, as quoted by J. Blanc, this volume.
17. C. J. Han, C. R. Helms, in "Silicon Nitride & Silicon Dioxide Thin Insulating Films," Electrochem. Soc. Proc, 87-10, 315 (1987).
18. The form involving β which is mathematically identical to the other form has been suggested by Demirlioglu and Plummer (to be published) and appears to provide more physical meaning to the parameters.
19. F. Rochet, B. Agius, S. Rigo, J. Electrochem. Soc., 131, 914 (1984).
20. C. J. Han, C. R. Helms, J. Appl. Phys., 59, 1767 (1986).
21. J. M. de Larios, C. R. Helms, D. B. Kao, B. E. Deal, Appl. Surf. Sci., 30, 17 1987).
22. R. W. Rendall, K. L. Ngai, this volume.
23. L. M. Landsberger, W. A. Tiller, this volume.
24. E. A. Taft, J. Electrochem. Soc., 125, 968 (1978).
25. C. J. Han, C. R. Helms, J. Electrochem. Soc., 132, 402 (1985).

THE ROLE OF SiO IN Si OXIDATION

AT A Si-SiO_2 INTERFACE

S.I. Raider

IBM T.J. Watson Research Center
Yorktown Heights, N.Y. 10598

Deal and Grove (1) modeled thin oxide growth by a Si-SiO_2 interfacial reaction whose rate is expected to vary linearly with time and exhibit a first order dependence on oxidant pressure. Deviations from this model, which increase with a decrease in temperature, are observed during oxidation in dry O_2 for oxides $\leq$ 30 nm thick. To account for the unexplained thin oxide growth process, it was proposed (2) that molecular SiO is formed in parallel with SiO_2. SiO formation and transport from the Si-SiO_2 interface into SiO_2 are identified as as a parallel oxidation process that originate at the interface and that affect SiO_2 growth. Experimental studies supporting this previously proposed model show a) evidence of significant SiO transport for oxide films that are $\lesssim$ 30 nm thick and b) that oxidation to SiO_2 is inhibited by blocking the intermediate step in which SiO forms.

A. INTRODUCTION: Si OXIDATION MODELS

The Deal-Grove linear-parabolic model of Si oxidation divides SiO_2 growth into two thickness regimes (1). Thick oxide growth is limited by oxidant diffusion through the oxide to the interface. Thin oxide growth is limited by the reaction rate at the interface and is expected to vary linearly with time and have a first order dependence on O_2 pressure. The sensitivity of the thin SiO_2 growth rate to the crystalline surface orientation (3) indicates that growth is limited by reaction at the Si-SiO_2 interface in agreement with the Deal-Grove model. During Si oxidation in H_2O, or in dry O_2 at a temperature of 1200°C, a linear thin oxide growth rate is approached (1). However, deviations from a linear model are reported for oxide films $\leq$ 30 nm thick grown in dry O_2 at lower temperatures. (1, 4). Growth rates increase nonlinearly with decreasing oxide thickness (5). Thin oxides exhibit an O_2 pressure power law dependence that is nearly first order at 1200°C (1, 6) but decreases towards a square root pressure dependence with decreasing temperature (4).

Oxygen tracer studies (7, 8) and oxidation rate crossovers with substrate orientation (9-11) suggest that the deviations from linear behavior are interdependent and originate from a common cause. During oxidation of Si-$Si^{16}O_2$ structures with $^{18}O_2$ at 930°C, the oxygen isotope ^{18}O unexpectedly accumulates at the outer SiO_2 surface (7, 8) although Si oxidation occurs at the Si-SiO_2 interface. What is most significant is that the ^{18}O concentration at the outer $Si^{16}O_2$ surface increases with the deviations from the linear growth model, i.e., as SiO_2 film thickness or $^{18}O_2$ ambient pressure decreases (8). The Si oxidation chemistry appears to be a source of these deviations. Secondly, a crossover occurs in the relative oxide growth rates with substrate orientation as O_2 pressure (9, 10) or as oxidation temperature is decreased (11).

Many chemical models have been suggested to derive thin oxide growth kinetics at the Si-SiO_2 interface. Deal and Grove (1) used the overall reaction, Si + SiO_2 → SiO_2, to describe the rate-limiting interfacial oxidation step. From low pressure oxidation studies, Van der Meulen (4) found that the temperature dependence of the linear rate constant is not described by a single activation energy and, consequently, the kinetics of the interfacial reaction could not be attributed to a single reaction. Ghez and Van der Meulen (12), and Blanc (13), discussed the oxide growth rate dependence on O_2 partial pressure in terms of an O_2 dissociation reaction competing with an overall reaction. From data fitting, Hopper et al. (14) proposed that thin oxide growth proceeds by two parallel rate-limiting processes which involve molecular and ionic oxygen species. From oxygen tracer studies, Rochet et al. (7) attributed the increase in ^{18}O concentration at the outer oxide surface to oxygen vacancies diffusing from the Si-SiO_2 interface. An intrinsic stress model (15) was proposed for the temperature-dependent crossover in oxidation rate with substrate orientation (10). This model is not applicable to the corresponding pressure-dependent crossover with substrate orientation (9, 10).

Recently, Han and Helms (16) extended the two parallel rate-limiting process approach using a much larger data base than Hopper et al. (14) They modeled oxide growth by a diffusion rate-limiting step in parallel with a (Deal-Grove) linear-parabolic growth rate. In the thin oxide limit, the diffusing species, which they suggest is either atomic oxygen or an oxygen vacancy, has a diffusion coefficient-solubility product about 10 times less than molecular O_2 at 1000°C but the reaction rate is greater. Most important, they (16) relate the oxide growth rate for the diffusion rate-limiting process with oxygen tracer accumulation at the outer SiO_2 surface when oxidizing Si-$Si^{16}O_2$ with $^{18}O_2$.

A limitation of crystalline Si-based systems is their tendency to form a gaseous suboxide, SiO. At elevated temperatures and in O_2-deficient ambients, SiO_2 oxidizes Si at a Si-SiO_2 interface to form SiO. To account for thin oxide growth at the Si-SiO_2 interface in dry O_2, Raider (2) proposed that molecular SiO and SiO_2 both form from the same intermediate species during Si oxidation. One of two parallel processes that originate at the Si-SiO_2 interface and limits thin SiO_2 growth rates is affected by molecular SiO transport from the interface. A linear, rate-limiting oxidation step, possibly O_2 chemisorption on the Si substrate surface, is assumed as the second oxide growth rate-limiting step.

SiO is formed during Si oxidation at a Si-SiO_2 interface as a byproduct of an intermediate, SiO*, adsorbed at the Si surface. The mechanism proposed for concurrent formation of both silicon oxides involves chemisorption of an O_2 molecule on the Si substrate surface followed by dissociation to form an adsorbed intermediate, SiO*. The SiO* species can either a) oxidize further to form SiO_2 or b) form a secondary product, molecular SiO. The SiO* and SiO concentrations at the interface are related according to the reaction SiO* ←→ SiO. A possible scheme for the thermal oxidation steps, showing the secondary path for SiO*, is given in Figure 1.

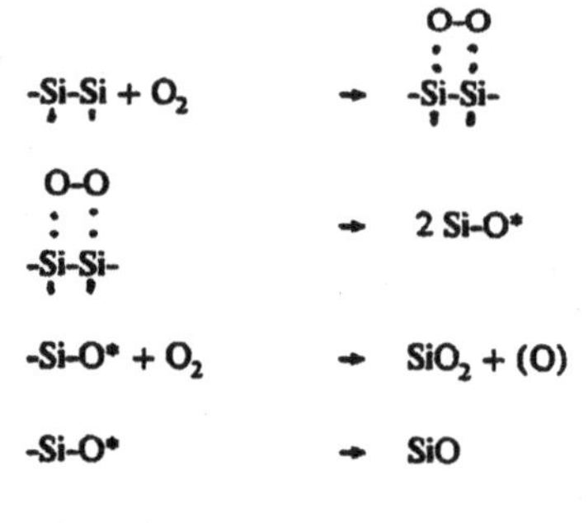

(1)

(2)

(3a)

(3b)

Figure 1. A reaction scheme with some of the elementary steps that occur during oxidation at the Si-SiO_2 interface. Two reaction paths (3a and 3b) are shown for the adsorbed intermediate suboxide species, SiO*. The formation and transport of SiO from the Si-SiO_2 interface influences SiO_2 growth by affecting the SiO* concentration available for reaction (3a).

The SiO species, which has a bond length about 25% longer than O_2, is metastable (17). Partial decomposition of SiO molecules into Si and (O) atoms is expected during transport through SiO_2 at elevated temperatures. SiO decomposition within an SiO_2 film yields free Si atoms which react with the SiO_2 within the bulk oxide according to the reaction Si + SiO_2 → 2 SiO. The bulk oxide reaction consumes SiO_2 and generate paths, or channels, in oxide films ≤30 nm thick through which SiO transport is enhanced. Using TEM, Gibson and Dong (18) have observed pores in 9 nm thick oxides grown in dry O_2. From analyses of O_2 oxidation rates, Revesz et al. (19) conclude that structural channels are present in thin oxides grown in dry O_2.

The deviations from a linear rate of thin oxide growth with decrease in oxidation temperature, in oxide thickness, and in O_2 pressure, are associated with enhanced SiO formation and transport from the interface during thin oxide growth. The oxygen isotope accumulation at the outer SiO_2 surface is caused by increased formation of SiO at the interface and transport during thin oxide, or low pressure, growth. The oxidation rate crossover with substrate orientation is determined by a change in relative contribution to SiO_2 growth of the two interfacial, rate-limiting, parallel processes.

To test this model, it is first necessary to show that SiO formation or transport is enhanced in the same oxide thickness region, i.e., for films $\leq$ 30 nm thick, in which deviations from linear oxide growth are observed. It is further necessary to demonstrate the existence of an SiO* intermediate. In section B, experimental studies are outlined describing the dependence of SiO transport on thermal SiO_2 thickness (20). In section C, evidence for SiO* as an intermediate during the thermal oxidation process is obtained from studies of oxidation barrier formation by local nitridation of Si with low concentrations of nitrogen (20). The formation and transport of SiO from the Si-SiO_2 interface provides the basis for a single, self-consistent explanation for deviations from a linear, thin oxide growth model (20). An explanation for these deviations is outlined in section D.

B. SiO TRANSPORT THROUGH SiO_2

The Interfacial Reaction Between Si and SiO_2

We have studied the interfacial reaction, Si + SiO_2 $\rightarrow$ 2 SiO, to determine the dependence of SiO transport on SiO_2 film thickness. This reaction is made visible by annealing Si-SiO_2 structures in an inert ambient at elevated temperatures. During annealing, oxide-free holes, or voids, form in the SiO_2 layer and etch pits form in the substrate. We have found that significant change in the rate of void formation occurs for oxide films $\lesssim$ 30 nm thick. Examples of voids formed across oxidized Si wafers with different oxide thicknesses, after annealing in He gas at 1030°C for 30 hours, are shown in Figure 2 (20). Void densities increase with decreasing oxide film thickness. Several size voids are seen in thin oxides with smaller voids forming at later stages of annealing (20). Once a group of voids with a particular size is detected, the void density in that group remains constant. Voids form towards the edge of each wafer, including those with thick oxides, after annealing for only ~ 4 hours in He at 1030°C. However, we detect a very low density of voids over the remainder of the wafer with a 150 nm thick oxide film after annealing for 30 hours and very little change in void density is observed after annealing for a total of 108 hours.

A void density dependence on oxide thickness may have several causes. These include a) an SiO transport thickness dependence, b) variations in substrate defect density associated with the thermal oxidation rate, c) variations in detection of the initial reaction stages, and d) changes in interfacial stress.

Figure 2. Void formation for different oxide thicknesses after annealing in He for 30 hours at 1030°C. SiO_2 thicknesses: a) 14.2 nm, b) 25.2 nm, c) 38.8 nm, d) 142 nm.

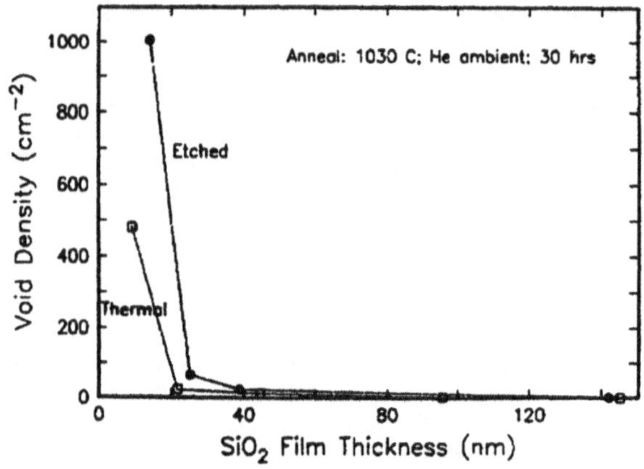

Figure 3. Void densities vs. SiO_2 thickness. SiO_2 films grown a) thermally or b)etched back from thick thermal oxide in buffered HF (9:1). Annealed at 1030°C for 30 hours in He gas.

We have studied the void densities formed after annealing 30 hours in He at 1030°C for thermal oxides, grown in dry O_2 at 1000°C to thicknesses between 10 nm and 150 nm, and for 150 nm thick thermal oxides that were wet chemically-etched in buffered (9:1) HF to different thicknesses. From Figure 3, it is apparent that void densities fall into two oxide thickness regimes which appear similar for thermally grown, and for etched, oxides. For oxides $\lesssim$ 30 nm thick, the void densities depend strongly on SiO_2 thickness. For oxides $\gtrsim$ 50 nm thick, the detected void densities are very low. The similarity between the curves for thermally-grown and for etched oxides (Figure 3) indicates that the void density is not directly related to substrate defects induced during high rates of thermal oxidation (20). We observe a direct correspondence between oxide void densities and etch pit densities in the substrate (20) which indicates that the observed thickness dependence is not due to an inability to detect the initial stages of void formation. At the anneal temperature of 1030°C, the intrinsic stress in the oxide films is near zero (21) and is not expected to induce substrate defects with varying oxide thickness. We conclude that the SiO_2 thickness dependence is related to SiO transport from the interface.

The void densities are a measure of the extent of interfacial reaction and of the ability of Si atoms to consume local paths through an oxide film. The appearance and size of voids is apparently dependent on the defect site size and the SiO_2 film thickness that must be consumed. Liehr et al. (22) associated void size with the size of stacking faults in the substrate. Presumably, a hole in SiO_2 first develops above a defect site due to reaction between SiO_2 and Si. Once initiated, voids grow laterally with time at a rate inversely dependent on oxide thickness (23).

The mechanism by which void formation is initiated is unknown since initial stage measurements are difficult to make. Oxide voids form when Si atoms reduce SiO_2. The Si atom source is either the substrate or from decomposition of the metastable SiO molecule. There are several indications that the latter occurs. From void cross-sections, Bradley (24) suggests that more SiO_2 than Si is consumed during this reaction in the void regions although a direct measurement has not been reported. Electrical measurements (25, 26) indicate that the SiO_2 film is degraded during inert ambient annealing of Si-SiO_2 structures prior to void formation. It is suggested that SiO transport is a source of Si in bulk SiO_2 but its relative importance remains to be determined. If SiO decomposes during transport through the oxide film, each Si atom reacts with more than one SiO_2 species. The oxygen atom attached to Si is therefore not the same atom in the transported SiO molecule as in the SiO initially formed at the interface.

An attempt was made to obtain direct evidence for molecular SiO transport through the oxide film into the gas ambient during in-situ oxidation studies (27, 28). Gaseous SiO molecules were excited by laser-induced fluorescence during low pressure O_2 oxidation and the signal was focused on a monochromator connected to a multichannel analyzer (29). Using this technique, a detectable flux of 3×10^{-5} SiO molecules/s from the Si surface was achieved (27). The SiO(g) signal increases with O_2 pressure as the O_2 reacts with a clean Si surface. When SiO_2 is formed on the Si surface, the SiO signal drops to the background level. Similar results were obtained during oxidation of Si/SiO structures (28). SiO molecules are not evaporated from SiO_2 into the ambient during oxidation. Escape of SiO from the SiO_2 surface into the ambient is inhibited either by the adsorbed O_2 molecules or by an instability of SiC molecules during transport through an SiO_2 film.

C. SiO - AN INTERMEDIATE IN THE OXIDATION PROCESS

When Si-SiO_2 structures are annealed in a low-reactivity, N_2 ambient at elevated temperatures, nitridation and SiO formation both occur at the interface. Lin (30) identified SiO by mass spectrometry as the only oxygenated species in the gas phase while thermally nitriding Si with N_2 at 1200°C. If Si or Si-SiO_2 is annealed in N_2 for sufficiently long times, or N_2 is implanted into the Si, a barrier to oxidation forms (31-33). The barrier acts to delay the normal rate of thermal oxidation of Si (31-33). At 1000°C, oxidation delays of 1 to 2 hours were obtained after Si is thermally annealed in N_2 at 1000°C for one week (31). The barrier region, <3 nm thick, contains 2 to 3 atomic percent N at the Si-SiO_2 interface (34). A barrier delay time of 20 hours for oxidation in O_2 at 1020°C was reported after N implantation into Si at a dose of $3 \times 10^{15}/cm^2$ (33). For reference, the concentration of N in a Si_3N_4 film is $5.9 \times 10^{22}/cm^3$ and the Si_3N_4 thermal oxidation rate at 1000°C is ~ 5 nm/hour (35). The number of implanted N atoms is equivalent to a 0.5 nm thick Si_3N_4 layer but the oxidation barrier effectiveness is equivalent to a 100 nm thick Si_3N_4 film. Clearly, the relationship between the long oxidation delay times and the low nitrogen concentrations indicate that barrier formation is not explained by formation of uniform Si_3N_4 films.

If a larger N ion dose, 10^{16} N ions/cm^2, is implanted into Si, only ~ 20% of the implanted N accumulates at the Si-SiO_2 interface after annealing (32). As the interfacial barrier N is consumed during oxidation, additional N diffuses from the bulk Si to the interface until the bulk N is consumed. A limited number of sites are apparently available at which N atoms accumulate at the Si-SiO_2 interface. This is not unusual, since thermal nitridation in N_2 at temperatures $\geq$ 1100°C yield small nitride crystallites rather than continuous nitride films (36, 37).

Oxidation barrier formation by reaction between Si-SiO_2 and N_2 is oxide thickness dependent, as is shown in Figure 4 (31). After thermal nitridation, an oxidation barrier forms on SiO_2-Si structures with oxides $\leq$ 45 nm thick but not on a similar structure with an oxide 100 nm thick. The cause for this thickness dependence is not immediately apparent since the N_2 anneal times are $\geq$ one week whereas the time for N_2 diffusion through a 100 nm oxide film at 1000°C is < 10 minutes (38, 39). Two diffusants, N_2 and SiO, are present in this system. The thickness dependence of barrier formation is therefore attributed to SiO transport from the interface through the SiO_2 layer. The observed thickness dependence indicates that the oxidation barrier depends on SiO formation and transport from local sites. Apparently, N_2 reacts either with the SiO at the reaction sites, or directly with the reactive sites. Nitridation and barrier formation occurs when these sites are present. We conclude that barrier formation by nitridation occurs only at the chemically reactive SiO defect sites rather than uniformly across the Si surface. Consequently, a relatively low N concentration is sufficient to form a barrier to further oxidation of Si to SiO_2. Since blocking the sites at which SiO forms prevents oxidation, these sites on the Si surface are unique and must be available for oxidation to occur. These sites are necessary for the intermediate SiO* to form in order that the oxidation process go to completion and form SiO_2.

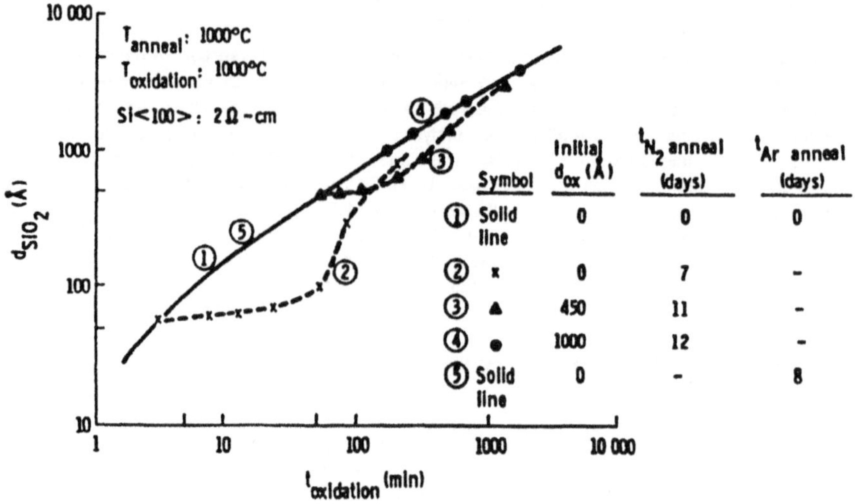

Figure 4. The dependence of SiO_2 film thickness on oxidation barrier formation by annealing in N_2 gas at 1000°C for 7 to 12 days (31). The solid line is the non-annealed control. Curve 2 forms an oxidation barrier which delays normal oxidation 1 to 2 hours.

D. DEVIATIONS FROM LINEAR, THIN SiO_2 GROWTH

From the nitridation studies, it is apparent that the oxidation rate is very sensitive to the availability of the intermediate SiO*. The SiO* forms SiO_2 according to reaction (3a), Figure 1, with the O_2 available at the interface. SiO forms by reaction (3b) at a much higher rate than SiO_2 is formed by reaction (3a) (40). The SiO concentration at the interface affects SiO* formation at the Si surface. Oxide growth is limited by the rate at which molecular SiO forms and escapes from the interface.

SiO transport from the interface rapidly increases with decreasing oxide thickness for oxide films $\leq$ 30 nm thick. Two consequences of the enhanced SiO transport for oxide films $\leq$ 30 nm occur. Oxide growth rates increase and SiO accumulation in the oxide layer increases. Transport of SiO from the interface also reduces interfacial stress that arises from a volume expansion that occurs during oxidation of Si to SiO_2.

The SiO transport rate from the interface through the SiO_2 film and the rate of formation of SiO* by reaction (2) in Figure 1 both increase with temperature. The influence that SiO transport has on the SiO* concentration at the interface plays in affecting the oxidation rate decreases at higher temperatures and the oxide growth rate is limited primarily by the rate of the linear, parallel reaction path. An inverse effect is observed as temperature decreases. The concentration of SiO* at the interface is affected by SiO transport and by the rate of SiO* formation.

The pressure dependence is directly related to the tendency to form SiO at a Si surface at reduced O_2 pressure. Since the O_2 concentration at the interface is low, SiO* is consumed by the faster reaction (3b) in which SiO is formed. The SiO* concentration for SiO_2 growth is reduced. The SiO_2 growth process is more limited by formation of the intermediate, SiO*, as O_2 ambient pressure decreases. The observed change with O_2 pressure from a linear to a square root dependence (4) is related to formation of the SiO* intermediate either by a) O_2 dissociation (12, 13) $O_2 \rightarrow 2\,(O)$, or b) by the reaction, $Si + 1/2\ O_2 \rightarrow SiO^*$. Three consequences of the reduced O_2 pressure are observed. As noted above, the SiO_2 growth rate is no longer first order with O_2 pressure. Excess SiO accumulates in the oxide layer. The formation of SiO* at the interface by reaction (2), plays a more dominant role at lower O_2 pressures than the parallel, linear oxide growth process.

The thickness and pressure dependence of the ^{18}O isotope accumulated at the outer SiO_2 surface are directly related to the quantity of $Si^{18}O$ formed at the interface and transported through the oxide. The SiO concentration at the outer oxide surface increases with decreasing oxide thickness and with decreasing oxide pressure. Observation of ^{18}O at the outer oxide surface is related either to a) formation of $Si^{18}O$ at the interface and transport to the oxide surface or to b) the formation of $Si^{18}O$ at the outer oxide surface by reaction or exchange between $^{18}O_2$ from the ambient and the transported Si species. The conclusion that Si is not transported through an SiO_2 film during oxidation to SiO_2, which is based on sequential tracer oxidations (7), must be reexamined in light of the thickness dependence of SiO transport (20).

As temperature or pressure decreases, the importance of the SiO* formation process limiting thin oxide growth rate increases relative to the linear process. From void formation studies, SiO growth appears to occur laterally along a crystalline Si surface. This has a different orientation dependence than a linear process, which presumably grows normal to the Si surface. As the relative contribution of the two reactions limiting oxide growth reverses, a crossover is observed in the orientation dependence. This provides a simple basis for the crystalline substrate orientation dependent crossover. Stress, steric hindrance, etc., remain as parameters that can further affect the relative oxidation rates on the different crystalline surfaces.

E. SUMMARY

A simple, chemical model for thin oxide growth at a Si-SiO_2 interface is derived by identifying one of two rate-limiting interfacial processes as the formation of SiO* at the Si-SiO_2 interface. The SiO* intermediate either forms SiO or reacts with O_2 to form SiO_2. SiO formation and transport from the interface affects the interfacial concentration of SiO*. Mechanisms for deviations from a linear model, associated with the SiO transport process, provide a unifying basis for understanding deviations due to oxide thickness, oxidation temperature, or O_2 ambient pressure, as well as for orientation-crossover and oxygen tracer results. Experimental data are presented showing that SiO transport from the Si-SiO_2 is strongly thickness dependent for oxide films $\leq$30 nm thick. In addition, oxidation barriers formed with low nitrogen concentrations are attributed to blocking the formation of the oxidation intermediate, SiO*, rather than forming a continuous nitride barrier. Blocking sites at which SiO forms inhibits the subsequent oxidation process.

ACKNOWLEDGEMENTS

The author wishes to thank C.C. Chi, N. Chou, G. Das, W.J. Gallagher, R. Ghez, T. Hickmott, F. Stern, and R. Walkup for many helpful discussions. The assistance of P. Fryer, J. Tornello, and R. McIntosh is greatly appreciated.

REFERENCES

1. B.E. Deal and A.S. Grove, J. Appl. Phys., 36, 3770 (1965).
2. S.I. Raider, Bull. Amer. Phys. Soc., 32; Electrochem. Soc. Ext. Abstr., 87-1, Abstr. 243; to be published.
3. J.R. Ligenza, J. Phys. Chem., 65, 2011 (1961).
4. Y.J. van der Meulen, J. Electrochem. Soc., 119, 530 (1972).
5. A. Reisman, E.H. Nicollian, C.K. Williams and C.J. Merz, J. Electr. Matls. 16, 45 (1987).
6. D.W. Hess and B.E. Deal, J. Electrochem. Soc., 122, 579 (1975).
7. E. Rosencher, A. Straboni, S. Rigo, and G. Amsel, Appl. Phys. Lett., 34, 254 (1979).
8. F. Rochet, B. Agius, and S. Rigo, J. Electrochem. Soc., 131, 1914 (1984).
9. S.I. Raider and L.E. Forget, J. Electrochem. Soc., 127, 1783 (1980).
10. Y. Kamigaki and Y. Itoh, J. Appl. Phys., 48, 2891 (1977).
11. H.Z. Massoud, J. Plummer, and E.A. Irene, J. Electrochem. Soc., 123, 1745, (1985).
12. R. Ghez and Y.J van der Meulen, J. Electrochem. Soc., 119, 1100 (1972).
13. J. Blanc, Appl. Phys. Lett., 33, 424 (1978).

14. M.A. Hopper, R.A. Clarke and L. Young, J. Electrochem. Soc., 122, 1216 (1975).
15. E.A. Irene, E. Tierney, and J. Angillelo, J. Electrochem. Soc., 129, 1253 (1986).
16. C.-J. Han and C.R. Helms, J. Electrochem. Soc., 134, 1297 (1987).
17. S.I. Raider and R. Flitsch, J. Electrochem. Soc., 123, 174 (1976); S.I. Raider, to be published.
18. J.M. Gibson and D.W. Dong, J. Electrochem. Soc., 127, 2722 (1980).
19. A.G. Revesz, B.J. Mrstic, H.L. Hughes and D. McCarthy, J. Electrochem.Soc., 133, 586 (1986).
20. S.I. Raider, to be published.
21. E.P. EerNisse, Appl. Phys. Lett., 35, 8 (1978).
22. M. Liehr, G.B. Bronner, and J.E. Lewis, Appl. Phys. Lett., to be published.
23. M. Liehr, J.E. Lewis, and G.W. Rubloff, J. Vac. Soc. Technol., 1986.
24. R.M. Bradley, J. Appl. Phys., 61, 545 (1987).
25. F. Montillo and P. Balk, J. Electrochem. Soc., 118,1463 (1971).
26. K. Hofmann, D.R. Young, and G.W. Rubloff, to be published.
27. R.E. Walkup and S.I. Raider, J. Appl. Phys., to be published.
28. S.I. Raider and R.E. Walkup, Materials Research Society, S. Pantelides and G. Lucovsky, Eds. Proc. Vol., Boston, MA, 11/87.
29. R. E. Walkup, Ph. Avouris, R. W. Dreyfus, J. M. Jasinski, and G. S. Selwyn, Appl. Phys. Lett., 45, 372 (1984).
30. S.-S. Lin, J. Amer. Cer. Soc., 58, 271 (1975).
31. S.I. Raider, R.A. Gdula, and J.R. Petrak, Appl. Phys. Lett., 27, 150 (1975).
32. W.J.M.J. Josquin and Y. Tamminga, J. Electrochem. Soc., 129, 1803 (1982).
33. K. Schott, K.C. Hofmann, and M. Schulz, Appl. Phys. A45, 73 (1988).
34. J.K. Howard, R. Flitsch, and S.I. Raider, J. Vac. Sci. Technol., 14, 69 (1977).
35. S.I. Raider, R. Flitsch, J.A. Aboaf, and W.A. Pliskin, J. Electrochem. Soc., 123, 560 (1976).
36. T. Ito, S. Hijiya, T.Nozaki, H. Arakawa, M. Shinoda, and Y. Fukukawa, J. Electrochem. Soc., 125, 448 (1978).
37. R.G. Frieser, J. Electrochem. Soc., 115, 1092 (1968).
38. R.M. Barrer, J. Chem. Soc., 136, 378 (1934).
39. J. Johnson and R. Burt, J. Opt. Soc. Am., 6, 734 (1922).
40. E.A. Gulbransen, K.F. Andrew and F.A. Brassart, J.Electrochem. Soc., 113, 834 (1966).

UNCERTAINTY ANALYSIS OF ANALYTIC OXIDATION MODELS

Zhi-Min Ling, Luc H. Dupas and Kristin M.De Meyer*

IMEC v.z.w.
Kapeldreef 75, B-3030 Leuven, Heverlee, Belgium

Abstract

The purpose of this paper is to indicate the parameters uncertainty in most of the proposed analytical oxidation models [1]-[6]. At first, the ε -indifference range, i.e. $|R(\vec{\theta}) - R(\vec{\theta}^*)| \leq \varepsilon$, is defined. Then the determination of the parameters uncertainty is converted into the determination of ε -indifference range. The methods are introduced to determine this ε -indifference range, and afterwards also the parameter uncertainty. The correlation matrices of the parameters for each model are given in this paper. The results of this paper show that most of the analytical oxidation model's parameters are interdependent within a widely used data fitting and parameter extraction range. Most of the extracted parameters for these proposed oxidation models [1]-[6] show a quite large uncertainty. Therefore, some physical mechanisms of oxidation which were concluded from or proven by the data fitting only could be questioned and debated.

1 Introduction

To improve the modelling accuracy of the classical Deal-Grove model [1] in the thin oxide regime (up to 300Å), many Si oxidation models with [2]-[6] or without analytical expression have been proposed. Most of the authors assert that their proposed mechanism is a predominant factor in the oxidation and relegate the others to a higher order or even negligible effect. It is possible however that several physical phenomena e.g. stress and viscous flow [2][3], internal field [4], parallel oxidizing species [6], structural channels [5], ... do play a role during the oxide growth. However, the question is what the predominant one in the oxide growth is, and which one(s) has to be included in the oxidation modelling, especially in the early stages.

It is rather difficult, at this moment, to demonstrate directly and experimentally which factor e.g. stress [2][3] or structural channels [5] is predominant for the oxidation mechanism. Therefore, the 'goodness' of data fitting and extracted parameters plays an important role in the oxidation modelling. Unfortunately, all of these models [2]-[6] have a good data fitting and sound reasonable parameter dependence. However, the question of "how much can the extracted parameters be changed without altering the fit quality significantly?" or "how large is the uncertainty in these extracted parameters, and therefore in the model itself?", is frequently ignored in most of the oxidation modelling papers.

---

*Professor at the Katholieke universiteit, Leuven. Research Associate of the Belgian National Found for Scientific Research

In contrast, it was indicated in our previous paper [7] that the parameters of the Deal-Grove model are quite insensitive which causes a serious problem of ambiguity in the oxidation kinetics study. Moreover, in this paper, we indicate that because of the non-negligible correlation between the model parameters, a significant uncertainty exists for most of the proposed analytical oxidation models [2]-[6]. At first, in sec.2, the techniques used to determine the parameter uncertainty are introduced. By using these techniques as such and after some modifications, analysis results related to the different proposed analytical oxidation models [2]-[6] were obtained and included in sec.3. Finally, in sec.4, some proposed and well known oxidation models in literature, which are basically supported and/or rejected by data fitting and parameter extraction only, are discussed, and critically questioned.

2 Determination of Parameter Uncertainty

Because of the imprecise nature of measurements we can never determine the real values with absolute certainty. Also, due to the random nature of the error in measurements, the parameter value $\vec{\theta}^*$ which is the best fitting of one series of measurements differs from the parameter value that fits another series, even though both series are made using the same equipment and experimental conditions. In our previous paper [7] we already indicated that there is no absolute reason to prefer $\vec{\theta}^*$ over any other value of $\vec{\theta}$ for which $\vec{\theta} \neq \vec{\theta}^*$ both satisfying the equation,

$$|R(\vec{\theta}) - R(\vec{\theta}^*)| \leq \varepsilon \tag{1}$$

where, ε is defined as the experimental error. For all of these parameter vectors $\vec{\theta}$ with a residual $R(\vec{\theta})$ less than the experimental error ε are acceptable. These acceptable parameter vectors form a ε -indifference region in $\vec{\theta}$ space (parameter space). The ε -indifference region is bounded by a n-1 hypersurface whose equation is,

$$R(\vec{\theta}) = R(\vec{\theta}^*) + \varepsilon \tag{2}$$

Direct method: Suppose we know the 'best fitting' parameter vector $\vec{\theta}^*$, according to eq.2, by changing the different values $\vec{\theta}$ surrounding the minimum point i.e. vector $\vec{\theta}^*$, the residuals related to these changed $\vec{\theta}$ are determined. Fig 1 shows contour lines of equi-fiterror ranges in the (B/A,B) plane for the classical Deal-Grove model. The data set is obtained for dry oxides grown at T=950°C and oxide thickness in the range from 200Å to 1000Å.

Ellipsoid approximation: It can be seen from fig.2 that, for a given ε which is very close to the minimum, the ε -indifference region can be approximated by an ellipsoid. The length of the ellipsoid axes reflect the parameters uncertainty. A model with n parameters will form a n-dimensional ellipsoid with n axes. In a sufficiently small neighborhood of $\vec{\theta}^*$, the objective function R($\vec{\theta}$) i.e. residual could be approximated by a Taylor series expansion with first two terms.

$$R(\vec{\theta}) = R(\vec{\theta}^*) + \mathrm{q}^{*T}\delta\vec{\theta} + \frac{1}{2}\delta\vec{\theta}^T\mathrm{H}^*\delta\vec{\theta} \tag{3}$$

where, $\delta\vec{\theta} \equiv \vec{\theta} - \vec{\theta}^*$, q* and $\mathbf{H}^*$ are, respectively, the gradient and Hessian of R($\vec{\theta}$) at $\vec{\theta}$ = $\vec{\theta}^*$. Since $\vec{\theta}^*$ is the optimum of R($\vec{\theta}$), then q* =0.

It is proven [8] that the eigenvalues of the $\mathbf{H}$ matrix are related to the length of the elliposid axes, and as such related to the parameter uncertainty. The correlation coefficients $\rho_{i,j}$ between parameters $\vec{\theta}_i$ and $\vec{\theta}_j$ can be obtained by:

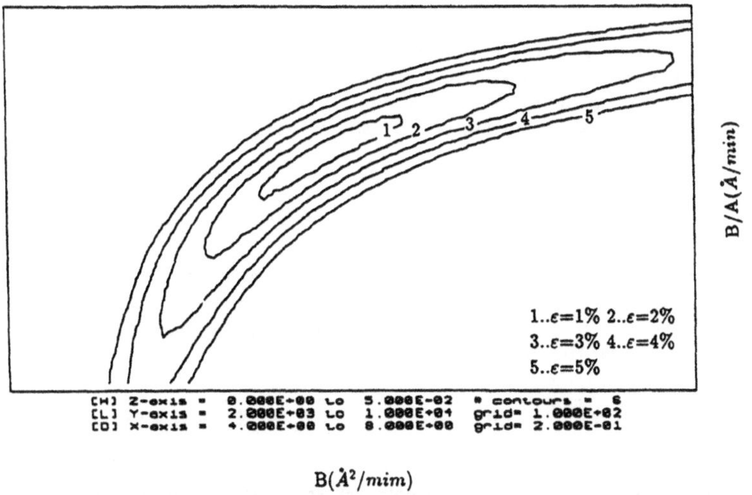

fig.1 ε-indifference range of the Deal-Grove model. The x-axis and y-axis are B/A and B respectivily

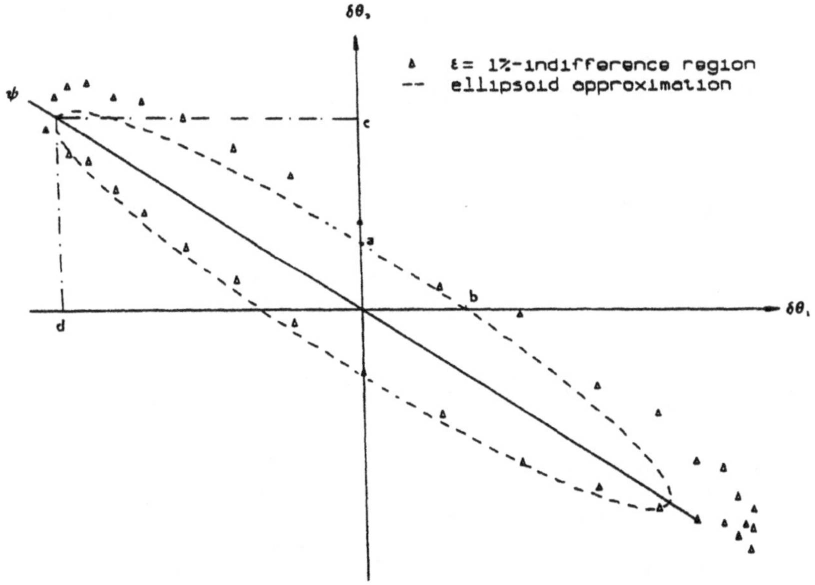

fig.2 The ε = 1%-indifference range in fig.1 is approximated by an ellipsoid

$$\rho_{i,j} \equiv \frac{V_{i,j}}{\sqrt{V_{i,i}V_{j,j}}} \quad V \approx 2\varepsilon H^{*-1} \tag{4}$$

Therefore, the study of the parameter uncertainty leads to the searching of the ε - indifference region defined by eq.2, then, to the determination of the length of the ellipsoid axes, and finally to the eigenvalues calculation of the $\mathbf{H}$ matrix. The interdependence of the parameters, which plays an important role in the parameter uncertainty analysis, is determined by the correlation matrix.

3 Parameter Uncertainty of Analytical Oxidation Model

Massoud et al [9] proposed a model composed of a classic Deal-Grove model with two extra exponential terms. Without introducing a significant error for X> $50\mathring{A}$ one of the exponential terms can be omitted. In this case the model reduces to a four parameters problem and can be written as below:

$$X = \sqrt{(A/2)^2 + Bt + K\tau(1 - e^{-t/\tau}) + X_n^2 + AX_n} - A/2 \tag{5}$$

Massoud's dry (100) Si oxidation (X,t) data set at T=1000°C [9] was used as an example. The oxide fitting range is from $52\mathring{A}$ to $680\mathring{A}$. The best fitting parameters $\vec{\theta}^*$ are extracted by the SIMPAR package [10]. By using the ellipsoid approximation approach, the eigenvalues of the $\mathbf{H}$ matrix are:

λ_1=1.5393 x 10^{-13} λ_2=8.2282 x 10^{-11} λ_3=1.0610 x 10^{-4} λ_4=8.9184 x 10^{-3}

correlation matrix:

	B	K	L	B/A
B	1.0000			
K	0.99362	1.0000		
L	0.76452	0.72054	1.0000	
B/A	-0.99026	-0.98577	-0.82327	1.0000

The smallest eigenvalue λ_{min} is related to the largest parameter variation. $|\psi|_{max} = \sqrt{2\varepsilon/\lambda_{min}}$, then the largest variation of each parameter $\delta\vec{\theta}$ can be obtained by the definition of $\Psi = U^T\delta\vec{\theta}$. Here, $\mathbf{U}$ is the unitary matrix whose colums are normalized eigenvectors of $\mathbf{H}$. For a given ε =1.%, the largest parameter variation region, i.e. $\vec{\theta} = \vec{\theta}^* \pm \delta\vec{\theta}$ can be expressed by:

$$B^* \pm \Delta B = 1.84 \times 10^4 \pm 1.487 \times 10^4 \quad K^* \pm \Delta K = 2.714 \times 10^4 \pm 3.48 \times 10^4$$
$$\tau^* \pm \Delta\tau = 5.24 \pm 1.933 \quad B/A^* \pm \Delta B/A = 10.2 \mp 3.988$$

However, this technique is valid only as ε is very close to the minimum point, otherwise, a larger error will be introduced i.e. out of the 'real' non elliptical ε -indifference region. If it happens, modification for some parameter variation ranges defined by the ellipsoid technique is necessary. In the above example, the largest parameter variation ranges are modified as ε is clamped at 1%. They are: B=(3.327~1.13)x10^4, B/A=8.09~14.2, K=(6.195~1.24)x10^4 and τ=(7.17~3.31), the fitting error is about 1% respectively.

By the same approach and for the same (X,t) data set used above, we are going to analyse other well known proposed oxidation models [2]-[6]. For each model, the eigenvalues, correlation matrix and largest parameter variation range have been obtained. However, in this paper, we only list the modified parameter variation range as ε is clamped at 1% .

Parallel Oxidizing Species: Based on the parallel oxidizing species, C.J.Han et al proposed an analytical model [6] with three parameters. The parameter uncertainty range determined by the direct method is plotted in fig.3. By using the ellipsoid technique and modification, the parameter uncertainty range is defined as A_2=(1.22~18.45)x10^3, B_1=(1.26~2.347)x10^3, B_2=(1.122~7.545)x10^4, and the integration constant t_o=(-0.488~0.068). By using these two modified parameter sets in the model, the fitting error is 1.8% and 1.4% respectively.

Structural channels: Based on the structural channels assumption, A.G.Revesz et al proposed an analytical oxidation model [3]. The parameter uncertainty range defined by the direct method is plotted in fig.4. By using the ellipsoid technique and modification, the parameter uncertainty range is defined as K_s=(377. ~33.2), K_t=(1.045~2.316)x10^3 K_o=(4.82~3.14) and the integration constant t_o=(-0.197~-1.18), the fitting error is 1.7% and 1.5%, respectively.

Internal electric field: This model, taking into account the influence of an internal electric field which originates from the presence of an exponential distribution of total net charges during oxidation, was proposed by Y.Z.Lu et al [4]. Using the ellipsoid technique, the parameter uncertainty range is defined as K_l=(14.0~8.11), K_p=(1.168~3.518)x10^4, L=(45.4~66.1) and Q_f=(4.39~4.9)x10^{11}. If the integration constant t_o is chosen as -0.942 and -0.862, respectively, the fitting error is 1.1% and 0.6%.

Stress relaxation: Another analytical model including stress relaxation by viscous flow during the oxide growth was proposed by A.Fargeix et al [2]. In this case, as another example, we do not use the Massoud data set which is used in the above four models, but instead an Irene data set is used, namely Si (100) dry oxidation experiments at T=980°C from Irene's paper [11], the oxide thickness is from 100Å to 1700Å. By using our ellipsoid technique and modification, the parameter uncertainty is defined as K_l=(8.04~8.40)Å/min, K_p=(1.368~2.994)x10^4Å^2/min, η=1.0x10^{14} ~5.0x10^{16} poise, α=0.01~0.58 $<r>$=7.0 and the integration constant t_o =-7.9~-7.2 min. The fitting errors to these Irene data are 1.9% and 1.8% respectively.

4 Discussion and Conclusion

It is clear that, the extracted parameter values in most of the oxidation models are prone to a significant uncertainty. However, to estimate the parameter uncertainty, one possibility which is used in most cases is to keep all parameters fixed except one and see how much deviation there is. Using this approach will produce the parameter uncertainty values at point **a** and **b** shown in fig.2. This approach is valid in case the model parameters are independent, which will result more or less in circular ε -indifference ranges. However, the parameters of the analytical oxidation models are very interdependent, and one has to change the parameters simultaneously across the ellipsoid axes, so much larger variations in parameter values have to be considered like obtained in points c and **d** in fig.2. Therefore, the key point in the parameter uncertainty analysis of this work, is the understanding of the parameter correlation i.e. interdependence.

It can be expected that by increasing the data fitting range a parameter set with a somewhat smaller uncertainty can be extracted. As an example we used the Irene data set for (100) Si dry oxidation at T=893°C [11] with the oxide thickness from 45Å to 1925Å. But even within such a quite wide thickness range the uncertainty of B and B/A is still about ±50%, of K and τ about ±20%. The correlation coefficient between B/A and B is -0.96783, between K and τ 0.58441.

The model based on the microstructure channels was proposed by Revesz et al [5]. The oxygen pressure dependence of the parameters was determined by regression analysis of Hopper et al data. The results indicate that K_s and K_o are proportional to $P^{0.6}$, whereas K_t depends linearly on oxygen pressure. The relationship between the oxygen transport in

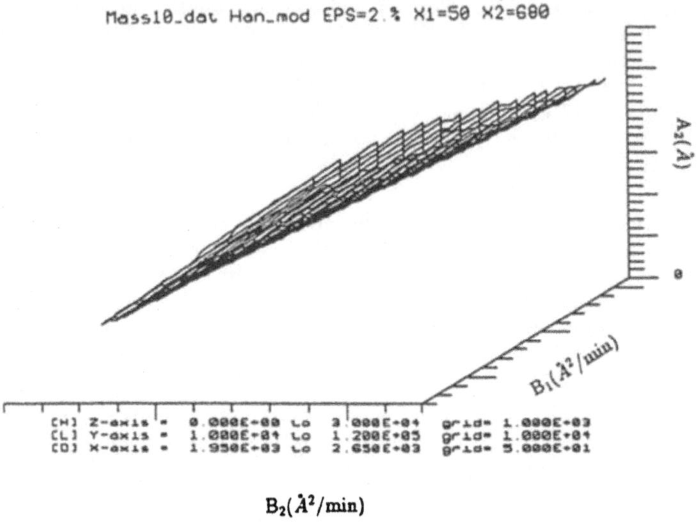

fig.3 3-D ε-indifference region plot of Han's model i.e. eq.12, Here, ε=2% and determined by direct method as described in sec.3

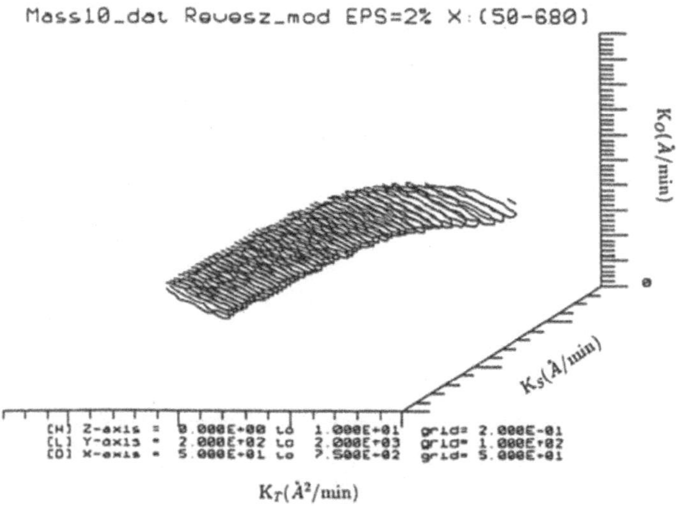

fig.4 3-D ε-indifference region plot of Revesz's model i.e. eq.13. Here, ε=2% by direct method as described in sec.3

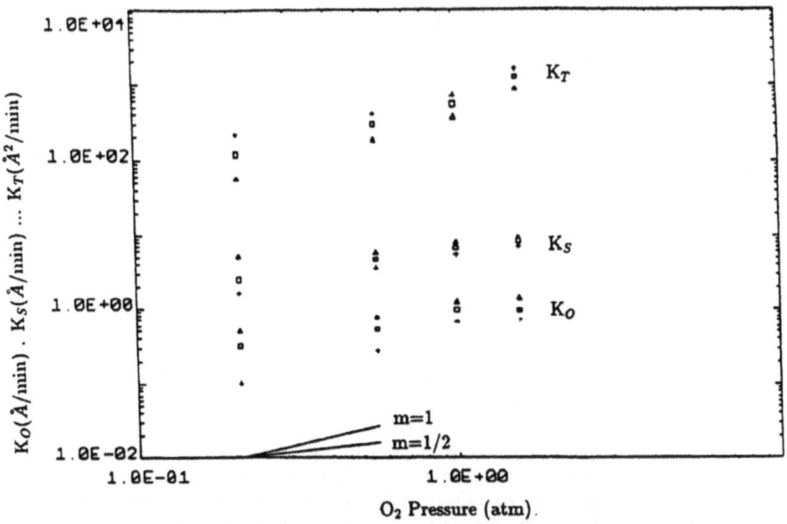

fig.5 Comparison of the oxygen pressure dependence of K_s, K_o and K_t in Revesz's model i.e. eq.13. The original parameter set (square) was from Ref.[5], the farest away point (cross and triangle) related to Ref.[5] parameters (square) determined by the ellipsoid technique as described in sec.3

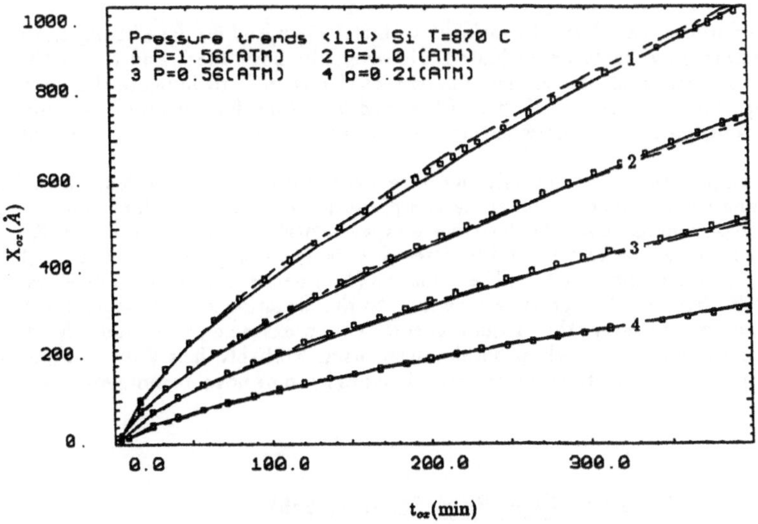

fig.6 Comparison of the oxidation data set on the different oxygen pressure. The original data are regenerated from the Revesz's model with their best fitting parameter set [5] i.e. square marks in fig.6. The fitting curves are calculated by Revesz's model with the parameter sets listed in in table 1 which are defined by ellipsoid technique as described in sec.3

the channels (K_o) and the reaction (K_s) at the interface is a strange phenomenon, and one seems to be encouraged to look for a (new) physical mechanism to explain this [5]. However, as one understands the parameter uncertainty property of this model, the above problem can be answered easily. The original data sets used by Revesz et al in the pressure dependence studies [5], i.e. Si (111) dry oxidation at T=870°C , with oxygen pressures as 0.21, 0.56, 1.0, 1.56 atm., are regenerated through their model with the parameters and the thickness range indicated in their paper [5]. The maximum parameter variation range i.e. the outer points in the Revesz's parameter set are obtained by the technique introduced in sec.3. These possible parameter variations are plotted in fig.5. Since the difference of the oxide thickness predicted by these different parameter sets and the Revesz's model is less than 2%, which are plotted in fig.6, it is difficult to distinguish, in the view of the data fitting, which parameter set is the 'real' one, and conclude from that the 'real' pressure dependence.

The model, based on the Deal-Grove process and a Maxwellian stress relaxation approach, was proposed by Fargeix et al [2]. However, it has been assessed [12] that the viscoelastic stress relief models required unrealistically low viscosity values, and have a number of conceptual difficulties. In this paper, we are not going to debate about the conceptual difficulties of the viscoelastic stress relief models. However, we don't think the requirement of the unrealistically low viscosity in this model [2] is an evidence to reject the viscoelastic stress relief models, since the viscosity extracted by this model [2] is quite unreliable. As shown in sec.3, for the Irene data set, the extracted viscosity η, in the first parameter set, is about 1.0×10^{14} poise which is close to the value in Fargeix's paper [2] and was criticized in Ref.[12] due to this low viscosity value extracted. However, the extracted viscosity value η, in the second parameter set, is 5.0×10^{16} which is close to the viscosity value of the dry SiO_2 suggested in Ref.[12]. Obviously, for the same experimental data set, the extracted viscosity η could be 1.0×10^{14} or 5.0×10^{16} poise. Since both of them cause a small fitting error, it is imposible to determine the 'real' viscosity values associated with this model by the data fitting only. Therefore, any assessment on the viscoelastic stress relief models which is based only on the extracted viscosity values is not convincing.

More examples, i.e. physical conclusions which were based on data fitting and parameter extraction only, could be evaluated. However, the basic ideas and the methods are the same as these used above. The purpose of this work is to indicate the uncertainty of the extracted parameters and thus of the models and mechanisms proposed and based on the data fitting and parameter extraction approach, rather than to reject some of them.

Since the parameters are strongly interdependent, it is impossible, at least not efficient, to determine the parameters separately and precisely only from the (X,t) data fitting. If some of the parameters could be determined experimentally independent on the (X,t) data measurement, it will be very helpful to determine the remaining parameters by (X,t) data fitting with a smaller uncertainty. The oxidation (X,t) data measurements by ellipsometry in situ [3][13] have made a great contribution to the oxidation kinetics studies in the last ten years or so. However, this is quite unsufficient to examine i.e. exclude the different possibilities to reveal the 'real' oxidation mechanism. Probably, it is a time to encourage more and more efforts to study the oxidation kinetics by some novel techniques, even in situ.

References

1. B.E.Deal and A.S.Grove, J.Appl.Phys., 36, 3770(1965).

2. A.Fargeix and G.Ghibaudo, J.Appl.Phys., 54(12), 7153(1983).

3. H.Z.Massoud, J.D.Plummer and E.A.Irene, J.Electrochem. Soc., 132, 1745(1985).

4. Y.Z.Lu and Y.C.Cheng, J.Appl.Phys., 56(6), 1608(1986).

5. A.G.Revesz, B.J.Mrstik, H.L.Hugher, and D.McCarthy, J.Electrochem. Soc., 133, 586(1986).

6. C.J.Han and C.R.Helms, J.Electrochem. Soc., 134, 1297(1987).

7. Z.M.Ling, L.Dupas and K.DeMeyer, this conference

8. Y.Bard, "Nonlinear Parameter Estimation", New York and London, 1974.

9. H.Z.Massoud and J.D.Plummer, J.Appl.Phys., 62(8), 3416(1987).

10. W.Maes, K.M.DeMeyer and L.H.Dupas, IEEE Trans. on CAD, CAD-5(2), 320,(1986).

11. E.A.Irene and Y.J.vanderMeulen, J.Electrochem.Soc., 123, 1380(1976).

12. B.J.Mrstik, A.G.Revesz, M.Ancona and H.L.Hughes, J.Electrochem.Soc., 134, 2020(1987).

13. M.A.Hopper, R.A.Clarke, and L.Young, ibid, 122, 1216 (1975).

MODELING OF THE OXIDE GROWTH IN A CHLORINE AMBIENT

Zhi-Min Ling, Luc H. Dupas and Kristin M.De Meyer*

IMEC v.z.w.
Kapeldreef 75, B-3030 Leuven, Heverlee, Belgium

Abstract

The kinetics of the oxide growth in TCA/O_2 a ambient is investigated. The oxide growth rate is related to the behaviour of the Cl concentration (as measured by SIMS) near the SiO_2/Si interface. A double layer structure for the SiO_2 was assumed in this work. One layer is the bulk oxide, the other one is a Cl peak containing layer near the SiO_2/Si interface (about 200Å thick). Because of the large amount of the nonbridging bonds in this Cl peak layer, the oxide species diffusivity enhancement is not the same in these two layers. Also because of the large amount of unstable reaction products and the catalytical action of the incorporated Cl in this Cl peak layer, the oxide reaction process takes place not only at the SiO_2/Si interface itself, but also near the SiO_2/Si interface within this Cl peak layer. It is observed that the Cl peak layer moves together with the SiO_2/Si interface but the width of this layer is nearly constant (about 200Å). On the other hand the total amount of incorporated Cl concentration increases during the oxide growth in chlorine ambient. Obviously, the effect of the Cl peak layer on the diffusion and the reaction fluxes should be thickness dependent. For the experimental data the plots of dt/dx *vs.* x also show a tendency to bend upwards for increasing x in certain growth conditions. Based on the above considerations, the behaviour of the Cl concentration near the SiO_2/Si interface seems to be responsible for this unusual growth rate. Furthermore, a model for the oxide growth in a chlorine ambient is proposed in this work. The experimental data are fitted by this model and the model parameters are extracted.

1 Introduction

The use of a chlorine additive during the growth of the gate oxide of a MOS device is a well established technique [1]. The characteristics of the oxide layer resulting from this technique are discussed in [2]. It was reported also that the oxide growth rate was enhanced by the chlorine additive in the gas phase [1][2][3]. However, the kinetics of the oxide growth in chlorine ambients is still unknown. Some studies emphasized the observation and discussion of the relation between the growth rate enhancement and the chlorine concentration in the gas phase [4]-[6]. It is believed that the chlorine species is decomposed during the oxidation, and diffuses through the oxide and, at the same time, it affects the oxide growth rate. Previous models and earlier studies apparently are merely descriptions and do not consider the effect of the incorporated Cl on the physical characteristics of the oxide layer and its growth rate.

---

*Professor at the Katholieke universiteit, Leuven. Research Associate of the Belgian National Found for Scientific Research

In this work, a model for the oxide growth in a TCA/O_2 ambient is proposed. First of all, in sec.2, the experimental results of the oxide growth in a chlorine ambient are summarized, the oxide growth rate is correlated with the incorporated chlorine which is measured by SIMS. In sec.3, the model is described qualitatively and derived mathematically. This model is based on the double layer structure due to the incorporated chlorine peak layer, and the catalytical effect of the chlorine in a two step reaction. Finally a model for the incorporated chlorine concentration at the SiO_2/Si interface is proposed in sec.4.

2 Experiments and Results

Sample Preparation: The oxide samples were grown in TCA/O_2 ambients at T=1000, 1025, 1050, 1075 and 1100°C, with TCA%=0.1, 0.2 and 0.3. Thicknesses ranged from 300$\AA$ to 2000$\AA$. The cleaned <100> wafers were pushed into the furnace in pure oxygen in 5 min to grow a native oxide to protect the gas phase chlorine etch effect on the bare Si surface. This pre-oxidation procedure is also carried out for safety reasons, since it is suggested that an excess of oxygen must be maintained to prevent phosgene formation in the lines or the tube [12]. The addition of chlorine to the oxidation ambient was carried out by sending a controlled amount of O_2 through a bubbler containing liquid TCA, i.e., $C_2H_3Cl_3$. The ellipsometer was used to measure the grown oxide thickness.

Oxide Growth Rate: The inverse differential of the experimental data (x, t) is a common way to investigate the thickness dependence of the oxide growth rate [13]. According to the Deal-Grove model [14], $dt/dx = 1/K_l + x/K_p$, the slope and the intercept of the dt/dx *vs.* x plot is related to the parabolic-rate constant K_p and the linear-rate constant K_l respectively. For dry oxidation, it was observed that the dt/dx *vs.* x plot is a straight line for X>300$\AA$. This means that the growth rate parameters K_l and K_p are independent of the oxide thickness. The dt/dx *vs.* x plots for our experimental (x,t) data for oxides grown in TCA/O_2 ambients were obtained by a conventional spline interpolation method. However, as shown in fig.1, these plots show an up-bending tendency. The slope (intercept) of the dt/dx *vs.* x plots increases (decreases) as oxide thickness increases. This seems to indicate that the diffusion flux of the oxidizing species decreases as oxide thickness increases more than expected from a $1/x$ relationship, or is increased more than expected for thinner oxides.

Chlorine Profile Measurement: To understand the relation between the oxide growth rate and the incorporated Cl in the oxide, the Cl profile in oxide was measured by SIMS using O_2^+ ions at 6 keV. Two different sample sets were used. The first set included oxides grown at $T = 1025°C$ with 0.3%TCA in the gas phase. The oxide thicknesses for these samples were 500$\AA$, 1100$\AA$, 1500$\AA$, 1700$\AA$ and 2000$\AA$ respectively. The second set was grown at T=1075°C with 0.2%TCA in the gas phase. The oxide thicknesses for these samples were 700$\AA$ and 2000$\AA$. The maximum Cl levels in the first set were about 10^{18} to 10^{19} cm^{-3}. As shown in fig.2, several conclusions can be drawn from the SIMS measurements of the first set: i) the incorporated Cl peak always exists at the SiO_2/Si interface, although the SiO_2/Si interface is moving during the oxide growth; ii) the FWHM (full width at half maximum) is 180$\AA$ to 220$\AA$ near the SiO_2/Si interface. iii) the concentration of incorporated Cl increases as the oxide thickness increases. These SIMS measurement results are comparable to those of ref.[7]-[11][15].

The measured Cl profiles were correlated with the oxide growth rate at the same oxide thickness. As shown in fig.2, the dt/dx *vs.* x curve could be approximated by two straight lines with a break point around X=1000$\AA$. It is interesting to see that the maximum Cl level increases much more slowly as the oxide thickness gets larger than 1000$\AA$.

3 Modelling of the Oxide Growth Rate Enhancement

Under thermal oxidation conditions, the TCA/O_2 system will be decomposed into Cl_2, HCl, O_2, and H_2O..., based on a thermodynamic analysis [1][16]. These active species diffuse through the oxide to react with the Si and to be incorporated into the network. It was suggested [10][11] that the incorporated Cl is unstable in the network and is replaced by diffused oxidizing species. The replaced free chlorine species will diffuse to the SiO_2/Si interface and will again be incorporated into the network. The chlorine species in the process seems to act as a catalyst. Meanwhile, the diffused oxidizing species react with Si directly to form SiO_2. As shown in fig.2, the incorporated Cl always shows a peak at the SiO_2/Si interface of about 20nm profile width. Moreover, the incorporated OH also seems to be increased near the SiO_2/Si interface [9][10]. To simplify, the total oxide layer can be subdivided into two layers: one (layer-1) is the bulk oxide, where only bonded Si-OH and Si-O exist, another (layer-2) is the Cl peak layer near the SiO_2/Si interface, where not only bonded O and OH but also bonded Cl are present. The schematic presentation of this two layer structure is shown in fig.3.

The diffusivity of oxidizing species in layer-1 is independent on the oxide thickness, but in layer-2 the diffusivity depends on the oxide thickness, because the incorporated Cl concentration increases towards the SiO_2/Si interface [9][10]. Therefore the effective diffusivity can be written as below:

$$\frac{1}{D_{eff}} = \frac{1}{X_1 + X_2}\left[\int_0^{X_1} \frac{dx}{D} + \int_{X_1} \frac{dx}{D(x)}\right] \tag{1}$$

Where, D is the diffusivity of the oxidizing species in layer-1. During the oxidation, the total oxide thickness (X_1+X_2) increases, whereas the profile width of the chlorine peak (layer-2) X_2 nearly stays the same. Hence, the effective diffusivity decreases because layer-2 contributes less to the total diffusivity enhancement. Since the actual relationship between the diffusivity enhancement and incorporated Cl concentration is unknown, a straightforward linear thickness dependence of the effective diffusivity is adopted.

On the other hand, the incorporated chlorine species are unstable and are replaced by the diffused oxidizing species, i.e. the Si-Cl network is transferred into the silicon oxide. Therefore, the formation of SiO_2 not only happens at the SiO_2/Si interface, but also within the clorine peak layer. During the oxide growth, the total incorporated Cl concentration increases [8][17], due to the catalytical effect of chlorine, so the contribution of the reaction enhancement from the Cl peak layer increases, *i.e.*, the apparent reaction rate increases. To simplify in the model derivation, a straightforward linear thickness dependence of the incorporated chlorine profile and diffusing oxidizing species profile are adopted. Therefore, the linear rate can be derived as below:

$$K_l(X_o) = K_{lo}\left[1 + \frac{(k_2/2k_3)L + (k_2L^2/6D)}{(1/C_f) - (k_2L^2/6D)}\right] \tag{2}$$

where L is the width of layer-2, which is supposed to be the value of FWHM (full width at half maximum), k_2, and k_3 are the reaction constants which are related to the reaction of the oxidizing species with Si-Cl and with Si; C_f is the concentration of the incorporated chlorine at the SiO_2/Si interface. It was suggested in Ref.[8][17] that the value of C_f shows a linear oxide thickness dependence. We adopted the expression $C_f = \lambda X_o$ in the following model simplification.

In order to keep the model as transparent as possible, a global oxide growth rate model, which is very complex from a mathematical point of view, is simplified as below:

$$\frac{dt}{dX} = \frac{X}{K_{po}}\{(1 - \frac{L}{X})[1 + \frac{1+\alpha}{\alpha}ln(\frac{1+\alpha\frac{X}{X-L}}{1+\alpha})]\} + [K_{lo}(1 + \frac{1}{1/(\eta X) - \xi})]^{-1} \tag{3}$$

Due to the lack of data for the physical constants (k_2, k_3, λ...) which are used in eq.(2), at this moment, the parameters η and ξ used in eq.(3) are just fitting parameters. α is also a fitting parameter which is related to the diffusivity enhancement from the chlorine peak layer.

This simplified model is used as input for the SIMPAR package [18] to fit our experimental data and extract the parameters. As shown in fig.1 (line), this model can fit our dt/dx $vs.$ x data set very well. The extracted parameters are listed below:

Temp($^\circ C$)(TCA%)	K_l(Å/min)	K_p(Å^2/min)	$\alpha \times 10^{-3}$	$\eta \times 10^{-4}$	$\xi \times 10^{-3}$
1000°C(0.2%)	6.72	5.7×10^3	230.	43.5	39.61
1025°C(0.2%)	15.34	1.22×10^4	102.	8.644	1.32
1025°C(0.3%)	16.5	1.30×10^4	199.	6.22	9.89
1050°C(0.1%)	17.0	1.46×10^4	2.91	8.101	372.
1075°C(0.2%)	35.1	2.80×10^4	1.20	0.312	1.23
1100°C(0.2%)	48.9	3.57×10^4	0.21	0.041	0.233

4 Modelling of The Interface Chlorine Concentration C_f

It is indicated in sec.3 that the thickness dependence of the parabolic and linear rate constant depends on the C_f value of the incorporated Cl at the SiO_2/Si interface. During the model derivation, the expression $C_f = \lambda X_o$ i.e. linear thickness dependence is adopted approximately In this section, we will present a model for the relationship between the incorporated chlorine concentration at the SiO_2/Si interface and the grown oxide thickness.

The basic ideas are the same as those described above. The replaced chlorine will diffuse to the SiO_2 Si interface to be incorporated again [11]. Therefore, the mobile chlorine species flux at the SiO_2/Si interface is composed of the diffused chlorine species from the gas phase and replaced chlorine species from the chlorine peak layer. The unstable incorporated chlorine can be taken as a generation center for the mobile chlorine species. We supposed that the maximum concentration of the incorporated chlorine species is located at the SiO_2/Si interface, and that all mobile chlorine species at the SiO_2/Si interface react with Si to be incorporated. The mobile chlorine species diffuse through the oxide layer fast enough [10], so that the surface reaction is the limiting step. Under steady state conditions, by using a linear profile for the mobile oxidizing species and a linear incorporated chlorine profile, the total amount of mobile chlorine at the SiO_2/Si interface can be deduced by solving the continuity equation. Under equilibrium conditions, at the SiO_2/Si interface, the concentration ratio of the mobile chlorine $[Cl]$ and the available silicon $[Si]$ to the incorporated chlorine $[Si\text{-}Cl]$ is a constant, Thus, the concentration of the incorporated chlorine at the SiO_2/Si interface C_f can than be obtained as below:

$$C_f = \frac{C_o^{Cl}}{[\frac{k_1}{D^{Cl}K} - \frac{k_2 L C_i}{2D^{Cl}}]X_o + [\frac{k_2 L 3(C_o - C_i)}{12 D^{Cl}}]/X_o + [\frac{1}{K} - \frac{k_2 L^2}{6D^{Cl}}(C_o - 2C_i)]} \quad (4)$$

All the symbols used here are the same as used in the above sections, and k_1 is the reaction constant of the reaction equation $Si + Cl = Si\text{-}Cl$, which takes place at the SiO_2/Si interface. D^{Cl} is the diffusivity of the mobile chlorine species in the oxide. C_o and C_i are the oxidizing species (mostly OH) concentration at the gas/SiO_2 interface and the SiO_2/Si interface. C_o^{Cl} is the concentration of the mobile chlorine at the gas/SiO_2 interface. K is a chemical equilibrium constant.

The linear thickness dependence of C_f is a rough approximation which is valid as long as $[B]/X_o >> ([A]X_o + [C])$ in eq.(4), $C_f \approx C_o^{Cl} X_o/[B]$. Because the lack of knowledge of the C_f value under the different conditions, in this work, we only present this

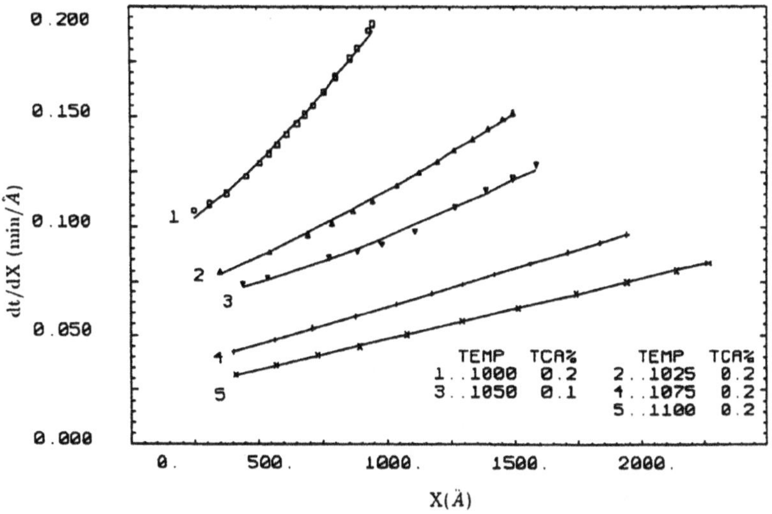

fig.1 dt/dx vs. x plots. These points are obtained by the spline interpolation method from our oxidation data grown in TCA/O_2 ambient. The lines are the simulation results from the model derived in this work eq.3

model as being a possible model and we are not going to extract and discuss in detail the model parameters. It should be mentioned that the Cl clustering, and out-diffusion of the replaced free Cl and some more unknown phenomena are not considered in this simplified model.

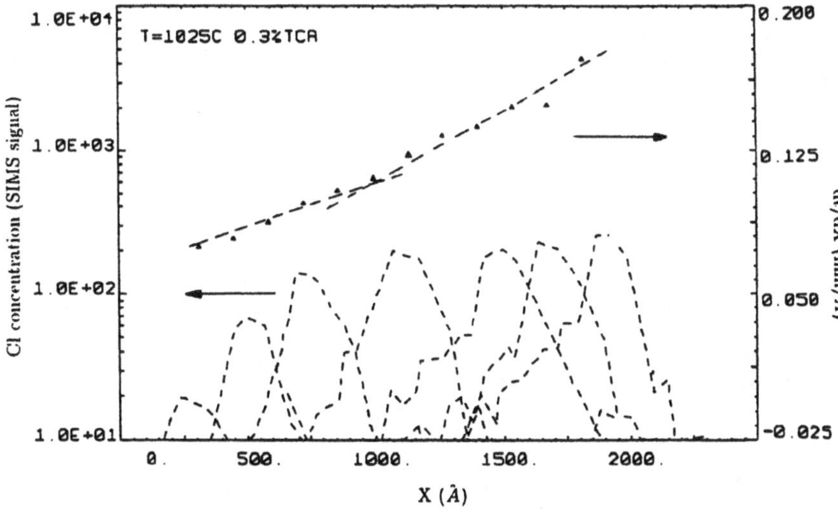

fig.2 Inverse oxide growth rate correlated with chlorine peak layer at different oxide thicknesses. These oxide samples were grown at $T = 1025°C$ and 0.3%TCA with different oxidation time (t=30min, 90min, 150min, 180min and 210min)

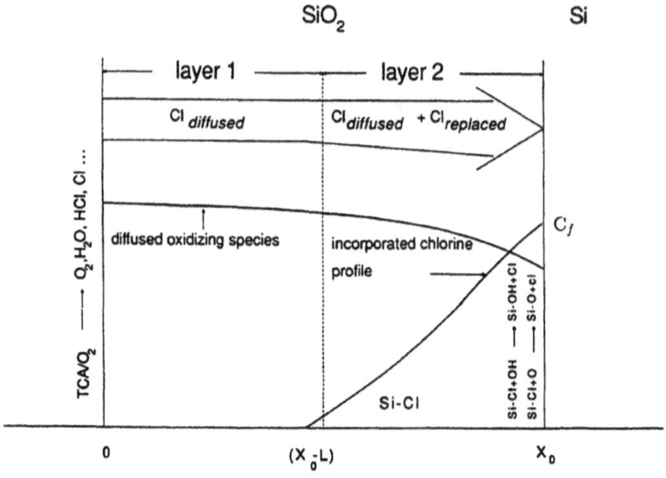

fig.3 Schematical representation of the double layer structure and the species diffusion flux

5 Discussion and Conclusion

As seen in fig.1 the dt/dx $vs.x$ behaviour is different at T>1050^oC. The plots of dt/dx $vs.x$ at T=1075^oC and T=1100^oC seem to be straight lines. This means that the diffusion enhancement from the chlorine peak layer is less when the oxidation temperature is higher. It is suggested that, for the oxide growth in a chlorine ambient, T=1050^oC is a critical temperature above which the growing oxide is essentially stress-free [19]. If the oxide network structure and diffusivity are correlated with the oxide stress, the temperature dependence of the parameter α is reasonable. It is indicated in our SIMS measurements (set 2) that, at T=1075^oC TCA%=0.2 the value of C_f increases slowly within the 700$\mathring{A} < X <$2000$\mathring{A}$ range. (The oxidation rate is too high to get data less than 500$\mathring{A}$ at these higher temperatures.) Therefore the reaction enhancement from the chlorine layer does not so strongly depend on the oxide thickness in this range. By the combination of these factors, it can be understood that the dt/dx $vs.x$ plots are nearly a straight line as T>1050^oC.

Since there are many simplifying assumptions used in the derivation of this data fitting model eq.(3), we are not going to discuss physically these extracted parameters. Moreover, this model is still a Deal-Grove-like model [20]. It was discussed in another [21] paper that the parameters of the Deal-Grove model are unsensitive to the simulation results. In other words, many other kinds of parameter combinations for this model could be extracted, with a fitting error smaller than the experimental error. The above extracted parameters are necessary but not sufficient to be used as a proof for the physical assumptions.

The understanding of the Cl incorporation process is a key point for the chlorine oxidation modeling. Once the parameters of this process are well known, this model will be a bridge between the oxidation conditions (*e.g.* temperature, TCA%,...) and the oxide growth rate enhancement. However, this model doesn't try to predict the thin oxide (X<300$\mathring{A}$) behaviour. The assumption of a double layer structure, which was adopted in this model, should be proven by some independent experiments. The relationship between the diffusivity enhancement and the concentration of the incorporated Cl and OH also needs further investigation.

References

1. E.Janssens and G.Declerck, J.Electrochem. Soc., 125, 1696(1978).
2. J.Monkowski, Solid State Technology, July 1979, pp.58
3. B.R.Singh and P.Balk, J.Electrochem. Soc., 126, 1288(1979).
4. K.Ehara, K.Sakuma and K.Ohwada, ibid, 126, 2249(1979).
5. K.Hirabayashi and J.Iwamura, ibid, 120, 1595(1973).
6. D.W.Hess and D.E.Deal, ibid, 124, 735(1977).
7. M.D.Monkowski, J.R.Monkowski, I.S.T.Tsong, J.Stach and R.E.Tressler, J.Non-Cryst.Solids, 49, 201(1982).
8. J.W.Rouse, C.R.Helms, B.E.Deal and R.R.Razank, J.Electrochem.Soc., 131, 887(1984)
9. W.A.Bosenberg, C.W.Magee and E.A.Botnick, ibid, 131, 2397(1984).
10. H.L.Tsai, S.R.Butter, D.B.Willians, H.W.Kraner and K.W.Janes, ibid, 131, 411(1984).
11. Y.D.Sheu, S.R.Butter and F.J.Feigl, ibid, 133, 2136(1986).
12. R.G.Cosway and Schyi-yi (C.E.) Wu, ibid, 132, 151(1985).
13. M.A.Hopper, R.A.Clark and L.Young, ibid, 122, 1216(1975).
14. B.E.Deal and A.S.Grove, J.Appl.Phys., 36, 3770(1965).
15. W.F.Banholzer and Ghezzo, J.Electrochem.Soc., 134, 415(1987).
16. R.E.Tressler, J.Stach and D.M.Metz, ibid, 124, 607(1977).
17. A.Rohatgi, S.R.Butter, F.J.Feigl, H.W.Kraner and K.W.Jones, J.Electrochem.Soc., 126, 143(1979).
18. W.Maes, K.M.DeMeyer and L.H.Dupas, IEEE Trans. on CAD, CAD-5 (2), 320(1986).
19. M.D.Monkowski, G.H.Fallouh and J.R.Monkowski, to be published (submitted to J.Appl.Phys)
20. J.Blanc, J.Electrochem.Soc., 133, 1982(1986).
21. Z.M.Ling, L.Dupas and K.DeMeyer, to be published in the J.Electrochem. Soc.

SILICON OXIDATION STUDIES: A REVIEW OF RECENT STUDIES ON THIN FILM SILICON DIOXIDE FORMATION

Eugene A. Irene

Department of Chemistry CB 3290
University of North Carolina
Chapel Hill, N.C. 27599-3290

ABSTRACT

The formation of thin SiO_2 films via thermal oxidation on single crystal Si substrates has been found to depend on the method of Si cleaning, impurities on the Si surface, the Si crystal orientation, film stress, and the availability of electrons at the Si surface. Recent studies on these topics are recounted along with a framework for understanding. No fully acceptable model for thin SiO_2 formation yet exists, but recent studies lead in new directions towards this goal.

INTRODUCTION

From studies of the Si oxidation process in the 1960's(1-3), has come the application of the Linear-Parabolic, LP, oxidation model of Evans(4) to the SiO_2 film growth kinetics. A goodly measure of understanding was afforded to the process as a result of the application of the LP model, but even from the earliest studies(1) it was recognized that the LP model did not account for the initial oxidation regime at the outset of the oxidation of Si and extending to several tens of nm. Furthermore, even prior to oxidation and after some initial cleaning, a native oxide was observed on the Si surface. This oxide was and still is difficult to study; it forms relatively instantly, somewhat independent of ambient conditions and extends to less than 2nm in thickness. In order to account for the mysterious initial regime which included the native oxide, an offset was built into the LP model(1). This entire offset regime extends up to about 30nm in SiO_2 film thickness, L, and for ease of discussion we herein use the label L_o for this regime, and similarly for the SiO_2 film thickness from 0nm to the native oxide thickness of less than 2 nm, we use the symbol L_n. The L_o regime was of little technological significance in the 1960's and early 1970's, but as device area was reduced, in order to increase device density on a chip, the SiO_2 thickness was also decreased. Now most advanced MOSFET's have gate oxides with L < 30nm, and recently IBM(5) showed an operating research device with L ≅ 5nm. Thus the L_o regime is no longer merely a scientific curiosity, and it becomes important to understand the thin film SiO_2 formation kinetics and the resultant thin film SiO_2 physical properties. This review is aimed at the present understanding of the L_o regime.

After the usual cleaning of a Si surface (by a procedure outlined in ref 6), the Si wafer undergoes an aqueous HF dip to remove any oxide that is formed during the cleaning process with concomitant impurities, and results in a hydrophobic surface(7) upon which a native oxide grows to about 1-2nm. Typical thermal oxidation experiments commence with the native oxide already formed, i.e. at oxidation time, t, and thickness, L : $t=0$ and $L=L_n$, and with oxidation temperatures above 800°C and usually near 1000°C, and in dry O_2 for the highest quality gate oxide or steam to accelerate oxide growth for field oxide applications. The rate of growth is observed to decrease as oxidation proceeds indicating that the growing oxide is providing a barrier to further oxidation. The SiO_2 film growth from L_n to L_o is characterized by faster oxidation rates than the LP regime which obtains above L_o. Before proceeding with a detailed discussion of recent studies on the two initial regimes: 0 to L_n and L_n to L_o, we briefly formulate the LP model for later use for comparison. Since this has been done many times previously(1-4,8), only the essential features are herein repeated. Firstly, we need consider the two essential processes that must occur during the thermal oxidation of Si. One process is the transport of oxidant across the growing oxide to the Si-SiO_2 interface where the oxidation reaction has been established to take place(9). This oxidant flux labeled F_1 is often approximated by a diffusive flux which at steady state is given by Fick's first law:

$$F_1 = D(C_1 - C_2)/L \quad [1]$$

where D is the oxidant diffusivity in SiO_2, C_1 is the oxidant solubility in SiO_2, and C_2 is the oxidant concentration at the Si-SiO_2 interface. In series with F_1 is the actual reaction between arriving oxidant and Si atoms at the Si-SiO_2 interface. This reaction is represented in its simplest form by a flux, F_2, of formation of SiO_2 as:

$$F_2 = kC_2 \quad [2]$$

where k is a reaction rate constant and F_2 is linear in oxidant concentration. It is easy to show that a steady state will obtain for the series fluxes F_1 and F_2, as these fluxes are self regulating. This steady state flux situation is represented in Figure 1 (10,11) in which the fluxes are plotted as linear functions of C_2. $F_1(max)$ is at $C_2 = 0$, and $F_2(max)$ is at $C_2 = C_1$. The intersection of these flux lines yields the observed oxidation flux, F, which can be converted to an oxidation rate, dL/dt by:

$$dL/dt = F/\Omega \quad [3]$$

where Ω is the conversion factor for the number of O_2's in a mole of O_2 gas to the number in a mole of solid SiO_2 which yields $\Omega = 2.3 \times 10^{22}$. The resulting integrated rate equation has linear-parabolic form as:

$$t - t_o = (L - L_o)/k_l + (L^2 - L_o^2)/k_p \quad [4]$$

where t_o, L_o represent the initial regime offset to the LP model up to L_o of 30nm. The linear and parabolic rate constants k_l and k_p respectively are defined by the following:

$$k_l = kC_1/\Omega = k'C_1[Si]/\Omega; \; k_p = 2DC_1/\Omega \quad [5]$$

where the linear rate constant is written in terms of C_1 and is explicit in the Si atom concentration(12).

Using the LP model as the basis for discussion, several points are made which form the basis for the discussion of the Si oxidation kinetics

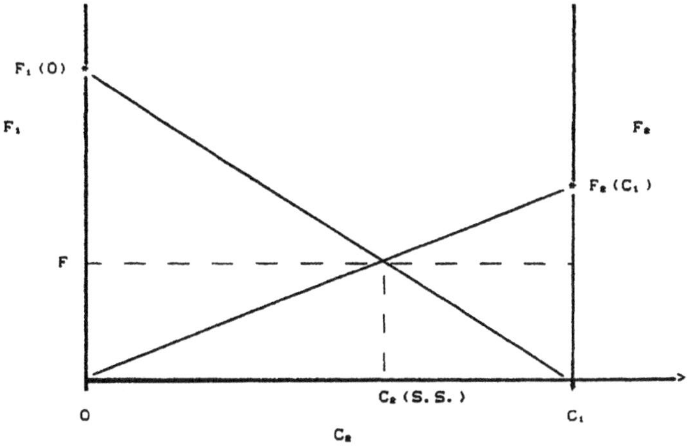

Figure 1. A representation of the diffusive flux, F_1, and reaction flux F_2 as a function of the concentration at the Si-SiO_2 interface, C_2. F is the steady state point (adapted from ref 11).

in the initial regime of oxidation. First it must be recognized that there is no commonly accepted model for SiO_2 growth up to L_o. Secondly, the growth regime below L_o is comprised of two regimes: 0 → L_n and L_n → L_o. Thirdly, for thin SiO_2 films it is likely that the transport of oxidant is less important than the interface reaction. From Figure 1 with $F_1 >> F_2$, the steady state point will lie on the F_2 line at a much lower value than $F_1(max)$. Since the observed rate can never exceed the smaller flux, which is thought to be F_2 for the initial regime, F_2 is said to be rate limiting. Therefore, experimental studies are aimed at parameters that affect F_2. From this, one may be tempted to argue that the dependence of the observed oxidation rate on D is nil for the initial regime, where the interface reaction is the slow step. However, Figure 1 teaches that there is a coupling between the series fluxes F_1 and F_2 as given by:

$$F = F_1 = F_2 \quad [6]$$

from which the observed rate F depends on both fluxes even if the variation of one of the fluxes is dominant. Notice from Figure 1 that a change in either F_1 or F_2 will shift the position of F irrespective of which flux is larger. With F_2 rate limiting as determined above, we now focus on specific studies of the parameters which affect F_2, but without forgetting entirely about transport.

THE VERY INITIAL REGIME: 0 TO L_n

Since most Si oxidation experiments commence with the native oxide already present as a result of the cleaning process, and since oxidation proceeds at the Si-SiO_2 interface, thus the native oxide is replaced on the Si surface with the first formation of grown oxide, the grown oxide is much more technologically relevant than the native oxide. However, the formation of the first oxide on the Si surface represents a regime rich in information about the mechanism of oxidation.

Several UHV studies(13-15) using optical techniques, AES and EELS have determined that the surface electronic states on the Si surface greatly decrease in number as a result of exposure to O_2. The first step at low temperatures and exposure is the adsorption of O_2 which then converts to atomic species with increased temperature and exposure. These studies utilize Si surfaces with the oxide removed by vacuum annealing. Actual oxidation processes utilize a Si surface that has been cleaned using various but quite similar aqueous cleaning procedures(6) involving

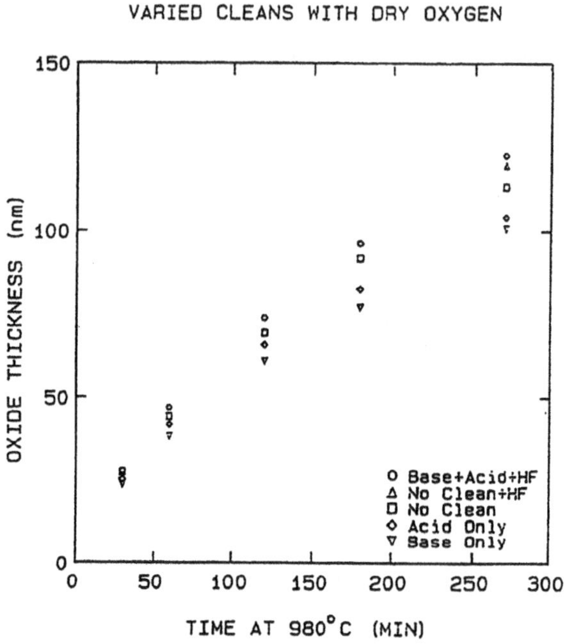

Figure 2. Comparison of Si oxidation data for various wet chemical cleaning procedures (data from ref 18)

a sequence of acidic and basic H_2O_2 exposures followed by an HF exposure to remove any contaminated native oxide. It has been confirmed that these cleaning steps have a measurable effect on the oxidation kinetics(16-19). Figure 2 shows some oxidation data(18) that demonstrates that the greatest difference in oxidation rates is seen between the HF treated samples which show the largest oxidation rate and NH_4OH treated samples which show the smallest rate. Recent studies(20,21) utilized the techniques of contact angle measurement and ellipsometry, both employed in-situ in the cleaning solution media, in order to directly monitor the changes that take place during wet chemical treatments. It was found that the contact angle for a SiO_2 covered Si surface in aqueous HF changed from being hydrophilic with SiO_2 exposed to the solution, to strongly hydrophobic near the Si surface. This behavior, viz. the production of a strongly hydrophobic Si surface, has been reported(7) and is concordant with the common experience of anyone who has treated a Si surface with HF. However, based on simple thermodynamics this kind of behavior cannot be due to the bare Si surface as it has been commonly portrayed. In fact both SiO_2 and Si have high surface energies greater than 1000 dynes-cm(22), and therefore both should be strongly hydrophilic when in contact with a low energy liquid such as aqueous media with surface energies about 30 dynes-cm. The usual observation that the Si surface is hydrophobic in aqueous solutions after HF exposure is clear evidence that the HF is profoundly altering the surface of the Si. In order to investigate this alteration, the limiting contact angle technique pioneered by Zisman(23) was employed. This technique uses a homologous series of solutions with differing surface tension, and the contact angle is measured with the surface under study. The resulting data is usually found to be linear, and is extrapolated to a contact angle value of zero, which is indicative of complete wetting. The surface tension corresponding to this contact angle is called the critical surface tension and has been found to be representative of the surface structure of the solid. For HF treated Si a critical contact angle indicative of either a fluorocarbon or hydrocarbon covered Si surface was found(20). There is analytical evidence for C, H, and F on the Si surface(24-26), but at the present time we cannot unambiguously determine the chemical nature of the surface layer.

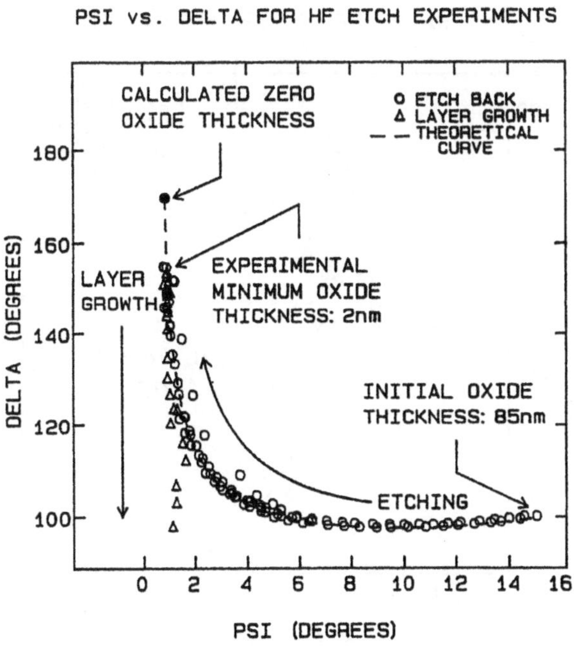

Figure 3a. In-situ ellipsometric analysis of the $Si-SiO_2$ system exposed to $HF-H_2O$ solutions.

In order to further elucidate the HF interaction with Si, in-situ ellipsometry was performed in HF-H_2O media on SiO_2 on Si and with some of the results shown in Fig. 3a(20). The initial SiO_2 thickness of 85nm is etched by the HF-H_2O, and the thinning is easily followed by in-situ ellipsometry with the data falling nearly identically on the theoretical Δ, ψ curve calculated for SiO_2 on Si. However, rather than the SiO_2 being etched to the bare Si surface, a minimum of about 2 nm is attained whereupon evidence is seen for layer growth, viz. Δ decreases and ψ increases (slightly). Presumably the HF removes the SiO_2, but it is replaced with a new film which renders the Si surface strongly hydrophobic. This result is contrasted with the effects of NH_4OH-H_2O on SiO_2 on Si as shown in Fig. 3b(21). In this case, the SiO_2 etching, much slower than in HF-H_2O, results in an apparently bare Si substrate with no film formation. It is believed that the scatter in the Δ, ψ data about the bare Si surface point is due to the observed roughening of the Si surface when exposed to NH_4OH. It then appears that the hydrophobic Si surface, for reasons not yet established, enhances the Si oxidation rate over the bare surface.

The very initial oxidation regime is fast, i.e., the native oxide forms virtually instantly at room temperature and to a thickness of about 1 nm. After this formation, further growth to 2 nm is relatively slow. It appears that the native oxide film formation may be due to surface electronic effects and related to the charge available in surface electronic states, but a discussion of this is delayed until electronic effects are discussed in the next section. Recently, a number of studies(16-21) have revived earlier work(27-29) that showed that

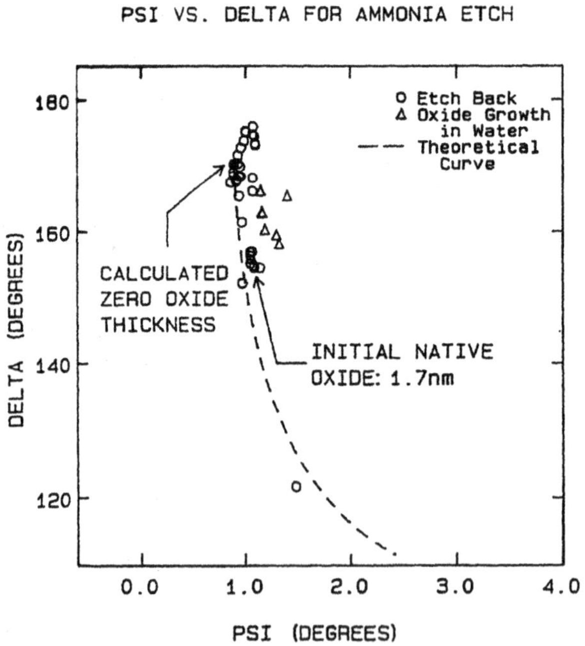

Figure 3b. In-situ ellipsometric analysis of the Si-SiO_2 system exposed to NH_4OH-H_2O solutions.

impurities have a profound effect on the initial oxidation kinetics. Indeed most impurities increase the propensity and speed of oxidation. The older work showed that Na, the ubiquitous impurity that plagued the early development of the technology, increased the oxidation rate. Even trace amounts by chemical analysis standards had measurable effects. No real mechanism was deduced but effects on the interface reaction and the diffusion of oxidant were deemed likely from the analysis of the oxidation data in terms of the LP model. Systematic studies with H_2O additions to an O_2 oxidation ambient(30) have demonstrated that trace amount effects on the oxidation rate were proportionately larger than for substantial H_2O additions to O_2, and that both the interface reaction and the transport of oxidant through the film were enhanced by the present of H_2O as obtained from applications of the LP model. More recent UHV studies with sub monolayers of Ag and Au on Si(31) report that the electronic nature of the impurity and surface interaction is important. In particular both Ag and Au were found to be deposited on (111) Si in a disordered state, and appeared to form a metal like surface on Si, as evidenced by a high density of electron states at the Fermi level. This kind of surface was found to oxidize readily when exposed to O_2. On the other hand, when heated, these metal-like disordered surfaces became ordered and semiconductor-like with a low density of states at the Fermi level, and under these conditions hardly oxidized at all. Similar oxidation enhancement effects were found for Cs and Na(32) and were attributed to metallization of the surface and a tendency towards producing atomic oxygen. Other work with K(33) also shows an oxidation enhancement, but the effect was thought to be due to an efficient transfer of O atoms to the Si surface, since the formation of K oxides was found and the SiO_2 formation was found to occur below the K layer. Si oxidation enhancement effects were also recently reported for Cu(34), Pd(35), Cr(36) and Cs(37), and other older reports are in the literature. We return to these ideas later after other enhancement effects are introduced.

THE INITIAL REGIME: L_a TO L_o

This regime extends to about 30 nm in SiO_2 film thickness, and due to device scaling considerations has recently become the focus of technological importance for gate oxide applications in MOS technology (38). Also, this regime is characterized by fast oxidation rates, relative to the L-P regime which follows, but not so fast as to be difficult to measure, as is the case for the very initial regime, $0 \rightarrow L_a$. Hence, there exists a goodly number of careful oxidation kinetics investigations yielding copious film growth data for study(8,39-42).

From the data, it is clear that this regime, which is sometimes referred to as the "linear" regime never displays purely linear film growth kinetics. The oxidation rate decreases in time similar to the L-P regime, but more slowly. There have been numerous attempts to fit the data to one kind of oxidation model or another. Rather than review these attempts which are technologically important for process models, but which are not usually scientifically rewarding, some of the important physical processes are identified through the use of a variety of experimental techniques.

Figure 4a shows oxidation data for the three low index Si orientations and for the entire L_o regime. The non linearity is clear as is a pronounced orientation dependence(12,40-42) over the temperature range studied of 600 to 1100°C. From a comparison of Figs. 4a and 4b

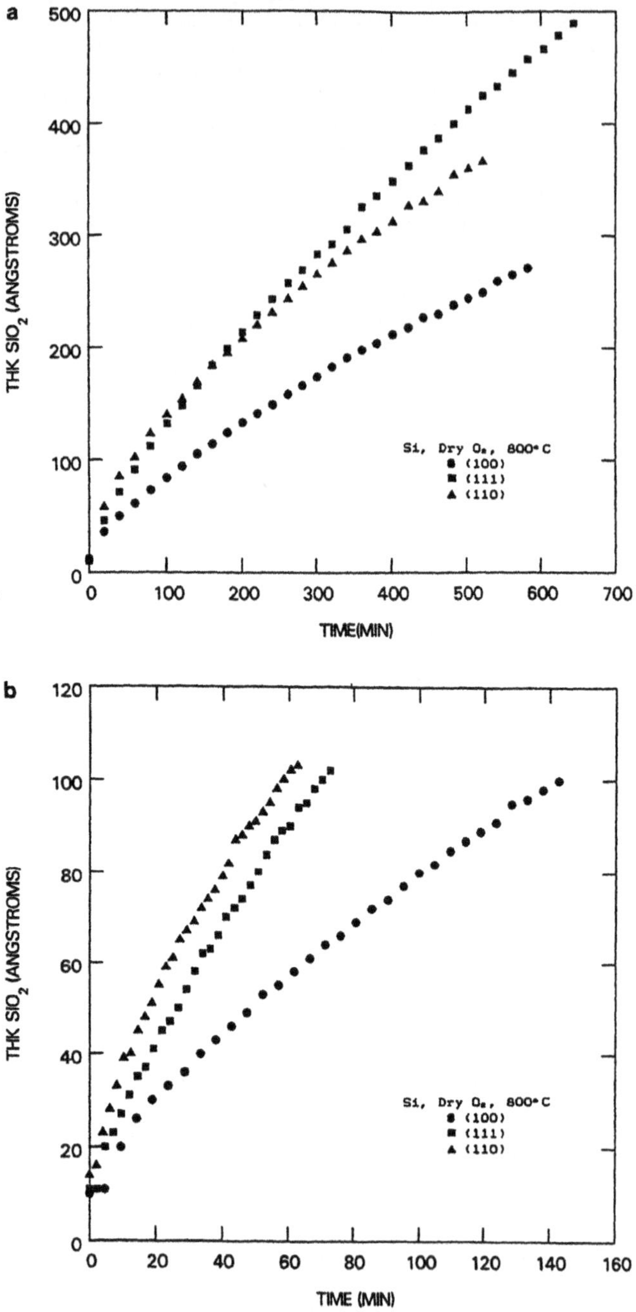

Figure 4. Comparison of Si oxidation data for the (111), (110) and (100) Si orientation for dry O_2 at 800°C for a) the range of 0 to 500A SiO_2 thickness and b) the same data with the 0 to 100A region expanded (from ref 12 and with permission of the Electrochemical Soc. Inc.).

which is the same data but with only the first 10 nm plotted in 4b, we observe the following orientation behavior for the oxidation rates, R:

0 to 15 nm: R(110) > R(111) > R(100)
15 nm upwards: R(111) > R(110) > R(100)

From this data, two questions arise: why the initial order and why the crossover between the (111) and (110) orientations as the oxide grows? The first question is answered by reference to equation [5] for k_1, which is a function of the Si surface concentration. It has been established(41-43) that the order for the Si atom areal density is as follows:

(110) > (111) > (100)

Thus the initial rates scale at least qualitatively according to eqn. [5] with [Si]. More complete recent work(41,42) on the Si orientation effect which included the (311) and (511) as well as the three major orientations also show that the initial oxidation rates up to at least 10nm scale with the Si surface atom density, [Si]. The second question is answered with less confidence. If it is considered that there is a compressive stress in the SiO_2 which can reduce D, then the orientation which exhibits the lowest SiO_2 compressive stress might be expected to exhibit the largest oxidation rate when the SiO_2 is sufficiently thick for the stress to be efficacious(42).

Figure 5 shows the SiO_2 intrinsic film stress as a function of oxidation temperature and for four orientations(44). Germane to the orientation dependence of the oxidation rate discussed above, it is seen that the (111) Si exhibits the lowest stress at all oxidation temperatures and should be expected to yield the largest oxidation rate, as is observed. However, while this interesting correlation provides a starting hypothesis, it remains to be conclusively established whether intrinsic stress really effects oxidation. One recent study suggests such a relation(45) by comparing the oxidation rate for 100nm SiO_2 covered samples, half of which have an intrinsic and half have had the stress relieved, with the finding of a higher rate for the stress relieved samples.

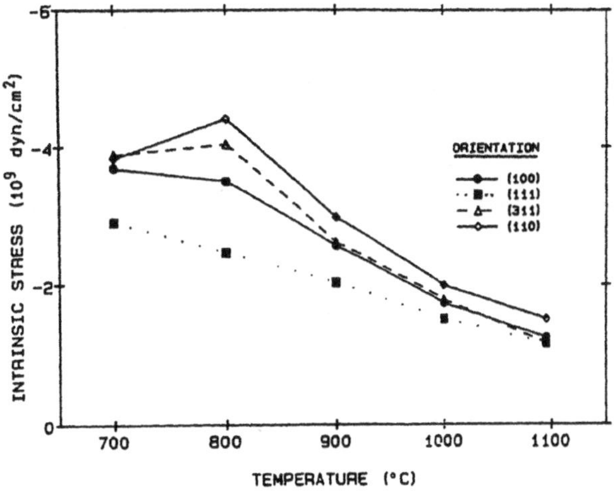

Figure 5. Measured intrinsic SiO_2 film stress as a function of oxidation temperature for four Si orientations (from ref 44 and with permission of the Am. Inst. Physics).

Two additional factors relative to stress effects on oxidation are worthy of further attention. One is the temperature dependence of the stress as is seen in Figure 5, and the other is the possible effect on the interface reaction. The considerable increase in intrinsic stress for lower oxidation temperatures suggests that the oxide viscosity which also increases sharply with decreasing temperature may be implicated. A viscous flow model for oxidation was proposed(46) based on earlier similar research(47). Figure 6 shows that according to the viscous flow model the SiO_2 formed flows normal to the Si surface in response to the lateral compression with a rate of flow proportional to the ratio of stress to viscosity. This relationship arises from considering the SiO_2 to be a Maxwell solid. Also, an oxidation model was proposed(12,48) for the initial regime where the rate of oxidation may scale with the rate at which the SiO_2 can relax. Very recent stress measurements(49) which compared as-grown SiO_2 film from 10 to 100 nm with similar thicknesses of SiO_2 but etched back from a starting 100 nm SiO_2 film revealed the same stress distribution, stress levels, and the same limiting value for stress at zero oxide thickness. The stress distribution shows a modest increase of less than a factor of two in intrinsic stress near the SiO_2-Si interface. The fact that the same values for the intrinsic stress are obtained from the two experiments means that once the oxide is formed, there is only very slow SiO_2 relaxation even with the vastly different oxidation rates for the formation of the thin SiO_2 films in the two experiments. The limiting stress value of between 4-5 X 10^9 dynes/cm^2 is more than an order of magnitude less than the maximum stress(46), and strongly suggests that there exists a low viscosity for the initially grown SiO_2, but this low value relaxes to a much higher viscosity value when the SiO_2 network is well formed. In this way a rapid initial relaxation lowers the stress, but then is followed by a slower relaxation rate in conformity with the equilibrium SiO_2 viscosity.

The possible effect of this stress on the interface reaction has been modeled(12,46,48) under the assumption that SiO_2 is a Maxwell solid. The essential idea is that the large molar volume change which attends the conversion of Si to SiO_2, and is responsible for the intrinsic stress, also couples into the interface reaction rate constant. To accomplish this, equation [5] is rewritten as follows:

$$k_i = k' C_i [Si]/\Omega = k' C_i [Si^*]/\Omega \qquad [6]$$

where $[Si^*]$ is the effective Si concentration and given by [Si], the areal density of Si atoms reduced by multiplying by the rate at which the SiO_2 flows away from the Si surface during oxidation. For a Maxwell solid this fraction is the strain rate, $\dot{\varepsilon}$, which is given as:

$$\varepsilon = \sigma/\eta \qquad [7]$$

where σ is the intrinsic stress as usual, and η is the oxide viscosity. Then k_i becomes:

$$k_i = k' C_i [Si] \sigma/\eta \qquad [8]$$

and the initial regime dominant rate process is a function of the rate of relaxation of the as-forming SiO_2.

There appears to be a growing body of experimental evidence that indicates that the electronic nature of the Si surface determines the initial Si oxidation rates. The evidence to be discussed herein is based on photonic effects, metallic Si surfaces and impurities.

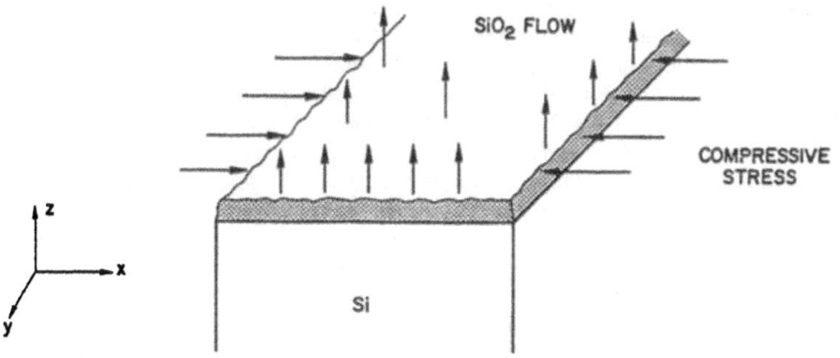

Figure 6. Viscous flow model for Si oxidation (from ref 46 and with permission of the Electrochemical Soc. Inc.

A number of studies(50-55) attest to the Si oxidation enhancement effects of intense light, usually laser light, on the oxidation of Si. Very recent results(53) estimate a 20-25% enhancement of the oxidation rate for 800°C oxidation using the 514.5 and 488 nm lines from an Ar ion laser. A number of models have been proposed, in order to explain the results. In one model the flux of excited electrons in Si over the Si-SiO_2 barrier is compared with the enhanced oxidation rate(54) and reasonable agreement is found. The photonically excited electrons are thought to assist in the decomposition of molecular oxygen and thereby accelerate the reaction. A similar mechanism but without photon enhancement of electrons was considered as a possibility to occur for the thermal oxidation of Si(56). In order to carefully test this model for thermal oxidation, oxidation data in the very initial oxidation regime of up to 5 nm was obtained and compared with the calculated thermionic flux of electrons(56). The agreement was found to be quite good for oxide growth less than 5 nm and for oxidation temperatures from 600 to 1000°C. Essentially, the thermionic oxidation model for the very initial thermal oxidation regime is as follows:

1. In the Si conduction band: $Si \rightarrow Si^{+} + e^{-}$

2. At the Si-SiO_2 barrier: e^{-}(Si C.B.) $\rightarrow$ e^{-}(SiO_2 C.B.)

3. In SiO_2: O_2(at the Si-SiO_2 interface) + $e^{-} \rightarrow O^{-} + O$

4. The Interfacial Reaction: $Si^{+} + O^{-} + O \rightarrow SiO_2$

with step 2. being rate limiting. We will return to this model to attempt correlation with the other oxidation effects which seem to effect the barrier, and thus affect the oxidation rate. For thicker films the transport of oxidant and other parameters discussed above cannot be ignored, as was discussed above. Also in this study(56) the role of electrons in intrinsic Si surface electronic states was found to be about the correct number to explain the initial very rapid formation of 1nm of native SiO_2 on Si, the L_n layer with the assumption that one electron in a surface state could be available for the oxidation reaction without the need for thermionic emission over the Si-SiO_2 barrier. Other models have used the idea of surfaces states or dangling bonds but these models have usually been used to explain the formation of oxide thicker than L_n(55,57).

The previously mentioned reports(31-37) show that metallic impurities of less than a monolayer accelerate the Si oxidation rate, strongly support the notion that the metallicity of the Si surface is crucial for the observed oxidation enhancements. In further support of this idea are recent reports(58,59) that a wide variety of metal silicides exhibit decidedly different oxidation behavior according to the electronic nature of the silicide surface. The oxidation of the variety of metal silicides was performed such that there was an ample supply of Si from the Si substrate underneath the silicide, and under oxidations conditions that yielded only SiO_2 on the silicide surface and the thickness of the silicide was preserved. In effect it is Si oxidation but on a silicide surface. Figure 7 summarizes these experimental results in which three bands of oxidation behavior is identified which correlate with the optical absorbtion of the silicide surfaces. The fastest oxidizing silicides exhibit the largest optical absorbtion as measured by ellipsometry, and the slowest oxidizing silicides oxidize at about the same rate as Si and these silicides are thought to be semiconducting silicides(58). The optical absorbtion in the visible is another measure of the surface metallicity with the greatest absorbtion attributed to a greater free electron concentration. These silicide oxidation experiments enable the oxidation of Si to occur on surfaces with decidedly different electronic properties.

Impurities on the Si surface usually increase the initial Si oxidation rate. Without impurities the initial rate is accounted for by the thermionic emission of electrons over the Si-SiO_2 interface. For consistency with this model, it is argued that the impurities alter the barrier, usually resulting in a lower barrier as would be expected for impurities on any surface and some precedent for this exists(60).

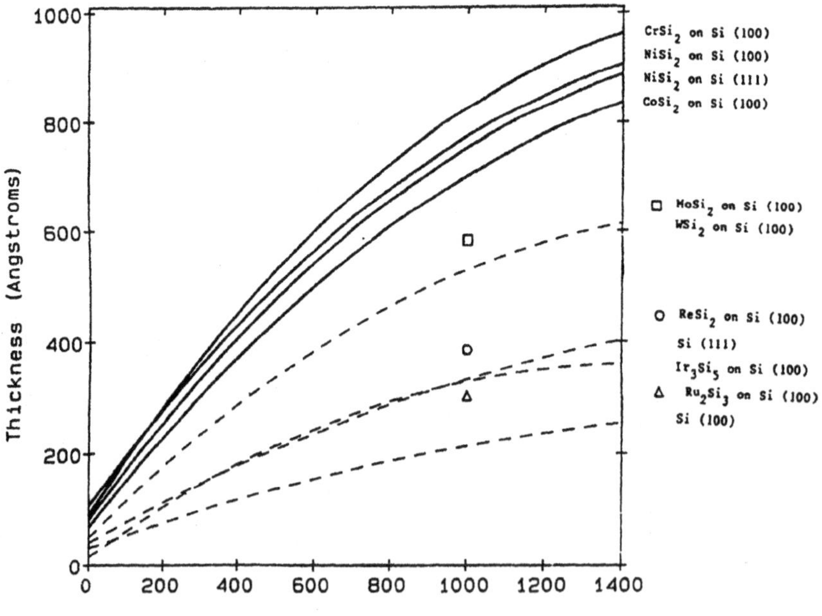

Figure 7. Oxidation behavior of a variety of metal silicides on Si oxidized at 750°C in dry O_2 and compared with pure Si (data from ref 59).

CONCLUSIONS

The initial Si oxidation regime remains without an accepted physical model. New experimental studies implicate impurities, cleaning procedures, the electronic nature of the Si surface, the crystallography of the surface, and mechanical stress as possible phenomena bearing on the physical processes that take place for the initial thin SiO_2 oxidation regime. Both the number and barrier for surface electrons appear to be important. These recent studies provide new paths for future Si oxidation research.

ACKNOWLEDGEMENTS

This work was supported in part by the Office of Naval Research, ONR, and the IBM corporation.

REFERENCES

1. B.E. Deal and A.S. Grove, J. Appl. Phys., 36, 3770 (1965).
2. W.A. Pliskin, IBM J. Res. and Develop., 10, 198 (1965).
3. A.G. Revesz, K.H. Zaininger and R.J. Evans, J. Electrochem. Soc., 113, 706 (1966).
4. U.R. Evans, "The Corrosion and Oxidation of Metals", Arnold, London (1960).
5. R. Rosenberg, Abstract DG-TUM2, presented at 34th AVS meeting Anaheim, CA, Nov. 2-6, 1987.
6. W. Kern and D.A. Poutinen, RCA Rev., 31, 187 (1970).
7. R. Williams and A.M. Goodman, Appl. Phys. Lett., 25, 531 (1974).
8. E.A. Irene and Y.J. van der Meulen, J. Electrochem. Soc., 123, 1380 (1976).
9. J.R. Ligenza and W.G. Spitzer, J. Phys. Chem. Solids, 44, 131 (1960).
10. V.G. Levich, "Physicochemical Hydrodynamics", Prentice Hall, Englewood Cliffs, N.J. (1962), pp. 72-78.
11. E.A. Irene and R. Ghez, Appl. Surface Sci., 30, 1 (1987).
12. E.A. Irene, H.Z. Massoud and E. Tierney, J. Electrochem. Soc., 133, 1253 (1986).
13. P. Chiaradia and S. Nannarone, Surface Science, 54, 547 (1976).
14. J. Derrien and M. Commandre, Surface Science, 118, 32 (1982).
15. H. Ibach, H.D. Bruckmann, and H. Wagner, Appl. Phys. A., 29, 113 (1982).
16. F.N. Schwettman, K.L. Chiang and W.A. Brown, 153rd Electrochemical Society Meeting, Abstract 276, May 1978.
17. F.J. Grunthaner and J. Maserjian, IEEE Trans. Nucl. Sci., NS-24, 2108 (1977).
18. G. Gould and E.A. Irene, J. Electrochem. Soc., 134, 1031 (1987).
19. J.M. DeLarios, C.R. Helms, D.B. Kao and B.E. Deal, Appl. Surface Sci., 30, 17 (1987).
20. G. Gould and E.A. Irene, J. Electrochem. Soc., accepted for publication (1988).
21. G. Gould and E.A. Irene, J. Electrochem. Soc., submitted 1988.
22. W.A. Zisman, "Contact Angle: Wetability and Adhesion", in Advances in Chemistry Series, Vol. 43, Ed. F.M. Fowkes, Am. Chem. Soc., Washington, D.C. (1964), Chap. 1.
23. H.W. Fox and W.A Zisman, J. Colloid Sci., 5, 514 (1950).
24. G.B. Larrabee, K.G. Heinen and S.A. Harrell, J. of Electrochem. Soc., 114, 867 (1967).
25. E. Yablonovitch, D.L. Allara, C.C. Chang, T. Gmitter and T.B. Bright, Phys. Rev. Lett., 57, 249 (1986).
26. R.E. Benenson, L.C. Feldman and B.G. Bagley, Nuclear Inst. and Methods, 168, 547 (1980).

27. A.G. Revesz and R.J. Evans, J. Phys. Chem. Solids, 30, 551 (1969).
28. E.A. Irene, J. Electrochem. Soc., 121, 1613 (1974).
29. Y.J. van der Meulen and J.G. Cahill, J. Electronic Materials, 3, 371 (1974).
30. E.A. Irene and R. Ghez, J. Electrochem. Soc., 124, 1757 (1977).
31. A. Cros, J. Physique, 44, 707 (1983).
32. A. Franciosi, P. Soukiassian, P. Philip, S. Chang, A. Wall, A. Raisanen and N. Troullier, Phys. Rev. B, 35, 910 (1987).
33. M.C. Asensio, E.G. Michel, E.M. Oellig, and R. Miranda, Appl. Phys. Lett., 51, 1714 (1987).
34. P.J. Moller and J. He, J. Vac. Sci. Technol. A, 5, 996 (1987).
35. G. Abbati, L. Rossi, L. Calliari, L. Braicovich, I. Lindau and W.E. Spicer, J. Vac. Sci. Technol., 21, 409 (1982).
36. P. Wetzel, C. Pirri, J.C. Peruchetti, D. Bolmont and G. Gewinner, J. Vac. Sci. Technol. A, 5, 3359 (1987).
37. P. Soukiassian, T.M. Gentle, M.H. Balshi and Z. Huryck, J. Appl. Phys., 60, 4339 (1986).
38. E.A. Irene, "Semiconductor International", p. 99, April 1983 and p. 92, June 1985.
39. M.A. Hopper, R.A. Clarke, and L. Young, J. Electrochem. Soc., 122, 1216 (1975).
40. H.Z. Massoud, J.D. Plummer and E.A. Irene, J. Electrochem. Soc., 132, 1745, 2685 (1985).
41. E.A. Lewis, E. Kobeda and E.A. Irene, in "Semiconductor Silicon 1986." H.R. Huff, T. Abe and B. Kolbeson, Eds. p. 416, The Electrochemical Soc., Pennington, N.J. (1986).
42. E.A. Lewis and E.A. Irene, J. Electrochem. Soc., 134, 2332 (1987).
43. J.R. Ligenza, J. Phys. Chem., 65, 2111 (1961).
44. E. Kobeda and E.A. Irene, J. Vac. Sci. Technol., B5, 15 (1987).
45. J.K. Srivastava and E.A. Irene, J. Electrochem. Soc., 132, 2815 (1985).
46. E.A. Irene, E. Tierney and J. Angillelo, J. Electrochem. Soc., 129, 2594 (1982).
47. E.P. EerNisse, Appl. Phys. Lett., 30, 290 (1977); 35, 8 (1979).
48. E.A. Irene, J. Appl. Phys., 54, 5416 (1983).
49. E. Kobeda and E.A. Irene, J. Vac. Sci. Technol. B, accepted for publication 1987.
50. R. Oren and S.K. Ghandı, J. Appl. Phys., 42, 752 (1971).
51. S.A. Schafer and S.A. Lyon, J. Vac. Sci. Technol., 19, 494 (1981); 21, 422 (1982).
52. Ian W. Boyd, Appl. Phys. Lett., 42, 728 (1983).
53. F. Micheli and Ian W. Boyd, Appl. Phys. Lett., 51, 1149 (1987).
54. E.M. Young and W.A. Tiller, Appl. Phys. Lett., 42, 63 (1983); 50, 46 (1987); 50, 80 (1987).
55. P. Quenon, M. Wautelet, and M. Dumont, J. Appl. Phys., 61, 3112 (1987).
56. E.A. Irene and E.A. Lewis, Appl. Phys. Lett., 51, 767 (1987).
57. S.A. Shafer and S.A. Lyon, Appl. Phys. Lett., 47, 154 (1985).
58. F.M. d'Heurle, A. Cros, R.D. Frampton and E.A. Irene, Philos. Mag. B., 55, 291 (1987).
59. R.D. Frampton, E.A. Irene and F.M. d'Heurle, J. Appl. Phys., 62, 2972 (1987).
60. M. Liehr, P.E. Schmid, F. K. Le Goues, and P.S. Ho, J. Vac.Sci. Technol., A4, 855 (1986).

Si OXIDATION MECHANISMS AS STUDIED BY OXYGEN TRACER METHODS

S. Rico

G.P.S. de l'E.N.S. - Universite Paris 7
Tour 23 - 2, Place Jussieu - 75251
Paris Cedex 05, France

ABSTRACT

The use of ^{18}O tracing to investigate atomic transport mechanisms during oxide growth is described. Experimental results for thermal oxidation of silicon are presented and discussed. It is shown that the growth in pure water vapor occurs by an interstitial diffusion of water molecules reacting with the silica network. It is shown that the growth in dry oxygen exhibits two main features :

- an interstitial transport of O_2, which is largely dominant for thick oxide films
- step by step motion of network oxygen atoms whose contribution increases as oxide thickness decreases.

INTRODUCTION

The thermal oxidation of silicon in dry oxygen, as well as in water vapor, is commonly described by the theory of DEAL and GROVE (1). This theory is based on the hypothesis that the oxidizing molecules (O_2 or H_2O) are dissolved in interstitial position and then, even though they react with the silica network diffuse in the molecular form through the oxide layer to the oxide/silicon interface where they oxidize the silicon substrate. This leads to a linear parabolic growth law which can written as follows : $dt/dX = 2X/k_p + 1/k_l$ (1)

where X is the oxide thickness, t the oxidation time, $k_p = 2\, D_c N^*_c/N_i$ and $k_l = k_r N^*_c/N_i$ are the parabolic and linear constant respectively ; N^*_c, N_i, D_c and k_r being respectively the solubility, the number of interstitial molecules needed to form of volume unit of oxide ($N_i = N_n$ for H_2O and $2N_n$ for O_2 where N_n is the concentration of network oxygen atoms), the diffusion coefficient and the reaction rate coefficient of the oxidizing molecule at the Si/SiO_2 interface.

Owing to the free volume model the solubility N_c^* is proportional to the oxidizing gas pressure. In the case of water vapor oxidation the kinetics well match with this theory, but for dry oxygen oxidation this theory only applies for thicknesses greater than 20-30 nm. Furthermore various mechanisms can give linear parabolic law.

Before concluding about the mechanism of growth which really occurs, it is necessary to answer the following questions:

i) what is the species transported through the oxide film: O or Si?

ii) what is the atomic transport mechanism: interstitial, vacancy or interstitialcy type? (For example for an interstitial transport of oxygen it could be O_2, O, neutral or charged). Previous results obtained by using inert marker and isotopic labelling have been reviewed (2). The aim of this paper is to describe the main features of the use of ^{18}O tracing for studying the mechanisms of oxide growth and to review the results and conclusions on thermal silicon oxidation including new ones.

PRINCIPLES

Principles of the use of ^{18}O labelling are the following : a SiO_2 layer, first grown up to a thickness X in ^{16}O oxidizing gas ($^{16}O_2$ or $H_2{}^{16}O$) undergoes further growth to a thickness X + Δ X in ^{18}O gas, then the depth profiles of the oxygen isotopes in the oxide film are determined. We will describe below the ^{18}O depth profiles corresponding to the various possible growth mechanisms.

A) Only silicon is mobile : (Fig. 1) whatever the silicon transport mechanism is, the newly formed oxide grows at the external surface of the oxide film where the reaction between silicon and the oxidizing gas occurs. Thus the ^{18}O atoms end up near the external oxide surface with an abrupt $^{16}O/^{18}O$ boundary since oxygen atoms are immobile.

B) Only oxygen is mobile : two limit mechanisms are possible :

Ba) oxygen moves interstitially (as an O_2 or H_2O molecule, or another oxygenated species formed by the decomposition of the O_2 or H_2O molecule at the external surface) through the silica network.

Ba i) if the oxygenated species migrates without reacting with the silica network, the newly formed oxide grows at the oxide/substrate interface. Thus the ^{18}O atoms end up at the oxide/silicon interface and there exist an abrupt $^{16}O/^{18}O$ boundary (Fig. 2) since oxygen atoms of the silica network are immobile.

Ba ii) if a chemical reaction occurs between the mobile oxidizing species and the network, this can lead to a replacement of the oxygen atoms of the network by those of the interstitial molecules i.e. an isotopic exchange of oxygen atoms occurs. Let us consider the typical case of water which is known to react with silica in forming SiOH groups (also called silanol groups). It is commonly admitted that the following equilibrium reaction occurs.

$$H_2O + Si\text{-}O\text{-}Si \underset{k_{nc}}{\overset{k_{cn}}{\rightleftharpoons}} SiOH + SiOH \qquad (2)$$

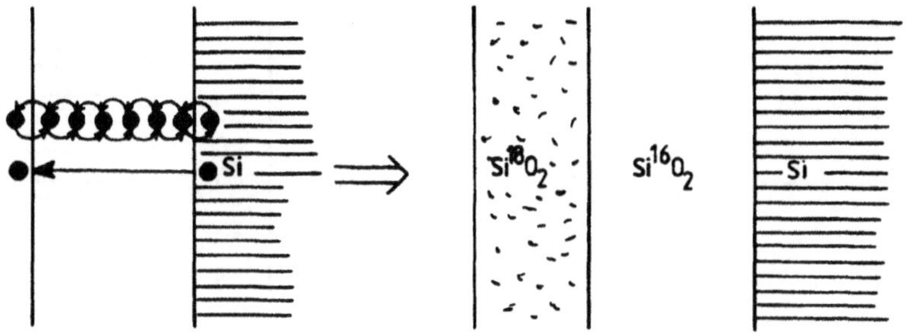

Fig. 1. ^{18}O profile when only silicon is mobile.

If an $H_2{}^{18}O$ molecule reacts with a $Si-{}^{16}O-Si$ network the two silanol groups is $Si^{18}OH$. The reverse reaction has a 50% probability to form a $H_2{}^{16}O$ molecule and $Si-{}^{18}O-Si$ network. This corresponds to an oxygen exchange between the water molecule and the network oxygen atoms given by $\gamma_{cn} = k_{cn}/2$ where k_{cn} is the reaction rate constant. This leads to a transport of the oxygen atoms of the network, by the water molecules which act as an oxygen carrier.

An exchange phenomenon means that the network oxygen atoms are free to move out of the oxide and can be replaced by atoms coming from the gas phase. This can occur with a total oxygen flux equal to zero (no oxide growth). In any case, this means that the amount of fixed ^{18}O atoms is always greater than that corresponding to the thickness increase of the oxide. The differential equations of this oxygen exchange have been stated in Ref. (3) and have been solved at equilibrium i.e. when the concentration N_c of the carrier is uniform in depth and no oxide growth occurs. It is shown that, in this case, the ^{18}O labelling $C_n{}^{18}$ of the oxide network gradually decreases from the gas/oxide interface to the oxide/silicon interface with a slope which decreases when the ratio between the exchange frequency γ_{cn} (i.e. the reaction rate) and the diffusion coefficient D_c of the carrier decreases.

Let us consider now what should be expected when further growth occurs in the two simplest cases where the differential equations simplify. We will name $[^{18}O_c]$ and $[^{18}O_n]$ respectively the concentrations of ^{18}O labelled carrier and of ^{18}O network ; the corresponding isotopic labelling $C_c{}^{18} = [^{18}O_c]/N_c$ and $C_n{}^{18} = [^{18}O_n]/N_n$ are fractions smaller or equal to the gas labelling $C_g{}^{18}$.

Ba ii.1) Isotopic quasi-equilibrium between carrier and network oxygen $C_n{}^{18} \sim C_c{}^{18}$ (Fig. 3a). This corresponds to an exchange frequency γ_{cn} much greater N_c, we obtain

$$\frac{\partial}{\partial t} C_n{}^{18} = \frac{\partial^2}{\partial x^2}(D^* C_n{}^{18}) \qquad (3)$$

where $D^* = D_c N_c/N_n$.

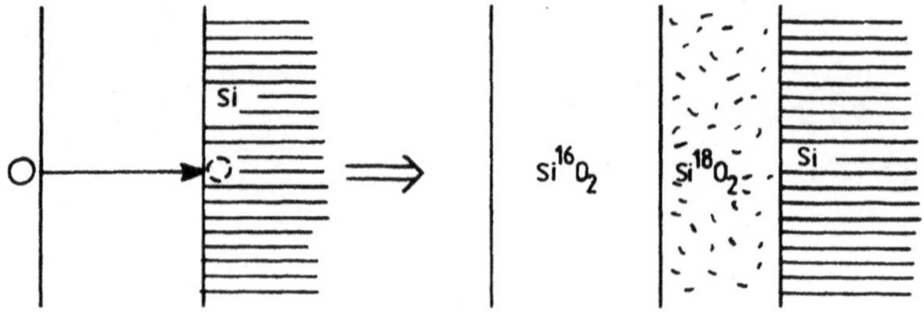

Fig. 2. ^{18}O profile when oxygen moves interstitially without reacting with the silica network.

Knowing that the transfer coefficient between the gas on the oxide is high, the solution will be $C_n^{18} = C_g^{18}$ erfc $x/2\sqrt{D^*t}$ with $D^* = D_c N_c^*/N_n$ if D^*(i.e. N_c) can be considered as constant through the diffusion depth $\sqrt{D^*t}$. This will be a reasonable approximation if the oxide thickness X is higher than $\sqrt{D^*t}$. Taking $N_c = N^*_c$ and $N_i = N_n$ we obtain $D^* = k_p/2$. The amount ^{18}Q of ^{18}O exchanged atoms will be equal to $C_g^{18} N_n(2/\sqrt{\pi})\sqrt{k_p\ t}/2$ Taking $C_g^{18} \simeq 1$, ^{18}Q can be much higher than the total amount of atoms corresponding to the thickness increase Δ X which is equal to $N_n\ t/(1/k_1 + 2X/k_p)$.

Ba ii.2) C_c^{18} much larger than C_n^{18} (Fig.3(b)) : this corresponds to an exchange frequency γ_{cn} much smaller than D_c and to times t sufficiently short to have little isotopic exchange. The equations for the initial oxide of thickness X reduce to

$$\left\{\begin{array}{ll} \dfrac{\partial}{\partial t}[O_c^{18}] = D_c\, \partial^2 \dfrac{[O_c^{18}]}{\partial x^2} - \gamma_{cn}[O_c^{18}] & (4) \\ \\ \dfrac{\partial}{\partial t}[O_n^{18}] = \gamma_{cn}[O_c^{18}] & (5) \end{array}\right.$$

We reach rapidly a steady state for $[O_c^{18}]$ and thus we obtain :

$$\left\{\begin{array}{ll} [O_c^{18}] = A\, e^{-x/x_o} + B\, e^{x/xo} & (6) \\ \\ [O_n^{18}] = \gamma_{cn}\, t\, [O_c^{18}] & (7) \end{array}\right.$$

where $x_o = (D_c/\gamma_{cn})^{1/2}$

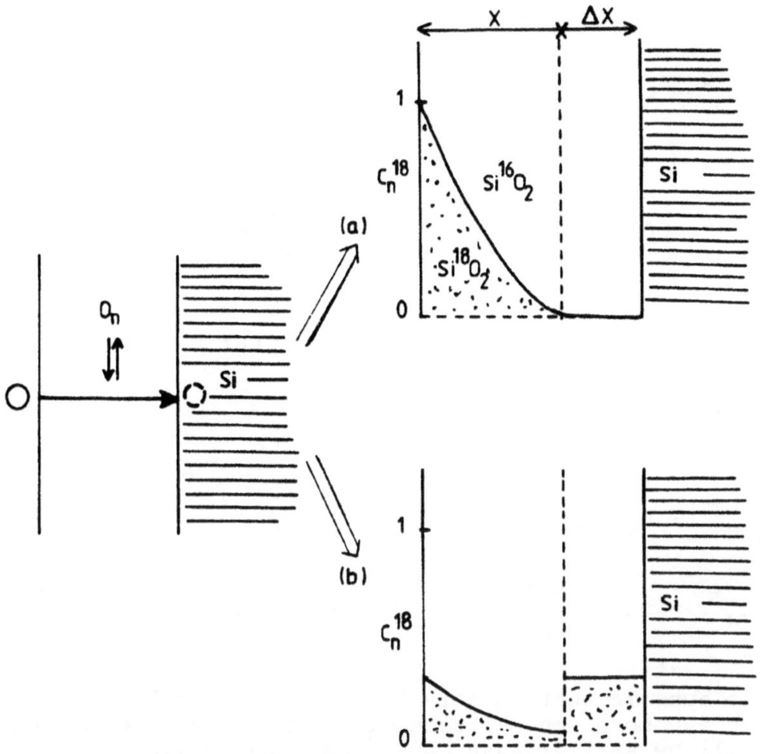

Fig. 3. ^{18}O profile when oxygen moves interstially with reacting with the silica network : (a) isotopic equilibrium (high γ_{cn}/D_c ratio (b) far from isotopic equilibrium (low γ_{cn}/D_c ratio).

We take the boundary condition

$$C_c^{18}\Big|_{x=0} = C_g^{18}$$

In the parabolic regime, considering that $[O_c^{18}]_{x=X} \simeq 0$, one obtains

$$[O_c^{18}] = C_g^{18}\, N_c^* \frac{\mathrm{Sh}\,\frac{X-x}{x_o}}{\mathrm{Sh}\,X/x_o} \qquad (8)$$

From this relationship we deduce

$$\begin{cases} C_c^{18} = C_g^{18} \dfrac{1}{1-\dfrac{x}{X}} \dfrac{Sh\left((X-x)/x_o\right)}{Sh\ (X/x_o)} & (9) \\ \\ C_n^{18} = C_g^{18} \dfrac{N_c^*}{N_n} \quad cn \ t \ \dfrac{Sh\left((X-x)/x_o\right)}{Sh\ (X/x_o)} & (10) \end{cases}$$

In the linear regime considering that $N_c \simeq N_c^*$ and $\left.\dfrac{\partial\,[O_c^{18}]}{\partial x}\right|_{x = X} \simeq 0$ one obtains

$$\begin{cases} C_c^{18} = C_g^{18} Ch\left((X-x)/x_o\right)/\ Ch(X/x_o) & (11) \\ \\ C_n^{18} = C_g^{18} \dfrac{N_c^*}{N_n} \ \gamma_{cn} \ t \ Ch\left((X-x)/x_o\right)/\ Ch(X/x_o) & (12) \end{cases}$$

It appears that in both linear and parabolic regime and thus in the more general case C_n^{18} remains smaller than C_c^{18}. C_n^{18} will be smaller than C_c^{18} as long as γ_{cn} t remains smaller than N_n/N_c^* whose value is $\sim 10^6$ at atmospheric pressure. This leads to the following conclusions :

1) the local ^{18}O labelling C_n^{18} of the initial oxide network (i.e. up to thickness X) increases linearly with the time and exhibits a depth distribution homothetic as a function of time and which becomes smoother when x_o increases. The corresponding amount of exchanged ^{18}O atoms increases linearly with the time.

2) the local ^{18}O labelling $^{18}C_c$ of the carrier remains constant as time increases. Thus the labelling of the further grown oxide (located between X and X + Δ X) which is that of the oxidizing carrier will be higher than that of the initial oxide layer for $\gamma_{cn}\ t < N_n/N_c^*$.

Bb) Oxygen is transported by a vacancy mechanism (or an interstitialcy mechanism) : the growth is due to the picking up by the silicon substrate of oxygen atoms of the oxide network leading to the injection of oxygen vacancies at the oxide/silicon interface. A step by step movement of these vacancies and thus of associated network oxygen atoms occurs. When a vacancy reaches the external oxide surface, it can be filled with a new oxygen atom coming from the gas phase. Thus the ^{18}O ends up near this external oxide surface. The net results are identical for an interstitialcy mechanism.

Bb i) if the vacancy migration occurs by a simple diffusion mechanism (Fig. 4), the vacancies jump in all directions,

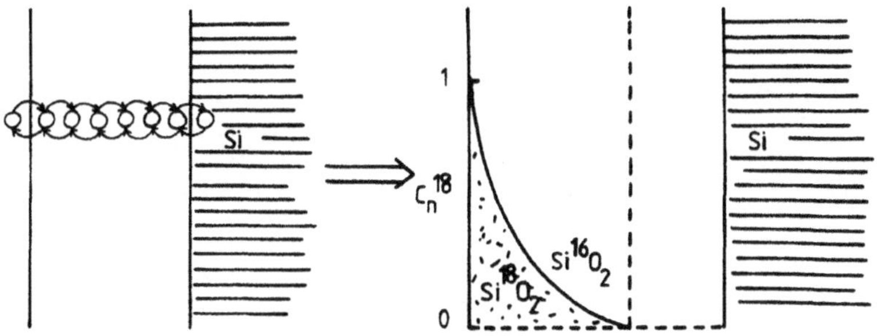

Fig. 4. ^{18}O profile : when oxygen moves by a simple diffusion of oxygen vacancies (or interstitialcies).

this means that vacancies are also created at the external oxide surface by the departure of oxygen network atoms. This means that ^{16}O atoms leaving the oxide network are replaced by ^{18}O atoms coming from the gas i.e. isotopic exchange occurs. This ^{18}O fixation by oxygen exchange adds to that one due to oxide growth.

In an oxide at or near equilibrium with the gas phase (i.e. without any oxide growth) the concentration of vacancies (or other defects) is homogeneous in depth and fixed by the thermodynamical conditions (external pressure, temperature). If the vacancies are mobile enough a very strong oxygen exchange can be observed while the vacancy concentration remains very small in comparison to that of network oxygen atoms. As a matter of fact the effective self diffusion coefficient D^* of ^{18}O is equal to $D_\square \, C_\square / N_n$ where $D_\square$ and $C_\square$ are respectively the diffusion coefficient and the concentration of vacancies. If the rate of atomic transfert between the gas and the oxide surface is high this gives an erfc ^{18}O profile for an oxide at equilibrium. For a growing oxide film the mathematics become very intricate. This mathematics given in (4) show that for $X > \sqrt{D^*}\, t$ the ^{18}O depth distribution is very near an erfc function and corresponds to an amount of exchanged atoms very near $C_g^{18}\, N_n\, 2/\sqrt{\pi}\, \sqrt{D^* t}$ and much more bigger than one corresponding to the oxide thickness increase.

Bb ii) if the vacancies (or interstitialcies) migrate in only one direction (Fig. 5) (this is the case of a charged defect migrating under a strong electric field for example in CABRERA and MOTT' mechanism), the O atoms jump only in the direction of the growth. In this case no oxygen atom leaves the oxide network and thus no isotopic exchange occurs. The $^{18}O/^{16}O$ frontier exhibits a small mixing due to the jumping statistics which give a poissonian depth distribution of ^{18}O atoms (5). The theory of this phenomenon is the same that the one of energy loss fluctuation of ions in matter described in (6).

EXPERIMENT

The thermal treatments in ^{18}O were carried out using set up allowing the getting back of the labelled gas (3,9). Further-

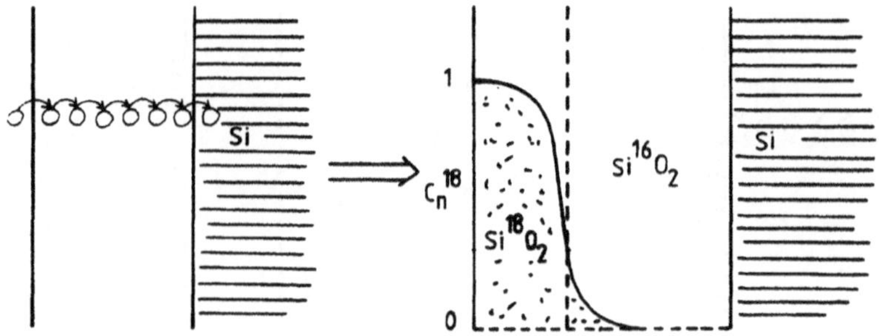

Fig. 5. ^{18}O profile when oxygen moves by a migration of oxygen vacancies (or interstitialcies) in only one direction.

more treatments in $^{18}O_2$ must be carried out in very clean and dry atmosphere in order to avoid the interfering contribution of water vapor traces (7,8,9). The ^{18}O depth profiles in silica were determined by using either the nuclear microanalysis technique (3,7,8,9,13,14) or the SIMS Analysis (10,11,12).

RESULTS AND INTERPRETATION

A) Wet thermal oxidation : ^{18}O experiments have been carried out in wide ranges of temperature and pressure. For pressures ranging from 0.3 to 13 Torr and temperatures ranging from 400°C to 930°C the present author obtained the following results (3,13,14). The shape of the ^{18}O profiles corresponds to the mechanism Ba ii) i.e. to a diffusion of water molecules reacting with the silica network. Using this model we deduced the values of the effective self-diffusion coefficient of ^{18}O values $D^* = D_C N_C^*/N_n$ and of the effective exchange coefficient $\gamma_{nc} = (k_{cn}/2) N^*_c/N_n$. The values found for D^* are in agreement with that of the parabolic constant k_p which should be equal to $2\ D^*$.

Furthermore they exhibit the same temperature dependance and the same linear proportionality to the water pressure (i.e. to N_C). This proves the validity of the DEAL-GROVE theory for oxidation in water vapor. The ^{18}O profiles become smoother when the temperature decreases. This corresponds to a higher activation energy for γ_{nc} (i.e. of k_{cn}) than for D^* (i.e. of D_C). In In contrast to D^*, γ_{nc} was not found proportional to the pressure, this departure means a decrease of k_{cn} when pressure increases. This is supported by the results of treatments in deuteriated water (15) in the same pressure range which show that the Deuterium concentration in oxide at equilibrium is not proportional to the square root of pressure as found by MIKKELSEN (16) in the 0.05-8 atm range and at 800°C. This decrease of k_{cn} in the low pressure range (< 13 Torr) can be due to a water clustering. For higher pressures it can be assumed that a structure relaxation due to a higher SiOH content occurs. As a matter of fact at very high pressures (35-160 atm) in the 500-700°C range SPITZER and LIGENZA (17) reported a flat ^{18}O profile, which means values of k_{cn} higher at high pressure than at low pressure.

B) <u>Dry oxidation</u> : We have studied the thermal growth in dry oxygen ($H_2O < 10^{-1}$ ppm) at temperatures ranging from 600 to 1090°C, pressures ranging from 10^{-1} to 100 Torr and for oxide thicknesses ranging from some monolayers to 300 nm (7,8,9,18). By making complete growth in $^{18}O_2$ we carried out also precise measurements of the kinetics since the first steps of growth, as ^{18}O content of the films can be measured by nuclear microanalysis without any backgrown and with a great sensitivity ($\sim 10^{-2}$ monolayer). When the O_2 gas was sufficiently dry, two regimes of oxide growth were clearly exhibited : an initial one, slow and of direct logarithmic type and where the oxide thickness never exceeds 6 nm, is succeeded by a faster one of roughly linear or linear parabolic law.

For PO_2=10 Torr, the logarithmic regime was the only one observed at 600°C and 810°C, while in 930-1090°C range its duration decreases as temperature and/or the content in water vapor traces incerases. After $^{16}O/^{18}O$ sequential oxidation during this logarithmic regime, ^{18}O is found essentially near the oxide/surface. For oxidation temperatures ranging from 810°C to 1090°C oxygen exchange is observed while at 600°C no oxygen exchange is observed. This shows that at 600°C oxide growth is due either to silicon migration or to a vacancy type migration of oxygen network atoms owing to mechanism Bb ii) which corresponds to the transport of charged defects under an electric field. From energetical arguments silica defects corresponding to the migration of silicon alone are not likely to be present. Thus the migration of oxygen network atoms seems the more likely. The exchange of oxygen atoms observed in the 810-1090°C range shows that migration of oxygen network atoms occurs owing to the mechanism Bb i) which corresponds to the single diffusion of the network oxygen defects (vacancies or interstitialcies).

During the faster regime $^{16}O/^{18}O$ sequential oxidation leads to the fixation of ^{18}O near both interfaces of the oxide film i.e. near the Si/SiO_2 interface (^{18}O FNI) and near the external surface of the oxide (^{18}O FNES). A very steep transition in isotopic concentration is observed between an ^{16}O intermediate layer and the ^{18}O FNI region where ^{18}O labelling is found equal to that of the $^{18}O_2$ gas. This ^{18}O FNI is due to growth by mechanism Ba i) i.e. an interstitial migration of the O_2 molecule without any reaction with the silica network in agreement with D.G. theory. In the ^{18}O FNES region the ^{18}O isotopic labelling progressively decreases from the surface to the bulk of the oxide film. Furthermore a large fraction of ^{18}O FNES is lost during a further growth in ^{16}O, showing that oxygen exchange occurs. Let us notice that this oxygen exchange between the oxide and the O_2 molecule occurs only at the oxide/gas interface showing the particular role of this external oxide surface. Thus ^{18}O FNES corresponds to a transport of oxygen atoms network by the mechanisms Bb i) i.e. a vacancy type diffusion mechanism. The amount N_{in} of ^{18}O FNI is found proportional to the duration Δ t of the $^{18}O_2$ treatment if the thickness increases Δ X is small enough. This is also in agreement with the D.G. theory which gives $\Delta X/\Delta t = k_p/(2X + k_p/k_1)$. For different initial oxide thicknesses equal to 50, 265 and 1300 Å (9,18) and oxygen pressures ranging from 10^{-1} to 100 Torr, N_{in} was found proportional to $PO_2{}^{\alpha}$ with α = 0.6 - 0.8 while it should be equal to 1 owing to D.G. theory. The amount N_S of ^{18}O FNES is found roughly proportional to the square root of time and to vary as a log PO_2 law. This shows that the contribution of growth to N_S is smaller than that due to exchange. The ratio N_S/N_{in} decreases as the initial oxide thickness increases. This shows that growth in this faster regime occurs essentially by a D.G. mechanism and that the contribution of a vacancy-type mechanism is strongly limited by the supply of network defect and/or electronics species. Results of ^{18}O tracer studies carried out by Han and HELMS (11) at 1000°C and COSTELLO and TRESSLER (10) at 1000°C and 1300°C are in qualitative agreement. An H_2O contamination might explain the higher N_S values reported in (10).

ACKNOWLEDGMENTS

The author gratefully acknowledges F. ROCHET for fruitful discussions and his participation in most of the reported experiments.

This research was supported by the C.N.R.S. (GRECO 86) and G.C.I.S.

REFERENCES

1. B.E. DEAL and A.S. GROVE, J. Appl. Phys. 36, 3770 (1965).
2. S. RIGO, "Instabilities in silicon devices" Vol. 1, Edited by G. BARBOTTIN and A. VAPAILLE, (Amsterdam, North Holland) (1986).
3. S. RIGO, F. ROCHET, B. AGIUS and A. STRABONI, J. Electrochem. Soc. 129, 867 (1982).
4. F. ROCHET, Thèse de Doctorat de 3e Cycle PARIS (1981).
5. G. AMSEL and D. SAMUEL, Anal. Chem. 39, 1689 (1967).
6. D. DIXMIER, A. L'HOIR and G. AMSEL, Nucl. Inst. Methods, 197, 537 (1982).
7. F. ROCHET, S. RIGO, M. FROMENT, C. D'ANTERROCHES, C. MAILLOT, H. ROULET and G. DUFOUR, Advances in Physics, 35, 237 (1986).
8. E. ROSENCHER, A. STRABONI, S. RIGO and G. AMSEL, Appl. Phys. Lett. 34, 254 (1978).
9. F. ROCHET, B. AGIUS and S. RIGO, J. Electrochem. Soc. 131, 914 (1984).
10. J.A. COSTELLO and R.E. TRESSLER, J. Electrochem. Soc. 131, 1944 (1984).
11. C.J. HAN and C.R. HELMS, J. Vac. Sci. Technol. A4, 853 (1986).
12. J.C. DUPUY, G. PRUDON, P. PINARD, S. RIGO, "Proceedings of SIMS", Versailles, 13-18 Sept. 1987, John Wiley and Sons Publishers, In Press.
13. S. RIGO, F. ROCHET, A. STRABONI and B. AGIUS, "The Physics of MOS Insulators" G. LUCOWSKY, S.T. PANTALIDES and F.L. GALEARNER, Editors, P. 167, Pergamon Press, Clumsford, NY (1980).
14. F. ROCHET and S. RIGO, Philosophical Magazine B, 55, 747 (1987).
15. F. ROCHET and S. RIGO, Philosophical Magazine B, 57, 123 (1988).
16. J.C. MIKKELSEN, Appl. Phys. Lett. 39, 601 (1981).
17. W.G. SPITZER and J.R. LIGENZA, J. Phys. Chem. Solids, 14, 131 (1960) and 17, 196 (1961).
18. S. RIGO, F. ROCHET, I. TRIMAILLE, Unpublished.

THE OXIDATION OF SILICIDES ON SILICON

F. M. d'Heurle

IBM T. J. Watson Research Center, PO 218, Yorktown Heights, NY 10598, USA
and
Institutionen för Mikrovagsteknik, Kungliga Tekniska Högskolan
100 44 Stockholm, Sweden

ABSTRACT

Under proper care the oxidation of silicide films formed over silicon substrates results in the formation of metal-free layers of silicon oxide. The silicide layers themselves appear to be unaffected by the oxidation process. The overall kinetic and thermodynamic conditions which make this possible are reviewed. The growth of the oxide is always faster over the silicides than over (100) silicon, but semiconducting silicides with large band gaps oxidize almost as slowly as silicon, whereas silicides with a strongly metallic character oxidize rapidly. Other factors such as whether the formation of the oxide layers occurs from the direct diffusion of silicon atoms or through the decomposition (followed by the reconstitution) of the silicide are considered. Some attention is also paid to such problems as that of the formation and behavior of point defects generated by the silicidation and oxidation processes.

INTRODUCTION

The issue of the formation of silicon oxide over silicides was first considered as a means of protection against the oxidation of parts made of metals such as molybdenum and tungsten (see for example: Perkins 1964, Gage and Bartlett 1965, Kofstad 1966, Abba et al. 1982). In the last twenty years with the initial suggestions (Mochizuki et al. 1978, Crowder and Zirinsky 1979), and then actual use, of silicides for silicon device interconnections there has been renewed interest in the properties of these materials including their behavior in oxidizing environments. This latter topic has been the subject of two recent and rather complementary reviews (Jiang et al. 1986, d'Heurle et al. 1987); the first of these offers a good summary of the accumulated experimental data, the second contains a longer discussion. (No attempt will be made here to duplicate the bibliography to be found in these two articles. The best source of general information about silicides remains Nicolet and Lau 1983). One of the main technological advantage of silicides is their relative inertness during processing at high temperatures in oxidizing environment. This results from the formation of a protecting layer of silicon oxide. From a practical point of view the oxide layer itself is not extremely important in the current electronic industry with the result that with few exceptions (Bartur and Nicolet 1984, Frampton 1987) one finds little published information about the electrical properties of the oxide layers grown over silicides. The silicides, however, remain quite fascinating since under the proper conditions they offer one the possibility of growing metal-free silicon oxide over a variety of substrates with greatly different properties, not only silicon, but semiconductors with band gaps as large as silicon, Ir_3Si_5 (Petersson et al. 1982) or Ru_2Si_3, semiconductors with smaller band gaps, $CrSi_2$ (about 0.4 eV, Bost and Mahan 1988), $ReSi_2$ (about 0.2 eV, Krontiras et al. 1988), semimetals such as $MoSi_2$ and WSi_2 (Laborde et al. 1986, Badoz et al. 1987), or materials with strong metallic character such as NiSi, $NiSi_2$ and $CoSi_2$ (Tersoff and Hamann 1983). Thus it is hoped that a proper understanding of the mechanism of oxide formation over the silicides would throw some light on the remaining unsolved problems of silicon oxidation. There are other means of oxidizing silicides, e.g. plasma anodization (Perrière et al. 1987), but this work will be limited to thermal oxidation.

OVERVIEW

a) Mass Transport

One is normally concerned with the oxidation of a silicide layer grown over a silicon substrate from the reaction of a metal film with that substrate. This insures that there exists no, or only a residual, layer of impurities at the silicide-silicon interface which could act as a diffusion barrier. With silicide layers formed by some codeposition process, such as sputtering, special care must be excercised to minimize the effects of any residual oxide over the substrates, possibly by prolonged high temperature treatment in a neutral atmosphere prior to oxidation (d'Heurle 1982). Figure 1 presents a series of superimposed backscattering spectra from layers of WSi_2 that have undergone oxidation at 800°C for different lengths of time. The gradual recess of the tungsten part of the spectra towards lower energy is due to the increasing thickness of SiO_2 with increasing oxidation treatments. Magnification of the spectra in the region of high backscattering energy would reveal that the oxide is free of detectible metal inclusions: given the high atomic weight of tungsten and the resulting high sensitivity this means a metal concentration lower than 0.1 at %.

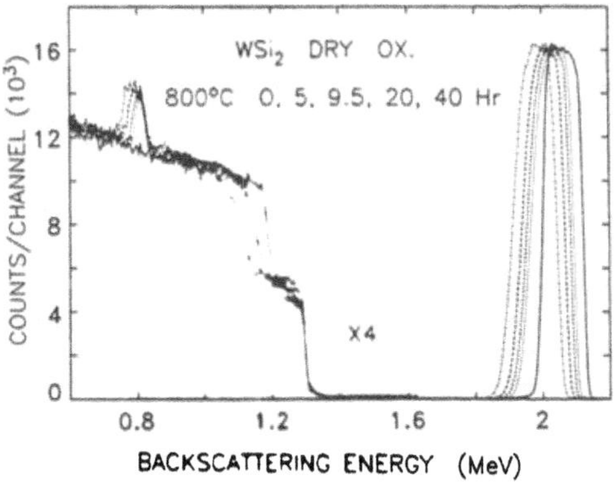

Figure 1 - Backscattering spectra of a series of WSi_2 layers having been oxidized at 800°C for periods of time from 0 to 40 hours.

The growth of the silicon oxide at the surface without any noticeable modification of the silicide layer indicates that there exists a mechanism for the net transfer of silicon from the substrate to the oxide, across the silicide. This can occur from the direct diffusion of silicon across the silicide, or from a more complex process which involves the decomposition of the silicide at the silicide- oxide interface, the "reverse" difffusion of the liberated metal atoms towards the substrate, and recombination of these atoms with silicon to reconstitute the decomposed silicide. These conditions are depicted in Fig. 2. One may determine which one of these processes occurs during oxidation by means of marker experiments. In Fig. 3 (from d'Heurle 1987) one observes that the position of a rhodium marker placed in the middle of a layer of Ir_3Si_5 does not change with respect to the iridium as a result of the oxidation treatment, thereby proving that oxidation occured via the diffusion of silicon atoms. This is in agreement with the "ordered Cu_3Au" rule which states that in compounds one anticipates the majority atoms to be more mobile than those of the minor component (d'Heurle and Gas 1986). With NiSi the contrary condition obtains, silicide formation and oxidation both occur via the motion of nickel atoms. The important point is that the apparent inertness of the silicide layer is insured by conditions of temperature and oxidant pressure such that the net transport of silicon across the silicide occurs at a rate equal to the rate of oxide formation. The rate of formation of the oxide shall be considered again later. One notes that the flux, J_{difsi}, of silicon atoms across the silicide will be given by the Nernst-Einstein equation as proportional to the mobility of the atoms and to the force acting on them (the gradient of chemical potential from one side to the other):

$$J_{difsi} = c.D/kT.\Delta G/t \tag{1}$$

Where c is the concentration of silicon in the silicide (e.g. 0.66 in WSi_2, or the equivalent in at/cm³), D is the diffusion coefficient of silicon in the silicide, and t is the thickness of the silicide. ΔG is the free energy change driving the diffusion of the silicon atoms from the silicon-silicide to the silicide-oxide interface. Were one to assume as usual that equilibrium prevails at these interfaces, ΔG would be the free energy of formation of SiO_2 at the temperature T of oxidation. It will be seen later on that the formation of metal-free SiO_2 excludes the possibility of such an equilibrium existing, so that Δ G should be relatively small. (For a study of the problems involved in the use of Eq. 1 and that of Fick, see d'Heurle et al.

1986). In the case of silicide decomposition and reverse motion of metal atoms the value of ΔG would appear to be the free energy of formation of the silicide (nearly equal to the enthalpy of formation - per metal atom). One direct consequence of these considerations is that the formation of metal free SiO_2 requires temperatures sufficiently high to allow relatively free motion of silicon in the silicides, and therefore to be safe a temperature above that at which a given silicide forms by metal-silicon reaction. A corollary therefrom is that in the oxidation process it is best to heat the samples in an inert atmosphere, and to introduce the oxidant only after the desired temperature has been reached.

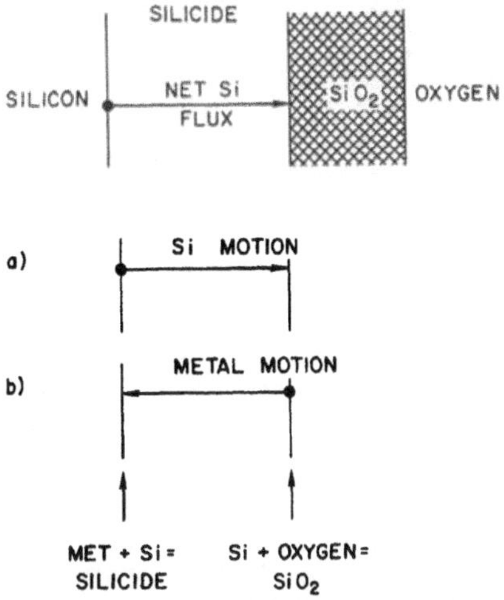

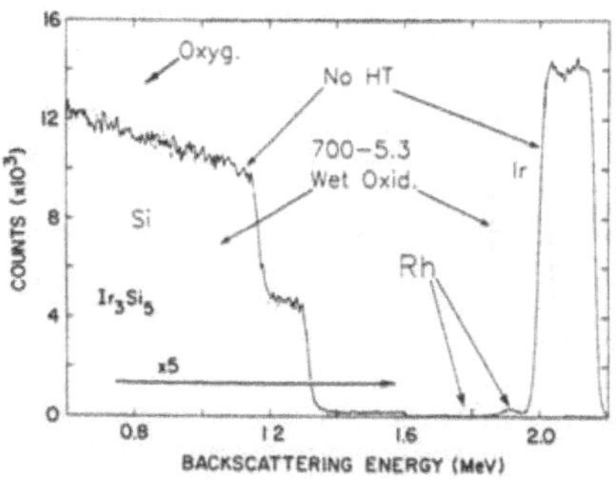

Figure 2 - (left) A schematic representation of the net transport of atoms during the oxidation of a silicide deposited, or formed, over a silicon substrate.

Figure 3 - (right) Backscattering spectra from a marker experiment showing that Ir_3Si_5 oxidizes via the transport of silicon atoms across the silicide layer.

b) Thermodynamic

It is perhaps perplexing that SiO_2 forms exclusively, without any formation of metal oxide, at least provided that the conditions considered above are fulfilled. The interpretation of these results is based on thermodynamic equilibrium (Bartur1983, Beyers 1984, Beyers et al. 1984). For the oxidation of $TiSi_2$ one must compare Eqs. 2 and 3 (below). The first of these, Eq. 2, corresponds to the oxidation of silicon that has diffused through the silicide and is being oxidized at the surface. Equation 3 corresponds to the simultaneous oxidation of both silicon and titanium, from the silicide itself. The main difference between the two, is that for the second reaction (Eq. 3) the driving force, which is the free energy change upon completion of the oxidation process, is reduced as compared with Eq. 1, by a quantity which corresponds to the decomposition of the silicide ΔG_f. When one considers the respective free energy changes, calculated per mole of reacting oxygen (this is important in considering the competition of various reactions for the available oxygen), one finds that equations of type 2 are favored with respect to equations of type 3. One encounters a limit with materials which form tightly bound oxides, such as zirconium and hafnium. The disilicides of these two metals cannot be oxidized to form metal free SiO_2, and actually titanium represents the upper (on a scale of increasing energy of oxide formation) limit of a silicide which can still be oxidized to pure SiO_2. Because the free energies of formation of the oxides as a function of temperature follow nearly parallel paths, the above argument in term of free energy remains valid if one compares only the various standard enthalpies of formation.

$$Si + O_2 = SiO_2 \qquad (2)$$

$$TiSi_2 + 3O_2 = 2SiO_2 + TiO_2 \qquad (3)$$

These considerations can be expressed graphically in the form of isothermal cross sections of ternary equilibrium diagrams for systems that include silicon, oxygen and the various metals. Figure 4 offers an example of such a diagram for the silicide of tungsten. Each of the triangles present in this diagram indicates that the three phases that correspond to the three apices of these triangles can coexist in equilibrium. However, as seen on the right side of this figure the conditions that prevail at the silicide-oxide

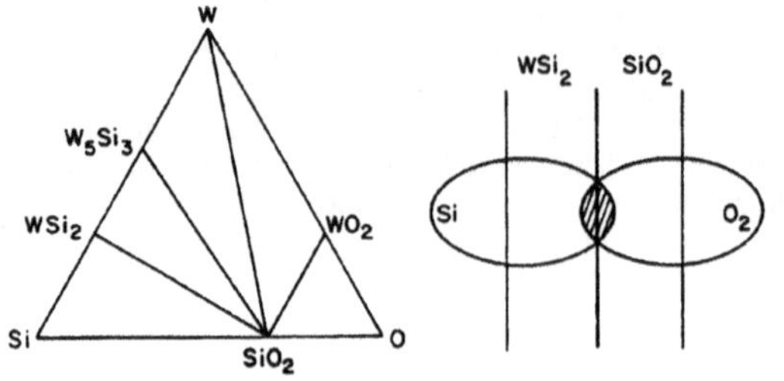

Figure 4 - Isothermal cross section of the ternary equilibrium diagram for the system silicon-tungsten-oxygen. (From Byers 1984). What happens at the silicide-oxide interface, however, is a function of what happens in the two adjoining layers.

interface depend on the kinetics of what happens both in the silicide and the oxide layers. If the oxide for some reason should cease to grow one is no more in the presence of a reaction (or lack therefrom) between a silicide and SiO_2 but, the oxide being now saturated with oxygen, between a silicide and oxygen. In theory, according to Fig. 4, if one had a very thick layer of WSi_2 resulting in a very slow diffusion of silicon from the substrate, it should be possible to reduce WSi_2 to W_5Si_3 and form SiO_2 from the silicon liberated in that process. Further growth of the oxide would then depend on diffusing silicon, from WSi_2 through a growing layer of W_5Si_3. However, if the diffusion of silicon through WSi_2 was too slow, the diffusion of silicon through W_5Si_3, is likely to be even slower since the respective activation energies are 2.2 eV and 3.5 eV (Bartlett et al. 1964, Gage and Bartlett 1965 a), although Wakita et al. (1984) report such a process taking place. It is this author's experience that, at temperatures below 1000°C, when the oxidation of WSi_2 does not proceed properly the most likely event is the production of some volatile tungsten oxides, that blister through already formed SiO_2, leading to a catastrophic failure of the oxidation process. However, the reported results are not unanimous on this point. Some authors (Rouse et al. 1980) have shown, at 1100°C, a relatively homogeneous formation of W_5Si_3 occuring through the whole thickness of the WSi_2 film, together with the growth of an adherent oxide film containing some tungsten; it is thought that a relatively high level of carbon in the silicide film stabilizes the W_5Si_3 phase, although the high temperature probably plays a determining role.

Actually the selective oxidation of silicides is similar to the selective oxidation of alloys in which one of the elements form a more stable oxide than the other. Because of the wide practical implications of this process it received attention quite early and was analyzed in detail by Wagner in 1952 (for a more recent treatment see: Gibbs 1981). For an alloy of two elements that form oxides of very different stabilities the condition for the formation of a protective film of the most stable oxide is that the concentration of that element should exceed a critical value N_{crit}; N_{crit} increases as the rate of oxide growth increases, and decreases as the diffusion coefficient of that element in the alloy increases. Gibbs (1981) underlines that this criterium which appears to be purely kinetic, is in fact a thermodynamic condition. The point can be made with reference to both Eq. 2 and Fig. 5. For Eq. 2 to be valid in the environment being considered here the activity coefficient of the silicon in the silicide must be one, namely as seen in

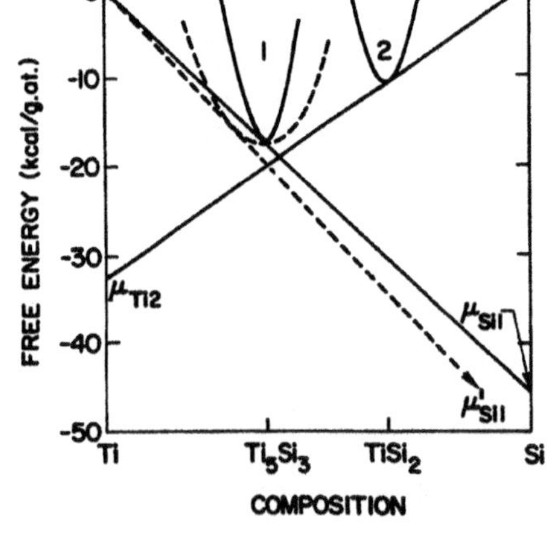

Figure 5 - Partial representation of free energy for the system titanium-silicon with emphasis on equilibrium between titanium and Ti_5Si_3, and between silicon and $TiSi_2$. Any decrease in the chemical potential of silicon at the silicide-oxide interface mandates a corresponding increase in that of the metal (Gibbs-Duhem relation). Thus there exists a precise equivalence between the driving force for silicon diffusion (Eq. 1) and that for the diffusion of metal in the opposite direction, as illustrated in Fig. 2.

Fig. 5, equilibrium between the silicide and silicon mandates that the chemical potential of the silicon in the silicide should be the same as that of pure silicon, while the chemical potential of titanium μ_{Ti2} is much lower than that of pure titanium. Were one to consider a silicide grown over the metal (Abba et al. 1982), one would most likely encounter a metal-rich silicide, e. g. Ti_5Si_3, that would be in equilibrium with the metal, a situation also described in Fig. 5. Instead of considering Eq. 2, one would have to consider the direct formation of the metal oxide. The exact values of the chemical potentials are a function of the shape of the free energy curves versus composition for the different phases (e.g. μ_{Si1} or μ'_{Si1}), but this is a secondary matter. The oxidation of "bulk" $MoSi_2$ at temperatures of the order of 1500°C was analyzed by Gage et al. (1965 b).

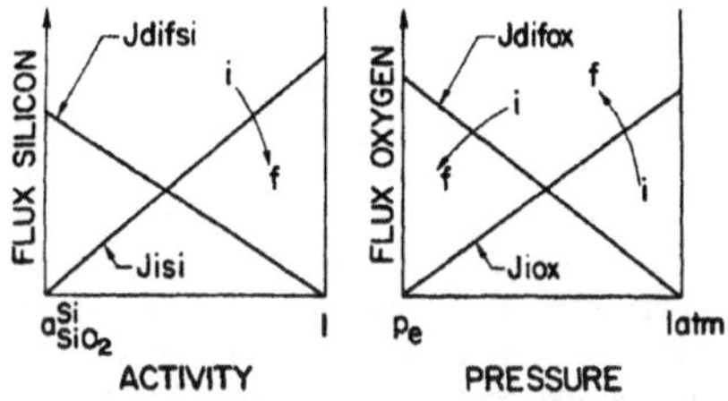

Figure 6 - Graphical representation of the oxygen (right) and silicon fluxes (left) during silicide oxidation. In both cases it is necessary to distinguish the flux across the layers, oxide and silicide, and the two fluxes of atoms across the silicide-oxide interface, that are proportional to the reaction rate.

The equilibrium condition which has just been described can exist only if there is no consumption of silicon at the surface of the silicide; the moment silicon is withdrawn from that surface the chemical potential of silicon is bound to drop, that of titanium to increase, so that the the standard thermodynamic values cease to be valid for Eqs. 2 and 3. The thermodynamic potentials of silicon and titanium are then described by a clockwise rotation of the tangent to the free energy curve for $TiSi_2$. Figure 6 is an attempt to be more quantitative about this question by means of a graphical scheme (Levich 1962) that has been used (Irene and Ghez 1987) to describe the linear parabolic law of oxidation kinetics. On the right hand side of Fig. 6, one distinguishes two fluxes of oxygen molecules both of which are plotted against the pressure (or concentration) of oxygen at that interface. The usual diffusive flux, J_{difox} through the oxide layer decreases linearly as the interface oxygen pressure p increases up to the effective pressure of oxygen (assumed here to be 1 atmosphere); it is represented by a line whose slope is proportional to the diffusion coefficient and inversely proportional to the thickness L of the oxide. One must also consider a second flux, that corresponds to the rate of oxygen consumption at the interface, J_{iox}, or to the reaction rate (the linear term in oxidation kinetics); it is given by the equation below:

$$J_{iox} = k(a - a^{si}_{sio2})(p - p_e) \tag{4}$$

In silicon oxidation the silicon activity term would be 1, but this cannot be the case for the oxidation of a silicide. The reaction rate constant k is not the usual one, valid for silicon (ignoring differences due to orientation changes), but will assume different values for each silicide. The actual rate of oxide growth, and the oxygen interface pressure are given by the intercepts of the two lines representing the two fluxes. Similarly one has on the left side of Fig. 6 two fluxes of silicon atoms, defined now as a function of the silicon activity at the silicide-oxide interface, one of which corresponds to the "net" diffusion of silicon through the silicide (Fig. 1), and a second interface flux which is also given by Eq. 4 (given that the diffusing species are silicon atoms on the one hand and oxygen molecules on the other). Again steady state is given by the intercept of the two silicon flux lines which defines the silicon interface activity and the flux of silicon at a given moment. Obviously the two fluxes obtained on the left and right hand side of Fig. 6, of silicon atoms and oxygen molecules, should be equal. One could have stated these considerations in the form of three equations which would not yield a linear-parabolic solution, but a complex version thereof. Let us note that the diffusive flux of oxygen decreases as the oxide layer increases so that the flux line becomes more horizontal from the intial to the final condition, i to f, so that for an infinitely thick oxide layer the flux line becomes identical with the x axis; the corresponding line for the diffusive flux of silicon remains fixed since the silicide thickness is constant. Some important consequences ensue from this. As oxidation progresses: 1) The interface pressure of oxygen tends towards the equilibrium value, where no growth takes place. 2) Because of this, the term $k(p - p_e)$ in Eq. 4 tends towards zero, so that the line for the interface flux of silicon becomes more horizontal; finally the activity of silicon at the interface becomes 1 (it exhibits a behavior which is exactly the opposite of that displayed by oxygen). 3) The line for the interface flux of oxygen becomes more vertical because the term $(a - a^{si}_{sio2})$ tends towards unity.

The most critical period during the oxidation of a silicide is the intial one since the rapid influx of oxygen decreases the silicon activity while increasing in proportion the activity of the metal, thereby making it much more likely to form metal oxides. In failed oxidation it is frequent to find inclusions of metal oxides in the top layers of an oxide film, while the inside layers can be quite pure. One remedy is to increase the slope of the diffusive silicon flux line in Fig. 6, either by selecting a thinner layer of silicide, or by increasing the silicon diffusion coefficient with a higher oxidation temperature. This is valid only if the flux of oxygen is not increased in proportion; fortunately the kinetic parameters of silicon oxide growth are characterized by relatively low activation energies, so that an increased temperature usually improves the possibility of forming oxide layers that are metal-free. The conditions of initial oxidation are not quite as unfavorable as shown above, for two different reasons. The first is directly related to what has already been said. The picture which has presented is one that corresponds to a steady state during some intermediate step in the oxidation process. However, this does not quite apply to the initial condition when the silicide, because of the equilibrium that prevails, is saturated with silicon. Thus the silicon being oxidized does not have to diffuse from the substrate and the flux can be higher than anticipated on basis of the model presented. A second reason is that there is a general tendency for silicide surfaces to be silicon-rich, presumably because of an adsorption process that decreases the surface energy. Examples of this effect relating to $NiSi_2$ and $CoSi_2$ can be found in the literature (Chang and Erskine 1982, Wang et al 1983, Chambers et al. 1986). Once the surface silicon has been oxidized the flux of oxygen at the interface is already reduced and so is the effect on the silicon activity. Man can help nature by depositing on top of the silicide a thin layer of silicon, or silicon oxide.

KINETICS

The question of silicide oxidation is not quite the proper point to raise the problem of the precise validity of the linear-parabolic law of oxide formation. In the previous part of the paper the law was assumed to be valid for silicon, and to require modification when applied to silicides. With rapidly oxidizing silicides, such as $NiSi_2$, oxide growth is nearly parabolic. Which is to say that with the silicides the linear term is higher than with silicon. On Fig. 7 (from Frampton 1987) the thicknesses of oxide layers grown in oxygen (about 20 ppm moisture) over various silicides and on both (111) and (100) silicon have been plotted as a function of time for two oxidation temperatures (700°C. and 900°C). It is seen that for thick layers the different curves are more or less parallel, indicating that the rate of diffusion in the various oxide layers are quite similar. There is much more discrepancy in the various slopes near the origin, especially at 700°C, symptomatic of a wide range of reaction rates. It is also seen that there is a greater divergence of oxidation rates at low temperatures (even after the small mismatch in the vertical scales on Fig. 7 is factored out). The silicide with the lowest resistivity, NiSi, oxidizes the fastest; it could not be oxidized at 800°C because at this temperature it reacts to form $NiSi_2$. The semiconducting silicides, Ru_2Si_3 and Ir_2Si_5, oxidize quite slowly, at rates close to those that obtain for pure silicon. WSi_2 also oxidizes slowly; this can be attributed to its electronic structure, since it has been proven to be a semimetal (Badoz et al. 1987), with a very low density of states at the Fermi level. It is possible, therefore, to correlate an increasing reaction rate with an increasing metallic character of the silicide (or silicon) substrate. This conclusion finds support in different sets of observations. 1) Surface spectroscopic studies showing that silicon surfaces modified with a few monolayers of metallic atoms oxidize or do not oxidize as a function of heat treatments that change the surface character from metallic to semiconducting (e.g. Cros et al. 1981, see also d'Heurle et al. 1987). 2) A strong correlation of the rate of silicide oxidation with the imaginary part of the silicide index of refraction. Since this measures the adsorption of light in the silicide it is a measure of the density of charge carriers (Frampton, and Frampton et al. 1987). 3) The present ideas about silicide oxidation concord with the thermionic model (Irene and Lewis 1988) where the rate of oxidation depends on the density of electrons excited in the conduction band of the oxide, and therefore also on the density of electrons at the Fermi surface of the substrates. That the oxidation rates should assume a common value at high temperatures would fit with these observations, since at these temperatures the semiconductors become increasingly metallic.

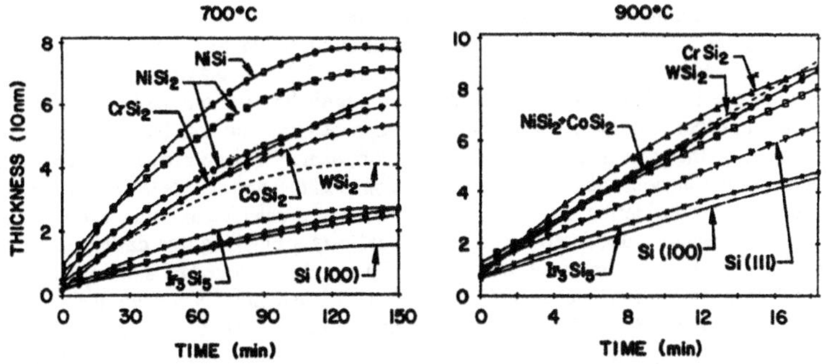

Figure 7 - The growth of oxide films over NiSi, $NiSi_2$, $CoSi_2$, $CrSi_2$, WSi_2, Ir_3Si_5, Ru_2Si_3, Si (111) and Si (100). The two sets of values at 700°C for $NiSi_2$ correspond to (111) and (100) substrates. The fastest rate is for the (100) substrate; this possible orientation effect disappears even at 750°C. (From Frampton 1987).

It has also been proposed (Lien et al. 1984) that the rate of oxidation depends on the availability of silicon atoms, so that those silicides that decompose during oxidation, NiSi and $CrSi_2$ are two of them, oxidize faster than the silicides where silicon atoms diffuse from the substrate, as is the case for Ru_2Si_3 and Ir_3Si_5 (Fig. 2, d'Heurle 1987). The present author does not see why silicide decomposition should necessarily increase the silicon supply since there is no decomposition followed by oxidation. The two phenomena are presumed to be simultaneous, possibly decreasing the driving force for oxidation by a quantity equal to the heat of formation of the silicide (small with respect to the heat of formation of SiO_2). In any case, to go back to the discussion above, how could the activity of silicon assume a value much greater than one as a result of decomposition? Thus even if this idea should allow one to compare different silicides it would remain to explain why k in Eq. 4 should be greater for the silicides than for silicon, requiring also some other mechanism similar to the proposed electronic one.

POINT DEFECTS AND POSSIBLY RELATED PROBLEMS

Two recent reviews about silicide were mentioned earlier. Since these were published some interesting observations were made about silicides and point defects which merit attention. It has been shown that the formation of $TaSi_2$ causes silicidation-enhanced diffusion of dopants in the underlaying silicon (Hu 1987), and that of the closely related $TiSi_2$ causes the disappearance of substrate dislocation loops through the injection of vacancies, the inverse of oxidation-enhanced stacking faults (Wen et al. 1987). Thus observations of point defects pertaining to the oxidation of silicon and of metals (Leroy 1987, Adda and Philibert 1987) have been generalized to other solid state reactions. It is possible that such creation and injection of point defects are related to some hitherto unexplained aspects of silicide oxidation.

NiSi oxidizes via decomposition and nickel motion, which is not too suprising since during the formation of that silicide nickel atoms constitute the dominant mobile species. However, things become more involved with the oxidation of the isomorphous $NiSi_2$ and $CoSi_2$ both of whom form via metal atom motion. One anticipates, therefore, that oxidation would also occur via metal atom motion, as verified by experiments. However, one observes not only the reverse motion of the metal atoms, but these are accompanied by a large quantity of silicon atoms, representing some 25 % of the silicon available at the silicide-oxide interface (one silicon atoms for every two metal atoms). Since this occurs against any possible equilibrium gradient in silicon concentration or chemical potential an explanation must have recourse to some instantaneous desequilibration similar to what occurs during silicon oxidation. Desequilibration of some sort seems to be required not only for the initial injection of silicon atoms as point defects, but also for their diffusion since under equilibrium germanium atoms, implanted in $CoSi_2$ in lieu of silicon atom tracers, diffuse extremely slowly even at 800°C (Thomas et al. 1988). In this context one must also consider the case of $CrSi_2$. With respect to crystallographic structure it belongs to the same category of silicides as $TiSi_2$, WSi_2, and $TaSi_2$ (quite distinct from $NiSi_2$ and $CoSi_2$ with their widely open CaF_2 structure). As such it forms as the others by silicon atom motion, presumably with the same injection of point defects as reported for the formation of $TiSi_2$ and $TaSi_2$. However, it oxidizes via the motion of chromium atoms (Lien et al. 1984), a fact so surprising as to require verification, now provided (Hudner et al. 1987). The use of vanadium atom tracers in equilibrated $CrSi_2$ (Finstad 1988) indicates that chromium atoms are mobile in $CrSi_2$ at temperatures as low as 800°C, much lower than the 1100°C required to provide mobility to the metal atoms in $MoSi_2$ or WSi_2, thus the motion of chromium at usual oxidation temperatures does not seem to depend on extraordinary interventions. In silicon oxidation one is well accustomed to deal with the problems of interface reactions, but these are often neglected in other reactions. During the formation of $CrSi_2$ the possible reactions at the silicon surface appear to be the same as during silicide oxidation, namely, silicon extraction from the substrate and "injection" in the silicide, and/or chromium injection in silicon and silicide formation. However, this parallelism fails when considering the other dissimilar interfaces, silicide-chromium and silicide-oxide. Is it legitimate to ask whether one interface reaction could play a blocking role during oxidation? Is decomposition leading to chromium sursaturation and diffusion characterized by a lower activation energy than direct silicon consumption? Silicon diffusion cannot be rate limiting, since the silicide forms at very low temperatures, about 500°C. The problem remains unresolved. Before leaving this topic of point defects one notes that experiments (Fahey 1988) conducted to determine whether the extraction of silicon atoms from the substrate during the oxidation of WSi_2 leads to the injection of point defects and enhanced dopant diffusion failed to reveal any such effects.

OXIDE PROPERTIES

A review of the limited amount of information on the properties of the oxide films grown over silicides is found in Jiang et al. (1086). We repeat here the earlier remark that kinetic studies (Fig. 7) indicate that the diffusion of oxygen through these oxides occurs at about the same rate as in oxides grown over silicon, implying that their structures are nearly identical. SiO_2 appears to be SiO_2 whether grown over silicon or over silicides. This is confirmed by the results below from etching studies on oxides grown over silicides and on silicon at 750°C and 800°C (Frampton 1987). The etchant was nominally 15 parts HF (49%) - 10 parts HNO_3 (70%) - 300 parts H_2O. There were two etching experiments apparently with slight variations in composition, or temperature, accounting for some difference in rates (see Si (100) a, grown at 1000°C). However, vertically the rates may be compared. Except for $CrSi_2$ the results are confined to narrow limits. There is a strong correlation between the rate of oxide growth and the rate of etching, but otherwise there is little difference between oxides grown over silicon or silicides at nearly similar rates. Experiments on oxide breakdown may reveal low breakdown voltages for oxides grown over silicides, of the order of 1 MV/cm. However, this is probably caused by the substrates rather than being an intrinsic property of these oxides, since higher values are also reported (7 MV/cm, Göltz and Ferrieu 1987).

Table I - Etch rates for oxides grown at 750°C and 800°C. (From Frampton 1987).

750°C, 1000 minutes	Å/min.	800°C, 480 minutes	Å/min.
$CrSi_2$	264	$CrSi_2$	237
		$CoSi_2$	181
Si (111) a	208		
Si (100) a	208	Si (100) a	170
$CoSi_2$	200	Ir_3Si_5	161
$NiSi_2$	183	WSi_2	157
Ir_3Si_5	169	Si (111)	155
Si (111)	164	Si (100)	154
WSi_2	161	Ru_2Si_3	140
Ru_2Si_3	158		
Si (100)	156		

a) oxides grown at 1000°C.

SUMMARY AND CONCLUSIONS

The growth of metal-free oxides over silicides depends on fulfilling both kinetic requirements about the supply of silicon from the substrate and thermodynamic requirements. The interdependance between these two sets of apparently unrelated conditions has been analyzed.

The kinetics of oxide growth are seen to be a function of the electronic character of the substrate. Strongly semiconducting substrates, including silicon, oxidize slowly, while metallic ones oxidize much faster. The decomposition, or the direct silicon oxidation of the silicides may help in comparing growth rates over different silicides; it does not explain high rates of silicide oxidation, as compared with those that obtain with pure silicon.

The role of point defects during silicidation and oxidation has been discussed. It remains to be understood.

The common rate of oxygen diffusion in oxides grown over silicides and over silicon, and comparable etching rates indicate that the characteristics of the oxide films are not a strong function of the substrates over which they are grown.

ACKNOWLEDGEMENTS

The author is particularly indebted to R. Frampton; without his diligent labor much of this work would not have been possible. The collaboration with both R. Frampton and E. Irene over the past three to four years has been a great source of pleasure and intellectual stimulation. The permission granted by R. Frampton and R. Beyers to reproduce published and unpublished results is acknowledged with gratitude.

REFERENCES

Abba, A., Galerie, A., and Caillet, M., 1982, High-temperature oxidation of titanium silicide coatings on titanium, Oxidation of Metals 17, 43.

Adda, Y., and Philibert, J., 1987, Tendances actuelles des études de diffusion en métallurgie physique et en science des matériaux, Mém. Sci. Revue de Métallurgie, 84, 293.

Badoz, P. A., Rosencher, E., Torres, J. and Fishman, G., 1987, Insulating, metallic and semimetallic electronic nature of XSi_2 compounds: Application to WSi_2, J. Appl. Phys. 62, 890.

Bartlett, R. W., Gage, P. R., and Larssen, P. A., Growth kinetics of intermediate silicides in the $MoSi_2$/Mo and WSi_2/W systems, Trans. Met. Soc. AIME 230, 1528.

Bartur, M., 1983, Thermal oxidation of transition metal silicides: the role of mass transport, Thin Solid Films 107, 55.

Bartur, M., and Nicolet, M.-A., 1984, Properties of SiO_2 grown on Ti, Co, Ni, Pd and Pt silicides, J. Electron. Mater. 13, 81.

Beyers, R., 1984, Thermodynamic considerations in refractory metal-silicon-oxygen systems, J. Appl. Phys. 56, 147.

Beyers, R., Sinclair, R., and Thomas, M. E., 1984, Phase equilibria in thin-film metallizations, J. Vac. Sci. Technol. B 2, 781.

Bost, M. C. and Mahan, J. E., 1988, An investigation of the optical constants and band gap of chromium disilicide, J. Appl. Phys. 63, 839.

Chambers, S. A., Anderson, S. B., Chen, H. W., and Weaver, J. H., 1986, High-temperature nucleation and silicide formation at the Co/Si (111)-7x7 interface: A structural investigation, Phys. Rev. B 34, 913.

Chang, Y.-J., and Erskine, J., 1982, Electronic structure of $NiSi_2$, Phys. Rev. B 26, 7031.

Crowder, B. L., and Zirinsky, S., 1979, 1 μm MOSFET VLSI technology: Part VII - Metal silicide interconnection technology - A future perspective, IEEE J. Solid State Circuits SC-14, 291.

Cros, A., Derrien, J., and Salvan, F., 1981, Cu-Si (111) interfaces: oxidation properties in relation with their structural properties, Surf. Sci. 152/153, 1239.

d'Heurle, F. M., 1982, in *VLSI Science and Technology/1982* edited by C. Dell'Oca and W. M. Bullis, The Electrochemical Society, Pennington, N. J., p. 194; 1987, Formation and oxidation mechanisms in two semiconducting silicides, Thin Solid Films, 151, 41.

d'Heurle, F. M., and Gas, P., 1986, Kinetics of formation of silicides: A review, J. Mater. Res. 1, 205.

d'Heurle, F. M., Cros, A., Frampton, R. D., and Irene, E. A., 1987, Thermal oxidation of silicides on silicon, Philos. Mag. B 55, 291.

Fahey, P., and Dutton, R. W., 1988, Investigation of point defect generation in silicon during oxidation of a deposited WSi_2 layer, Appl. Phys. Lett. 52, 1092.

Finstad, T., Thomas, O., and d'Heurle , F. M., 1988, unpublished results.

Frampton, R., 1987, *Thermal Oxidation Kinetics of Metal Silicides on Silicon*, Thesis, University of North Carolina, Chapel Hill.

Frampton, R. D., Irene, E. A., and d'Heurle, F. M., 1987, A study of the oxidation of selected metal silicides, J. Appl. Phys. 62, 2972.

Gage, P. R., and Bartlett, R. W., 1965 a, Diffusion kinetics affecting formation of silicide coatings on molybdenum and tungsten, Trans. Met. Soc. AIME 233, 832; 1965 b, Oxidation of molybdenum silicides at high temperatures and low pressures, Trans. Met. Soc. AIME. 233, 968.

Gibbs, G. B., 1981, On the influence of metal lattice diffusion on oxidation of metals and alloys, Oxidation of Metals 16, 147.

Göltz, G., and Ferrieu, F., 1987, Thermal oxidation of WSi_2-polycide and the characterization of the oxide layer, Le Vide-Les Couches Minces, 42-236, 183.

Hu, S. M., 1987, Point defect generation and enhanced diffusion in silicon due to tantalum silicide overlays, Appl. Phys. Lett. 51, 2368.

Hudner, J., Jiang, H.,and Petersson, C. S., 1987, Marker experiments in the Cr/Si-system-formation and oxidation, Le Vide -Les Couches Minces, 42-236, 63.

Irene, E. A. and Ghez, R., 1987, Thermal oxidation of silicon: new experimental results and models, Applied Surface Science 30, 1.

Irene, E. A., and Lewis, E. A., 1987, Thermionic emmission model for the initial regime of silicon oxidation, Appl. Phys.Lett. 51, 767.

Jiang, H., Petersson, C. S., and Nicolet, M.-A., 1986, Thermal oxidation of transition metal silicides, Thin Solid Films, 140, 115.

Kofstad, Per, 1966, *High-Temperature Oxidation of Metals*, John Wiley, New York, p. 310.

Krontiras, Ch., Grönberg, L., Suni, I., d'Heurle, F. M., Tersoff, J., Engström, I., Petersson, C. S., and Karlsson, B., 1988, Some properties of $ReSi_2$, Thin Solid Films,

Laborde, O., Thomas, O., Senateur, J. P., and Madar, R., 1986, Resistivity and magnetoresistance of high-purity monocrystalline $MoSi_2$, J. Phys. F: Met. Phys. 16, 1745.

Leroy, B., 1987, Stresses and silicon interstitials during the oxidation of a silicon substrate, Philos. Mag. 55 B, 159.

Levich, V. G., 1962, *Physicochemical Hydrodynamics*, Prentice-Hall, Englewood Cliffs, N. J., pp. 72-78.

Lien, C.-D., Bartur, M., and Nicolet, M.-A., 1984, Marker experiments for the moving species in silicides during solid phase epitaxy of evaporated Si, Mater. Res. Soc. Symp. Proc. 25, 51.

Mochizuki, T., Shibate, K., Inoue, T., and Ohuchi, K., 1978, A new MOS process using $MoSi_2$ as a gate material, Jpn. J. Appl. Phys. 17 Suppl. 17-1, 37.

Nicolet, M.-A., and Lau, S. S., 1983, Formation and characterization of transition metal silicides, in *VLSI Electronics: Microstructure Science, vol. 6* edited by N. G. Einspruch and G. B. Larrabee, Academic Press, New York, p. 330.

Perkins, R. A., 1964, in *The Science and Technology of Molybdenum, Tungsten, Niobium, and Tantalum and their Alloys* edited by N. E. Promisel, Pergamon, London.

Perrière, J., Pelloie, B., and Siejka, 1987, J., Ionic movement during oxide growth by plasma anodization, Philos. Mag. B 55, 271.

Petersson, C. S., Reimer, J. A., Brodsky, M. H., Campbell, D. R., d'Heurle, F. M., Karlsson, B., and Tove, P. A., 1982, $IrSi_{1.75}$ a new semiconducting compound, J. Appl. Phys. 53, 3342.

Rouse, J., Mohammadi, F., Helms, C. R., and Saraswat, K. C., 1980, Studies of steam-oxidized WSi_2 by Auger sputter profiling, Appl. Phys. Lett. 37,305.

Tersoff, J., and Hamann, D. R., 1983, Bonding and structure of $CoSi_2$ and $NiSi_2$, Phys. Rev. B 28, 1168.

Thomas, O., Gas, P., Charai, A., d'Heurle, F. M., LeGoues, F. K., Michel, A., and Scilla, G., 1988, Diffusion of elements implanted in films of $CoSi_2$, submitted to J. Appl. Phys..

Wagner, C., 1952, Theoretical analysis of the diffusion processes determining the oxidation rate of alloys, J. Electrochem. Soc. 99, 369.

Wakita, A. S., Sigmon, T. W., and Gibbons, J. F., 1984, Effects of impurities on the oxidation of $MoSi_2$ on silicon, Appl. Phys. Lett. 45, 140.

Wang, W. S., Jona, F., and Marcus, P. M., 1983, Formation and structure of epitaxial nickel silicide on Si (111), Phys. Rev. B 28, 7377.

Wen, D. S., Smith, P. L., Osburn, C. M., and Rozgonyi, G. A., 1987, Defect annihilation in shallow p^+ junctions using titanium silicide, Appl. Phys. Lett. 51, 1182.

OXIDATION KINETICS OF SI IN DRY CO_2

Tadashi Sakon and J.B. Wagner, Jr.

Center for Solid State Science
Arizona State University, Tempe, AZ 85287-1704

Oxidation kinetics of p-type (100) silicon in 1 atm dry carbon dioxide-carbon monoxide-argon gas mixture were examined using thermogravimetry. The kinetics follow the parallel diffusion model originally proposed by Han et al. for oxidation in dry oxygen. The oxidation rates are governed by only the partial pressure of carbon dioxide, and they are independent of the partial pressures of carbon monoxide and oxygen. The possible reaction mechanism is proposed, where the dominant diffusion species and less dominant species are the interstitial oxygen molecule and oxygen vacancy.

INTRODUCTION

Recently special techniques to form a thin and uniform gate oxide have been required especially in advanced MOS (Metal-Oxide-Semiconductor) technology, and various techniques have been developed.[1-6] Thermal oxidation in 1 atm dry O_2 is the most simple and reliable one among such techniques. The oxidation is done at a relatively low temperature (e.g. 950°C) in order to reduce the oxidation rate, and it results in the linear-law-like film growth rather than the parabolic one.[2] However, the parabolic-law-like film growth is preferable for obtaining higher dielectric strength of the films produced. The low P_{O2} (oxygen partial pressure) technique[4,6] seems to overcome such a difficulty, since the kinetics follow the parabolic law from the very beginning of oxidation.[4] It, however, essentially requires a strict P_{O2} control resulting in higher production cost.

In the present paper the oxidation kinetics of p-type (100) Si at 1418 K in 1 atm dry CO_2 and CO_2-CO-Ar were measured by using thermogravimetry in order to develop an alternative oxidation procedure. The authors believe that htis is the first experimental work on the kinetics in the gases. The two probable diffusion species in SiO_2 film and possible oxidation mechanism are proposed based on the parallel diffusion model.[7] The dominant and less dominant diffusion species are the interstitial oxygen molecule and oxygen vacancy, respectively. The parabolic rate constant of the dominant diffusion species in 1 atm dry CO_2 at 1418 K is only 3% of that in 1 atm dry O_2. Moreover, it is verified that only the CO_2 partial pressure, P_{CO2} governs the oxidation rates in CO_2-CO-Ar gases. The rate is independent of the CO partial pressure, P_{CO}. In other words, the oxidation rates are not proportional to an equilibrium P_{O2}. Thermodynamic considerations explain the above experimental results as shown later.

EXPERIMENTAL

Sample Preparation

Specimens used for the present study were commercially available p-type (100) CZ Si wafers with a resistivity of 8.5-11.5 Ω·cm and about 0.5 mm in thickness. The face side of all wafers was already mechano-chemically polished. For a comparison, a both side polished wafer corresponding to wafers was oxidized, and it was found that the oxidation rates were consistent with those of the one side polished specimens. The wafer was cut into 20 x 50 mm^2 and cleaned by the RCA type sequence.[8]

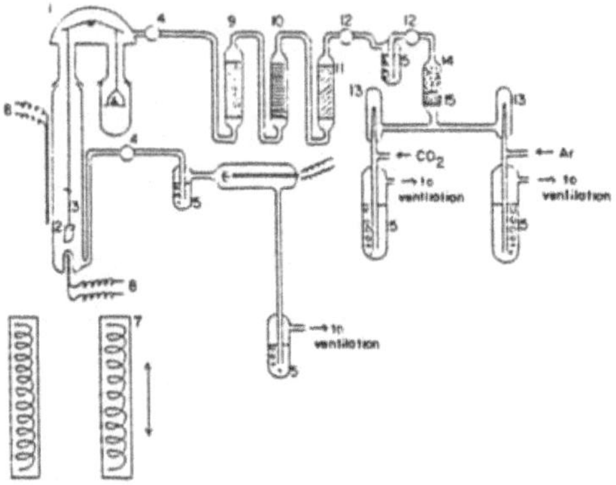

Fig. 1 Oxidation Apparatus and Gas Flowing System

1. microbalance, 2. specimen
3. quartz glass string
4. glass stopcock, 5. oil trap
6. zirconia oxygen sensor
7. electric furnace
8. thermocouple
9. phosphorus pentoxide column
10. heated Cu-turning column
11. calcium sulfate column
12. needle valve
13. glass capillary (exchangeable)
14. glass beads, 15. glass wool

Oxidation. The oxidation equipment with gas flowing system is illustrated in Fig. 1. Industrial grade CO_2 and CO and highly pure Ar were used. The impurity H_2O and O_2 in the gas mixture were removed by an anhydrous calcium sulfate column and by a column of heated copper turnings followed by a phosporous pentoxide column. Detailed explanation on the gas flowing system is available elsewhere.[9] The fused silica tube and the silica suspension string were degreased and heat-treated at around 1150°C for more than 100 hours in dry O_2 before the oxidation measurements. Prior to each oxidation, the entire oxidation chamber was evacuated three times. Then the reaction gas flowed in the chamber for at least 24 hours prior to the oxidation. The measured P_{O2} in pure CO_2 at 1418 K was 7×10^{-5} atm and consistent with the calculated equilibrium P_{O2} by supposing that the impurity levels of CO, O_2, and H_2O in the inlet gas were significantly low. From the above fact it is considered that the impurity O_2 of a inlet gas is quantitatively removed by the copper turning column and a chemical equilibrium is achieved in the gas phase of the reaction tube at our experimental conditions.

The furnace was pre-heated at a certain temperature, and was moved up to surround the reaction tube. Temperature distribution around the specimen along the vertical axis was ±3 °C/50 mm or less. The weight gain was measured by Cahn type 1000 microbalance. The pseudo-weight gain caused by a convection in the reaction tube was measured after long time oxidation where oxidation weight change was negligibly small for several hours, and it was found that the pseudo-gain becomes constant within 10 minutes from moving the furnace up. The oxidation weight gain was calculated by subtracting the pseudo-gain from the measured weight gain. The noise level of the measurements was $\pm 1.2 \times 10^{-5}$ [$kg \cdot m^{-2}$] (≈ ±100 Å) or better, and it was constant during the entire measurement.

RESULTS

Weight gain curves per unit area are shown in Figs. 2-4. The oxidation rates gradually decrease with time. As shown in Figs. 2 and 3, the slope, $d\ell nW/d\ell nt$ is close to 1.0 in the thinner oxide film regime, and it approaches 2.0 as the film thickness increases. This suggests that the kinetics follow the linear-parabolic law[10], or other more extensive film growth laws proposed by Han[7], Hopper[11], Irene[12], Massound[13,14], Revesz[15] et al. In the present work the parallel diffusion model proposed by Han et al[7] was adopted for two reasons. Firstly, our data fit the model better than the linear-parabolic model. Secondly, the previous work in our laboratory[16,17] has shown that two parallel diffusion species, O_2 molecule and oxygen vacancy or oxygen ion are likely in transport through silica.

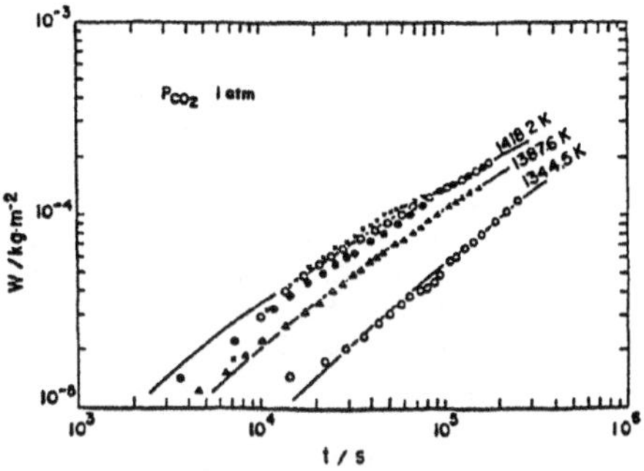

Fig. 2 Weight Gain Curve in 1 atm CO_2

dot: measured value
solid line: fitted curve

In Figs. 2 and 3, dots show measured values, and solid lines show fitted curves using the parallel diffusion model. The solid line in Fig. 4 is a calculated curve supposing that the oxidation rate is independent the P_{O2} and the CO partial pressure, P_{CO}. The agreement of the curve with the measured values suggests that the oxidation kinetics are governed by only the CO_2 partial pressure, P_{CO2}, even if the P_{CO} is rather large.

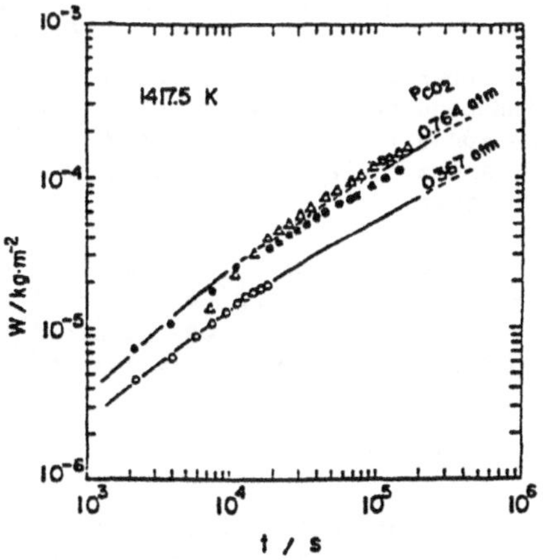

Fig. 3 Weight Gain Curve in 1 atm CO_2-Ar at 1418 K

dot: measured value
solid line: fitted curve

Data Analysis. Film growth expressed in terms of weight gain per unit area, $W[kg\cdot m^{-2}]$, is written by eq. (1) based on the parallel diffusion model.[7]

$$W^2+(2W_o+C)W - G\,\ell n\left(1+\frac{2EW}{2EW_o+F}\right) = Et \quad (1)$$

$$C = A_2B_2/(B_1+B_2)$$

$$E = B_1+B_2$$

$$F = A_2B_1$$

$$G = B_1B_2A_2^2/(2E^2)$$

where $t[s]$, $B_2[kg^2\cdot m^{-4}\cdot s^{-1}]$ and B_2/A_2 $[kg\cdot m^{-2}\cdot s^{-1}]$, B_1 $[kg^2\cdot m^{-4}\cdot s^{-1}]$, and W_o $[kg\cdot m^{-2}]$ are oxidation time, parabolic and linear rate constants of the dominant diffusion species, parabolic rate constant of the less dominant diffusion species, and an equivalent weight per unit area of an initial oxide film before oxidation. 1.2×10^{-6} [$kg \cdot m^{-2}$] (≈ 10 Å) was chosen as the W_o, since silicon wafer treated by similar procedure to our sample preparation has an 10-12 Å SiO_2 film on it.[18]

The following procedure was used for fitting the experimental data; 1. assuming a value of E in eq. (1), 2. determining the coefficients of the second and the third term in eq. (1) by the least square method supposing a value of F, 3. calculating B_1, B_2, and A_2 from the coefficients, 4. repeating procedure 2 and 3 until resulting A_2B_1 = F, 5. calculating a standard deviation using the set of B_1, B_2, and A_2 which satisfies A_2B_1 = F, 6. Determining the most appropriate set of B_1, B_2, and A_2 by repeating procedures 1 to 5.

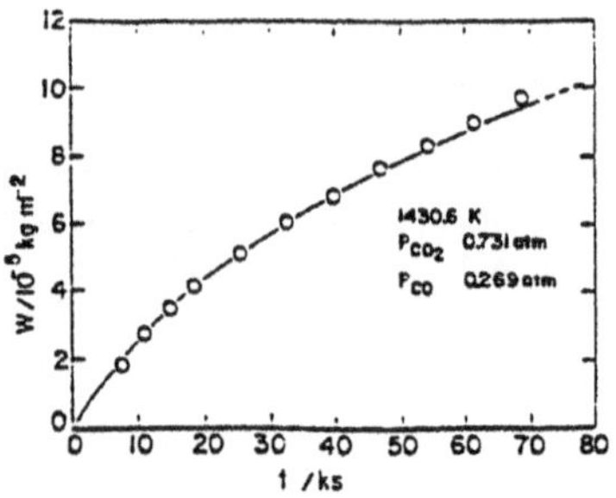

Fig. 4 Weight Gain Curve in 1 atm CO_2-CO at 1431 K

dot: measured value
solid line: calculated from eqs. (2)-(4) supposing that CO does not affect the oxidation rate

Rate Constants. Results of the data analysis are compiled in Table 1. The temperature and P_{CO2} dependencies of the rate constants are given in the following equations;

$$B_2/P^2_{CO_2} = 1.27 \times 10^{-6} \exp(-\frac{1.80\times10^5 \text{ J/mol}}{RT}) \; [kg^2 \cdot m^{-4} s^{-1} \cdot atm^{-2}] \quad (2)$$

$$\frac{B_2/A_2}{P^2_{CO_2}} = 3.56 \times 10^8 \exp(-\frac{4.55\times10^5 \text{ J/mol}}{RT}) \; [kg \cdot m^{-2} \cdot atm^{-2}] \quad (3)$$

$$B_1 = 1.1 \exp(-\frac{4.0\times10^5 \text{ J/mol}}{RT}) \; [kg^2 \cdot m^{-4} \cdot s^{-1}] \; (T>1388K) \quad (4)$$

As shown in Fig. 5, B_2 and B_2/A_2 are proportional to P^2_{CO2}, but B_1 is independent of P_{CO2}. The activation energy of B_2 is comparable to those measured in 1 atm O_2.[7,10,13,18-22] The activation energies of B_2/A_2 and B_1 are much larger than those[7,13,18] in 1 atm O_2. We have no explanation for that.

The ratio of B_2 (2.8×10^{-13} [$kg^2 \cdot m^{-4} \cdot s^{-1}$]) in 1 atm CO_2 to the average B_2 (10^{-11} [$kg^2 \cdot m^{-4} \cdot s^{-1}$] of Ref. 7, 10, 13, 18-22 in atm O_2 is 0.03 at 1418 K. Thus, we may consider that the B_2 in CO_2 is not proportional to the equilibrium P_{O2} (≈ 7×10^{-5} atm).

DISCUSSION

The oxidation kinetics are governed by only P_{CO2}, and independent of P_{CO} and P_{O2}. This suggests that CO molecule does not penetrate the SiO_2 film formed. In addition, the O_2 molecules in the reaction gas does not produce the dominant diffusion species. Moreover, it is shown by eqs. (2) and (3) that two moles of CO_2 molecules produce one mole of the dominant diffusion species. Thus, the only probable dominant diffusion species is the interstitial oxygen molecule, O_{2I}, since CO_2-dimer will not penetrate the SiO_2 film considering a diameter of the CO_2 molecule (3.24-4.60 Å[24]) and an average hole size in the SiO_2 network (2.16-3.35 Å). This was estimated from the size[22,23] of a six-sided SiO_4^{2-} ring pattern taking account of the ionic radius of O^{2-} ion and the fact that the density of crystalline quartz is larger than that of amorphous silica by about 20%.

As shown in Fig. 5 and eq. (4), B_1 is independent of P_{CO2}. This suggests that oxygen vacancy, V_O^{Z+} is more probable than interstitial oxygen ion. This consideration is in agreement with that of Srivastava et al.[16]

Table 1 Rate Constant in 1 atm CO_2-Ar

T/K	P_{CO2}/atm	$B_1/10^{-17}kg^2 \cdot m^{-4} \cdot s^{-1}$	$B_2/10^{-13}kg^2 \cdot m^{-4} \cdot s^{-1}$	$A_2/10^{-5}kg \cdot m^{-2}$	$(B_2/A_2)/10^{-10}kg \cdot m^{-2} \cdot s^{-1}$
1418.2	1.00	159	2.80	4.14	67.7
1387.6	~	75.9	2.19	8.73	25.1
1344.5	~	5.53	1.19	14.4	8.29
1417.5	0.764	234	1.59	4.14	38.3
	0.367	252	0.308	1.21	25.4

Reaction Mechanism. A probable mechanism consistent with the experimental results is as follows; for the dominant diffusion species,

$$2CO_2\ (gas) \underset{k_{-2}}{\overset{k_2}{\rightleftarrows}} O_{2I}(SiO_2/gas) + 2CO\ (gas) \tag{5}$$

$$O_{2I}(SiO_2/gas) \underset{k_1}{\overset{k_{-1}}{\rightleftarrows}} O_2\ (gas) \tag{6}$$

$$O_{2I}(SiO_2/gas) \xrightarrow{D_I} O_{2I}(SiO_2/Si) \tag{7}$$

$$O_{2I}(SiO_2/Si) + Si \xrightarrow{k_i} SiO_2 \tag{8}$$

where k_2, k_{-2}, k_1, k_{-1}, and k_i are reaction rate constants, and D_I is independent of P_{O2}, since the activity coefficient would be constant for a dilute solution of O_{2I} in SiO_2.

For the less dominant diffusion species,

$$S_i \underset{k_{-3}}{\overset{k_3}{\rightleftarrows}} Si_{Si}(SiO_2/Si) + 2V_O^{Z+}(SiO_2/Si) + 2Ze^- \tag{9}$$

$$V_O^{Z+}(SiO_2/Si) \xrightarrow{Dv} V_O^{Z+}(SiO_2/gas) \tag{10}$$

$$CO_2\ (gas) + V_O^{Z+}(SiO_2/gas) + Ze^- \underset{k_4}{\overset{k_{-4}}{\rightleftarrows}} O_o(SiO_2/gas) + CO\ (gas) \tag{11}$$

$$O_2\ (gas) + 2\ V_O^{Z+}(SiO_2/gas) + 2Ze^- \underset{k_5}{\overset{k_{-5}}{\rightleftarrows}} 2O_o(SiO_2/gas) \tag{12}$$

where k_3, k_{-3}, k_4, k_{-4}, k_5, k_{-5} are reaction rate constants. Dv is the chemical diffusion constant of V_O^{Z+}, which is independent of P_{O2} because of a dilute solution of V_O^{Z+}. Si_{Si} and O_o denote Si and O on normal sites in the SiO_2. In the above two series of reaction mechanisms, an interaction of O_{2I} with V_O^{Z+} was not considered, since Rosenscher et al.[25] have shown from their $^{18}O_2$ experiment that the $^{18}O_2$ transports through SiO_2 without exchange with O_o in the SiO_2 network and the ^{18}O reacts with Si at Si/SiO_2 interface.

The production rate of O_{2I} at SiO_2/gas interface can be written by the following equation;

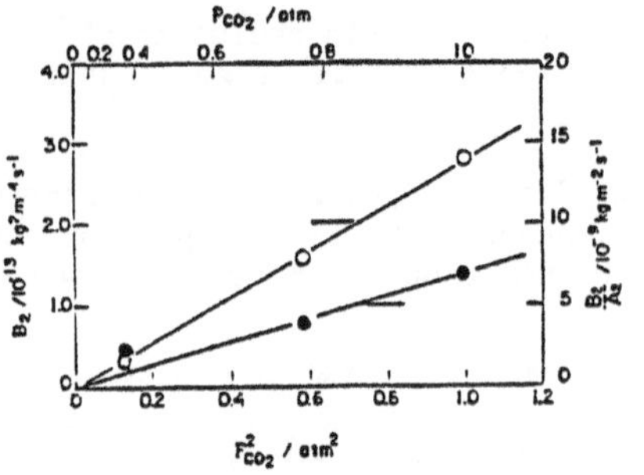

Fig. 5 P_{CO2} Dependences of the Rate Constants for the Dominant Diffusion Species in 1 atm CO_2-Ar at 1418 K

$$d[O_{2I}]_s/dt = k_2 P^2_{CO_2} - k_{-2}[O_{2I}]_s P^2_{CO} + k_1 P_{O_2} - k_{-1}[O_{2I}]_s$$

$$- \text{(diffusion term)} \qquad [mol \cdot m^{-3} \cdot s^{-1}] \qquad (13)$$

where [ ] and the suffix "S" denote concentration $[mol \cdot m^{-3}]$ and SiO_2/gas interface, respectively. Under steady state condition, $[O_{2I}]_s$ can be written by eq. (14) considering that the diffusion term is significantly small.

$$[O_{2I}]_s = (k_2 P^2_{CO_2} + k1 P_{O_2})/k_{-2}P^2_{CO} + k_{-1}) \quad [mol \cdot m^{-3}] \qquad (14)$$

According to Deal et al[10], B_2 and B_2/A_2 are

$$B_2 = 2D_I[O_{2I}]_s(10^{-3}M_{O_2})^2(10^3\rho_{SiO_2})/(10^{-3}M_{SiO_2}) \; [kg^2 \cdot m^{-4} \cdot s^{-1}] \qquad (15)$$

$$B_2/A_2 = k_i[O_{2I}]_s(10^{-3}M_{O_2}) \; [kg \cdot m^{-2} \cdot s^{-1}] \qquad (16)$$

where $M[g \cdot mol^{-1}]$, $\rho[g \cdot cm^{-3}]$, and $k_i[m \cdot s^{-1}]$ are molecular weight, density, and interface reaction rate constant of reaction (8), respectively. In the case of pure CO_2 we may consider that P_{CO} and P_{O2} are much smaller than P_{CO2}. Thus, B_2 and B_2A_2 are proportional to P_{CO2} squared.

We can rewrite eq. (14) as the follows;

$$[O_{2I}]_s = \frac{(k_2/k_{-1})P^2_{CO_2} + (k_1/k_{-1})P_{O_2}}{\{(k_2/k_{-1})/(k_2/k_{-2})\}P^2_{CO} + 1} \quad [mol \cdot m^{-3}] \qquad (14\text{-b})$$

In the above equation, $(k_1/k_{-1})^{-1}$ is equal to Henry's constant for O_2 dissolution into SiO_2, which can be calculated from the average parabolic rate constant of Ref. 7, 9, 13, 18-22 in dry O_2 using D_I measured by Norton[20] (cf. eq. (15)). Similarly, k_2/k_{-1} is available from B_2 in 1 atm dry CO_2. k_2/k_{-2} can be calculated by using the following thermodynamic relationship;

$$-RT\,\ln(k_2/k_{-2}) = 2G°(CO) - 2G°(CO_2) + G°(O_{2I}) \quad [J \cdot mol^{-1}]$$

$$= 2G°(CO) - 2G°(CO_2) + G°(O_2) - RT\,\ln(k_1/k_{-1}) \qquad (16)$$

where G° denotes a standard free energy of formation. Using free energy data of Ref. 27, $[O_{2I}]_s$ at 1418 K was calculated;

$$[O_{2I}]_2 = \frac{P^2_{CO_2} + 31.4\, P_{O_2}}{3.3 \times 10^{-2} P^2_{CO} + 5 \times 10^5} \cdot 10^3\ [\mathrm{mol \cdot m^{-3}}]\ (P:[\mathrm{atm}]) \qquad (17)$$

Thus we may conclude that B_2 and B_2/A_2 are independent of P_{CO} even if P_{CO} is fairly high. This is in agreement with the experimental result shown in Fig. 4.

Using Wagner's theory[28], B_1 can be written in the following form;

$$B_1 = \frac{(1+Z)(10^3 \rho_{SiO_2})(10^{-3} M_{O_2})^2}{10^{-3} M_{SiO_2}} \int_{[V_O^{Z+}]_s}^{[V_O^{Z+}]_{Si/SiO_2}} D_v t_e\, d[V_O^{Z+}]\ \ [\mathrm{kg^2 \cdot m^{-4} \cdot s^{-1}}] \qquad (18)$$

where t_e is the tranport number of electrons in SiO_2. Considering that D_v and t_e are constant at a given temperature and $[V_O^{Z+}]_{Si/SiO_2} \gg [V_O^{Z+}]_s$, eq. (18) becomes

$$B_1 = \frac{(1+Z)\rho_{SiO_2} M^2_{O_2}}{M_{SiO_2}} D_v t_e [V_O^{Z+}]_{Si/SiO_2}\ \ [\mathrm{kg^2 \cdot m^{-4} \cdot s^{-1}}] \qquad (19)$$

$$[V_O^{Z+}]_{Si/SiO_2} = \left(\frac{k3/k{-3}}{Z^Z}\right)^{\frac{1}{1+Z}} (P_{O_2(Si/SiO_2)})^{\frac{1}{2(1+Z)}}\ [\mathrm{mol \cdot m^{-3}}] \qquad (20)$$

where $P_{O2}(Si/SiO_2)$ denotes the dissociation pressure of SiO_2. Thus, B_1 must be independent of P_{CO2}, P_{O2}, and P_{CO}.

As far as the dominant diffusion species is the interstitial oxygen molecule, the SiO_2 film formed in CO-CO_2-Ar mixture is equivalent to that formed by the low P_{O2} technique. Therefore, we expect that this oxidation technique would be applicable for forming gate oxide on Si wafers.

CONCLUSIONS

Oxidation kinetics of Si in CO_2-CO-Ar were measured, and the following facts were found;

1. The oxidation rates are governed by only P_{CO2}, and CO acts as if an inert gas.
2. The kinetics can be expressed by the parallel diffusion model where the dominant and less dominant diffusion species are the interstitial oxygen molecule and oxygen vacancy, respectively.
3. The proposed reaction mechanism suggests that the SiO_2 film formed in CO_2-CO-Ar mixture is equivalent to that formed by the low P_{O2} technique.

REFERENCES

1. L. E. Katz, Oxidation, in: "VLSI Technology," S. M. Sze, ed., McGraw-Hill International Book, New York (1983).
2. E. A. Irene, J. Electrochem. Soc., 125, 1708 (1978).
3. C. Hashimoto, S. Muramoto, N. Shiono, and O. Nakajima, ibid., 127, 129 (1980).
4. A. C. Adams, T. E. Smith, and C. C. Chang, ibid, 127, 1787 (1980).
5. E. A. Irene, An Overview of the Kinetics of Oxidation of Silicon, in: "Passivity of Metals and Semiconductors," M. Froment, ed., Elsevier Sci. Pub., Amsterdam (1983).
6. K. K. Ng, W. J. Polito, and J. R. Ligenza, Appl. Phys. Lett., 44, 626 (1984).
7. C. Han and C. R. Helms, J. Electrochem. Soc., 134, 1297 (1987).
8. W. Kern and D. A. Puotinen, RCA Rev., 31, 187 (1970).
9. G. M. Crosbie, J. Solid State Chem. 25, 367 (1978).
10. B. E. Deal and A. S. Grove, J. Appl. Phys., 36, 3770 (1965).
11. M. A. Hopper, R. A. Clarke, and L. Young, J. Electrochem. Soc., 122, 1216 (1975).
12. E. A. Irene, J. Appl. Phys. 54, 5416 (1983).
13. H. Z. Massoud, J. D. Plummer, and E. A. Irene, J. Electrochem. Soc., 132, 1745 (1985).
14. H. Z. Massoud, J. D. Plummer, and E. A. Irene, ibid., 132, 2685 (1985).
15. A. G. Revesz, B. J. Mrstik, H. L. Hughes, and D. mcCarthy, ibid., 133, 586 (1986).
16. J. K. Sriastava, M. Prasad, and J. B. Wagner, Jr., ibid., 132, 955 (1985).
17. J. K. Srivastava, V. B. Tare, and J. B. Wagner, Jr., ibid., 132, 310 (1985).
18. E. A. Irene, ibid., 121, 1613 (1974).
19. P. J. Burkhardt and L. V. Gregor, Trans. Metall. Soc. AIME, 236, 299 (1966).
20. A. G. Revesz and R. J. Evans, J. Phys. Chem. Solids, 30, 551 (1969).
21. R. B. Fair, Impurity Diffusion and Oxidation of Silicon, in: "Silicon Integrated Circuits, Part B," Dawon Kahng, ed., Academic Press, New York (1981).
22. R. M. Burger and R. P. Donovan, "Fundamentals of Silicon Integrated Device Technology, vol. 1, Oxidation, Diffusion and Epitaxy," pp. 48, Prentice-Hall, Englewood Cliffs (1968).
23. E. H. Nicollian, J. R. Brews, "MOS (Metal Oxide Semiconductor) Physics and Technology," pp. 665, Wiley International Pub., New York (1982).
24. W. J. Moore, "Physical Chemistry (4th edition)," pp. 157 (1972), Prentice-Hall, Englewood Cliffs (1972).
25. E. Rosencher, A. Straboni, S. Rigo, and G. Amsel, J. Appl. Phys. Lett., 34, 259 (1979).
26. F. J. Norton, Nature, 171, 701 (1961).
27. D. R. Stull and H. Prophet, "JANAF Thermodynamical Tables (2nd edition)," NSRDS-NBS 37 (1971).
28. O. Kubaschewski, B. E. Hopkins, "Oxidation of Metals and Alloys (2nd edition)," pp. 82, Butterworths, London (1962).

A NOVEL SILICON OXIDATION METHOD

— HF ENHANCED OXIDATION

Long Wei, Xu Yuan-sen and Zheng Yang-shu

Shanghai Institute of Metallurgy, Academia Sinica
Shanghai 200050, China

ABSTRACT

In this paper, we report a new silicon oxidation method, at least one order magnitude higher in oxide growth rate can be achieved by only addition of a few thousand PPM HF to the pure oxygen gas. After rapid HF doped oxidation and subsequent annealing at low temperature (800°C) in pure oxygen, high quality gate oxide for VLSI can be obtained, which exhibits a tight distribution in thickness and breakdown voltage measurements leading to a defect densities about $2/cm^2$, and intrinsic breakdown events more than 99%. The interface charge densities are very low, and the Si/SiO_2 interfaces are nearly atomically flat as revealed by HREM at 200K magnification. These superior properties of the SiO_2 thin films would very promise for VLSI application.

INTRODUCTION

There is currently increasing demand to low temperature fabrication of large scale integrated circuits in order to reduce difficulties such as substrate warpage, dopant impurity redistribution and defect generation. Oxidation is a key process in semiconductor device fabrication. Oxidation of silicon is presently carried out at temperatures over 950°C and mostly at about 1000°C. But the oxide growth rate of silicon in dry oxygen becomes extremely low at temperature below 850°C and the quality of the Si/SiO_2 interface gets worse[1].

To significantly enhance the oxide growth rate, several novel oxidation techniques have been reported. These techniques, such as high-pressure oxidation[2], plasma oxidation[3], photon-assisted oxidation[4,5] and NF_3-enhanced thermal oxidation[6], have still not found industrywide application, they have some problems which should be resolved. A major problem of high-pressure oxidation is the apparent inadequacy of its equipment to overcome the temperature nonuniformity arised due to the severe gas convection that occurs at high gas pressures. The use of plasma enhanced oxidation, however, always encounters the problem of ion bombardment damage. The report on electrical properties of photon-assisted grown oxide is far rather scant. As for NF_3 enhanced oxidation, because of the high concentration of very reactive species in the oxidation system, sometimes undesirable products are deposited on the substrate via gas phase reactions of the active species and the resulting oxide films are inferior when compared to thermally oxidized silicon with oxidation temperature of the order of 1000°C.

In this paper, the authors report for the first time that at least one order of magnitude higher in the oxide growth rate can be achieved in the temperature range of 800°C to 950°C by adding a very small quantity of HF vapour into the oxygen gas. Therefore, it was attemped to form thin and high reliable oxide at 800°C by use of HF oxidation. After some investigation, however, it become evident that single HF enhanced oxidation step is not suitable for forming thin and high reliable oxides. This problem was solved by two step HF enhanced oxidation. First, an oxide of considerable thickness was formed using dry oxygen containing low HF concentration at low temperature of 700-900°C. Under this condition, the oxide growth rates are significantly high and resulting oxides are incorporated with fluorine atoms. Then fluorine atoms in the oxides was removed and high quality oxide was obtained by post annealing in dry oxygen without HF vapor. In this paper we report this new HF enhanced oxidation technique and the characteristics of oxides formed by two step HF enhanced oxidation.

EXPERIMENT

HF gas was introduced to the furnace tube by oxygen passing over the surface of HF acid in a plastic container. The temperature of the container was hold constant at 22°C throughout the experiments. P or N type CZ Si(100) wafers with resistivities ranging from 6 to 10Ω-cm were subjected to a standard cleaning process consisting of HCl/H_2O_2 clean, followed by a 1:40 HF dip to remove any native oxide. The wafers were then oxidized in a dry oxygen atmosphere containing HF up to 2600PPM and, were subsequently annealed in a pure oxygen ambient at the same temperature.

Oxide thickness and refractive index were measured physically by an automatic ellipsometry. The compositions of the thin oxide film were determined by Auger electron spectroscopy (AES) and chemical analytical method. The electrical properties of the oxide films and Si/SiO_2 interface charge density were evaluated by using polysilicon gate capacitors of $0.6mm^2$ in area. The breakdown voltage is defined as the voltage across a thin oxide when current level of 3 μA is conducted through this oxide. For breakdown measurements, a total of 300 devices per wafer were tested. A high resolution transmission electron microscopy (HREM) was used to study the cross section of the polysilicon/SiO_2/Si structure.

RESULTS

Kinetics of oxide growth. Remarkable enhancement of oxide growth rate by adding HF gas to dry oxygen is shown in Fig.1. It is clearly indicated that the oxide growth rates at 800°C reach the rate of pure dry oxygen at 1000°C. For instance, 30nm thick oxide layer can be obtained by HF enhanced oxidation in 35 min at 800°C, but for pure oxygen oxidation it takes more than 10 hr at the same temperature. In Fig.2, the kinetics of HF enhanced oxidation are plotted. It is evident that the experimental data are well modeled by Deal and Groves simple lineal-parabolic (L-P) relationship.

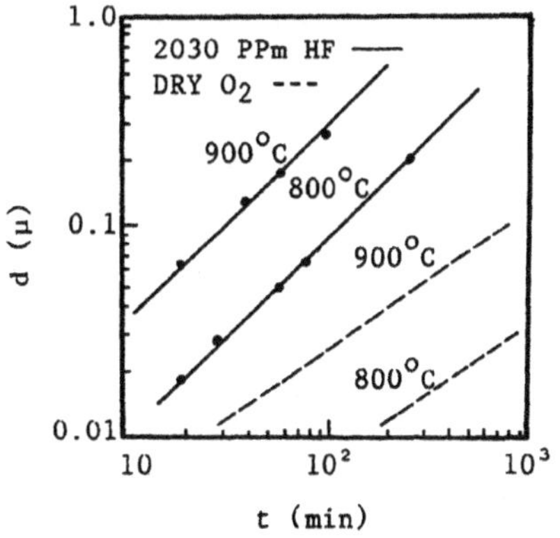

Fig. 1. Comparison of oxidation rate for HF enhanced oxidation and standard dry oxidation.

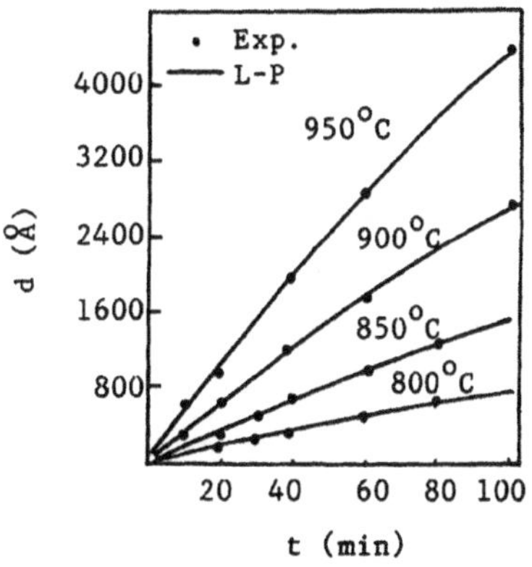

Fig. 2. Oxide thickness vs. oxidation time for (100)-oriented lightly doped silicon oxidized in O_2/HF gas mixture in the 800°C-950°C range and its fit to the linear-parabolic relationship. The HF concentration is about 2000PPM.

Table 1. Linear and parabolic rate constant for the oxidation of (100)-oriented silicon in O_2/HF mixture

Oxidation temperature (°C)	Linear-rate constant K_{LF} (μm/hr)	Comparison with dry O_2 K_{LF}/K_{LD}	Parabolic constant K_{PF} (μm²/hr)	Comparison with dry O_2 K_{PF}/K_{PD}
800	0.053	20.5	0.038	95.6
850	0.115	20.6	0.065	57.3
900	0.210	16.8	0.207	61.7
950	0.342	14.2	0.469	64.6

Table 1 shows the model constants obtained by fitting the data to L-P equation. It can be seen that, by adding a low concentration of HF gas, a large increase in linear and parabolic oxide growth rate constants for dry oxygen was brought. These results can not be interpreted simply in terms of small amount of H_2O containing in gas ambient, since our system contains less than 2500 PPM H_2O coming from HF acid, which can only increases the oxidation rate by 30% [7] while our experimental results are greater by a factor of 40 or more.

Oxide density. The SiO_2 density values were obtained from the measured refractive index values using the Lirentz-Lorenz relationship: $\rho=K(n^2-1)/(n^2-2)$ where n is the refractive index and K is a constant. K was evaluated by using values for n and ρ for SiO_2 films grown at 1000°C in dry O_2. The values, n=1.462 and ρ=2.212, were used to obtain K=8.0461. Our experimental data have been presented in table 2 which shows that the SiO_2 density decreases as the HF concentration increases. That is to say, by adding HF, the oxide will become less dense and contain more free volume.

Table 2. Variation of density and refractive index for HF oxidation

Temperature(^{o}C)	HF content(PPM)	SiO_2 thickness(Å)	Refractive index (n)	Density (g/cm^3)
950	1000	447	1.457	2.191
	2000	654	1.442	2.129
	2650	854	1.429	2.074

We have done several experiments involving the measurements of Auger electron spectroscopy (AES) and refractive index of oxide obtained with different annealing times. The results strongly suggest that, once a HF doped SiO_2 film is annealed, the film density increases and fluorine content decreases. For example, when a fluorine doped oxide was annealed at 900°C in pure oxygen for 30 min, it loss 60% of the fluorine content and its density increased 5.2%. These experiments demonstrate that fluorine atoms incorporated in the oxide can be removed and high dense oxides can be obtained during the post annealing in dry oxygen.

Oxide thickness uniformity. Fig.3 shows the relationship between oxidation temperature and oxide thickness uniformity. The experimental data throw some new light on the idea that the lower the oxidation temperature is, the better the uniformity of resulting oxide thickness.

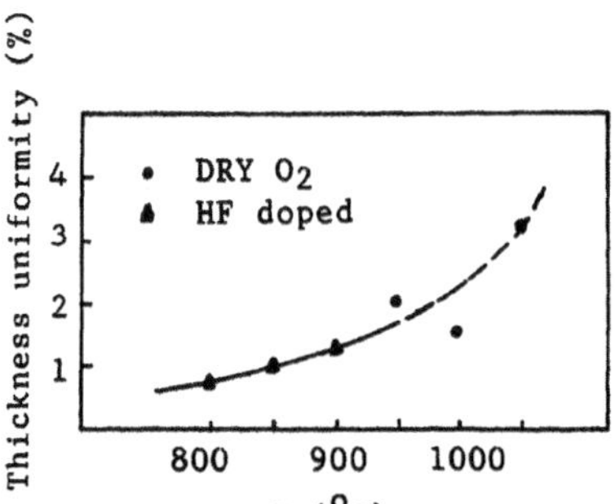

Fig. 3. Variation oxide thickness vs. oxidation temperature.

This finding seems to have a bearing on the characteristics of resistance heated furnace that the variations of temperature are larger at higher temperature. So we can conclude that the HF enhanced oxidation can achieve not only remarkable increased oxide growth rate, but also the better uniformity of oxide thickness as compared with the conventional dry oxidation.

C-V characteristics. High frequency (1MHz) capacitance-voltage measurements have been made on some thin (20nm) SiO_2 samples obtained by two step HF enhanced oxidation. The results are that: (1) typical flatband voltage for 20nm SiO_2 films grown on a p-type substrate is -0.85V; (2) no C-V hysteresis was detected; (3) the level of fixed positive charge is about $1\times10^{11}cm^{-2}$; (4) the mobile Na^+ density is less than $4\times10^{10}cm^{-2}$. These measurements suggest that the C-V stability of SiO_2/Si structures for HF enhanced oxide is good and equivalent to that high quality dry oxide grown at high temperature.

Post annealing. The effects of post-annealing on the electrical properties of HF enhanced oxide are presented in Fig.4. In case the annealing time was insufficient, such as 70 min, the percentage of breakdown events occurred at 6 to 8 MV/cm are high. As the annealing time increases to 100 min, the number of non-intrinsic breakdown events decrease to about only 1%. These is possibly because of fluorine atoms in the oxide form Si-F bonds and became physical weak spots or local trap sites which led catastrophic breakdown at low-field and, nevertheless, these Si-F bonds can be converted into Si-O-Si bonds by post-annealing in oxygen atmosphere.

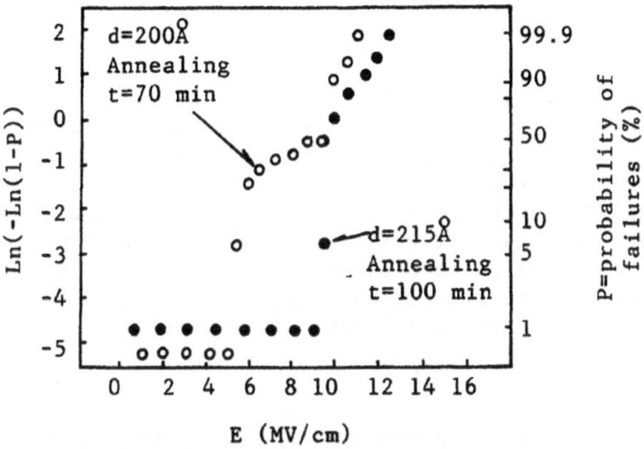

Fig. 4. Plot of cumulative probability of failures (percentage) vs. electric field for thin oxides grown under two step HF enhanced process at 800°V.

Breakdown field strength. HF oxide breakdown histograms for MOS capacitors are shown in Fig.5. The yields of those capacitors that failed at electrical field more than 9.5 MV/cm were 99%, which appear equivalent to oxide obtained by high temperature dry oxidation. (see Fig.6 and Fig.7).

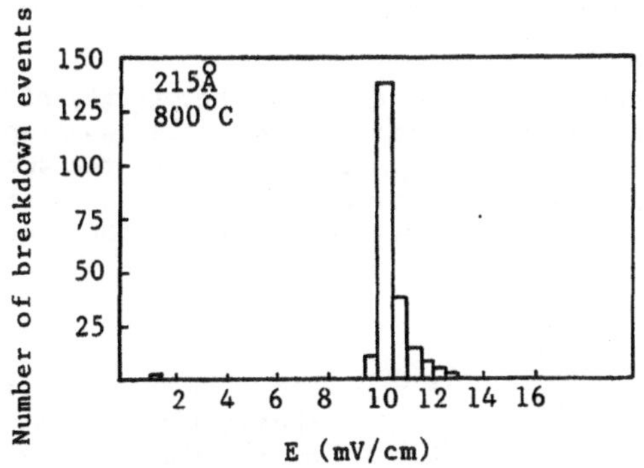

Fig. 5. Breakdown histogram for MOS capacitors with HF enhanced oxides.

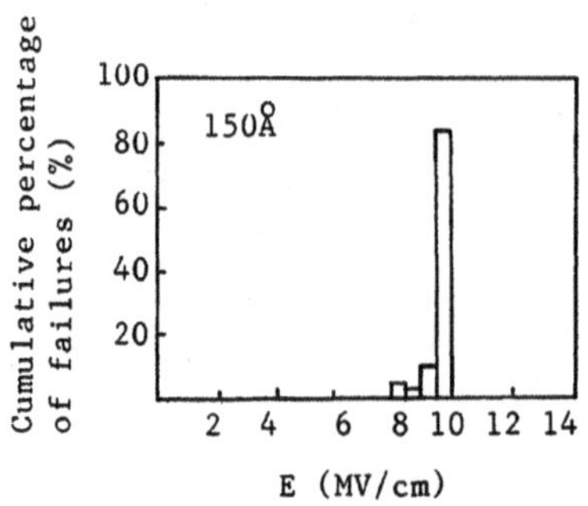

Fig. 6. Breakdown histograms for HCl oxidation MOS capacitors.

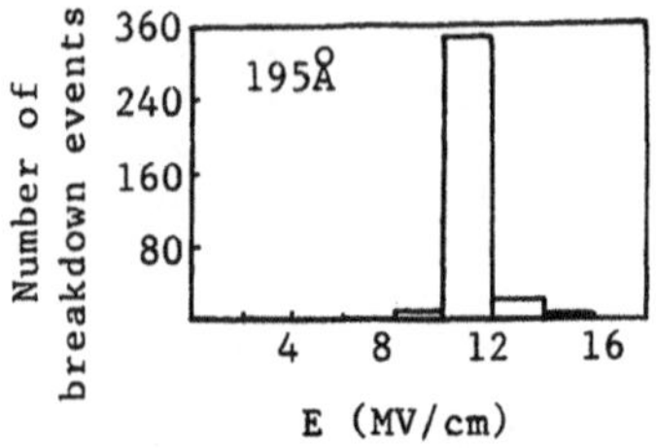

Fig. 7. Breakdown histograms for standard oxidation with long-time postannealing MOS capacitors.

Defect density. A value for the defect density Q was derived in the standard manner from the MOS capacitor breakdown data using the relationship $Q=(-Ln(1-P))/A$[8], where P is the cumulative failure probability and A is the capacitor area. This value serves as an indicator of the oxide film quality, although it is not necessarily directly interpretable in terms of actual physical defects. Table 3 summarizes the various parameters of thin oxides such as defect density as a function of oxide growth conditions. It is confirmed that the two-step HF oxidation process leads to oxides with much lower defect density than those from the dry oxygen process.

Table 3. Summary of the oxide breakdown voltage characteristics

Process variable	Oxide thickness (Å)	Defect density (Q/cm^2)
HCl oxidation (900°C), annealed at 900°C for 1hr.	150	20
Dry oxygen two-step oxidation (950°C), anneal at 1050°C for 1hr.	86	4
HF oxidation (800°C), anneal at 800°C for 100 min.	215	2

Si-SiO_2 interface. Fig.8 is a HREM photography of the thin oxide interface for the sample that had undergone the two-step HF oxidation at 800°C. It is evident that the interface between the substrate and the oxide is atomically flat. No any silicon protrusions are observed, and the degree of interface roughness never exceeds the width of a single atomiclayer.

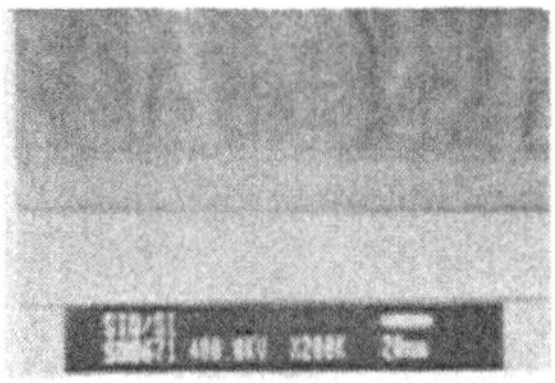

poly-Si
SiO_2
Si

Fig.8 High resolution electron microphotographs, illustrating the thin oxide interface morphology for samples that had undergone the two-step HF enhanced oxidation process.

DISCUSSION

1. It is well known, the reaction of fluorine with the silica network[9] (see Fig.9) is:

$$H\text{-}F + \text{-}\overset{|}{\underset{|}{Si}}\text{-}O\text{-}\overset{|}{\underset{|}{Si}}\text{-} = \text{-}\overset{|}{\underset{|}{Si}}\text{-}OH + F\text{-}\overset{|}{\underset{|}{Si}}\text{-}$$

So, as shown in Fig.10, some Si-O bonds are broken by HF and voids in SiO_2 network are enlarged and then many micropores are formed. These micropores led to lower SiO_2 density, higher percentage of free volume to dissolve more oxygen from the gas phase, and with fast transportation rate of oxygen to the SiO_2/Si interfaces, to supply the silicon oxidation reaction. Furthermore, the fluorine and OH radicals have great effect on the lowering of SiO_2 viscosity, accelerate the stress relaxation at SiO_2/Si interface and in SiO_2 network.

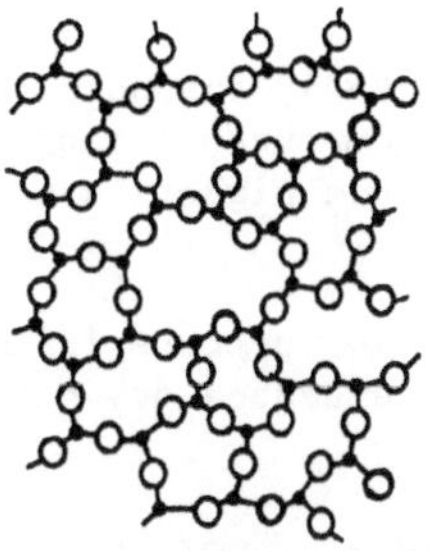

Fig. 9. Two-dimensional representation of perfect vitreous silica (black dots represent Si atoms, circles represent O atoms).

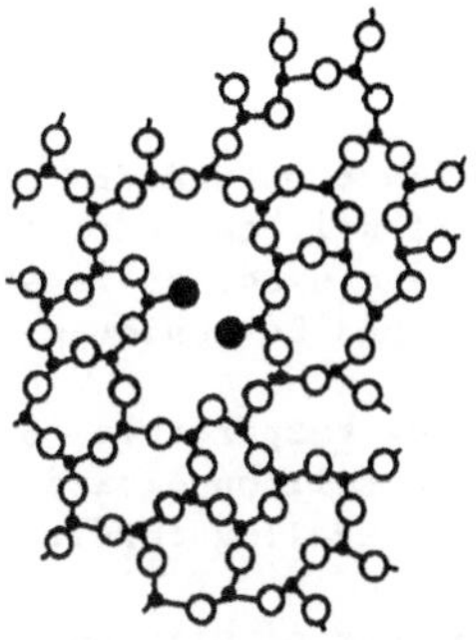

Fig. 10. Two-dimensional representation of fluorine doped SiO_2. (larger black dots represnet F atom or "OH").

It is generally believed that the oxide growth rate is dependent upon the behavior of the oxidant diffusion through the growing oxide and the oxidant reacts with silicon at the Si/SiO_2 interface. It has been demonstrated above that there are large amount micropores which can accelerate diffusion and great stress relaxation that allow rapid interface reaction to take place in HF doped SiO_2 film, thus, oxide growth rates are remarkably enhanced by the generation of micropores and by the reduction of viscosities in SiO_2 films.

2. In two-step HF enhanced oxidation the mechanism of the post annealing step is somewhat different from conventional high-temperature annealing step. The conventional annealing are performed at high temperature about 1050°C allowed viscous flow and stress relaxation to take place, in order to reduce the interface roughness[1], while post annealing of HF enhanced oxidation at low temperature about 800°C in order to reduce broken bonds in HF doped oxide as following reaction:

$$\text{-}\overset{|}{\underset{|}{Si}}\text{-}F + \text{-}\overset{|}{\underset{|}{Si}}\text{-}OH = \text{-}\overset{|}{\underset{|}{Si}}\text{-}O\text{-}\overset{|}{\underset{|}{Si}}\text{-} + HF ,$$

and restore to ordinary SiO_2 network structure. Note that the oxide thickness is slighty increased by the pure oxygen annealing and a intrinsic compressive stress may be developed in the Si/SiO_2 interface. Therefore, the oxide films would be densified by this compressive stress and a much denser oxide can be obtained. This conception is in good agreement with our experimental findings.

3. As seen from the results of the electrical measurements, the two-

step HF enhanced oxidation is as effective as conventional dry oxygen high temperature oxidation methods in reducing the oxide defect density. Moreover, thin oxide grown with HF oxidation compared to other low temperature oxidation methods produce lower defect density. There seem to be a definitive correlation between the electrical properties of oxides and the cleanness of oxide growth environment. In other low temperature techniques such as CF_4 or NF_3 enhanced oxidation, due to the high concentration of very corrosive gas in the oxidation system, contamination may be unavoidable. Thus, similar to other low-temperature techniques, the SiO_2 films resulting from CF_4 or NF_3 enhanced oxidation are inferior in electrical properties as compared with high temperature dry oxidation films. Nevertheless, the superior SiO_2 films are obtained with the HF enhanced oxidation. The reason is that this new method adoptes a rather clean system in which only a few thousand PPM of very pure HF vapour is presented and harmful etching effects are avoidable.

CONCLUSION

1. A few thousand PPM of HF vapour can remarkably enhances the oxidation rate of silicon wafer in dry oxygen at the temperature range (800-950°C) we have used, at least one order of magnitude higher in oxide growth rate can be achieved.

2. A low-temperature (800°C) two-step oxidation process for high quality thin silicon dioxide film for VLSI by HF enhanced dry oxidation and a post annealing condition is devised.

3. The use of high purity hydrofluoric acid as fluorine source for enhanced oxidation of Si wafer has been shown to be rather attractive, since it is easier to handle, less corrosive, and without special equipments.

ACKNOWLEDGEMENT

The authors are grateful to the support of many members at the IC laboratory of our Institute. They are also indebted to Ni Ru-shan for performing HREM on the Si/SiO_2 interface.

REFERENCES

[1] A. Bhattacharyya, and C. Vorst, J. Electrochem. Soc., 132, 1900(1985).
[2] E.A. Irene, D.W. Dong and R.J. Zete, J. Electrochem. Soc., 127, 396 (1980).
[3] A.K. Ray and A. Reisman, J. Electrochem. Soc., 128, 2424 (1981).
[4] E.M. Young, and W.A. Tiller, Appl. Phys. Lett., 42, 63 (1983).
[5] I.W. Boyd, Appl. Phys. Lett., 42, 728 (1983).
[6] M.Morita, S. Aritome, M. Tsukde and M. Hirose, IEDM, 144 (1984).
[7] E.A. Irene, J. Electrochem. Soc., 125, 1708 (1976).
[8] N.J. Chou and J.M. Eldridge, J. Electrochem. Soc., 120, 1140 (1973).
[9] Wells, "Structure Inorganic Chemistry", Oxford Press, 3rd Edition, 786-808 (1962).

THERMAL OXIDATION OF SILICON IN AN AFTERGLOW GAS

Andrew M. Hoff and Jerzy Ruzyllo

Center for Electronic Materials and Devices
Department of Electrical Engineering
The Pennsylvania State University
University Park, PA 16802

Low temperature thermal oxidation of silicon is desired for many reasons in device manufacture. Thermal activation alone at temperatures below 700°C is ineffective. The proposal that other oxidants might act via parallel mechanisms instigated this study in which the thermal process is enhanced by the presence of non-thermally produced oxidants in the afterglow of a microwave gas discharge. Film growth in atmospheric O_2, in an O_2 afterglow at 1 Torr, and in an O_2/NF_3 afterglow at 1 Torr is studied. Fluorine containing afterglows are found to enhance film growth at 700°C compared with that obtained in O_2 afterglows, however, at lower temperatures the apparent enhancement vanishes. The effect is attributed to the existence of a concurrent etching reaction. Further, the electrical properties of films grown in the various regimes are discussed.

INTRODUCTION

Conventional silicon thermal oxidation at temperatures below 700°C results in little film growth even at extended times. Such processes are therefore unreasonable from a manufacturing point of view. However, an obvious need exists motivating the development of such low temperature processes. A parallel mechanism by which other oxidants than molecular oxygen, such as atomic oxygen might play a role in thermal oxidation has already been proposed (1-3). Molecular oxygen may undergo decomposition near the Si-SiO_2 interface to form atomic oxygen which then can act along with other oxidants in the surface reaction controlled oxidation mode. In conventional thermal oxidation systems, this process of decomposition may be controlled only by temperature, and is not very efficient at temperatures of oxidation below about 700°C. Hence, the possible enhancement of oxide growth rate at low temperature due to atomic oxygen formation can not be achieved in furnace oxidation, unless atomic oxygen generated by a process independent of the thermal decomposition of O_2 will be added to the oxidizing ambient.

In our recent work (4) we report the results of a study of the role of atomic oxygen in enhancing oxide growth kinetics. The addition of atomic oxygen to the thermal oxidizing ambient was achieved through the use of a microwave discharge in oxygen outside of the conventional oxidation chamber. The presence of a few percent of atomic oxygen in an oxidizing ambient consisting of pure molecular oxygen was found to substantially increase the oxide growth rate in the temperature range of oxidation below 850°C.

In this work we are expanding the investigation into the area of thermal oxidation of silicon in an afterglow gas produced by a microwave discharge in either oxygen only or oxygen with a small addition of nitrogen trifluoride (NF_3). The reason for adding NF_3 to the ambient is twofold. First, the incorporation of NF_3 in the thermal oxidizing ambient has previously been found to increase the oxide growth rate (5), and hence, the time of low temperature oxidation needed to grow a 100Å to 200Å thick gate oxide may be shortened. Second, the role of fluorine in the oxide, although not yet entirely clear, is thought to improve some of its properties such as dielectric strength and radiation hardness. In any case no chemically caused degradation of the oxide from fluorine should be expected (6).

In this paper, the results of an investigation of the thermal oxidation of silicon in both oxygen and oxygen / NF_3 afterglows are reported. The discussion of oxide growth characteristics is accompanied by an evaluation of the basic electrical parameters of these oxides.

EXPERIMENTAL

Figure 1 shows a schematic diagram of the oxidation system used in this experiment. It consists of a standard thermal oxidation apparatus to which a microwave atom source at the inlet of the furnace tube and vacuum pumping capability at its outlet are added. A unique feature of this system is the fact that, in contrast to other systems implementing the concept of plasma enhanced oxidation (7-9), with the microwave generator off it acts as a conventional thermal oxidation apparatus. With the microwave power on, a plasma is created by the 2.45 GHz excitation at 400 watts and molecular oxygen is dissociated to produce atomic oxygen. In both cases the oxidant gas travels through the microwave cavity, past the wafer in the furnace, and to the pump port beyond the furnace, entirely within high purity fused silica. Therefore, the silicon surface may be exposed to both thermally and non-thermally generated reactants, but not to contaminants or species normally found in a plasma environment.

The silicon wafers used in this study were p-type with 6-10 Ω-cm resistivity and (100) orientation. Following a standard RCA clean (10), each sample was loaded into the oxidation system with N_2 flowing. Oxidation was carried out either in (a) undissociated O_2 at atmospheric pressure in which case the process is conventional thermal oxidation or , (b) the afterglow of a microwave discharge in oxygen at 1 Torr or, (c) the afterglow of a microwave discharge in O_2/NF_3 at 1 Torr. Oxidation times ranged from 5 to 120 minutes at

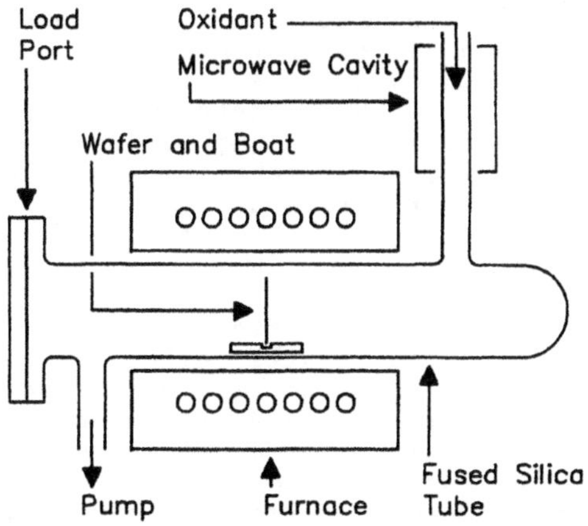

Fig. 1. Schematic of the afterglow thermal oxidation system.

temperatures from 400°C to 700°C. The thickness of the grown film was measured by ellipsometry. Subsequent to oxidation, aluminum gate MOS capacitors were fabricated using filament evaporation and a forming gas anneal. Capacitance voltage measurements were used to evaluate the electrical properties of the grown oxides.

Additional experiments were performed to assess the role of plasma and thermal processes in the O_2/NF_3 oxidation process. In the first group, wafers with steam grown oxide approximately 0.7μm in thickness were used. The second group used samples with 150Å initial oxide grown in dry O_2 at 1000°C. Individual samples from each group were processed in the NF_3 containing afterglow for 30 minutes each at temperatures ranging from 180°C to 700°C to study the balance between etching and oxidation in the NF_3 containing ambient.

RESULTS AND DISCUSSION

Although the results obtained for the atomic oxygen afterglow were interesting from the viewpoint of studying the mechanisms of thermal oxidation under vacuum (4) and clearly showed an enhancement over thermal oxidation at temperatures below 850°C, this process requires two to four hours to produce a useful gate oxide in the thickness range of 100Å to 200Å. However, when 50ppm of NF_3 was added to the oxygen flowing through the microwave cavity a substantial increase in growth was observed. Isothermal growth results for the three conditions of oxidation studied in this work are presented in Fig. 2a. This data shows the large increase in film thickness obtained at 700°C in the NF_3 containing afterglow, achieving 100Å in as little as 10 minutes. It should be noted that thermal activation alone (5) required three times the concentration of NF_3 used in this work to produce a similar thickness at 30 minutes. Another reason for using a low concentration of NF_3 in this study was that afterglow mixtures containing significantly higher (2-3 times) concentrations than 50ppm of NF_3 were found to etch SiO_2 rather than enhance its growth.

An important practical aspect of the discussed oxidation technique is its similarity to conventional thermal oxidation. In the three modes studied, O_2 at atmospheric pressure, O_2 afterglow at 1 Torr, and O_2/NF_3 afterglow at 1 Torr, when multiple wafers were loaded in the apparatus and all were at the same process temperature, equivalent film thickness was achieved. Further, the orientation of the front surface of the silicon to the oxidant flow through the system has no effect on the process result.

In the previously reported case of the atomic oxygen afterglow thermal oxidation of silicon (4), an enhancement was observed over the conventional thermal oxidation process. When the two processes were compared it was concluded that in the initial oxidation regime the thermal process exhibited a strong temperature dependence, whereas when an independent, non-thermal, source of atomic oxygen was made available to the process environment little if any thermal activation was evident. In other words, in the initial thickness regime, temperature over the range from 400°C to 700°C had little effect on the thickness of film achieved and only exhibited an effect as the film thickness increased. Hence, in the presence of atomic oxygen, the process appeared to operate in a reactant limited mode.

Previous authors (5,9) have noted the "catalytic" effect which fluorine may have on the oxidation process. Auger depth profiles of plasma grown oxides where fluorine was added to the ambient (9) have shown that fluorine has its highest concentration at the Si-SiO_2 interface. In a thermal process alone (11) the fluorine concentration throughout the film was found to be essentially uniform by XPS profile analysis. Also from the XPS data it was concluded that as a result of the thermal process the fluorine was bonded to silicon throughout the film. In the O_2/NF_3 afterglow thermal process studied in this work, similar activation of the growth process has been observed. However, as previously noted, the concentration of NF_3 required to achieve this activation was substantially smaller.

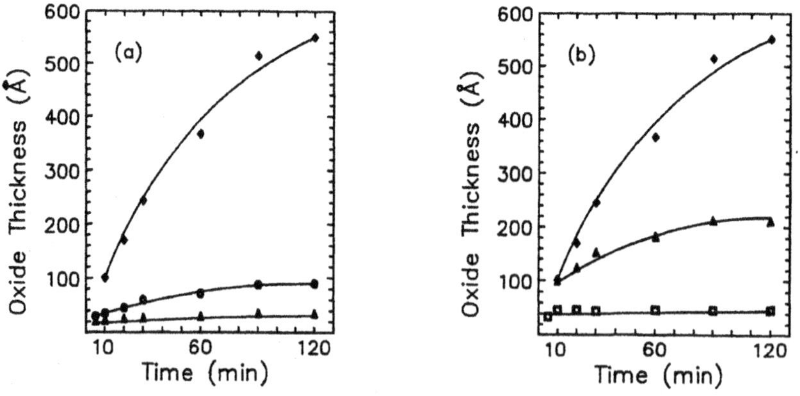

Fig. 2. Thickness (Å) vs. time (min) plots for (a) three modes of oxidation at 700°C: ▲ O_2 thermal, 760 Torr; ● O_2 afterglow, 1 Torr; ◆ O_2/NF_3 afterglow, 1 Torr and (b) O_2/NF_3 afterglow, 1 Torr: ◆ 700°C, ▲ 550°C, ◘ 400°C.

The afterglow process chemistry between NF_3 discharge products and atomic oxygen is not well understood at this time. Atomic oxygen has been reported to react with NF, NF_2, NO, and OF (12) to produce such species as free F, NO_2, more NO, O_2, and FNO. In particular, Clyne (12) has noted that when the concentration of O is higher than the concentration of NF_2, the yellow-green air afterglow emission is present. In this study when 50ppm of NF_3 in O_2 is passed through the microwave discharge a very faint air afterglow was found to be present in the downstream region. This observation indicates that a similar chemistry to that found in afterglow chemical kinetic studies on NF (12) may exist.

The data depicted in Fig. 2b show the temperature dependence of the O_2/NF_3 afterglow oxide growth process. A strong temperature enhancement is noted between the 550°C and 700°C growth conditions. Further, by comparison with Fig. 2a the film growth rate at 550°C is roughly double that achieved in a pure O_2 afterglow at 700°C. The data at 400°C however, points to a particularly interesting feature of the studied process. At this temperature the thickness of film achieved is independent of time. Comparison with data for O_2 afterglow oxidation suggests that in the NF_3 containing afterglow two competing mechanisms are present at this temperature, film growth and etching. The competition between growth and removal processes in the O_2/NF_3 afterglow may be illustrated by the group of experiments where an initial oxide 150Å thick was grown in a conventional furnace system and then processed in the afterglow thermal system for 30 minutes at various temperatures. The results of these experiments are shown in Fig. 3. These data indicate that at temperatures below approximately 525°C the etching reaction dominates in the O_2/NF_3 afterglow whereas at temperatures higher than this value the growth process dominates the reaction. Although the chemistry of this effect is not entirely clear, it is apparent that under the process conditions used in this experiment, temperatures higher than about 500°C are needed to achieve continuous growth of an oxide in an O_2/NF_3 afterglow thermal oxidation process. There are several possible explanations for this phenomenon, one being related to enhanced diffusivity of oxidizing species in the oxide, or the creation of the species responsible for enhanced oxidation at temperatures exceeding 500°C.

It should be strongly emphasized that the possibility of film growth as a result of the redeposition of silica species from the gas phase was excluded by the results of experiments utilizing 0.7μm thick oxides. At 0.7μm, any further oxidation was precluded, and hence, the observation of any increase in measured oxide thickness after processing could only result from the existence of a redeposition process. However, over the range of process conditions used in this experiment this film only exhibited a loss in thickness and never an increase.

In order to evaluate the performance of the thin oxide grown at low temperature in afterglow gases, electrical measurements on the fabricated MOS capacitors were performed to determine interface trap density (D_{it}) and fixed oxide charge (Q_f). Figure 4 shows the variation of current flowing through the oxide as a function of electric field in the oxide. As may be seen, very limited leakage current was detected over a wide range of electric field variation which indicates a

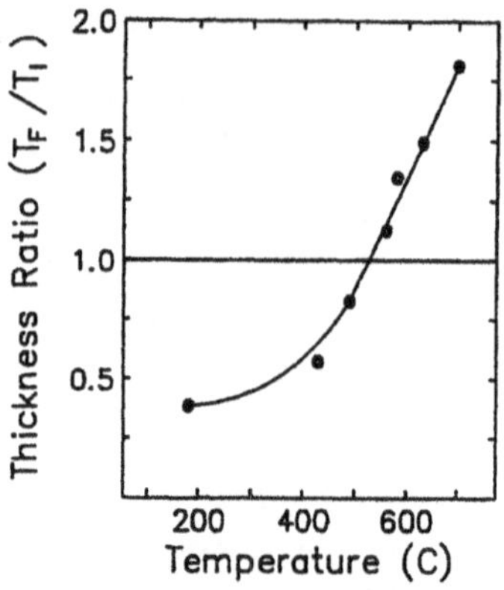

Fig. 3. Ratio of final oxide thickness T_F to initial thickness T_I vs. temperature after 30 min O_2/NF_3 afterglow at 1 Torr.

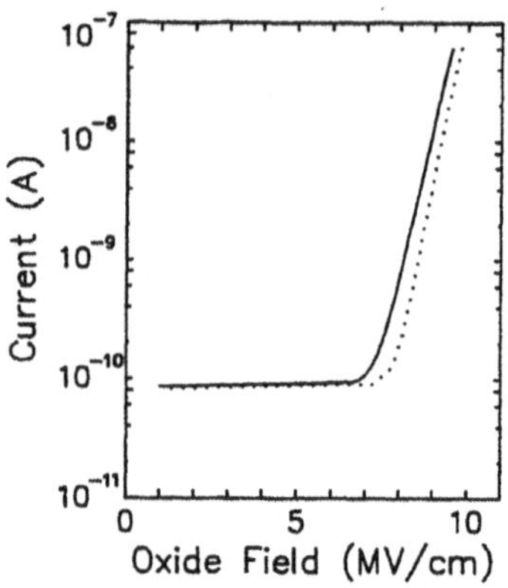

Fig. 4. Leakage current vs. oxide field for: ——conventional oxide grown at 1000°C, and --- oxide grown in O_2/NF_3 afterglow at 700°C and 1 Torr.

relatively dense, defect free structure of the afterglow oxides comparable with conventional thermal oxide. Also, similarly satisfactory results were obtained in terms of interface trap density which has varied, without any apparent relation to the afterglow oxidation conditions, from $5x10^{10}$ eV^{-1} cm^{-2} to $8x10^{10}$ eV^{-1} cm^{-2}. Somewhat inferior values were recorded however, for fixed oxide charge density. In this case oxidation temperature dependence was observed when compared with films grown under high temperature conditions. All afterglow oxides grown at 700°C and 550°C exhibited Q_f values of $2x10^{11}cm^{-2}$ to $3x10^{11}cm^{-2}$. In addition, equivalent values were obtained whether or not NF_3 was present in the afterglow gas mixture. We note that this result was expected from low temperature oxidation.

SUMMARY

We have presented the results of a study of the low temperature oxidation of silicon using a combined afterglow and thermal process. Oxide films with reasonable properties have been grown at temperatures at or below 700°C and times of oxidation compatible with circuit manufacturing requirements. The process system utilized consists of a conventional, resistance heated oxidation process furnace, to which low pressure capability and a microwave atom source have been added. The electrical properties of the films are promising, however more effort will be required to lower the fixed charge values obtained. It is believed that this might be achieved through the use of a rapid annealing technique. Finally, the concept of using afterglow gas mixtures in a thermal environment should allow increased chemical and process flexibility in the future.

REFERENCES

1. R. Ghez and Y. J. van der Meulen, J. Electrochem. Soc., 119, 1100 (1972).
2. J. Blanc, Appl. Phys. Lett., 33, 424 (1978).
3. E. A. Irene, Appl. Phys. Lett., 40, 74 (1982).
4. A. M. Hoff and J. Ruzyllo, Appl. Phys. Lett., accepted for publication.
5. M. Morita, T. Kubo, T. Ishihara, and M. Hirose, Appl. Phys. Lett., 45, 1312 (1984).
6. P. J. Wright, M. Wong, and K. T. Saraswat, Int. Elect. Dev. Mtg. 1987, Technical Digest, 574 (1987).
7. J. R. Ligenza, J. Appl. Phys., 36, 2703 (1965).
8. A. K. Ray and A. Reisman, J. Electrochem. Soc., 128, 2467 (1981).
9. R. P. H. Chang, C.C. Chang, and S. Darack, Appl. Phys. Lett., 36, 999 (1980).
10. W. Kern, Semicond. Int., 7(4), 94 (1984).
11. M. Morita, S. Aritome, M. Tsukude, and M. Hirose, Int. Elect. Dev. Mtg. 1984, Technical Digest, 144 (1984).
12. M. A. A. Clyne and I. F. White, Chem. Phys. Lett., 6, 465 (1970).

DEPOSITION OF SiO_2 THIN FILMS BY REMOTE PLASMA ENHANCED CHEMICAL VAPOR DEPOSITION (REMOTE PECVD)

D.V. Tsu, S.S. Kim and G. Lucovsky
Departments of Physics, and
Materials Science and Engineering
North Carolina State University
Raleigh, NC 27695-8202

ABSTRACT

This paper discusses the growth of thin films of silicon dioxide (SiO_2) by the process of remote plasma enhanced chemical vapor deposition (RPECVD). For comparison, we also discuss the deposition of silicon nitride and oxynitride films by the same RPECVD technique. We describe; (1) the deposition process chamber and reaction pathways; (2) the characterization of the local atomic structure in the deposited thin films: and (3) the applications of dielectric thin films in metal insulator semiconductor (MIS) structures.

INTRODUCTION

There is considerable interest in the low temperature (300 to 500°C) formation of dielectric thin films for semiconductor device technology applications. This is driven primarily by two considerations: (1) the available thermal budget for silicon device processing; and (2) the fabrication of device structures using compound semiconductors, e.g., GaAs, (Cd,Hg)Te, etc., which because of volatile atomic constituents, As and Te, respectively, require low temperature processing. There have been many different approaches to low temperature film growth and/or deposition; e.g., plasma and photo assisted CVD, plasma and anodic oxidation and nitridation, etc. In this paper we restrict the discussion to the application of the RPECVD technique to silicon based dielectric materials; the oxides, nitrides and oxynitrides.[1-3] We first discuss the RPECVD process, comparing it with conventional or direct PECVD.[3] We discuss selected properties of the deposited films[1-5], and conclude with a discussion of device applications. In this final section of the paper, we compare the chemical bonding and device properties of the RPECVD silicon oxide films with thermally grown SiO_2 films.[6,7]

THE REMOTE RPECVD PROCESS

We first describe the ***conventional or direct*** PECVD process[3] which is generally carried out within a single chamber with the substrate immersed in the plasma. All reactant gases, as well as any noble gas diluents, are fed into the chamber and then subjected to either an inductively or capacitively coupled RF plasma discharge. The purpose of the plasma excitation is to create chemically active (and generally short-lived) species, e.g., ions, radicals, metastables, etc., which serve to increase the reaction rates for the chemical processes involved in the thin film deposition. The primary consequence of this increased chemical reactivity is to reduce the temperature for CVD deposition relative to what it would be for an ***unassisted or thermal*** CVD process. However, the simultaneous plasma excitation of all reactant gases generally opens multiple and parallel deposition pathways, and as a consequence the deposited films usually contain ***undesirable*** chemical bonding groups.[8] This is particularly true for the silicon dielectrics where films grown by conventional PECVD are generally non-stoichiometric, i.e., the are Si-rich with respect to the compound compositions, SiO_2, Si_3N_4, etc., and also usually contain significant amounts of bonded hydrogen. This hydrogen incorporation occurs in SiH, SiOH, and SiNH groups, with concentrations about 5-10 at % in the oxides, and as high as 20-30 at % in the nitrides.[8] The direct PECVD dielectric films are not suitable for the insulating layers in MIS structures, but have been used in other ways in semiconductor device technology, e.g., as field insulators, and as "caps" for annealing processes involving compound semiconductors. Recently, Batey and coworkers[9] have significantly modified the direct process by diluting the reactant gases with He by factors between 100 and 10,000:1; they have been successful in producing gate quality SiO_2 films on ***carefully prepared*** silicon substrates. The high dilution rates have modified the deposition process chemistry in a way that minimizes any direct plasma excitation of the silane reactant[3], and thereby limits the number of parallel deposition reaction pathways. The heavy dilution also produces a significant decrease in the deposition rate, from Angstroms/s to tenths of Angstroms/s. The remote PECVD process, we discuss in this paper accomplishes the same objective of limiting the deposition reaction pathways and specifically avoiding plasma excitation and decomposition of the silane reactant, but in a qualitatively different manner.[1-3]. It is interesting to note that the deposition rates for device quality SiO_2 films by the modified direct PECVD process, and the RPECVD process are about the same.[1-3,9]

The main of objective of the RPECVD process steps is to control and limit the reaction pathways leading to thin film deposition. In particular the process is specifically designed to prevent direct plasma excitation and decomposition of the silane reactant. The overall reactions in the RPECVD process for the deposition of silicon oxide, nitride and oxynitride thin films are given below[1-3]:

$[O_2,He]^* + SiH_4 \text{ ----> } SiO_2;$
$[N_2,He \text{ (or Ar)}]^* + SiH_4 \text{ ----> } Si_3N_4;$
$[NH_3,He]^* + SiH_4 \text{ ----> } Si_3N_4;$ and
$[NH_3,O_2,He]^* + SiH_4 \text{ ----> } (SiO_2)_x(Si_3N_4)_{1-x};$

where the [*] identifies the gas mixture that is plasma excited. The deposition process includes in four separate steps. Figure 1 indicates the design of our newest processing chamber[10] in which these steps have been implemented. The four steps in the deposition process are described below, where we have illustrated the process as it is applied to the deposition of SiO_2 thin films:

A. ***Plasma excitation*** of the appropriate gas mixture, $[O_2,He]$, in the fused silica tube at the top of the chamber.

B. ***Extraction of active species*** from the plasma region and transport into the main chamber. This is accomplished by a relatively high gas flow rate for the O_2/He mixtures (at least 100 sccm). Back streaming of the plasma excited species, and more importantly the SiH_4 reactant (coming from the dispersal ring) into the plasma region is prevented by the relatively short mean free path associated with the main chamber pressure of 300 mTorr, and the high flow rate of gases out of the fused silica tube.

C. ***Mixing*** of the extracted active species with silane, SiH_4. The chamber pressure of 300 mTorr insures a relatively high number of bimolecular collisions between the active species and the silane. This high collision rate serves to promote the ***formation of deposition precursors*** involving either: (1) the chemical consumption; and/or (2) the selective fragmentation of the silane at by the active species delivered from the plasma region into the main chamber. Recent experiments have shown that processes (1) and (2) take place predominantly at the growth substrate rather than in the gas phase. as we had originally proposed.[1-3]

D. ***Film deposition*** in the main chamber, and hence outside of the plasma region by a CVD reaction process at a heated substrate (generally 100 to 550°C).

The RPECVD deposition process steps and reaction chemistry proceed in a ***serial*** manner, with the number of reaction pathways limited by the sequenced approach described above. This is the aspect of the process that produces the capability: (1) to control the stoichiometry, and thereby limit the formation of Si-Si bonds in the compound materials, SiO_2 and Si_3N_4; (2) to insure that oxynitride alloys are on the join line between the end two member compounds; and (3) to reduce bonded hydrogen incorporation in the films to concentrations below the levels of ir detection (to at least about 0.5 to 1 at %). Figures 2 and 3 illustrate, respectively the spectra of oxides and nitrides grown by the conventional direct PECVD, and the remote PECVD processes. The major differences in these spectra are in the stoichiometry, and hydrogen incorporation.[3] The deviations from stoichiometry show up in shifts to lower wavenumber of the dominant Si-O and Si-N bond-stretching absorption bands[1-5]; and the bonded hydrogen is readily identified via discrete bond-stretching and bending vibrations; e.g., for the SiH group in SiO_2 these occur, respectively at 2250 and 875 cm^{-1}.[1-3]

CHEMICAL BONDING IN THE RPECVD FILMS

We have been able to control the stoichiometry and hydrogen content of ***silicon oxide alloy*** films by controlling: (1) the ratio of O_2/He in the feed gas mixture; and (2) the substrate temperature for film deposition. We have thereby been able to grow three different types of silicon oxide films:

A. ***sub-oxides*** in which the Si/O ratio is greater than 0.5, and in which any bonded hydrogen is present only in SiH groups (no SiOH);

B. ***stoichiometric oxides*** in which the Si/O ratio equal 0.5, and in which there is no ir detectable bonded hydrogen; and

C. ***super-oxides*** in which the Si/O ratio is less than 0.5, and in which any bonded hydrogen in present only in SiOH groups (no SiH).

Representative spectra for the three types of oxides are shown in Fig. 4. We have also grown silicon oxide films using deuterated silane, SiD_4 in place of the SiH_4. These experiments have established that: (1) the bonded hydrogen in SiH groups in the ***sub-oxides*** derives from the SiH_4 feed gas (as expected) whilst (2) the bonded hydrogen in the SiOH groups in the ***super -oxides*** comes for the most part from impurities (CH_4, H_2O) in the oxygen feed gas (not as expected!).

We have grown ***nitride films*** using both N_2 and NH_3 as the source of nitrogen atoms. We limit our discussion to the films grown from NH_3/He plasma excited mixtures. Figure 5 represents the range of nitride films we have grown:

A. ***stoichiometric*** silicon nitride, Si_3N_4, with little or no detectable SiNH bonding, and no detectable SiH bonding. The determinant factors for hydrogen incorporation as SiNH are the substrate temperature and the NH_3/He ratio in the feed gas mixture;

B. ***stoichiometric*** silicon diimide, $Si(NH)_2$, an analog of SiO_2 in which the bridging two fold-coordinated oxygen atoms of SiO_2 are replaced by isoelectronic bridging NH groups; and

C. silicon nitride/diimide ***pseudobinary alloys***. These are analogs of the silicon oxynitrides ($[SiO_2]_x[Si_3N_4]_{1-x}$) and have compositions on a join-line between $Si(NH)_2$ and Si_3N_4.

We have grown nitride/diimide alloys (which alternatively can be described as silicon nitride films with NH bonding groups) using deuterated ammonia, ND_3, and have established that the hydrogen atoms in the NH groups derive from the ammonia rather than the silane reactant. Films grown from N_2/He and N_2/Ar mixtures are discussed in Ref.4. Using the N_2 source gas mixtures, we have grown ***sub-nitrides*** ; these contain no SiNH groups, but display ir evidence for Si-Si bonds, and also can contain bonded hydrogen in the form of SiH groups only.[4]

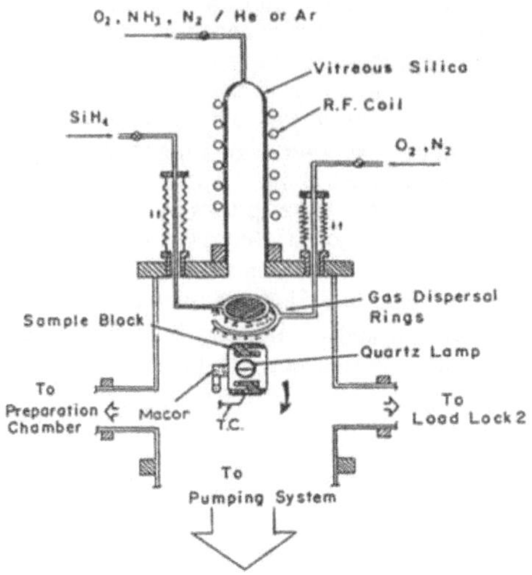

FIG. 1. Deposition chamber for RPECVD.

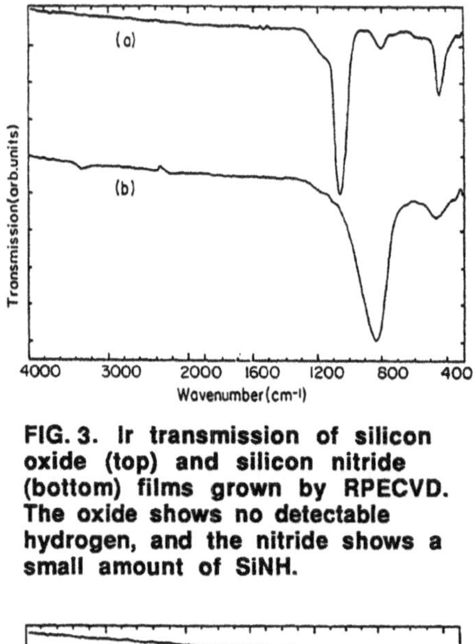

FIG. 3. Ir transmission of silicon oxide (top) and silicon nitride (bottom) films grown by RPECVD. The oxide shows no detectable hydrogen, and the nitride shows a small amount of SiNH.

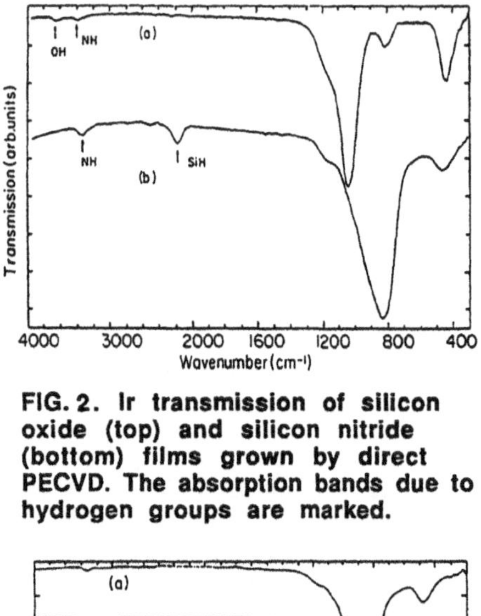

FIG. 2. Ir transmission of silicon oxide (top) and silicon nitride (bottom) films grown by direct PECVD. The absorption bands due to hydrogen groups are marked.

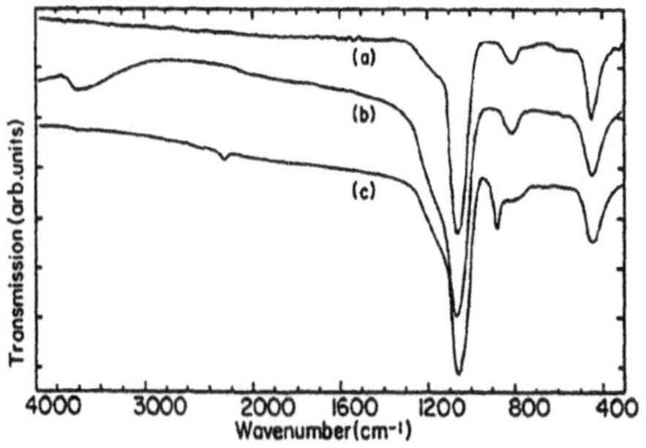

FIG. 4. Ir transmission spectra of silicon: (a) dioxide; (b) super-oxide with SiOH; and (c) sub-oxide with SiH.

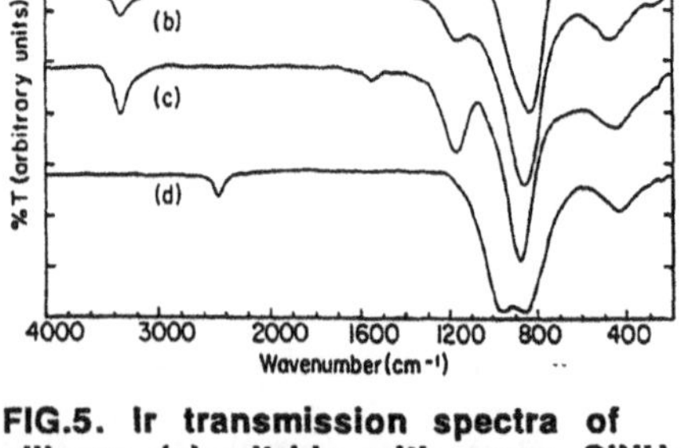

FIG.5. Ir transmission spectra of silicon: (a) nitride with some SiNH; (b) nitride with a large concentration of SiNH; (c) diimide with additional absorption due to $SiNH_2$; and (d) nitride grown from ND_3 showing SiND bands.

We have also grown ***silicon oxynitride alloy films*** from source gas mixtures of $O_2/NH_3/He$. By controlling the ratio of O_2 to NH_3 for a fixed He dilution, and the substrate temperature, we have been able to control the oxide/nitride ratio and the hydrogen incorporation in the deposited films. We have been able to grow films along the join-line between SiO_2 and Si_3N_4, and maintain bonded hydrogen concentrations, mostly as SiNH, close to, or below the levels of ir detection. We have used Auger Electron Spectroscopy (AES) to establish the O/N ratios, and ir to determine the extent and nature of any bonded-hydrogen incorporation.[5]

PROCESS CHEMISTRY

A system designed specifically to investigate the process chemistry is shown in Fig. 6.[11] This deposition/analysis system allows for: (1) analysis of gas phase components by optical spectroscopy and mass spectroscopy; and (2) film deposition on a movable heated substrate. This system is just becoming operational, and the following initial results have been obtained.

We have studied the emission spectra, in the visible and near UV, from O_2/He discharges, and have established empirical relationships between spectral features, associated with either He or O species in the plasma and the compositions of the deposited films. We find that stoichiometric oxides and also super-oxides can be produced under conditions in which all of the He emission lines are quenched. Such conditions exist when the relative O_2 in He flow rate is between about 2% and 100 % (total flow rate ≈100 sccm). We have established that the oxide film deposition rate is not proportional to either the atomic oxygen or O_2^+ emission intensities.[11] Future studies will center around oxygen species which do not have strong emission bands, e.g., metastable molecular oxygen and ozone. For relative oxygen flow rates below about 2 %, the deposited oxide films depart from SiO_2 stoichiometry with ir evidence for both Si-Si and SiH bonding groups. The Si-Si bonding is evident from shifts to lower wavenumber in the absorption maximum of the 1075 cm^{-1} stretching band, and the SiH bonding from distinct features at 2250 cm^{-1} and 875 cm^{-1}. This transition to sub-oxide deposition with SiH bonding is correlated with the emergence of strong He emission bands. The particular bands we have identified have terminal states in the singlet and triplet He metastable electronic configurations. When the oxygen is completely eliminated from the feed gas these strong He emission bands persist, and thin films of hydrogenated amorphous silicon (a-Si:H) are deposited.[12,13] The implication here is that the same mechanism, presumed to be a ***fragmentation and/or excitation*** of the SiH_4 by ***active*** He species, that is operative for the a-Si:H deposition, also leads to Si-Si and SiH bonding in the sub-oxide films. It should be noted further that the evidence for generation of He metastables in the plasma does not mean that the active species transported out of the plasma are necessarily He metastables; in fact our most recent experiments suggest that He+ ions, which we can not detect optically due to the band-pass limitations in our optical detection system, many indeed play an important role in the deposition process for the a-Si:H films.

We have found three different types of emission spectra from NH_3/He plasmas: (a) for low to intermediate He dilutions the dominant emission lines are for NH and H; (b) for high dilutions, we find predominantly N and H emission lines; whilst for (c) very high dilutions, we find essentially only He emission.[11] We have correlated the deposition of $Si(NH)_2$ and $Si(NH)_2/Si_3N_4$ alloy films with the occurrence of NH (and H) plasma emission lines, the deposition of stoichiometric Si_3N_4 films with the occurrence of no NH, but only N (and H) emission lines; and finally the deposition of sub-nitrides with SiH bonding with the onset of strong He emission.

We are just beginning complementary studies of the process reactions using mass spectrometry. This is being done to correlate features in the optical emission spectra with both neutral and ionic groups that are extracted from the plasma, and hence are available for mixing and reaction with the SiH_4. For example, as noted above, our preliminary results suggest that the observation of He emission lines, associated with transitions into the singlet and triplet metastable states is associated with He ion extraction from the plasma region.

FORMATION AND CHARACTERIZATION OF DEVICE STRUCTURES

We have used the multi-chamber surface preparation/insulator deposition/in-situ analysis system shown in Fig. 7 to form MOS device structures. This system contains: (1) two substrate introduction and extraction load locks; (2) a substrate processing chamber; (3) an analysis chamber with reflection high energy electron diffraction (RHEED) and AES; and (4) an insulator deposition chamber. We at North Carolina State University, and our colleagues at the Research Triangle Institute[14], have found that a key aspect of MOS device fabrication based on deposited silicon dielectrics is in the processing and characterization of the semiconductor substrate. For silicon substrates, using a modified RCA, and subjecting the chemically processed surface to a short hydrogen plasma discharge promotes surface reconstructions that are observable via RHEED. Depositing SiO_2 films on these reconstructed surfaces at temperatures in the range of 350°C correlates with the formation MOS structures with low defect densities (typically less than $3 \times 10^{10}/cm^2$) and high break down field strengths (typically greater than about $8\text{-}10 \times 10^6$ V/cm). These devices additionally have flat-band voltage shifts less than 0.2 V, and do not show ionic motion in the dielectrics under stress bias testing. Similar gate-quality dielectrics have been formed by Batey and coworkers[9] using a modified direct PECVD process with high He dilution: $SiH_4/N_2O/He$ = 1/100/10,000. As we have mentioned above, this modified direct process prevents direct excitation of SiH_4 in the plasma through the high He dilution, thereby cutting off unwanted deposition pathways and producing deposited SiO_2 thin films that are also stoichiometric and hydrogen free.

We have not studied nitride or oxynitride films as extensively in device structures, but preliminary studies have shown: (1) that RPECVD nitride films can be combined with SiO_2 in tri-layer structures to form gate quality dielectrics for (In,Ga)As field effect transistor structures[1-3]; and (2) that gate quality silicon oxynitride layers can also be grown with defect states densities $<5 \times 10^{10}/cm^2$, high breakdown fields $>6 \times 10^6$ V/cm, and small flat band offset voltages <0.5 V.

COMPARISONS BETWEEN THE PROPERTIES OF RPECVD OXIDES AND THERMALLY GROWN OXIDES

We have compared selected properties of RPECVD SiO_2 films with thermally grown and annealed SiO_2.[6,7] We have emphasized in these comparisons: (1) the frequency and width, ν and $\Delta\nu$, respectively of the Si-O infrared active bond-stretching vibration; (2) the optical index of refraction, n, at 632.8 nm; and (3) the etch rate of the oxide films in buffered HF. We observe that:

A. Comparisons between the properties of low temperature thermal oxides, formed by both wet and dry oxidation procedures, and RPECVD oxides indicate that the frequency and half width, ν and $\Delta\nu$, of the bond-stretching mode obey the same universal relationship shown in Fig. 8 with the same value of the fit parameter, $\Delta\theta/\theta = 0.21$; 2θ is the bond-angle at the oxygen atom site.[6,7] This means that the width of the Si-O-Si stretching mode derives from a sum of narrow modes, each associated with a specific bond-angle in a statistical distribution. The range of Si-O-Si bond-angles in the RPECVD oxides, based on the value of ν is 140 to 144 degrees, whilst the range of bond-angles in thermal oxides is from 147 to 150 degrees.[7]

B. The RPECVD oxides also display a linear relationship between n and ν, but with an intercept that is different from the thermal oxides (see Fig.8).[7] The particular values of n and ν for the RPECVD oxides are consistent with these films being less densified, by a factor of about 1.4 % than the low temperature thermal oxides. This is supported by the higher etch rates for the RPECVD oxides in buffered HF relative to the etch rates for the thermal oxides.

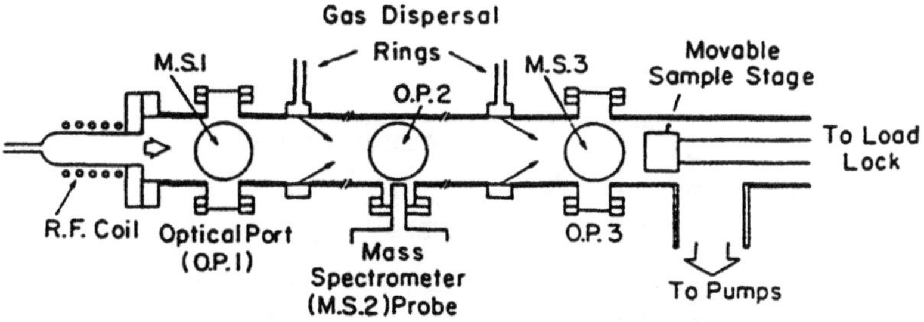

FIG. 6. Schematic representation of deposition/analysis system with provisions for optical spectroscopy, mass spectrometry and film deposition on a movable substrate.

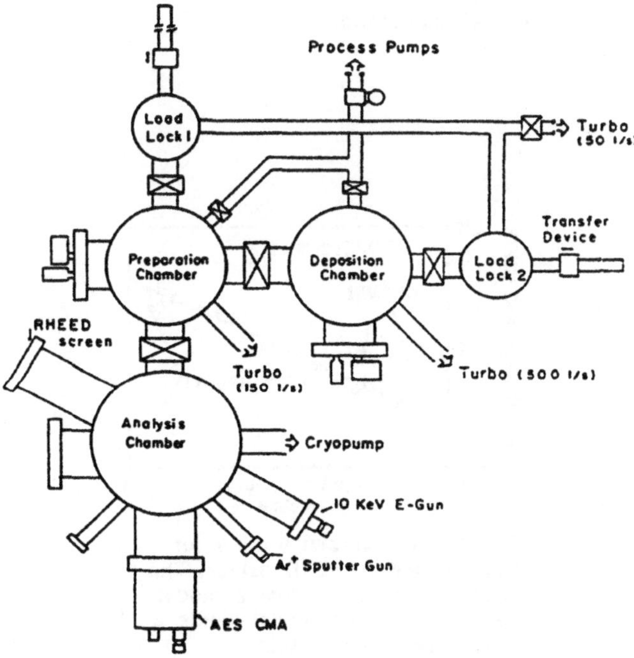

FIG. 7. Schematic representation of multi-chamber deposition system with in situ analysis capabilities.

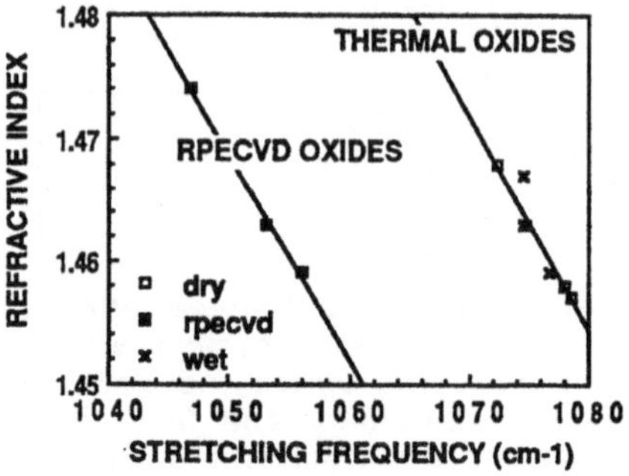

FIG. 8. Index of refraction at 632.8 nm versus ν for wet and dry thermal oxides and for oxides deposited by RPECVD. The solid lines derive from model calculations.

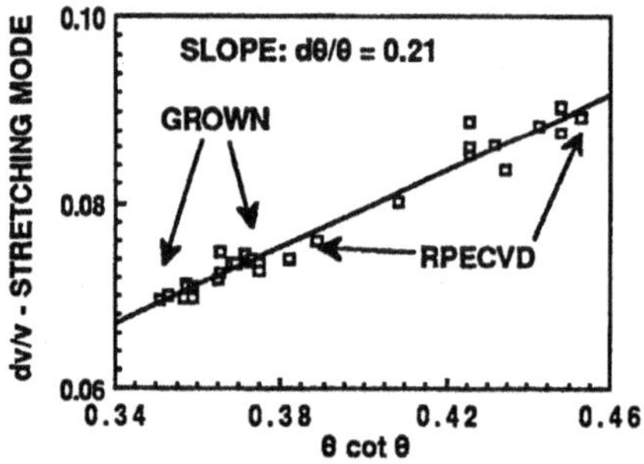

FIG. 9. Δν/ν versus θ cot θ for thermal and RPECVD oxides. The solid line derives from a model calculation. The model yields a fit parameter dθ/θ = 0.21.

C. Low temperature (700 to 800°C) thermal oxide films are less suitable for gate oxide applications than the RPECVD oxides and thermal oxides grown at temperatures >1000°C. However, annealing low temperature thermal oxides at temperatures in excess of 1000°C improves their properties considerably; in fact sufficiently long anneals (> 10 - 100s) at temperatures in excess of 1100°C generally render low temperature oxides comparable in optical, vibrational and electronic properties to oxides grown at 1100°C. In contrast, and depending mostly on Si surface preparation RPECVD oxides grown at temperatures of about 350°C can be of gate quality.[1-3,10] This means that the local atomic structure of the oxide, e.g., the Si-O-Si bond-angle is not a determinant factor for defining device quality oxide films.

CONCLUSIONS

We observe the following with respect to the formation of silicon dielectric thin films:

A. Thin films of silicon oxide, nitride and oxynitride alloys can be grown by RPECVD. The RPECVD process affords greater control over stoichiometry, and/or alloy composition than the conventional direct PECVD process, and can also be used to grow hydrogen free silicon based dielectric thin films as well.

B. The process chemistry is being studied, and the identification of specific gas phase species in the plasma discharge have been observed to correlate with the occurrence of specific bonding groups in the films. This supports the assertion made in (1), and also the general observation relative to the ***inherent controllability*** of the RPECVD ***serial*** reaction process chemistry.

C. Device quality films, mostly oxides have been deposited by RPECVD and provide an interesting and potentially important technology option particularly in: (1) silicon device structures where the thermal budget is an issue; and (2) in other device structures using compound semiconductors which require low temperature processing for all of the fabrication steps.

ACKNOWLEDGEMENTS

This research is supported by the Office of Naval Research, the Semiconductor Research Corporation and the National Science Foundation. We wish to acknowledge the contributions of our collaborators at the Research Triangle Institute, R.J. Markunas, R.A. Rudder and G.G. Fountain.

REFERENCES

1. P.D. Richard, R.J. Markunas, G. Lucovsky, G.G. Fountain, A.N. Mansour and D.V. Tsu, J .Vac. Sci. Technol. **A3**, 867 (1985).
2. G. Lucovsky, P.D. Richard, D.V. Tsu, S.Y. Lin and R.J. Markunas, J .Vac. Sci. Technol. **A4**, 681 (1986).
3. G. Lucovsky and D.V. Tsu, J .Vac. Sci. Technol. **A5**, 2231 (1987).
4. D.V. Tsu and G. Lucovsky, J .Vac. Sci. Technol. **A4**, 480 (1987).
5. D.V. Tsu, G. Lucovsky, M.J. Mantini and S.S. Chao, J .Vac. Sci. Technol. **A5**, 1998 (1987).
6. J.T. Fitch and G. Lucovsky, Mater. Res. Soc. Proc. **92**, 89 (1987); and Mater. Res. Soc. Proc. (Spring 1987 Meeting), in press.
7. J.T. Fitch, G. Lucovsky, E. Kobeda and E.A. Irene, these proceedings.
8. A.C. Adams, Solid State Technol. **26**, 135 (1983).
9. J. Batey and E. Tierney, J. Appl. Phys. **60**, 3136 (1986).
10. S.S. Kim, D.V. Tsu and G Lucovsky, Sang, J .Vac. Sci. Technol. **A7**, (1988), in press.
11. D.V. Tsu and G Lucovsky, J .Vac. Sci. Technol. **A7**, (1988), in press.
12. G.N. Parsons, D.V. Tsu and G Lucovsky, J. Non-Cryst. Solids 97 & 98, 1375 (1988).
13. G.N. Parsons, D.V. Tsu and G Lucovsky, J .Vac. Sci. Technol. **A7**, (1988, in press.
14. G.G. Fountain, R.A. Rudder, S.V. Hattangadi, R.J. Markunas and P.S. Lindorme, J. Appl. Phys. **63** (1988), in press.

ANODIC SiO_2 FOR LOW TEMPERATURE GATE DIELECTRICS

Keyvan Sayyah

Hughes Research Laboratories
3011 Malibu Canyon Road
Malibu, CA 90265

ABSTRACT

Electrical properties of anodic SiO_2 grown at room temperature and annealed at 450°C have been shown to closely approach those of thermal SiO_2. Midgap interface state densities in the low-10^{10} $cm^{-2}.eV^{-1}$, oxide surface charge at flatband in the low-10^{10} cm^{-2}, average breakdown fields in excess of 10 MV/cm, and dc resistivities above 10^{16} Ω.cm at 1MV/cm have been measured. The oxide surface charge at flatband is thickness dependent. The dependence of electrical and structural properties on various annealing conditions will be discussed.

INTRODUCTION

The trend towards shrinking the dimensions of silicon integrated circuit devices for improved speed and packing density necessitates the reduction in both the thickness and processing temperature of MOS gate oxides. Temperatures of 900-1100°C used in growing conventional thermal gate oxides give rise to several undesirable effects such as broadening of dopant profiles, generation of stacking faults, wafer warpage, and oxidation enhanced diffusion which hinder the performance of small dimension MOS devices. Furthermore, the availability of a good quality, low temperature gate oxide is of utmost importance for amorphous and polycrystalline silicon thin film transistors on glass substrates used for display applications. Therefore, there is a substantial need for the development of low temperature, gate quality SiO_2.

During recent years, there has been a wide range of activities in the fabrication of thin layer of silicon dioxide at low temperatures. A variety of techniques such as plasma-enhanced CVD (PECVD) (1-3), photo-assisted CVD (4-6), laser-enhanced oxidation (7), plasma oxidation (8-10), low temperature CVD (11,12), and sputtering (13) have been employed. Most of these processes result in oxides which are less dense, slightly deficient in oxygen, and have poor electrical and interfacial properties in comparison with the state-of-the-art thermal oxides, even under optimum low temperature anneal conditions. Among these techniques, plasma-enhanced and photo-assisted CVD have resulted in the best quality low temperature oxides. Annealed midgap interface state and flatband oxide surface charge densities in the range of mid-10^{10} to mid-10^{11} $cm^{-2}.eV^{-1}$ and cm^{-2}, respectively, and mean breakdown fields of 9-10 MV/cm have been reported in these two low temperature oxides (1-6). The PECVD process, however, has the disadvantage of heavy substrate damage from charged particles and UV light created in the plasma (8), while the photo-assisted CVD oxides suffer from annealed interface state densities in the low to mid-10^{11} $cm^{-2}.eV^{-1}$ range (4,6) which is more than an order of magnitude higher than those measured in thermal oxides. Furthermore, it is difficult to deposit oxides with thickness uniformities as good as those obtained in thermal oxides using these two techniques.

Besides some early work in the 60's and 70's (14-16), little attention has recently been paid to anodic silicon dioxide as a low temperature dielectric (17-19). The purpose of this paper is to investigate the structural, electrical, and interfacial properties of thin layers of SiO_2 grown by anodic oxidation of Si at room temperature with a subsequent low temperature anneal, as well as to examine their suitability for MOS gate oxides. We report here midgap interface state densities as low as $1.7x10^{10}$ $cm^{-2}.eV^{-1}$ together with mean breakdown fields in excess of 10 MV/cm at a maximum processing temperature of 450°C. To our knowledge, this is the lowest reported value of midgap interface state density in any low temperature SiO_2 with a maximum processing temperature below 500°C.

EXPERIMENTAL

Two inch p-type (100) Si wafers with boron doping levels of $1\text{-}2x10^{15}$ cm^{-3} were used in this study. The oxides were grown in two different electrolytes: a 0.04M solution of KNO_3 in N-methylacetemide (NMA), and a solution of 0.02M KNO_3 and 0.02M KNO_2 in ethylene glycol (EG). The potassium nitrate (KNO_3) used had a purity of 99.999%, the potassium nitrite (KNO_2) powder was 97% pure, and the two solvents (NMA and EG) were 99% pure. The anodic oxides were grown at room temperature at a constant current density of 2.5 mA/cm^2 until the desired thickness was reached, and then maintained at the forming voltage for an additional 10 minutes. This procedure has been shown to result in improved interfacial properties (15). For electrical measurements, MOS capacitors with thermally evaporated aluminum gates 635 μm in diameter were constructed. Most of the capacitors received a post-metallization anneal (PMA) step at 450°C in forming gas for one hour before defining the Al dots by photolithography.

The thickness of the anodic oxide films increases linearly with the forming voltage, defined as the difference between the final and initial cell voltages, as previously reported (14,17). Anodic oxide layers were grown on Si in both solutions in the thickness range of 325-1600 Å, corresponding to forming voltages between 35 and 255 V. The thickness uniformities obtained across the 2 inch wafers were better than 1% for both growth solutions, and better than 5% from run to run. This uniformity and reproducibility, which is an inherent consequence of the anodization process, is one the major attributes of this growth technique.

A combination of high frequency (1 MHz) and quasistatic CV measurements were used to extract the interface state and the flatband oxide charge densities. The breakdown fields, leakage currents, and current-voltage characteristics were evaluated using stepped and ramp IV measurements. The structural properties of these oxides as a function various annealing conditions were examined using infrared absorption spectroscopy. The thickness of the oxides grown were measured using Nanospec assuming a refractive index of 1.46. Secondary ion mass spectrometry (SIMS) was used to determine the purity of the anodic oxides.

STRUCTURAL PROPERTIES

Stoichiometric SiO_2 exhibits 3 characteristic absorption bands near 1075, 800, and 450 cm^{-1} which are due to the stretching, bending, and rocking motions of the Si-O bond, respectively (20). These band shift towards lower frequencies as the oxide moves off stoichiometry and becomes deficient in oxygen (21), or

Table 1. Position and FWHM of Si-O stretching band in anodic SiO_2 as-grown and annealed at 300 - 900°C in forming gas.

ANNEAL TEMP (°C)	PEAK (cm^{-1})	FWHM (cm^{-1})
Anodic SiO_2: as grown	1050 ± 5	89 ± 2
Anodic SiO_2: 300	1049 ± 4	81 ± 2
Anodic SiO_2: 450	1061 ± 4	93 ± 2
Anodic SiO_2: 600	1062 ± 4	91 ± 2
Anodic SiO_2: 700	1066 ± 3	84 ± 2
Anodic SiO_2: 800	1071 ± 4	80 ± 2
Anodic SiO_2: 900	1072 ± 4	80 ± 2
Thermal SiO_2: 450	1075 ± 4	85 ± 2

becomes more porous (22). The shift of the Si-O stretching band with oxygen concentration in SiO_x has been found to be linear (21). The width of this band is a manifestation of the statistical distribution of different bonding arrangements at each Si atom site. In general, the narrower this width the more structurally ordered the oxide film. The IR transmission spectra of as-grown and annealed anodic oxide films in the frequency range of 200-1400 cm^{-1} is shown in Fig. 1. Also shown in this figure is the IR transmission spectrum of a thermal oxide sample annealed at 450°C for comparison. In obtaining these spectra, a Si substrate was inserted in the reference beam in order to eliminate the effects of substrate absorption . The position and full width at half maximum (FWHM) of the Si-O stretching band of anodic oxide samples annealed in the temperature range of 300-900°C are listed in Table 1. These data clearly shows that the as-grown anodic oxide is slightly deficient in oxygen, and is more porous and less structurally ordered than thermal oxide. However, as the oxide is annealed in forming gas from 300 to 900°C, the Si-O stretching band shifts towards higher frequencies and its FWHM decreases, both approaching thermal oxide values and

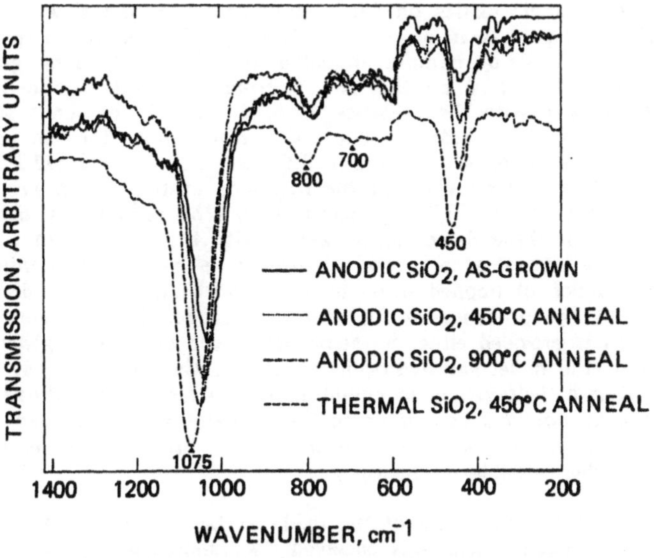

Fig. 1. IR transmission spectra of anodic SiO_2 as-grown and annealed at 450 and 900°C, and thermal SiO_2 annealed at 450°C in forming gas.

indicating denser, more stoichiometric and more structurally ordered films. Also, as the annealing temperature increases, the absorbance of this band increases indicating an increase in the Si-O stretching mode oscillator strength. A close look at the spectra of the anodic oxide reveals small absorption bands at around 540, 610, and 700 cm^{-1} present at all various annealing conditions. The latter two bands are also observed in the thermal oxide samples annealed at 300 and 450°C in forming gas. The small absorption peak near 610 cm^{-1} observed in all the samples is most probably due to the strong absorption of the Si substrate at this frequency (23) which was not completely compensated for by the reference sample. The weak absorption band near 700 cm^{-1} was not observed in the as-grown thermal oxide as well as in the thermal oxide sample annealed at 450°C in forming gas but covered with aluminum. Thus, it appears that this band might be related to the incorporation of hydrogen in the films since nitrogen, which constitutes 85% of the forming gas, is very stable at temperatures of only 450°C. The absence of the Si-N absorption band near 900 cm^{-1} in the annealed thermal oxides further supports this conclusion. The origin of the weak band near 540 cm^{-1} only observed in the anodic oxide samples is not known.

INTERFACIAL AND BULK ELECTRICAL PROPERTIES

Besides being porous and nonstoichiometric, the as-grown anodic oxide films have very poor interfacial properties. Very large densities of positive surface oxide charge at flatband (N_{ss}) and interface states (D_{it}) are present in the as-grown oxides. However, a post-metallization anneal (PMA) step in forming gas at 450°C significantly improves these interfacial properties. The high frequency and quasistatic CV curves of an anodic SiO_2 sample after receiving the 450°C PMA step are shown in Fig. 2. The resulting density of interface states calculated as a function of energy within the Si bandgap using these two curves is shown in Fig. 3. The midgap interface state density of this oxide is about 1.7×10^{10} $cm^{-2}.eV^{-1}$. To our knowledge, this is the lowest reported value of midgap interface state density in any low temperature SiO_2 with a maximum processing temperature below 500°C.

The dependence of the midgap interface state and the surface charge densities of the anodic oxides on 450°C post oxidation anneal with and without the presence of Al gates was examined. The high frequency and quasistatic CV curves of an anodic oxide sample annealed at 450°C in forming gas with and without the presence of Al on the oxide during annealing is shown in Fig. 4. The sample annealed with the presence of Al has an N_{ss} of about 5.4×10^{10} cm^{-2} and midgap D_{it} of about 5.3×10^{10} $cm^{-2}.eV^{-1}$, while those for the one annealed without the presence of Al are about 2.9×10^{11} cm^{-2} and 3.5×10^{11} $cm^{-2}.eV^{-1}$, respectively. It is well known that the interface states, which are a result of dangling Si bonds at the Si/SiO_2 interface, can be reduced by attachment of atomic hydrogen at these defect sites (24). With the presence of Al during the low temperature anneal step, atomic hydrogen is generated within the oxide through the reaction of trapped water molecules with Al at the oxide/Al interface (24). Without the presence of Al, atomic hydrogen for the neutralization of interface states is provided either by some other type of internal generation or by dissociation of the molecular hydrogen present in the forming gas. Very little amount of atomic hydrogen is in equilibrium with molecular hydrogen at 450°C. Thus, the results obtained here indicate that generation of hydrogen through the reaction of water molecules trapped in the anodic SiO_2 with Al is by far more efficient than other possible mechanisms.

An anomalous dependence of the oxide surface charge at flatband on the thickness of the anodic oxides was observed. As shown in Fig. 5, this charge is negative for thinner oxides, becomes less negative as the oxide thickness increases, crosses zero somewhere between 1050-1200 Å, and turns abruptly positive above 1200 Å. This phenomenon , which has not been observed in thermal oxides, was discovered in samples grown in both types of growth solutions. The oxide thickness dependence of the midgap interface state density for the same set of samples is also shown in Fig. 5. Below a thickness of 1050 Å, the midgap D_{it} is relatively constant while above 1200 Å it abruptly increases by more than an order of magnitude. For a uniformly doped substrate, which is the case in the low temperature anodic SiO_2, the oxide surface charge (Q_{ss}) obtained from the measurement of the flatband voltage is a lumped sum of two charge contributions (24):

$$Q_{ss} = -C_{ox} \cdot [\ V_{FB} - W_{MS}\ ] = Q_f + Q_{it}(0)$$

where C_{ox} is the oxide capacitance, V_{FB} is the measured flatband voltage, W_{MS} is the known metal-semiconductor work function difference, Q_f is the fixed oxide charge, and $Q_{it}(0)$ is the charged interface traps at flatband. In practice, it is very difficult to separate these two charges due to the uncertainty in obtaining the total charge contribution from the interface traps. This charge is obtained by adding the contributions of both the donor and acceptor-like interface state densities integrated over the Si bandgap. One possible explanation for the strange dependence of Q_{ss} on the thickness of the anodic oxide is that the fixed oxide charge is negative, relatively independent of the oxide thickness, at least for thicknesses below 1050 Å where the contribution of the interface state charges is the smallest. This implies that the rapid positive increase in the magnitude of Q_{ss} with oxide thickness, specially above 1200 Å, is due to a positive increase in the charged interface states at flatband. This explanation would only be valid if the interface states contributing to this charge are donor-like since acceptor-like states can not result in an effective positive charge. The fact that both the interface state and oxide surface charge densities at flatband increase rapidly in the thickness range between 1050 and 1200 Å supports the above model. Negative fixed oxide charges have been reported at the Si/Ta_2O_5 interface (25). It should also be mentioned that no hysteresis has been observed in the CV curves of the anodic SiO_2 samples biased to ±10 V. This indicates that very

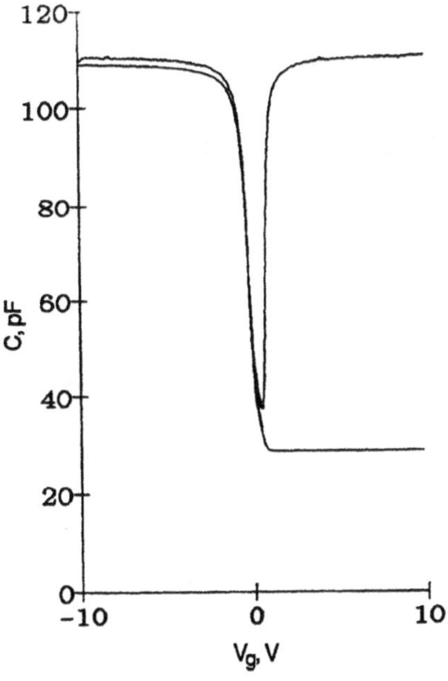

Fig. 2. High frequency (1 MHz) and quasistatic CV curves for a 1015Å anodic SiO_2 sample after 450°C post-metallization anneal.

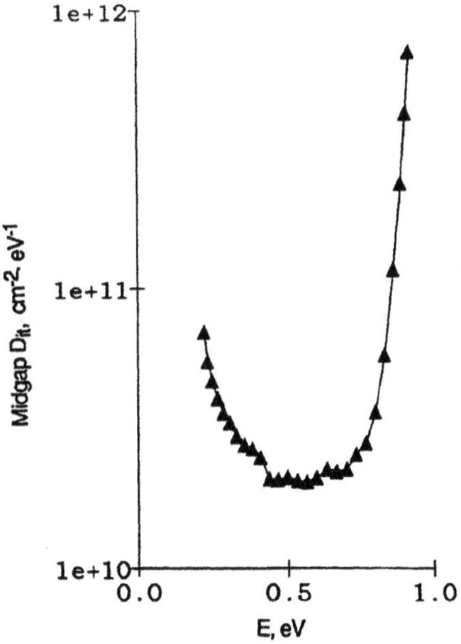

Fig. 3. Midgap density of interface states as a function of energy within Si bandgap for sample in Fig. 2.

little trapping of electrons and holes by traps located in the oxide adjacent to Si occurs at these bias levels.

The breakdown field histograms of two anodic oxide samples using 50 dots per sample statistics are shown in Fig. 6. Average breakdown fields of 11.1, 10.4 and 8.8 MV/cm were measured for oxides with thicknesses of 325, 407 and 1006 Å, respectively. These values are as good as those obtained in thermal oxides with the same thickness. Anodic oxides grown on polycrystalline Si have been shown to have a breakdown field strengths better than thermal oxides grown on the same material by a factor of two (26). This difference has been speculated to be caused by the smoothing effect of the anodization process on the interface. It is known that both the leakage current and breakdown field strength of poly-Si thermal oxides are due to locally enhanced fields at the rough poly-Si/SiO_2 interface. Anodization is also a self-healing growth process which results in very low densities of pinholes (15). This effect results in improved mean breakdown fields due to the absence of pinhole shorts compared to other low temperature SiO_2 processes.

Stepped IV measurements were used to determine the leakage currents and dc resistivities of the anodic oxide films. Leakage current densities as low as $3x10^{-11}$ A/cm^2, and dc resistivities of about $4x10^{16}$ Ω.cm at 1 MV/cm have been measured in anodic SiO_2 films. These values are as good, if not better than those obtained in thermal oxides. The ramp IV measurements of an anodic and a thermal SiO_2 sample, 712 and 730 Å thick, respectively, are shown in Fig. 7. The Fowler-Nordheim (FN) conduction mechanism in the thermal oxide starts at a slightly higher field and the current increases more rapidly with the field than the anodic oxide. Furthermore, the thermal oxide breaks down at less than 10 MV/cm, while at this field the conduction mechanism in the anodic oxide changes and the rate of current increase with field increases. The differences in the conduction mechanism between these two oxides is under investigation.

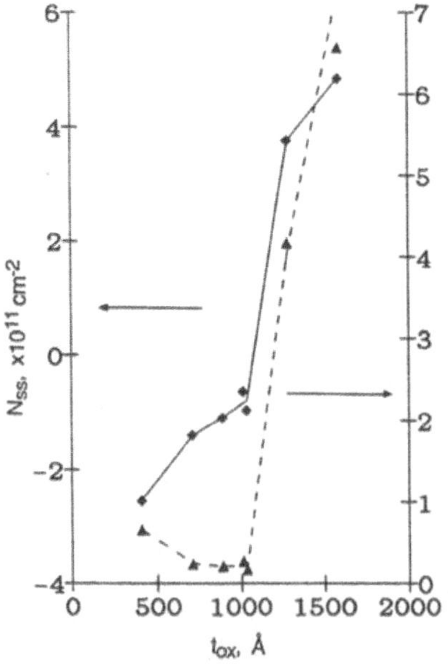

Fig. 4. High frequency (1 MHz) and quasistatic CV curves for an anodic SiO_2 sample annealed at 450°C with (solid) and without (dotted) Al.

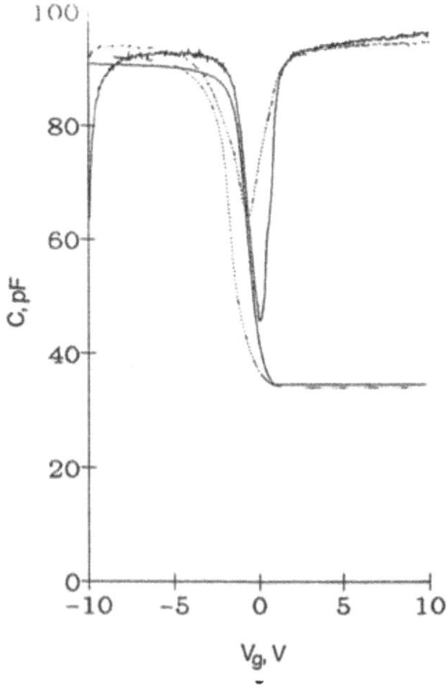

Fig. 5. Dependence of anodic SiO_2 surface charge at flatband and midgap interface state density on oxide thickness.

Finally, bias-temperature-stress (BTS) measurements were used to measure the mobile charge content of these oxides. Positive mobile ion densities of 1-2x10^{11} cm^{-2} and 4-7x10^{11} cm^{-2} were measured for the EG and NMA growth solutions, respectively. SIMS measurements were used to find the impurity content of these films. Other than various Si and O related peaks, these measurements only revealed the presence of traces of potassium, sodium, calcium and carbon. However, these measurements were inconclusive in terms of determining the source(s) of the mobile ions since reference oxides with the same amount of detected sodium and potassium showed no bias-temperature-stress instability. The main source of the mobile ions is most probably in the EG and NMA growth solutions. These solutions were not electronic grade and were only 99% pure. Although the KNO_2 powder used in the EG growth solution was only 97% pure, the anodic oxide samples grown using this solution showed less mobile ion content than the ones grown using the NMA growth solution which only contained the 99.999% pure KNO_3 powder. This fact supports the above supposition. Thus, purifying the anodic growth solutions could remove the source of the mobile ions. Anodic oxides grown in EG solutions containing only KNO_3 have been reported to have mobile ion contents of about 2-3x10^{10} cm^{-2} (19), a rather respectable value. It is therefore possible to obtain anodic oxide films with acceptable BTS instabilities.

CONCLUSION

The results obtained here indicate that the electrical and interfacial properties of anodic oxides grown at room temperature closely approach those of state-of-the-art thermal oxides after receiving a post-metallization anneal step at 450°C, with the exception of the mobile charge content. There is, however, a strong possibility that purifying the non-electronic grade growth solutions might remove

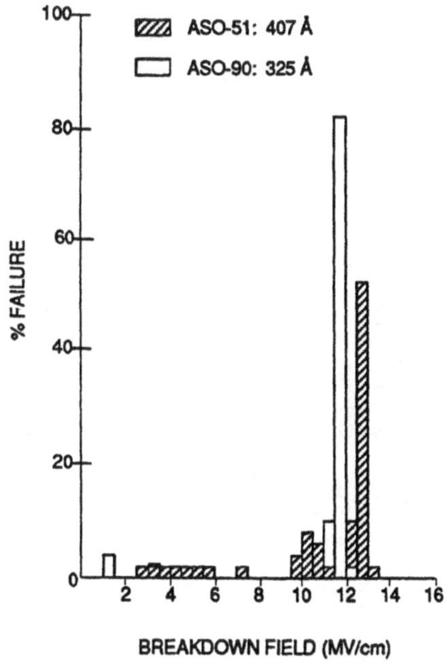

Fig. 6. Breakdown field histogram of 2 anodic SiO_2 samples.

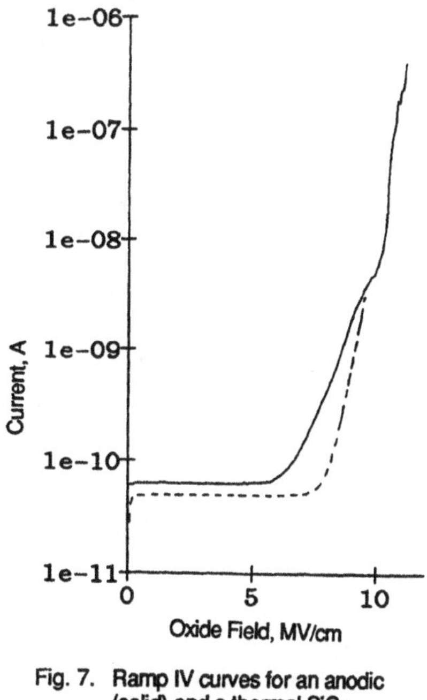

Fig. 7. Ramp IV curves for an anodic (solid) and a thermal SiO_2 (dashed) sample. Ramp rate is 0.4 V/s.

the source(s) of this mobile charge. A maximum processing temperature of 450°C, excellent oxide thickness uniformity and control, very low interface state and oxide surface charge densities below thicknesses of 1100-1200 Å, very high breakdown fields and low leakage currents, lack of any substrate damage, and the ease in the implementation of the process are major attributes of the anodic SiO_2. Until the source(s) of the mobile charge are removed, the anodic oxide can be used as low temperature gate dielectrics in applications where bias-temperature-stress instability is not of critical importance.

ACKNOWLEDGMENTS

The author is thankful to V. Thompson and N. Goodwin for their technical assistance.

REFERENCES

1. J. Batey and E. Tierney, J. Appl. Phys., 60, 3136 (1986).
2. J. Batey, E. Tierney, and T.N. Nguyen, IEEE Electron Device Lett., EDL-8, 148 (1987).
3. G. Lucovsky, P.D. Richard, D.V. Tsu, S.Y. Lin, and R.J. Markunas, J. Vac. Sci. Technol. A, 4, 681 (1986).
4. U. Sharma, RCA Review, 47, 551 (1986).
5. P.K. Boyer, G.A. Roche, W.H. Ritchie, and G.J. Collins, Appl. Phys. Lett, 40, 716 (1982).
6. K. Inoue, M. Michimori, M. Okuyama, and Y. Hamakawa, Jpn. J. Appl. Phys., 26, 805 (1987).
7. H. Richter and T.E. Orlowski, J. Appl. Phys., 56, 2351 (1984).
8. S. Taylor, E. Eccleston, and P. Watkinson, Electron. Lett., 23, 732 (1987).
9. S. Kimura, E. Murakami, K. Miyake, T. Warabisako, H. Sunami, and T. Tokuyama, J. Electrochem. Soc., 132, 1460 (1985).
10. A.K. Ray and A. Reisman, J. Electrochem. Soc., 128, 2466 (1981).
11. B.R. Bennett, J.P. Lorenzo, and K. Vaccaro, Appl. Phys. Lett., 50, 197 (1987).
12. B.R. Bennett, J.P. Lorenzo, and K. Vaccaro, Electron. Lett., 24, 172 (1988).
13. S. Suyama, A. Okamoto, and T. Serikawa, J. Electrochem. Soc., 134, 2260 (1987).
14. P.F. Schmidt and W. Michel, J. Electrochem. Soc., 104, 230 (1957).
15. A.G. Revesz, J. Electrochem. Soc., 114, 629 (1967).
16. J.D.E. Beynon, G.G. Bloodworth, and I.M. Mcleod, Solid-State Electronics, 16, 309 (1973).
17. H. Hasegawa, S. Arimoto, J. Nanjo, H. Yamamoto, and H. Ohno, J. Electrochem. Soc., 135, 424 (1988).
18. H. Yamamoto, T. Sawada, S. Arimoto, H. Hasegawa, and H. Ohno, Electron. Lett., 19, 607 (1983).
19. G. Mende, K.D. Butter, and B. Schmidt, Thin Solid Films, 102, 65 (1983).
20. I.W. Boyd and J.I.B. Wilson, J. Appl. Phys., 62, 3195 (1987).
21. P.G. Pai, S.S. Chao, Y. Takagi, and G. Lucovsky, J. Vac. Sci. Technol. A, 4, 689 (1986).
22. W.A. Pliskin and H.S. Lehman, J. Electrochem. Soc., 112, 1013 (1965).
23. I.W. Boyd and J.I.B. Wilson, J. Appl. Phys., 53, 4166 (1982).
24. E.H. Nicollian and J.R. Brews, MOS Physics and Technology, Wiley, New York, 1982.
25. S. Seki, T. Unagami, and B. Tsujiyama, J. Electrochem. Soc., 131, 2621 (1984).
26. G. Mende and J. Wende, Thin Solid Films, 142, 21 (1986).

II. THERMAL AND STRUCTURAL PROPERTIES OF SiO_2

INTRODUCTION

D. L. Griscom

Naval Research Laboratory
Washington, DC 20375

Two major difficulties in quantitatively understanding the Si-SiO_2 system relate to the structure of amorphous SiO_2 and the role of small amounts of hydrogen. The purpose of this chapter is to review the state of our understanding of the thermal and structural properties of SiO_2 with these two specific factors in mind.

In the first paper, Lucovsky, Fitch, Kobeda and Irene studied the IR spectra, index of refraction n, and in-plane film stress for a variety of SiO_2 films thermally grown or deposited by plasma assisted chemical vapor deposition (PECVD) on silicon substrates.It was suggested that IR bond stretching frequency is related to the Si-O-Si bond angle 2θ through a force-constant model and that the experimental linewidth results mainly from the statistical distribution in 0. An analysis of the data under these model assumptions gave $2\Delta\theta \cong 30°$, consistent with the results of x-ray and neutron diffraction. An independent test of the model was carried out by plotting the measured oxide stress against the calculated strain using d_{Si-Si} as a strain parameter, resulting in a linear relationship with a slope (the in-plane elastic modulus) virtually equal to the literature value. The authors concluded by contrasting their results for thermal and PECVD oxides, noting that the model interpretation of the data implies very different Si-O-Si bond-angle distributions for the two cases, even though both are gate quality oxides.

Rendell and Ngai discussed the potential applicability of a theoretical formalism which has exhibited considerable success in explaining structural relaxations in glasses to SiO_2 films on Si. Near structural equilibrium, the so-called "two-state" relaxations commonly reported for such parameters as density, refractive index, shear modulus, and viscosity can often be described by a "stretched exponential" relaxation function, $\phi(t) = \exp[t/\tau^*)^{1-n}]$. Physically, it is supposed that in an amorphous material there exist local structural groupings (e.g., SiO_4 tetrahedra) with a local relaxation time τ_0 governing times t short compared to a characteristic time $1/\omega_c$. On the longer term, cooperative interactions among local groups become important so that the overall measured relaxation rate becomes time dependent. The stretched exponential formalism is the result of solving the first order rate equation with a rate constant of the form $\tau_0^{-1}(\omega_c t)^{-n}$. The authors went on to describe a general approach valid for significant departures from structural equilibrium when both τ^* and n may depend on thermal history.

Landsberger and Tiller reviewed previous work and presented more recent results relating to structural relaxation and oxygen transport in SiO_2 films on Si. Systematic inert thermal anneals of dry 800°C-grown SiO_2 films demonstrating a "two-stage" relaxation (c.f., Rendell and Ngai) wherein the film thickness swelled in proportion to the decrease in refractive index. Kobeda and Irene (1987) showed that intermediate refractive index levels correlate both structural relaxation (density) and residual stress levels with processing temperature. These correlations provide the opportunity to study the impact of such subtle structural effects on oxidation rates during second-step oxidations. Extensive new data were presented which showed that high-temperature-grown films exhibit enhanced second-step oxidation rates at lower temperatures due to the thermal history of the sample. Many data of these types were analyzed to extract transport parameters as a function of the relaxation state of the film. The higher the initial density (the less relaxed the network) the slower the transport of oxidant and the higher the activation energy for transport.

Pfeffer presented the latest results of his ongoing electron spin resonance (ESSR) study of radiation-induced defects in oxygen-stuffed low-water silica. Following γ irradiation, all samples exhibit the ESR spectra of E' centers (≡Si·) and superoxide radicals (≡Si-O-O·). Pfeffer's data provide an important window on the supposed diffusion-limited reaction: ≡Si· + O_2 → ≡Si-O-O·. Since $[O_2] >> [E']$ for all of the O_2-stuffed samples, quasi-first-order kinetics were expected. But instead, the author found that his isochronal anneal curves were accurately fitted by the form: $[E'(t)]/[E'(0)] = \{1 + t/\tau\}^{-(2/3)}$, where the value of τ^{-1} all display the same temperature dependence as the diffusivity of O_2 in silica. Thus, while the experiments verify the reaction of E' centers and O_2 molecules, the kinetics present a puzzle.

In the final paper of this chapter, Gale, Chew, Feigl, and Magee employed secondary-ion mass spectrometry (SIMS) and C-V methods to explore the influences of the low-temperature incorporation of water in ultra-dry thermal oxides on silicon. Surprisingly, lower temperature treatments (predicted to affect only the surface) resulted in significant changes in hydrogen isotope concentration throughout the bulk of the 80-90 nm films. The samples then subjected to avalanche electron injection from the silicon substrate showed (for the 200° diffused samples) a buildup of hydrogen isotope at the SiO_2/Si interface accompanied by buildups of negative charge-trapping sites in the bulk oxide (with strongly isotope dependent cross sections), positive trapping sites at or near the interface, a large increase in interface traps D_{it}, and the deactivation of acceptors in the depletion region of the substrate. The latter phenomenon suggested that the silicon substrate acts as a sink for hydrogen released from the Al /SiO_2 interface during electron injection. The sensitivity of the low-temperature water-diffused oxides to electron-injection-induced degradation was almost totally suppressed by a standard post-metallization anneal at 450 C in a hydrogen-containing ambient.

LOCAL ATOMIC STRUCTURE OF THERMALLY GROWN SiO_2 FILMS

G. Lucovsky and J.T. Fitch

Departments of Physics and
Materials Science and Engineering
North Carolina State University
Raleigh, NC 27695-8202

E. Kobeda and E.A. Irene

Department of Chemistry
University of North
Carolina at Chapel Hill
Chapel Hill, NC 27514

ABSTRACT

We have grown thin films of SiO_2 by thermal oxidation of crystalline silicon and have determined as functions of the growth and annealing conditions: (1) the frequency, ν, and line-width, $\Delta\nu$, of the Si-O infrared (ir) active bond-stretching absorption band; (2) the index of refraction, n, at 632.8 nm; and (3) the intrinsic in-plane film stress, $\sigma_x(SiO_2)$. Linear relationships between; (1) $\Delta\nu$ and ν; and (2) n and ν, are explained in terms of a microscopic model in which the determinant atomic scale variable is the bond-angle, 2θ, at the oxygen atom site. Using this model, we have been able to obtain an approximate value of the in-plane Young's modulus for SiO_2 from a combination of stress and ir measurements.

INTRODUCTION

This paper discusses changes in the properties of thermally grown films of SiO_2 with processing conditions, e.g., the oxidation temperature (T_o), the oxidizing species, the annealing time and temperature, etc., in terms of the local atomic structure of the oxide material. We are able to interpret systematic changes in: (1) the frequency and width of the Si-O bond-stretching ir absorption band; (2) the index of refraction at 632.8 nm; and (3) the intrinsic film stress in terms of a model in which changes in the bond-angle at the oxygen atom sites are the determining factor in the local atomic structure.[1-4] We discuss the relevance of these studies to the question of low temperature/low thermal budget processing, and include comparisons between the properties of low temperature thermally grown and annealed oxides, and oxides deposited by low temperature plasma enhanced chemical vapor deposition (PECVD).[5-7]

EXPERIMENTAL PROCEDURES

We have grown thin SiO_2 layers on crystalline silicon substrates which have been prepared by standard (and modified) ***RCA clean*** procedures.[2-4] Several different oxidation procedures have been employed: (1) oxidation in dry oxygen at atmospheric and elevated pressures;[1,2] (2) wet oxidation using pyrolytic steam;[3] and (3) oxidation accomplished by ***sequences*** of dry and wet processing procedures.[4] The majority of the oxidation runs were for times that yielded a film thickness of about 1200 Angstroms; this corresponds to one half an ellipsometric period and is the optimal thickness for accurate determination of both the index of refraction and the film thickness.[1] This thickness is also convenient for the ir absorption measurements. We have studied selected properties of the thermally grown and annealed SiO_2 films, specifically: (1) the ir absorption in the spectral range of the three first order vibrational modes (approximately 300 to 1500 cm^{-1}) using a Perkin-Elmer Model 983 Spectrophotometer;[1-4] (2) the optical index of refraction at 632.8 nm via ellipsometry using a Gaertner L115A Automatic Ellipsometer;[1-4] and (3) the mean stress using an apparatus developed by Kobeda and Irene.[8]

IR Measurements

Figure 1 is a representative ir absorption spectrum for an SiO_2 film grown in dry oxygen at a temperature of about 1100°C. There are three first order features in the ir absorbance corresponding to bond-stretching (at about 1075 cm^{-1}), bond-bending (at about 800 cm^{-1}), and bond-rocking (at about 460 cm^{-1}) motions. These vibrations can be characterized by the oxygen atom motions which are also shown in the figure. We emphasize the behavior of the bond-stretching feature. We have indicated in the figure the center of this band, ν, and the full width at half maximum, $\Delta\nu$. The data in Figs. 2-5 represent averages over measurements made on many films grown in the same way.[2-4] Figures 2(a) and (b) includes plots of ν and $\Delta\nu$, respectively, versus oxidation temperature (T_o) for films grown in dry oxygen at atmospheric pressure. We observe that ν increases, and $\Delta\nu$ decreases with increasing T_o. Figure 3 presents ν versus T_o for films grown by dry and wet oxidation procedures. We observe that the values of ν are very similar for T_o >950°C, but that for T_o <950°C, the dry oxides have smaller ν. Figures 4(a) and (b) indicate the effects of annealing on ν and $\Delta\nu$, respectively. The oxide films were grown either with T_o equal to 850°C in dry oxygen, or 800°C in steam, and the annealing was accomplished in a time period of 100s over a temperature range that extended to 1200°C.[3] The annealing was done with a rapid thermal annealing (RTA) instrument, an AG Heatpulse 210M, in an argon ambient. The effect of annealing at a temperature in excess of the growth temperature, T_o, was always the same; to increase ν and to decrease $\Delta\nu$.

We will not discuss the results of experiments in which films of SiO_2 were grown via sequences of oxidation runs in wet and dry environments; these are discussed in detail in Ref. 4. All of the observed behavior are correlated with the results presented for films grown either entirely in wet or dry oxidizing environments, subject to the condition that the last environment generally determined the film properties. Finally, in Fig. 5, we present the results of ir studies on a series of oxides which were grown to a range of thicknesses from 60 to 700 Angstroms.[9] We note that as the thickness increases, ν increases, and further that as the oxide thickness approaches 700 Angstroms, the values of ν begin to approach a values characteristic of thick oxides (> 1000 Angstroms) grown at the particular T_o and oxidation condition.[9] The variation of vibrational frequency with thickness is a manifestation of a non-homogeneous thermal history in the sense that material close to the Si/SiO2 interface has had less time to ***equilibrate*** than material at the oxide surface. The details of this relaxation process are determined by the viscoelastic properties of the SiO_2; in particular the rate of viscoelastic relaxation as a function of the temperature.

Ellipsometry Measurements

We have performed ellipsometry measurements on thermally grown and annealed oxide films.[1-4] As in the ir studies, the data shown in Fig. 6 were derived by averaging over measurements made on many similarly prepared films. We display n versus T_o for oxides grown at atmospheric pressure, and note that as T_o increases, the index of refraction at 632.8 nm decreases. The manner in which n varies is ***complementary*** to the variation of ν with T_o shown in Fig. 2 in the sense that the variation is linear in the low temperature range, from about 700 to 900°C, and then the variation of n (and ν) with higher T_o tend toward a ***limiting high temperature value.*** In general, the index of refraction

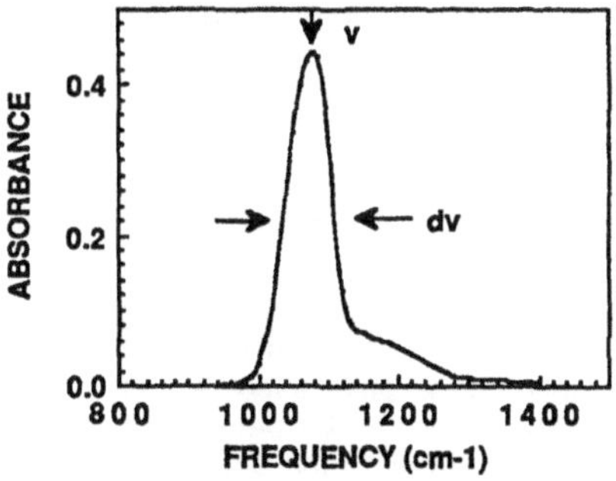

FIG. 1 Ir absorbance for the bond-stretching vibration in SiO_2. Also shown are the frequency, ν, and width , Δν, of this band.

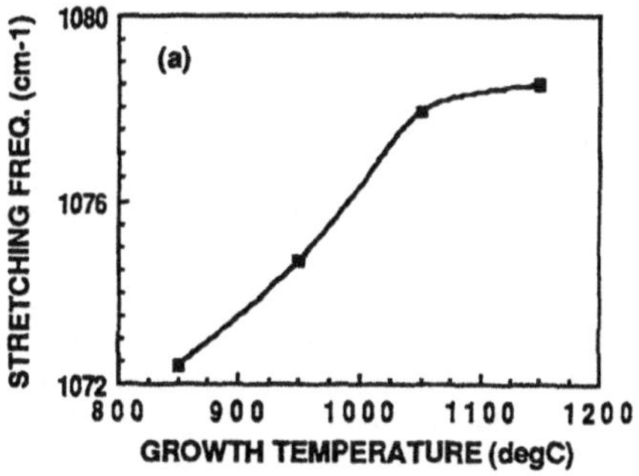

FIG. 2(a) ν versus growth temperature for films grown in dry oxygen.

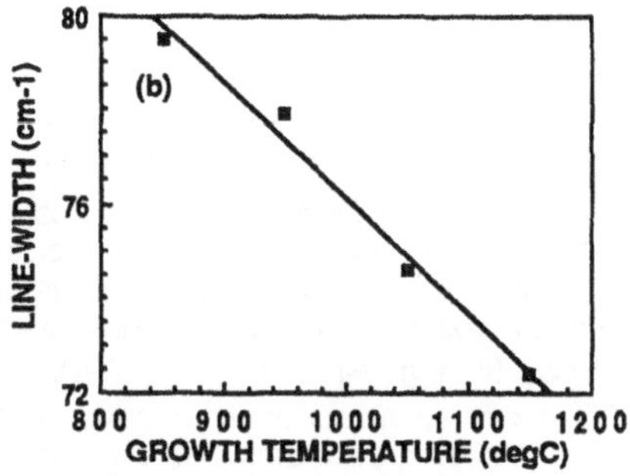

FIG. 2(b) Δν versus growth temperature for films grown in dry oxygen.

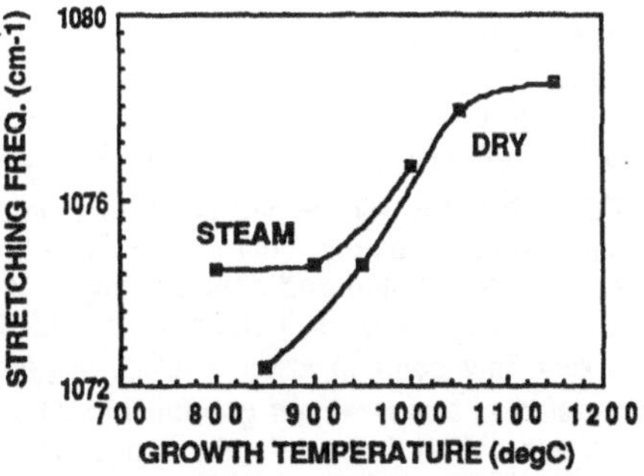

FIG. 3 ν versus growth temperature for films grown in dry oxygen and in steam.

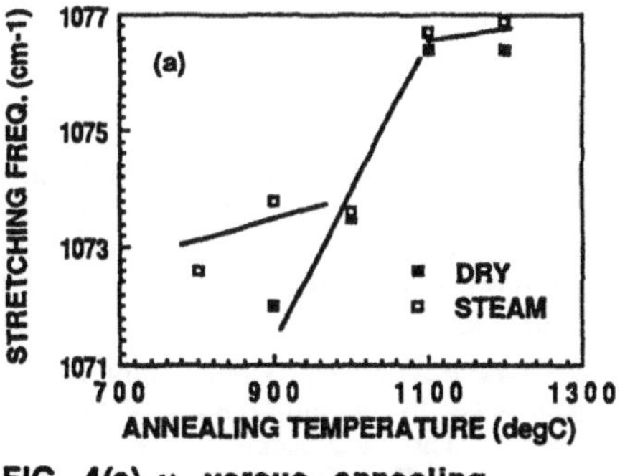

FIG. 4(a) ν versus annealing temperature for films grown in dry oxygen and in steam.

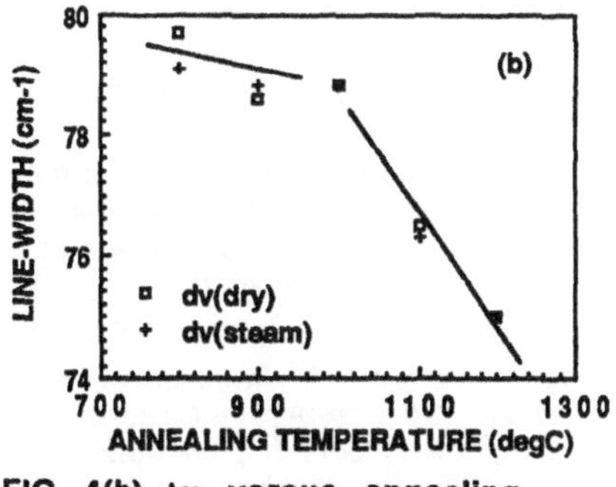

FIG. 4(b) Δν versus annealing temperature for films grown in dry oxygen and in steam.

decreases in oxide films that are grown at low temperature, 800 to 850°C, and then subsequently annealed at higher temperatures, 900 to 1200°C.

Stress Measurements

We have measured the average intrinsic stress in the SiO_2 film using the method of Kobeda and Irene.[8] The oxide film on the 'back side' of the oxidized silicon wafer is removed prior to this measurement to induce curvature. The Kobeda and Irene method employs two parallel laser beams incident on the curved sample. From the deviation from parallelism of the reflected laser beams, measured at the end of a long optical path, a radius of curvature of the sample can be calculated. The strain in the silicon wafer can be determined from this radius. The elastic constants of the silicon are used to determine the stress in the silicon from which the stress in the oxide can be deduced. The average intrinsic stress in the SiO_2 film is then determined by subtracting from the measured stress, the stress that is associated with the different thermal expansion coefficients of crystalline silicon and SiO_2.[3,8] Figure 7 displays the intrinsic stress versus film thickness for films grown in dry oxygen. We find that the stress is compressive, and that it decreases as the film thickness is increased. We have also studied the intrinsic stress of SiO_2 films grown in pyrolytic steam and find that the stress in these films is always less than the stress in an oxide film of similar thickness grown in dry oxygen at about the same temperature.[2]

DATA REDUCTION

Figure 8 displays plots of n versus ν for oxide films grown at atmospheric pressure in dry oxygen and in pyrolytic steam[2,3], and for SiO_2 films that are deposited by remote PECVD.[5-7] We will not discuss the relationship between the data for thermal oxides and the remote PECVD oxides until a later point in the text. Data from thermally grown (wet or dry oxidation conditions) and annealed oxide films also approach the line drawn in this figure as the annealing time is increased, provided that the annealing temperature is above the viscous flow point of about 950°C. In particular, oxide films grown at 800 or 850°C, and annealed at temperatures greater than about 1000°C for times in excess of 10s have values of n and ν that fall within the range of the data for the as-grown films. We observe an essentially linear dependence between n and ν, with n decreasing as ν increases. The solid lines in the diagram derive from a model described in the next section of the paper.[1]

Figure 9 presents a plot of $\Delta\nu$ as a function of ν. The data points include deposited oxides, formed by remote PECVD (hereafter RPECVD)[5-7], as well as the thermally grown (and annealed) oxides described above. The data exhibit an approximately linear dependence with $\Delta\nu$ decreasing as ν increases.[2,3] We will show later on that this behavior can also be understood in terms of a model for the local atomic structure of SiO_2 in which stress is taken up in the Si-O-Si bond-angle.

LOCAL ATOMIC SCALE STRUCTURE OF SiO_2

Before we present the models developed to interpret trends in the data identified above, e.g., the linear relationships between $\Delta\nu$ and ν, and n and ν, we discuss some important aspects of the local atomic structure in non-crystalline, or as it is alternatively designated amorphous, vitreous, or glassy SiO_2. There is a ***general consensus*** that the structure of the non-crystalline SiO_2 thin films is best described as a chemically ordered, covalent random network[10-12] in which short range order, determined by bonding chemistry is maintained at the respective silicon and oxygen atom sites. In this sense each Si atom has four oxygen neighbors in a near-tetrahedral arrangement. The Si-O bond-lengths show very little variation, but the bond-angles at the Si atom sites (O-Si-O) may show deviations from the ideal tetrahedral value of 109.47 degrees of up to about 10 degrees.[10-12] Each oxygen atom has two Si neighbors and the spread in the oxygen atom bond-angle (Si-O-Si) is much greater, of the order of +/- 30 degrees.[10-12] The experimental support for this model comes from a variety of measurements, with the most important being X-Ray radial distribution functions[10-12] and vibrational spectroscopy.[13] We use this structural picture as a basis for developing model relationships for the changes in the frequency of the ir-active bond-stretching vibration and the index of refraction as a function of the stress induced changes that occur with different thermal processing cycles.

We first discuss the ir vibrations. The vibrational mode that we have tracked is the bond-stretching vibration, and in the context of a central force model[13], the frequency, ν, of this bond-stretching vibration is given by:

$$(1) \qquad \nu = \nu_0 \sin\theta,$$

where $\nu_0 = 1117\ cm^{-1}$, and where 2θ is the average bond-angle at the oxygen atom sites (the value for ν_0 is determined from the condition that $\nu = 1078.5\ cm^{-1}$ for $2\theta = 150$ degrees[2,3]). The additional consideration of non-central forces, as in valence force field representation of the short range forces does not change any of the qualitative conclusions that can be drawn using the simpler central force picture we have employed. The only effects are quantitative, and are in effect a renormalization of the model parameters. We further assume that the width of the bond-stretching vibration derives from a distribution of Si-O-Si bond-angles. In this case, we can take the derivative of Equation (1) to obtain the variation of $\Delta\nu$ with $\Delta\theta$, i.e., $d\nu = \cos\theta\, d\theta$, so that dividing by ν, we can then write:

$$(2) \qquad \frac{\Delta\nu}{\nu} = \theta \cot\theta \frac{\Delta\theta}{\theta}.$$

The assumption used in deriving Equation (2) is consistent with the RDF studies of glassy SiO_2 in the sense that these studies indicate a considerable spread in the bond-angle at the oxygen atom sites[10-12] approximately +/- 30 deg. We further assume that the vibrational mode associated with each bond-angle in the distribution is narrow with respect to the observed ir line-width (75-100 cm^{-1}), so that the observed line shape is then approximated as a sum over these discrete and narrow absorption features.

We use a Clausius-Mossotti model[1] for the optical frequency dielectric constant, $\varepsilon = n^2$, where ε is the real part of the optical frequency dielectric constant in a transparent region of the absorption spectrum of the SiO_2; e.g., at 632.8 nm. If V is the molar volume of SiO_2, then:

$$(3) \qquad \frac{[\varepsilon - 1]}{[\varepsilon + 2]} = \frac{A}{V},$$

where A is a normalized sum of silicon and oxygen atom polarizabilities. It has been established that relationships of this type hold in glassy compounds and their alloys.[14] The normalization factor, A, can be determined by noting that for the high temperature SiO_2 films, n approaches a limiting value of about 1.458.[2,3]

A connection between the vibrational frequency, ν, as described by Equation (1), and n can be established by assuming that the density of the SiO_2 films is inversely proportional to the cube of the silicon-silicon distance, $d_{Si\text{-}Si}$.[1] This distance is determined by the Si-O-Si bond-angle through the relation:

$$(4) \qquad d_{Si\text{-}Si} = 2\, r_0 \sin\theta,$$

where r_0 is the Si-O bond length, nominally, 1.6 Angstroms.[1,10-12] Equation (4) neglects small changes in r_0 that occur when the bond-angle, 2θ, is changed;[15] the inclusion of these changes in r_0 in the calculations does not alter any of the interpretations or conclusions that follow. Combining Eqns. (1), (3) and (4), we obtain:

(5) $$\frac{\varepsilon_{eff}[1]}{\varepsilon_{eff}[2]} = \frac{v[1]^{3/2}}{v[2]^{3/2}},$$

where $\varepsilon_{eff} = \frac{[\varepsilon-1]}{[\varepsilon+2]}$, and where [1] and [2] designate two different Si-O-Si bond-angles.[1]

The final modeling relates to the determination of the average intrinsic stress in the SiO_2 film, obtained from the radius of curvature, R, of the oxidized silicon wafer using the method of Kobeda and Irene.[8] This curvature provides a measure of the strain in the silicon wafer. Using the elastic constants of silicon, this in turn allows a determination of the in-plane stress in the silicon water, $\sigma_x(Si)$. If t_s is the thickness of the silicon wafer, and t_f the SiO_2 film thickness, then the in-plane stress in SiO_2 film, $\sigma_x(SiO_2)$ is related to, t_f and t_s by:

(6) $$\sigma_x(SiO_2)\, t_f + \sigma_x(Si)\, t_s = 0 .$$

Using Equation (6), we can then develop an expression, usually designated as Stoney's "beam bending" relation[8] for the average intrinsic in-plane stress in the SiO_2 film;

(7) $$\sigma_x(SiO_2) = \frac{E}{6[1-P]} \frac{t_s^2}{t_f} \frac{1}{R}.$$

We can relate the average intrinsic in-plane compressive stress in the SiO_2 film to the strain in the SiO_2 film. This is done using a microscopic model for the strain; the validity of this approach will be demonstrated by the degree to which this assumption allows us to estimate the macroscopic elastic properties of the SiO_2 material. We ***assume*** that any strain in the SiO_2 film is taken up in the Si-O-Si bond-angle, and that an appropriate strain variable is then the Si-Si distance. The strain, ε_α, is then written as:

(8) $$\varepsilon_\alpha = \frac{\Delta d_{Si\text{-}Si}}{d_{Si\text{-}Si}}.$$

Using the relationship between $d_{Si\text{-}Si}$ and sin θ Equation (4,) and the relationship between ν and sin θ, we can write the strain as:

(9) $$\varepsilon_\alpha(2) = \frac{[2 \cos \theta(+)\ \theta(-)]}{\sin \theta_1},$$

where $\theta(+) = \frac{[\theta_1 + \theta_2]}{2}$, $\theta(-) = \frac{[\theta_1 - \theta_2]}{2}$, θ_1 is a reference Si-O-Si bond-angle, and $\varepsilon_\alpha(2)$ is the strain for $\theta = \theta_2$. Therefore, the slope of a plot of $\sigma_x(SiO_2)$ versus $\varepsilon_\alpha/6$ will yield Young's modulus, E, divided by [1-P], where P is Poison's ratio for SiO_2. The factor of 1/6 is used to maintain consistency with Stoney's equation.

INTERPRETATION OF EXPERIMENTAL DATA FOR SiO_2 FILMS

Figure 10 presents a plot of Δν/ν versus θ cot θ, where Δν and ν are determined directly from experiment and θ is determined from the relationship between ν and sin θ. We find that the data show an essentially linear dependence . A linear regression analysis of the data yields a "y" intercept very close to zero, and an R^2 coefficient of determination of 0.98. In addition we find a slope of approximately 0.21. In the context of our model this slope is equal to Δθ/θ. Since the average value of 2θ is 150 degrees, this means that the full width half maximum of the angular spread (2Δθ) at the oxygen atom sites, as determined from this data is about 30 degrees. This is in excellent agreement with the analysis of X-Ray data; i.e.,

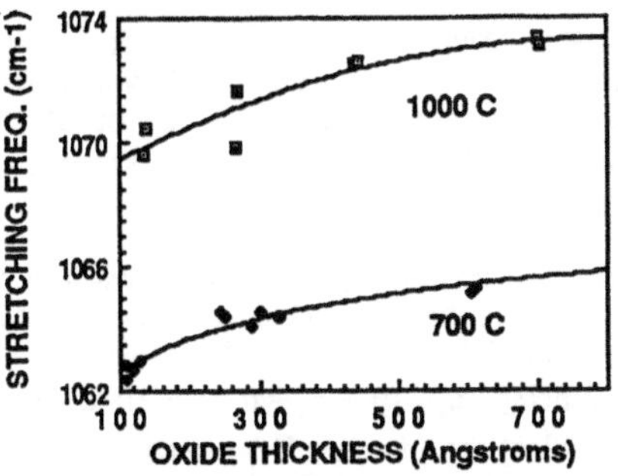

FIG. 5 ν versus oxide thickness for films grown in dry oxygen at 700°C and 1000°C.

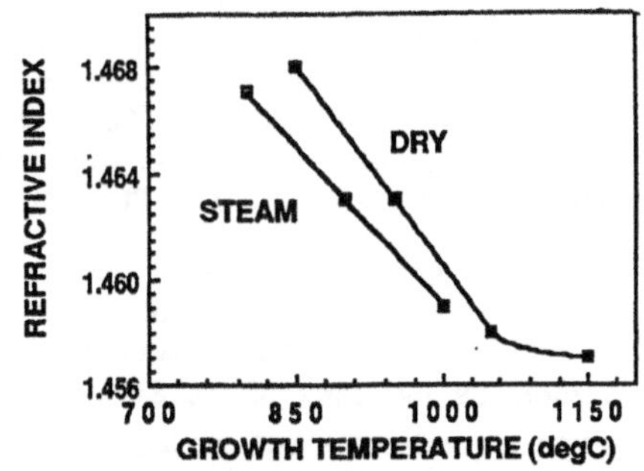

FIG. 6 Refractive index at 632.8 nm versus growth temperature for films grown in dry oxygen and in steam.

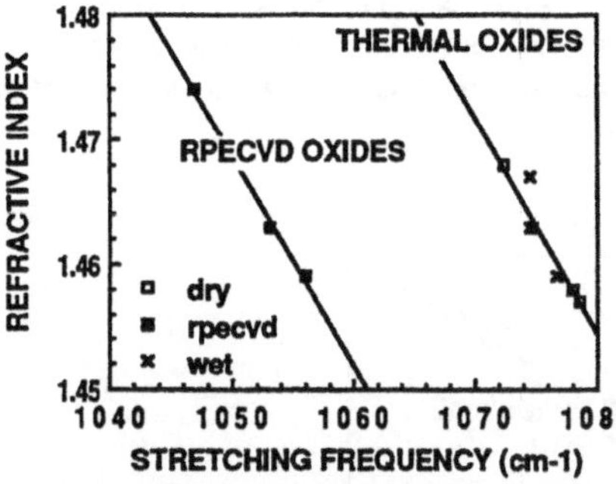

FIG. 8 Index of refraction at 632.8 nm versus ν for wet and dry thermal oxides and for oxides deposited by RPECVD. The solid lines derive from a model calculation.

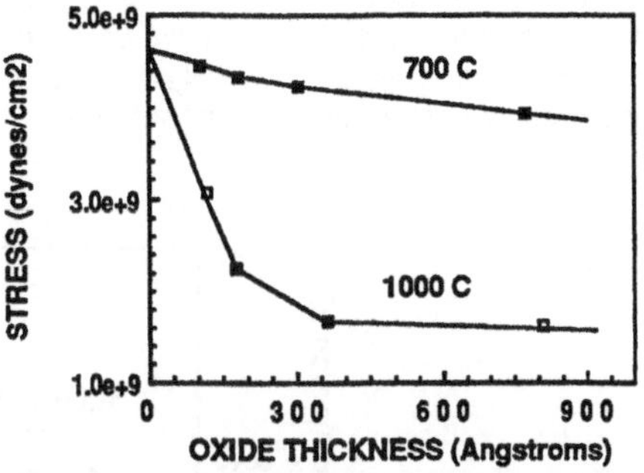

FIG. 7 Intrinsic stress versus oxide thickness for films grown in dry oxygen at 700°C and 1000°C.

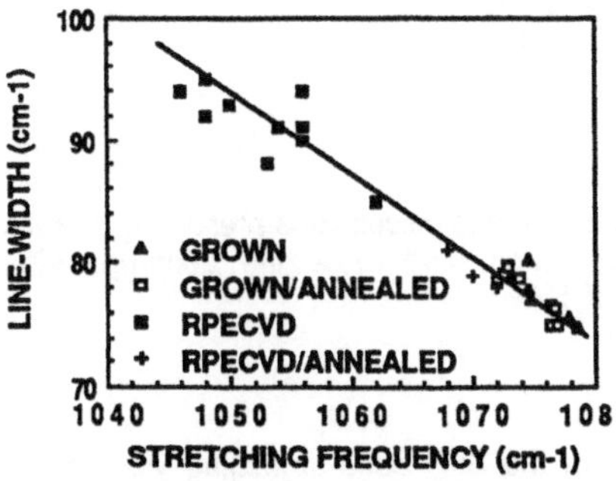

FIG. 9 Δν versus ν for thermal and RPECVD oxides, and films that have been subjected to RTA.

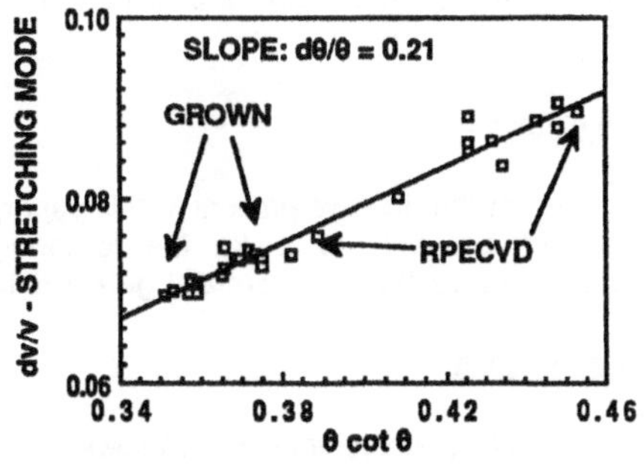

FIG. 10 Δν/ν versus θ cot θ for thermal and RPECVD oxides. The solid line derives from a model calculation. The model gives a fit parameter $d\theta/\theta = 0.21$.

the interpretation of RDF curves derived from X-Ray scattering data.[10-12] This in turn means that the assumption made in the derivation of the expression for $\Delta\nu/\nu$ is valid; i.e., that the width of the bond-stretching band in SiO_2 derives from a sum over a distribution of narrow modes, each associated with a particular Si-O-Si bond-angle in the statistical distribution, and centered around the average value.

In discussing the data displayed in Fig. 8, we noted that the straight line shown in the diagram derived from a model that related n to ν. That model has been discussed above, and the important aspects of the model are two fold: (1) that the density of the SiO_2 film varies inversely as the cube of the Si-Si interatomic spacing, $d_{Si\text{-}Si} = r_o \sin\theta$; and (2) that systematic and correlated changes in Fig. 8 then reflect the effects of densification.[1] We have tested this densification model by comparing the experimentally determined stress with a model for the strain based on the use of $d_{Si\text{-}Si}$ as a strain parameter. The results of this model are displayed in Fig. 11, where we have plotted the measured stress as a function of the calculated strain. The data have been subjected to a linear regression analysis, and the slope of the line has been determined. This analysis yields a value for the in-plane elastic modulus, E/[1-P], for SiO_2 of 1.3×10^{12} dynes/cm^2, that is approximately equal to the literature value of 1.6×10^{12} dynes/cm^2.[16] This close correspondence between the experimentally determined value of E/[1-P] and the reference value closes the loop in our interpretation of the data we have so far obtained on low temperature thermally grown SiO_2 thin films. In particular, we have used a microscopic model for the strain to extract a macroscopic characterization parameter, E/[1-P].

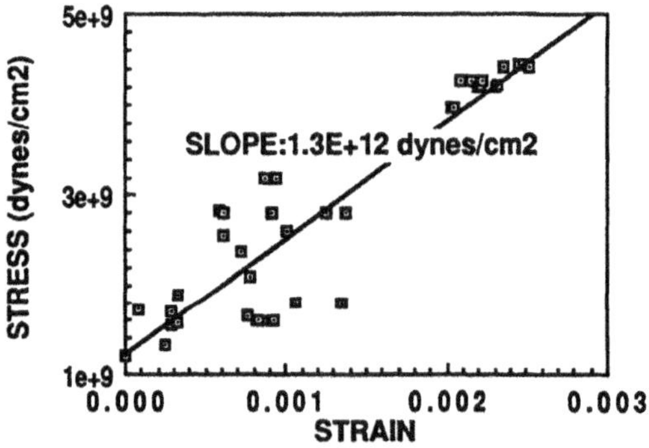

FIG. II Experimentally determined intrinsic stress versus calculated strain. The slope of the this line gives a value for E/[1-P].

SUMMARY

From the data we present in this paper, and the data and discussions presented in Refs. 1-9, we are able to make the following observations, and draw the accompanying conclusions relative to thermally grown thin films of SiO_2:

<u>Observations</u>

1. We observe that the frequency and width of the Si-O bond-stretching vibration, ν and $\Delta\nu$, respectively, and the index of refraction, n, at 632.8 nm vary with oxidation temperature, oxidation ambient, annealing temperature and time, and for a fixed set of processing conditions, also with the oxide thickness.

2. We further observe that thermal oxides grown at temperatures below about 1000°C are under a relatively high degree of compressive stress, and that under any set of growth

conditions, this stress is larger at the Si/SiO_2 interface than at the outer surface of the oxide film.

3. We find correlations between n, ν, and Δν in the form of linear relationships between n and ν, and Δν and ν.

Interpretation in Terms of Atomic Structure

We have developed a microscopic model for the temperature dependence of ν, Δν and n, and specifically;

1. We have explained the correlations between n and ν, and Δν and ν in terms of a model in which the correlated changes are related to stress induced changes in the bond-angle, 2θ, at the oxygen atom sites.

2. The model predicts a linear and universal relationship between the ratio of the ir line-width and frequency, Δν/ν, and θ cot θ. This is observed to hold not only for the thermally grown and annealed oxides, but also for oxides deposited by Remote PECVD and additonally, subjected to thermal annealing. The analysis of the data in Fig. 10 yields a value of $\Delta\theta/\theta = 0.21$, which for an average bond-angle of $2\theta = 150$ degrees, gives $2\Delta\theta = 30$ degrees. This is essentially the same value of the bond-angle spread at the oxygen atom sites that is derived from X-Ray RDF studies.[10-12]

3. In addition the model explains the linear relationship between n and ν, if we make the assumption that the film density is inversely proportional to the cube of the Si-Si atom distance. This means that the index of refraction, n, depends on θ through the density, and that ν depends on θ through a force constant model; hence the dependence of n on ν.

CONCLUSIONS

The observations and modelling discussed above are brought to fruition in terms of a model in which the observed effects derive from stress, and stress relief associated with various aspects of the growth and thermal history of the SiO_2 thin films. Specifically:

1. The driving force for the atomic structure dependent changes in ν, Δν and n, with processing variables is the relief of intrinsic stress originating at the growth interface between the silicon substrate and the SiO_2 film. We observe that the ***extrapolated intrinsic stress*** (see Fig. 7) at this Si/SiO_2 interface is independent of the growth temperature, and is approximately equal to about $4.5 - 5.0 \times 10^9$ dynes/cm^2.

2. The changes in ν, Δν and n, with processing are further accounted for in a quantitative way by assuming that the intrinsic stress orginating at the growth interface is relieved in part primarily through systematic changes in the bond-angle at the oxygen atom sites between the corner-connected SiO_4 tetrahedra.

3. The local atomic structure model is self-consistent in explaining these structurally driven changes; e.g., the correlations we find between Δν and ν, and n and ν.

4. The local atomic structural model is also consistent with one macroscopic property, the in-plane value of Young's modulus for SiO_2. This parameter, E/[1-P] can be obtained from a plot of the measured stress versus the calculated strain. The strain is assumed to be taken up in the bond-angle at the oxygen atom sites (or equivalently the Si-Si distance), which can in turn be determined from the analysis of the ir data.

COMPARISONS BETWEEN THERMALLY GROWN OXIDES AND RPECVD OXIDES

1. Comparisons between the properties of low temperature thermal oxides (grown by both wet and dry oxidation procedures) and RPECVD oxides indicate the following: (a) the frequency and half width, ν and Δν , of the bond-stretching mode obey the same universal

relationship, with the same value of the fit parameter , $\Delta\theta/\theta$ = 0.21. This in turn means that the width of the Si-O-Si stretching mode in all instances derives from a sum of narrow modes, each associated with a specific bond-angle in a statistical distribution. The range of bond-angles in the RPECVD films is from 140 to 144 degrees, whereas, the range in thermal oxides grown between about 800 and 1150°C is from 147 to 150 degrees [these values are determined relative to an assumed value of 2θ =150 degrees for an oxide grown at 1150°C].

2. The RPECVD oxides display a linear relationship between n and ν, but with an intercept that is different from the thermal oxides. The particular values of n and ν for the RPECVD oxides are consistent with these films being less densified than the low temperature thermal oxides by a factor of about 1.4 %. This density deficiency in the RPECVD oxides is supported by the enhanced etch rates of the RPECVD oxides, relative to the thermal oxides, in buffered HF acid.

3. Low temperature thermal oxide films [T_o = 700-800°C] are not as suitable for gate oxide applications as some of the RPECVD oxides.[17] However, annealing the low temperature thermal oxides at temperatures in excess of 1000°C improves their properties by relieving stress, primarily at the growth interface. In fact, sufficiently long anneals at temperatures in excess of 1100°C generally render many of the vibrational, optical and electronic properties of the low temperature oxides indistinguishable from oxides grown at higher temperatures, e.g., 1000-1150°C. In contrast, and depending mostly on the silicon surface preparation, RPECVD oxides grown at temperatures of about 350°C can be of gate quality.[5-7] This means that since the Si-O-Si bond-angle is very different in the bulk of ***both gate quality thermal and RPECVD oxides,*** this aspect of the local atomic structure of bulk SiO_2 is not a determinant factor in the specification of device quality oxide films.

ACKNOWLEDGEMENTS

This work has been supported in part by the Office of Naval Research (at both UNC-CH and NCSU), and the Semiconductor Research Corporation, through sub-contracts from the Research Triangle Institute to NCSU.

REFERENCES

1. G. Lucovsky, M.J. Mantini, S.K. Srivastava and E.A. Irene, J. Vac. Sci. Technol. **B5**, 530 (1987).
2. J.T. Fitch and G. Lucovsky, Mater. Res. Soc. Proc. **92**, 89 (1987).
3. J.T. Fitch and G. Lucovsky, Amer. Inst. of Phys. Conf. Proc., AVS Sub-series (1988), in press.
4. J.T. Fitch and G. Lucovsky, Mater. Res. Soc. Proc. (Spring 1987 Meeting), in press.
5. P.D. Richard, R.J. Markunas, G. Lucovsky, G.G. Fountain, A.N. Mansour and D.V. Tsu, J. Vac. Sci. Technol. **A3**, 867 (1985).
6. G. Lucovsky and D.V. Tsu, J .Vac. Sci. Technol. **A5**, 2231 (1987).
7. D.V. Tsu, S.S. Kim and G. Lucovsky, these proceedings
8. E. Kobeda and E.A. Irene, J .Vac. Sci. Technol. **B4**, 720 (1986).
9. E. Kobeda, J.T. Fitch, G. Lucovsky and E.A. Irene, private communication.
10. R.J. Bell and P. Dean, Nature **212**, 1353 (1966).
11. R.L. Mozzi and B.E. Warren, J. Appl. Cryst. **2**, 164 (1969).
12. I. Simon, in *Modern Aspects of the Vitreous State*, ed. by J.D. McKenzie (Butterworths, London, 1960), p. 120.
13. F.L. Galeener, Phys. Rev. **B19**, 4292 (1979); F.L. Galeener and P.N. Sen, ibid. **B17**, 1928 (1978).
14. J.E. Stansworth, *Physical Properties of Glasses* (CLarendon, Oxford, 1950), p. 56.
15. M.D. Newton and G.V. Gibbs, Phys. Chem. Miner. **6**, 221 (1980).
16. A.K. Sinha, H.J. Levinstein and T.E. Smith, J. Appl. Phys. **49**, 2423 (1978).
17. B.E. Deal, M. Sklar, A.S. Grove and E.H. Snow, J. Electrochem. Soc. **114**, 266 (1967).

STRUCTURAL RELAXATION AND GROWTH OF SiO_2 FILMS ON Si

R. W. Rendell and K. L. Ngai

Naval Research Laboratory
Washington, DC 20375-5000

Recent works have noted that the structural relaxation of thermally grown SiO_2 films and of pressure compacted silica glass both consist of an initial rapid change to a level which depends on temperature followed by a slow approach to equilibrium. In this work we discuss structural changes of thermally grown SiO_2 using concepts and models previously developed for quenched glasses. These naturally exhibit the observed features when applied to thermal oxides. They are based on a time-dependent rate coefficient which leads to a Kohlrausch stretched exponential form $\exp[-(t/\tau^*)^{1-n}]$. This is generalized to include a dependence on thermal and annealing history through structural variations of n and τ^*. The physical interpretation of these models is described, numerical simulations of relaxation of a thermally grown oxide are presented, and the relations between different relaxation times and activation energies are discussed.

INTRODUCTION

The index of refraction and density of a thermally grown SiO_2 film are sensitive to both the growth temperature and the amount of annealing which has taken place (1,2). The refractive indices decrease with increasing growth temperature until they reach the approximate value 1.460, corresponding to the index of a completely relaxed film (1). At higher temperatures, T>950 C for wet-grown and T>1150 C for dry-grown oxides, the index remains constant at this value. Densified SiO_2 films grown in dry O_2 have been observed to decrease in density with time when annealed at 1000 C (2). Recent works have suggested that the oxide as grown near the interface is similar to that of pressure compacted silica glass (3,4). It has been pointed out (3,4) that the volume expansion of compacted silica (5,6) consists of an initial rapid increase (t<2 hr) to a level which depends on temperature followed by a slow approach to equilibrium (t>100 hr) and that this behavior is consistent with the annealing kinetics of thermal oxides. Recent measurements (4) have found such a two-stage process in the index of refraction and thickness of dry-grown oxides in inert ambients. Models with a single relaxation time cannot account for such behavior. Although a correspondence between the timescale for oxidation and that of the viscosity has been indicated (4) at the high temperatures where the index of refraction becomes constant, no methods have been available for quantitatively describing and interpreting these structural variations. Some of these thermal and annealing features are similar to those exhibited by quenched glasses. However, there may also be differences between the relaxation of the structure of a glass formed by quenching and the structure of an oxide grown on a silicon surface. There have been recent suggestions (7) that the expansion of volume occuring during the oxidation process may differ in some aspects from that in a glass. In this work we discuss how theoretical methods previously developed for quenched glasses (8-10) can be adapted to the situation of thermal oxides. Characteristics of relaxation near structural equilibrium are first summarized. This is discussed in terms of a connection between short-range spatial order and time-dependent rate coefficients. A relation is obtained between a local rate constant and the relaxation time observed in long-time relaxation. Situations with significant departure from equilibrium are then discussed by incorporating the dependences of the rate parameters on structure or fictive temperature. The dependence on thermal and annealing histories is illustrated by a numerical simulation of relaxation of a dry thermal oxide. The relation between several different activation energies in the problem is discussed.

RELAXATION NEAR STRUCTURAL EQUILIBRIUM

Relaxation showing behavior similar to the two-stage behavior described above is commonly found in glasses and other complex material systems. Near structural equilibrium they are often well described by the Kohlrausch (11) stretched exponential relaxation function:

$$\phi(t)=\exp\left[-(t/\tau^*)^{1-n}\right] \tag{1}$$

where τ^* is the relaxation time and the coupling parameter n lies in the range $0\leq n<1$. If n is significantly greater than zero, part of the relaxation occurs for $t<\tau^*$ followed by a slow long-time decay. If Eq.(1) is plotted on a linear time scale, it exhibits the behavior described as "two-stage relaxation" by several groups (3-5). Note, however, that Eq.(1) does not actually have two distinct stages of relaxation. Rather the approach to equilibrium at long times is "stretched" by the fractional exponent (1-n). Materials which relax according to the Kohlrausch function include silica glass. For example, the longitudinal strain in suprasil W silica glass was studied by Bucaro et al (12) in structural equilibrium for 1305 C<T<1750 C and it conforms well to Eq.(1) with n= 0.3. Shear modulus measurements by Mills (13) (see discussion in ref. 12) on infrasil at lower temperatures over the range 1000-1200 C also obey Eq.(1) but n in this range has increased to 0.37. In this high temperature range, the average relaxation time is related to the viscosity by $\langle\tau^*\rangle=\eta/G_\infty$ where G_∞ is the infinite frequency shear modulus. Many other materials near structural equilibrium are observed to relax according to Eq.(1), including oxide and flouride glasses B_2O_3, GeO_2, BeF_2, As_2O_3; alkali silicate and borate glasses; salts such as $Ca(NO_3)_2$ and $ZnCl_2$; glassy amorphous polymers and many others (11).

Although Eq.(1) can be regarded as a useful parameterization of data near structural equilibrium, attempts to understand its physical origin have been successful in identifying the physical meaning of the relaxation time τ^*. Eq.(1) can be regarded as the solution of a rate equation with a time-dependent rate coefficient. This can be understood physically as follows. Well-defined short range order is known to exist within silicate glasses. Suppose relaxation of a structural unit in a glass has a simple rate constant τ_o^{-1} within each domain of short-range order. Relaxation will proceed according to the rate constant τ_o^{-1} as long as the structural unit is not influenced by other structural units outside the domain of short-range order. However, larger scale structural changes will eventually involve cooperative interactions with other structural units. Physically, we expect that such cooperative movements will slow down the relaxation process and modify the rate constant τ_o^{-1} for times longer than a characteristic time ω_c^{-1} at which cooperativity becomes important. Thus $\phi(t)$ obeys a rate equation with rate coefficient W(t):

$$d\phi/dt=-W(t)\phi(t), \tag{2}$$

where $W(t)=\tau_o^{-1}$ for $\omega_c t<1$ and $W(t)=\tau_o^{-1}f(t)$ for $\omega_c t>1$. The function $f(t)<1$ represents the slowing down of the rate constant due to cooperative interactions with other structural units. The specific choice $f(t)=(\omega_c t)^{-n}$ will result in the solution given by Eq.(1) along with an additional expression which relates τ^* to τ_o. Accordingly, with:

$$W(t)=\begin{cases}\tau_o^{-1} & ,\omega_c t<1\\ \tau_o^{-1}(\omega_c t)^{-n} & ,\omega_c t>1\end{cases} \tag{3}$$

the solution of Eqs.(2) and (3) becomes $\phi(t)=\exp[-t/\tau_o]$ for $\omega_c t<1$ and

$$\phi(t)=\exp[-(t/\tau^*)^{1-n}] \qquad ,\omega_c t>1 \tag{4}$$

$$\tau^*=[(1-n)\omega_c^n\tau_o]^{1/(1-n)} \tag{5}$$

Here we have assumed that n and τ^* both remain constant during the relaxation. As discussed below, this may not be the case for significant departures from structural equilibrium.

The time-dependent rate coefficient W(t) predicts not only the Kohlrausch relaxation but also a relation between the local rate constant τ_o^{-1} and the observed relaxation time τ^*, Eq.(5). This relation has consequences which can be checked experimentally. If the value of n is determined by fitting Eq.(4) to experimental data, Eq.(5) can be used to predict the behavior of τ^*. For example if $\tau_o=\tau_\infty\exp(E_a/kT)$ has an Arrhenius temperature dependence, then Eq.(5) implies that $\tau^*=\tau_\infty^*\exp(E_a^*/kT)$ where $E_a^*=E_a/(1-n)$. The local relaxation time τ_o describes the relaxation in the absence of cooperative interactions and in some applications it may have a direct microscopic interpretation. The coupling parameter n is a measure of the strength of the cooperative interactions. Thus E_a^* is an effective activated barrier which is observed for relaxation of a cooperatively interacting species and its value can be predicted by $E_a^*=E_a/(1-n)$. This as well as other consequences of Eq.(5) have been directly verified in many examples of structural relaxation, ionic conductivity, electronic transport and other relaxations in glasses (11). The crossover between single exponential and Kohlrausch relaxation occurs approximately at the timescale ω_c^{-1}. However ω_c also determines, in part, the magnitude of τ^* by Eq.(5). This allows the value of ω_c to be determined experimentally (11). The value of ω_c for several glasses has been found to be in the range 10^{11}-10^{12} s^{-1}. The value for silica glass is not presently known. Examples such as these, which simultaneously verify the predictions of Eqs.(4) and (5), provide evidence that the time-dependent rate coefficient in Eq.(3) plays a basic role in glassy systems with short range order. In this sense, the form of Eq.(3) could be regarded as an ansatz which produces several predictions supported by experiment. There is considerable experimental support for this approach. However, we point out that Eq.(3) can be derived from the dynamical properties of complex Hamiltonian systems interacting with a thermal reservoir (11). Thus there is also theoretical support for the time-dependent rate coefficient. Eq.(5) has been verified (11) for electronic transport in SiO_2 but the experimental information has not been available to test it for structural relaxations in SiO_2. However, as discussed above, Eq.(4) does describe structural relaxations in silica glass for T>1000 C. It is expected that the rate coefficient Eq.(3) should also be relevant for structural relaxation in thermally grown oxides.

STRUCTURAL NONEQUILIBRIUM

The Kohlrausch form Eq.(4) is known to be a good approximation near structural equilibrium. The rate parameters n and τ^* had been assumed to remain constant in deriving Eq.(4) from Eqs.(2) and (3). In general, they are expected to vary if there is a significant departure from structural equilibrium. In this situation, Eqs.(2) and (3) lead instead to:

$$\phi(t) = \exp\left[-\int_0^t dt'(1-n)\tau^{*-1}(t'/\tau^*)^{-n}\right] \qquad (6)$$

where the dependences on structure or fictive temperature T_f are now included: $n=n[T_f]$, $\tau^*=\tau^*[T_f,T]$. The time-dependence of T_f following a given thermal and annealing history has been found to be accurately described by Eq.(6) for many glasses (8,11). For example, following a sudden temperature increase from T_o to T:

$$T_f(t) = T - (T-T_o)\phi(t) \qquad (7)$$

The fictive temperature T_f changes from T_o to T in parallel with the relaxation of a physical quantity such as volume or index of refraction. For example, the index of refraction N(t) would be related by $\left(N(t)-N(\infty)\right)/\left(N(o)-N(\infty)\right) = \left(T_f(t)-T\right)/\left(T_o-T\right)$. When n and τ^* are functions of T_f, Eqs.(6) and (7) must be solved self-consistently for $T_f(t)$. For different thermal and annealing histories, different solutions for $T_f(t)$ will be obtained. The fictive temperature concept first introduced by Tool (14), has previously been useful for understanding such memory effects in quenched glasses (8-10). These methods should also be helpful in studying thermal oxides. Thermal and annealing history dependences for the index of refraction and density of thermal oxides have already been demonstrated (1,15).

Formulations involving fictive temperature studied previously by several groups (9-10) have some similarities to Eqs.(6) and (7). These have been successful in making qualitative and semiquantitive predictions for relaxations of enthalpy, volume, index of refraction and other quantities. However, there are several phenomena which these older formulations completely fail to describe. A notable example occurs during volume expansion in glasses following an increasing temperature jump (10). Small differences (e.g. 2°C) in the initial temperature of the experiment result in large differences (e.g. 0.5 decades) in the rate of change of volume as the equilibrium volume is approached. This anomalous dependence on thermal history could not be qualitatively reproduced by previous formulations (9,10) even with arbitrary adjustment of the model parameters. However, Eqs.(6) and (7) have been able to quantitatively predict this feature with all parameters determined by independent experimental measurement. This was recently studied in detail for the case of glassy polyvinylacetate (8). For this material, the dependences $n=n[T_f]$ and $\tau^*=\tau^*[T_f,T]$ were independently determined by shear measurements using the same sample and thermal history as used for volume measurements. This was done by monitoring the dynamic mechanical frequency response as a function of the time after quench. With this information, Eqs.(6) and (7) were able to quantitatively reproduce the detailed time dependences of the volume expansion at different temperatures (8). The temperature dependence of the underlying local relaxation time τ_o was also successfully predicted using Eq.(5). Insight into the physical meaning of the relaxation times and effective activation energies was thus obtained. This gives further confidence in the time-dependent rate coefficient since Eqs.(5)-(7) were all obtained from Eqs.(2) and (3). Notice that Eq.(6) reduces to the Kohlrausch form Eq.(4) in the absence of structural variations. The results of experiments performed with different thermal and annealing histories can accurately be predicted by these methods.

These same methods are expected to be useful for understanding structural relaxation of thermal oxides. In particular, volume expansion is known to occur in the compacted structure of thermally grown oxides. At present, experimental information about $n[T_f]$ and $\tau^*[T_f,T]$ is not available for thermal oxides (or even densified silica glass) at the low temperatures, T=600-1000 C, relevant to oxidation. In the following sections, an illustrative simulation will be presented to discuss possible structural changes in thermal oxides. However, we first discuss some features of Eqs.(6) and (7) using silica and boron oxide glasses as examples. The time-dependent volume expansion of densified silica and boron oxide glasses was investigated by MacKenzie (5) for different thermal and annealing histories. Nonlinearity was observed in the deviations from equilibrium and MacKenzie stressed that the annealing kinetics could not be described by a single relaxation time. In the absence of a self-consistent description of this process, he arbitrarily defined a time τ' at which the system relaxed to 1/e of its original value. For silica glass, τ' was found to have an apparent energy of activation of $E_a'=2.4$ ev. However, the physical interpretation of such an activation energy is certainly problematical. A similar situation occurs in B_2O_3 where more experimental information is available. Bucaro et al (16) have measured both pressure-jump volume relaxation and dynamic light-scattering spectroscopy (which is a measure of the longitudinal stress modulus) for B_2O_3. Their samples had been well annealed to near equilibrium and the relaxation could be accurately described by the Kohlrausch function, Eq.(4), with n=0.4 and $\tau^*=A^*\exp(E_a^*/kT)$ with $E_a^*=3.6$ ev. This is not unexpected since Eq.(6) reduces to the Kohlrausch form for situations near structural equilibrium. Shelby (17) also measured volume relaxation in B_2O_3 but with significant departures from structural equilibrium. He noted that the relaxation is nonexponential and also defined a 1/e time τ'. From his volume data it is found that $\tau'=A'\exp(E_a'/kT)$ with $E_a'=0.86$ ev. This differs greatly from the result of Bucaro et al since $E_a'=0.86$ ev << $E_a^*=3.6$ ev. In addition to possible differences in samples, a large part of this difference might be attributed to the importance of structural evolution in Shelby's case. If we describe his data using Eq.(6), then τ' corresponds to the time at which the integral in Eq.(6) equals one. With this definition of τ', it can be shown directly from Eq.(6) that:

$$E_a'=E_a^*\left[\tau^*[T_f(\tau'),T]/\tau'(T)\right]^{1-n} \tag{8}$$

For simplicity, n has been assumed constant in the derivation of Eq.(8). There will be further modification if n changes with T_f. The relation between E_a' and E_a^* can be studied by using explicit solutions of Eqs.(6) and (7). Given $n[T_f]$ and $\tau^*[T_f,T]$, Eqs.(6) and (7) are solved numerically by a self-consistent interactive procedure. Numerical examples which have been used to evaluate Eq.(8) show that E_a' can be both independent of T and significantly smaller than E_a^* if τ^* has a sufficiently strong dependence on T_f. The 1/e time τ' can appear to be thermally activated with an apparent activation energy $E'_a<E^*_a$. Independently measured information on $\tau^*[T_f,T]$ is not available to test the B_2O_3 case in detail. However, as an illustration, we consider n=0.4, E^*_a=3.6 ev, and log $\tau^*[T_f]/\tau^*[T]$ decreasing linearly from 2.5 to 0 as T_f approaches T. This is consistent with the roughly exponential decrease of τ^* with T_f by several orders of magnitude found in many glasses (8-10). Numerical solution of Eqs.(6) and (7) with these parameters finds $E'_a \cong$1ev from Eq.(8). Thus this is a plausible explanation for at least part of the difference between the Bucaro et al and Shelby data and it illustrates the importance of having a proper self-consistent description of structural relaxation. Note that in this case we have an apparent activation energy E^*_a=3.6 ev which is observed near structural equilibrium, and additional apparent activation energy E'_a=0.86 ev which differs from E^*_a due to time-dependent structural changes, and a local activation energy $E_a=(1-n)E^*_a$=2.2 ev which differs from E^*_a due to cooperative interactions between structural units. This relation between E^*_a and E_a has been studied and verified in several different glasses (11). The exercise presented here demonstrates the appearance of an additional apparent activation energy E'_a for relaxation accompanied by structural changes.

ANNEALING OF A THERMAL OXIDE

The methods developed for describing thermal and annealing dependences of structural relaxation in quenched glasses (8-10) have not previously been used for thermal oxides. However, thermal and annealing history dependences have already been demonstrated for thermal oxides (1,15). Recently, Landsberger and Tiller (4) have also measured time-dependent changes of the refractive index N_{ox} at various temperatures during anneals in an inert ambient of dry-grown SiO_2 initially at 800 C. These are the only data presently available for structural changes of a thermal oxide as a function of time. They found that N_{ox} relaxed to an intermediate value in less than about 2 hours while approximately 100 hours or more is then required to completely relax the films except at the highest temperatures (T>1100 C). This is similar to the situation found for compacted bulk silica (5). We now consider how these features can be understood by Eqs.(6) and (7). This will include only the structural relaxation (i.e. volume expansion) and not the oxidation and growth processes. The oxide growth process also involves volume expansion and the framework of Eqs. (6) and (7) could also be incorporated into a model of oxidation. However this would be considerably more complicated. Several workers have presented evidence that the oxides have generally uniform densities throughout their thickness (4,18) except for a thin dense interface layer (i.e. 6 Å). Therefore the annealing of a 1000 Å thick dry-grown SiO_2 in an inert ambient by Landsberger and Tiller should be adequately represented by Eqs.(6) and (7). For the inert anneals of the thermal oxides, there is the possibility that some of the observed changes in refractive index result not only from volume expansion but also from processes such as chemical degradation. These possibilities do not occur for quenched glasses. For purposes of illustration we will assume that only volume expansion occurs during the annealing.

In the absence of direct experimental information for n and τ^*, we will assume n=0.7 and $\tau^*[T_f,T]=a(T_f)\ A^*\exp(E^*_a/kT)$ with $A^*=10^{-10}$ s and E^*_a=3 ev. Values in the range of n=0.7-0.8 and E^*_a=3-4 ev produce relaxation curves with approximately the time and temperature dependence observed in the lower temperature curves in the initial stage of relaxation. The coupling parameter n is a measure of the strength of cooperativity between structural units. The structural dependences of n and τ^* depend on how cooperativity is affected by the volume expansion. As discussed previously, some authors (3-4) have pointed out similarities between relaxation in quenched glasses and thermally grown oxides. Other recent work (7) on oxidation kinetics has suggested that volume expansion during oxidation may differ from a glass in important ways. Expansion of a growing oxide as it moves away from the Si surface may require reconfigurations of Si atoms into larger ring structures (7). This would result in increasing cooperativity between the neighboring atoms as the volume expands. If this is the situation during annealing, we would expect n and τ^* to increase as volume increases. In contrast to this, motion of structural units in a quenched glass involves higher cooperativity when it is in a dense state. If a thermal oxide were similar to a glass in this respect, we would expect n and τ^* to decrease as volume increases. Experimental information is not presently available to tell whether or not a thermal oxide differs from a glass in this way. Either possibility can be simulated using Eqs. (6) and (7). However, each of these possibilities would require different dependences of n and τ^* on structure. Independent experimental information for n and τ^* would then allow us to distinguish which situation is appropriate for a thermal oxide. This can give insight into the relation between molecular motion and bulk structure. The present example will simply illustrate how a simulation is carried out using Eqs. (6) and (7). For concreteness, we will assume that an oxide is similar to a glass in that cooperativity decreases as volume increases. For the T_f dependence of τ^* we will assume that log $a(T_f)$ decreases linearly between 2.0 and 0.0 as T_f increases between 800 C and 1150 C. This is similar to that used in the B_2O_3 example. For simplicity, we will assume n is constant. The oxide refractive index is related to T_f by $\left(N_{ox}(t)-N_{ox}(\infty)\right)/\left(N_{ox}(0)-N_{ox}(\infty)\right) = \left(T_f(t)-T_o\right)/\left(T_o-T\right)$. For an oxide dry-grown at 800 C, we have (4) $N_{ox}(0)=1.472$ and $N_{ox}(\infty)=1.460$. For various temperatures T, Eqs.(6) and (7) were numerically solved self-consistently for $T_f(t)$ and the result then used to obtain $N_{ox}(t)$. The results for the intermediate stage of relaxation (t<10 hrs) are shown in Fig. 1. At each T, the curves drop from the initial value of $N_{ox}(0)=1.472$ to an intermediate value which depends on temperature. This occurs within about 2 hours. This is followed by a slow long-time approach to equilibrium. For the T=900 and 950 C curves, there is only a moderate decrease of τ^* as T_f goes from 800 C to T. These curves are not too different from the Kohlrausch function Eq.(4). The approach to equilibrium extends for several hundred hours and the spacing between the curves is determined mostly by E_a^*. For the T=1000-1100 C curves, there is a progressively stronger decrease of τ^* during the relaxation. The approach to equilibrium is thus more rapid so that at the highest temperature, equilibrium is reached within 10 hours. The precise form of the time decay is determined self-consistently from Eqs.(6) and (7). The relaxation of $T_f(t)$ corrects the value of $\tau^*[T_f,T]$ which in turns corrects the subsequent relaxation of $T_f(t)$. It is this nonlinear self-consistent effect which produces memory effects. For instance, if the temperature was changed back to T=800 C at t=10 hours, the subsequent relaxation of each curve would depend on its past history. If the $n[T_f]$ and $\tau^*[T_f,T]$ dependences could be determined, the result of experiments such as this could be predicted.

We have demonstrated how the observed features of the thermal oxide data can emerge from Eqs.(6) and (7) with structural changes of the rate parameters n and τ^*. Calculations such as these have been previously verified for quenched glasses for which $n[T_f]$ and $\tau^*[T_f,T]$ could be independently measured (8). It would be useful to consider whether more direct experimental probes of n and τ^* could be devised. This type of information about $n[T_f]$ and $\tau^*[T_f,T]$ will allow us to predict the rich nonlinear behavior of the oxide films under a variety of thermal and

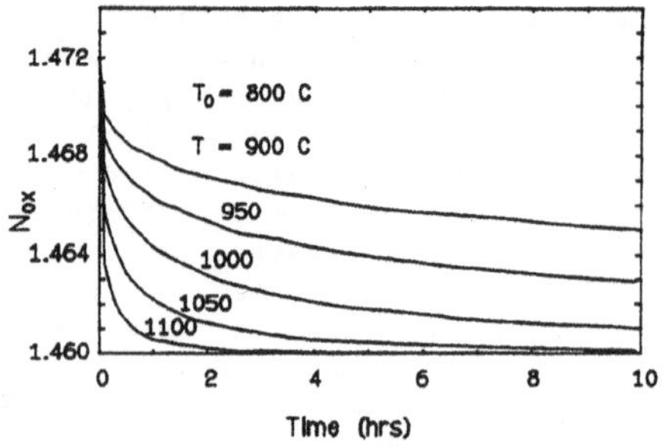

Fig. 1 Simulation from Eqs.(6) and (7) of the annealing at different temperatures T of a dry-grown oxide initially at T_o=800 C.

annealing histories. However, we would also like to be able to physically interpret the n and τ* dependences in terms of the known properties of SiO_2. This will allow a comparison of relaxation in a thermally grown oxide with that occuring in glasses. In addition, the several different but related activation energies found in the B_2O_3 example in the previous section could also be discussed for the thermal oxide films. The illustration in Fig. 1 has a τ* with activation energy E_a^*=3.0 ev. However, the 1/e time τ' of the curves is activated with an energy smaller by about 1 ev due to the time-dependent structural changes. This has been determined from Eq.(8) to be E_a^*=2.1 ev. The underlying local activation E_a is related to E_a^* via Eq.(5) as E_a = (1-n)E_a^*=0.9 ev. Fig. 1 was only an illustration. If the actual values of n and τ* appropriate to the thermal oxides were available, we could attempt to interpret these activation energies in terms of known physical properties of SiO_2. Such interpretations have previously been successful and provided physical insight for quenched glasses (8,11). The methods and concepts presented here have potential for a quantitative understanding of structural relaxation in thermal oxides. Development will require an interplay between both theory and experiment.

Supported in part by ONR Contract No. N0001488WX24074

REFERENCES

1. E.A. Taft, J. Electrochem. Soc. 125, 968 (1978).
2. E.A. Irene, E. Tierney and J. Angilello, J. Electrochem. Soc. 129, 2594 (1982).
3. B.J. Mrstik, A.G. Revesz, M. Ancona and H.L. Hughes, J. Electrochem. Soc. 134, 2020 (1987).
4. L.M. Landsberger and W.A. Tiller, Appl. Phys. Lett. 151, 1418 (1987).
5. J.D. MacKenzie, J. Am. Ceram. Soc. 46, 470 (1963).
6. J. Arndt and D. Stoeffler, Phys. Chem. Glasses 10, 117 (1969).
7. E.H. Nicollian and A. Reisman, J. Elec. Mat. 17, 263 (1988).
8. R.W. Rendell, K.L. Ngai, G.R. Fong and J.J. Aklonis, Macromolecules 20, 1070 (1987).
9. C.T. Moynihan, P.B.Macedo, C.J. Montrose, P.K. Gupta, M.A. DeBolt, J.F. Dill, B.E. Don, P.W. Drake, A.J. Easteal, P.B. Elterman, R.P. Moeller, H. Sasabe, and J.A. Wilder, Ann. N.Y. Acad. Sci. 279, 15 (1976).

10. A.J. Kovacs, J.J. Aklonis, J.M. Hutchinson, and A.R. Ramos, J. Polym. Sci., Poly. Phys. Ed. 17, 1097 (1979).
11. K.L. Ngai, R.W. Rendell, A.K. Rajagopal and S. Teitler, Ann. N.Y. Acad. Sci. 484, 150 (1986).
12. J.A. Bucaro and H.D. Dardy, J. Non-Cryst. Solids 24, 121 (1977).
13. J.J Mills, J. Non-Cryst. Solids 14, 255 (1974).
14. A.Q. Tool, J. Amer. Ceram. Soc. 29, 240 (1946).
15. E.A. Irene, D.W. Dong and R.J. Zeto, J. Electrochem. Soc. 127, 396 (1980).
16. J.A. Bucaro, H.D. Dardy and R.D. Corsaro, J. Appl. Phys. 46, 741 (1975).
17. J.E. Shelby, J. Non-Cryst. Solids 14, 288 (1974).
18. E.A. Taft and L. Cordes, J. Electrochem. Soc. 126, 131 (1979).

STRUCTURAL RELAXATION EFFECTS IN DRY THERMAL SILICON DIOXIDE FILMS ON SILICON

L.M. Landsberger* and
W.A. Tiller**

*Department of Electrical Engineering
**Department of Materials Science and Engineering
Stanford University, Stanford, CA, 94305

Abstract

In recent papers, work was presented on (i) inert thermal anneals of low-temperature-grown (highly-stressed) SiO_2 films, (ii) two-step oxidations to relate the second-step oxygen transport characteristics to the density of the existing film, and (iii) negative-point oxygen corona-discharge relaxation of low-temperature-grown SiO_2 films. This paper will review some of the previous work, present some more comprehensive results, and highlight the relationship between the SiO_2 film structure and transport of oxidant through the film. Oxidant transport through an SiO_2 film is found to depend strongly on the state of relaxation of the film.

INTRODUCTION

Silicon oxidation has been modelled as a simple linear-parabolic [1] growth process, and, for many years was thought to be well enough understood. However, in recent years, fuelled by the drive to smaller devices and lower temperature processing, oxidation has come under closer scrutiny. The original understanding has been found to be insufficient.

In response, a variety of studies have tried to match the existing oxidation data with more complicated models [2]-[7]. While these models in general have reasonable physical bases and fit the data well, other work [8] has shown that further experimentation is still necessary to distinguish between the mechanisms. In particular, experiments are necessary which sensitively test the actual physics of the models. The original assumptions and parameters related to the interface and surface reactions, diffusion and solubility of oxidant, must be better understood or modified.

Previous work [9]-[12] on bulk silica has shown that both stress and thermal history can have subtle effects on the properties of SiO_2. Since the ground-breaking work of EerNisse [13] established that significant stresses existed during thermal growth of SiO_2 films, and since Taft [14]-[15] showed that the refractive index of such films varied as a function of growth temperature, the arguably interrelated bulk properties of stress, density, refractive index and viscosity have emerged as significant [16]-[19]

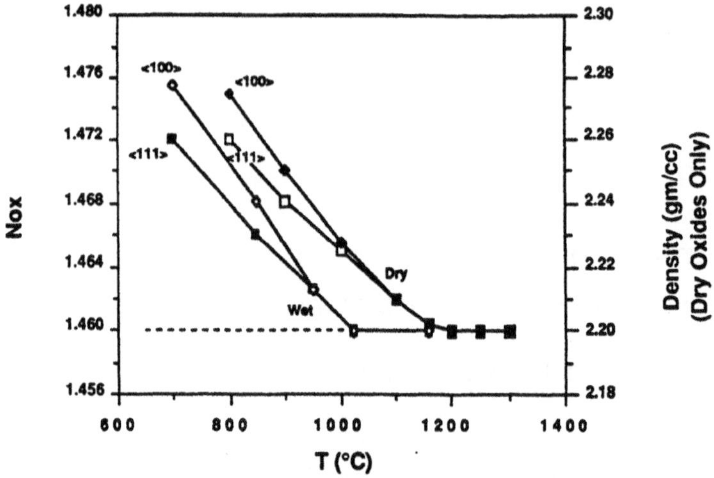

Figure 1: Refractive index vs. growth temperature for wet [15] and dry [14] oxides, and density [17] for dry-grown SiO_2 films on Si.

topics in the study of SiO_2 films on Si. Because the existing film growth models are insufficient, such subtle structural effects are good candidates for further study because they could impact oxidation rates.

In recent papers [20]-[24] work was presented on the following topics:

1. structural relaxation of dry low-temperature-grown (highly stressed) SiO_2 films by inert thermal anneals [20],

2. two-step oxidations to relate the second-step oxygen transport characteristics to the density of the existing film [21], [22],

3. negative-point oxygen-corona relaxation at 800°C of SiO_2 films grown at low temperatures [21]-[23].

4. conformal SiO_2 films grown by oxygen-corona at 900°C in dry O_2 on two-dimensional Si structures [24].

This paper will review some of this previous work, present some more comprehensive results, and highlight the relationship between the SiO_2 film structure and transport of oxidant through the film.

THERMALLY-INDUCED STRUCTURAL RELAXATION

Fig. 1 shows the relationship between the refactive index and the growth temperature of SiO_2 films. The refractive indices for both wet- and dry-grown films decrease with increasing growth temperature until they near 1.460. Wet oxides reach this fully-relaxed refractive index level at about 1000°C, while dry oxides reach it at 1150°C-1200°C. Above these breakpoint temperatures, the refractive index is constant, and this level can be assumed to be the same as the index of pure fused quartz under no stress [25],[20].

Previously reported work [20] showed that these breakpoint temperatures could be explained [25], [20] by comparing the oxidation times with the viscoelastic relaxation times calculated from literature viscosity values [12] and corroborated by thermal annealing experiments. It was concluded that the simple linear viscoelastic model [16], [17] is sufficient to quantitatively explain the breakpoints in refractive index, but is unable to explain the variations in refactive index below the breakpoint temperature.

Systematic inert thermal anneals [20] of dry 800°C-grown SiO_2 films demonstrated two distinct stages of relaxation. The film relaxes quickly to an intermediate level, then progresses more slowly toward full relaxation. The oxide thickness swelled during annealing in proportion to the refractive index changes, confirming that the refractive index changes reflected bulk structural relaxation of the film. The maximum thickness swelling of 3% correlated well with predicted (see Fig. 1) density differences between 800°C-grown and fully-relaxed films.

The intermediate levels were found to be a function of anneal temperature. An 800°C anneal barely changed the properties of the 800°C-grown film, while a higher temperature (900°C-1180°C) anneal caused an immediate (less than 30mins) drop in refractive index to a temperature-dependent level. For a given temperature, these intermediate levels were slightly lower than the characteristic N_{ox} levels for as-grown dry oxide in Fig. 1, suggesting that the relaxing oxide seeks a particular density as a function of temperature.

When a high-temperature-grown (1180°C) film was annealed in N_2 at 900°C, the oxide thickness and refractive index remained unchanged for anneals of up to 160 hours. The intermediate levels are therefore believed to be metastable, preferred over less relaxed levels but not preferred over more fully relaxed states.

The above scenario is in accord with the recent intrinsic stress measurements by Kobeda & Irene [27]. In a careful study measuring the total residual film stress and subtracting off the thermal expansion mismatch stress, they found the intrinsic residual stress as a function of growth temperature for 700°C-1100°C for different substrate orientations. Their results showed that the intrinsic stress is at a maximum for the lowest growth temperatures, and decreases with increasing growth temperature. The intrinsic stress is still greater than 0 at 1100°C, consistent with the dry oxide refractive index data in Fig. 1. Furthermore, they found residual stresses in ⟨100⟩-grown films to be higher than those in ⟨111⟩-grown films, again in agreement with Fig. 1. Their work thus supports the use of the refractive index as an indicator of residual stress and structural strain in SiO_2 films.

Kobeda & Irene [27] also measured the intrinsic stress in films grown at 700°C-800°C and annealed for 1 hour at 1000°C in N_2, finding that the stress is reduced to a level near (slightly below) the 1000°C as-grown stress level. This is in remarkable agreement with the behaviour of the refractive index during annealing described above. The intermediate refractive index levels correlate both structural relaxation (density) and residual stress levels with processing temperature.

TWO-STEP OXIDATION EXPERIMENTS TO DETERMINE FILM HISTORY EFFECTS

The above correlations provide the opportunity to explore the impact of such subtle structural effects on oxidation rates. Since, by the above experiments and analysis, an oxide film carries the signature of a previous processing temperature, one can use a two-step oxidation strategy to vary physical parameters which might depend on such stress or structural variations.

Hamasaki [28] was the first to perform two-step oxidation experiments on thermally-grown films. He found considerable history effects when the growth temperature of the initial oxidation was different from that during a second oxidation step. In 1982, Irene [29] reoxidized thick films at lower second-step temperatures. His work was interpreted as a history effect by Han & Helms [30]. More recently, Srivastava & Irene [31] and Taft [32] have similarly related the extra thickness grown during a second oxidation step to the prior thermal history of the sample.

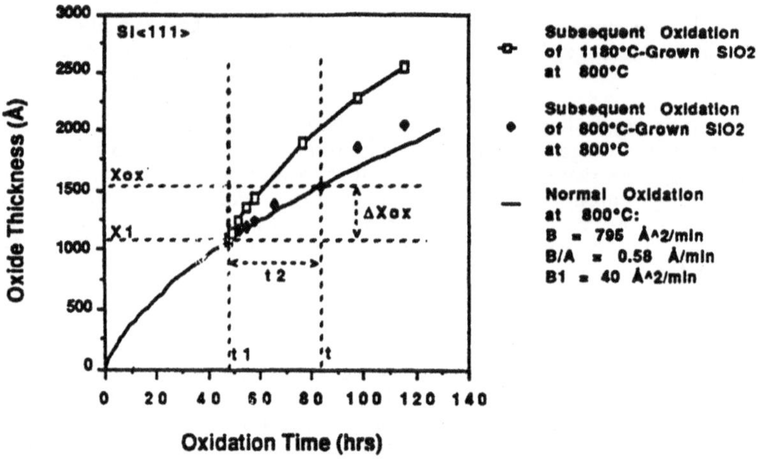

Figure 2: Oxide thickness vs time for 1-step and 2-step oxidation on Si⟨111⟩ at 800°C. The two sets of data points represent subsequent oxidation of films with initial thickness $X_1 \sim 1070$ Å for a second-step time t_2, to grow an extra thickness ΔX_{ox}.

Further, by studying stress and strain relaxation in SiO_2 films, several other researchers [2]-[5] have hoped to better model dry thermal oxidation. They hoped to explain the rapid initial oxidation regime by the effects of strain variations on the diffusion of oxidant.

The remainder of this paper will use a two-step oxidation strategy to relate the structural state of relaxation of the film prior to the second oxidation step to the transport of oxidant through the film during the second oxidation step. For most of the experiments described, the initial oxide films were grown to a thickness of $X_1 \approx$ 1070 Å at a variety of temperatures T_1. Considering Fig. 2, these initial films are represented by the first (leftmost) data point at 1070 Å for each data set. The two sets of data points are for subsequent oxidations of such initial films. In other words, in a second-step time t_2, an extra thickness of $\Delta X_{ox}=X_{ox}-X_1$ was grown on top of the initial 1070 Å. The smooth linear-parabolic curve from represents normal X_{ox} vs t data for dry oxidation of a bare Si⟨111⟩ wafer at 800°C with $X_1=0$.

The second-step oxidations in Fig. 2 were performed at $T_2=800$°C. The lower set of points, for initial films grown at 800°C, can be seen to follow near the normal linear-parabolic kinetics expected for a one-temperature oxide, as if we had not taken them out of the furnace at t_1. The upper set of points, however, is for initial films grown at 1180°C, and the difference in behaviour is obvious. For a given second-step time t_2, the 1180°C-grown films grew a substantially larger extra thickness $\Delta X_{ox}=X_{ox}-X_1$. The higher-temperature-grown film experiences an enhanced second-step oxidation rate due to the thermal history of the sample.

Fig. 3, on the other hand, investigates extra oxidation behaviour for T_1 less than T_2, showing results for 800°C-grown films reoxidized at 900°C. As is readily apparent, the extra oxidation data follow closely along the linear-parabolic curve, indicating that the 800°C-grown films are behaving as if they were grown at 900°C with no significant thermal history effect. Since the thermal annealing studies described in the preceding section show that an 800°C-grown film will equilibrate at the refractive index characteristic of 900°C in a time much shorter than the extra oxidation times presented here, it is reasonable to expect no deviation from the normal oxidation rate. Accordingly, the enhancements in Fig. 2 are maintained because the higher-temperature films are the more stable structures.

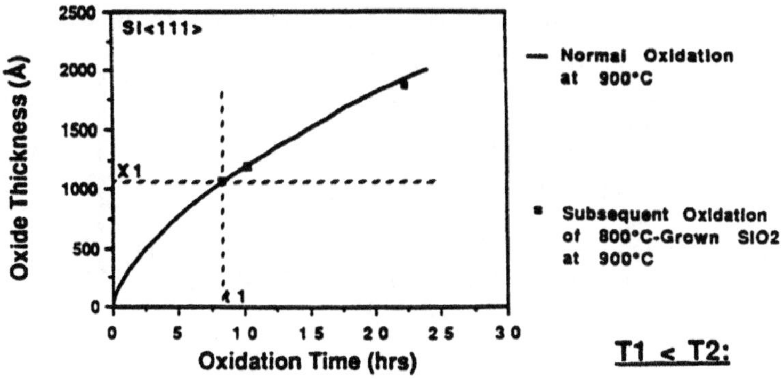

Figure 3: Extra oxidation at 900°C of 800°C-grown films with initial thickness $X_1 \sim$ 1070 Å. The extra oxidation data follows the linear-parabolic curve normal for 900°C, indicating that there is no history effect.

Analysis of extra oxidation data such as in Fig. 2 will require a more complicated mathematical formalism than the one-layer linear-parabolic [1] model. If the linear-parabolic formalism is expanded to include the effects of an initial layer [33], [21], [34], the resulting rate constants for the extra growth ΔX_{ox} during the second oxidation step are as follows:

$$B_\Delta = 2D_2\frac{C^*}{M_{ox}} \tag{1}$$

$$\frac{B}{A}_\Delta = \frac{\frac{C^*}{M_{ox}}}{\frac{1}{k_i}+\frac{k_c}{h}+\frac{X_1 k_c}{D_1}} \tag{2}$$

where D_1, D_2 are the diffusivity of oxidant through the existing and newly forming oxide, respectively, X_1 is the thickness of the initial oxide layer, M_{ox} is the number density of SiO_2 molecules, k_i is the interface reaction constant, h is the gas transfer coefficient, and $k_c = C^*/C^{*'}$. C^* is the normal solubility of oxidant in SiO_2, while $C^{*'}$ is the solubility in the initial layer of different structure.

Then, in order to use this formalism, ΔX_{ox} vs t_2 data sets were fitted to a linear-parabolic curve to find B_Δ and $(B/A)_\Delta$. In all cases a good fit was obtained by a two-parameter model, indicating that there was no rapid initial oxidation regime in the second-step oxidation. B_Δ were in all cases close enough to the B normal for the second-step temperature to support the application of the above formalism.

The interface reaction constant, k_i, is not expected to change, because, as soon as the second oxidation step begins, the SiO_2/Si interface is no longer in contact with the 1180°C-grown oxide. Since the second-step oxide is growing at 800°C, we therefore expect that the interface will assume the properties characterstic of 800°C. Then,

$$\Delta A/B = (A/B)_\Delta - (A/B)_{normal} = \frac{X_1 k_c M_{ox}}{D_1 C^*} \tag{3}$$

and

$$D_1 C^{*'} = \frac{X_1 M_{ox}}{\Delta A/B} \tag{4}$$

It is important to note that we can find only the product $D_1 C^{*'}$ from this: a change in oxidant diffusivity cannot in general be distinguished from a change in oxidant solubility.

Table 1. Parameters extracted from ΔX_{ox} vs t_2 for extra oxidations at 800°C

Initial Thickness X_1	Initial Refr. Index N_{ox}	Temp. Initial Oxide Grown	Fitted B_Δ (Å²/min)	Fitted $(B/A)_\Delta$ (Å/min)	B/A_{normal} (Å/min)	$(D_1C^{*'})\frac{2}{M_{ox}}$ (Units of B) (Å²/min)
1070 Å	1.460	1180°C	980	0.78	1.0	7645
1070 Å	1.463	1100°C	980	0.60	1.0	3195
1070 Å	1.472	800°C	1000	0.28	0.58	1163

All initial and subsequent oxidations in dry O_2.
All Si⟨111⟩.

Table 2. Parameters extracted from extra oxidations in dry O_2 for 19 hours at 800°C

Initial Processing	Initial Refr. Index N_{ox}	Initial Thickness X_1	ΔX_{ox}	Fitted $(B/A)_\Delta$ (Å/min)	$(D_1C^{*'})\frac{2}{M_{ox}}$ (Units of B) (Å²/min)
800°C-grown	1.472	1070 Å	320 Å	0.28	1165
1000°C-grown	1.466	950 Å	515 Å	0.58	2635
1000°C-N_2-1hr	1.466	1080 Å	510 Å	0.57	2860
1100°C-grown	1.463	1070 Å	550 Å	0.60	3195
1000°C-N_2-16hrs	1.462	1090 Å	575 Å	0.71	5100
1150°C-grown	1.461	1070 Å	580 Å	0.72	5350
1180°C-grown	1.460	1070 Å	630 Å	0.78	7645
800°C-grown	1.470	1760 Å	255 Å	0.23	1037

Any initial oxides prior to annealing were grown at 800°C in dry O_2.
All Si⟨111⟩.

Table 1 shows the transport parameter $2(D_1C^{*'})/M_{ox}$ calculated from extra oxidation data sets for initial oxides of three different growth temperatures. In order to supplement the data in Table 1, further extra oxidations were performed on a variety of initial films. The initial films were all prepared as shown in Table 2 and subjected to a constant extra oxidation in dry O_2 at 800°C for 19 hours. Because the second-step B_Δ's in Table 1 were constant over the whole range of initial growth temperatures (1180-800°C), the ΔX_{ox} in Table 2 can be conveniently analyzed in spite of the fact that there is only one data point for each initial condition. B_Δ can be held constant at 980 Å^2/min, and $(B/A)_\Delta$ varied until the data point is fitted. The ensuing analysis yields the $D_1C^{*'}$ shown for comparison with the previous results. The last row of Table 2 shows that essentially the same $D_1C^{*'}$ is found if the initial thickness of the film is varied.

The parameters calculated in Tables 1 and 2 form a consistent picture when related to the refractive indices of the initial films prior to the second oxidation step in Fig. 4. Fig. 4 also includes extracted parameters from the extra oxidation at 800°C of 2.2μm thick oxides grown at 1155°C. The extracted $2(D_1C^{*'})/M_{ox}$ increase monotonically as the refractive index decreases. Of particular interest is the behaviour of the 1000°C- N_2-annealed cases. The oxidant transport properties of the films are a direct function of the state of relaxation of the film, whether the particular state has been reached by normal growth in dry O_2 or by subsequent thermal processing in an inert ambient. Furthermore, by the correlations presented in the previous section, the oxidant transport properties are therefore directly related to the residual intrinsic stress in the film.

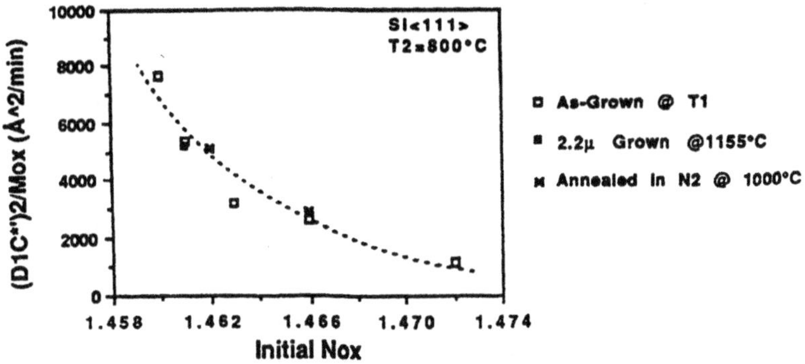

Figure 4: Extracted $\frac{2(D_1C^{*'})}{M_{ox}}$ are plotted as a function of initial refractive index prior to the second oxidation step. All second-step oxidations were carried out at 800°C.

Many extra oxidation experiments were performed for a variety of different initial oxides and a variety of different second-step temperatures. The extra oxidation data sets were analyzed as described above, and transport parameters $2(D_1C^{*'})/M_{ox}$ calculated. Fig. 5 compiles the results for all of the different T_1, T_2 data sets studied. The two-step lines can be modelled as simple Arrhenius dependences of the form $\frac{2(D_1C^{*'})}{M_{ox}} = (K_o)e^{-E_a/kT}$, These lines intersect the curve for normal oxidation at or near the initial growth temperature, yet diverge above it for lower second-step oxidation temperatures. While the fitted [6] activation energy for B_{normal} is 1.6eV, the activation energy for $\frac{2(D_1C^{*'})}{M_{ox}}$ varies slightly in the vicinity of 1eV, from 0.85eV to ~1.3eV.

As also seen in Fig. 4, the history effects are monotonically related to the refractive index of the initial oxides. The higher the initial refractive index (the higher the initial density, the less 'relaxed' the network) the slower the transport of oxidant and the higher the activation energy for transport. These simple relationships are intuitively sensible in view of the interstitial nature of solubility and diffusion in SiO_2. In a less dense network with a less tightly packed ring structure, one would naturally expect both that larger interstitial spaces would accomodate a higher concentration of oxidant and that less constricted channels would require a lower activation energy for diffusive jumps between interstitial sites.

EXTRA OXIDATION OF CORONA-RELAXED OXIDES

Previous work [21]-[24] showed that a negative-point oxygen corona-relaxation process was able to fully structurally relax low-temperature-grown SiO_2 films at low temperatures (600°C-900°C). The top solid line in Fig. 5 represents extra oxidations of 800°C, oxygen-corona-relaxed films. As can be seen, the oxidant transport in such a corona-relaxed film is even more enhanced over the one-temperature case than in a 1180°C-grown film.

ACKNOWLEDGEMENTS

This work was supported by the Defense Advanced Research Projects Agency (DARPA) under contract No. MDA 903-79-C-0257, by DARPA/VHSIC under contract No. DAAL 01-86-K-0101, and by the Semiconductor Research Corporation under contract No. SRC 85-08-062.

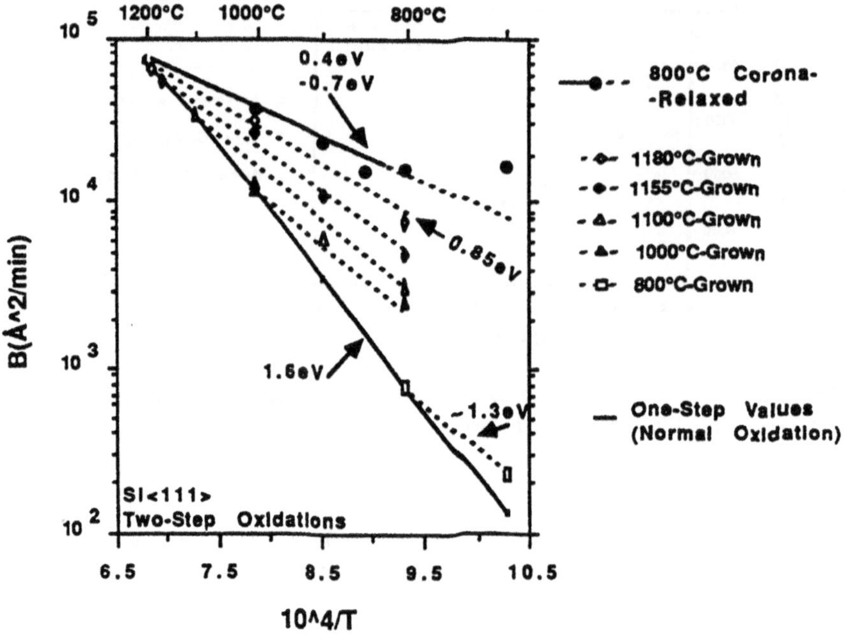

Figure 5: Combined results showing history effects in $\frac{2(DC^{*'})}{M_{ox}} = \mathrm{B}$ for SiO_2 grown at T_1 and reoxidized at a lower $T=T_2$.

References

[1] B.E. Deal, A.S. Grove, *J. Appl. Phys.* **36**, 3770 (1965).

[2] A. Fargeix, G. Ghibaudo, *J. Appl. Phys.* **54**, 7153 (1983).

[3] G.C. Sarti, F. Santarelli, and G. Camera Roda, *Chem. Eng. Sci.* **41**, 2699 (1986).

[4] R.H. Doremus, *Thin Solid Films* **122**, 191 (1984).

[5] R.H. Doremus, *J.Electrochem. Soc.* **134**, 2001 (1987).

[6] C.J. Han, and C.R. Helms, *J. Electrochem. Soc.* **134**, 1299 (1987).

[7] A. Reisman, E.H. Nicollian, C.K. Williams, and C.J. Merz, *Journal of Electronic Materials* **16**, 45 (1987).

[8] J. Blanc, presented at the Symposium on Silicon Nitride and Silicon Dioxide Thin Insulating Films at the Electrochemical Society Meeting inm San Diego, CA, Oct. 20-24, 1986.

[9] H.M. Cohen and R. Roy, *Phys. Chem. Glasses* **6**, 149 (1965).

[10] J. Arndt and D. Stoeffler, *J. Am. Ceram. Soc.* **10**, 117 (1969).

[11] J.D. Mackenzie, *J. Am. Ceram. Soc.* **46**, 470 (1963).

[12] G. Hetherington, K.H. Jack, and J.C. Kennedy, *Phys. Chem. Glasses* **5** 130 (1964).

[13] E.P. EerNisse, *Appl. Phys. Lett.*, **35**, 8 (1979).

[14] E.A. Taft, *J. Electrochem. Soc.*, **125**, 968 (1978).

[15] E.A. Taft, *J.Electrochem. Soc.*, **127**, 993 (1980).

[16] W.A. Tiller, *J. Electrochem. Soc.* **127**, 619 (1980).

[17] E.A. Irene, E. Tierney, J. Angilello, *J. Electrochem. Soc.* **129**, 2594 (1982).

[18] E.A. Taft, *J. Electrochem. Soc.* **134**, 475 (1987).

[19] B.J. Mrstic, A.G. Revesz, M. Ancona, and H.L. Hughes, *J. Electrochem. Soc.* **134**, 2020 (1987).

[20] L.M. Landsberger, and W.A. Tiller, *Appl. Phys. Lett.* **51**, 1416 (1987).

[21] L.M. Landsberger, and W.A. Tiller, *Appl. Phys. Lett.* **49**, 143 (1986).

[22] L.M. Landsberger, and W.A. Tiller, in "Symposium on Silicon Nitride and Silicon Dioxide Thin Insulating Films" at the Electrochemical Society Meeting in San Diego, CA, Oct. 20-24, 1986, Proc. Vol. 87-10, p.333.

[23] L.M. Landsberger, and W.A. Tiller, in "Symposium on Photon, Beam, and Plasma Stimulated Chemical Processes at Surfaces" at the Materials Research Society Meeting in Boston, MA, Dec. 1-6, 1986, Mat. Res. Soc. Symp. Proc. Vol. 75, p. 803.

[24] L.M. Landsberger, D.B. Kao and W.A. Tiller, *J. Electrochem. Soc.* **135**, No. 7, (1988).

[25] I.H. Malitson, *J. Opt. Soc. America*, **55**, 1205 (1965).

[26] D.J. Chin, S.M. Hu, R. Dutton, and J. Moll, *IEEE Trans. Electron Devices* **ED-30**, 744 (1983).

[27] E. Kobeda, and E.A. Irene, *J. Vac. Sci. Technol. B* **5**, 15 (1987).

[28] M. Hamasaki, *Solid State Electronics*, **25**, 479 (1982).

[29] E.A. Irene, *J. Electrochem. Soc.*, **129**, 413 (1982).

[30] C.J. Han, C.R. Helms, *J. Electrochem. Soc.*, **132**, 516 (1985).

[31] J.K. Srivastava, E.A. Irene, *J.Electrochem. Soc.*, **132**, 2815 (1985).

[32] E.A. Taft, *J. Electrochem. Soc.*, **132**, 2486 (1985).

[33] W.A. Tiller, *J. Electrochem. Soc.* **130**, 501 (1983).

[34] L.M. Landsberger, Ph.D. Thesis, Stanford University, June 1988.

MOLECULAR DIFFUSION IN a-SiO_2: ITS ROLE IN ANNEALING RADIATION-INDUCED DEFECT CENTERS

Robert L. Pfeffer

US Army Electronics Technology and Devices Laboratory (LABCOM)
Fort Monmouth, New Jersey 07703-5000

The reversible annealing of E' centers (paramagnetic bridging oxygen vacancies) in a-SiO_2 at temperatures lower than about 300°C was studied by electron paramagnetic resonance (EPR) spectroscopy in samples of bulk dry a-SiO_2 which were specially prepared with widely differing amounts of dissolved O_2. It was found that all decreases in E' concentrations were exactly matched by increases in corresponding $SiOO^-$ (superoxide radical) concentrations, and that there was a pronounced persistence of E' centers at higher temperatures in oxygen-deficient samples. The annealing process was modelled by a diffusion-limited quasi-unimolecular reaction between dissolved O_2 and fixed trapping centers in the context of the continuous time random walk (CTRW) formalism.

INTRODUCTION

Buildup of trapped positive charge in gate and field oxides is an important cause of ionizing radiation induced degradation of MOS integrated circuit devices. The charge trapping sites in the oxides have by now been identified principally as E' centers (paramagnetic bridging oxygen vacancies).[1] These are formed during irradiation with energetic photons or electrons mainly by ionization processes,[2] i.e., bond rearrangements after electron-electron collisions, rather than displacements from electron-atom collisions.

The annealing of E' centers follows different routes in different temperature regimes. Above 700°C the annealing process is irreversible, and involves strictly local thermally activated bond rearrangement.[3] However, at temperatures lower than about 300°C, an alternate process occurs which is reversible.[4] This process appears to be nonlocal, involving the thermally activated diffusion of some species to the E' site and its subsequent reaction to form a distinct metastable entity rather than to merely involve a change in the vacancy's electrical state such as hole detrapping. Edwards and Fowler[5] have suggested that the process is mediated by the diffusion of dissolved O_2 and its subsequent capture to form superoxide radicals (E' + O_2 $\rightarrow SiOO^-$). Various annealing studies have been performed to test this idea[6]; these have been either EPR observations of E' concentrations in MOS oxides or bulk SiO_2 or of oxide trapped charge densities in MOS structures. Results have been consistent with it in general but have failed to agree in detail with the predictions of a diffusion-limited bimolecular reaction model.[7] A distribution of activation energies[8] which purportedly characterizes O_2 diffusion in SiO_2 has frequently been invoked to ameliorate the discrepancy. We have sought a physical basis for the seemingly ad hoc assumption that a distribution of activation energies exists through a computer simulation of the diffusion process based on the Anderson-Stuart model.[9] We found that even a wide distribution of doorway and interstice radii in the SiO_2 network yielded diffusion characterized by a single effective activation energy.

To improve our understanding of the low-temperature diffusion and annealing process we have conducted annealing experiments involving samples of bulk dry a-SiO_2 which were specially prepared with widely differing amounts of dissolved O_2. Our intention was to modify the annealing kinetics by controlling the average path length of a diffusing molecule. In addition to tracking the evolution of E' center concentrations we also monitored that of $SiOO^-$ centers by means of EPR. Our studies involve samples which we pre-loaded with different gas concentrations, irradiated to different doses, and isochronally and isothermally annealed.

In this paper we review this work as well as report on our latest experiment in the series, in which the low-temperature isothermal annealing kinetics of E' centers were determined. We then interpret our data by invoking models analyzing diffusion in the presence of fixed traps in the context of the continuous time random walk (CTRW) formalism.

REVIEW OF ISOCHRONAL ANNEALING EXPERIMENTS

All our studies were done on samples cut from bulk Suprasil W1 (Heraus-Amersil, Somerville, NJ), a high-purity fused silica with very low OH content. For the first of these experiments,[10,11] we prepared three samples. One was degassed by a 900°C vacuum bake at 7 mTorr for 5 days, one was oxygen stuffed at a pressure of 8000 psi for the same time at the same temperature, and one was left untreated. These durations were sufficient to establish equilibrium distributions of dissolved O_2 throughout the samples at temperatures well below the devitrification point of fused silica. All samples were then exposed at room temperature to gamma rays from a ^{60}Co source. After a 6-month period at room temperature, a sequence of cumulative isochronal anneals was conducted in air on all samples, with E' and $SiOO^-$ spin concentrations measured after each by EPR at -170°C. Our results qualitatively supported the Edwards-Fowler model:

(1) All three samples showed increases in $SiOO^-$ densities as temperatures were increased. These increases exactly matched corresponding decreases in E' densities. In every sample, the total defect density (i.e., [E']+[$SiOO^-$]) remained approximately constant from room temperature up to 600 K, above which it decreased steadily.

(2) There was a pronounced persistence of E' centers at higher temperatures in the oxygen-deficient (degassed) sample. The onset of annealing occurred 50 K higher than in the other samples, and the E's were still evident even over 800 K, having largely vanished by 500 K in the others.

Quantitatively, however, the data provided a poor fit to Waite's diffusion-limited bimolecular reaction model, upon which Edwards and Fowler based their analysis. Furthermore the E' anneal curves in the untreated and oxygen-stuffed samples were quite similar, suggesting that the as-received Suprasil W1 contained a good deal of excess oxygen.

In our second isochronal annealing study,[11] we degassed all our samples by a 900°C vacuum bake at 10^{-6} Torr for 8.5 days before any further treatments, guaranteeing a uniform starting material. Samples were then backfilled with oxygen at pressures of 1, 22 and 500 atm. at similar temperatures and times, and all were ^{60}Co-irradiated as before. After a 3-week period at room temperature, a sequence of cumulative isochronal anneals was again conducted in air, and the fraction of E' centers remaining after each was recorded via EPR at -135°C. We observed the following:

(1) There was a remarkable temperature dependence of the remaining fraction of E' centers in every sample that had been backfilled with O_2. Above a critical temperature, below which the fraction remained virtually unchanged, the [E]' fraction after each successive 10-minute anneal turned out to be a constant fraction of its value preceding that anneal. This corresponded to exponential decreases in [E'] with time, independent of temperature. The exponentials' rate constants depended only weakly on the initial value of $[O_2]$. However, the critical temperatures were quite sensitive to dissolved $[O_2]$, decreasing as $[O_2]$ increased. In a sample that had been degassed but not backfilled, the [E'] fraction never fell below 0.2, even at 500°C; those of the other samples had diminished by about 3 orders of magnitude by 400°C.

(2) All increases in $[SiOO^-]$ again tracked corresponding decreases in [E']. All the $SiOO^-$ centers again annealed at the same temperature (much higher than the E's), suggesting a local rather than an oxygen diffusion mediated process for $SiOO^-$ annealing.

(3) Each of the pre-anneal defect concentrations, i.e., both initial [E'] and initial $[SiOO^-]$, varied directly with dissolved oxygen content.

ISOTHERMAL ANNEALING EXPERIMENT

Procedure

All samples were 4 x 23 x 0.5 mm slabs of Suprasil W1. They were all degassed by a 900°C vacuum bake at 10^{-6} Torr for 8.5 days. Then one group was oxygen-stuffed at 875°C for 8 days at 14.7 psi, another at 850°C for 7 days at 370 psi, a third at 850°C for 8 days at 7580 psi, and a fourth received no further treatment. All then received room-temperature irradiations of 10^8 rad(Si) by gamma rays of 1.17 MeV and 1.33 MeV from a ^{60}Co source at a dose rate of 0.7 Mrad/hr. After a 10 month period at room temperature, a sequence of isothermal anneals was conducted in air on various samples, and the concentration of E' centers in each sample was determined by EPR measurement of spin concentrations after each anneal. The spectrometer was a standard Varian E-9 type operating near 9.1 GHz in the absorption mode at a modulation frequency of 100 kHz, with the samples at -135°C in a (011) cylindrical cavity. The E' signals were recorded using a 0.4 G modulation amplitude and a 0.2 uW power level (switching to 1 mW out of phase for the weakest signals). No measurements were made of $[SiOO^-]$.

The E' concentration was determined in one sample (which was irradiated with all the others but which had undergone no heat treatments or annealing) by numerical integration and comparison with a strong pitch standard, with microwave saturation effects taken into account. The relative [E'] for all other samples were obtained by comparing the peak-to-peak amplitudes of each to the one in that sample.

Results

The fraction of E' centers remaining in each type of sample after each anneal is shown in Figs. 1a-1d, in which the symbols indicate data obtained from samples identified in the caption. The sample identifiers are intended as mnemonic acronyms denoting the sequence of treatments they had received; e. g., $VO_{22}R_8$ denotes samples which had been vacuum-degassed, oxygen-

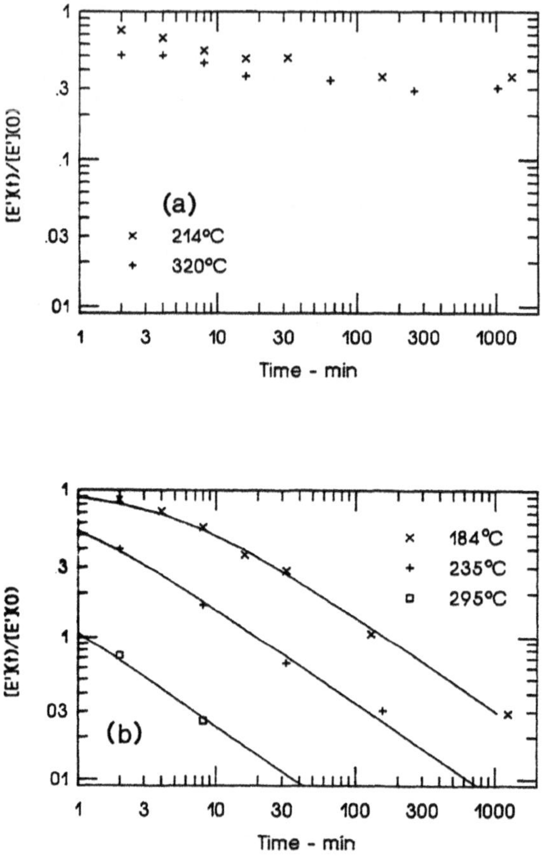

Fig. 1. Observed fraction of E' centers remaining in Suprasil W1 samples after isothermal anneals in air at indicated temperatures. Continuous curves represent fits to the data as discussed in the text.
(a) Sample group VR_8
(b) Sample group VO_1R_8

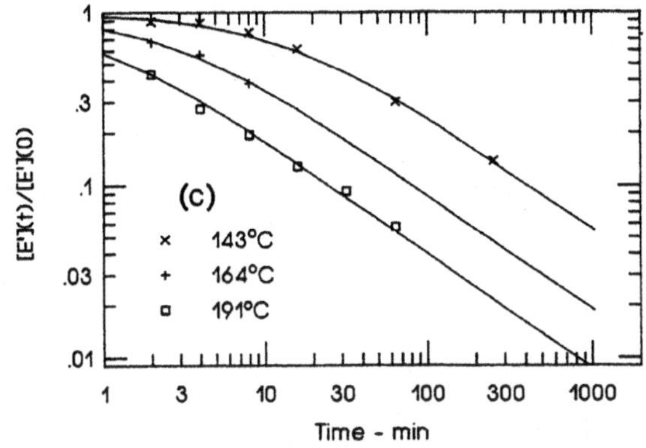

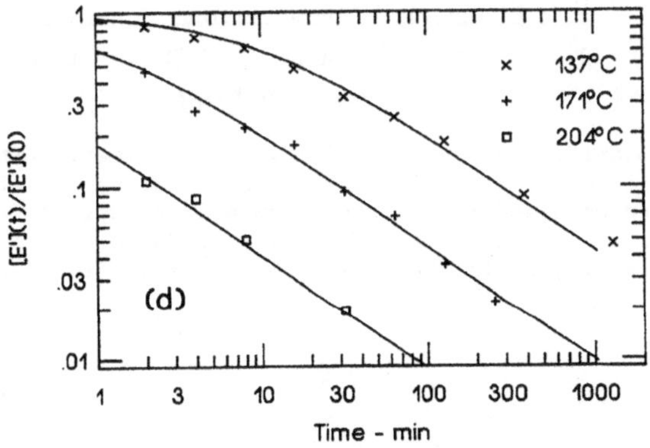

Fig. 1. (Cont'd) Observed fraction of E' centers remaining in Suprasil W1 samples after isothermal anneals in air at indicated temperatures. Continuous curves represent fits to the data as discussed in the text.
(c) Sample group $VO_{22}R_8$
(d) Sample group $VO_{500}R_8$.

backfilled at 22 atm and exposed to a gamma-ray dose of 10^8 rad(Si). Examination of the figures shows that:

(1) Very little E' annealing occurred in the degassed samples; even at 320°C, the highest temperature used, E' annealing ceased altogether within 2 hrs., leaving almost half the original number of defect centers intact.

(2) In every sample which was backfilled with O_2 the fraction of remaining E's remained relatively steady for a while, then dropped off with time according to a power-law. For every value of $[O_2]$, the initial relaxation time shortened drastically with increasing temperature. However, all the samples shared the same power-law exponent.

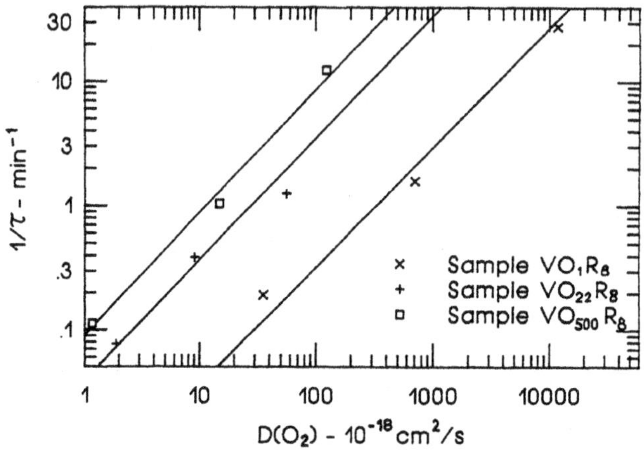

Fig. 2. Temperature dependence of relaxation times. Symbols denote reciprocals of relaxation times appearing in Table 1, which were obtained by fitting E' annealing data to C(t) given by Eq. 1. Abscissa is diffusivity of O_2 in SiO_2 at temperatures listed in the table. Solid lines are of unit slope.

Discussion

The strikingly similar behavior of all the anneal rates led us to attempt a universal fit to the data from the backfilled samples. This was accomplished with the fitting function

$$C(t) \equiv [E'](t)/[E'](0) = (1 + t/\tau)^{-(2/3)} \quad (1)$$

where the relaxation time $\tau(T,[O_2])$ is a function of both temperature T and initial oxygen concentration $[O_2]$. The value of τ for each sample was determined by eyeball fit (easily done by sliding a calculated curve over identically-scaled figures showing the data). The values appear in Table 1. The closeness of fit can be judged by inspection of Fig. 1b - 1d, in which the continuous curves were calculated using those values in Eq. 1.

The temperature dependence of τ reflects the process by which E' annealing occurs. To establish what its temperature dependence was, we calculated the diffusivity of O_2 in SiO_2 for every temperature listed in

Table 1. Summary of Results

Sample Type	Presumed[a] $[O_2]$ (cm^{-3})	Initial [E'] (cm^{-3})	Anneal Temp. (°C)	Relaxation Time τ (min)
VR_8	$6.5(10)^{12}$	$4.7(10)^{15}$	214	b
			320	b
VO_1R_8	$1.5(10)^{16}$	$4.1(10)^{15}$	184	5.2
			235	0.63
			295	0.036
$VO_{22}R_8$	$3.8(10)^{17}$	$6.6(10)^{15}$	143	13.2
			164	2.6
			191	0.80
$VO_{500}R_8$	$7.7(10)^{18}$	$10.5(10)^{15}$	137	8.9
			171	0.96
			204	0.081

[a]Oxygen concentrations in backfilled samples are based on gas-phase solubilities quoted by Norton.[12] That in degassed samples is based on Fickian desorption from SiO_2 presumed to initially contain 1 atm; however, Suprasil W1 seems to contain considerably more O_2, making this value questionable.

[b]The extent of observed annealing was insufficient to establish relaxation times in these cases.

Table 1 by extrapolating those measured by Norton[12] with a simple Arrhenius law. Against these values of $D(O_2)$ we plotted in Fig. 2 the reciprocal of τ at those temperatures. As can be seen, there is a direct proportionality between $1/\tau$ and $D(O_2)$ for each of the sample groups which extends over 3 orders of magnitude, clearly indicating that the time scale for the low-temperature conversion of E' to $SiOO^-$ centers is governed by the diffusion of molecular oxygen in SiO_2.

We have been less successful at establishing a plausible (quantitative) concentration dependence of τ, which could reasonably supposed to go as $[O_2]^{(-2/3)}/D(O_2)$; this would imply a ratio between lines in Fig. 2 of about 1:8:64, clearly not the case. Although the specific values of $[O_2]$ listed in Table 1 are thus questionable, the data do indicate the correct ordering among sample groups. And beyond that, the lack of annealing in the degassed samples suggests a lack of O_2 in that sample group; on the other hand, the similarity in form among all the other anneal curves suggests $[O_2]$ so exceeded [E'] in all the backfilled samples (c.f. Table 1) that E' annealing did not significantly reduce its initial value.

Interpretation of the data in terms of a model for the diffusion process at low temperatures requires an analysis of the kinetics of diffusion and subsequent reaction. This has recently been done by several groups.[13-15] In the caes treated, diffusing "scavenger" particles consume randomly distributed "target" particles. (When the target particle concentration is tracked, the reaction is referred to as scavenging or target annihilation; when that of the random walkers is tracked, it is termed trapping. When the scavengers far outnumber the targets, as they do in this case, the reaction is generally denoted quasi-unimolecular.) In the scavenger reaction, the target particle density for short times is given by $C(t) = \exp(-[O_2]\,S(t))$, where $S(t)$ is the mean number of sites visited by a scavenger after time t.

The behavior of S(t) is determined by the properties of a given system. If transport in a random medium is involved, S(t) can be calculated from a "waiting-time" distribution $\psi(t)$ according to the continuous time random walk (CTRW) formalism introduced by Montroll and Weiss.[16] We are currently attempting to determine an appropriate form of $\psi(t)$ based on the data and to compare it to our calculations of $\psi(t)$ based on the Anderson-Stuart model; the results will be reported shortly.

CONCLUSIONS

We have performed isochronal and isothermal annealing experiments which demonstrated that the low-temperature annealing kinetics of E' centers in dry a-SiO_2 are consistent with the Edwards-Fowler model of an E' to $SiOO^-$ transformation mediated by the diffusion of dissolved molecular oxygen.

ACKNOWLEDGMENTS

The author is grateful to M. Harmatz, P. Caplan, D. Brocking, E. Potenziani, and J. Conrad for sample preparations; to J. Rodgers, S. Morris and the staff of ATT Bell Laboratories (Murray Hill) for irradiations; to G. Gerardi for help with spin concentration determination; to D. R. Griscom for invaluable guidance on EPR, and to J. Klafter for enlightening discussions of CTRW analysis.

REFERENCES

1. P. M. Lenahan and P. V. Dressendorfer, J. Appl. Phys. 55, 3495 (1984).
2. R. L. Pfeffer, J. Appl. Phys. 57, 5176 (1985) and refs. cited.
3. R. A. B. Devine and C. Fiori, J. Appl. Phys. 58, 3368 (1985).
4. A. Golanski, R. A. B. Devine, and J. C. Oberlin, J. Appl. Phys. 56, 1572 (1984).
5. A. H. Edwards and W. B. Fowler, Phys. Rev. B26, 6649 (1982).
6. D. B. Brown, D. I. Ma, C. M. Dozier, and M. C. Peckerar, IEEE Trans. Nucl. Sci. NS-30, 4059 (1983); A. Golanski and T. Nicolle, in Induced Defects in Insulators, ed. by P. Mazzoldi (Materials Research Society, 1984), p.255.
7. T. R. Waite, Phys. Rev. 107, 463 (1957).
8. R. A. B. Devine, J. Appl. Phys. 58, 716 (1984); A. Golanski and T. Nicolle, op cit.
9. O. L. Anderson and D. A. Stuart, J. Am. Ceram. Soc. 37, 573 (1954).
10. R. L. Pfeffer and D. L. Griscom, "Annealing Kinetics of Radiation-Induced Defect Centers in Pressure-Treated SiO_2: an EPR Study", presented at the Conference on Radiation Effects in Insulators 3, Guildford, Surrey, UK, 15-19 July 1985 (unpublished).
11. R. L. Pfeffer, "Transformation of Radiation-Induced Defect Centers as a Probe of Molecular Diffusion in a-SiO_2", in Proceedings, Conference on the Physics and Technology of Amorphous Silica, Les Arcs, France, 29 June - 3 July 1987, ed. by R. A. B. Devine (in press).
12. F. J. Norton, Nature 191, 701 (1961).
13. A. Blumen, G. Zumofen, and J. Klafter, Phys. Rev. B30, 5379 (1984).
14. S. Redner and K. Kang, J. Phys. A: Math. Gen. 17, L451 (1984).
15. A. Blumen, J. Klafter, and G. Zumofen, "Models for Reaction Dynamics in Glasses", in Optical Spectroscopy of Glasses, ed. by I. Zschokke (D. Reidel Publishing Co., 1986) p. 199.
16. E. W. Montroll and G. H. Weiss, J. Math. Phys. 6, 167 (1965).

CURRENT-INDUCED CHARGES AND HYDROGEN SPECIES DISTRIBUTIONS IN MOS SILICON DIOXIDE FILMS

R. Gale, H. Chew, and F.J. Feigl

Department of Physics and Sherman Fairchild Center
Lehigh University, Bethlehem, PA. 18015

C.W. Magee

David Sarnoff Research Center
Princeton, NJ 02954

The effect of low temperature incorporation of H_2O and D_2O on ultradry thermal silicon dioxide films has been studied using chemical analysis by secondary ion mass spectrometry and oxide capacitor current flow by avalanche injection of electrons from the silicon substrate. Hydrogen was transported from the metal interface to the silicon interface during current flow and diffused into the silicon to form electrically inactive complexes with boron donors. The presence of hydrogen near the silicon interface enhanced the production of traps at this interface during current flow. Also during current flow, electrons were captured by oxide defects containing hydrogen or deuterium impurities. The capture cross sections for defects involving the two isotopes differed by an order of magnitude.

INTRODUCTION

A number of investigators who have studied the electrical properties of metal-oxide-silicon (MOS) devices have concluded that the passage or production of electron currents within the silicon dioxide layer induces a redistribution of hydrogen impurities within the device (1-5). Although the association of mobile hydrogen species with electrical phenomena remains controversial, an accumulation of hydrogen at the silicon dioxide-silicon interface during avalanche electron current injection has been directly demonstrated (6,7). The present study has extended this work, using refined sample preparation and analytical techniques.

EXPERIMENTAL DETAILS

Boron-doped (100) silicon wafers subjected to ultra dry thermal oxidation were used exclusively in these experiments. The Si wafers were oxidized at 1000°C to a thickness of 80-90 nm in a double walled oven tube with dry N_2 continuously flowed through the outer chamber. The water vapor concentration in the wafer ambient was below 1 part per million by volume at all times that the wafers were at temperatures in excess of 100°C. The ultra-dry oxidation procedures used in this study, as well as the water diffusion and electrical measurement procedures described below, have been described in detail (8).

The controlled, reproducible water contamination of the ultra-dry oxides was accomplished using the technique developed by Nicollian, et al. (1). The "bulk diffused" oxides were exposed to a 150 Torr partial pressure of H_2O or D_2O vapor at 200°C for 10 minutes. The "surface exchanged" oxides were treated with H_2O or D_2O vapor at 150 Torr and 60°C for 10 minutes. After cooling 15 minutes, the wafers were removed to either the aluminum evaporator or a storage box. Oxide films were also prepared with all possible combinations of 200°C diffusion treatment (H_2O or D_2O) followed by 60°C surface exchange treatment (H_2O or D_2O).

Immediately following the final treatment to be received, wafers were metallized with aluminum vapor deposited from a low sodium filament source through a circular electrode contact mask. The completed MOS capacitors were tested for mobile ion contamination and general oxide film quality by bias-temperature-stress measurements (1.5 MV/cm, 200°C, 300 sec) using 1 MHz capacitance versus voltage (C-V) characteristics. The oxides are identified by a two letter code indicating the diffusion species (first letter, 200°C treatment) and the exchange species (second letter, 60°C treatment). "H" in the sample code denotes H_2O, "D" denotes D_2O, and "X" denotes no treatment. The samples are listed in Table I below.

Selected capacitors on each of the wafers were avalanche injected using a high voltage constant current network based on that described by Nicollian and coworkers (1). On a given capacitor, injection was continued at a current density of 68 $\mu A/cm^2$ to a predetermined fluence (integrated current density). The 1 MHz C-V flatband voltage was automatically recorded at pre-set time intervals throughout a given avalanche injection run. A full 1 MHz C-V characteristic was obtained before and after each run. A number of capacitors on each wafer were injected to different fluences. On selected capacitors, additional electrical characterization was obtained.

Uninjected capacitors and the avalanche-injected capacitors were profiled for hydrogen. Hydrogen concentration vs. depth profiles were obtained by secondary ion mass spectrometry (SIMS) performed using Cs^+

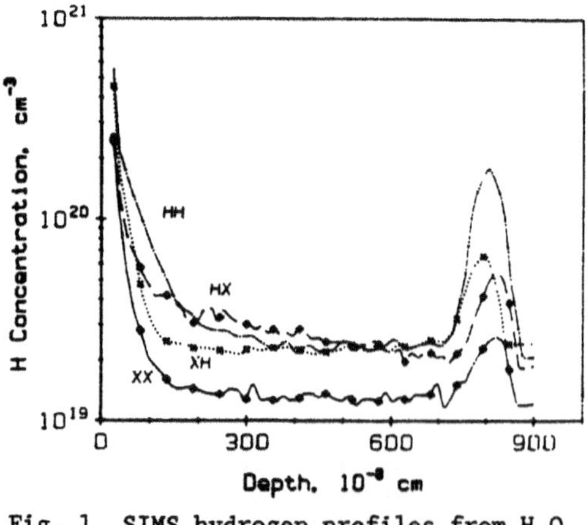

Fig. 1 SIMS hydrogen profiles from H_2O diffused and exchanged oxides.

primary ions and monitoring H^- and D^- secondary ions. In order to eliminate profile artifacts and degradation of depth resolution, the Al electrodes were removed chemically before SIMS profiling. The SIMS measurements have been described in detail (8) and are generally similar to those reported previously in which Ar^+ primaries and H^+ secondaries were used (6,7). The present procedures resulted in significantly less scatter in the profile data.

EXPERIMENTAL RESULTS

The basic experimental results were hydrogen and deuterium profiles within the silicon dioxide layer of Al-gate MOS capacitors, as shown in Figs. 1 and 2. In each of these figures, several profiles (individual curves) are shown together for comparison purposes. The origin of the depth scale is the Al-SiO_2 interface of the original capacitor. The SiO_2-Si interface is located at a depth of 80-90 nm. This interface was determined from oxygen concentration profiles, and differs for each sample. Each curve, representing a specific sample, was drawn through approximately fifty data points. For clarity, a small number of individual points are shown. SIMS data within the 2.5 nm nearest the aluminum electrode has been discarded to avoid surface contamination artifacts. The profiles shown are representative of those obtained from several capacitors on a given oxide film.

Water Diffusion And Exchange

Figure 1 shows the effect of H_2O diffusion and/or exchange on the hydrogen (^1H) concentration in silicon dioxide films. The diffusing species was H_2O. Figure 2 shows deuterium (^2D) profiles for the oxides diffused with D_2O at 200°C. The samples exchanged with D_2O at 60°C (XD, HD) are not presented in Fig. 2 because deuterium levels in these samples were on the order of the SIMS background signal. The ultra-dry hydrogen profile for sample XX shown in all figures may be background limited. Throughout this report, therefore, the XX data will be used as a baseline profile in evaluating the effects of low temperature water contamination and electron current injection on ultra-dry high temperature thermal oxides.

The hydrogen and deuterium profiles were integrated numerically to

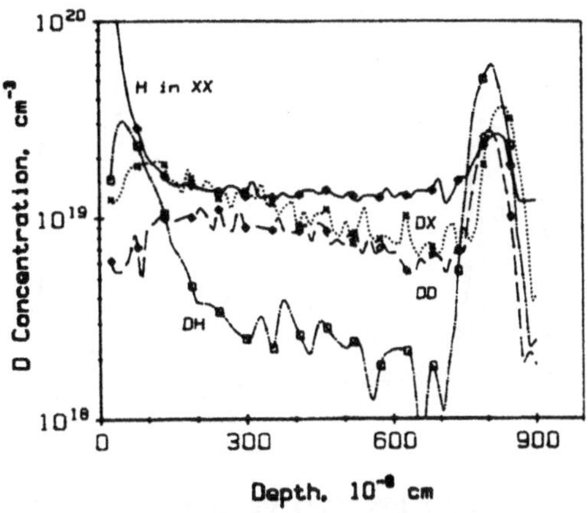

Fig. 2 SIMS deuterium profiles from D_2O diffused oxides DX, DD, and DH.

TABLE I. The effects of low temperature post-oxidation water diffusions on the oxide hydrogen content and on hydrogen redistribution.

SAMPLE	TREATMENT		OXIDE BULK H AND D CONCENTRATIONS[a]			INTERFACE
	$200^\circ C$	$60^\circ C$	$H_{ox} - H_{ox}^{XX}$	D_{ox}	TOTAL	dH_{os}/dF_a
			(units of $10^{13} cm^{-2}$, error ± 20%)			(± 20%)
XX	--	--	--	--	--	0.5×10^{-6}
XH	--	H_2O	6	--	6	4×10^{-6}
XD	--	D_2O	7	--	7	3×10^{-6}
HX	H_2O	--	8	--	8	30×10^{-6}
HH	H_2O	H_2O	10	--	10	43×10^{-6}
HD	H_2O	D_2O	6	--	6	41×10^{-6}
DX	D_2O	--	5	6	11	13×10^{-6}
DD	D_2O	D_2O	2	4	6	10×10^{-6}
DH	D_2O	H_2O	11	1	12	18×10^{-6}

a. Baseline Data: $H_{ox}^{XX} = (7 \pm 1) \times 10^{13} cm^{-2}$

determine areal densities of hydrogen (H_{ox}) or deuterium (D_{ox}) within the mid-film region of the oxide films. The integration region extended from depth 20 nm to depth (d - 15 nm), where d is the measured oxide film thickness. The results of these analyses are summarized in Table I. The areal densities for hydrogen listed in Table I are corrected by subtracting the corresponding baseline densities determined for the ultra-dry sample XX. These results indicate that both $60^\circ C$ and $200^\circ C$ H_2O diffusion treatments were effective in producing H species within the silicon dioxide film bulk. The $200^\circ C$ treatment was somewhat more effective, as indicated

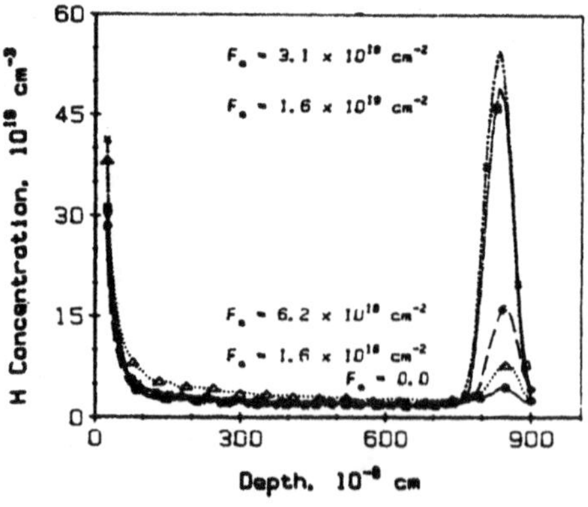

Fig. 3 SIMS hydrogen profiles from D_2O diffused oxide DX after electron injection to different fluences.

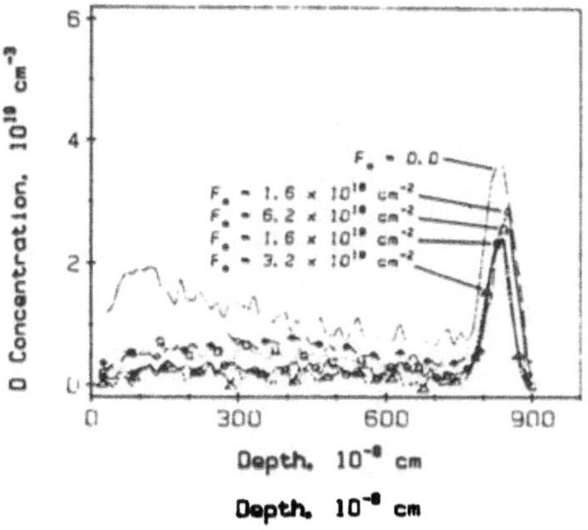

Fig. 4 SIMS deuterium profiles from D_2O diffused oxide DX after electron injection to different fluences.

by a direct comparison of XH, HX, and HH profiles in Fig. 1. The DH sequence (D_2O at 200°C followed by H_2O at 60°C) removed 5×10^{13} cm^{-2} of deuterium from the oxide bulk (see D_{ox} for samples DX and DH in Table I). Most of this deuterium appears to have been redistributed within the oxide film (see the DX and DH profiles in Fig. 2). Control of the oxide surface species by the D_2O diffusion and exchange treatments was not successful. Hydrogen and deuterium profiles taken through the aluminum electrodes indicated a net surface deuterium concentration only two orders of magnitude over the 0.01 percent natural abundance in D_2O diffused or exchanged oxide films, and thus only one percent of the surface hydrogen.

Avalanche Electron Injection

Figure 3 is representative of the results obtained from the SIMS analyses of the avalanche injected capacitors. In this figure are plotted

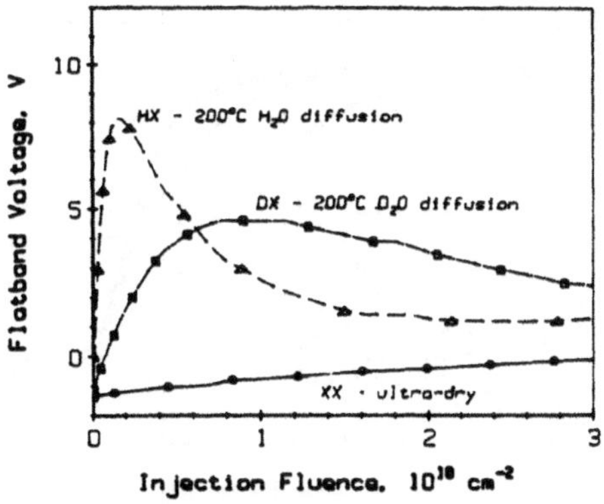

Fig. 5 Oxide charging curves for ultra-dry, D_2O diffused, and H_2O diffused oxides in MOS capacitors.

the hydrogen concentration versus depth profiles from several capacitors on a given oxide film. The several capacitors were injected to different fluences of electron current, as indicated. F_a is a parameter specifying the fluence in units of electrons/cm^2. The results shown in Fig. 3 are similar to those reported previously (6,7): a significant change of the hydrogen concentration profile is produced by avalanche-injected electron current. In all cases, a peaked distribution approximately triangular in shape and with a full width at half maximum amplitude of 7 nm develops. This peak is located at the silicon dioxide-silicon interface, and, as discussed below, both the peak height above background and the integrated areal density above background increase linearly with F_a.

Figures 3 and 4 show hydrogen and deuterium concentrations versus depth, respectively, for injected capacitors on the D_2O diffused oxide DX. Relative to the large change observed in the H concentration profiles in the same samples (Fig. 3), there is little change produced in the deuterium concentration profiles by avalanche injection (Fig. 4). If anything, there are indications of a net reduction in the D-species concentration. Similar results were obtained for samples DH and DD.

For each sample, the interfacial hydrogen areal density, H_{os}, has been computed as follows: A baseline extended from the H concentration data measured in the oxide bulk was subtracted from the data in the vicinity of the silicon dioxide-silicon interface. The remaining volume concentration vs. depth profile was integrated to yield H_{os}, the areal density of hydrogen at that interface. H_{os} has been related to the avalanche injection fluence, F_a, in a linear fashion as $H_{os} = a + bF_a$ and the coefficients a and b giving the least squares best fit determined. The transport efficiencies (dH_{os}/dF_a, which is the coefficient b in the linear fit above) are listed in Table I above.

Electrical Characterization

The oxide film charging characteristics were obtained from the variation of the 1 MHz flatband voltage with the injection fluence F_a. Figure 5 shows the measured flatband variation for the ultradry and the diffused oxide films. The ultradry charging curve XX shows that the initial oxide was relatively trap free. The surface exchanged oxides XH and XD exhibited similar characteristics in the fluence regime shown in the

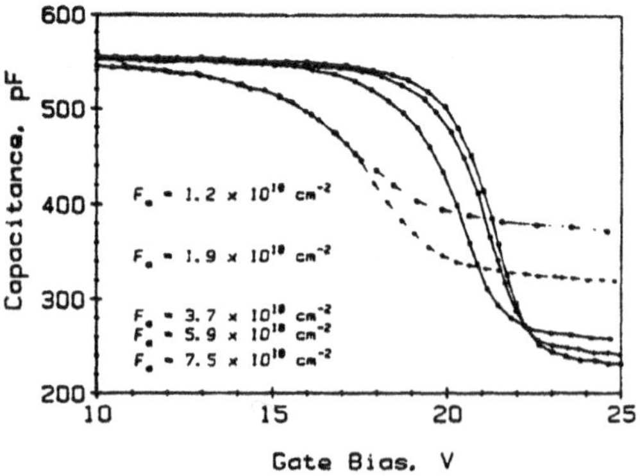

Fig. 6 High frequency C-V curves from an H_2O-diffused oxide injected to different fluences.

TABLE II. The effects of low temperature post-oxidation water diffusions on oxide charge trapping and silicon acceptor neutralization.

SAMPLE	OXIDE BULK AND INTERFACE ELECTRON AND HOLE TRAPS						SUBSTRATE
	Type [a]	Density (cm-2)	Cross Section (cm2)	Type [a]	Density (cm-2)	Cross Section (cm2)	$-dN_a/dt$ ($cm^{-3}s^{-1}$)
XX	--	--	--	i;e	$7x10^{11}$	$2x10^{-19}$	$0.3x10^{11}$
XH	--	--	--	i;e	$8x10^{11}$	$4x10^{-19}$	$2.3x10^{11}$
XD	--	--	--	i;e	$9x10^{11}$	$1x10^{-19}$	$2.0x10^{11}$
HX	b;e	$4x10^{12}$	$1x10^{-17}$	i;h	$3x10^{12}$	$2x10^{-18}$	$15x10^{11}$
HH	b;e	$4x10^{12}$	$1x10^{-17}$	i;h	$4x10^{12}$	$2x10^{-18}$	$16x10^{11}$
HD	b;e	$5x10^{12}$	$1x10^{-17}$	i;h	$4x10^{12}$	$2x10^{-18}$	$17x10^{11}$
DX	b;e	$5x10^{12}$	$2x10^{-18}$	i;h	$4x10^{12}$	$8x10^{-19}$	$10x10^{11}$
DD	b;e	$4x10^{12}$	$2x10^{-18}$	i;h	$3x10^{12}$	$7x10^{-19}$	$7x10^{11}$
DH	b;e	$4x10^{12}$	$2x10^{-18}$	i;h	$4x10^{12}$	$6x10^{-19}$	$11x10^{11}$

a. Legend for traps: b = bulk, i = interface; e = electron, h = hole

figure. The 200°C diffused oxides showed greatly enhanced trapping (see data for samples HX and DX in Fig. 5).

The shifts in the 1 MHz C-V characteristic can be related to the total centroid-weighted charge in the oxide, Q, using the relation $Q = -C_{ox}\Delta V$, where ΔV is the shift in the flatband voltage and C_{ox} is the oxide capacitance (9). Q was then analyzed using a standard multiple-trap model with first-order trapping and trap generation kinetics (9). In this model, the oxide charging curves (individual curves in Fig. 5) are

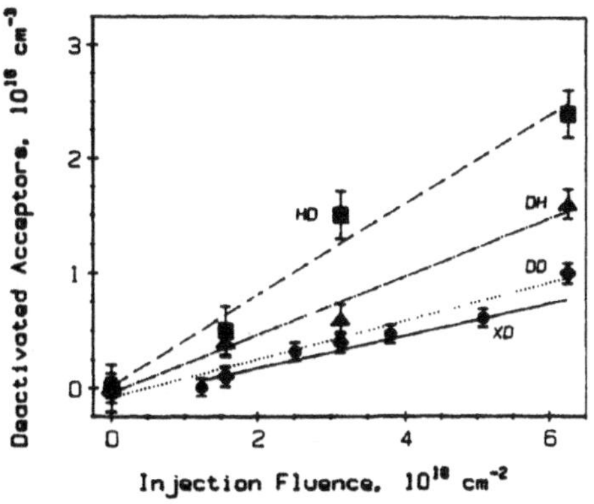

Fig. 7 N_{ad} as a function of F_a for diffused and exchanged oxides. The lines are fitted to data points.

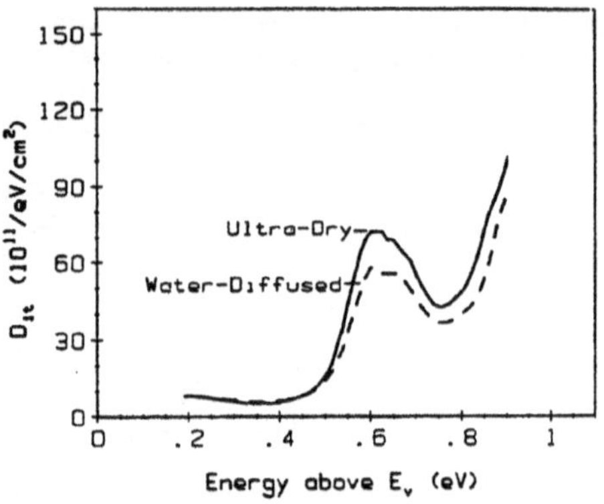

Fig. 8 Interface trap density D_{it} vs. energy for as-prepared XX (ultra dry) and HH (diffused) oxides.

interpreted as a sum of exponential curves representing a negative charge Q_{ot-b} (or positive ΔV) due to trapping of electrons in the oxide bulk and a generally positive charge Q_i due to generation of trapping centers at the silicon dioxide-silicon interface or within approximately 3 nm of this interface. The results of these analyses are presented in Table II.

Variations of the effective acceptor doping density with F_a were determined from the maximum and minimum capacitance values on C-V curves measured before and after each injection run. This analysis yields active acceptor concentrations N_a within the MOS depletion region of the silicon substrates (8). Figure 6 shows capacitance-voltage curves measured on an H_2O diffused oxide film. The deactivated acceptor concentration N_{ad} at a given fluence F_a was calculated by the equation $N_{ad}(F_a) = N_a(0) - N_a(F_a)$, where $N_a(0)$ and $N_a(F_a)$ were obtained at fluences 0 and F_a, respectively. For each sample, a straight line was fitted to such data (N_{ad} vs. F_a).

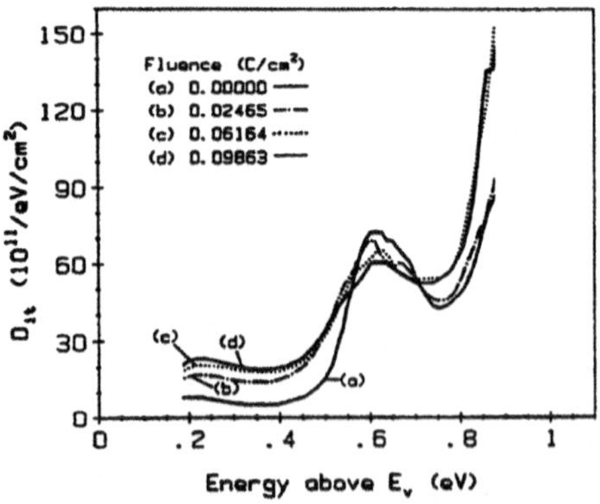

Fig. 9 Interface trap density D_{it} vs. energy for an XX oxide. Fluence varies from 0 to 6.2×10^{-17} cm^{-2}.

These data, along with the fitted curves, are shown in Fig. 7 for a selection of samples. The slope of the fitted curves are presented in Table II as dN_a/dt, where $t = F_a/eJ$ is the time of injection at current J.

Finally, Figs. 8-10 are density-of-states curves for electron traps at the silicon dioxide-silicon interface. These results were obtained using standard high frequency and low frequency capacitance measurements (9). The data for as-prepared oxides (Fig. 8) indicate that the 200°C H_2O diffusion did not effectively passivate the large concentration of interface traps produced by ultra dry oxidation. These traps are presumably the amphoteric dangling silicon interface defects described by Poindexter and his collaborators (10). However, the H_2O diffused oxides (Fig. 10) are much more susceptible than the ultra dry oxides (Fig. 9) to interface degradation, and specifically to the production of interface traps, during avalanche electron injection.

CONCLUSIONS

A striking result is the difference in the kinetics of bulk electron trapping between H_2O-diffused oxides and D_2O-diffused oxides. In all cases, 200°C H_2O diffusion produces a dominant trap with an electron capture cross section of 10^{-17} cm^2 (HX, HH, and HD in Table II). The cross section characteristic of D_2O-diffused oxides is 10^{-18} cm^2 (DX, DP, and DH in Table II). This large isotope effect argues against a simple model for electron capture based on defect energy levels, and supports the ideas of Young and Lai that capture involves a metastable state from which thermal detrapping and relaxation may occur (11). In a different context, Emin has argued that thermal rearrangement of hydrogen containing complexes may show large isotope effects (12).

The results on hydrogen distributions and redistribution in the oxide films (Figs. 1-4 and dH_{os}/dF_a in Table 1) and on boron acceptor deactivation in the silicon substrate (Fig. 7 and dN_a/dF_a in Table II) indicate that the source of the redistributed hydrogen must be, as Sah and co-workers suggested (4), the region immediately adjacent to the aluminum-silicon dioxide interface (0-2.5 nm in the depth scale on Figs. 3 and 4). First, there is a very strong correlation between hydrogen redistribution (dH_{os}/dF_a) and acceptor neutralization (dN_a/dt). This

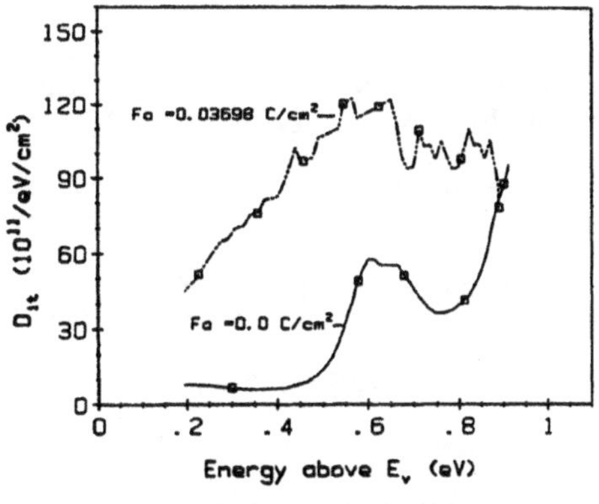

Fig. 10 Interface trap density D_{it} vs. energy for an HH oxide film at $F_a = 0$ and $F_a = 2.3x10^{-17}$ cm^{-2}

indicates that the silicon substrate is a sink for hydrogen released during electron current injection, since boron acceptor neutralization has been shown to involve formation of B-H complexes (13). Second, the SIMS profiles are sufficiently precise to indicate directly that the large amounts of interfacial hydrogen H_{os} produced by injection of water-diffused oxides to high electron fluences could not have originated within the measured part of the oxide (depth 2.5 nm to approximately 75 nm in Figs. 1-4).

Finally, the interface trap measurements (Figs. 8-10) indicate that either the presence of hydrogen-containing species at a water-diffused silicon dioxide-silicon interface, or the transport of such species to that interface during electron current injection, enhances interface trap formation during the avalanche injection process. In the course of this study, both ultra dry oxides and 200°C H_2O-diffused oxides were subjected to standard post-metallization anneal at 450°C in a hydrogen-containing ambient. This process resulted in (1) an oxide charging curve for both films which was indistinguishable from that reported above for an ultra dry oxide (XX in Fig. 5) and (2) a reduction of interface trap densities by one to two orders of magnitude over those reported above (Fig. 8).

Acknowledgements. This research was supported by the Office of Naval Research. Present addresses: R. Gale, Central Research Laboratories, Texas Instruments, Dallas, Texas 75265; H. Chew, AT&T Bell Laboratories, Allentown, Pa. 18103; C. W. Magee, Evans East, Plainsboro, N. J. 08536.

REFERENCES

1. E. H. Nicollian, C. N. Berglund, P. F. Schmidt, and J. M. Andrews, J. Appl. Phys., 42, 5654 (1971).

2. F. B. McLean, IEEE Trans. Nucl. Sci., NS-27, 1651 (1980).

3. F. J. Feigl, D. R. Young, D. J. DiMaria, S. Lai, and J. Calise, J. Appl. Phys., 52, 5665 (1981).

4. C. T. Sah, J. Sun, and J. Tzou, Appl. Phys. Lett., 43, 204 (1983).

5. D. L. Griscom, J. Appl. Phys., 58, 2524 (1985).

6. R. Gale, F. J. Feigl, C. W. Magee, and D. R. Young, J. Appl. Phys., 54, 6938 (1983).

7. F. J. Feigl, R. Gale, H. Chew, C. W. Magee, and D. R. Young, Nucl. Inst. and Meth. in Phys. Res., B1, 348 (1984).

8. R. Gale, Ph. D. Dissertation (Lehigh University, Bethlehem, Penna., 1984), unpublished. Available from University Microfilms.

9. F. J. Feigl, VLSI Microelectronics: Microstructure Science, Vol. 6, N. Einspruch and G. Larrabee, eds. (Academic, New York, 1983), p. 147.

10. E. H. Poindexter and P. J. Caplan, Prog. Surf. Sci., 14, 201 (1983).

11. D. R. Young, private communication.

12. D. Emin, M. Baskes, and W. Wilson, Phys. Rev. Lett., 42, 791 (1979).

13. G. G. DeLeo and W. B. Fowler, Phys. Rev. B, 31, 6861 (1985).

III. THE ATOMIC AND ELECTRONIC STRUCTURE OF THE Si-SiO$_2$ INTERFACE

INTRODUCTION

C. R. Helms

Department of Electrical Engineering
Stanford University, Stanford, CA 94305

Many of the properties that make the Si-SiO$_2$ system so unique relate to a region only tens of Angstroms thick at the interface between Si and SiO$_2$. Although there is still considerable uncertainty in our knowledge of the details of the interfacial region, some of the key factors are clear. First, it is possible and, in the last few years, quite a simple task to manufacture an interface with a negligible number of Si atoms with broken bonds. This property leads to device structures where the semiconductor surface potential is determined only by doping, oxide capacitance, and applied voltage. It has been generally accepted that this property is related to the small free energy differences accompanied by Si-O-Si bond angle changes from 120° to 180°. This is thought to allow a matching between amorphous SiO$_2$ and crystalline Si without the need for broken bonds. This idea may seem to be contradicted by the first paper of this chapter which shows strong evidence for a few layers of crystalline SiO$_2$ matched to a strained Si substrate lattice. However, the possible Si-O-Si bond angle variation is still required to allow for the structural relaxations that occur as the oxide grows (at least for high temperature grown thermal oxides). The possibility of strained Si bonds in the substrate has been nearly neglected over the years even though their presence was strongly suggested by high energy ion channeling results which are discussed by Feldman in the second paper. The possible effects of such strained Si bonds in the substrate have not been extensively investigated.

Cross-section TEM results as discussed by Ourmazd and Bevk give us a nearly atomic level view of the interface. However, it is important to recall that such a technique is based on the reconstruction of scattering data which is very sensitive to diffraction effects and can be difficult to interpret. A new technique, scanning tunneling microscopy (STM), in its various forms has the potential for viewing bonding at the Si-SiO$_2$ interface (at least for thin oxides) directly. Recent progress in this area is discussed in the next paper by Hamers and Koch.

The next five papers provide one of the most comprehensive discussions of the results from electron spectroscopy measurements of thin SiO$_2$ layers on Si assembled in one volume. These techniques have demonstrated that SiO$_x$ bonding configurations are found at or near the interface in concentrations ranging from 1 to 2 monolayers. The exact value of this concentration, the distribution over the various possible bonding configurations, and the effect of oxide growth methods and conditions have yet to be agreed upon. In addition, the origin of the "chemical shift" of the near interface SiO$_2$ related feature in the silicon spectra which has been interpreted as due to SiO$_x$, strained SiO$_2$, as well as an artifact of the spectroscopy is still not agreed upon. It would appear that further theoretical as well as experimental work (especially on ideal interfaces as discussed by Ourmazd and Bevk) will be necessary to establish the chemical nature of the near interface region.

One property related to both interfaces of the insulator which greatly affects MOS electrical properties is the built in potential between electrode and semiconductor. The paper by Massoud and Plummer discusses a dipole mechanism that is suggested to be responsible for variations as a function orientation and processing.

The final two papers of this chapter discusses the state-of-the-art of theoretical calculations of the atomic and electronic structure of the Si-SiO_2 interface and SiO_2 defects which can be compared to the experimental work discussed earlier and serve as a bridge to the next chapter on defects.

THE STRUCTURE OF THE Si/SiO_2 INTERFACE: A REVIEW

A. Ourmazd
AT&T Bell Laboratories, Holmdel, NJ 07733
and
J. Bevk
AT&T Bell Laboratories, Murray Hill, NJ 07974

ABSTRACT

We briefly review the structure of the Si/SiO_2 interface, so far as it is known, and present new results on the atomic structure of the interface, when atomically flat Si(100) surfaces are oxidised under a variety of conditions. In all cases we observe a crystalline interfacial layer. On the basis of diffraction, lattice imaging and photoemission spectroscopy data, we discuss the possible atomic configurations of the interface.

INTRODUCTION

The technological drive towards the fabrication of ever smaller and faster Si-based devices has resulted in extensive and sustained effort to determine the properties of the Si/SiO_2 system on an increasingly fine scale. The structure of this system in the close proximity of the interface is now reasonably well-established; the transition from c-Si to stoichiometric a-SiO_2 is complete within ~10 Å. The way this transition occurs on an atomic scale, and the structure of the Si/SiO_2 interface itself still remain to be established.

Often in order to explain the results of a particular set of experiments, a number of models for the structure of the Si/SiO_2 interface have been proposed. Broadly speaking, these models can be divided into three categories. In the first, the c-$Si \rightarrow a$-SiO_2 transition is thought to proceed via an ordered, stable, bulk phase of c-SiO_2, which bears an epitaxial relationship to the Si substrate. Since cristobalite is obtained by the simple insertion of oxygen atoms between the Si-Si bonds of c-Si, it has been the most popular candidate for an epitaxial oxide [1]. While structurally similar to Si, cristobalite has a lattice parameter 40% larger, and thus an epitaxial relationship between these two materials results in severe bond distortions. The second class of models rests on the proposal that the c-$Si \rightarrow a$-SiO_2 transition is affected through a "sub-stoichiometric oxide" layer, which, although absent from the bulk phase diagram, is nevertheless stabilised by the special conditions pertaining at the Si/SiO_2 interface [2]. Finally, it has been pointed out that it is structurally feasible to achieve an abrupt transition from c-Si to a-SiO_2 with no intervening layer [3].

Two general experimental approaches have been adopted to investigate the structure of the Si/SiO_2 interface. First, probes have been used which provide direct *structural* information. Electron diffraction and imaging, and most recently X-ray scattering with synchrotron sources, are examples of such structural probes. Diffraction techniques in general provide superior resolution and sensitivity, but imaging techniques have the important advantage of revealing directly the region from which the signal emanates. The second class of experiments, such as magnetic resonance and X-ray Photoemission Spectroscopy (XPS), provide *electronic* information, which are then used to deduce the structure of the interface. Since in the final analysis it is the electronic properties of the interface which are of primary interest to the device technologist, these latter techniques directly provide some of the information needed for device design. The use of electronic data, such as those obtained from XPS, to deduce the interfacial *structure*, however, often involves a further stage of interpretation and analysis, which strictly speaking requires complex electronic calculations of the type rarely undertaken.

In this paper, we attempt a brief review of the structure of the Si/SiO_2 interface, so far as it is known. Due to the overwhelming volume of the literature, our treatment is somewhat arbitrarily selective. Moreover, we discuss "electronic" probes such as XPS only in as much as they illuminate the data obtained by direct structural techniques.

X-RAY SCATTERING

The investigation of the structure of the Si/SiO_2 interface requires the ability to obtain clearly resolved signal from the interface in the presence of large contributions from the neighbouring c-Si and a-SiO_2. The recent emergence of bright synchrotron sources has made possible the X-ray study of interfaces in a number of ways. The grazing incidence geometry, where the incident beam is totally externally reflected from an interface, is capable of yielding information from Si/SiO_2 interfaces buried under varying thicknesses of differently prepared oxide layers [4]. The results reveal the presence of crystallites in the close vicinity of the interface, whose orientation is highly correlated with that of the Si substrate. This orientational order, which decays with distance from the interface, may be indicative of the way in which the crystalline to amorphous transition takes place. However, the question remains whether the transition to the oxide with its crystallites occurs directly from the c-Si substrate, or via an epitaxial c-SiO_2 layer. Nevertheless, these results are important, because many techniques, such as high resolution transmission electron microscopy have difficulty in characterising the decay of crystalline order into the amorphous state.

ELECTRON DIFFRACTION AND IMAGING

Electron probes are in many ways idealy suited for the investigation of Si/SiO_2 systems. The necessarily thin samples reduce the effect of non-interfacial material, and the strong interaction of electrons with matter ensures sufficient signal from the interface. Electron diffraction and imaging of the Si/SiO_2 system have traditionally been carried out in one of two configurations. In "plan-view" experiments, the electron beam impinging on the sample is perpendicular to the interface plane. This arrangement has the advantage of producing a two dimensional view of the interface, but cannot easily distinguish between signals emanating from the interface, and those from the neighbouring c-Si and a-SiO_2 layers. In "cross-sectional" experiments, the electron beam is parallel to a direction lying *in* the interfacial plane. Particularly when an image is formed, the position from which a particular signal originates can be easily determined, i.e. it is immediately possible to establish whether the signal comes from the interface, its close vicinity, or the neighbouring layers. On the other hand, only a one-dimensional projection of the interface is obtained, which can complicate data analysis.

Electron scattering experiments can also be conducted in less familiar configurations. For example, Fresnel fringe contrast has been used to study the change in the mean potential across the Si/SiO_2 interface [5]. While these techniques sometimes provide valuable information, one is often attempting to characterise the technique and the sample at the same time.

Electron Diffraction

Electron diffraction, in the conventional "plan-view" transmission [6], in the grazing reflection (RHEED) [7], and the low energy diffraction (LEED) [8] modes have been used to study the Si/SiO_2 interface. In several instances features characteristic of microcrystallinity have been clearly observed, and interreticular distances measured. Dark-field images, obtained by allowing only a few of the diffracted spots to pass through a small aperture, show them to arise from small, presumably crystalline regions ~100Å in size [6]. The nature and presence of these features in the diffraction pattern depend on the oxidation procedure. The deduced interreticular distances have not been consistently assigned to a particular oxide phase, and it has not been possible to deduce the exact location of the ordered regions with respect to the interface. It is thus possible that at least some of the diffracted spots correspond to ordered regions in the oxide layer itself, and may in part arise from Si inclusions in the a-SiO_2.

Lattice Imaging

Lattice imaging can, in principle, directly reveal the atomic structure of the Si/SiO_2 interface. In practice, great care is required in image formation, analysis and interpretation. Since the initial applications of lattice imaging to the study of the Si/SiO_2 system [9], the instrumental resolution has improved by more than a factor of two and now approaches 1.6Å, which is well beyond the level needed to resolve individual atom-columns in semiconductors in certain projections [10]. This, together with the availability of accurate and well-established algorithms for image simulation [11] have in part alleviated some of the hazards involved in image interpretation. Nevertheless, as with any technique, an uncritical analysis of lattice images can lead to erroneous conclusions, either through over-interpretation, or by missing vital clues encased in the wealth of data in the image.

Fig. 1 is typical of published lattice images of the Si/SiO_2 interface. The sample is viewed in cross-section, with the electron beam parallel to a <110> direction lying in the plane of the interface. Each blob on the image represents a *pair* of atom-columns, which, depending on the sample thickness and imaging conditions, can appear black or white. Both the c-Si substrate and the a-SiO_2 layer are clearly observed, but no evidence of an intervening interfacial layer is present. Such images have usually been taken to mean that no ordered crystalline phase is present at the Si/SiO_2 interface. Since it is not possible to determine by lattice imaging the atom positions in a highly disordered material such as a-SiO_2, lattice imaging has traditionally remained mute regarding the exact atomic configuration at the interface. Depending on the sample preparation conditions, small crystallites have been observed both

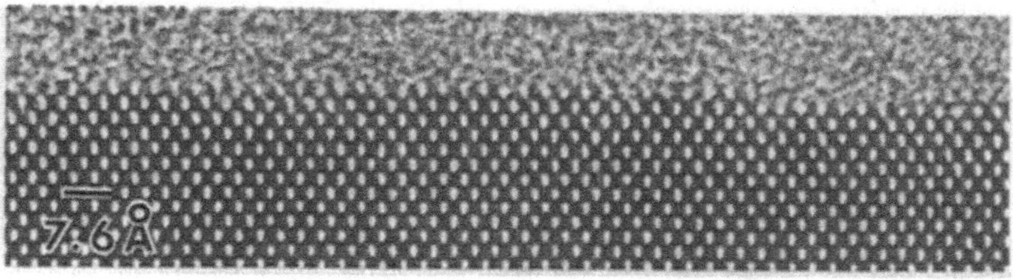

Fig. 1 Typical <110> lattice image of Si/SiO_2 interface.

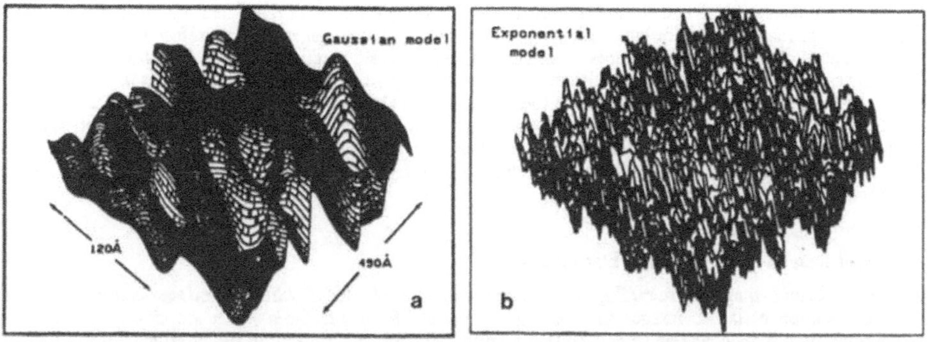

Fig. 2 (a) Two-dimensional random surface generated from a Gaussian model, with an rms surface fluctuation of 3Å, and "mean spacing between steps" of 20Å [12]. (b) Surface generated assuming an exponential model with the same parameters. The exponential model provides a better description of the observed roughness.

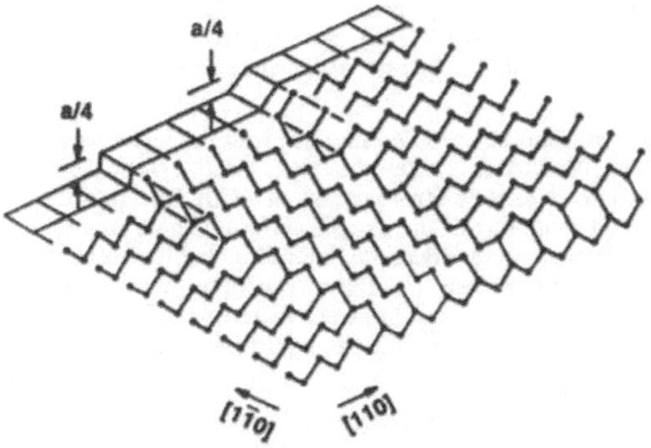

Fig. 3 Schematic representation of the Si(100) surface. Note that the back bonds, and hence the dangling bonds, rotate by 90° on crossing a step an odd number of monolayers high.

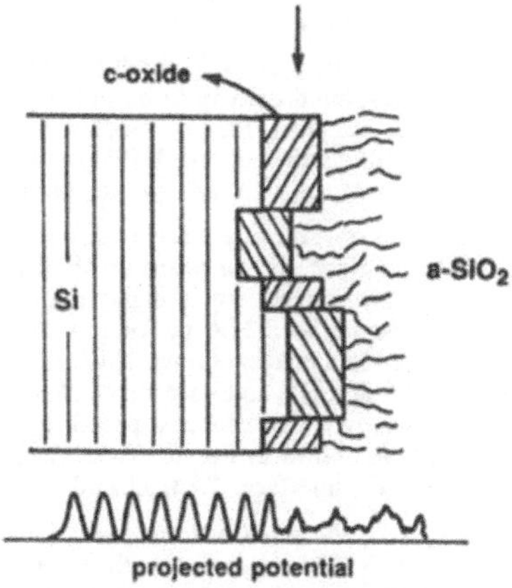

Fig. 4 Schematic representation of $c\text{-}Si/c\text{-}SiO_2/a\text{-}SiO_2$ interface, viewed in cross-section. Monolayer interfacial steps cause 90° rotations in the Si bonds and hence the $c\text{-}SiO_2$ grains. Thus, unless the $c\text{-}SiO_2$ is four-fold symmetric, the atomic columns in the difference $c\text{-}SiO_2$ grains do not lie on top of each other.

at the interface and in the amorphous oxide layer [6]. The significance of these isolated observations for the atomic structure of the interface is not clear.

The problem of interfacial *roughness*, however, itself of importance for device performance, has been characterised after a variety of oxidation treatments [12,13]. In one of the most sophisticated applications of quantitative image analysis to lattice imaging, Goodnick et al. have obtained statistical information regarding interfacial roughness on an atomic scale [12]. Such sophisticated approaches are necessary, because one is attempting to reconstruct a three-dimensional picture of the interface from a finite sample of its cross-sectional view, and because the absence of order on one side of the interface removes many of the geometric constraints which help narrow the range of possibilities for the interfacial atomic configuration in epitaxial crystalline structures [14]. Goodnick's analysis shows the "standard" Si/SiO_2 interface to be extremely rough. This roughness is analysed in terms of its autocovariance, which is shown to be non-Gaussian and characterised by an rms fluctuation of ~3Å, with a correlation length of ~20Å. In less precise but physically more transparent terms, the interface has a mean spacing of ~20Å between ~3Å high steps. Fig. 2 is a two-dimensional view of such an interface, generated by a computer assuming an exponential model for the autocovariance [12]. Note that the dimensions of interfacial segment shown are comparable with the thickness of typical cross-sectional specimens used for lattice imaging. The resulting lattice images must therefore be considered projections of such rough interfaces. The effect of such extreme roughness on the formation of lattice images, where multiple scattering (dynamical) effects dominate, is far from clear. That is to say, while lattice images have been used to deduce the interfacial roughness shown, the final check necessary for self-consistency, namely the simulation of the effect of such roughness on lattice images has not been conducted. The exact effect of substantial roughness on lattice images of Si/SiO_2 interfaces is thus unknown. Simplistic interpretations of the effect of interfacial roughness on lattice images [15] must therefore be treated with caution.

Recent Results from Atomically Flat Interfaces

An important question regarding the atomic structure of the Si/SiO_2 interface concerns the interpretation of lattice images such as Fig. 1, obtained from interfaces which are clearly far from smooth. Is it indeed correct to deduce that no crystalline phase is present at the interface? Fig. 3 is a schematic representation of the Si(100) surface, with the lines indicating the back-bonds the surface atoms make with their neighbours. The surface dangling bonds are at 90° to the back bonds, and it is these dangling bonds which determine the way an epitaxial phase could be attached to the Si substrate. As shown in Fig. 3, the crossing of a surface step an odd number of monolayers high causes a 90° rotation in the direction of the surface dangling bonds, and thus an epitaxial oxide would also have to rotate through 90° on crossing a step. Consequently, unless the crystalline oxide is four-fold symmetric, it would necessarily be poly-crystalline, with a grain size determined by the spacing between the surface steps. Fig. 4, depicts schematically the situation when a sample with a non-four-fold symmetric interfacial oxide is imaged in cross-section. The overlap of the differently oriented interfacial oxide grains is likely to make their clear observation difficult.

The recipe for a careful search for the presence of an interfacial epitaxial phase is now clear. The interfacial step density must be so low that a single oxide grain extends over the thickness of the cross-sectional sample (~100Å). Modern Molecular Beam Epitaxy (MBE) can produce surfaces of this quality. In particular, the MBE growth of Si on a Si(100) substrate under suitable conditions can produce samples with very low step densities.

Fig. 5 is a transmission electron diffraction pattern from the Si/SiO_2 interface with the beam parallel to the interface normal, when an MBE grown Si(100) surface is allowed to form a native oxide [16,17]. This diffraction pattern is highly unusual, because it displays strong peaks at (110) type positions, corresponding to a 3.8Å periodicity. A 20 second sputtering of the sample surface with an Ar beam, estimated to damage/remove ~20 Å of material destroys these reflections, indicating that they originate from the Si/SiO_2 interface or its close vicinity. The exact origin of these reflections, not previously observed from Si surfaces exposed to air, or oxidised, is at once apparent in cross-sectional lattice images, as shown in Fig. 6. The $c\text{-}Si/a\text{-}SiO_2$ interface is atomically flat and displays a remarkable phase at the interface ~7Å thick [16].

This phase, characterised by its strong 3.8Å periodicity, is observed under essentially all imaging conditions and sample thicknesses, unless the interface has been damaged by ion-milling in very thin regions of the foil (<50 Å), or by extended electron beam irradiation. The possible role of imaging artefacts, such as lens defocus and dynamical effects, in giving rise to the observed interfacial images has been discussed in detail in a previous publication [17]. Briefly, it can be shown on quite general grounds, that such effects can only produce artefacts whose spatial frequencies are the same or *higher* than the fundamental frequency present in the sample. The new phase observed at the Si/SiO_2 interface has a frequency of $(3.8Å)^{-1}$, which is *half* that of the Si lattice, $(1.9Å)^{-1}$. From such considerations, and also because the phase is observed in electron diffraction, artefacts can be ruled out as the origin of the observed interfacial layer.

Fig. 5 Montage showing transmission electron diffraction pattern of MBE *Si* (100)/SiO_2 sample in plan-view. Note the presence of sharp spots at arrowed positions, corresponding to a 3.8Å periodicity.

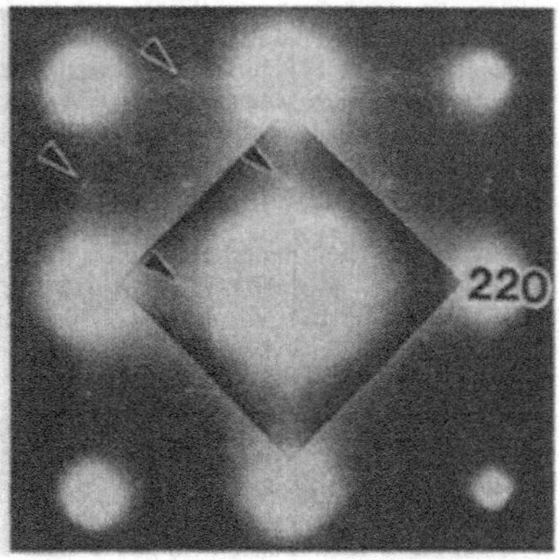

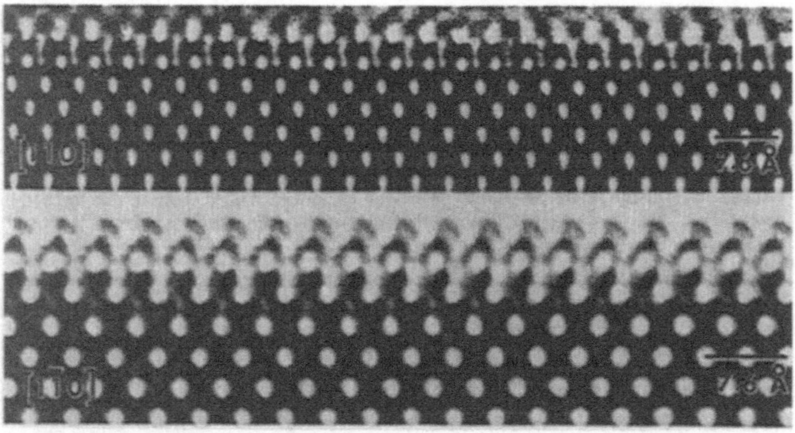

Fig. 6 [110] and [$\bar{1}$10] lattice images of an atomically flat Si/native oxide interface. Note the presence of a new phase at the interface, displaying 3.8Å periodicity in both projections. However, the [110] and [$\bar{1}$10] images are different.

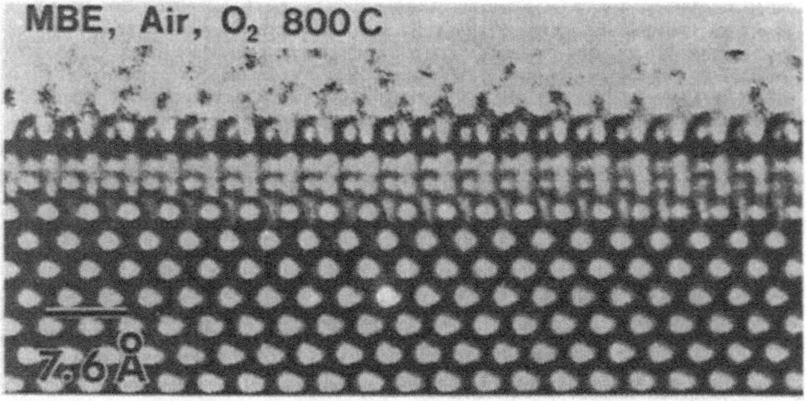

Fig. 7 [$\bar{1}$10] lattice image of Si/native oxide sample annealed in oxygen at 800 C.

The lattice images appear different in the [110] and [$\bar{1}$10] projections. This indicates the interfacial phase not to be four-fold symmetric and, as discussed earlier, helps explain the previous lack of success in its detection by HRTEM. The crystalline phase displays a characteristic common to the oxides of Si, namely that its lattice images change (but do not disappear) with changing sample thickness and lens defocus. Nevertheless, this phase can always be identified by its strong 3.8Å periodicity, which acts as its hallmark both in diffraction and lattice imaging, and helps its ready distinction from the neighbouring c-Si and a-SiO_2 layers.

The question arises whether the crystalline interfacial phase is peculiar to the MBE Si(100)/native oxide interface, or it occurs more generally. Figs. 6 - 9 show the interfacial structure formed when MBE grown Si(100) is oxidised at different annealing temperatures, and after a variety of steps characteristic of device processing. These lattice images establish a number of interesting facts. (For details see [18]).

1. The interfacial crystalline phase is present even when the sample is oxidised at temperatures as high as 900 C (the highest temperature we have so far investigated). At least in the case of samples where the native oxide is not removed before high temperature oxidation, the crystalline phase covers the entire interface.
2. The non-four-fold nature of the oxide can sometimes be directly observed. As pointed out above, crossing a monolayer step causes a 90° rotation in the Si bonds at the interface, but the bonds across a bilayer step retain the same orientation. This implies that the epitaxial oxide, which is tied to the substrate dangling bonds, must undergo a 90° rotation across a monolayer step, but not across a bilayer step. Fig. 8 shows the behaviour of the crystalline phase on crossing a monolayer and a bilayer step. Note that each blob represents a pair of atom-columns. In Fig. 8b, the image of the interfacial oxide is the same on both sides of the bilayer step; it consists of vertical pairs of white dots, separated from each other by 3.8Å along the interface. In Fig. 8c, however, the lattice image of the interfacial phase changes abruptly across a monolayer step, but retains its 3.8Å periodicity. This behaviour is repeated in reverse further along the interface, with the phase reverting to its original image, and the net result is a change in the interface position of one bilayer. These results directly show that a 90° rotation of the Si bonds at the interface changes the image of the interfacial phase, thus demonstrating the phase not to be four-fold dymmetric. This had been previously surmised from the different images in the [110] and [$\bar{1}$10] images, obtained form the Si/native oxide sample (Fig. 6).
3. High temperature oxidation may induce a slight increase in the interfacial roughness. This conclusion must be treated with caution due to the small density of steps and the consequent difficulty in obtaining reliable statistics.
4. A significant increase in the interfacial roughness is produced when the native oxide is removed in HF before oxidation. The roughest interface is produced when the Si/native oxide sample is etched in a 10% solution of HF/HNO_3 before oxidation. At the roughest interfaces, the crystalline phase is observed only in thicker regions of the cross-sectional sample and appears somewhat disturbed (Fig. 9). This and the larger width of the interfacial image in Fig. 9 may be due to the cross-sectional projection of a rough interface, where the overlap of several differently oriented grains, each on a different interface terrace, can give rise to an apparently thicker and less well-ordered image of the interfacial phase. The apparent absence of the crystalline phase in thinner regions of the sample may also indicate genuine interfacial disorder, or lack of complete coverage of the interface.
5. In all samples we observe a remarkable uniformity in the total oxide thickness. Even at rough interfaces, the oxide surface closely follows the undulations of the Si/SiO_2 interface. Thus we do not observe the substantial changes in the interfacial roughness sometimes reported when polished Si surfaces are oxidised [13]. This emphasises the advantage of starting from a surface which is known to be atomically perfect and free from damage prior to oxidation, thus eliminating possible uncertainties due to initial surface damage and roughness.

These results establish that when atomically flat Si(100) is oxidised after a variety of treatments, an epitaxial phase is present at the interface, but that the ease with which it can be observed deteriorates with increasing interfacial roughness. It is now important to attempt to identify the structure of this interfacial phase.

The Atomic Structure of the Interface

Due to the difference in the atomic numbers of Si and O, oxides of Si display complex lattice images, whose features do not in general bear a simple relationship to the atomic positions. The determination of the structure of the interfacial phase must therefore proceed with the matching of experimental images with simulated images of trial structures. Assessment of the reliability of the resultant structural models is complicated by the absence of quantitative means for judging the success of a fit, and the problem of uniqueness. In the following, we describe the procedure we have used to determine the structure of the Si/SiO_2 interface. We cannot be certain that the resultant model is unique.

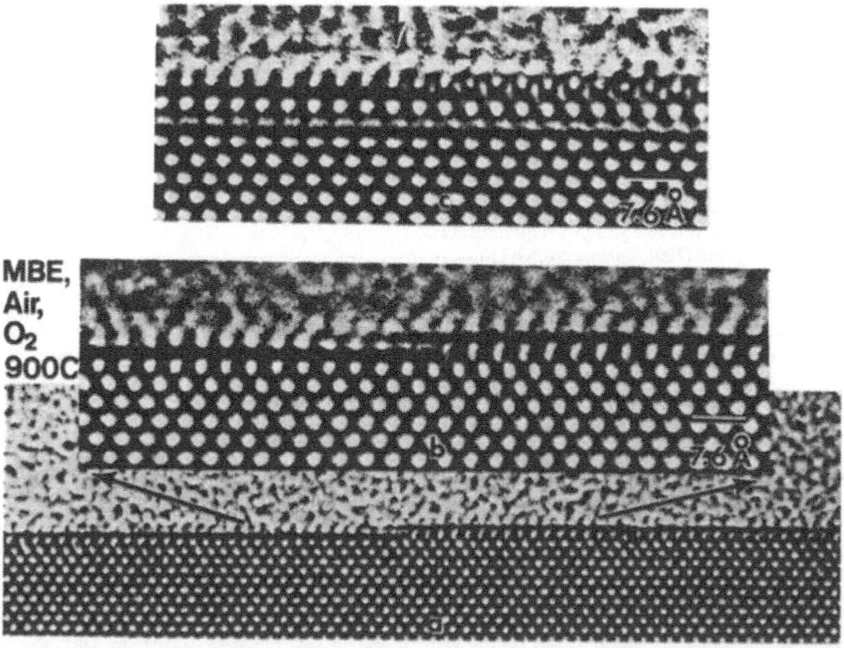

Fig. 8 (a) <110> lattice image of Si/native oxide sample annealed in oxygen at 900 C. Each blob represents a pair of atom-columns. Note the bilayer step in the middle of the image. (b) The bilayer step shown at higher magnification. The interfacial phase produces the same image on both sides of the bilayer step. (c) Lattice image of the same interface around a monolayer step (arrowed). Note the change in the image of the interfacial phase across the monolayer step.

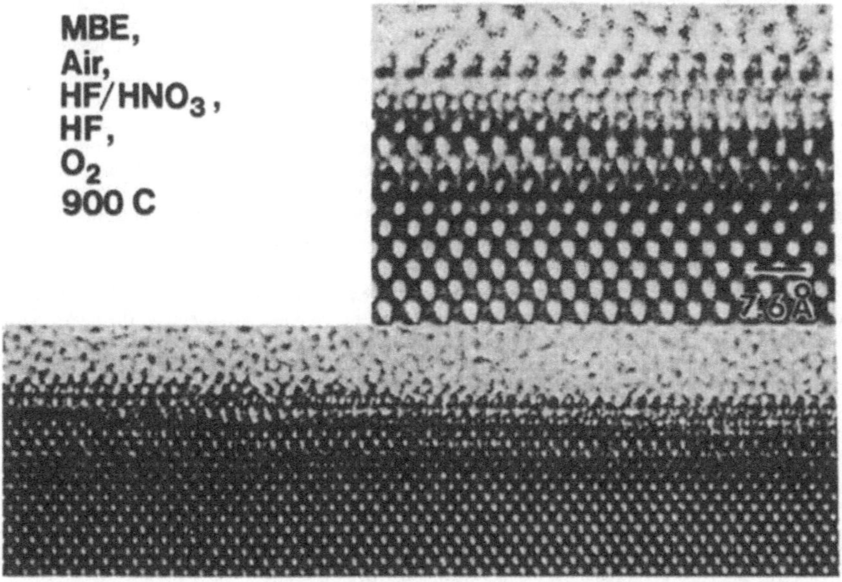

Fig. 9 Lattice image of sample annealed at 900 C after etching in HF/HNO_3 and oxide removal in HF. Note the rough interface and the presence of the interfacial phase, which appears more disordered.

To determine the structure of the phase observed at the interface, we have conducted a systematic search of all known phases of Si (including possible but undiscovered high pressure analogues of Ge), as well as all known phases of SiO_2, as follows. To establish the candidacy of a particular phase, we required that it exhibit two planar spacings within 20% of 3.8 Å, with the angle between these planes lying within 15° of a right angle. In this way we were seeking phases which could, with or without distortion, reproduce the observed diffraction pattern. When such a candidate was found, we attempted to obtain a match between its simulated images and our experimental lattice images. It is noteworthy that, while every known phase of SiO_2 satisfied the candidacy requirements on the basis of diffraction information alone, only one phase, tridymite, gave rise to simulated images which even resembled our experimental images. Tridymite is a stable form of SiO_2 with a hexagonal structure (a=5.046, c=8.23 Å) [19,20]. The orientational relationship between the Si substrate and the tridymite satisfying the diffraction and imaging requirements is:

Si (110) || tridymite (100), mismatch 13.2%;
Si (100) || tridymite (002), mismatch 6.7%;
Si [001] || tridymite [010].

Starting with the atom positions of perfect, high tridymite [20], we have refined the interfacial structure by comparing experimental and simulated images over the thickness range 50 - 250Å and the defocus range -100 - -800Å, seeking fits in the two orthogonal [110] and [$\bar{1}$10] projections. Fig. 10 shows the deduced three-dimensional map of atom positions at and near the interface.

Assuming the oxide indeed to have the tridymite structure, the following observations can be made. The crystalline oxide is 5 monolayers (ML) thick, where each ML denotes a layer of atoms at a particular height above the interface plane. The in-plane lattice spacings of tridymite are compressed to match those of the Si substrate and provide perfect epitaxy. Detailed matching of experimental and simulated images suggests that the tridymite lattice sites in the third layer have an occupancy of ~3/4, while the fourth and fifth layer lattice sites are only half-occupied and are sheared with respect to the interface. These features indicate loss of perfect epitaxy and order, most probably due to stress relief. The Si surface lattice vector 1/2[110] is 3.8Å long, while the parallel (compressed) tridymite lattice vector [001] is 7.6Å in length. Thus tridymite can grow epitaxially on the Si substrate in two configurations, which are identical but shifted with respect to each other by 1/2Si[110]. Our lattice images in the [110] projection show that, over the thickness of our sample (~100 Å), we observe both configurations in superpositions. Thus the tridymite overlayer contains planar defects, which have been extensively observed in oxides in general and tridymite in particular [19,20].

On the basis of the tridymite model, the oxidation of Si leads to the formation of a crystalline, epitaxial oxide phase which is in compression, and the process can be simply regarded as a particular example of a strained-layer epitaxy. Since the overriding constraint is that of epitaxy with the substrate as the oxidation proceeds, and as the crystalline oxide can grow only to a certain critical thickness before stress relaxation, its extent perpendicular to the interface is limited by simple strain considerations. In this way one can understand the small thickness of the crystalline oxide and the formation of a-SiO_2 as simple consequences of stress relaxation. Indeed our results provide strong indication that the relaxation begins after 2 MLs and is complete after 5MLs. The relaxation of a crystalline layer by amorphisation is well-known in metallic systems and also appears to be at work in the Si/SiO_2 system.

While tridymite is the most likely candidate for the observed interfacial phase, the question still remains as to the details of the epitaxial arrangement at the interface between the Si substrate and the tridymite. In the simplest arrangement consistent with the lattice images, each substrate Si atom at the interface bonds to one oxygen atom presented by the tridymite, and half the interfacial Si dangling bonds are saturated (Fig. 10). In this case the remaining dangling bonds may dimerise, thus saturating all the interface bonds. Another possibility, suggested to us by Fowler and Chu [21], consists of replacing some, or all the interfacial dimers by Si-O-Si bridges (Fig. 10). Lattice imaging cannot directly detect these bridges at the interface, and thus the dimer and Si-O-Si arrangements are both consistent with the TEM data.

A possible clue to the exact interfacial arrangement may come from XPS experiments, which can measure the oxidation state of the Si atoms at the interface; in the dimer arrangement, the interfacial oxidation state is Si^{+}, while in the Si-O-Si case the oxidation state becomes Si^{2+}. Several sets of XPS data from Si/SiO_2 interfaces are available in the literature [22,23], but the results from different groups are in qualitative disagreement. This has been, at least on occasion, ascribed to differences in sample preparation [22], and thus implies significant variation in the details of the interfacial atomic configuration depending on the preparation procedure used. To arrive at a consistent picture of the interface, it is therefore essential that the *same* samples are examined by HRTEM and XPS.

Prelimenary examination of our samples by soft X-ray spectroscopy [24] shows the Si^{+}, Si^{2+} and Si^{3+} oxidations states all to be present at atomically flat Si/SiO_2 interfaces, with Si^{+} being the most prominent, and the Si^{3+} having the smallest concentration. It thus appears that both Si dimers and Si-O-Si bridges may be present at the Si/SiO_2 interface. However, it is likely that the exact concentration of each species will depend on the details of the sample preparation.

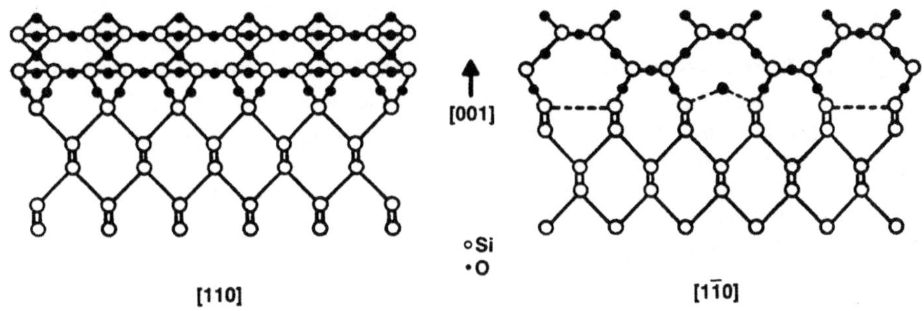

Fig. 10 Schematic representation of the Si/SiO_2 interfacial atomic structure, deduced from diffraction and lattice imaging data. The dashed lines indicate possible dimerisation of the unsaturated bonds. The dangling bonds may also be saturated by forming Si-O-Si bridges as indicated.

CONCLUSIONS

It is clear that, while the exact atomic structure of the Si/SiO_2 interface is still under investigation, certain features are known with confidence. Roughness at typical interfaces has been quantitively characterised. There is extensive, and now persuasive, diffraction evidence for the presence of crystallinity in the vicinity of the interface. The examination of atomically flat interfaces has directly revealed the presence of an epitaxial crystalline phase at the interface between the c-Si and the a-SiO_2.

The most likely candidate for the interfacial crystalline phase is tridymite. However, this identification is subject to the usual problem of uniqueness, generic to all simulation and matching techniques. This problem arises because it cannot be ruled out that another model, not thought of by the investigator, may prove to be an equally good, or better candidate. To overcome such difficulties and distinguish between the different interfacial arrangements admitted by the HRTEM data, it is necessary to obtain complimentary information from a variety of other techniques such as XPS, and XAFS.

However, a characteristic of the Si/SiO_2 literature is that different sets of data obtained by different experimenters using the same technique are often in qualitative disagreement. It is most likely that many of these disagreements indicate genuine differences in the samples used. To make further progress, it is essential that the *same* samples are examined by a variety of techniques, rather than attempting to transpose results between different samples. Finally, while the *identification* of the interfacial phase is subject to the usual uniqueness problems, it is important to recognize that the *existence* of this phase is a fact based on substantial experimental evidence.

FUTURE DIRECTIONS

The structure of the Si/SiO_2 interface is now established in sufficient detail that it may be realistic to contemplate using this knowledge to manipulate the properties of the Si/SiO_2 system. In particular, it may be possible to exploit the atomic perfection of MBE Si/SiO_2 interfaces to achieve superior electronic characteristics. The epitaxial nature of the oxide also offers the intriguing possibility of fabricating Si/epitaxial oxide superlattices, which may possess interesting properties. However, these avenues remain to be explored, and at present can be regraded only as interesting possibilities.

We are grateful to S. M. Goodnick for valuable discussions and for Fig. 2.

REFERENCES

[1] F. Herman and I. P Batra, *Physics of SiO_2 and its Interfaces*, ed. S. Pantelides (Pergamon, New York, 1978) p. 333.

[2] A. M. Stoneham, C. R. M. Grovenor and A. Cerezo, Phil. Mag. B 55, 201 (1987).

[3] S. Pantelides and M. Long, *Physics of SiO_2 and its Interfaces*, ed. S. Pantelides (Pergamon, New York, 1978), p. 339.

[4] P. H. Fuoss, L. J. Norton, S. Brennan and A. Fischer-Colbrie, Phys. Rev. Lett. **60**, 600 (1988).

[5] F. M. Ross and W. M. Stobbs, Proc. of MRS Fall Meeting, Boston (1987), in press.

[6] F. Rochet, S. Rigo, M. Froment, C. d'Anterroches, C. Maillot, H. Roulet and G. Dufour, Advances in Phys. **35**, 237 (1986).

[7] B. Agius, S. Rigo, F. Rochet, M. Froment, C. Maillot, H. Roulet and G. Dufour, Appl. Phys. Lett. **44**, 48 (1984).

[8] P. O. Hahn and M. Henzler, J. Vac. Sci. Technol. A **2**, 574 (1984).

[9] O. L Krivanek, T. T. Sheng and D. C. Tsui, Appl. Phys. Lett. **32**, 437 (1978).

[10] A. Ourmazd, K. Ahlborn, K. Ibeh and T. Honda, Appl. Phys. Lett. **47**, 685 (1985).

[11] M. A. O'Keefe, P. R. Busek and S. Iijima, Nature **274**, 322 (1978).

[12] S. M. Goodnick, D. K. Ferry, C. W. Wilmsen, Z. Liliental, D. Fathy and O. L. Krivanek, Phys. Rev. B **32**, 8171 (1985).

[13] A. H. Carim and R. Sinclair, J. Electrochem. Soc. **134**, 741 (1987).

[14] A. Ourmazd, W. T. Tsang, J. A. Rentschler and D. W. Taylor, Appl. Phys. Lett. **50**, 1417 (1987).

[15] A. H. Carim and R. Sinclair, Materials Lett. **5**, 94 (1987).

[16] A. Ourmazd, D. W. Taylor, J. A. Renstchler and J. Bevk, Phys. Rev. Lett. **59**, 213 (1987).

[17] A. Ourmazd and J. Bevk, Proc. of MRS Fall Meeting, Boston (1987), in press.

[18] A. Ourmazd, J. A. Renstchler and J. Bevk, to be published.

[19] V. G. Hill and R. Roy, Trans. Br. Ceram. Soc. **57**, 496 (1958).

[20] W. A. Dollase, Acta Crystallogr. **23**, 617 (1967).

[21] W. B. Fowler and A. X. Chu, private communication.

[22] P. H. Grunthaner, M. H. Hecht, F. J. Grunthaner and N. M. Johnson, J. Appl. Phys. **61**, 629 (1987).

[23] G. Hollinger and F. J. Himpsel, Appl. Phys. Lett. **44**, 93 (1984).

[24] J. F. Morar, private communication.

THE STOICHIOMETRY AND STRUCTURE OF THE Si/SiO_2 INTERFACE: ION SCATTERING STUDIES

L. C. Feldman

AT&T Bell Laboratories
600 Mountain Avenue
Murray Hill, N.J 07974-2070

High energy ion scattering combined with particle channeling is an interfacial probe which reveals information on the stoichiometry and structure of an amorphous/crystalline interface.[1,2] In early studies the technique was extensively applied to probing the Si/SiO_2 interface in a standard channeling geometry.[3-6] The original interpretation of these results was somewhat confused by a lack of understanding of the basic surface interactions associated with MeV channeling.[3] Theory and experiments on atomically clean surfaces quickly resolved the difficulty.[1] The final results of these early studies were as follows: The oxide is stoichiometric to within ~5Å of the interface and there is either: 1) one monolayer of reconstructed Si and ~5Å of nonstoichiometric oxide or 2) two monolayers of reconstructed Si. These results held for Si(100) and Si(110).

Further insight into the structure of the oxide came from "thin-crystal transmission" experiments which provided a still greater sensitivity to atomic displacements at the interface.[7,8] This experiment showed that the two monolayers of reconstructed Si is more correctly described as a strained layer with relatively small displacements from bulk-like sites.

Recently additional work has revealed some new characteristics of the interface; namely transmission electron microscopy has revealed the existence of an epitaxial oxide layer at the interface.[9] In light of these findings we can reinterpret the ion scattering results to take into account this possibility. In particular we can place a limit on the thickness of the epi-oxide layers of <5Å.

Further aspects of the original work can now be related to progress in surface science which has clearly shown the principal structural characteristics of the Si(100) and Si(111) surfaces.[10] In effect the (100) surface contains in-plane displacements which can extend a few monolayers into the bulk of the substrate. The main structural feature of the Si(111) surface is a stacking fault which gives rise to a anomalously large off-normal ion scattering, but a "bulk-like" scattering yield in the normal direction. The results for the interfacial strain in the Si/SiO_2 system follows the same pattern, leading to the result that the anisotropy of the interfacial strain follows the anisotropy of the clean surface reconstruction.

More recently ion scattering has been applied to other aspects of the Si/SiO_2 interface. Using transmission channeling Kahn and Eades have studied the atom site location of Sb when segregated at the growing Si/SiO_2 interface.[11] The Sb is found at a substitutional site, within ~100Å from the interface. Additional new applications of ion scattering to the Si/SiO_2 structure include measurements of the decomposition of an oxide[12] and studies of buried oxides formed by ion implantation.[13]

REFERENCES

1. L. C. Feldman, J. W. Mayer, S. T. Picraux, *Materials Analysis by Ion Channeling*, (Academic Press, N.Y. 1982).
2. L. C. Feldman and J. W. Mayer, *Fundamentals of Surface and Thin Film Analysis*, (North Holland, N.Y. 1986).
3. T. W. Sigmon, W. K. Chu, E. Lugujjo and J. W. Mayer, Appl. Phys. Let. *24*, 105 (1970).
4. L. C. Feldman, I. Stensgaard, P. J. Silverman and T. E. Jackman, in *The Physics of SiO_2 and Its Interfaces*, ed. by S. T. Pantelides (Pergamon, New York, 1978).
5. T. E. Jackman, J. R. MacDonald, L. C. Feldman, P. J. Silverman and I. Stensgaard, Surf. Sci. *100*, 35 (1980).
6. R. Haight and L. C. Feldman, J. Appl. Phys. *53*, 4884 (1982).
7. L. C. Feldman, P. J. Silverman, J. S. Williams, T. E. Jackman and I. Stensgaard, Phys. Rev. Lett., *41*, 1396 (1978).
8. N. W. Cheung, L. C. Feldman, P. J. Silverman, I. Stensgaard, Appl. Phys. Lett. *35*, 859 (1979).
9. A. Ourmazd, D. W. Taylor, J. A. Rentschler and J. Bevk, Phys. Rev. Lett. *59*, 213 (1987).
10. R. J. Culbertson, Y. Kuk and L. C. Feldman, Surf. Sci. *167*, 127 (1986).
11. A. D. F. Kahn and J. A. Eades, Phys. Rev. B, *36*, 6532 (1987).
12. R. Tromp, G. W. Rubloff, P. Balk, F. K. LeGoves and E. J. van Loenen, Phys. Rev. Lett. *55*, 2332 (1985).
13. A. E. White, K. T. Short, J. L. Batstone, D. C. Jacobson, J. M. Poate and K. W. West, Appl. Phys. Let. *50*, 19 (1987).

SCANNING TUNNELING MICROSCOPY AND SPECTROSCOPY OF SILICON DANGLING BOND DEFECTS

R.J. Hamers and R.H. Koch

IBM Research Division
Thomas J. Watson Research Center
Yorktown Heights, N.Y. 10598

ABSTRACT

Scanning Tunneling Microscopy (STM) and tunneling spectroscopy have been used to study localized dangling bond defects at the Si/SiO_2 interface and in metal overlayers on Si(111). On mildly oxidized silicon, electron traps are observed which give rise to telegraph noise on the tunneling current. In tunneling I-V measurements at constant gap spacing, these traps appear as regions of negative resistance or as discontinuous jumps. A wide variety of trapping behavior is observed, indicating that there is no unique trap on mildly oxidized Si(001). A second kind of dangling bond defect arises at sub-monolayer coverages of Al on Si(111). Atomic-resolution STM measurements demonstrate that these Si p_z-like dangling bonds are very localized. Simultaneous tunneling spectroscopy measurements show that they give rise to a defect state at -0.4 eV and a minimum in the density of states at E_F due to many-body effects.

INTRODUCTION

Defects at surfaces and interfaces play crucial roles in many practical applications. Yet, since most surface science techniques sample surface areas which are large compared to the dimensions of these defects, it has not generally been possible to study the geometry or electronic properties of individual defects. Scanning Tunneling Microscopy (STM) and Current Imaging Tunneling Spectroscopy (CITS) have recently emerged as atomic-scale tools for the study of surface "topography" and electronic properties, respectively. We have applied these techniques to the study of silicon "dangling bond" defects in two prototypical systems. Dangling bonds at the Si-SiO_2 interface have long been recognized to act as electron trapping centers and have been widely studied using a variety of techniques. [1-6] Here, we use the STM to characterize individual electron trapping defects.[7, 8] A second application of STM to "dangling bonds" is a study of Si substitutional defects occurring in low-coverage aluminum overlayers on Si(111).[9] These three-fold coordinated Si atoms constitute localized "dangling bonds" at the surface; tunneling measurements permit a direct assessment of the spatial localization of the wavefunctions associated with the defect.

In order to understand the nature of defects at the Si-SiO_2 interface, we have investigated the in-situ oxidation of Si(111) and Si(001) surfaces. The samples were commercial Si wafers with a resistivity of $\simeq$ 50 mΩcm, N-type. The samples were cleaned by careful degassing and finally annealing to 1325 Kelvin while maintaining the background pressure at less than $1X10^{-9}$ torr. This procedure produces the characteristic Si(111)-(7X7) or Si(001)-(2X1) reconstructions and leads to large, well-ordered terraces as observed in the STM. After characterization of these clean surfaces, the vacuum chamber was filled to $\simeq 10^{-7}$ torr O_2 for approximately 20-60 sec to partially oxidize the silicon surfaces.

Tunneling microscopy is ordinarily restricted to conductive materials. However extremely thin layers of insulating materials can also be studied if they are on a conductive substrate and care is taken to ensure that true vacuum tunneling is taking place (so that sample and tip are not in physical contact). In our experiments this was confirmed by modulating the piezoelectric transducer attached to the tip to change the tunneling gap Z, and measuring the induced modulation of the tunneling current I. The rapid decay of the surface wavefunctions in the vacuum gap leads to $d(\log_{10}I)/dZ \simeq 1$ $Å^{-1}$ for true vacuum tunneling; much smaller values are observed when the tip and sample are in contact. In these measurements we found that while tunneling measurements can be performed on clean Si surfaces with less than 1 V between sample and tip, on the oxidized surfaces a minimum voltage of approximately 2 V is usually required to establish vacuum tunneling with a 1 nA current. In the experiments reported here on oxidized Si, the feedback loop was always stabilized at a voltage of greater than 3 volts under conditions where true vacuum tunneling occurs.

The Current Imaging Tunneling Spectroscopy[10] technique was used to obtain tunneling I-V spectra simultaneously with STM "topographic" images, while maintaining a constant gap spacing. In the CITS technique, the feedback loop (normally used to maintain a constant tunneling current) is gated in such a way that the feedback loop is only active $\simeq$ 10% of the time. During the 10% active period, a constant voltage is applied to the sample and the feedback system corrects the tip position in order to maintain a constant tunneling current at this "stabilization" voltage. The feedback system is then disabled for a period of $\simeq$ 500 microseconds, and the sample bias is ramped linearly while simultaneously recording $\simeq$ 512 points on the I-V curve. The sample bias then returns to the "stabilization" value and the feedback loop is activated again. By recording such I-V curves synchronously with the STM topographs, it is possible to directly correlate the STM "topographic" features with the local tunneling I-V characteristics.

Figure 1 shows a typical STM image of the Si(001) surface after oxidation. The main feature of these images is the presence of circular areas approximately 20-50 Å diameter which appear to be lower by 2 Å or more than the surrounding surface regions. Unlike the clean Si(111)-(7X7) and Si(001) surfaces where the STM resolution is evident from the atomic corrugations themselves, STM images of oxidized surfaces generally do not exhibit obvious atomic-sized features. This is partially attributed to the fact that the SiO_x species are amorphous so that there is no recognizable pattern corresponding to an atomic reconstruction. Additionally, local height changes associated with the atoms themselves may be overwhelmed by *electrostatic* effects due to charged defects within the oxide layer and at the Si-SiO_x interface. These electrostatic effects have characteristic length scales larger than the size of a typical atom, making it difficult to resolve the atoms themselves.

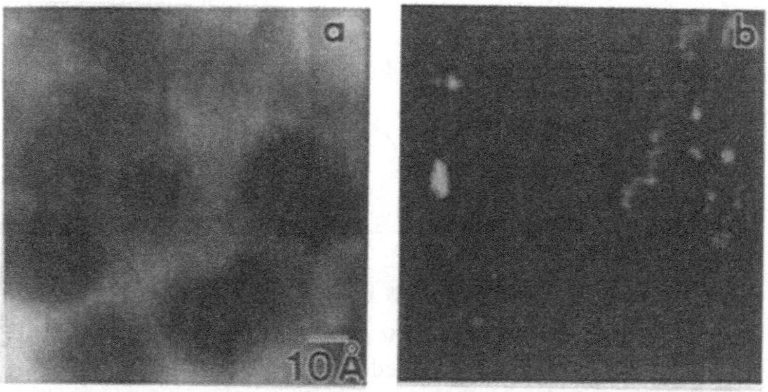

Figure 1. Grayscale image of surface topography (a) and tunneling current variance (b) of a 100 X 100 Å region of in-situ oxidized Si(001) surface. Regions of higher surface height (a) or high noise (b) are light. The vertical height scale in (a) spans 3 Å.

When tunneling on these oxidized surfaces, localized regions of ≃10 Å diameter are observed where the tunneling current quickly switches between two discrete values, as would be expected for a trap switching between two charge states.[1] Although the STM is nominally in a "constant current" mode of operation, the characteristic switching times are faster than the feedback response time and so appear as fluctuations in the tunneling current. In order to measure the spatial dependence of this noise, we have measured the variance of the tunneling current while simultaneously scanning the surface topography. Fig. 1b shows the variance of the tunneling current measured simultanously with the topography shown in fig. 1a. At the region indicated by the arrow, the tunneling current noise is approximately 1000 times larger than in the surrounding regions as a result of the trap-induced switching. The fluctuations amount to ≃ 30% of the total tunneling current and are much larger than the fluctuations observed in MOSFET devices (where the drain current typically changes by less than 1%) due to the close proximity of the tip to the trap. [1, 3] Comparing the STM topograph with the simultaneous noise measurement shows that the electron traps are usually observed at the edges of the circular oxidized regions, where unsaturated Si danging bonds should be more prevalent. The lateral extent of the "noisy" regions (≃10-15Å) is larger than the physical size anticipated for a trap and is determined by electrostatic screening at the surface. In addition to the lateral dependence shown in fig. 1b, the switching noise is also strongly dependent on the sample-tip bias voltage.

The electronic characteristics of these traps can be understood more completely from measurements of the tunneling current as a function of voltage. Although these

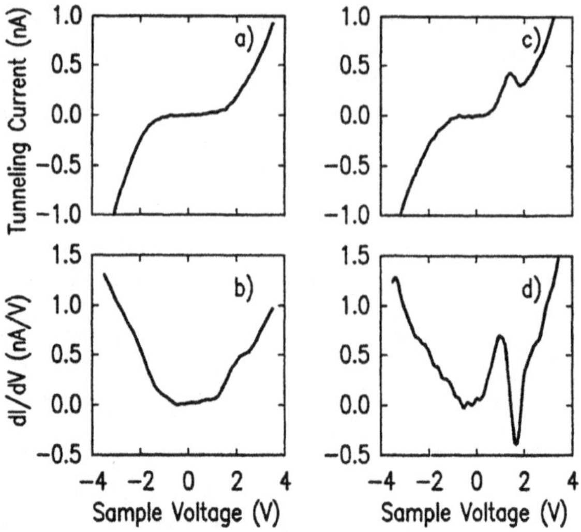

Figure 2. Tunneling I-V and dI/dV measurements on oxidized Si(001) surfaces. On most areas of oxidized surface, I vs. V (2a) and dI/dV vs. V (2b) show little structure. Over electron trap, I vs. V (2c) shows a pronounced feature near 1.5 V and a region of negative resistance is observed in the derivative spectrum (2d).

measurements reveal a wide variety of oxidation-related defect states, Figure 2 shows representative tunneling I-V characteristics obtained on an oxidized Si(001) surface using the CITS technique.[10] In most regions of the oxidized surfaces, the tunneling I-V curve (2a) and its derivative (2b) are both relatively featureless. Occasionally, atomic-sized defects are observed which give rise to drastically different tunneling I-V properties. Typical tunneling I-V characteristics at an oxide-related defect (here, on Si(001)) are shown in fig. 2c and 2d. As on the unoxidized surface, when tunneling into the sample (positive sample bias) the tunneling current first increases with voltage. The most distinctive feature of many oxide defects is that at slightly higher voltage, the tunneling current decreases strongly and a region of *negative resistance* is observed. At still higher voltage, the tunneling current again increases. The spatial extent of this I-V feature is limited to approximately 10 Å diameter. As the tip is moved laterally away from the trap center, the conductance peak can be observed to shift in energy by ≃0.2 eV before disappearing.

We have observed a variety of trapping defects both in noise measurements and I-V measurements. In our earlier work we concentrated on trap states which were near the Fermi energy of the sample and relied on local band-bending to change the trap energy and thermal activation to cause the switching. In that case, voltage-dependent band-bending will shift the energy of the trap with respect to the Fermi energy of the sample, leading to switching noise when they are nearly equal. This type of behavior was reported both in our earlier work [7] and in previous experiments on MOSFETS. [1, 3] Other traps (such as that shown in fig. 2) appears to lie significantly above the Fermi energy, and trapping is induced by the tunneling electron themselves. These two cases can be readily distinguished from the I-V curves and trapping behavior. At positive sample (or negative tip) bias, traps near E_F should show a decreasing occupation with

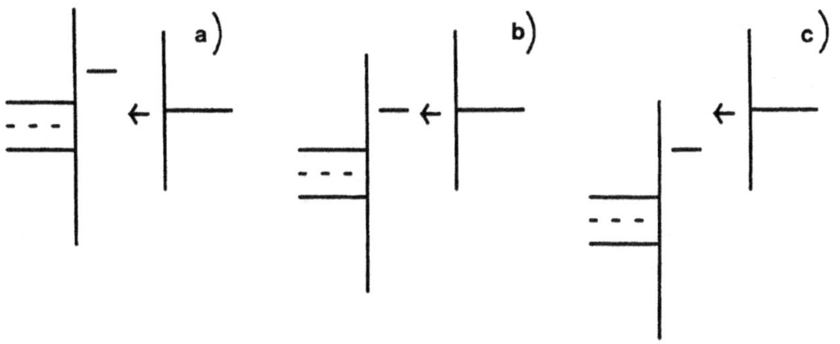

Figure 3. Schematic illustration of the origin of negative resistance and switching noise in oxidized silicon at positive sample bias (as in figures 1 and 2). At low voltage (a) the trap is always empty and electrons tunnel from Fermi level of tip (right) to sample (left). At slightly higher voltage (b) the trap become occupied, leading to negative resistance and switching noise. At higher voltages yet (c) the trap is always occupied and electrons again have sufficient energy to tunnel.

increasing voltage and give rise to a discontinuous increase in the tunneling current as the trap energy sweeps through E_F. Traps directly populated by the tunneling electrons and lying far from E_F show an *increasing* occupation with increasing voltage and give rise to negative resistance (as in fig. 2). At negative sample bias, the situation reverses.

Our measurements show that there is not a "unique" dangling bond defect on mildly-oxidized silicon. Our noise measurements and I-V measurements detect only those defects with trapping times and energies in certain windows. Thus, there are likely oxidation-related defects which escape our detection. However, localized regions of showing negative resistance such as those in fig. 2c and 2d are quite common on oxidized surfaces and are *not* observed on clean silicon surfaces, leading us to associate them with oxidation-related defects. These spatially-localized defect states in oxidized silicon can be observed at positive and negative sample bias, corresponding to tunneling into and out of the silicon sample, respectively. Although the I-V curves and noise measurements cannot be performed simultaneously, the switching and negative resistance are both attributed to charge trapping at these defects, which are located either at the Si-SiO_2 interface or within the oxide itself.

While the the switching noise on trap states near E_F is often thermally activated, it is also possible to study trapping in states lying well above E_F. For example, the tunneling I-V curve shown in figure 2 shows a negative resistance near +1.5 eV, when electrons tunnel into the sample. Such negative resistance features cannot be attibuted to occupation changes of traps lying near E_F resulting from band-bending effects, since these would produce a steep increase in the tunneling current as the positive sample voltage was increased. (Discontinuous jumps in the tunneling current with increasing positive sample voltage are observed but are not discussed here). Instead, the I-V characteristics shown in fig. 2 likely arise from normally unoccupied states trap states lying well above E_F, as indicated in figure 3. At small positive sample bias (3a) the electrons are unaffected by the trap and tunnel normally through the vacuum into the oxidized silicon. When the tunnel bias is increased enough so that energy of the trap and the tip are equal (fig. 3b), the trap occasionally becomes occupied by the tunneling electrons. When this occurs, the local negative charge creates a large Coulombic repulsion, which pinches off the tunneling current until the trapped electron dissipates into the sample. [8] Thus, the average tunneling current *decreases* with increasing voltage when the energy of the tip and trap are nearly equal. In this same range, switching noise is observed as the trap occupation fluctuates. In this case, the switching noise shows that the trap occupation increases with increasing tunnel bias. At higher sample voltages yet, electrons are able to overcome the Coulombic repulsion and again tunnel through the barrier (fig. 3c) and the resistance again becomes positive. In this case, the lifetime of the occupied trap state must be relatively long and it must have a high probability of being filled by the tunneling electrons.

Further information about the dynamics of charge trapping can be obtained from measurements of the tunneling current in the time domain, since the rate of transitions between trapped and untrapped states contains information about the activation barriers for electrons to fill and empty the trap. [7] Measurements of the the temporal distribution of "up" and "down" times show that it is often Poisson in form, as expected for a thermally activated process.[1, 8] A detailed analysis of these switching times[7] for traps lying close to E_F has shown that the characteristic activation barriers for trapping and untrapping are on the order of 0.3-0.7 eV and the preexponential factor is approximately 10^{10} $seconds^{-1}$. These numbers are representative of those traps which can be measured with the STM but may not be representative of electron traps in general. They are consistent with values obtained previously using a similar analysis on the switching noise observed in MOSFETS. [3]

The time-domain analysis of the switching times provides a measure of how strongly the trap energy is affected by the tunnel bias. The energy of the trap relative to the tip can be expressed as $E_{trap}=E_0 + \alpha eV_{sample}$ where V_{sample} is the voltage on the sample relative to that of the tip (here, $V_{sample}>0$). An analysis of the switching times typically yielded $\alpha = 0.14 \pm 0.04$. This small value of α means that the energy of the trapping level with respect to the Fermi level of the bulk silicon is only weakly dependent on the voltage on the tip, as would be expected for a defect at the surface. Essentially the same value of α can be determined from the spatially-dependent I-V measurements as in fig. 2, where it was noted that the electronic feature near +2V shifted in energy by $\simeq$ 0.2 eV when the tip was moved away from the center of the defect; this leads to an estimate of $\alpha \simeq 0.1$.

The wide variety of possible trapping centers precludes making generalized statements about their nature. Nevertheless, tunneling measurements like those reported here and in our previous work[7, 8] make possible measurements of the energies and dynamic behavior of individual electron trapping defects. From the voltage-dependence of the telegraph signals and from the tunneling I-V curves, we are able to distinguish traps lying close to E_F whose occupation is changed due to band-bending from those lying further from E_F whose occupation is changed due to the tunneling electrons themselves.

Our time-dependent measurements have also sometimes observed three-level switching, believed to arise from either traps with three charge states or else two traps which interact. Future work will be aimed at accessing "faster" traps and understanding the origin of the more complicated behavior of three-level traps.

LOCALIZED DANGLING BOND DEFECTS IN METAL-SEMICONDUCTOR OVERLAYERS

Silicon "dangling bond" defects are also of interest in understanding the electronic properties of metal-semiconductor interfaces. We have recently used STM to characterize the atomic structure, charge state, and density of electronic states of localized defects naturally occurring in Al overlayers on Si(111). Depositing 1/3 monolayer of Al onto the Si(111) produces an extremely stable overlayer with ($\sqrt{3}$ X $\sqrt{3}$) symmetry. This $\sqrt{3}$ -Al overlayer consists of a bulk-like Si(111) surface with Al atoms adsorbed on the three-fold sites. This configuration provides four-fold coordination of all Si atoms in the outermost Si layer, thereby saturating all Si dangling bonds.

As shown in fig. 4, STM images of this surface show that although much of the surface is like the "ideal" surface described above, there are also two types of characteristic defects which are frequently found. One type of defect, labeled "S", appears to be a substitutional defect as it still has a protrusion in the same location where (for the ideal surface) an Al atom would be expected. A second type of defect, labeled "V", is a vacancy defect. At negative sample bias, the "S" defects appear *higher* than the ideal $\sqrt{3}$ -Al layer, while the "V" defects still appear to be $\simeq$ 2 Å lower. These voltage-dependent differences are due to the different electronic state density of the ideal $\sqrt{3}$ -structure and the "S" defects. Studying the number of "S" defects as a function of Al coverage shows that the "S" defects are actually Si atoms (instead of Al atoms) adsorbed on the Si(111) top layer. Since they are only three-fold coordinated, these Si atoms have an unsaturated "dangling bond".

In order to study the energy spectrum of these defects, the tunneling I-V characteristics were analyzed on "S"-type defect sites and on the "ideal" $\sqrt{3}$ -Al regions. Figure 4 shows these results in the form of plot of (V/I)dI/dV vs. V; this quantity is approximately proportional to the local density of states but is *normalized* to unity at V=0. [11] While the $\sqrt{3}$ -Al structure has a surface state bandgap of $\simeq$ 2 eV, figure 4 shows that the "S" defects give rise to a strong defect state at 0.4 eV below E_F. For comparison, in fig. 4d we show the ultraviolet photoemission results of Kinoshita, et al. [12], which also show a defect state at this energy. Our STM results allow us to assign the atomic origin of this feature to Si atoms substituting for Al adatoms in the outermost surface layer.

On of the most striking features of these subsitutional defects is that they are very localized. For example, measurements of the apparent height of Si adatoms in individual defects and 2-3 atom clusters of defects relative to the "ideal" $\sqrt{3}$ -Al surface show only small changes as a function of cluster size. Additionally, tunneling spectroscopy measurements show no significant difference between the state density at Al adatoms in "ideal" regions and those adjacent to the Si defects. [12, 13] These measurements show that the nature of the "S" defect is independent of the surrounding adatoms, so that it can be thought of as an isolated Si adatom. In many respects, such a localized defect bears a strong resemblance to the well-studied P_b center at the Si-SiO_2 interface. [4, 5] In both cases, the Si atoms have a half-filled Si p_z-like dangling bond. Also in both cases

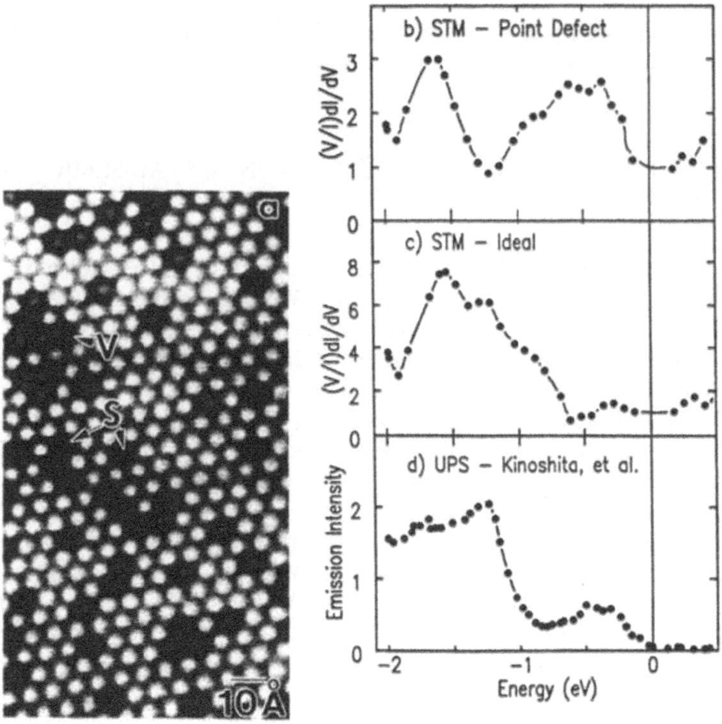

Figure 4. STM grayscale image of Si dangling bond defects imbedded in $\sqrt{3}$ Al/Si(111) overlayer (a) and corresponding tunneling state density measurements on Si dangling bond defects (b) and on Al adatoms (c) compared with photoemission results of Kinoshita, et al. (d, from Ref. 12).

the strong spatial localization of this state results in strong many-electron effects, so that the energy of the dangling bond state is strongly dependent on its electron occupation.

One-electron band theories predict that such a Si "dangling bond" defect should have a half-filled p_z-like state at E_F.[14, 15] We instead see a state at -0.4 eV; this energy suggests that the state is completely filled and the defect is therefore negatively charged. In order to determine whether these defects are charged or not, measurements were made of the apparent height of the Al adatoms at varying distances from the defect. If the "S" defects were charged, local band-bending would be expected to produce an apparent difference between those Al atoms close to the defect and those further away. While such band-bending revealed charging around *other* types of defects, the Si "S"-type defects observed here appeared to be electrically neutral, both for n- and p-type bulk doping.

The observed dangling bond energy of -0.4 eV and its apparently neutral charge state can be understood by considering a many-electron picture of "dangling bond" defects instead of the single-electron band picture presented above. Such many-electron calculations predict that the three possible charge states for a trivalent defect such as a

Si dangling bond produces two observeable electrical levels, one a donor and one an acceptor, separated by an effective interaction energy U_{eff}, with the Fermi energy between these levels. These calculations predict a *minimum* in the state density at E_F, which is confirmed by bulk tunneling junction measurements in α-SiH.. [16] The peak we observe at -0.4 eV on the three-fold coordinated Si adatoms then corresponds to the energy of the donor state relative to E_F We do not observe a well-defined acceptor state above E_F due to the pronounced p_z-like state on the nearby Al adatoms at +0.9 eV. However, we do observe a clear minimum in the state density at E_F.

Our results indicate that U_{eff} for these Si dangling bond defects is greater than 0.4 eV. This value is in agreement with both theoretical and experimental studies of other, similar Si "dangling bond" defects. For example, Fowler, et al.[17] calculated the energies of the donor and acceptor states of an isolated Si dangling bond and obtained values of U_{eff}=0.4-0.5 eV. Deep level transient spectroscopy measurements of the P_b center at the Si-SiO_2 interface clearly show donor and acceptor state separated by U_{eff}=0.7 eV, and similar measurements for dangling bond defects in α-SiH give U_{eff}=0.35 eV. Given the strong geometric similarity between those dangling bond defects and the the one at the Al/Si(111) interface observed here, the value of $U_{eff} > 0.4$ eV implied by our tunneling spectroscopy measurements is quite reasonable.

SUMMARY

Combining the high spatial resolution of STM with the energy resolution of tunneling spectroscopy allows the electronic properties of individual atomic-sized defects to be directly studied. Here, we have reported the use of STM and tunneling spectroscopy to identify and characterize electron trapping defects in oxidized Si(001) and localized Si "dangling bond" defects in Al overlayers on Si(001). In both cases, the electronic states associated with the defect are strongly localized in space. This spatial localization leads to energy levels which are dependent on the defect charge state. The scanning tunneling microscope provides a means of simultanously studying the geometry and electronic structure of individual atomic defects.

This work is supported in part by the U.S. Office of Naval Research.

REFERENCES

1 K.S. Ralls, W.J. Skocpol, L.D. Jakcel, R.E. Howard, L.A. Fetter, and R.W. Epworth, and D.M. Tennant, Phys. Rev. Lett. **52**, 228 (1984).

2 D.K. Biegelsen, N.M. Johnson, M. Stutzmann, E.H. Poindexter, and P.J. Caplan, Applic. Surf. Sci. **22/23**, 879 (1985).

3 M.J. Kirton and M.J. Uren, Appl. Phys. Lett. **48**, 1270 (1986).

4 N.M. Johnson, D.K. Biegelsen, M.D. Moyer, S.T. Chang, and E.H. Poindexter, and P.J. Caplan, Appl. Phys. Lett. **43**, 562 (1983).

[5] N.M. Johnson, W.B. Jackson, and M.D. Moyer, Phys. Rev. B **31**, 1194 (1985).

[6] P.J. Caplan, E.H. Poindexter, B.E. Deal, and R.R. Razouk, J. Appl. Phys. **50**, 5847 (1979).

[7] R.H. Koch and R.J. Hamers, Surf. Sci. **181**, 333 (1987).

[8] M.E. Welland and R.H. Koch, Appl. Phys. Lett. **48**, 724 (1986).

[9] R.J. Hamers, J. Vac. Sci. Tech., in press.

[10] R.J. Hamers, R.M. Tromp, and J.E. Demuth, Phys. Rev. Lett. **56**, 1972 (1986).

[11] R.M. Feenstra, J.A. Stroscio, and A.P. Fein, Surf. Sci. **181**, 295 (1987).

[12] T. Kinoshita, S. Kono, and T. Sagawa, Phys. Rev. B **32**, 2714 (1985).

[13] R.I.G. Uhrberg. G.V. Hansson, J.M. Nicholls, P.E.S. Persson, and and S.A. Flodstrom, Phys. Rev. B **31**, 3805 (1985).

[14] A. Zur, T.C. McGill, and D.L. Smith, Phys. Rev. B **28**, 2060 (1983).

[15] J.E. Northrup, Phys. Rev. Lett. **57**, 154 (1986).

[16] J. Stuke, Phil. Mag. **B 52**, 225 (1985).

[17] W.B. Fowler and R.J. Elliott, Phys. Rev. B **34**, 5525 (1986).

THE MICROSTRUCTURE OF SiO_2-Si(100) INTERFACES INVESTIGATED BY XPS AND HRTEM

G. Hollinger and R. Saoudi

Laboratoire d'Electronique - Ecole Centrale de Lyon
B.P. 163 - 69131 Ecully Cedex - France

P. Ferret and M. Pitaval

Département de Physique des Matériaux - UCB Lyon 1
69622 Villeurbanne Cedex - France

ABSTRACT

Thin oxides (20-100 Å) grown on Si(100) surfaces using a variety of oxidation processes were examined by X-ray Photoelectron Spectroscopy (XPS) and cross sectional Transmission Electron Microscopy (TEM). Device quality oxides exhibit flat but not ideal interfaces and extended chemical transition layers with about 2 monolayers of suboxide Si atoms.

1. INTRODUCTION

Over two decades, intense research efforts have been concentrated on the SiO_2-Si interface both because of its practical importance and fundamental interest. Despite a large number of studies, a microscopic description of this interface has still to be established. This is mainly due to the non crystalline character of the SiO_2 overlayer and to the inability of usual experimental tools to investigate the atomic structure of disordered systems.

However, an extensive use of Transmission Electron Microscopy TEM[1,4], electron and X-ray diffraction techniques[5,6] and photoelectron spectroscopies[7-9] allowed to make some interesting progresses these last years. TEM can visualize the structure of the interface, diffraction techniques probe the crystallinity of the oxide and photoelectron spectroscopies give the number of interfacial Si atoms and define their chemical environments.

Results which are apparently contradictory have emerged. Epitaxial interfacial oxides were observed with TEM[4,5] whereas it was shown that very rough interfaces could be easily prepared by thermal oxidation[3]. An abrupt interface with about one monolayer of Si atoms in intermediate oxidation states was observed with XPS[8] whereas photoemission with Synchrotron radiation measurements have reported an extended interface with about two monolayers of Si[7,9].

The difficulties to reconciliate these results come from two main reasons. First, a large variety of oxides processed in different conditions was studied (thick and thin oxides, device quality thermal oxides, very thin oxides grown in an ultra high vacuum environment, etc...). Second, just one experimental technique was generally used which makes almost impossible to interpret unambiguously the results. For example, we do not really know how the XPS results obtained on device quality oxides have to be compared to the Synchrotron results obtained on very thin UHV oxides. It is not known whether these published photoemission data apply to "ideal" flat interface or to defected rough interfaces. Concerning more specifically the photoemission results, the part of the structural reality and the part of the authors interpretation have to be clearly established.

Consequently, to clarify these points it appears necessary to associate different complementary techniques in the characterization studies of SiO_2-Si interfaces. In this paper we present the first results of an extensive research program based on a combined use of XPS and TEM techniques. Possible correlations between the density of interfacial Si atoms and the interface roughness have been examined.

2. EXPERIMENT

The samples were (100) oriented Si wafers (P-type ≈ 1 Ω.cm) covered with thin (20-100 Å) SiO_2 overlayers.

Samples A,B,C,D and E were oxidized at 900°C in dry diluted O_2 + 2 % HCl during various times. Samples A and B were annealed at 1050°C in nitrogen. These oxides are device quality oxides (N_{ss} ≈ 2-4x10^{11} cm^{-2}) as checked by electrical measurements done on capacities processed on the same wafers[10]. Samples F and G were oxidized anodically in pure water[11] and annealed at 980°C in nitrogen. Sample H was grown in an UHV system by exposing an atomically clean Si(100) -2x1 surface to 1 atmosphere of oxygen at ≈ 700°C. Sample I was thermally grown in dry oxygen at 900°C. It is similar to sample (a) with a rough interface described in ref.12.

The XPS Si 2p spectra were measured with a Hewlett Packard 5950 A spectrometer and a monochromatized AlK_α radiation. The total instrumental resolution was equal to ≈ 0.7 eV. In order to measure the interfacial signals with enough accuracy, the oxides have been chemically thinned using a 0.1 % HF-ethanol solution and a spin etch procedure similar to that used by Grunthaner et al[8]. For clarity, the secondary electron background and the Si 2p spin orbit doublet were removed. Only the Si $2p_{3/2}$ components are shown in figure 1.

The high resolution transmission electron microscopy measurements were performed using a JEOL 200 cx electron microscope with a point to point resolution of 2.2 Å at 200 kV. Cross sectional samples with the <110> axis to be oriented along the electron beam direction were prepared using a standard method[13]. Although some samples were prepared with a capping layer of polysilicon, we found that the epoxy glue used in the cross sectional sample preparation was appropriate to observe a contrast at the outer boundary of the SiO_2 layer. This made possible to evaluate absolute SiO_2 thicknesses. The micrographs shown in the paper were obtained near the Scherzer defocussing distance (- 450 Å) using the transmitted and 6 diffracted beams (111 and 002) from the substrate. For samples with SiO_2 layers thicker than 30 Å we used defocused images (- 2500 Å) to increase the contrast at the Si-SiO_2 interface and at the SiO_2-epoxy interface, thanks to Fresnel oscillations. The cross section thicknesses were somewhere between ≈ 140 and ≈ 400 Å.

3. XPS RESULTS

Fig.1 illustrates the informations extracted from the analysis of the Si $2p_{3/2}$ core level spectra. The two main components separated by approximatively 4 eV are characteristic of the silicon substrate (peak position

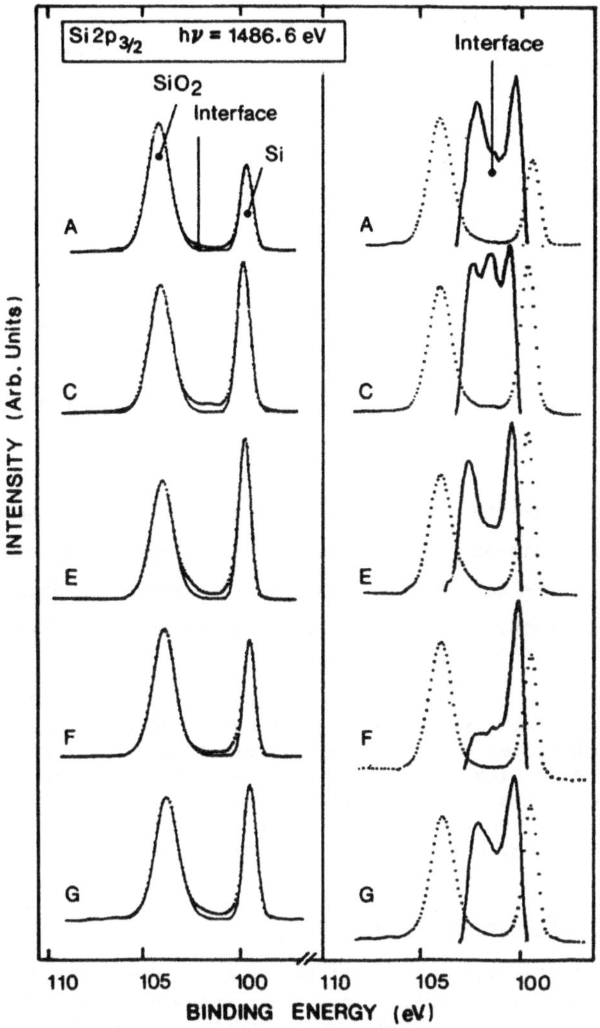

Fig.1. Si $2p_{3/2}$ XPS spectra for samples A,C,E,F and G with the fitted Si and SiO_2 components (left hand side). The suboxides contributions are zoomed on the right hand side.

at ≈ 99.5 eV) and of the SiO_2 overlayer. The weaker components correspond to silicon atoms in intermediate oxidation states Si^{1+}, Si^{2+} and Si^{3+}. To quantify the SiO_2 and Si intensities we have fitted, using analytical functions, the low binding energy side and the high binding energy side of the Si and SiO_2 peaks, respectively. The suboxide contributions were evaluated by subtracting the fitted SiO_2 and Si components to the experimental spectra.

The intensity of the SiO_2 peak relative to the Si signal is related to the thickness d of the SiO_2 overlayer and to the electron mean free path λ of 1380 eV electrons in SiO_2.

The relative intensity of the suboxide components gives the number of interfacial Si atoms or the thickness of the chemical transition layer.

The distribution of the oxidation states give some insight on the ideality of the interface.

a) <u>the oxide electron mean free path</u>

To determine the escape depth we use the following equation :

$$\lambda_{SiO_2} \operatorname{Sin}\theta = d/\ln\left[\frac{I_{SiO_2}}{I_{Si}}\;\frac{I^{\infty}_{Si}}{I^{\infty}_{SiO_2}} + 1\right] = d/\ln\left[\frac{R}{C} + 1\right] \qquad (1)$$

I_{SiO_2} and $I^{\infty}_{SiO_2}$ are the oxide core level intensities for the SiO_2 overlayer of thickness d and for a bulk SiO_2 sample, respectively.

I_{Si} and I^{∞}_{Si} are the substrate Si intensity for an oxidized sample and for a clean surface, respectively. θ is the electron emission angle with respect to the surface.

The thicknesses d were determined using TEM as discussed below. We found mean free path values ranging from 36 to 42 Å. This is in general agreement with values published by Hill et al[14]. λ is expected to increase when the oxide density is decreasing. We obtained the highest λ values for samples A and B. These two samples were annealed at 1050°C above the temperature for which the oxide structure is expected to relax and the density to decrease. Using C = 0.8 we find the λ value in silicon equal to 22 Å.

b) <u>the number of interfacial Si atoms</u>

Using a relation similar to (1) we can calculate the equivalent average thickness of the "SiO_x" transition layer. Assuming realistic parameters for SiO_x ($\lambda \approx 30$ Å, $\rho = 3.1\times10^{22}$ cm^{-3}) and taking into account the surface atom density for the Si(100) surface (6.8×10^{14} cm^{-2}), it is possible to estimate the number of interfacial Si atoms[9]. These numbers are presented in Table 1 for the different samples. Due to uncertainties on the SiO_x parameters these numbers are not absolute values and cannot be used to determine accurately the thickness of the interface. To give exact numbers it would be necessary to use reference samples where the number of surface or interfacial Si atoms is known. Himpsel et al[9] used the H_2O-exposed Si(100) 2x1 surface (0.5 monolayer of Si^{1+}) in their synchrotron radiation core level studies. CaF_2/Si (111) and As terminated Si surfaces are other candidates. Such calibration procedures are in progress in our laboratory. At present the calculated values are relative numbers to be multiplied by a factor k in order to get the exact values.

Table 1. TEM oxide thicknesses, oxide mean free paths λ, calculated interfacial Si atom density n SiO_x and calibrated interfacial Si atom density θ (in Si monolayers - see text) for a variety of thin oxide films grown on Si(100).

		A	B	C	D	E	F	G	H	I
TEM	d (Å)	32	42	63	81	96	66	29	≈17	
X	λ (Å)	40	45	38	39	36	42	42		
P	n SiOx ($x\ 10^{15}cm^{-2}$)	1.43	0.97	1.0	0.97	1.17	1.03	0.97	0.93	2.87
S	θ Si monolayers	2.9	2.06	2.06	2.06	2.35	2.06	2.06	1.9	5.9

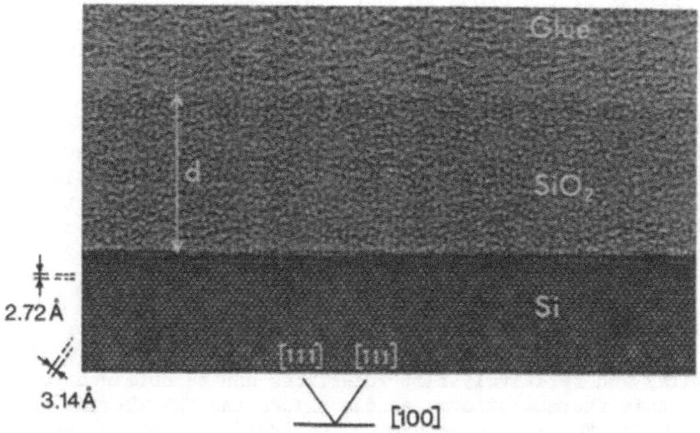

Fig.2. HRTEM image of a cross section through Si/SiO_2/epoxy glue interfaces.

From Table 1, it appears that most of the thermal device quality oxides (B,C,D,E,F,G) are characterized by the same number of interfacial Si atoms (≈ k $1x10^{15}$ cm^{-2}) it is interesting to note that we found the same value for the very thin UHV oxide H. For this sample after internal calibration Himpsel et al[9] found ≈ 2 monolayers of Si atoms (≈ $1.4x10^{15}$ cm^{-2}) at the interface. From this result, we can deduce k ≈ 1.4. It means that for our device quality oxides the density of Si atoms in intermediate oxidation states is equal to about 2 monolayers of Si atoms. These conclusions are at variance with the findings of Grunthaner et al[8] who found about 1 monolayer of Si atoms. A careful examination of their data indicates that these discrepancies could be due to differences in data processing (we always find higher Si^{+1} signals than they do) and in the parameters used (atomic densities, electron mean free paths) involved in the conversion of the XPS interface signals into atom densities.

Samples A and I have higher interface Si densities. This will be discussed in section 5.

4. TEM RESULTS

An example of a lattice resolution image for a SiO_2-Si(100) interface is shown in Fig.2. The image is oriented along the <011> direction. Lattice fringes of (111) type planes are clearly visible in the silicon substrate. The (111) plane spacing of 3.14 Å allows an accurate calibration for the magnification. The interface between Si and SiO_2 is clearly discerned. It is more difficult but still possible to distinguish SiO_2 from the epoxy glue. However it was impossible to measure the oxide thickness for very thin films (< 20 Å) due to the lack of contrast. The oxide thicknesses reported in Table 1 are estimated with an accuracy of ± 3 Å.

TEM allows to investigate almost directly the interface roughness[1-3]. However, the images have to be interpreted carefully. Some artefacts may lead to images which do not reflect the actual interface morphology. The ion milling thinning process may disturb the interface structure for very thin cross sectional samples (< 50 Å). The one dimensional image roughness is in fact the projection of the two dimensional surface roughness through the thickness of the cross section. This leads to underestimate the actual interface roughness in TEM images[2]. Steps are clearly visible in very thin cross sections[1]. The minimum visible lattice separation in the vertical direction is 2.72 Å.

Most of the samples investigated presented relatively smooth and abrupt interfaces (samples A,B,C,D,E,G,H) as seen in Fig.3. However these interfaces are not ideally flat and a few steps are visible. Some of them can be due to the compensation of the misorientation of the wafer. Sample H is flat with a few protusions. Sample F is characterized by long wavelength smooth ondulations. Sample I is described in ref.12.

5. DISCUSSION

Similar and relatively flat interfaces can be obtained for device quality thin thermal oxides, anodic oxides and UHV thermal oxides (samples B,C,D,E,G,H). For all these samples we found the same number of interfacial Si atoms -about 2 monolayers of Si if we use the calibrated values of Himpsel et al[9]- this last number is in agreement with the non ideal distribution of oxidation states shown in Fig.2. Only one oxidation state is expected for ideal atomically abrupt interfaces. This confirms that even for device quality oxides the SiO_2-Si interface has an extended structure.

Sample A has the thinnest device quality oxide (32 Å). The interface was found to be flat whereas the number of interfacial Si is higher than for other device quality thermal oxides : about 3 monolayers of Si. This result tends to indicate that wide chemical transition layers could exist for some thin oxides. Note however that we cannot exclude the presence of surface suboxides on the SiO_2 layer which would surestimate the apparent interface signal.

Two samples, F and I exhibit interface roughness. Sample I is extremely rough with large (≈ 20 Å) protusions and wavelengths of ≈ 40 Å[12]. This sample is characterized by a high number of interfacial suboxide Si atoms -about 4 monolayers-. In contrast sample F has a smaller roughness (10 Å) with longer wavelengths (200 Å). This sample has a 2 monolayer thick chemical interface similar to that of flat samples.

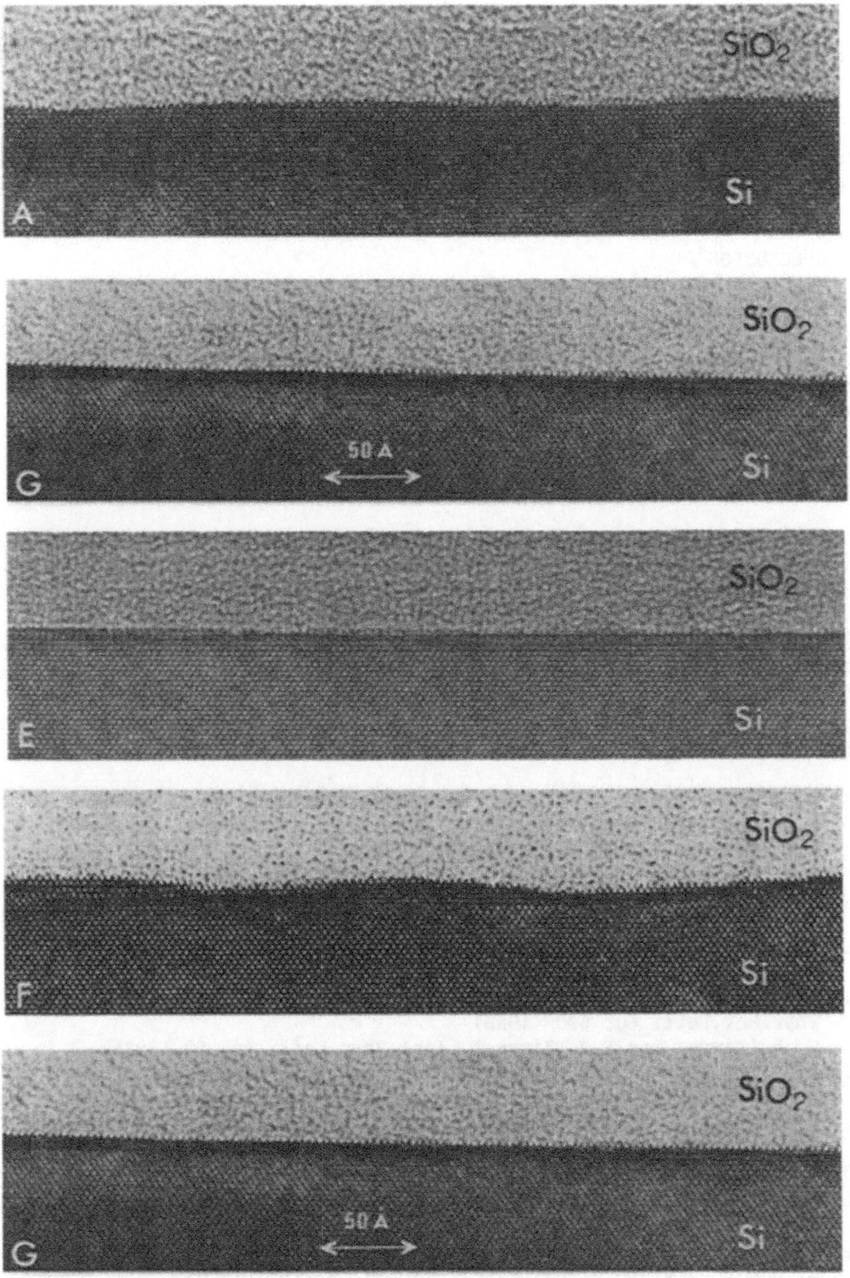

Fig.3. High resolution TEM images of SiO_2-Si interfaces for samples A,C,E,F and G.

These results can be understood as follows :

- for rough interface with narrow protusions and low wavelengths, the oxide regions between protusions are made up with SiO_x species which give a thick chemical transition layer,

- for rough interfaces with large ondulations the behavior is similar to that of flat interfaces and the chemical transition layer is locally equal to about 2 Si monolayers.

6. CONCLUSION

A large variety of processes, including device quality oxidation, produce similar SiO_2-Si interfaces. The best oxides display flat but not ideal interfaces with about 2 monolayers of suboxide Si atoms. Rough samples with low wavelengths are characterized by a much higher number of suboxide Si atoms.

ACKNOWLEDGEMENT

We acknowledge G. Sarrabayrouse, F. Gaspard, A. Carim, A. Cros and F. Himpsel for providing oxidized Si wafers and for helpful and stimulating discussions.

This work was supported by the French "Groupement Circuits Integrés au Silicium".

REFERENCES

1 C. D'Anterroches, J.Microsc. Spectrosc.Electron. 9, 147 (1984)
2 S.M. Goodnick, D.K. Ferry, C.W. Wilmsen, Z. Liliental, D. Fathy, O.L. Krivanek, Phys.Rev.B 32, 8171 (1985)
3 A.H. Carim, and R. Sinclair, J.Electrochem.Soc. 134, 741 (1987)
4. A. Ourmazd, D.W. Taylor, J.A. Rentschler and J. Bevk, Phys.Rev.Lett. 59, 213 (1987)
5. F. Rochet, S. Rigo, M. Froment, C. D'Anterroches, C. Maillot, H. Roulet, G. Dufour, Adv.Phys. 35, 237 (1986)
6. P.H. Fuoss, L.J. Norton, S. Brennan, A. Fischer-Colbrie, Phys.Rev.Lett. 60, 600 (1988)
7. G. Hollinger and F.J. Himpsel, Appl.Phys.Lett. 44, 93 (1984)
8. P.J. Grunthaner, M.H. Hecht, F.J. Grunthaner, N.M. Johnson, J.Appl.Phys. 61, 629 (1987)
9. F.J. Himpsel, F.R. Mc Feely, A. Taleb-Ibrahimi, J.A. Yarmoff and G. Hollinger, Phys.Rev.B (to be published)
10. G.Sarrabayrouse and M. Prom (unpublished)
11. F. Gaspard, A. Halimaoui and G. Sarrabayrouse, Revue Phys.Appl. 22, 65 (1987)
12. A.H. Carim, A. Bhattacharyya, Appl.Phys.Lett. 46, 872 (1985)
13. M. Dupuy, J.Microsc.Spectrosc.Electron. 9, 163 (1984)
14. J.M. Hill, D.G. Royce, C.S. Fadley, L.F. Wagner and F.J. Grunthaner, Chem.Phys.Lett. 44, 225 (1976)

MICROSCOPIC STRUCTURE OF THE SiO_2/Si INTERFACE

F. J. Himpsel, F. R. McFeely, A. Taleb-Ibrahimi, and J. A. Yarmoff*

IBM Research Division, T. J. Watson Research Center
Yorktown Heights, NY 10598

G. Hollinger

Laboratoire d'Electronique Automatique et Mesures Electriques
Ecole Centrale de Lyon
F 69131 Ecully Cedex, France

The bonding of Si atoms at the SiO_2/Si interface is determined via high-resolution core level spectroscopy with synchrotron radiation. For oxides grown in pure O_2, the SiO_2/Si interface is found to contain Si atoms in intermediate oxidation states with a density of $1.5 \pm 0.5 \times 10^{15}$ cm^{-2}. From the density and distribution of intermediate oxidation states, models of the interface structure are obtained. The interface is not abrupt, as evidenced by the non-ideal distribution of intermediate oxidation states and their high density (about 2 monolayers of Si). The finite width of the interface is explained by the bond density mismatch between SiO_2 and Si. Annealing in H_2 is found to influence the electrical parameters by removing the P_b centers that pin the Fermi level. The distribution of intermediate oxidation states is not affected.

INTRODUCTION

The structure of the SiO_2/Si interface has been elusive despite many efforts to come up with models. Two distinct interface regions have been identified. The near interface is 5Å wide (or less) and consists of Si atoms in oxidation states intermediate between Si and SiO_2. A second region extends about 30Å into the into the SiO_2 overlayer. The SiO_2 in this region is compressed because the bond density at the Si surface is too high to be matched by the bond density of SiO_2. Several structural models have been proposed (1-8) for the SiO_2/Si (100) interface. Most of them assume an atomically abrupt interface. We focus onto the near interface region and study the distribution of intermediate oxidation states (Si^{1+}, Si^{2+}, Si^{3+}) using core level spectroscopy (9-11). From our data, we can exclude abrupt interface models, since they cannot explain the large portion of Si^{3+} observed at the interface. Recent calculations by Ohdomari, et al. (8) have arrived at an extended interface model for $SiO_2/Si(100)$ by minimizing the strain energy (8). This model is found to be in good agreement with our data. It is not unique, though. A new model is proposed for $SiO_2/Si(100)$ based on the observed distribution and intensity of intermediate oxidation states. Both models are characterized by an extended interface, with protrusions of Si^{3+} reaching about 3Å into the SiO_2 overlayer. In our work, we consider mostly SiO_2/Si interfaces grown in pure O_2, i.e., free of hydrogen or OH. By covering a broad range of oxidation temperatures and pressures we ascertain that the interface properties hold for a wide class of oxides. The preparation conditions play only a minor role as long as one starts with clean and smooth Si surfaces.

EXPERIMENTAL

Thin SiO_2 films were obtained by oxidizing atomically-clean Si surfaces in pure O_2. Clean Si(111) 7x7 and Si(100) 2x1 surfaces were prepared by ohmic heating of Si wafers to 1050°C or by dipping in 10% HF and heating to 850°C. Using an ultrahigh vacuum transfer system, the clean Si samples were transferred from the spectrometer (pressure 10^{-10} Torr) to a bakeable chamber which could be filled with gases up to a pressure of 1 atmosphere and subsequently pumped down to ultrahigh vacuum. During all oxygen exposures, the sample was cooled while still in oxygen. Thereby, the formation of holes in the SiO_2 layer is prevented, which occurs when a thin SiO_2 film is heated in vacuum. Vacuum-annealed films with pinholes exhibit a larger signal from bulk Si. The density of Si atoms in intermediate oxidation states is underestimated when using such data.

The experimental technique employed in this work is core level spectroscopy. It has the advantage of resolving the four oxidation states (9,10) of Si via the chemical shift of the Si2p core level. The depth resolution is optimized by using tunable synchrotron radiation. At a photon energy of 130 eV, the escape depth (11) of photoelectrons from the Si2p core level has a minimum of about 3Å in Si (7 Å in SiO_2) compared with about 15Å (25Å) in conventional XPS. The short escape depth makes it easy to detect monolayer quantities of Si atoms at the interface. For clarity, all Si2p core level spectra are decomposed into the $Si2p_{1/2}$ and $Si2p_{3/2}$ spin-orbit partner lines using a splitting of 0.60 ± 0.01 eV and the statistical $2p_{1/2}$:$2p_{3/2}$ intensity ratio of 1:2.

DISTRIBUTION OF OXIDATION STATES

In the $Si2p_{3/2}$ core level spectra shown in Fig. 1, one can clearly identify three peaks between the Si and the SiO_2 lines. They are due to intermediate oxidation states, which are labelled Si^{1+}, Si^{2+}, Si^{3+}. One should keep in mind that these symbols refer to the oxidation state, not to the actual charge transfer. The charge transfer is only about half an electrons per oxidation state (9). The energy positions and the widths of the various core lines are given in Table 1. Fig. 1 compares $Si2p_{3/2}$ core level spectra for SiO_2 films of different thickness. The intensity distribution of intermediate oxidation states $I^{3+}:I^{2+}:I^{1+}$, does not vary much with the oxide overlayer thickness, showing that this is a universal property of the SiO_2 interface. It depends only on the crystallographic orientation of the Si substrate. Si(111) produces less Si^{2+} than Si(100). Results for the intensities obtained from least squares fitting of the data in Fig. 1 are given in Table 1. The intensity of intermediate oxidation states relative to that of the Si substrate (I^0) is a measure of their density. It is generally independent of the layer thickness. However, it can be affected both ways by poor sample preparation. Pinholes due to vacuum annealing of thin SiO_2 films cause the Si signal to increase. Rough starting surfaces with poor surface core level signals produce more intermediate oxidation states.

Several processing parameters have been varied in order to test the universal nature of the SiO_2/Si interface. Pressure and temperature variations (from 10^{-5} Torr to 20 Torr and from 700°C to 1100°C rsp.) do not affect the distribution of oxidation states significantly. We have also looked into the effect of H_2 annealing on the distribution of intermediate oxidation states (Fig. 2). Such annealing is used to reduce the number of interface states in device oxides. There is no noticeable difference. This is understandable since the number of dangling bonds (P_b centers) that can be saturated by hydrogen reaches at most about $10^{12}cm^{-2}$ (12) i.e., 1/1000 of a monolayer. Such small changes are not detectable directly with our technique, but the unpinning of the Fermi level can be seen as an overall shift of the spectrum. The presence of H_2, e.g. in steam oxidation, has been reported (13) to give rise to an abrupt interface where 6%-17% of the interface bonds are saturated by H. In this case, the higher temperature may cause the strong Si-O-Si bridge bonds to break up. However, there are alternative explanations for the low density of intermediate oxidation states reported in (13). The back-etching of thick SiO_2 films by HF may have created openings with H-terminated silicon (14). Also, the intensity of Si^{1+} may have been underestimated due to the proximity of the dominant bulk Si line in XPS.

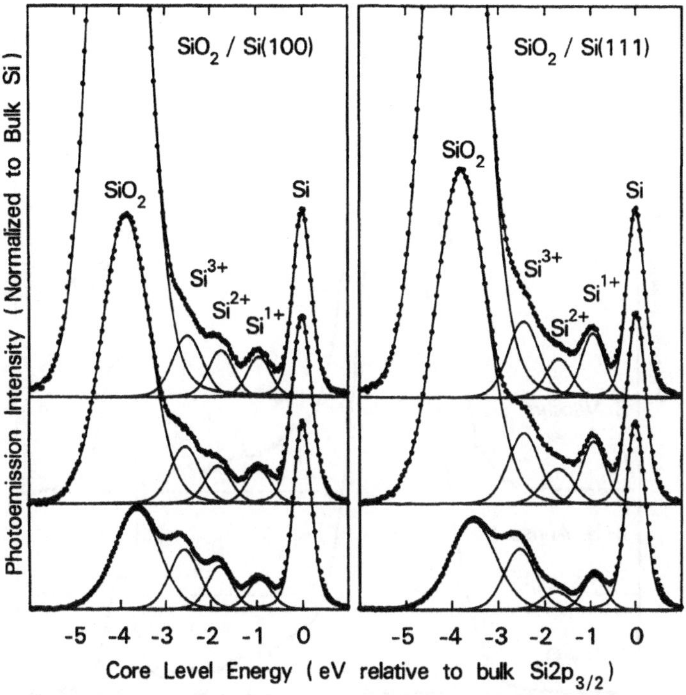

Fig. 1. Intermediate oxidation states at the SiO_2/Si(100) and SiO_2/Si(111) interfaces. All three intermediate oxidation states are seen. Si(111) exhibits less Si^{2+} than Si(100). The oxide thickness (5Å, 9Å, and 14Å for the three spectra) does not affect the distribution of intermediate oxidation states.

INTERFACE STRUCTURE

A coarse look at the distribution of oxidation states shows that the interface is not ideal, as evidenced by the coexistence of all three intermediate oxidation states Si^{1+}, Si^{2+}, Si^{3+}. For a truncated bulk structure one would obtain only Si^{2+} for Si(100) and Si^{1+} or Si^{3+} for Si(111). In order to come up with sensible models, we have to consider the possible driving forces for the interface structure, mainly the density mismatch between SiO_2 and Si. The SiO_2 lattice has a density of Si atoms that is 2.2 times lower than that in Si. Consequently, a SiO_2 surface has only half as many open bonds as a Si surface. In the following we will discuss how this difference can be accommodated for SiO_2 on Si(100).

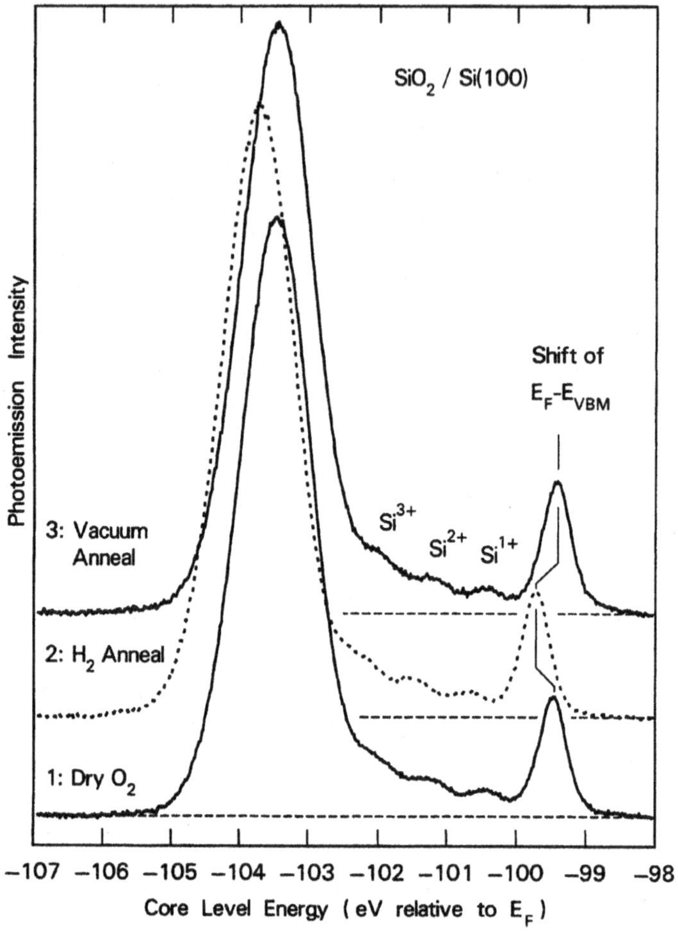

Fig. 2. Effect of H_2 annealing (1 atmosphere H_2, 850°C, 10 sec.) on the SiO_2/Si(100) interface. A rigid shift of the whole spectrum is caused by the Fermi level moving from a pinned position to its bulk position. The distribution of oxidation states does not change, however, since the number of pinning defects is too small to be detected. Subsequent annealing in vacuum (850°C, 1 min.) restores the pure interface by driving off hydrogen.

Table 1. Energies, widths, intensities, and photoionization cross sections for the Si2p core levels at a photon energy $h\nu = 130eV$. To obtain the relative abundance of oxidation states, the intensities I have to be divided by the cross sections σ.

		Si^{4+}	Si^{3+}	Si^{2+}	Si^{1+}	Si^{0}
Energy (eV)		-3.9^{a}	-2.48^{a}	-1.75^{a}	-0.95	0
FWHM (eV)^{b}		1.15	0.66	0.58	0.44	0.28
$\left(\frac{I}{I^{1+}+I^{2+}+I^{3+}}\right)$	for Si(111)		0.48	0.21	0.31	0.74
$\left(\frac{I}{I^{1+}+I^{2+}+I^{3+}}\right)$	for Si(100)		0.48	0.28	0.24	1.1
σ/σ_{Si}		2.2	1.7	1.1	1.0	1

a) The Si^{4+} core level moves up for thin films due to a smaller valence band offset and extra screening by the Si substrate. The Si^{3+} and Si^{2+} core levels move down for films thinner than 5Å due to the absence of dielectric screening by the SiO_2 overlayer.

b) Instrumental resolution subtracted.

The truncated bulk structure of Si(100) can be connected to an amorphous SiO_2 network despite the density mismatch, as shown by Pantelides et al. (1). Such a model has a full monolayer of Si atoms in the 2+ oxidation state. An epitaxial model by Herman et al (2,3) has the density difference between SiO_2 and Si built in. This model uses a diamond-like structure of SiO_2 which is obtained by straightening the Si-O-Si bonds in the β-cristobalite structure. The bond length ratio between Si-O-Si and Si-Si is close to $\sqrt{2}$. Therefore, an epitaxial $\sqrt{2} \times \sqrt{2}$ R 45° structure can be formed on Si(100). In this case, the bond density across the interface changes by a factor of 2. Consequently, only half of the atoms at the Si(100) surface are connected to oxygen in SiO_2. To absorb the remaining broken bonds, one has to introduce impurities such as H, F, and OH at the interface. Alternatively, oxygen atoms may form double bonds with the free Si surface atoms. The distribution of oxidation states expected at such interfaces is between 0.5 and 1.0 monolayer of Si^{2+} and up to 0.5 monolayer of Si^{1+}. Thereby, Si atoms bonding to hydrogen are counted as Si^{0}. An alternative way to take care of the density mismatch has been proposed recently by Ourmazd et al. (6). Only one of the two broken bonds of the Si surface atoms is connect to SiO_2. The other is presumed to pair up with a neighboring Si atom in a 2x1 structure, like on the clean Si(100) surface. This configuration matches tridymite, another crystalline configuration of SiO_2. The interface consists of a full monolayer of Si^{1+} for this model. A third possibility to fix the bond density mismatch has been proposed by Ohdomari et al (7,8). By terminating the Si(100) surface with Si(111) facets one is able to reduce the dangling bond density to about one per surface atom, resulting in about a monolayer of Si^{1+} at interface. All of the atomically abrupt interface models discussed so far give only Si^{2+} and Si^{1+} as intermediate oxidation states. A look at the bond topology of the Si(100) surface shows that the common defects (e.g. adatoms, vacancies, steps) also produce only Si^{2+} and Si^{1+}, but not Si^{3+}. In order to explain the observed Si^{3+} peak we have to resort to extended interface models.

A solution to the density mismatch problem is a graded interface which is about 2 atomic layers wide (see Fig. 3). Strain energy calculations by Ohdomari et al. (8) show that a properly-graded $SiO_2/Si(100)$ interface has a signficantly lower strain energy than an abrupt one. A specific model has been proposed based on strain energy minimization (Fig. 3 model A). We add a second structure (Fig. 3, model B), which agrees equally as well with our data. A peculiar feature common to these two models is the occurrence of Si^{3+} protrusions into the SiO_2 layer.

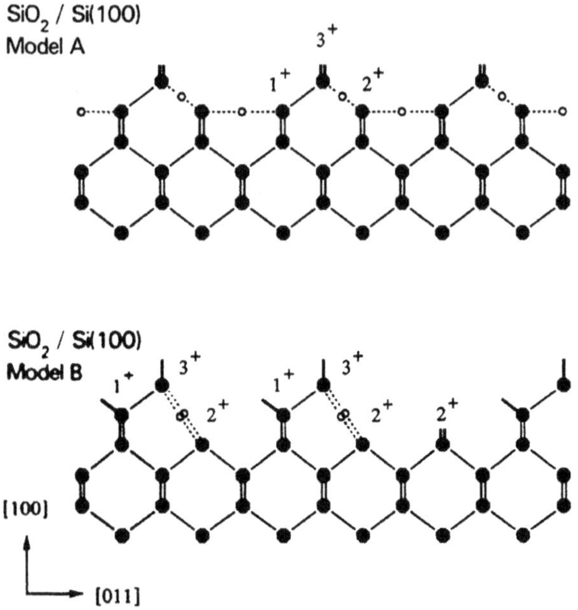

Fig. 3. Models of the $SiO_2/Si(100)$ interface that match the distribution and intensity of intermediate oxidation states. Model A is similar to a structure proposed by Ohdomari et al. (8) based on strain energy minimization. Open bonds are to be connected to a SiO_2 network via oxygen bridge bonds. The bond topology is given without taking relaxation into account.

Acknowledgement: We acknowledge M. Liehr for providing calibrated oxide films and B. Robinson for ellipsometry measurements. Research was carried out in part at the National Synchrotron Light Source, Brookhaven National Laboratory, which is sponsored by the U. S. Department of Energy, Division of Materials Sciences and Division of Chemical Sciences under contract number DE-AC02-76CH0016.

*Present address: National Bureau of Standards, Surface Science Division, Gaithersburg, MD 20899.

References

1. S. T. Pantelides and Marshall Long, in *The Physics of* SiO_2 *and its Interfaces*, ed. Sokrates T. Pantelides, (Pergamon Press, NY, 1978) p. 339.
2. F. Herman and I. P. Batra, *The Physics of* SiO_2 *and its Interfaces*, ed. Sokrates T. Pantelides, (Pergamon Press, NY, 1978) p. 333.
3. F. Herman, and R. V. Kasowski, J. Vac. Sci. Technol. **19(3)**, 395 (1981).
4. T. Hattori, M. Muto, and T. Suzuki, Proceedings of the 17th International Conference on the Physics of Semiconductors, ed. by D. J. Chadi and W. Harrison (Springer-Verlag, New York 1985), p. 229.
5. F.Rochet, S. Rigo, M. Froment, C. D'Anterroches, C. Maillot, H. Roulet and G. Dufour, Adv. in Phys. **35**, 237 (1986).
6. A. Ourmazd, D. W. Taylor, J. A. Rentschler, and J. Bevk, Phys. Rev. Lett. **59**, 213 (1987).
7. I. Ohdomari, H. Akatsu, Y. Yamakoshi, and Koji Kishimoto, J. of Non-Crys. Solids **89**, 239 (1987).
8. I. Ohdomari, H. Akatsu, Y. Yamakoshi, and K Kishimoto, J. Appl. Phys. **62**, 3751 (1987).
9. F.J. Grunthaner, P.J. Grunthaner, R.P. Vasquez, B.F. Lewis, J. Maserjian, and A. Madhukar, Phys. Rev. Lett. **43**, 1683 (1979) and J.Vac. Sci. Technol. **16**, 1443 (1979).
10. G. Hollinger and F. J. Himpsel, Appl. Phys. Lett. **44**, 93 (1984).
11. For a detailed account of our work see F. J. Himpsel, R. R. McFeely, A. Taleb-Ibrahimi, J. A. Yarmoff, and G. Hollinger, Phys. Rev. B, to be published.
12. N. M. Johnson, D. K. Biegelsen, M. D. Moyer, and S. T. Chang, Appl. Phys. Lett. **43**, 563 (1983).
13. P. J. Grunthaner, M. H. Hecht, and F. J. Grunthaner, J. Appl. Phys. **61**, 629 (1987).
14. It has been found that a Si surface etched in aqueous HF solution is terminated mainly by hydrogen, see E. Yablonovitch, D.L. Allara, C.C.Chang, T. Gmitter, and T.B. Bright, Phys. Rev. Lett. **57**, 149 (1986), and B. S. Meyerson, F. J. Himpsel, and J. A. Yarmoff, unpublished.

ARXPS ANALYSIS OF $Si-SiO_2$-INTERFACES

J. Halbritter

Kernforschungszentrum Karlsruhe GmbH
Institut für Kernphysik II, Postfach 3640
D-7500 Karlsruhe, Federal Republic of Germany

ABSTRACT

Angle resolved x-ray photo electron spectroscopy (ARXPS) results in a very detailed analysis of minor amounts ($\geq$ 0.3 nm) of interface compounds and their spatial distribution. First experimental results on the $Si-SiO_2$ interface used in microelectronics, are presented. The ARXPS results on plane, single cristalline (100) Si, oxidized to about 5nm SiO_2, indicate a planar Si surface connected by about one monolayer Si* to a compressed SiO_2^* layer coated by amorphous SiO_2. Seperated from this interface region Si cluster stabilized by a compressing SiO_2^* coating have been found in SiO_2. Dehydrogenation is showing up in binding energy changes in SiO_2 indicating that H or OH is not only saturating Si* but is bonded also to SiO_2. The O deficiency of the amorphous SiO_{2-x} coating is increasing towards the outer SiO_2 surface.

1. INTRODUCTION

The $Si-SiO_2$ interface is one of the most extensively studied object in solid state physics [1,2]. Despite many studies and the success of microelectronics based on the excellent quality of the $Si-SiO_2$ interface achieved by phenomenological recipes, there still exist many questions. As summarized in [3-8], XPS (x-ray photo electron spectroscopy) gives reliable information on the chemical bonds via their binding energies (B.E.) shown for the Si 2p line in Fig.1 [7]. As obvious from Fig.1a, a fit to one XPS line for one angle cannot resolve interface oxides with certainty. But in our case shown in Figs.1a and b, the Si2p line changes with take-off angle θ of the photo electrons yielding so an enhanced data base. Thus the simultaneous fit of the XPS lines between 5° and 80° yields reliable information on the interface oxides and their chemical shifts as summarized in Table 1. At the same token, the simultaneous fit of the XPS lines for different take-off angles adds a new dimension to the analysis, namely a destruction free depths analysis. [9] In Section 3, our ARXPS results [7-9] are summarized, which have been obtained for a Si (100) single crystal oxidized at 900°C in $Ar-O_2$ to 5 nm SiO_2 suited for microelectronics [10]. Thus, we present here the first ARXPS results on the $Si-SiO_2$-interface, which are also the first (AR)XPS measurements on SiO_2 barriers suited for microelectronics. XPS result presented before [3-6] are obtained on thin ($\leq$ 1 nm) SiO_2 coatings or on

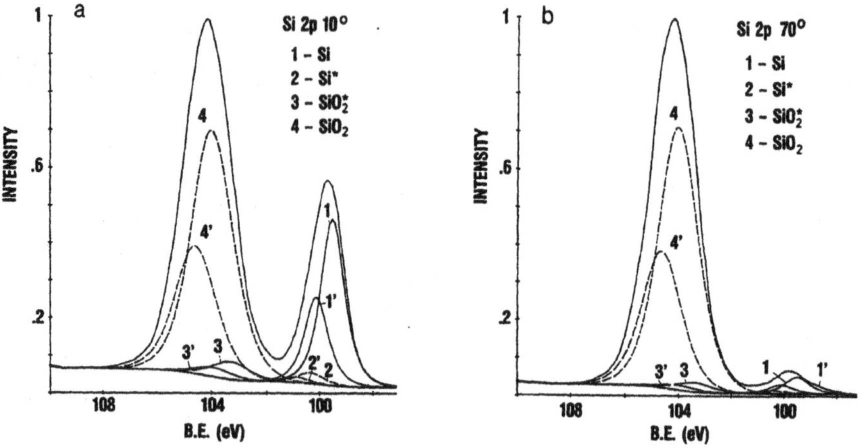

Fig. 1. a and b: Si 2p photoelectron peaks fitted by the minimal number of doublets allowing a simultaneous fit ($10 \leq \theta \leq 80°$) after background subtraction of the 100 plane of Si oxidized at 900° by an O-Ar mixture to about 5 nm and annealed in H_2. The fit started with the known doublets of Si and SiO_2 and then the intermediate compounds were identified. The Si 2p 3/2 positions are in [eV]: Si ≙99.4; Si* ≙100.3; SiO_2^* ≙103.2 and SiO_2 ≙103.9. (a) θ = 10° take-off angle and, (b) θ = 70° take-off angle.

T A B L E 1. Mean escape depths λ of Si 2p for Al K_a and binding energy shifts ΔE relative to Si and half widths FWHM for Si (100) oxidized in O_2 + Ar at 900°C and hydrogenated if not otherwise stated

		Si λ/nm 2.5		Si* 2.56		SiO_2* 2.66		SiO_2 2.7	
		ΔE/eV	FWHM/eV	ΔE/eV	FWHM/eV	ΔE/eV	FWHM/eV	ΔE/eV	FWHM/eV
5nm SiO_2	10°	0	1.04	0.9	1.2	3.8	1.6	4.5	1.82
	70°	0	1.1	0.9	1.2	3.9	1.5	4.55	1.80
5nm SiO_2 de-hydrogenated	10°	0	1.02	0.7	1.2	4.5	1.6	4.7	1.8
	70°	0	1.1	0.8	1.4	4.6	1.6	4.8	1.77
7nm SiO_2	10°	0						4.35	
10 nm SiO_2	10°	0						4.4	

chemically thinned (<5nm) SiO_2 coatings on Si. Both are not representative for the Si-SiO_2 interfaces used in microelectronics as shown by our ARXPS results.

Our enlarged ARXPS (angle resolved XPS) data base yields the following results (Table 1) discussed in Sect. 4: as interface oxides we found Si*, which may be assigned [3-6] to Si^{1+} or Si^{2+}, and SiO_2*. SiO_2* we assign to compressed SiO_2^* which recently has been identified as strained cristalline SiO_2 [11]. In situ dehydrogenation showed the growth of the amount of Si* and the growth of the binding energies of SiO_2* and

SiO_2 by more than 0.3 eV. The destruction free ARXPS depth analysis confirmed [1], that in oxidation (100) Si single crystals stay planar being connected by the planar Si* layer (-0.3nm) to the compressed SiO_2^* layer (~0.6nm) confirming the strained layer epitaxy [11] quantitatively. This strained layer epitaxy allows ordered strain relaxation at the interface via epitaxial SiO_2^* films which decay (≧1nm) into amorphous SiO_2. The amorphous SiO_2 contains in distances larger than 3.5 nm from Si Si cluster compressed by SiO_2^*. The O/Si^{4+} - ratio in amorphous SiO_2 declined towards the outside surface, which can only partly be accounted for by the Si-SiO_2^* clusters.

2. XPS-ANALYSIS [7-9]

The photoelectron spectra are taken with a AEI ES 200 ESCA spectrometer, where the photoelectrons are excited with the Al $K_{\alpha 1,2}$ lines. The samples were mounted on a rotatable sample holder enabling the collection of photoelectrons (aperture ≤ 10°) emitted at take-off angles θ between 5° and 85°. Each XPS-line is taken 2-8 times and averaged to reduce the noise. Before fitting the data with a Zenith Z-100 PC the background caused by inelastic processes and satelites is subtracted. These data are then fitted by a model function which is the sum of a Gaussian and a Lorentzian, where the Gaussian contribution increases with FWHM (full width half maximum). Then angle dependent line shapes shown Fig.1 are fitted simultaneously, yielding an enhanced sensitivity in stoichiometries and depths. This simultaneous fit makes ARXPS a sensitive, nondestructive method to analyze interfaces up to depths of about 10-20 nm. The stability of the fit is proven by measuring differently prepared surfaces.

To evaluate the stoichiometry and the distribution quantitatively the fitted peak area $F \propto n\sigma\lambda$ is used with n: atomic concentration of the element producing the peak; σ: photo ionization cross section of the atomic orbital and λ: mean escape depth of photoelectrons - see Table 1. Interfaces may show a sequence of stoichiometries corresponding to different depths and distributions. These sequences can be plotted as shape functions $d_i(\theta)$ [9]. These shape function for a stack of layers (i) on a substrate (s,i=1)

$$d_i(\theta) = \lambda \cos\theta \ln[1 + \frac{F_i}{n_i \lambda_i} \cdot \frac{n_s \lambda_s \cdot T_2 \cdot T_3 \cdots T_{i-1}}{F_s}] \qquad (1)$$

are constant for flat surfaces and uniform layers, being the actual layer thickness = $d_i(\theta)$. $T_i = \exp(-d_i/\lambda_i \cos\theta)$ are the transmission coefficients. $d_i(\theta)$ = constant = thickness is due to the fact that the shielding $T_i(\theta)$ for the substrate is compensated by the cosθ forfactor in Eq. (1). For serrated surfaces the "substrate" is coated by an ordered sequence of oxides, where I_i is now attenuated by all T_i's, i.e. I_{ox}/I_s is decreasing with θ more strongly than for the planar case. This causes the decrease of all shape functions $d_i(\theta)$ with angle. [9] As shown in Fig. 2 for Si-SiO_2 - interfaces only d(θ) of SiO_2 and of Si* is decreasing strongly, whereas d(θ) of SiO_2^* peaks.

In Fig. 3 a model is presented [8] for this scenario fitting the shape functions in Fig. 2. The decrease of d(θ) of Si* and SiO_2, in contrast to SiO_2^* peaking at θ≈60°, is explained by Si-SiO_2^* clusters in SiO_2. Such Si cluster not only shadow the Si* and the Si^{4+} but also shadow O resulting so in a (fig. 2c) decrease of the O/Si^{4+} ratio with θ. In our analysis we neglect the small [12] elastic mean free path effect (< 10%), channeling and refraction [13] because the decline of O/Si^{4+} with θ in amorphous SiO_2 cannot be explained by these effects.

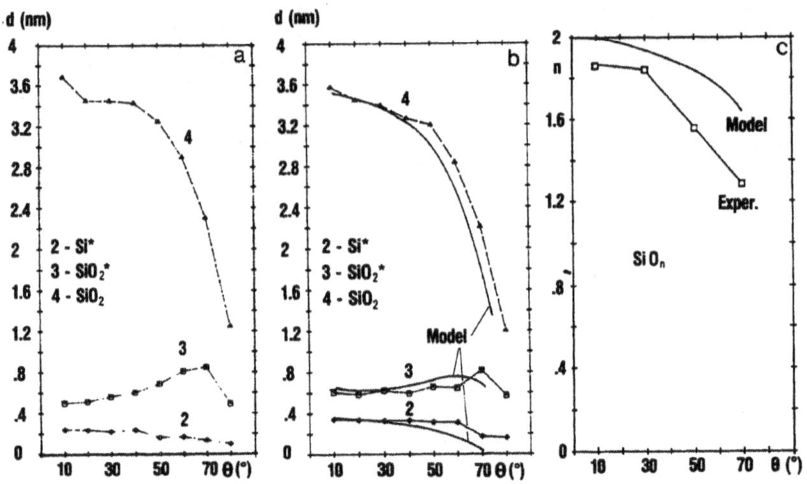

Fig. 2. Oxide thicknesses d(θ) on Si (100) oxidized at 900°C. The d(θ≥60°) dependencies indicate a fairly lateral oxide growth [7].

(a) annealed in H_2 (Fig. 1),

(b) UHV annealed for 2h at 690°C to drive H_2 out. This dehydrogenation changes the binding energies of SiO_2^* and SiO_2, as summarized in Table 1, indicative of bonding of H to SiO_2^* and to SiO_2 or due to strain relaxation. The dehydrogenation shows up also in the growth of the thickness of the Si* layer. As indicated, the shape functions can be fitted by the model of Fig. 3.

(c) increase of O deficiency of SiO_n ($n \leq 2$).

3. EXPERIMENTAL RESULTS ON THE Si-SiO_2-INTERFACES

The samples studied where cut from a Si single cristal (100) wafer, which was oxidized by Siemens at 900°C in a Ar-O_2 mixture followed by loading with hydrogen. [10] The ellipsometric SiO_2 thickness was stated as 5 nm. As discussed by the author [7,8], the fit of the usually assumed [6] interface oxides Si^{1+}, Si^{2+} and Si^{3+} and strained SiO_2 to our ARXPS data gave a fit worse than our fit assuming only the two lines Si* and SiO_2^* between Si and SiO_2. The fit with the two intermediates, Si* and SiO_2*, was also stable by UHV annealing the hydrogenated Si-SiO_2 samples in the XPS system . By this dehydrogenation the amount of Si* grows (Fig. 2) and the binding energies of SiO_2* and SiO_2 shifts to higher values and both became more similar in ΔE (Table 1). In Fig. 4 the shape function of the total oxide thickness d(θ) for thicker oxides [10] are plotted. In Fig. 2c also the ratio O/Si^{4+} is plotted. First we want to point out that the effect of the CH_x adsorption layer (~0.3 nm) have been substracted and that a similar θ dependence and 5% higher O/Si^{4+} ratios have been observed before the UHV anneal. [7] An oxygen deficiency increase with θ indicates that there is less O at the outside SiO_2-surface despite the oxidation by O_2 and despite a hydrocarbon adsorption layer containing additional oxygen not bonded to Si.

4. DISCUSSION OF Si-SiO_2-INTERFACES

Before analyzing our ARXPS results some commonly accepted properties are summarized:

- SiO_2 grows by oxygen and water diffusion tnrough SiO_2 towards the SiO_2-Si-interface, where
- SiO_2 is nucleated accompanied by a volume increase of a factor of 2.2, which
- compresses SiO_2 severely and strains Si, where a
- backflow of O containing species to the SiO_2 outside surface is observed also [14].

Till now, the thermal oxidation is usually assumed to proceed uniformly. But the volume increase by SiO_2 nucleation makes this unlikely, because the compression of SiO_2 reduces, e.g., O_2 and H_2O solution and diffusion. [8] Indication for a nonuniform oxide growth are the recently identified backflow of an O containing species away from Si towards the SiO_2-surface. [14]

Discussion of XPS-measurements

As pointed out above, the XPS studies usually concentrate on the initial stages of oxidation (≲1nm,) [4,5] or on thick oxides (≳2nm) [6], which have been thinned by chemical etching to thicknesses suited for XPS analysis. As shown by the cristalline interface layer [11] and by our successful fits with Si and SiO_2^* alone, both types of SiO_2 are quite distinct and not representative for Si-SiO_2 interfaces used in microelectronics. In the following the ARXPS analysis of Si-SiO_2 interfaces used in microelectronics is reported. The simultaneous fit of ARXPS results has already been discussed in Sec. 2 [7-9], the results of Si-SiO_2-interfaces are summarized in Table 1 and Figs. 1-4. Before comparing our fits with Refs. 4 and 5, the shifts of Si* SiO_2* and SiO_2 with dehydrogenation are discussed: Table 1 shows, that with dehydrogenation a 10% ΔE reduction occurs for Si* , whereas ΔE of SiO_2* increases from 3.8 to 4.6 eV, i.e. SiO_2* approaches SiO_2 in ΔE with ΔE(SiO_2) increasing from 4.5 eV to 4.8 eV. Because dehydrogenation corresponds to a relief of compression and because of the spatial position (Fig. 3) SiO_2* is identified by strained SiO_2. This identification has recently been confirmed quantitatively in thickness by the TEM result. [11] The binding energies for SiO_2 are typical for thick (> 2 nm), thermal oxides [4,6,15] but the shift from 4.5 to 4.8 eV is astonishingly large. The other astonishing fact is, that SiO_{2-x} is understoichiometric, where x is increasing with take-off angle θ (Fig. 2c).

To explain the ΔE enhancements by 690°C UHV anneal for SiO_2* and SiO_2, the larger ΔE for thicker oxides [6,7] as compared to thin SiO_2 layers (≲ 1 nm), [4,5,15] the author proposes, that in very thin oxides SiO_2 ≈ SiO_2^* is heavily strained, cristalline SiO_2 [11]. This heavy strain yields oxygen vacancies or bond angle changes causing the smaller ΔE values. This picture is confirmed by Table 1 where ΔE(SiO_2*) ≈ 3.8 eV agrees quite well with ΔE(SiO_2) ≈ 3.6 eV for thin oxides. [4,5,15] For thicker oxides (≳ 2 nm) the compression is relaxed by viscous flow in the amorphous SiO_2, so that only close to Si SiO_2* exists modelled in Fig. 3. Also very long room temperature relaxation yield this flow and the ΔE(SiO_2) increase. [6] With annealing H_2 out and strain relaxation, ΔE(SiO_2*) and ΔE(SiO_2) are growing in line with this proposal and ΔE(SiO_2*) ≈ ΔE(SiO_2) is approached. This is in line with the shape functions of Si*, SiO_2 and SiO_2* (Fig. 2) not changing with this anneal.

Si* (Table 1) fits [4-6] very well to Si^{1+} by ΔE and by the property, that its amount is growing with dehydrogenation as the interface states [1]. The small half width and the amount of Si* of about 1ML (Fig.3) fits well to the TEM [11] results of an ordered transition layer. The question if this ordered transition layer Si* is "Si^{1+} or, e.g., Si^{2+}" seems academic to the author, because Si* is a "transient" between Si and SiO_2. As shown by Table 1 by the unchanged FWHM's and by Fig. 2,

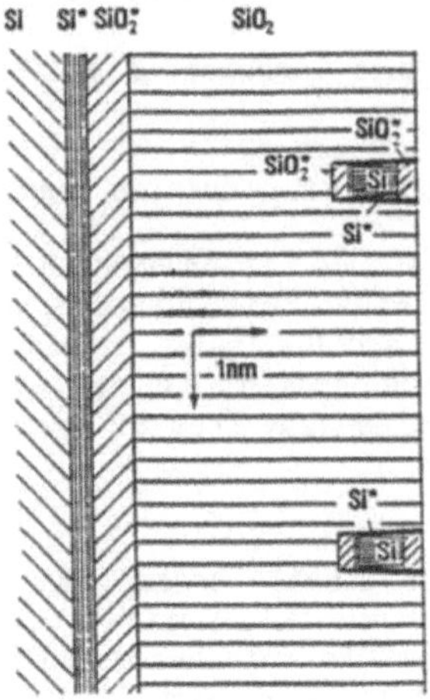

Fig. 3. Model of the $Si\text{-}SiO_2$-interface suited to describe the ARXPS results.[7] The shape function resulting from this model are plotted in Figs. 2b and c.

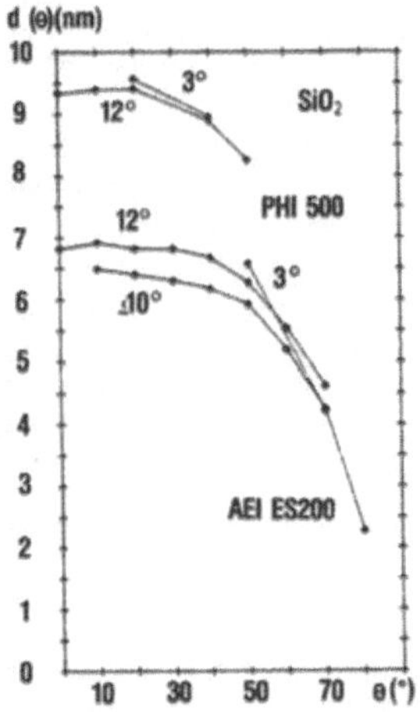

Fig. 4. Shape functions of SiO_2 oxidized to 7 and 10 nm and measured in different XPS systems.

the dehydrogenation starts with "homogeneous, amorphous SiO_2" and ends with "homogeneous , amorphous SiO_2". Also the $Si^*\text{-}H\text{-}SiO_2^*$ bonds are well defined as indicated by the identical FWHM's and by the negligible ΔE changes with dehydrogenation of Si*. Above SiO_2^* was identified as strained cristalline [11] SiO_2 where the uniform, cristalline layer explains well the ΔE and the halfwidth (Table 1) being smaller than FWHM of amorphous SiO_2. This strained, cristalline SiO_2^* connects epitaxially the Si-Si* surface to partly strain relaxed amorphous SiO_2 [11] and is thus quite different from thin (<1nm), in situ oxidized $Si\text{-}SiO_2$ interfaces. In the latter cases [4,5,15,18] the oxides are strained and cristalline, as discussed above, but neither uniform nor ordered, which explain the observed Si^{2+} and Si^{3+} contents [4,5].

The observed angle dependencies yield the following result on the spatial distributions at the $Si\text{-}SiO_2$-interface (Fig. 2): SiO_2 is the outermost layer with $d(\theta)$ decreasing with θ, as confirmed by Carim [16]; SiO_2* underneath shows a uniform thickness of $d \approx 0.6$nm, where $d(\theta)$ is peaking at $\theta \approx 60°$; and Si* on the Si surface has a uniform thickness around 0.3 nm. The weak angle dependencies of $d(\theta)$ of Si* and SiO_2* rule out serration found in the case of Nb or NbN [7,17]. As shown in Fig. 3, the shape functions are explained by a planar Si surface, which is coated by a Si* layer and a SiO_2* layer. To model the decreases of $d_i(\theta)$ of SiO_2 being similar to the O-deficiency [16] in Fig. 2c and the peak in $d_i(\theta \approx 60°)$ of SiO_2* simultaneously, $Si\text{-}SiO_2^*$ clusters are proposed in Fig. 3. These Si cluster may be remnants from the initial, thin (<2nm) cristalline oxide where excess Si has been found at the SiO_2 surface [18]. The cluster may carry the O back flow [14] and seem to decrease in cristallinity and size with further SiO_2 growth beyond 6nm. [16]

In Fig. 3 lateral distances between $Si\text{-}SiO_2^*$ clusters of about 5 nm are assumed. In the fits cluster size and distance enter as product, thus, 10 nm distance yields Si-SiO_2^* cluster of 0.8 nm size. The thickness of the oxide layer in Fig. 2 is with 4.5 or 4.2 nm thinner than the ellipsometric thickness [10]. This discrepancy that ellipsometric thicknesses are larger than XPS thicknesses is well known [4] and is easily explained by the Si-SiO_2^* cluster. The Si-SiO_2^* clusters explain part of the increased O deficiency with θ (Fig. 2c). The 10% O deficiency not explained by the clusters hint to O vacancies which may be viewed as "small Si cluster" being rather common in SiO_2 [3] or may be due to systematic errors discussed in Sect. 2 or Refs. 9 or 13. The O/Si^{4+} ratios of amorphous SiO_2 declining with θ in parallel to d(θ) of SiO_2 (Fig. 2) are strong hints for our model (Fig. 3).

5. SUMMARY

The above XPS analysis using an enhanced data base has shown, that ARXPS is a powerful method to study Si-SiO_2 interfaces in their make-up, stoichiometry and space wise. Our results for 4-5 nm thick SiO_2 coatings confirm the strained layer epitaxy for Si surfaces, where an epitaxial SiO_2 layer decaying into amorphous SiO_2 performs the strain relaxation. This yields a planar and ordered Si-Si*-SiO_2^* interface, which shifts defects away (> 1nm) from Si where the the actual strain relaxation involving O back flow happens.

Thin Si-oxides (≲ 1nm) may be microcristalline and irregular which in oxide growth can be converted to strained layer epitaxy at about 1-2 nm SiO_2. The study of this conversion and the shift of the residues from the thin initial oxide should clarify several important aspects of oxide growth and defects in SiO_2.

References

1. The Physics of SiO_2 and its Interfaces (S. T. Pantelides Ed., Pergamon, New York, 1978); Semiconductor Silicon (edited by H.R. Huff, T. Abe, B. Kolbesen Electrotechnical Society, N.J. 1986); Insulating films on Semiconductors (edited by M. Schulz, G. Pensl, Springer Series in Electrophysics, 7, Springer, Berlin, 1981)
2. M. Schulz, Surf. Sci. 132, 422 (1983)
3. N.S. Mc Intyre, W.M. Lau, J.R. Mycroft, S.I. Ingrey, to be publ.
4. M. Grundner, H. Jacob, Appl. Phys. A39, 73 (1986)
5. G. Hollinger, F.J. Himpsel, Appl. Phys. Lett. 44, 93 (1984); W. Braun, H. Kuhlenbeck, Surf. Sci. 180, 279 (1987)
6. F.J. Grunthaner, P.J. Grunthaner, Mat. Sci. Reports 1, 65 (1986);
7. A. Darlinski, J. Halbritter, J. Vac. Sci. Technol. A5, 1235 (1987)
8. J. Halbritter, J. Mat. Res. 2, May (1988)
9. A. Darlinski, J. Halbritter, Surf. Interface Anal. 10, 223 (1987)
10. Research Lab. Siemens-München
11. A. Ourmazd, D.W. Taylor, J.A. Rentschler, J. Bevk, Phys. Rev. Lett. 59, 213 ('87);F.M. Ross, W.M. Stobbs, Surf. Interf. Anal. 12 ('88)
12. A. Jablonski, H. Ebel, Surf. Interface Anal. 6, 21 (1984)
13. W. Huschka, E. Umbach, Diplom thesis (TU-München, 1988)
14. F. Rochet, B. Agius, S. Riger, J. Electroch. Soc. 131, 914 (1984); P. Collot, G. Gautherin, B. Agius, S. Rigo, F. Rochet, Phil. Mag B52, 1051 (1985) and to be published.
15. S.I. Raider, R. Flitsch in: The Physics of SiO_2 and its Interfaces (Ed. S.T. Pantelides, Pergamon, New York, 1978), 384
16. A.H. Carim, thesis (Stanford University, 1987)
17. J. Halbritter, Appl. Phys. A43, 1 and J. Less-Common Mat. 139, 133 (1987)
18. L.B. Hazell, I.S. Brown, F. Freisinger, Surf. Interface Anal.8, 25 (1986)

DEPENDENCE OF SiO_2/Si INTERFACE STRUCTURES ON OXIDATION PROCESS

Takeo Hattori, Hiroaki Yamagishi and Noboru Koike
Department of Electrical and Electronic Engineering
Musashi Inst. Technol., Setagaya-ku, Tokyo 158, Japan

Keitaro Imai and Kikuo Yamabe
VLSI Research Center, Toshiba Corp., Kawasaki 210, Japan

ABSTRACT

The XPS depth profiling of oxide films grown on Si(100) surface has been used to investigate the dependence of SiO_2/Si interface structures on oxidation process. Analysis of the XPS data indicates that almost abrupt and atomically flat interface is formed if oxidation time is shorter than stress relaxation time. The stress dependent oxide structures near the interface deduced from the measurements on chemical etching rate of the oxide films is also discussed.

INTRODUCTION

As a result of significant progress in microfabrication technology, approximately 10 nm thick silicon oxide film is now used as a gate insulator of Si MOSFET. The oxide structures with such thickness are expected to be affected by the interface structures. However, two important problems are not solved yet. Firstly, controversy exists on the orientation dependence of interface structures. Namely, orientation dependent interface structures were deduced from XPS studies(1-4), while orientation independent interface structure was deduced from surface sensitive XPS studies(5). However, it was found recently that suboxide distributions deduced from these studies(5) should be modified(6). According to our studies, orientation dependent abrupt and atomically flat interface can be realized only if the oxide film thickness increaes and approaches 10 nm. Therefore, the orientation dependence of interface structure should be discussed on oxide films with such thickness. Secondly, several studies were reported concerning the effect of oxidation process on the interface structures(4,7). However, the systematic studies on oxidation process to form abrupt and atomically flat interface were not performed yet. It is the purpose of the present paper to investigate the oxidation process to form abrupt and atomically flat interface on Si(100) surface.

INTERFACE STRUCTURES

The Si2p photoelectron spectra were analyzed by the same method as that described previously(2). Firstly, backgroud signal is considered to depend on material^{2+}. Secondly, the observed spectra were separated into three spectra originating from silicon substrate, silicon in silicon dioxide and silicon in silicon suboxides. The spectrum for suboxides is further separated into three spectra originating from suboxides designated by Si^{x+}(x=1,2,3) with bonding configuration of $Si\text{-}Si_{4-x}O_x$. The integrated spectral intensity for silicon in silicon substrate, that for silicon in silicon dioxide and that for silicon in silicon suboxides is expressed in the following by NS, NO and NI, respectively. Among these suboxides, Si^{2+} is related with interface structure on Si(100) surface(1-4). If the atomically flat interface is formed, transition layer at the interface consists of monolayer of Si^{2+} as shown in Fig. 1(2). In this case, NI is equal to numbers of photoelectrons emitted from Si^{2+}. Then, NI devided by NS is approximately equal to the areal density of Si^{2+} atom devided by $n_s \Lambda_s$. Here, n_s is the density of silicon atom in the substrate and Λ_s is the electron escape depth in silicon substrate. The value of (NI/NS) calculated for atomically flat interface is 0.057. According to the previous studies(2), the amounts of Si^{3+} distributed in the oxide film is less than one monolayer. Based on these considerations, the typical thickness dependence of (NI/NS) in Fig. 2 is calculated by assuming one monolayer suboxide at the interface and in the oxide film, respectively. Here, L expresses the range of suboxide distribution shown in Fig. 2. Therefore, the suboxide distribution can be deduced from thickness dependence of (NI/NS).

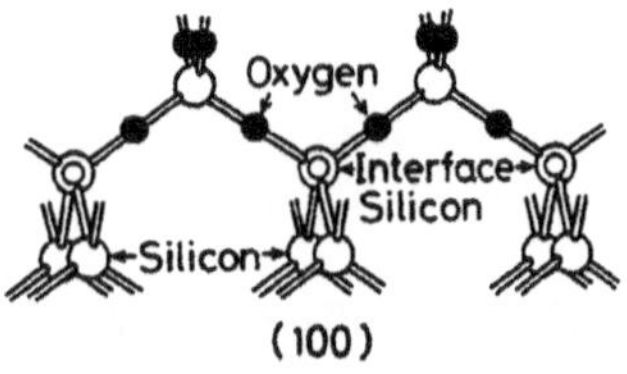

Fig. 1 Atomically flat interface on Si(100) surface

Figs. 3 and 4 show dioxide film thickness dependence of (NI/NS) obtained for oxide films grown on Si(100) surface in dry oxygen at 800 °C(2). Here, silicon dioxide film thickness is calculated from spectral intensity ratio (NO/NS). Figs. 3 and 4 were obtained from the nondestructive depth profiling measurements on oxide films having different thickness prepared at the same oxidation condition. Figs. 3 and 4 are obtained from Mg Kα radiation excited photoelectron spectra and Si KLL Auger electron spectra, respectively. In these figures, Si^{4+} spectral intensity NO normalized by NS is also shown as a function of dioxide film thickness. The difference between Figs. 3 and 4 arises from the difference in escape depth and leads to the following understanding. Namely, if suboxides is assumed to exist only at the interface, dotted lines in Fig. 4 is obtained from the experimental value of (NI/NS) in Fig. 3. On the other hand, if suboxides is assumed to exist not only at the interface but also in the oxide film, calculated curves for Si^{2+} and Si^{3+} in Fig. 3 and solid lines in Fig. 4 is obtained. From the comparison of these calculated curves with experimental values in Fig. 4, it is concluded that Si^{2+} exists mostly at the interface and Si^{3+} exist mostly in the oxide film. Therefore, Figs. 3 and 4 leads to the important aspect that the amount of Si^{2+} at the interface increases with the increase in oxide film thickness and approaches to atomically flat interface at dioxide film thickness of approximately 10 nm. Furthermore, destructive depth profiling of the oxide film grown at the same

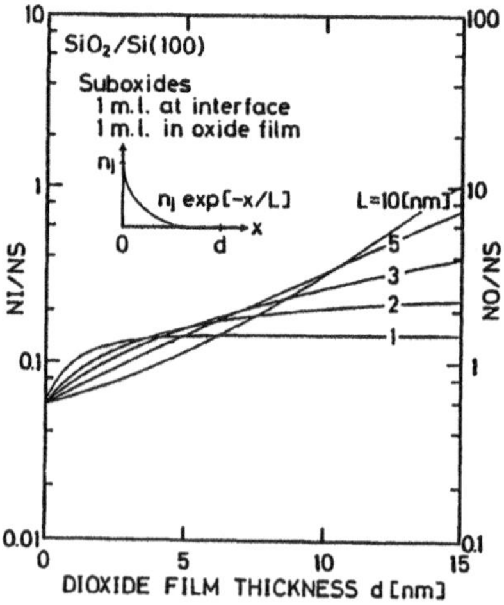

Fig. 2 Calculated SiO_2 film thickness dependence of (NI/NS) obtained for Mg Kα excited Si 2p spectra

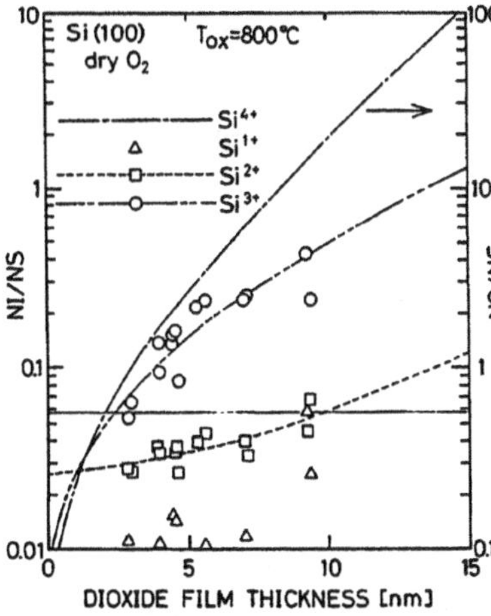

Fig. 3 Dependence of (NI/NS) on SiO_2 film thickness obtained from nondestructive measurement of Mg Kα excited Si 2p spectra

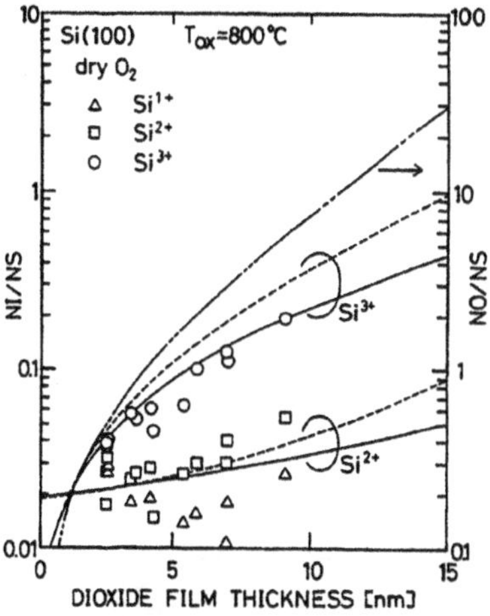

Fig. 4 Dependence of (NI/NS) on SiO_2 film thickness obtained from nondestructive measurement of Si KLL Auger electron spectra

oxidation condition with that used for the growth of oxide films discussed in Figs. 3 and 4 results in the thickness dependence of (NI/NS) shown in Fig. 5. Here, the depth profiling was performed using chemical etchinng. This figure confirms the existence of atomically flat interface for starting oxide film thickness of 9.3 nm.

Therefore, the flatness of the interface can be probed by XPS depth profiling. In the following, this technique is applied to investigate the dependence of flatness on oxidation and annealing conditions. Namely, from XPS depth profiling of Si^{2+} spectral intensity, thickness dependence of (NI/NS) is obtained to determine the flatness of the interface. Here, the flatness can be determined by the deviation from atomically flat interface, for which the value of (NI/NS) is 0.057 as discussed in the preceding section.

EXPERIMENTAL DETAILS

In order to clarify the flatness of the interface in atomic scale, silicon surface used for ultrathin oxide film growth should be step free surface. According to Hahn and Henzler(8), Si(100) surface with little atomic step densities can be prepared, if the oxide film with film thickness of more than 30 nm was formed in dry oxygen at oxidation temperature of more than 800°C and then removed. Based on these observation, the (100) surfaces with little atomic step densities were prepared by removing 100 nm thick oxide films grown in dry oxygen at 1000 °C. Oxide films studied were prepared by, 1) oxidation in 100% dry O_2 at 800 and 1000° C, 2) oxidation in dry atomosphere(10% O_2 + 90 %

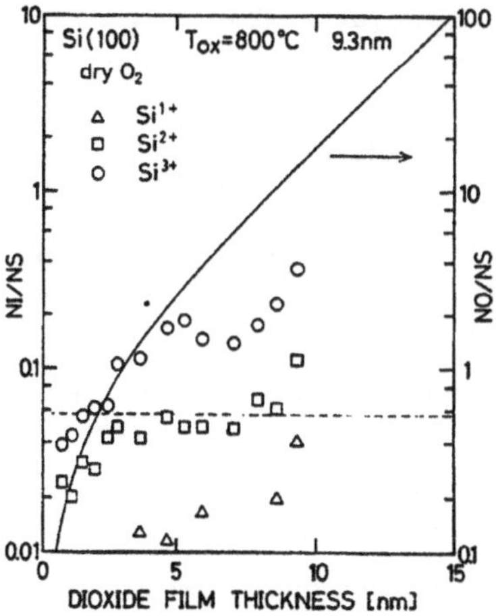

Fig. 5 Dependence of (NI/NS) on SiO_2 film thickness obtained from destructive depth profiling measurements.

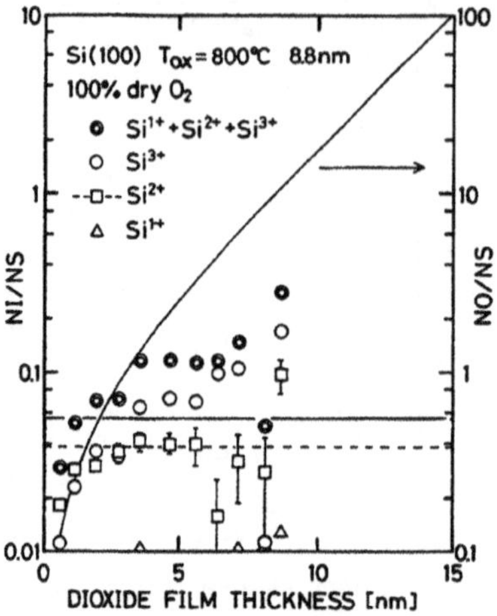

Fig. 6 Dependence of (NI/NS) on SiO_2 film thickness obtained from depth profiling measurements on 8.8 nm thick oxide film grown in 100% dry O_2 at 800°C

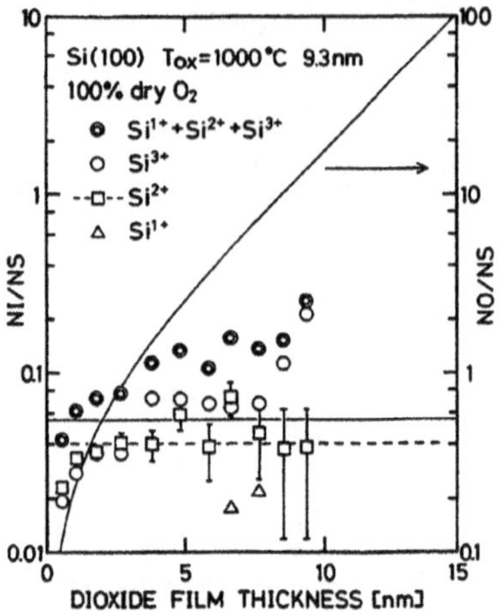

Fig. 7 Dependence of (NI/NS) on SiO_2 film thickness obtained from depth profiling measurements on 9.3 nm thick oxide film grown in 100% dry O_2 at 1000°C

argon) at 1000 ° C, 3) oxidation in dry atomosphere(10% O_2 + 90 % Ar) at 1000° C followed by one hour annealing in 100% dry argon at 1000° C. Depth profiling was performed using chemical etching. In this case a dilute HF solution ($HF:CH_3OH:H2O$ = 1:18:1) was used while maintaining solution temperature at 25°C and with supersonic vibration of the solution. Other experimental details were the same as those described previously(2).

A. DEPENDENCE OF INTERFACE STRUCTURES ON OXIDATION TEMPERATURE

In Figs. 6 and 7 dioxide film thickness dependence of (NI/NS) are shown for oxide films grown in 100% dry O_2 at 800 and 1000 ° C, respectively. The thickness of dioxide film studied in Figs. 6 and 7 is 8.8 and 9.3 nm, respectively. The value of (NI/NS) for Si^{2+} is 0.038 and 0.040, respectively.

B. DEPENDENCE OF INTERFACE STRUCTURES ON SiO_2 FILM THICKNESS

In Fig. 8 dioxide film thickness dependence of (NI/NS) are shown for oxide film grown in 100% dry O_2 at 1000 ° C. The dioxide film thickness studied is 22.5 nm. The value of (NI/NS) for Si^{2+} is 0.047.

C. DEPENDENCE OF INTERFACE STRUCTURE ON OXIDATION ATOMOSPHERE

In Fig. 9 dioxide film thickness dependence of (NI/NS) are shown for oxide film grown in dry atomosphere(10% O_2 + 90 % argon) at 1000 ° C followed by the slow cooling in dry atomosphere(5% O_2 + 95 %

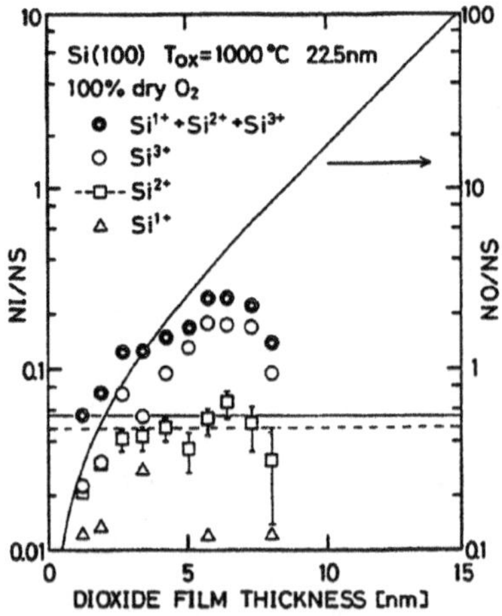

Fig. 8 Dependence of (NI/NS) on SiO_2 film thickness obtained from depth profiling measurements on 22.5 nm thick oxide film grown in 100% dry O_2 at 1000°C

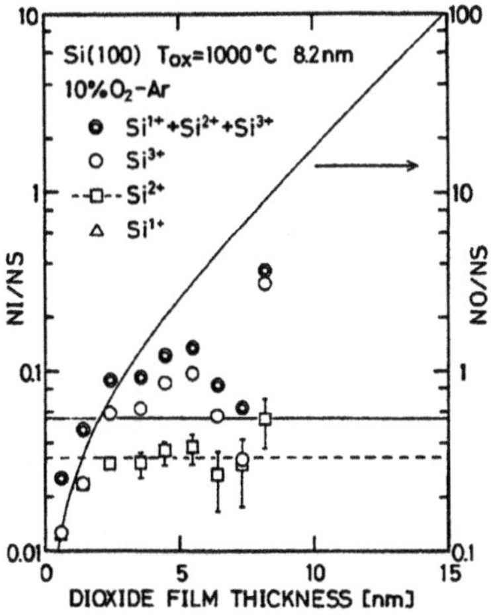

Fig. 9 Dependence of (NI/NS) on SiO_2 film thickness obtained from depth profiling measurements on 8.2 nm thick oxide film grown in (10% O_2/90% argon) at 1000° C.

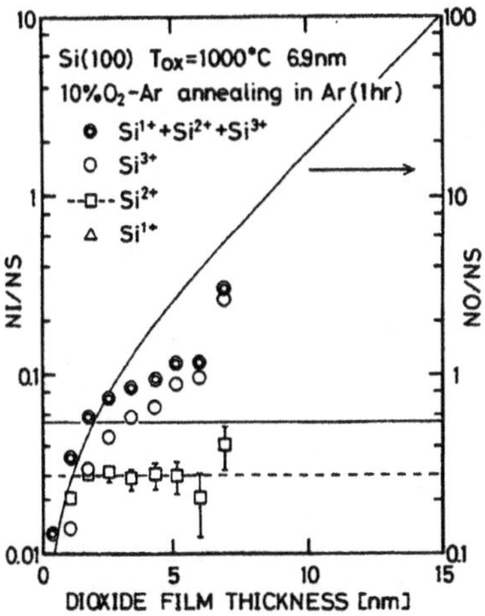

Fig. 10 Dependence of (NI/NS) on SiO_2 film thickness obtained from depth profiling measurements on 6.9 nm thick oxide film grown in (10% O_2/90% argon) at 1000 ° C followed by annealing in 100% dry argon at 1000 ° C.

argon) for 10 minutes. The dioxide film thickness studied is 8.2 nm. The value of (NI/NS) for Si^{2+} is 0.034.

D. DEPENDENCE OF INTERFACE STRUCTURES ON ANNEALING

In Fig. 10 dioxide film thickness dependence of (NI/NS) are shown for oxide film grown in dry atmosphere(10 % O_2 + 90 % argon) at 1000°C followed by annealing in dry argon at 1000 ° C for one hour. The dioxide film thickness studied is 6.9 nm. The value of (NI/NS) for Si^{2+} is 0.027. Almost same result was obtained for the same oxidation followed by annealing in dry nitrogen at 1000 ° C for one hour.

E. CHANGE IN CHEMICAL ETCHING RATE IN DEPTH DIRECTION

In Fig. 11 dioxide film thickness is shown as a function of chemical etching time. Etching interval is 60 seconds. In the case of oxide films grown in 100 % oxygen, etching rate is constant down to the thickness of 5.5 nm, then increases in the thickness range from 5.5 to 1.5 nm, and then decrease below the thickness of less than 1.5 nm. In the case of oxide films grown in 10 % dry O_2 diluted with argon followed by slow cooling in 10 minutes, the thickness range where etching rate increase, decrease. Furthermore, in the case of oxide films grown in dry (10 % O_2 + 90 % argon) followed by annealing in argon at 1000 ° C for one hour, the thickness range where etching rate increase, disappears.

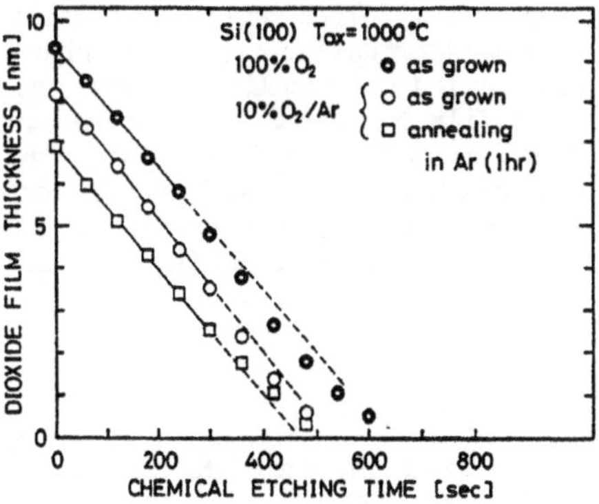

Fig. 11 Dependence of dioxide film thickness on chemical etching time

FLATNESS OF INTERFACE

From the depth profiling of Si^{2+} spectral intensities in oxide films prepared by various oxidation processes, following understanding were obtained.

(1) In the case that oxide films were grwon in 100 % oxygen at 800 and 1000° C shown in Figs. 6 and 7, the Si-SiO_2 interface is close to flat interface. In these cases, oxidation time of 5400 and 180 sec is shorter than stress relaxation time(9) of 1.9 x 10^7 and 720 sec, respectively.

(2) In the case of thick oxide film shown in Fig. 8, the interface is more close to flat interface as compared with thin oxide film case in Fig. 6. In this case oxidation time of 840 sec is comparable to stress relaxation time of 720 sec. However, in this case oxidation rate is faster than stress relaxation rate. This leads to atomically flat interface for thick oxide film.

(3) In the case of oxide film grwon in diluted oxygen shown in Fig. 9, interface is not close to flat interface. This can be attributed to the fact that oxidation time of 660 sec is comparable to stress relaxation of 720 sec.

(4) In the case of oxide films annealed after oxidation shown in Fig. 10, the flatness is further decreased as compared with fig. 8. Therefore, stress relaxation caused by annealing results in the decrease in flatness.

(5) The existence of stress can be probed by measuring the change in chemical etching rate as shown in Fig. 11. Namely, the increase in etching rate near the interface for oxide film grown in 100 % oxygen at 1000°C can be attributed to the existence of stress near the interface. The decrease in thickness range where etching rate increases is observed for the oxide film grown in dilute oxygen. This can be attributed to the stress relaxation caused by long oxidation time. The thickness range where etching rate increases disappears by annealing. This can be attributed to the disappearance of intrinsic stress. The chemical etching rate of oxide film grown at 800° C was larger than the chemical etching rate of oxide film grown at 1000° C.

CONCLUSION

Depth profiling of oxide films prepared by various oxidation process leads to the following conclusion. Firstly, from the dioxide film thickness dependence of (NI/NS) indicates that Si^{2+} suboxides detected exist mostly at the interface. Secondly, the amount of Si^{2+} depends strongly on the oxidation processes. Thirdly, Etching rates of oxide films change in distance from the interface especially for oxide films grown in 100 % dry oxygen at low oxidation temperature. These observation leads to following understanding: (1) In the case that oxidation time is shorter than stress relaxation time, interface approaches to atomically flat interface. The flatness increases with increas in oxide film thickness. In other words, in the case of 22.5 nm thick oxide film formed at 1000° C in 100% dry O_2, the largest value of (NI/NS), therefore the maximum flatness was obtained. (2) Stress relaxation caused by annealing at 1000° C results in the decrease in flatness. (3) The chemical etching rate of oxide film is affected by the existence of stress near the interface.

REFERENCES

1. T. Hattori and T. Suzuki, Appl. Phys. Lett., 43, 470 (1983).
2. T. Suzuki, M. Muto, M. Hara, K. Yamabe and T. Hattori, Jpn. J. Appl. Phys., 25, 544 (1986).
3. P. J. Grunthaner, M. Hecht, F. J. Grunthaner and N. M. Johnson, J. Appl. Phys., 61, 629 (1987).

4. F. J. Grunthaner and P. J. Grunthaner, Mater. Sci. Rep., 1, 65 (1986).
5. G. Hollinger and F. J. Himpsel, Appl. Phys. Lett., 44, 93 (1984).
6. W. Braun and H. Kuhlenbeck, Surf. Sci., 180, 279 (1987).
7. T. Hattori, M. Muto, T. Suzuki, K. Yamabe and H. Yamauchi, AIP Conf. Proc. No. 122, Proc. Int. Conf. on the Physics of VLSI, J. C. Knights, ed., p.45 (Am. Inst. of Physics, New York, 1984).
8. P. O. Hahn and M. Henzler, J. Vac. Sci. & Technol., A2, 574 (1984).
9. E. A. Irene, E. Tierney, and J. Angilello, J. Electrochem. Soc., 29, 2594(1982)

THE EFFECT OF ELECTROSTATIC SCREENING ON ENERGY POSITIONS OF SPECTRA NEAR SiO_2/Si INTERFACES

R.Browning, M.A.Sobolewski and C.R.Helms

Stanford Electronics Laboratory
Dept. of Electrical Engineering
Stanford University, Stanford, Ca 94305

ABSTRACT

XPS and Auger spectra of thin SiO_2 films grown on Si substrates are shifted when compared with those from thick oxide films. The spectral features associated with the SiO_2 films are shifted towards lower binding energies as the thickness of the film decreases. Using an electrostatic image charge calculation it can be shown that a large part of these spectral shifts may simply be due to the proximity of the semiconducting substrate. This brings into question any model of the SiO_2/Si interface based on these shifts.

INTRODUCTION

It has been recognized for some time that the X-ray photoelectron spectra(XPS) and Auger electron spectra (AES) from thin SiO_2 films grown on Si substrates show differences compared to those from similarly grown thick oxide films. The most well established of these differences are: the presence of additional spectral features, which are believed to be associated with SiO_x, and the shifts in energy of the transitions associated with SiO_2. These shifts, which are seen in both XPS and AES, are towards lower binding energies as the thickness of the oxide film decreases[1-3]. The oxygen KLL Auger transitions and the KLL and the LVV transitions from the silicon dioxide all shift a similar amount. The shifts are quite large, and are almost always observed for thermally grown oxide films. For films a few Ångstroms thick grown on Si substrates the oxide related spectral shifts are found to be in the 2-3eV range for Auger features, and 0.5-1.5eV for the XPS transitions. Thin films prepared from thicker thermally grown oxide films, either by ion bombardment[3] or by chemical etching[4], show similar effects.

The electron spectral shifts appear to correlate with other measurements on thermally grown oxide films and the interface between the oxide and the Si substrate. Ellipsometry[5], ion scattering[6], and measurements of electron mean free paths[7], all appear to show that there is good evidence for an extended SiO_2/Si interface region in the oxide, between 5-20Å in thickness. This interface layer has different physical and electronic properties to the bulk oxide, including a much increased effective density. Although there is apparently good evidence for this interface layer, and there has been considerable discussion of the spectral shifts, no consensus as to the actual origin of the spectral shifts has emerged, and the implications for properties of thin dielectric films are unclear.

The XPS and Auger spectral shifts have alternatively been related to chemical modifications of the SiO_2 near the Si/SiO_2 interface[8], to extended strain fields in the near interfacial region, the bond strain model[9], and in the absence of any definite evidence for the above, to relaxation effects[3]. However, as all the oxide related features move towards lower binding energies in the proximity of the substrate, electronic screening of the charges by the semiconductor is also a possibility that must be considered[10].

The influence of the local electronic enviroment is an important contribution to the measured electron energies in both XPS and Auger spectroscopy. These relaxation processes are normally considered to be very local, with the nearest neighbor atoms providing the major extra-atomic screening of the final state holes. Calculating the electronic screening for an extended inhomogeneous enviroment, such as in a thin oxide, would be expected to be very complex, and the suggestion that the binding energy shifts are just an artifact of the electronic enviroment has not been considered in detail. However we believe it is possible to produce a reasonable estimate of these effects, and in this paper we show that the direction, and particularly the magnitude, of the observed binding energy shifts are consistent with electronic screening by the proximity of the semiconductor substrate. If this assertion is correct, then one of the main strands of evidence for the existence of an extended SiO_2/Si interface layer is undermined, and the chemical and strain models of the interface based on these shifts should be reexamined.

CALCULATION OF SUBSTRATE SCREENING EFFECTS

The method chosen to model the effects of substrate screening was to investigate the set of electrostatic image charges that could be said to exist after the hole and electron pair are created. Two large assumptions are made in the calculation presented here. Firstly it is assumed that the substrate and oxide are individually homogeneous on the atomic scale; that is, the different atomic sites, oxygen, silicon, intersitial, have similar electronic environments. The second assumption is that if the interactions are treated classically, there will only be a quantitative error compared to a quantum mechanical treatment[11].

The main part of the image charge calculation is the determination of the change in the binding energy of the electron due to a series of images, mirrored around the two interfaces. Figure 1 shows the first set of reflected images formed for a single charge located within the oxide. The image charges form an infinite series of decreasing magnitude with reflection coefficients determined by the oxide and semiconductor dielectric constants, ε_1 and ε_2 respectively. The potential along the z axis inside the oxide due to the image charges is[12]:

$$V=(q/4\pi\varepsilon_1)\Sigma_n(k_1k_2)^n[k_1/(z-2a-2nd)+k_2/(z-2b-2nd)+2k_1k_2/(z-2d-nd)] \quad (1)$$

Where $n=0$ to ∞,

and where the two reflection coefficients k_1 and k_2 are given by:

$$k_1=(\varepsilon_1-\varepsilon_0)/(\varepsilon_1+\varepsilon_0) \text{ and } k_2=(\varepsilon_1-\varepsilon_2)/(\varepsilon_1+\varepsilon_2) \quad (2)$$

Images of both the emitted electron and of the final hole (a singly charged hole for XPS, or a doubly charged hole for Auger) are considered. The binding energy shift from any depth in the oxide is calculated by reference to the composite binding energy from a thick oxide. The change in binding energy is calculated for a particular depth in the oxide, and this is used to shift a Gaussian line shape which represents the spectral feature. The line width is chosen to be similar to that measured experimentally. All the shifted Gaussians from the different depths within the oxide thickness are then summed. The sum includes factors for the degree of attenuation due to the electron mean free paths in the oxide. The summed shift in the binding energy for a thin oxide is referred to the similarly calculated shift in the thick oxide, not only because this is the experimental comparison, but also to avoid the infinite energy artifacts at the vacuum interface. It is possible to do this in a sensible way, because the intensity contribution to the summed Gaussian from very near the (theoretical) surface, is small compared with that from the rest of the oxide. The calculation of the shift can therefore be terminated near the surface without significantly changing the intensity or the peak position. A similar argument holds for the oxide/semiconductor interface. This argument of course breaks down for the thinnest oxides where there can be considerable shifting and smearing of the peak due to the high image charge interaction energies. The analog of these effects are indeed seen experimentally in monolayer growth systems where the contribution of first monolayer and second monolayers are similarly shifted[13].

There is an element of artificiality about the choice of a distance to terminate the summation at the oxide/semiconductor interface. However the effects of this choice become rapidly unimportant at oxide thicknesses above 8Å (~4 monolayers). The minimum distance chosen was based on the oxygen ion radius ~1.4Å. Comparison with jellium metal-vacuum calculations[14] suggest that the distance should be adjusted by a factor representing the overlap of the charge density at the interface, ~0.5Å. The numerical termination was therefore rounded to 1Å. The jellium metal-vacuum calculations also indicate that the image charge calculation should

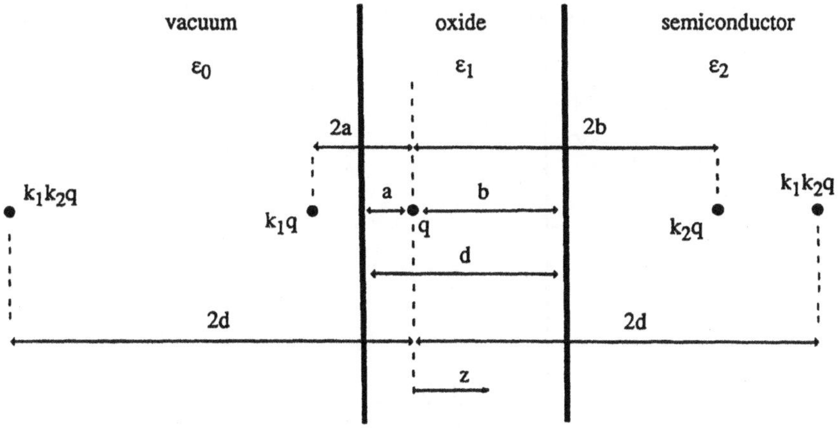

Fig 1. The first image charge reflections for the potential due to a charge within an oxide film on a semiconductor substrate. The charge q is reflected in an infinite series with reflection coefficients k_1 and k_2. The images shown are from the n=0 term of equation (1). The dielectric constants are ε_0, ε_1, and ε_2.

overestimate the energy shift for the first monolayer. However above 1.5Å the classical image charge theory is remarkably well obeyed. The thinnest oxide for which the shift was calculated was 3Å. With the strong influence of the two near interfaces, and the validity of the simple image charge model suspect at this point, only an estimate can be claimed. However the calculation for films thicker than 8Å should be better than 10%, given homogeneous dielectric constants and an electrostatic calculation.

An important part of the image charge calculation is determining the effective dielectric constants while the dynamic events are in progress. Indeed this is the critical difficulty with the classical calculation. In essence the problem is that the electronic relaxation times are similar to the interaction times; the dielectric constants are therefore complex and time dependent. However the core hole initiating the Auger process exists for a time comparable to the optical relaxation time. This suggests that the image charges of the initial Auger hole can be treated separately as a quasi-static charge, using the optical dielectric constants. Assuming no other effects, the shift in the binding energy of the Auger electron due to the inital hole image charges is the shift of the XPS Auger parameter. The Auger parameter, besides its easier calculation within this model, also has the experimental advantages that it is independent of charging or referencing methods, such as to adventious carbon, which will also be shifted for thin oxides.

CALCULATED RESULTS AND DISCUSSION

The calculated shifts in the Auger parameter for oxygen KLL-1s using Mg Kα radiation are plotted in figure 2.

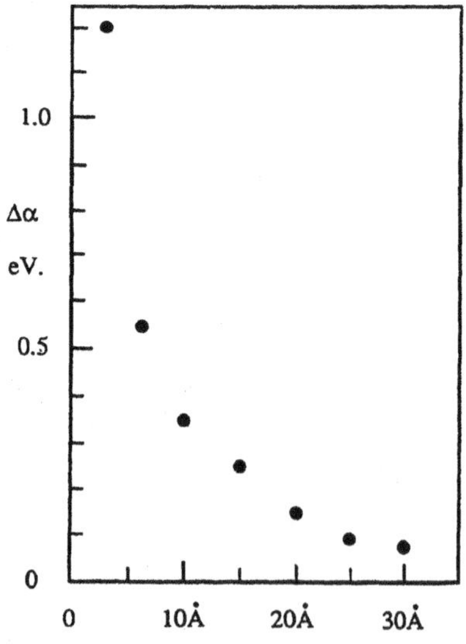

Fig 2. Calculated shift in the Auger parameter, Δα, with SiO_2 film thickness, for O KVV-1s, Mg Kα radiation.

The optical dielectric constants chosen were 2.25 for the oxide and 12 for the Si substrate. The mean free path for the oxygen KLL and 1s electrons using Mg Kα radiation are similar, and an average value of 14Å was used[15]. The detection direction was normal to the surface. The change in the Auger parameter is negative for oxygen and Mg Kα photons. The higher energy 1s photoelectrons are shifted less distance towards lower binding energy than the lower energy Auger electrons. The effect of the interface extends out to 30Å where the change in the Auger parameter compared to the thick oxide is 0.1eV. For thinner oxides the shift rapidly becomes larger and is approximately 1eV for a 3Å film. The influence of the interface is dependent on the mean free path of the electrons so that comparing two very different energies will not produce as simple a curve as figure 2.

Oxygen was chosen as an example here because it does not overlap with any substrate peak. Therefore the experimental determination of the Auger parameter shift for thin oxides should be more reliable. While this particular experimental measurement does not appear to have been published, the trend and magnitude is similar to existing evidence in the literature, where the Si transiinons have been concentrated on. These effects are sufficient to explain the magnitude of the observed shifts in the Auger parameter[16].

To make a comparision with published results we have calculated XPS and Auger shifts using the same optical dielectric constants used for the Auger parameter calculation. The shifts found for 3Å oxides were 0.9eV for XPS oxygen 1s, and 2eV for Auger Si LVV. These shifts compare with a shift of 1.3eV measured by Hollinger[1] for the oxygen 1s and a 4.9Å film, while the Auger shift is aproximately the same as that reported by Sobolewski and Helms[3] for the silicon LVV transitions. Both the magnitude of the effect and its extent out beyond 30Å are in agreement with experiment. Figure 3 shows the comparision of the data of Sobolewski and Helms[3] with the image charge model calculation. The experimental data is from four oxide films of different initial thicknesses that have been subsequently argon ion etched.

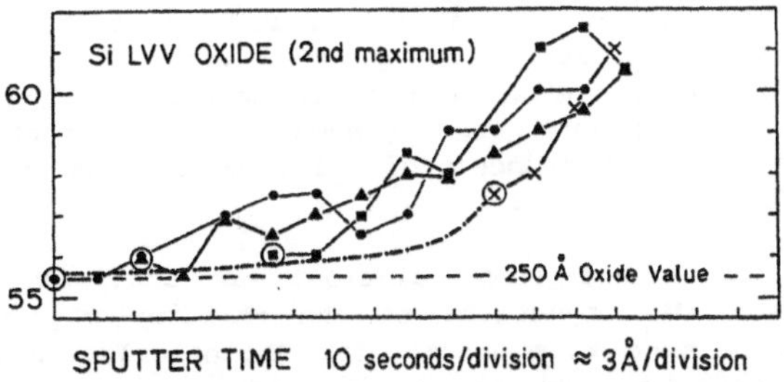

Fig 3. Comparison of the experimental and calculated shift in the Auger energy with oxide film thickness for the oxide related Si LVV transition. Experimental data after Sobolewski and Helms[3]. Four film thicknesses were used 30, 25, 17, and 3Å. The films were subsequently argon ion profiled. The initial peak energies of each sample are circled. The theoretical calculation, dots and dashes, gives the same result as the 3Å film initial value, and extends out to over 30Å.

It can be seen that the calculation gives a good fit for the shifts for the initial oxide thicknesses before depth profiling. As no additional parameter has been fitted to obtain this result, and as the parameters used are reasonable, the authors are quite confident that the calculated magnitude of the shift is realistic. The sensitive parameter is the dielectric constant of the oxide which is the divisor. This was chosen as 2.25 or $(1.5)^2$, an approximate value for cristobalite. Obviously this will change for an ion bombarded film, and the change in the binding energy with ion bombardment may be due to this effect, rather than a change in chemistry. The calculations of the XPS and Auger shifts must however still be regarded as estimates, as the calculation is very sensitive to assumptions about the optical constants.

Despite the agreement between theory and experiment shown here, comparing with the broad range of experimental data is not straightforward. There are considerable experimental disagreements between authors. Iqbal et al.[16] point out that there are measurement inconsistancies of 2-3eV in the change seen between the core levels of Si 2p and O 1s with film thickness. Iqbal et al.[15] believe there is only a small change in this measurement with film thickness, which would be consistant with the image charge model, while Hollinger[1] and Grunthaner et al.[17] have shifts in opposite directions. This is a measurement which should be independent of energy referencing and charging of the film. Other effects such as argon ion bombardment, both bond breaking and mixing of the interface, and the condition and cleanliness of the original oxide surface, all add to the difficulties in a direct comparison of theory to experiment over a wide range of experiments.

SOME IMPLICATIONS OF SUBSTRATE SCREENING EFFECTS ON INTERFACE MODELS

The XPS and Auger energy shifts found in thin SiO_2 films have been used to support models of the interface that include an extended transition region of up to 50Å between the Si substrate and the normal oxide. These results are accepted, and although there is some discussion on the details of the transition region, it is generally accepted that such a region is necessary to produce agreement with experiment.

The calculations reported here suggest that the XPS and Auger shifts seen in thin thermally grown SiO_2 films, and at the interface of thinned thicker films, are due to an artifact resulting from the proximity of the substrate. This explanation is in agreement with TEM results, which show that the SiO_2/Si interface is atomically abrupt, except for monoatomic steps[18] and some roughness[19]. This agreement between spectroscopy and TEM enhances the value of other models of the interface that do not require an extended interface region[20,21].

While the image charge calculations are consistant with electron spectroscopy and TEM, objections to the validity of the image charge calculation can be made from other data. The calculation is sensitive to the optical constants; the assumption made was a uniform dielectric constant in the oxide, equivalent to that of cristobalite. A high refractive index interlayer in the oxide at the interface would decrease the XPS and Auger binding energies, while at the same time reduce the effects of image charges in the Si substrate.

Ellipsometry[5] indicate the presence of a high refractive index phase of the oxide at the interface. These results from ellipsometry are well established, and the presence of the interface layer appears to be accepted in the literature. The properties of the interface layer found from ellipsometry are a layer of 5-10Å with a refractive index of 2-3, halfway between silica and silicon.

The ellipsometry and spectroscopy results are normally seen as mutually supporting. However there is no evidence from the properties of pure silicates that there are thermodynamically stable silicon oxide phases which meet both the ellipsometry criteria of high refractive index, and the SiO_2 stoichiometry found spectroscopically[22,23].

There are alternative explanations for the ellipsometery results, and interface roughness is one such mechanism. Interface roughness can be considered to produce a mixing of the properties of the Si and the oxide. The rule of mixtures, for two continuous phases of equal concentration[24], can thus be used to predict an intermediate refractive index of 2.6. To be consistent with ellipsometry results this rough layer would be 6Å thick. This argument can be put another way. If the ellipsometry results are used to claim there is an interface layer with different optical properties, then first it must be shown that the interface is smooth, to less than 6Å, over a lateral extent that is a large fraction of the wavelength of the light used. This has never been demonstrated.

Given the above arguments, the direct evidence for any interface interlayer models must rest on changes in binding energy, due to chemical bond effects, in a different medium to that normally associated with SiO_2 stoichiometry. The films are known to be constant in stoichiometry, even for very thin films <5Å[22,23]. Whether any change in binding energy is largely due to a change in the dielectric environment or a change in local electronic screening (atomic and chemical) is one of those points that inevitably excite considerable comment. However if Occam's razor is applied to this problem, it can be seen, that the screening of the XPS and Auger interactions by the substrate, is not only the simplest explanation, but the only explanation necessary at this point.

CONCLUSION

The calculations reported here, of the influence of image charge effects, show that the spectral shifts observed in electron spectroscopy of SiO_2 can be explained, in large part, as due to a spectroscopy artifact. This brings into question any chemical or physical model based on these shifts. Indeed it is necessary to calculate to a reasonable degree of accuracy the image charge peak shifts, before anything can be said about the chemical or physical environment at the Si/SiO_2 interface from such experiments. It follows that there does not appear to be sufficient spectroscopic evidence to suggest that there is anything special about the chemical or mechanical properties of thin oxide films in the range 3-30A grown on Si substrates.

ACKNOWLEGEMENTS

This work was supported in part by a grant from the Center for Integrated Systems, Stanford, and by a Semiconductor Research Corporation Contract No. 83-01-031.

REFERENCES

1. G.Hollinger, Appl. Surf. Sci. 8, 318 (1981)
2. F.J.Grunthaner, MRS Bul. 12(6), 60 (1987)
3. M.A.Sobolewski and C.R.Helms, J. Vac. Sci. Tech. A3, 1300 (1985)

4. F.J.Grunthaner and J.Maserjian, in 'The Physics of SiO_2 and its interfaces', ed. S.T.Pantelides, p389 (1979)
5. E.Taft and L.Cordes, J.Electrochem. Soc. 126, p131 (1979)
6. L.C.Feldman I.Stensgaard P.J.Silverman and T.E.Jackman, as ref. 4, p344 (1979)
7. R.P.Vasquez and F.J.Grunthaner, Sur. Sci. 99, p681 (1980)
8. S.I.Raider and R.Flitsch, IBM J. Res. Develop. 22, 294 (1978)
9. F.J.Grunthaner, P.J.Grunthaner and J.Maserjian, IEEE Trans. Nucl. Sci. NS29, 1462 (1982)
10. C.D.Wagner A.Joshi L.Gulbrandsen and B.E.Deal, J. Vac. Sci. Technol. B2, 107 (1984)
11. F.Sols and R.H.Ritchie, Sur. Sci. 194, p275 (1988)
12. M.Kleefstra and G.C.Herman, J. App. Phys. 51, 4923 (1980)
13. G.Kandl, T.C. Chiang D.E.Eastman and F.J.Himpsel, Phys. Rev. Lett. 45, 1808 (1980)
14. J.A.Appelbaum and D.R.Hamann, Phys. Rev. B 6, 1122 (1972)
15. M.P.Seah and W.A.Dench, Sur. Interface. Anal. 1, 2 (1979)
16. A.Iqbal C.W.Bates and J.W.Allen, Appl. Phys. Lett. 10, 1064 (1985)
17. F.J.Grunthaner P.J.Grunthaner R.P.Vasquez B.F.Lewis J.Maserjian and A.Madhukar, Phys. Rev. Lett. 43, p1683 (1979)
18. O.L.Krivanek D.C.Tsui T.T.Sheng and A.Kamagar, as ref. 4, p356 (1979)
19. A.H.Carim and R.Sinclair, J. Electrochem. Soc. 134, p741 (1987)
20. S.T.Pantelides and M.Long, as ref. 4, p339 (1979)
21. C.R.Helms N.M.Johnson S.A.Schwarz and W.E.Spicer, as ref. 4, p366 (1979)
22. R.S.Bauer J.C.McMenamin H.Petersen and A.Bianconi, as ref. 4, p401 (1979)
23. F.J.Grunthaner and J.Maserjian, as ref. 4, p389 (1979)
24. L.E.Nielsen, 'Predicting the Properties of Mixtures', Marcel-Dekker NY, chap. 4, p73 (1978)

A PHYSICAL MODEL FOR THE OBSERVED DEPENDENCE OF THE METAL-SEMICONDUCTOR WORK FUNCTION DIFFERENCE ON SUBSTRATE ORIENTATION

Hisham Z. Massoud

Department of Electrical Engineering
Duke University
Durham, North Carolina 27706

James D. Plummer

Center for Integrated Systems
Stanford University
Stanford, California 94305

The metal-semiconductor work function difference ϕ_{ms} in the Al–SiO_2–Si system has been measured using capacitance-voltage (C-V) measurements, and it was found to be orientation dependent, with the (100) orientation having a ϕ_{ms} larger than that of (111) by ~ 0.250 V. This result is in agreement with previous studies which showed that ϕ_{ms} is both orientation and processing dependent. These observations are interpreted in terms of the contribution of dipole moments resulting from partial charge transfer in interface bonds to the measured ϕ_{ms}. Such charge transfer takes place in bonds at the Si–SiO_2 and gate–SiO_2 interfaces and in their vicinity. The contribution of these dipole moments to the measured ϕ_{ms} has been calculated from first principles, and the results predict the orientation dependence of ϕ_{ms}, and provide a physical model for further investigating its processing dependence. Measurements and process modeling of charge components in metal-oxide-semiconductor (MOS) devices will be discussed.

INTRODUCTION

The metal-semiconductor work function difference ϕ_{ms} is an important electrical parameter in MOS device physics because it partially determines the flatband voltage V_{FB} and the threshold voltage V_T of MOS devices, and partially contributes as to whether the semiconductor surface is in accumulation, depletion, or inversion [1]. It is defined as the difference between the work function ϕ_m of the gate-electrode metal and the work function ϕ_s of the semiconductor. The metal work function depends on the type of gate material, while that of the substrate depends on the semiconductor electron affinity, the bandgap energy, the device operating temperature, and the substrate doping concentration N_{SUB} [1]. It has been experimentally measured by optical and electrical techniques, but C-V measurements have been the most widely used method because of its sensitivity and accuracy [1,2]. If the allowable error in calculating the oxide fixed charge density N_f for a 100 Å thick oxide is to be less than 1×10^{10} cm^{-2}, then V_{FB} must be measured and ϕ_{ms} must be known with an accuracy better than ~ 5 mV. The need for precise measurements of V_{FB} and accurate values of ϕ_{ms} when characterizing the electrical charges in ultrathin oxides used in VLSI and ULSI MOS devices motivated this study.

The above definition of ϕ_{ms} predicts that it should be solely dependent on the gate electrode material, the device operating temperature, the substrate doping concentration and type, the semiconductor bandgap energy, and the electron affinity. However, in addition to the the results obtained in this study, there is a large number of experimental investigations of ϕ_{ms} which show, beyond any experimental error, that it is both orientation and processing dependent [3–22]. The results of these studies and those obtained in this work are compiled in Fig. 1. Figures 1(a) and 1(b) show the dependence of ϕ_{ms} on N_{SUB} for n-type and p-type (100) silicon, respectively. Figures 1(c) and 1(d) show this dependence for n-type and p-type (111) silicon, respectively. It can be easily seen that there is a substantial spread in the experimental data which cannot be attributed to experimental error. The spread in these measurements makes it difficult to estimate the threshold voltage V_T of MOSFETs, perhaps the most important device design parameter. The Si–SiO_2 barrier energy was found by Deal *et al.* [3] to be independent of silicon orientation and type. Werner [4] measured ϕ_{ms} on n- and p-type silicon. Deviations of as much as 380 and 330 mV were observed between the two studies [3,4] on n- and p-type silicon substrates, respectively.

While early measurements of ϕ_{ms} were limited to one orientation, later studies compared the (100) and (111) orientations of silicon, and the results indicated that ϕ_{ms} has an orientation dependence [7–22], as well as a processing dependence [8–14,19,20,22]. For CVD oxides [8–10] where the oxides are deposited in a mainly H_2 atmosphere, it was proposed that the chemisorption of atomic hydrogen at the Si–SiO_2 interface affects the silicon work function and, consequently, the Al–Si work function difference [23,24]. Forming-gas anneals of the Al–SiO_2 interface also affect the internal photoemission characteristics [25]; the barrier height is raised by 250 mV and the electric-field dependence of the photocurrent is increased, and this, in turn, would also affect the ϕ_{ms} photoemission measurement. In a study using Al and Al–Si alloys sputtered on thermal oxides grown [15], it was found that ϕ_{ms} is dependent on the substrate orientation of p-type silicon, with (100) being 200 mV more positive than (111); no difference in ϕ_{ms} was observed between pure Al and Al–Si(1%). In a similar study but with reactively sputtered SiO_2 [11], a 400 mV difference in ϕ_{ms} between the two orientations was reported , with (100) again being more positive than (111). The dependence of ϕ_{ms} on substrate orientation and on the details of the processing and annealing conditions for oxides grown in steam were also observed [13,14].

The effects of processing parameters – particularly post-metallization H_2 anneals – on ϕ_{ms} were investigated by Razouk and Deal [19,20]. Variations, exceeding 500 mV for a given n- or p-type silicon substrate, were found to be dependent on processing conditions. A recent systematic study of processing parameters such as the effects of the oxidation temperature, pre- and post-metallization processing, and forming-gas anneal temperature and time, was done by Akinwande and Plummer [22]. The results showed again the orientation and processing dependence of ϕ_{ms}. These results emphasize the critical need for a physical model that would predict the dependence of ϕ_{ms} on substrate orientation and processing conditions.

EXPERIMENTAL PROCEDURES AND RESULTS

The structure used in this investigation of ϕ_{ms} in thin oxides is an MOS capacitor with a guard ring. The substrate is a lightly doped (100) or (111) silicon wafer. The samples were first cleaned with a sequence of H_2O_2-based solutions, followed by a dip in dilute HF, and a final rinse in deionized water. They were then dried and immediately introduced into a double-walled quartz oxidation furnace with Ar flowing.

They were then subjected to a pre-oxidation *in situ* cleaning/gettering cycle using a small concentration of HCl in Ar [21]. Oxidations were done at 1000°C in dry O_2 to thicknesses in the 200–2000 Å range. At the end of the oxidation, the samples were either pulled in O_2 and immediately quenched to room temperature, or annealed in Ar at 1200°C for 30 min to anneal N_f. After etching the back side oxide, the wafers were then metallized on both sides in a flash evaporation system. After defining the capacitors using optical lithography, a post-metallization anneal was done in N_2 for 40 min at 420°C.

The electrical characterization was done in an electrically shielded box under nitrogen purge after the probing leads were calibrated for leakage ($< 1 \times 10^{-14}$ A) and capacitance (< 0.1 pF). Capacitance and conductance measurements were made at 1 MHz and corrected for the sample series resistance at each bias point [2]. Plots of V_{FB} vs X_{ox} for all wafers resulted in straight lines. The intercept of these lines with the voltage axis yielded the measured ϕ_{ms}, and the results are plotted as a function of N_{SUB} in Fig. 1 (closed circles). It can be seen that ϕ_{ms} – measured on wafers which were processed simultaneously – is dependent on the substrate orientation, with $\phi_{ms(100)}$ being larger than $\phi_{ms(111)}$ by $\sim$ 0.250 V.

PHYSICAL MODEL

A first-order model has been developed to calculate the magnitude of the dipole moment at the Si–SiO_2 interface resulting from partial charge transfer that takes place upon the formation of interface bonds [26,27]. The charge transfer occurs because of the difference in electronegativity between the two species across that interface. At the Si-SiO_2 interface these bonding species are silicon atoms and SiO_2 molecules. At the gate–SiO_2 interface, it is the metal atoms and SiO_2 molecules. The partial transfer of charges in interface bonds results in the formation of relatively uniform dipole layers at these interfaces. The charge transfer calculation is based on the principle of electronegativity equalization [28,29]. The electronegativity S_m of a molecule or compound is given by the geometric mean of the electronegativities S_i of its constituent atoms [29]. The relative compactness electronegativity scale of Sanderson [29] was used because it provides a physical picture of electronegativity equalization in terms of the adjustment of the relative compactness of the electronic clouds of the individual atoms to a uniform value throughout the compound. It was found that most elements and compounds of interest have electronegativities larger than that of silicon, which means that electrons will be partially displaced towards the oxide, thus generating a dipole where the positive charge is on the silicon side of the interface and the negative charge on the oxide side.

The charge transfer ρ_i, expressed in electronic charges per bond, which affects the ith atom in a compound is given by $\rho_i = (S_m - S_i)/\Delta S_i$, where S_m is the molecule or compound electronegativity, S_i the electronegativity of the ith atom in the compound, and ΔS_i the change in the electronegativity of the atom that would result from the acquisition of a unit charge [29], and is given by $\Delta S_i = 2.08\sqrt{S_i}$. The amount of partial charge transfer in a bond between a substrate silicon atom and an SiO_2 molecule was found to be $\rho_{Si-SiO_2} = 0.182$ electrons/bond.

Charge transfer is not limited to interface bonds only, it also takes place in bonds beyond the interface both in the silicon substrate and in the silicon dioxide layer [27]. The charge transfer in the nth bond removed from the Si–SiO_2 interface was calculated and found to fit a relation of the form $\rho_n = 1/(\alpha + \beta n)$. For charge transfer in silicon bonds away from the interface, it was found that $\alpha_{Si} = 2.467$, and $\beta_{Si} = 3.053$, whereas in silicon dioxide, $\alpha_{ox} = 3.064$ and $\beta_{ox} = 2.486$ [27]. In this first-order model, the dependence of the dipole moments on substrate orientation was the main interest. Consequently, dipoles at the gate–SiO_2 interface were assumed to be identical for both orientations, and only Si–SiO_2 interface dipoles were considered [27]. The charge separation d in the partial charge transfer in interface bonds, equal to the sum of the covalent radius of Si and an equivalent radius of an SiO_2 molecule derived from the average oxide bulk density, was found to $d = 2.88$ Å [26,27].

Upon charge transfer in the interface bonds, two charge sheets of equal charge density $q\rho N_B$, where N_B is the bond density in bonds per unit area, and opposite polarity are separated by a distance δ perpendicular to the Si–SiO_2 interface, where $\delta = d \sin\theta$ and θ is the angle between the bonds and the interface. As shown earlier, the positive charge sheet is on the silicon side of the interface, and the negative charge sheet on the oxide side. It has also been shown that the flatband voltage V_{FB} of such a structure can be written as [26,27]

$$V_{FB} = \phi_{ms} - \frac{Q_f X_{ox}}{\epsilon_{ox}} - \frac{q\rho N_B \delta}{\epsilon}, \tag{1}$$

where Q_f is the oxide fixed charge density, C_{ox} the oxide capacitance per unit area, and ϵ is an average dielectric permittivity between that of Si and SiO_2, i.e., $\epsilon = (\epsilon_{Si} + \epsilon_{ox})/2$ [30]. It is important to note that the contribution of the dipoles to V_{FB} is independent of their distance from the interface. Consequently, the contribution of dipoles at the gate–SiO_2 interface is equally as important as that of dipoles at the Si–SiO_2 interface. The oxide-thickness-independent contribution of dipoles to V_{FB} can be added to ϕ_{ms} in defining an effective ϕ_{ms} by rewriting Eq. (1) as

$$V_{FB} = \phi_{ms}^{eff} - \frac{Q_f X_{ox}}{\epsilon_{ox}}, \tag{2}$$

where

$$\phi_{ms}^{eff} = \phi_{ms} - \frac{q\rho N_B \delta}{\epsilon}, \tag{3}$$

is the effective metal-semiconductor work function difference obtained by plotting V_{FB} vs X_{ox}, as suggested by Eq. (2), and finding the intercept at $X_{ox} = 0$. The experimentally observed dependence of ϕ_{ms} on substrate orientation can be explained by suggesting that the quantity measured experimentally is ϕ_{ms}^{eff} and not ϕ_{ms}. As can be seen in Eq. (3), ϕ_{ms}^{eff} is expected to be orientation dependent through N_B and θ, and processing dependent through ρ which is determined by the chemistry of the interfaces and their vicinity.

In the (100) orientation of silicon, there are 6.783×10^{14} bonds/cm^2 inclined at an angle $\theta = 35.26°$. If we assume that all silicon bonds on this orientation are bonded to SiO_2 molecules, then $-(q\rho N_B \delta)/\epsilon$ for (100) is −0.476 V. For (111) silicon, there are 7.833×10^{14} bonds/cm^2 that are perpendicular to the interface, i.e., $\theta = 90°$, and $-(q\rho N_B \delta)/\epsilon$ for (111) is −0.951 V. These calculations show that ϕ_{ms} measured on (100) could be as much as 0.475 V larger than that on (111) based on the assumptions made in this model [26,27]. This is, however, an upper limit because all surface bonds were assumed bonded to silicon dioxide molecules, which have the largest magnitude charge transfer when compared with those in suboxide or other bonds [26,27]. If the density of silicon surface bonds is distributed among the different suboxide species at the interface as observed by XPS [31], the difference in the dipole moments of (100) and (111) is expected to be smaller than 0.475 V depending on the distribution of suboxides at the Si–SiO_2 interface.

The nature of the bonds at the gate–SiO_2 and its chemistry will determine the amount of partial charge transfer in the bonds of that interface. In the Al–SiO_2 interface, it was found that Al first clusters about each surface oxygen atom, and then grows Al_2O_3 by reducing SiO_x ($x<2$) and leaving excess silicon atoms at the interface [32]. In contrast, Au was found to form islands on the oxide, with evidence of Au—Si bonding, causing an SiO_x layer beneath the contact [32]. Dipole layers at the Au–SiO_2 interface were proposed by Hickmott [13,14], in expaining the dependence of the apparent ϕ_{ms} on processing conditions which influence metal–insulator reactions. Partial charge transfer can also be calculated to compare the effect of different gate materials and post-metallization processing on ϕ_{ms} measurements.

Further refinements in this model, such as estimating the contribution of dipoles away from the interfaces would make the description of both interfaces more complete. Despite the many approximations and simplifications made in this model, its predictions for different orientations are in agreement with experimental results for (100) and (111), and the agreement in magnitude is thought to be good, taking into account that the interface dipole contribution is expected to be strongly process-dependent. This model also provides a framework for further studies of the partial charge transfer in bonds not only at the Si–SiO_2 interface, but also at the gate–SiO_2 interface, and away into the bulk of the gate material, the bulk of the oxide, and that of the substrate.

MOS CHARGE MEASUREMENTS AND MODELING

High-temperature annealing of thermally grown oxides is typically used to anneal the oxide fixed charge density N_f to its minimum value. Differences were observed, however, depending on the annealing ambient where it was observed that annealing in N_2 results in an initial decrease followed by a gradual increase for long anneal times [33]. In Ar, N_f reaches a minimum value and remains at that value with long annealing cycles [34]. This difference can be interpreted both qualitatively and quantitatively. Annealing has a thermal component which is represented by the behavior of N_f in Ar. When annealed in N_2, in addition to the thermal component in annealing, N_2 diffuses in the oxide and reacts with the surface, replacing Si—O bonds with Si—N bonds. The charge transfer in such bonds is 0.241 electrons/bond, which is larger than the 0.182 value in Si—SiO_2 bonds. The magnitude of the negative dipole moment term in Eq. (1) increases, which was assumed to date to be an increase in N_f. Observed dependencies with nitridation or ammonia treatments could be similarly interpreted. Modeling the exact amount of change in the dipole moment term will only be made possible after correlating spectroscopy and profiling studies of N_2 in SiO_2 and at the Si–SiO_2 and gate–SiO_2 interfaces with electrical measurements.

It becomes, therefore, necessary to identify changes in charge densities with annealing as partly due to changes in the intrinsic mechanism that results in N_f (the thermal component) and partly due to changes in the dipole moment contribution at both interfaces and throughout the oxide as a result of processing. The influence of charge transfer in bonds on electrical measurements in MOS devices can be a new starting point in revisiting the understanding of charges components in thermally oxidized silicon structures and their process dependence.

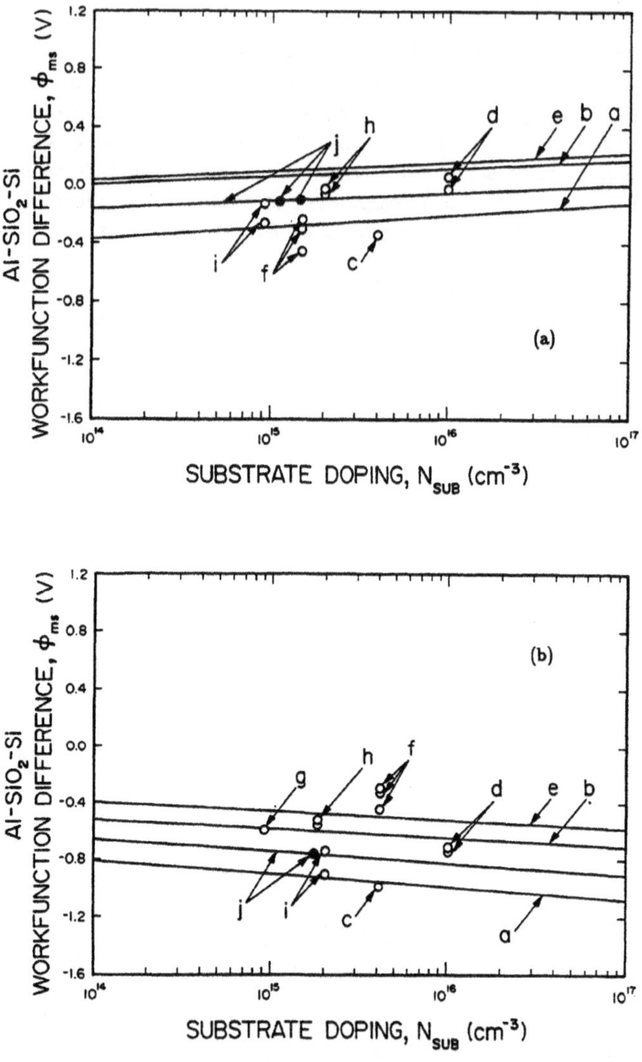

Fig. 1. Experimental measurements of the metal-oxide semiconductor work function difference ϕ_{ms} as a funstion of the substrate doping concentration N_{SUB} for (a) n-type (100) Silicon, (b) p-type (100) Silicon, (c) n-type (111) Silicon, and (d) p-type (111) Silicon: a [3], b [4], c [5,6], d [8–10], e [11], f [13,14], g [15], h [16–18], i [19,20], and j (this work).

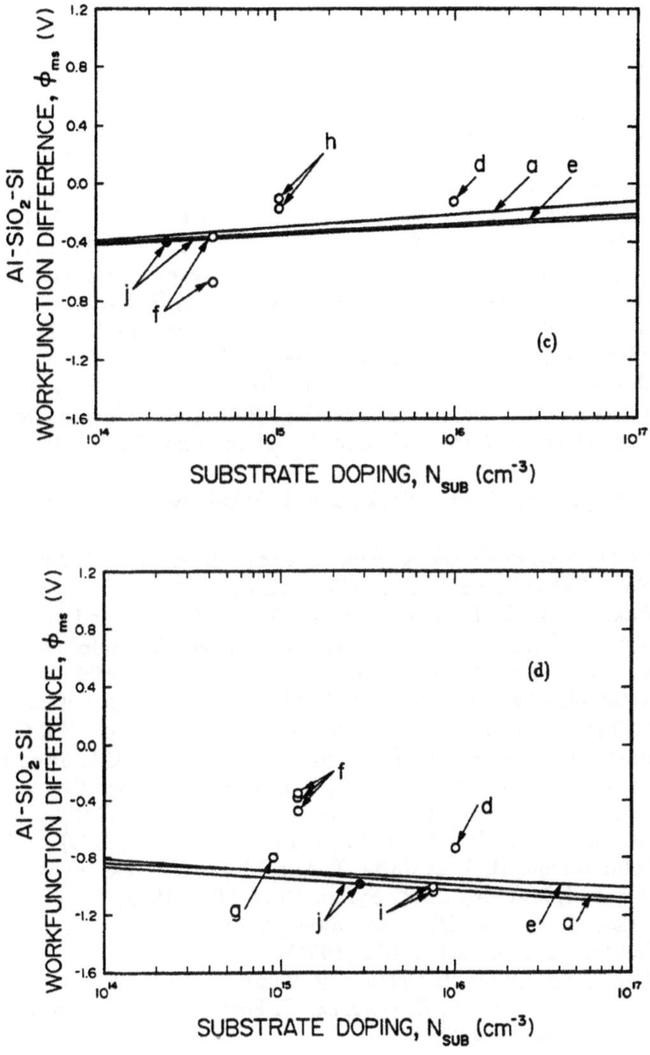

Fig. 1. Continued.

REFERENCES

1. S. M. Sze, *Physics of Semiconductor Devices*, 2nd Edition, Wiley-Interscience, John Wiley & Sons, New York, 1981.
2. E. H. Nicollian and J. R. Brews, *MOS (Metal Oxide Semiconductor) Physics and Technology*, John Wiley & Sons, New York, 1982.
3. B. E. Deal, E. H. Snow, and C. A. Mead, J. Chem. Phys. Sol. 27, 1873 (1966).
4. W. M. Werner, Solid-State Electron. 17, 769 (1974).
5. S. Kar, Solid-State Electron. 18, 169 (1975).
6. S. Kar, Solid-State Electron. 18, 723 (1975).
7. H. B. Michaelson, J. Appl. Phys. 48, 4729 (1977).
8. L. A. Kasprzak and A. K. Gaind, in *The Physics of SiO_2 and Its Interfaces*, S. T. Pantelides, Ed., Pergamon Press, New York, 1978.
9. A. K. Gaind and L. A. Kasprzak, Abstract No. 344, Extended Abstracts Vol. 77–2, The Electrochemical Society Fall Meeting, Atlanta, Georgia, Oct. 9-14, 1977.
10. A. K. Gaind and L. A. Kasprzak, Solid-State Electron. 22, 303 (1979).
11. K. Haberle and E. Fröschle, J. Electrochem. Soc. 126, 878 (1979).
12. L. A. Kasprzak and A. K. Gaind, IBM J. Res. Develop. 24, 348 (1980).
13. T. W. Hickmott, in *The Physics of MOS Insulators*, G. Luckovsky, S. T. Pantelides, and F. L. Galeener, Eds., Pergamon Press, New York, 1980.
14. T. W. Hickmott, J. Appl. Phys. 51, 4269 (1980).
15. N. W. Jones and M. R. Gulett, Abstract No. 523, Extended Abstracts Vol. 80–2, The Electrochemical Society Fall Meeting, Hollywood, Florida, Oct 5–10, 1980.
16. H. M. Przewlocki, S. Krawczyk, and A. Jakubowski, Phys. Stat. Sol. (a) 65, 253 (1981).
17. S. K. Krawczyk, H. M. Przewlocki, and A. Jakubowski, Revue Phys. Appl. 17, 473 (1982).
18. H. M. Przewlocki, in *Physics of Semiconductor Devices*, S. C. Jain and S. Radhakrishna, Eds., Halsted Press, John Wiley & Sons, 1982.
19. R. R. Razouk and B. E. Deal, Abstract No. 376, Extended Abstracts Vol. 81–2, The Electrochemical Society Fall Meeting, Denver, Colorado, Oct 11–16, 1981.
20. R. R. Razouk and B. E. Deal, J. Electrochem. Soc. 129, 806 (1982).
21. H. Z. Massoud, *Thermal Oxidation of Silicon in Dry Oxygen – Growth Kinetics and Charge Characterization in the Thin Regime*, Tech. Rep. No. G502-1, Stanford Electronics Laboratories, Stanford University, Stanford, CA, June 1983.
22. A. I. Akinwande and J. D. Plummer, J. Electrochem. Soc. 134, 2297 (1987).
23. T. Sakurai and H. D. Hagstrum, Phys. Rev. B12, 5349 (1975).
24. T. Sakurai and H. D. Hagstrum, Phys. Rev. B14, 1593 (1976).
25. P. M. Solomon and D. J. DiMaria, J. Appl. Phys. 52, 5867 (1981).
26. H. Z. Massoud, Mat. Res. Soc. Symp. Proc. 105, (1988).
27. H. Z. Massoud, J. Appl. Phys. 63, 2000 (1988).
28. R. T. Sanderson, Science 114, 670 (1951).
29. R. T. Sanderson, *Inorganic Chemistry*, Reinhold, New York, 1967.
30. P. Perfetti, C. Quaresima, C. Coluzza, C. Fortunato, and G. Margaritondo, Phys. Rev. Lett. 57, 2065 (1986).
31. F. J. Grunthaner and P. J. Grunthaner, Materials Science Reports 1, 65 (1986).
32. R. S. Bauer and R. Z. Bachrach, Appl. Phys. Lett. 37, 1006 (1980).
33. B. E. Deal, J. Electrochem. Soc. 121, 198C (1974).
34. M. Revitz, S. I. Raider, and R. A. Gdula, J. Vac. Sci. Technol. 16, 345 (1979).

THEORETICAL CALCULATIONS OF THE ELECTRONIC STRUCTURE IN THE Si-SiO_2 SYSTEMS

Michel Lannoo

Laboratoire de Physique des Solides (U.A. 253 CNRS)
I.S.E.N., 41, boulevard Vauban, 59046 Lille Cedex, France

ABSTRACT

The electronic structure of the Si-SiO_2 interface is discussed with emphasis on descriptions relating the density of states to the local chemical bonds. The bulk properties of Si and SiO_2 are first considered and then used to derive the essential features of the interface. It is shown that recent photoemission measurements can be interpreted by the existence of valence alternation pairs. Finally the electronic properties of the P_b center are reviewed.

INTRODUCTION

Although there is a wealth of experimental informations concerning the Si-SiO_2 interface its atomic and electronic structures are not yet completely understood. There is now good evidence that the interface is structurally abrupt with low roughness as shown by high resolution transmission electron microscopy [1]. There is then a transition layer of intermediate structure SiO_x and of thickness equivalent to two monolayers, i.e ~ 5 Å [2,3]. It is also likely that, at least in some cases, there is a strained layer of about 20 Å as discussed by Grunthaner et al [3]. However the general situation is more complex and the detailed atomic structure of the interface depends sensitively on the oxide growth conditions.

In this context the aim of the present work is to discuss some key features of the electronic structure of the Si-SiO_2 interface. This will be based on a very simple tight binding analysis, starting from molecular models and using them to understand the results of more complex numerical calculations. This is first tested in the case of bulk Si and SiO_2 and then applied to the study of the intermediate SiO_x composition where we analyze the results of cluster Bethe lattice calculations and compare with experimental data essentially obtained by photoemission [4]. In a second part we concentrate on recent photoemission measurements [5] on thin SiO_2 films and show that the relative amplitude of the peaks can only be understood on the basis of the existence of valence alternation pairs (VAP). We perform a cluster tight binding total energy calculation of such VAP and show that their stability is close to that of a normal Si_2O_4 unit explaining why they are likely to be formed under conditions of interface formation. In a final part we discuss the best known interface defect :

the P_b center and show that theory is able to account for most of its properties as determined from electron paramagnetic resonance (e.p.r.), optical spectroscopy and deep level transient spectroscopy (D.L.T.S.). We also comment on the possibility that this defect could behave as a negative U center.

ELECTRONIC STRUCTURE OF Si, SiO_2 AND SiO_x

Our analysis makes use of a tight binding description of these materials where the electron wave function is expressed as a linear combination of atomic orbitals. These are assumed to form an orthonormal basis set (i.e interatomic overlaps are neglected) and furthermore we limit ourselves to a "minimal" basis set consisting of one "s" and three "p" orbitals on each Si and O atom. The electronic states are then obtained by diagonalisation of the hamiltonian matrix $H_{i\alpha, j\beta}$ where i and j specify the atoms, α and β the orbitals. Such a diagonalisation must be performed numerically but here we proceed in a different way, making use of molecular models to discuss the formation of the band structure. For this we make a basis change chosen in such a way that the hamiltonian matrix becomes dominated by some interatomic terms, the other elements being much smaller. When this is possible the neglect of the smaller matrix elements corresponds to a molecular description, intimately connected to the nature of the chemical bonds, where each band reduces to a highly degenerate set of molecular levels. Inclusion of the smaller matrix elements will then broaden these levels into bands (first order effect) and eventually slightly shift these bands (second order).

The simplest illustration is for bulk Si. There, according to symmetry, one can build four orthonormal orbitals of the form [6]

$$\varphi_{ij} = 1/2\ (s_i + \sqrt{3}\ p_{ij}\ )$$

where s_i is the "s" orbital of atom i, p_{ij} the "p" orbital of i pointing towards its nearest neighbor j. These orbitals have strong directional properties and the hamiltonian matrix is dominated by the elements $\langle \varphi_{ij}\ |H|\ \varphi_{ji}\ \rangle = -\beta$ $(\beta > 0)$ connecting the two sp_3 hybrids involved in the same bond. If one neglects all other matrix elements the system becomes equivalent to a set of diatomic molecules formed by the pairs φ_{ij}, φ_{ji}. This results in bonding and antibonding states of energies $\bar{E} \pm \beta$ (where $\bar{E}$ is the sp_3 atomic energy) and wave functions $(\varphi_{ij} \pm \varphi_{ji})/\sqrt{2}$. The degeneracy of these two states is N, the number of bonds. With two electrons per bond, the bonding states are filled, the antibonding states empty. This bonding-antibonding picture is the crudest description of the electron states of bulk Si. It will be instructive for the following to discuss now the broadening of these states into bands to get an idea of the shape of density of states. For this we consider the bonding states (only valence bands will be studied in the following in relation to photoemission data) and call $-\ \Delta\ (\Delta > 0)$ the hamiltonian matrix element between two adjacent bonding states, which must be the next important interaction to include. To discuss the broadening effect we write the valence band wave function ψ_v as a combination of the bonding states

$$B_{ij} = \varphi_{ij} + \varphi_{ji}\ /\ \sqrt{2}$$
$$\psi_v = \sum_{\text{pairs } ij} \alpha_{ij}\ B_{ij} \tag{1}$$

Projecting Schrödinger's equation $(E - H)\,\psi_v = 0$ onto one basis state B_{ij} gives

$$(E - \bar{E} + \beta)\,\alpha_{ij} + \Delta \sum_{\substack{k \neq j \\ \ell \neq i}} (\alpha_{ik} + \alpha_{j\ell}) = 0 \qquad (2)$$

where k and ℓ are nearest neighbors of i and j. Introducing $S_i = \sum_k \alpha_{ik}$ on each atom we get

$$(E - \bar{E} + \beta - 2\Delta)\,\alpha_{ij} + \Delta(S_i + S_j) = 0 \qquad (3)$$

which, when summed over the four neighbors j, gives

$$(E - \bar{E} + \beta + 2\Delta)\,S_i + \Delta \sum_j S_j = 0 \qquad (4)$$

Equation (4) is formally equivalent to the nearest neighbors' tight binding problem of a "s" band on the silicon lattice. It gives rise to a broad band whose extrema correspond to all S_i equal or to alternant S_i leading to $\bar{E} - \beta - 6\Delta < E < \bar{E} - \beta + 2\Delta$. This band contains one state per atom. Other non trivial solutions correspond to all $S_i = 0$, i.e, from equ. (3) to $E = \bar{E} - \beta + 2\Delta$. These states have pure p character ; their number is one per atom and they correspond to a flat band. The resulting density of states is pictured schematically on Fig. 1. Inclusion of further interactions will mainly broaden the upper flat band as shown again on figure 1. Thus with no calculations we have obtained correct information about the valence band density of states of Si.

Let us now apply similar reasoning to the case of SiO_2. The molecular model is essentially the one developed by Harrisson and Pantelides [7]. The building Si-O-Si unit is shown on figure 2. Again one builds sp3 hybrids on the Si atoms while on the oxygen atom one keeps the natural sp basis. The oxygen "s" state is by far the lowest in energy and, to first order, its coupling with other states can be neglected. It will remain

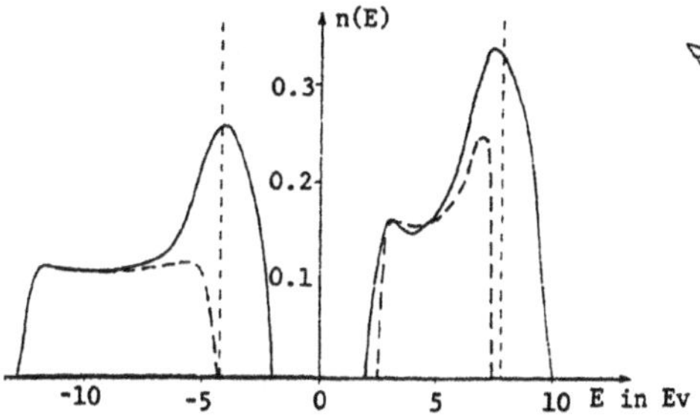

Figure 1 - Density of states for silicon calculated on a Bethe lattice :—— full nearest neighbors calculation --- simplified form corresponding to the broad and flat bands discussed in the text.

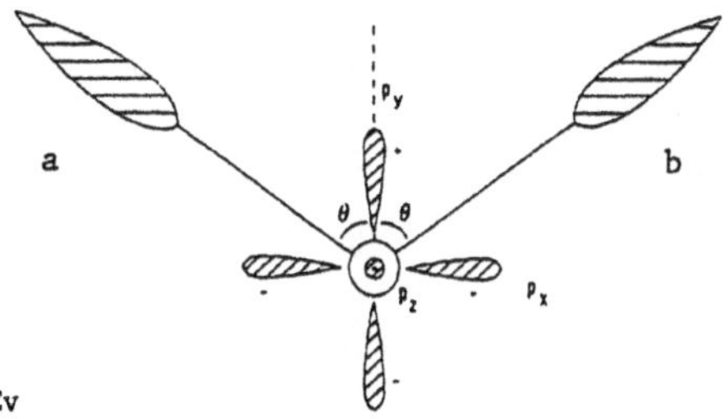

Figure 2 - Si-O-Si unit for building the molecular model.

atomic like at its atomic value E_s (0). On the other hand the oxygen p energy $E_p(0)$ is closer to the silicon sp3 energy $\bar{E}$ and the interaction of the corresponding states must be taken into account. The molecular states of the Si-O-Si unit of Fig. 2 are then built from the two sp3 hybrids a and b pointing towards the oxygen atom and the p_x, p_y and p_z oxygen "p" states. It is clear that p_z, being perpendicular to the S_i -O-Si plane, will remain uncoupled at this level of approximation giving one state at the atomic value E_p (0). Thus the sp3 states a and b will only couple to p_x and p_y via the projection of these states along the axis of the corresponding nearest neighbor direction. All interactions only reduce to one parameter β_p defined as the interaction between an sp3 orbital and the p orbital along the corresponding bond. By symmetry p_x only interacts with $(a - b)/\sqrt{2}$ giving rise to strong bonding and antibonding states at energies

$$E_{S_A^B} = \frac{\bar{E} + E_p\ (0)}{2} \mp \sqrt{\left(\frac{(\bar{E} - E_p\ (0)}{2}\right)^2 + \beta_p^2 \sin^2 \theta} \qquad (5)$$

while p_y and $(a + b)/\sqrt{2}$ lead to weak bonding and antibonding states

$$E_{W_A^B} = \frac{\bar{E} + E_p\ (0)}{2} \mp \sqrt{\left(\frac{\bar{E} - E_p\ (0)}{2}\right)^2 + \beta_p^2 \cos^2 \theta} \qquad (6)$$

where 2θ is the Si-O-Si bond angle.

The resulting valence band density of states per Si-O-Si unit is pictured on Fig. 3. It consists of delta functions at energies $E_s(0)$, E_{SB}, E_{WB} and E_p (0), the weight of each state per Si-O-Si unit being equal to

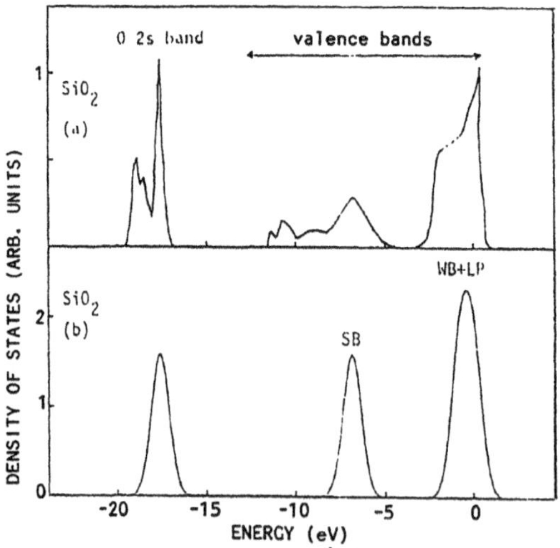

Figure 3 - a) Density of states for SiO_2 calculated in ref [11]. b) Density of states in the mole-cular model, with the same tight binding parameters, broadened by a Gaussian of half width at half maximum equal to 0.5 eV.

unity. The influence of further interactions can now be analyzed as for pure Si. If we call Δ_S and Δ_W the interaction between strong and weak bonding states belonging to adjacent Si-O-Si units then we can repeat the treatment previously applied to Si simply replacing Δ by Δ_S or Δ_W. This means that we get densities of states in the strong and weak bonding bands that have exactly the same shape as for Si, consisting of the superposition of a broad and a narrow almost flat band. This behavior is apparent on the calculated density of states of Fig. 3a. These results are in good qualitative correspondence with photoemission data [4,8,9] and more sophisticated numerical calculations [10,11]. An essential information provided by equations (5) and (6) is that the splitting between the strong and weak bonding bands is a very sensitive function of the Si-O-Si angle 2θ. Any cause of randomness in θ such as the existence of a strained SiO_2 layer is then likely to induce some broadening of these two bands and partially fill the gap between them. Note that in an extreme situation where $\theta \sim 90°$ like in $GeSe_2$ the strong and weak bonding energies become identical, leading to qualitative change in the shape of the density of states.

We are now in position to discuss qualitatively the behavior of the SiO_x layer as a function of composition. Numerical calculations have been performed that are all based on a tight binding treatment combined with a more or less refined cluster Bethe lattice approximation [12 to 14]. Again the molecular model gives precious informations that are confirmed by the full calculations. For this let us consider the results of ref. [12] that are pictured on Fig. 4. In the molecular model each Si-O-Si bond corresponds to a density of states as given on Fig. 3b while each Si-Si bond corresponds to one bonding state which, for the parameters corresponding to Fig.4, falls at an energy slightly higher than $E_p(O)$, the energy of the SiO_2 non bonding state. If we start from the SiO_2 limit each Si-Si bond acts like an isolated defect giving rise to a defect level at the energy of the bonding state of Si, i.e just above the SiO_2 valence band. When the concentration of Si-Si bonds increases this defect state will begin to broaden into a band with no defined structure at small concentration. This corresponds to $SiO_{4/3}$ on Fig. 4. It is only when the concentration is high enough for Si chains to appear that the situation changes qualitatively. If, as before, Δ characterizes the interaction between two adjacent Si-Si bonding states, then the D.O.S. of a Si-Si chain exhibits two divergences at $\pm\ 2\Delta$ from the bonding energy. At still higher Si concentrations there will be three and then four Si-Si bonds connected to each Si atom. For reasons discussed above

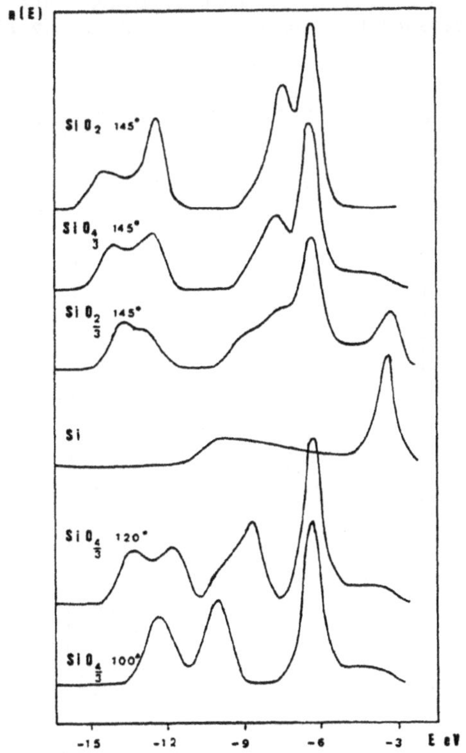

Figure 4 - Theoretical density of valence band states (arbitrary units).

such situations are characterized by a flat band at $+2\Delta$ and a broad band at lower energies. This regime will thus exhibit a peak of increasing height at the energy of the top of the pure Si valence band. This is exactly what happens on Fig. 4. Note also that the strong bonding band has exactly the reverse behavior : the height of its peak and its width both decrease. The last point that is clearly seen on the Figure is that, for smaller θ, the separation between the strong bonding states and the weak plus non bonding states is smaller.

All these conclusions are in qualitative agreement with experimental observations concerning the valence band of the SiO_x systems and of the transition layer at the Si-SiO_2 interface [4].

VERY THIN SiO_2 FILMS ON SILICON

The case of thin SiO_2 films thermally grown at low pressure (10^{-6} Torr) and low temperature (700°) reveals very interesting features. Fig. 5 shows the UPS valence band spectra corresponding to cases where the oxide layer is 11, 20 and about 100 Å thick. The last two layers are sufficiently thick for the spectra to correspond only to the SiO_2 layer. This is not true for the 11 Å layer where interface and bulk components have been substracted by a procedure described in ref. [5]. The thickness of the residual top SiO_2 layer is estimated to be ~ 6 Å. As for bulk SiO_2 the valence band spectra of this layer are characterized by three structures labelled I, II and III on Fig. 5 -. The most striking feature of these data is that the ratio R between the intensities of the bands II

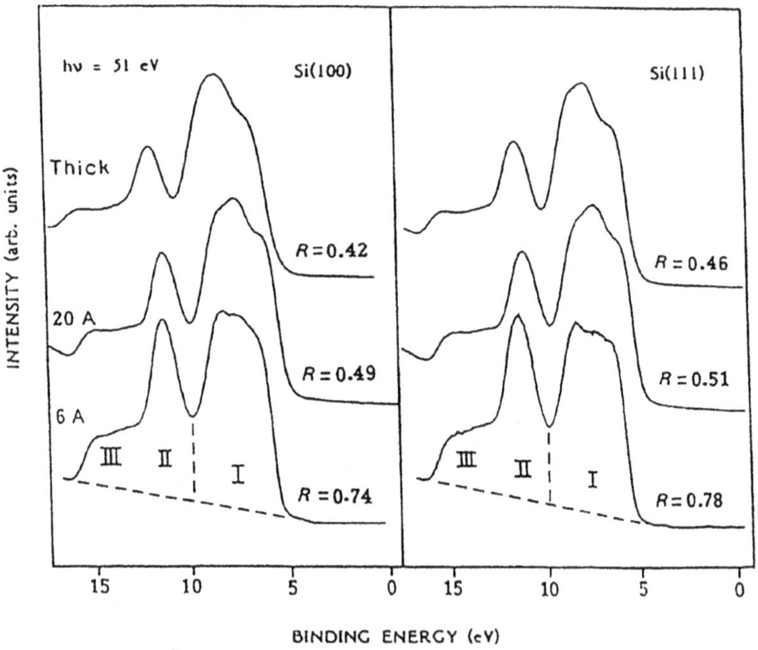

Figure 5 - Comparison of UPS valence band spectra for different thicknesses of SiO_2 thermally grown on Si. The corresponding R = (II+III)/I intensity ratios are reported.

+ III and of band I varies significantly between 0.42 and 0.74 for (100) orientation, 0.46 and 0.78 for (111) orientation. From the discussion of the last section bands II and III correspond to the strong bonding band of SiO_2 (one state per Si-O-Si unit) and band I to the sum of the weak and non bonding bands (two states per such unit). Thus the value characteristic of SiO_2, i.e of 4:2 coordination, is R = 0.5 which is not far from what is obtained for the thick layer (the actual value of R cannot be directly compared to 0.5 since the experimental bands overlap and the dip between I and II is arbitrarily chosen as their separation). Higher values of R cannot be explained by bond angle or bond length variations since, as shown in the first section, these only modify the band positions and spacings but not the value of R. We have therefore to admit that there are deviations from the classical 4:2 coordination.

One constraint that such deviations must respect is that valence bands should remain similar to the SiO_2 case but with only an increase in R. From this one can rule out the possibility of 6:3 coordinated SiO_2 stishovite which does not display the dip between bands I and II [15]. This is also the case for 3:3 coordinated Si-O whose band diagram is completely different [16]. The only hypothesis left seems to correspond to a high concentration of defects in a 4:2 SiO_2. Many types of defects have been proposed in crystalline and vitreous SiO_2 [11]. Among these one can reject peroxyl bridges or radicals which give oxygen states in the gap. Other defects such as oxygen vacancies or Si-Si bonds cannot be easily detected in UPS and should not strongly modify the spectra as observed.

We now discuss the defect model which we believe to be the most appropriate to explain these experimental observations. It corresponds to the valence alternation pair (VAP) [5,17 and references therein] which consists of one threefold overcoordinated oxygen O_3^+ and a one-fold under-coordinated oxygen O_1^-. In this model the fourfold coordination of Si, charge neutrality, number of Si-O bonds and therefore SiO_2 composition are preserved. To check quantitatively the validity of the VAP model, we have calculated the electronic structure of O_3^+ and O_1^- components using the simple molecular models of Fig. 6, with the tight binding (TB) approximation. We have used the TB parameters given by O'Reilly and

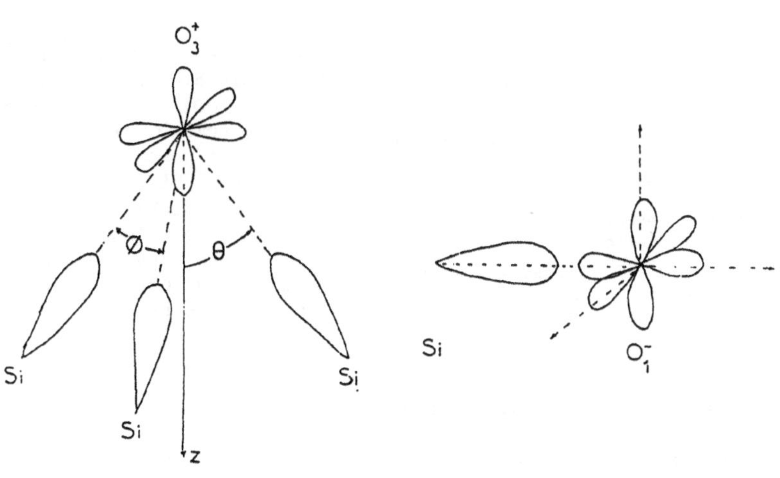

Figure 6 - Molecular units for O_3^+ and O_1^-.

Robertson [11] that reproduce well the UPS spectrum of SiO_2. We have also included the relaxation of the distance between the O_3^+ or O_1^- atom and its nearest neighbors (in the O_3^+ case we have also varied the θ angle) making a total energy calculation. We write the total energy for each molecular model as

$$E(R) = 2 \sum_{i}^{occ} \epsilon_i (R) + n f(R)$$

where R is the Si-O bonding distance, ϵ_i (R) are the one-particle energies, f(R) is the interatomic repulsion energy and n the coordination number. As in ref. [18] we consider that the TB parameters vary exponentially with distance as exp(-qR) while f(R) varies as exp(-pR), with $qR_0 = 2$, $pR_0 = 4$ for the normal Si-O distance $R_o = 1.61$ Å of α quartz. The prefactor of f(R) has been adjusted to get the correct R_0 for the molecular model of SiO_2. Applying the method to the two units of Fig. 6 the essential results of the approach are :

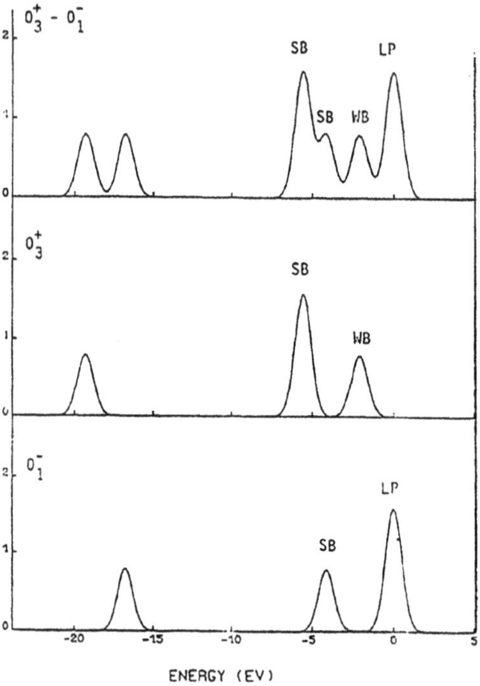

Figure 7 - Density of states in the molecular model of O_3^+, O_1^- and the VAP $O_3^+ - O_1^-$, broadened by a Gaussian of half width at half maximum equal to 0.5 eV.

i) The total energy of a VAP is within 2 % of the energy of two Si-O-Si units
ii) We find a reduction in Si-O bond length and angle for the VAP which is in agreement with the increase in atomic density expected at an abrupt interface.
iii) The densities of states of our simple models which are plotted on Fig. 7 - (note that for SiO_2 there is good agreement with the full calculation of ref. [11]) show that the VAP produces the right transformation of weak or non bonding states into strong bonding states. Each VAP corresponds to a ratio R = 1.

From these results it is likely that thin SiO_2 oxide films grown on Si (100) or (111) contain a large number of VAP. Their energy is comparable to that of two normal SiO_2 units and they could be stabilized near the interface to release the strain. Furthermore they can explain the qualitative change observed in the UPS spectra.

THE P_b CENTER AT THE Si-SiO_2 INTERFACE

Trivalent silicon atoms (P_b center) occur not only at the Si-SiO_2 interface [19 to 23], but also in amorphous silicon, grain boundaries in silicon and perhaps at some dislocations too. Each trivalent Si is characterized by an isolated dangling bond which has been identified mainly through electron spin resonance (e.s.r) [19], deep level transient spectroscopy (D.L.T.S.) [21,24], capacitance measurements versus frequency and, more recently, through optical experiments [25,26]. The qualitative picture which emerges is the following : i) the isolated dangling bond can exist in three charge states : positive D^+, neutral D^0 and negative D^- ; ii) the effective Coulomb term U, i.e the difference in energy between the acceptor and donor levels ranges from 0.4 eV in a - Si [25] to about 0.6 eV at the Si-SiO_2 interface [21] ; iii) in the paramagnetic state D^0 (one electron in the dangling bond state), the spin density is localized at 80% on the trivalent atom with 15 % s character.

There have been several calculations of the electronic properties of dangling bonds. Here we present the results of a Green's functions treatment [27] that gives a quantitative description of the following experimental data : i) order of magnitude and shape of the optical absorption versus frequency ii) the magnitude of the effective Coulomb term (often termed the correlation energy) and iii) the localization and character of the defect wave function as given by e.p.r. experiments.

To calculate the optical properties of a trivalent silicon atom at the Si-SiO_2 interface and this especially for transitions between the dangling bond level and the silicon valence band we need a model which correctly describes the top of the valence band. This eliminates the possibility of using a cluster Bethe lattice approximation which gives a bad density of states in this energy region. The calculation is based on the Green's function technique in the tight binding approximation. The perfect crystal is described in a second nearest neighbors approximation. We calculate the corresponding Green's functions by numerical integration over the Brillouin zone. We then remove an atom and add a perturbative potential which suppresses the interaction between the four dangling bonds of the resulting vacancy. This is done by use of Dyson's equation.

The first result is that we get a non self consistent localized dangling bond level at +0.14 eV from the top of the valence band. We find that the localization of this bound state on the trivalent atom is about 60% with 16% "s" character. This is in good correspondence with the e.p.r. data.

The determination of the Coulomb (or correlation) energy U requires a self consistent treatment. The simplest way to do this in tight binding is to put an intraatomic perturbation on the trivalent silicon atom and adjust its value to get a zero net charge on this atom, a procedure which is quite accurate for defects in silicon. This can be done for any occupancy N of the dangling bond level (N = 0,1, 2). The observable quantities are the electrical levels $\epsilon(0,+)$ and $\epsilon(-,0)$ [28] which can be calculated using Slater's transition state as the one - electron eigenvalues for N = 1/2 and N = 3/2 respectively. In this way we get $\epsilon(0,+)$ = 0.05 eV and $\epsilon(-,0)$ = 0.7 eV. The effective Coulomb U is the difference between these two values, so that we get U = 0.65 eV.

The value of U is quite good agreement with the experimental value (0.6 eV at the Si-SiO_2 interface, 0.4 eV in a - Si). Strictly speaking the local neutrality argument should be more valid for bulk material (a - Si for instance) where screening is more efficient than at the Si-SiO_2 interface. However the effective U can be reduced by relaxation effects [28], explaining the difference of 0.25 eV between our predicted value and the a-Si one. On the other hand the almost perfect agreement with the value at the Si-SiO_2 interface is the result of a cancellation between the decrease due to relaxation and the increase due to less efficient screening in that case.

We now consider optical transitions between the valence band and the level $\epsilon(0,+)$, for which there is experimental information at the $Si\text{-}SiO_2$ interface. We first calculate the electronic contribution to the optical cross section $\sigma_e(h\nu)$ versus frequency ν by using the Fermi Golden rule. Our calculated results are compared to the experimental curve of ref. [26] on Fig. 8 for the same polarization direction and the perpendicular one. The order of magnitude is clearly obtained but the shape near threshold is different. This represents evidence for the existence of electron-lattice coupling, leading to lattice relaxation. Such coupling causes a broadening of the lineshape. In this case $\sigma(h\nu)$ is obtained as the convolution of the purely electronic part $\sigma_e(h\nu)$ by a vibrational lineshape function [29]. This one can be approximated for relatively strong coupling by a Gaussian [28] whose position and width are functions of one parameter, the Franck-Condon shift d_{FC}. An interesting point is that the plot of the theoretical $h\nu\,\sigma_e(h\nu)$ near threshold is a straight line. If this is correct it is easy to show that convolution with a Gaussian also leads to a straight line with a Gaussian tail near threshold. This is precisely what is obtained experimentally on Fig. 9 which can be fitted by the theoretical curve for $d_{FC} = 0.27$ eV which is a manifestation of substantial electron-lattice interaction.

This conclusion is quite important and is confirmed by theoretical calculations of this parameter [30] which directly lead to $d_{FC} \sim 0.3$ eV. This however has the important consequence that the distance in energy between the two dangling bond levels $\epsilon(-,0) - \epsilon(0,+)$ is no more U but becomes equal to $U^* = U - 2d_{FC}$. If U is equal to 0.6 eV or smaller then one gets a negative U center ($U^* < 0$) for which the charge state 0 (with one unpaired electron) is not stable [31]. This contradicts experimental e.p.r. information. The way out of this internal contradiction is that the value found for the purely electronic contribution $U \sim 0.6$ eV is an

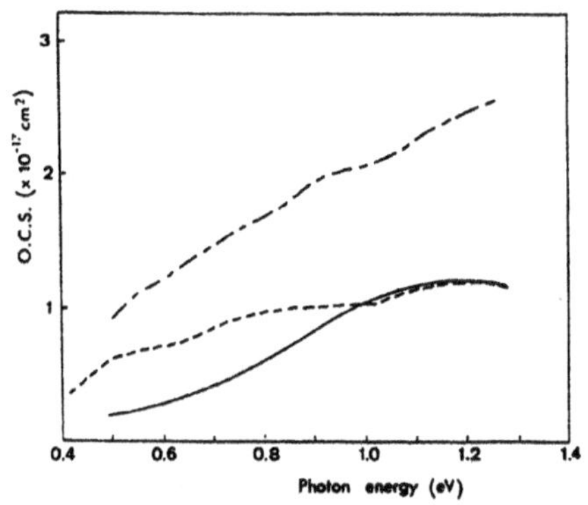

Figure 8 - Optical cross section versus photon energy : —— experiment, --- theory with the same polarisation direction, _._._. theory with perpendicular polarization.

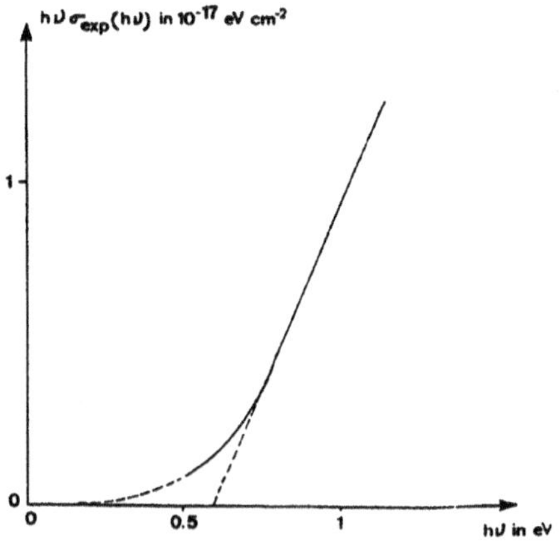

Figure 9 - Plot of experimental $\sigma(h\nu)$ showing the linear behavior and the tailing effect due to the electron-lattice interaction.

underestimation. Indeed it has been calculated with bulk silicon screening whereas the screening should be reduced at the $Si-SiO_2$ interface. Simple reasoning based on the replacement of ϵ by $(\epsilon+1)/2$ leads to an enhancement by a factor of 2, i.e $U \sim 1.2$ eV leading to the final result $U^* \sim 0.6$ eV which is coherent with the experimental situation.

REFERENCES

1. O.L. Krivanek, T.T. Scheng and D.C. Tsui, Appl. Phys. Lett., 32:439 (1978).
2. G. Hollinger and F.J. Himpsel, Appl. Phys. Lett., 44:93 (1984).
3. F.J. Grunthaner, P.J. Grunthaner, R.P. Vasquez, B.F. Lewis, J. Maserjian and A. Madhukar, Phys. Rev. Lett., 43:1683 (1979).
4. G. Hollinger, Y. Jugnet and Tran Minh Duc, Sol. State Comm., 22:277 (1977).
5. G. Hollinger, E. Bergignat, H. Chermette, F. Himpsel, D. Lohez, M. Lannoo and M. Bensoussan, Phil. Mag., 55:735 (1987).
6. C.A. Coulson, "Valence", Oxford University Press (1961).
7. S.T. Pantelides and W.A. Harrisson, Phys. Rev. B, 13: 2667 (1976).
8. T.H. Di Stefano and D.E. Eastman, Phys. Rev. Lett., 29, 1088 (1972).
9. H. Ibach and J.E. Rowe, Phys. Rev. B, 10:710 (1974).
10. J.R. Chelikowsky and M. Schlüter, Phys. Rev. B, 15: 4020 (1977).
11. E.P. O'Reilly and J. Robertson, Phys. Rev. B, 27: 3780 (1983).
12. M. Lannoo and G. Allan, Sol. State Comm., 28: 733 (1978).
13. E. Martinez and F. Yndurain, Phys. Rev. B, 24: 5718 (1981).
14. K. Hübner, A. Stern and E.D. Klinkenberg, Phys. Stat. Sol., 136: 211 (1986).
15. Y.P. Li and W.Y. Ching, Phys. Rev. B, 31: 2172 (1985).
16. M. Bensoussan and M. Lannoo, J. Phys. Paris, 40: 749 (1979).
17. G. Hollinger, S.J. Sferco and M. Lannoo, Phys. Rev., to be published.
18. M. Lannoo, J. Phys., 40: 461 (1979).
19. E.H. Poindexter and P.J. Caplan, Prog. Surf. Science, 14: 201 (1983).
20. P.J. Caplan, E.H. Poindexter, B.E. Deal and R.R. Razouk, J. Appl. Phys., 50: 5847 (1983).
21. N.M. Johnson, D.K. Biegelsen, M.D. Moyer, S.T. Chang, E.H. Poindexter and P.J. Caplan, Appl. Phys. Lett., 43: 563 (1983).
22. K.L. Brower, Appl. Phys. Lett., 43: 1111 (1983).
23. B. Henderson, Appl. Phys. Lett., 44: 228 (1984).
24. J.D. Cohen and D.V. Lang, Phys. Rev. B, 25: 5285 (1982).
25. W.B. Jackson, Sol. State Comm., 44: 477 (1982).
26. N.M. Johnson, W.B. Jackson and M.D. Moyer, Phys. Rev. B, 31: 1194 (1985).
27. J. Petit, M. Lannoo and G. Allan, Sol. State Comm., 60: 861 (1986).
28. J. Bourgoin and M. Lannoo, "Point Defects in Semiconductors II", ed. by M. Cardona, Springer Verlag, New York (1983).
29. S. Loualiche, A. Nouailhat, G. Guillot and M. Lannoo, Phys. Rev. B, 30: 5822 (1984).
30. M. Lannoo, unpublished.
31. Y. Bar-Yam and J.D. Joannopoulos, Phys. Rev. Lett., 56: 2203 (1986).

THEORY OF DEFECTS IN THE MOS SYSTEM

Arthur H. Edwards

U.S. Army E.T. and D. Laboratory
Ft. Monmouth, N.J. 07703-5000

INTRODUCTION

Without doubt, the single most important technology for electronic device fabrication relies on the metal-oxide-silicon (MOS) system. In recent years III-V semiconductors have promised great increases in device speed and, because they are direct-gap materials, have been very important in electro-optical applications (i. e. light emitting diodes). However, these semiconductors have very complicated surfaces with high densities of surface-states and, hence, very low surface electron densities. While bare silicon surfaces also have low electron densities due to surface-states, most of these states can be eliminated by growing a thermal oxide. For other semiconductors, no such passivating process exists, so that, for instance, inversion devices such as enhancement-mode MOSFET's (field effect transistors) are practical only in silicon. It is not surprising, then, that the MOS system has been studied intensely over the past thirty years. One of the most important areas of investigation has been the study of point defects. Defects in the oxide lead to shifts in threshold voltage in MOSFET's [1]. Also, the residual unpassivated defects at the Si/SiO_2 interface are responsible for decreased surface mobility, and for "soft" threshold characteristics. We should note that while some defects are inherent, i. e. arise from differences in the thermal expansion coefficients of silicon and silicon dioxide, and from lattice-network mismatch, most exist due to ionizing radiation or hot-electron injection. They can be created during fabrication (during X-ray lithography, or during the various plasma-assisted etching steps), or they can be a result of exposure to a space environment where even a modest flux of gamma rays can cause dramatic changes in device characteristics [2].

In this paper we will discuss the physics of some of the intrinsic defects in the MOS system [3]. We will present the current understanding of the important defects in bulk SiO_2 and attempt to draw comparison with those found in thin films. We will also discuss at length the set of dangling silicon orbital defects found at the Si/SiO_2 interface. This will include a discussion of our calculations of model defects at the <100> interface. We will not present an exhaustive history of MOS defects. Rather, we refer the reader to several reviews [4,5]. Furthermore, we will only give a cursory account of the experimental data, as this is covered in other papers in this proceedings [6,7]. We will instead concentrate on recent theoretical advances. Finally, we limit our attention to the paramagnetic defects. For discussions of the diamagnetic defects see ref. 8.

POINT DEFECTS IN BULK SiO_2

We can separate the intrinsic diamagnetic defects in bulk SiO_2 into two catagories: (1) oxygen vacancy-related defects, and (2) oxygen excess-related defects.

Oxygen vacancy-related defects

We first discuss the oxygen vacancy-related, or E′, defects. The invariant feature of all of these is a neutral silicon atom bonded to three oxygen atoms with an unpaired electron in the dangling orbital, shown in Fig. 1. The E_1', E_2', and E_4' centers have been observed in α-quartz, so that identification is fairly well established. The E_4' center, shown in Fig. 2a, is the best understood. The ESR analysis of Isoya *et al.* [10] shows clearly that the unpaired spin is shared unequally between two silicon atoms and one hydrogen atom. The strong temperature dependence of the hyperfine parameters is thoroughly explained by a double-well model with inequivalent wells. At low temperatures, the hydrogen atom is strongly localized in the lower well, i. e. it is attached to one silicon atom, so that the unpaired spin is localized strongly on the other silicon. As the temperature rises, the hydrogen atom nuclear wave-function is spread more evenly so the unpaired spin is delocalized onto the three atoms. Isoya and co-workers performed extensive *ab initio* molecular orbital calculations that are in excellent agreement with experiment. Rudra *et al.* [11] have also studied this center using a semiempirical technique applied to large atomic clusters with similar agreement.

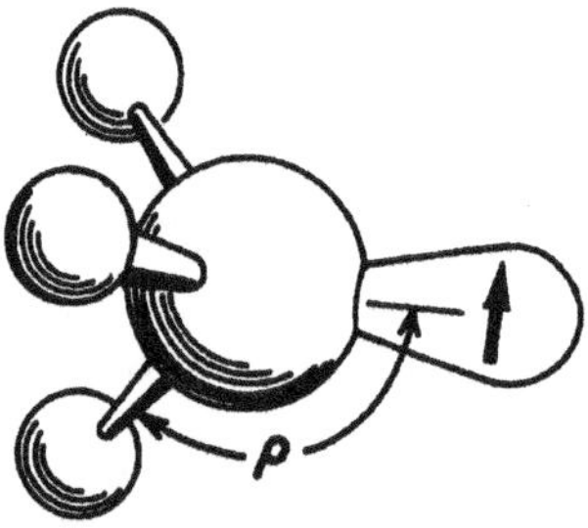

Fig. 1. The generic E′ center. Large sphere represents a silicon: smaller spheres are oxygens. Arrow indicates an unpaired spin. From ref. 9.

The E_2' is also an oxygen vacancy with an associated hydrogen atom. However, the proton hyperfine interaction is roughly an order of magnitude smaller than for the E_4' center, so that a specific microscopic model was, until recently, lacking. Rudra and Fowler [11] have proposed the model shown in Fig. 2b. While it will be difficult to attain the same level of certainty for this defect as exists for the E_4' center, the model agrees with all ESR data, i. e. the magnitudes and directions of the ^{29}Si and ^{1}H hyperfine tensors. We note that the E_4' and E_2' centers can transform into one another. In fact, the Rudra-Fowler-Feigl calculations predict that the E_2' center is more stable by about 0.1 eV and that the barrier to transformation is about 0.6eV. As a function of temperature, then, we would expect the ratio of densities for these two defects to vary in a predictable way.

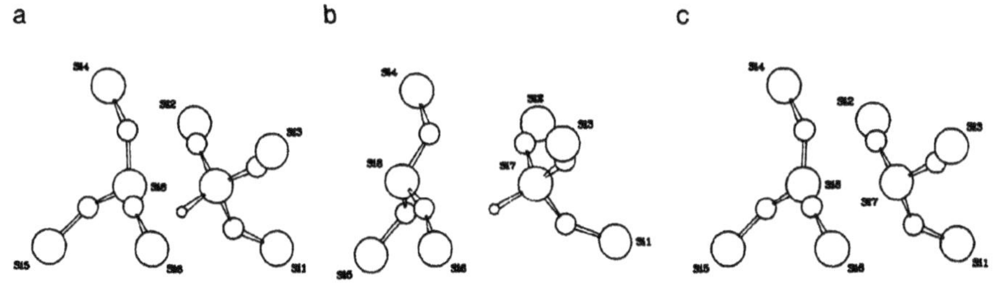

Fig. 2. a) The Isoya-Weil model for the E_4' center. b)The Rudra-Fowler model for the E_2' center. c) The Feigl-Fowler-Yip model for the E_1' center. The unpaired spin is on Si8 in a) and b), and on Si7 in c). The smallest circles in a) and b) are hydrogen atoms.

The simplest and, perhaps, most interesting member of the E' family of defects is the E_1' center. The accepted model for this defect is shown in Fig. 2c. In this case an oxygen vacancy has trapped a hole, and has undergone an asymmetric relaxation. The unpaired spin is again trapped predominantly on one silicon atom. The hyperfine data in α-quartz is fairly compelling, especially when combined with the E_4' data. Jani *et al.* have performed ENDOR on the E_1' center [12]. In agreement with the original Silsbee results [13], they found one strong (principal) and two weak (super-) hyperfine interactions. By analyzing the directions of the principal axes of the associated hyperfine tensors they concluded that: 1. the principal hyperfine interaction is directed along the short bond direction (recall that there are two inequivalent Si-O bond directions in α-quartz, labeled long and short). 2. the weak hyperfine interactions arise from two of the three second-nearest neighbors on the short bond side of the vacancy. Note that statement 1 implies that there is no observable spin density on the relaxed silicon atom (Si8 in Fig. 2c) [14].

The disposition of spin density, and the magnitudes and directions of the hyperfine interactions for the E_1' center are faithfully reproduced in the cluster calculations of Rudra and Fowler [15]. The inequivalence of Si3 in Fig. 2c (with respect to Si1 and Si2) led naturally to a suppression of spin density on that nucleus. Furthermore, the presence of an oxygen directly behind Si8 (not shown) led to a much larger relaxation and, hence, a much smaller spin density on Si8 than was estimated from calculations on smaller clusters.

It is interesting to consider the cause for the strong, asymmetric relaxation of Si8 that leads to the spin localization on one silicon atom. This has been reproduced in every atomic cluster calculation on the E_1' center [15-18]. It was originally attributed to a pseudo-Jahn-Teller effect. Simply stated, this is an extension of the Jahn-Teller theorem from cases of exact, group theoretical degeneracy to, cases where states are close enough in energy to interact through a suitable electron-phonon interaction. In the original Jahn-Teller effect, if the electronic quantum state is part of a degenerate manifold, and the total system has symmetry higher than linear, then the the system is guaranteed to spontaneously distort, lowering the symmetry, and splitting the degeneracy. For the pseudo-Jahn-Teller effect, the system *can* spontaneously distort to lower the symmetry, but this is not guaranteed. Rather an inequality must be satisfied. For the case of the E_1' center, where for simplicity, C_{2v}

symmetry was assumed, and where the interacting states were assumed to be the symmetric (bonding) and antisymmetric (antibonding) combinations of dangling sp^3 orbitals pointing toward the vacancy, the following inequality determined whether the system would relax asymmetrically

$$G^2/f > |E_{sp}|, \qquad (1)$$

where G is a linear electron-phonon coupling constant, E_{sp} is the energy difference between the bonding and antibonding states in the symmetrical, unrelaxed configuration, and f is the effective vibrational force constant (i. e. $2f = \mu\omega^2$, where μ is an effective mass). The vibrational mode of interest is an antisymmetric displacement of the two silicon atoms- one toward and one away from the oxygen vacancy.

Recently, we have re-examined this simple picture. We found that for clusters exhibiting true C_{2v} symmetry, the two silicon atoms relaxed symmetrically except for exceptionally large initial Si-Si separations [19]. The relaxation is then obviously a function of the initial asymmetry of the system. From analysis of a modified vibronic Hamiltonian, due to Schirmer [20], that includes an asymmetry parameter, and from cluster calculations wherein we systematically varied the degree of initial asymmetry, we found that the degree of distortion was a continuous function of initial asymmetry. Unlike, the simple pseudo-Jahn-Teller case, where the system can localize in one of two equivalent wells, the modified Hamiltonian can, and in the case of the E_1' center does, yield only one ground state minimum energy configuration. This theory explains the observation of E_1' centers with spin localization always on the short-bond side of the vacancy. Finally, we note that recent calculations by Bar-Yam and Pantelides [21] in β-cristobalite and Allan and Teter [22] in α-cristobalite have not obtained the asymmetrically relaxed state as the minimum energy configuration. Both use the local-density density approximation applied to super-cells. Both of these systems exhibit milder asymmetry in the unrelaxed configuration than seen in α-quartz, so that we would expect smaller (and in the case of β-cristobalite negligible) relaxation.

We now turn to vacancy-related defects in amorphous SiO_2. ESR has also been used here, although most of our fundamental ideas are derived from analogy with the well established crystalline models. The most common E' center in a-SiO_2, E_γ' is most likely the glass-analog of the E_1' center [23, 24]. To date, there have been no observations of the glass-analog of the E_2' or E_4' centers, although other hydrogen-related E' centers have been studied in a-SiO_2[25,26]. Models for these defects (from ref. 26), labeled by the size of the proton hyperfine splitting, are shown in Fig 3. Edwards and Germann have studied these models [8] using semiempirical molecular orbital techniques, briefly described below. Their results favor strongly the Tsai-Griscom model for the 74 G doublet, while lending some support to the 10.4 G model. The remaining E' variants, E'_α and E'_β are observed at low temperature.

Excess Oxygen-Related Defects

The oxygen related defects in a-SiO_2 have no analogs in α-quartz. Careful spin resonance studies have rendered the models shown in Fig. 4 unambiguous [27-29]. Of these two defects, only the superoxide radical has been observed in a non-neutral charge state, although, the NBOHC has been shown theoretically to have a negative charge state in the gap [8]. We will not discuss these at length,as they are not as important as the E' centers to the degradation of MOS device performance.

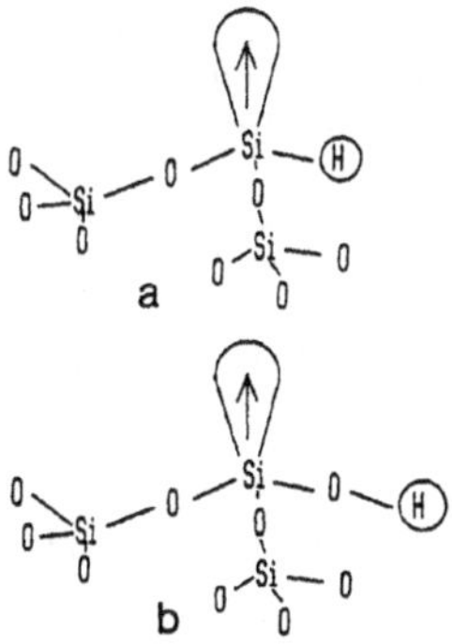

Fig. 3. Tsai-Griscom models for the hydrogen related E′ centers observed in a-SiO_2. (a) 74G doublet (b) 10.4 G doublet

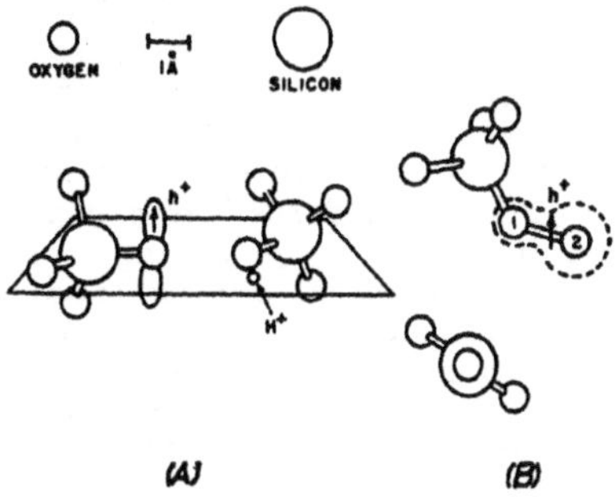

Fig. 4. Models for the two paramagnetic oxygen related defects in a-SiO_2. (a) The nonbridging oxygen-related hole center (NBOHC) (b) The superoxide radical.

DEFECTS IN THERMAL OXIDES

Before we turn to the interface defects, we should discuss the relevance of studies of bulk SiO_2 to thermal oxides. Several groups have investigated the defects in SiO_2 grown on silicon using ESR [30-33]. It turns out that, with the exception of the low temperature E′ centers alluded to above and the 10.4 G doublet, all of the paramagnetic defects observed in bulk a-SiO_2 have been observed in thermal oxides. However, the most important defect is E'_γ. For radiation doses occurring in fabrication, or in a space environment, this is the most abundant defect. Also, the evidence is strong that it is the defect responsible for the build up of positive charge near the Si/SiO_2 interface.

DEFECTS AT THE Si/SiO_2 INTERFACE

Our understanding of interface states has expanded dramatically since ESR was first used to probe their microscopic structure. Combined with more traditional electrical techniques (i. e. C-V, G-V, DLTS) ESR has shown us the most fundamental features of an important subset of the electrically active interface states, namely the dangling silicon orbital defects. These are conveniently segregated according to crystalline orientation of the Si/SiO_2 interface.

Dangling Orbitals at the <111> Si/SiO_2 Interface

The careful analysis of the g-tensor anisotropy of Poindexter *et al.*[34] combined with Brower's ^{29}Si hyperfine data [35] led to the positive identification of the P_b center at the <111> Si/SiO_2 interface as a trivalent silicon, shown in Fig. 5. The field-dependent ESR of Poindexter *et al.*[36] and of Lenahan and Dressendorfer [37] shows that this defect is amphoteric with electrical levels 0.3 eV (+/o transition) and 0.85 eV (o/- transition) above the silicon valence-band edge.

The theoretical understanding of the P_b center is also reasonably sound. Several groups have treated this system using self-consistent molecular orbital methods [38-41]. Two of these groups [39-41] calculated equilibrium geometries for a Si_4H_9 cluster (similar to the moiety in Fig.

5a, but with terminating hydrogen atoms on the three back-bonded silicon atoms) wherein the central silicon atom is allowed to move along the <111> direction. All of these calculations are in qualitative agreement, and have given us a good intuitive understanding of these relaxations. In the neutral charge state, the central atom can strengthen the three back bonds at the expense (in electron density) of the dangling orbital by relaxaing slightly toward its three nearest-heighbors. The other two charge states are understood by noting that the dangling orbital has anti-bonding character. Hence, adding (removing) an electron to (from) an antibonding orbital should lower (raise) the overall bonding and cause the central atom to seek a position more distant from (closer to) its three nearest neighbors. We note that these trends were preserved for all clusters considered by Edwards [38].

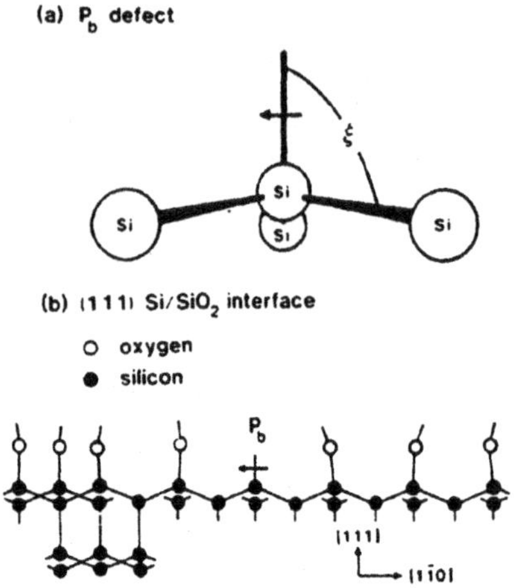

Fig. 5. Poindexter model for the P_b center at the <111> Si/SiO$_2$ interface.

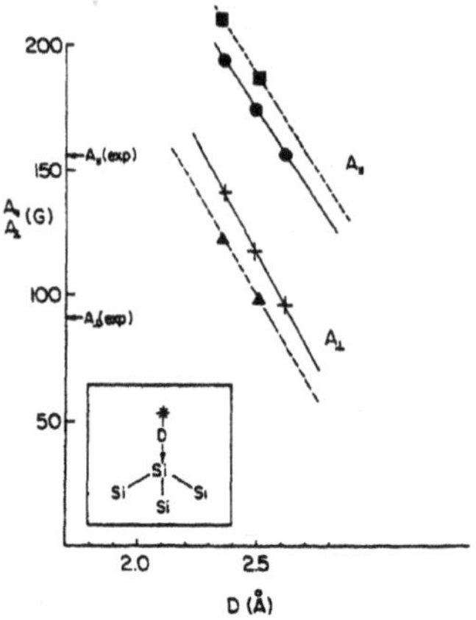

Fig. 6. Theoretical hyperfine interactions on the defect atom of the P_b center as a function of position (see insert). Dotted lines are from ref. 38, solid lines are from ref. 39.

Recent calculations by Cook and White [38] and by Edwards [39] have faithfully reproduced Brower's hyperfine data. In fact, both calculations showed the origin of superhyperfine interactions evident, though previously unexplained, in Brower's data. The superhyperfine lines result from spin density on second-nearest neighbors (see refs. 38 and 39). The agreement of these calculations with the experimental data, and with each other is particularly striking considering the basic differences between the two methods of solution. While both groups used clusters containing up to twenty-two silicon atoms, and both used hydrogen atoms to terminate unwanted surface bonds, Cook and White used the local-density approximation (X-α) while Edwards used a semiempirical technique. Both groups calculated hyperfine tensors for different positions of the defect atom. Their results are shown in Fig. 6. Note the agreement in magnitude and in functional dependence. Note also that both calculations are within ten percent of the experimental values.

Several groups estimated of level positions for the (+/o) and (o/-) transitions. From cluster calculations, Edwards predicts that the (+/o) transition is 0.06 eV from the valence band and that the (o/-) transition is 0.66 eV so that the effective correlation energy, U, is 0.6 eV. However, in their local-density calculation, Bar-Yam and Joannopoulos [42] obtain a U of -0.2 eV. We should point out that this is their predicted value for a dangling bond in crystalline silicon, and that it is a super-cell calculation. If this result were valid for the interface, it would imply the dangling bond would never be paramagnetic. It would indeed be surprising if this were the case. We point out that this is a spin-restricted calculation. The added variational freedom inherent in a spin-unrestricted calculation would lower only the total energy of the neutral state. This would favor a change toward a positive U. We also note that the super-cell size is surprisingly small- it seems possible that lattice distortions, if not defect wave-functions, in adjacent super-cells could interact.

Dangling Orbitals at the <100> Si/SiO_2 Interface

The picture for the <100> interface is less clear. This interface has not yielded as complete a set of data as did the <111> interface for several reasons. First, the density of interface states at the <100> interface is inherently smaller (by roughly an order of magnitude) than it is at the <111> interface. Hence, reliable ESR signals are more difficult to obtain. Second, the <100> interface state spectrum is more complex. That is to say, for <111> and <100> MOS capacitors that have undergone similar processing to maximize interface defects, the <111> interface state distribution, D_{it}, shows two peaks, corresponding to the two electrical level of the P_b center discussed above, while D_{it}(<100>) shows at least three separate, yet strongly overlapping, spectral features.

These difficulties notwithstanding, the following experimental picture has emerged:

i. The ESR experiments of Poindexter *et al.* [43] show that there two spectroscopically different paramagnetic defects on the <100> interface, labeled P_{bo} and P_{b1}.

ii. The g-tensor for P_{bo} is almost identical to that of the <111> P_b center. That is to say, it has approximately axial symmetry, its principal values are of nearly the same magnitude, and they point in the same crystallographic directions.

iii. The g-tensor for P_{b1} is clearly not axially symmetric. However, the principal values are similar to the other dangling orbital defects, and the principal directions are also approximately the same, so it is also identified as a dangling silicon orbital. Poindexter has suggested models for these defects, shown in Fig. 7. Note that these are simple yet they hold promise of explaining all of the observed ESR data.

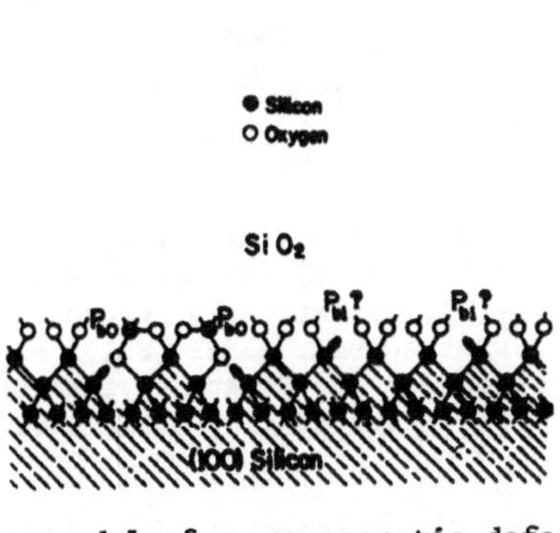

Fig. 7. Poindexter models for paramagnetic defects at the <100> Si/SiO_2 interface.

iv. Gerardi *et al.* [44] have performed field-dependent ESR studies using both capacitors (with aluminum gates) and bare Si/SiO_2 samples (they used corona discharge to generate an electric field). This study showed that the P_{bo} electrical levels were approximately the same as the <111> P_b center, that the P_1 center was also amphoteric, and that the P_{bo} levels bracket the P_{b1} levels, so that $U_{eff}(P_{bo}) > U_{eff}(P_{b1})$

v. Brower has performed some preliminary hyperfine measurements on these two defects and found that, for one orientation of the external magnetic field, the ^{29}Si hyperfine interaction for P_{b1} is about 30 G larger than that of P_{bo}. Furthermore, in oxides containing 52% ^{17}O, he found no ^{17}O hyperfine structure in either ESR spectrum (P_{bo} or P_{b1}) [45].

These experimental data comprise the set to which we should compare all theoretical calculations on models for P_{bo}, and P_{b1}.

Model Calculations

We report here a set of preliminary calculations on models for the P_{b1} defect. We chose this because, initially, P_{bo} seemed to be without interest. That is to say, the Poindexter model seemed almost inevitable. His model for P_{b1} also appeared plausible, so we performed an initial set of calculations on this defect model. We used a combination of *ab-initio* and semiempirical molecular orbital techniques applied to clusters of atoms used to represent various defects at the <100> interface. As discussed above, this has been a very fruitful approach for defects in SiO_2 and for the <111> P_b center. For the *ab-initio* calculations, we used Gaussian-82 [46], and for the semiempirical calculations we used MOPN, an unrestricted Hartree-Fock version of MINDO/3 [18,47,48]. The *ab-initio* calculations were performed only on the smallest cluster, Fig. 8b, to calibrate the semiempirical results. We passivated the dangling orbitals that originated from cluster termination with hydrogen atoms. We have calculated equilibrium geometries under a variety of constraints discussed in detail below. We also calculated hyperfine tensors from the full spin-density matrix using the method described in Ref. 39, and we estimate electrical level positions from the one-electron eigenvalues, appealing to Koopmans' theorem. In previous studies, we have investigated convergence of these observables as a function of cluster size. Rather than repeat that study here, we appeal to the result that third-nearest-neighbor coordination is enough to assure convergence of equilibrium geometries, as well as spin densities and one-electron level positions. Finally, we have studied all three charge states of interest.

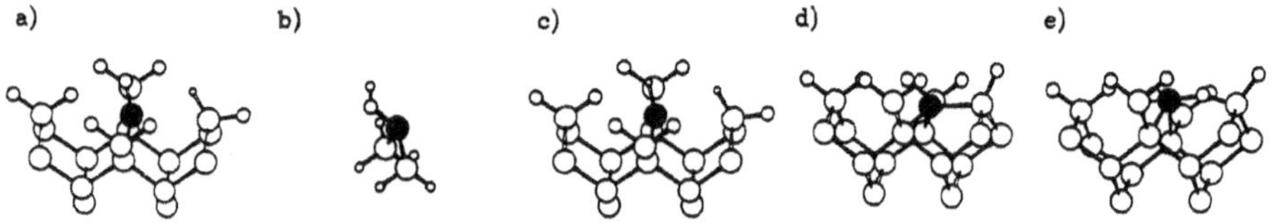

Fig 8. Clusters used for studying the P_{b1} center. Terms in parentheses refer to Fig. 9. a) Poindexter model ($P_{b1}(O)$); b) Small cluster $P_{b1}(O)$; c) Paritally hydrogenated dangling bond ($P_{b1}(H)$); d) Strained bond model (SB1); e) Strain relief model (SB2). Large circles are Si atoms, intermediate are O atoms, and the smallest are H atoms. Defect atom is black.

Poindexter Model for the P_{b1} Center: For these calculations we used the cluster shown in Fig. 8a. Before we discuss our results we should mention two special features of this cluster. First, we have represented the oxide by attaching OH groups. On the <111> surface, there was no appreciable improvement when we replaced these with SiO_4H_3 units. Also, OH groups lead to no inherent assumptions about local connectivity in the oxide. The second and more important feature is the hydrogen atom facing the dangling bond. This ensures that the defect will have dangling bond-character. When we replaced this with an OH group we obtained a three-fold coordinated oxygen atom (bonded to two silicon atoms and one hydrogen atom). The equilibrium position of the hydrogen atom in Fig. 8a was distant enough from the defect atom that there was negligible interaction.

We first consider the neutral charge state. A schematic representation of the defect level with respect to the valence band edge is given in Fig. 9 [49]. Note the the +/o transition is virtually degenerate with the conduction-band edge. This is in dramatic disagreement with experiment. We believe that this is a physical result and not a result of poor assumptions in our method of calculation. We note that Carrico *et al.* obtain a similar result from a tight-binding calculation[50]. The difference between the energy level positions for the Poindexter P_{b1} model and the <111> P_b center can be attributed to differences in interaction strengths between the defect state and orbitals on the first nearest neighbors. Specifically, the overlap, and hence the interaction, is stronger between the dangling orbital and the lone-pair orbital on a neighboring oxygen atom (P_{b1}) than between the dangling orbital and the symmetric combination of sp^3 orbitals on three neighboring silicon atoms (<111> P_b). This is reflected in the contributions to the defect wave-function. In summary, our calculations indicate that the Poindexter model for the P_{b1} defect is untenable, as the proximity of the o/+ electrical level to the conduction band precludes observation of an ESR signal. However, before we turn to alternative models for this defect we will complete our analysis, as there is no reason to doubt the existence of a defect similar to the poindexter model, even though it would not be paramagnetic for reasonable values of surface potential.

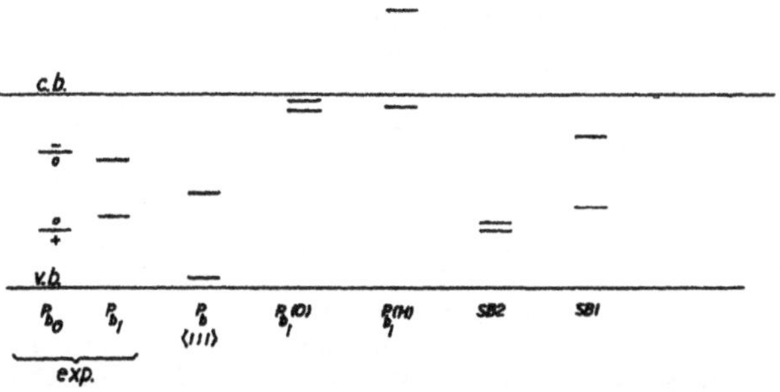

Fig 9. Schematic representation of the electrical level positions for various models for the P_{b1} center. We include the experimental values for P_{bo} and P_{b1}, as well as the theoretical results for the <111> P_b center. For definition of labels, see Fig. 8.

We have also calculated the equilibrium geometry and electrical level position for the negative charge state. The most interesting result is the negligibly small effective correlation energy we predict for this defect. This is most surprising considering our prediction of 0.6 eV for the <111> P_b center. There are two possible mechanisms for this decrease. First, if the defect wave-function were more delocalized, then the bare correlation energy (not including e-lattice interaction) would be smaller because the coulomb interaction would be reduced. Second, the e-lattice coupling could be appreciably stronger. To study these mechanisms, we consider a much smaller cluster, shown in Fig. 8 b. It turns out that even for this small cluster, U is ~ 0.1 eV. To uncouple the two mechanisms we have performed calculations with and without lattice relaxation. For comparison with the <111> P_b center, we performed the same analysis on the Si_4H_9 cluster. The results are summarized in Table I.

TABLE I. Decomposition of effective correlation energies for the P_b(<111>) center and the P_{b1}(<100>) center into coulombic and e-lattice contributions. All values are in eV.

	U_{bare}	U_{e-l} (U_{eff}-U_{bare})
P_b	1.25	0.76
P_{b1}	1.07	1.08
Δ	0.18	0.32

While both mechanisms give important contributions, the increased e-lattice interaction is the dominant effect. Again, this can be traced to the silicon dangling orbital-oxygen lone-pair orbital interaction.

Note that for this small cluster the Poindexter P_{b1} defect would actually be a negative-U system, while for larger clusters the U is slightly positive. However, for most experiments, this small uncertainty is unimportant, especially when we consider the role of inhomogeneous line broadening. This result implies that for all values of surface potential, there would be no ESR signal.

Finally, we note that the Poindexter P_{b1} model is a strong candidate for a radiation induced interface state. On <100> samples, both hole-injection and γ-radiation produce a D_{it} spectrum with a large peak in the upper half of the silicon band gap. However, even in heavily doped n-type MOS capacitors, there is no associated ESR signal. The Poindexter model would explain both of these observations.

Other Models for the P_{b1} Center: We have studied several other models for the P_{b1} center. The first we will discuss is a trivalent silicon atom with two silicon and one hydrogen nearest neighbors, shown in Fig. 8c. The electrical levels are shown in Fig. 9. Note that the +/o transition is also in the upper half of the silicon band gap, contrary to experimental results. This is also due to an increased interaction between the dangling orbital and its nearest neighbor- in this case the hydrogen 1s state. Unlike the Poindexter model, this defect has a positive-U of ~0.6 eV. Hence, in principle, it could be observable in ESR, so that it is useful to calculate the ^{29}Si hyperfine interaction. The principal values of the hyperfine tensor on the trivalent silicon atom are A_1 = 280. G, A_2 = 286. G, and A_3 = 300. G. These values are roughly twice the observed values for any of the paramagnetic interface defects so that if this defect were to exist in measurable densities, it would be easy to identify.

The final two models we have considered are shown in Figs. 8 d and e. Note that they are actually defect pairs. The first case, Fig. 8d, is a dangling bond adjacent to a reconstructed silicon-silicon bond. The second might be called a dangling bond adjacent to a strain-relief site. For brevity, we label these SB1 and SB2 respectively.

Before we present our results, it would be useful to discuss the origin of these two models. It is instructive to consider the oxidation process of <100> silicon surfaces. Several groups have built ball-and-stick models to study how an SiO_2 network can be attached to the crystalline silicon so that the transition from pure silicon to stoichiometric SiO_2 occurs within a monolayer, and so that all the silicon surface orbitals (assumed to be sp^3 orbitals pointing in the ideal <111> directions) can be saturated in Si-O bonds [51,52]. That is to say, is it possible to build an ideal Si/SiO_2 interface without surface states using only silicon and oxygen atoms? The answer is yes, as long as you allow for strain-relief sites in the form of Si-O-Si units, where both silicon atoms are on the <100> surface. The source of strain is easily understood. On a <100> surface, the surface sp^3 orbitals point in pairs toward missing silicon sites (This is in contrast to the <111> surface, where the surface orbitals all point in the same direction- perpendicular to the surface.) We cannot put a monolayer of oxygen atoms on this system unless we accept unphysically small O-O distances (0.24 A smaller than free O_2^-, the appropriate charge state for SiO_2), or unless we introduce bond strain. This strain was large enough to pull the ball-and- stick models apart [52]. However, if you included roughly an atomic per cent of the strain-relief sites, they can absorb the strain by forming smaller Si-Si distances (i. e. the single oxygen atom pulls the two surface silicon atoms together). If one allows some strained Si-Si bonds to form, similar to the 2X1 surface reconstruction on bare <100> silicon surfaces, then strain can be relieved even more efficiently. On the best practical interface, these are the natural interface states, as dangling bonds can form naturally next to them. There is no need to replace the oxide (the opposing OH group) with a hydrogen atom, as was done for the Poindexter model, and for the partially hydrogenated silicon atom.

The electrical levels for SB1 and SB2 are shown in Fig 9. Note that the SB1 has a level structure that is quite close to the experimental values for P_{b1}. Note also that SB2 has a very small U, similar to the Poindexter P_{b1}, but that the levels are shifted to mid-gap. We attribute this to on a weakening of the dangling bond-lone-pair interaction due to the added constraint imposed by the surface bonding geometry. Of the two centers, SB1 would appear to be a strong candidate for the P_{b1} defect. With this in mind, we calculated all the relevant hyperfine interactions. We present these in Table II, along with Brower's result for one orientation of the external magnetic field.

Table II. ^{29}Si hyperfine values for Si_1 (the central defect atom, black in Fig. 8d) and Si_2 (nearest neighbor on the surface). These were the only important interactions. Experimental value is taken from ref. 45. All values are in Gauss.

	A_1	A_2	A_3	$A_{exp.}$
Si_1	49.3	83.8	128.4	-186.0
Si_2	26.0	26.9	37.9	

It is clear that, while the electrical levels are in good agreement with experiment, the hyperfine results disagree strongly enough that we believe SB1 is not the P_{b1} center. This is reinforced by our <111> P_b results, where theory actually *over* estimated the hyperfine interaction (Brower's data clearly show that $A(P_{b1}) > A(P_{bo})$). We note in passing that this is an example of a strain-induced enhancement of nearest-neighbor hyperfine strength, as predicted by Biegelsen and Stutzmann in their work on the dangling bond in a-Si:H [53]. While Cooks' observations regarding the possible role of increased freedom of relaxation in amorphous materials are probably also to the point [54], strain should play an important role.

Discussion and Conclusion

While we have not yet performed a study of a suitable candidate for the P_{b1} defect, the negative results detailed above are not without value. The features of each lead us to speculate on the general nature of a viable model. We believe that, while not completely satisfactory, SB1 contains many of the necessary features. First, on the basis of electrical level spectra, we speculate that both P_o and P_{b1} have three nearest-neighbor silicon atoms. Furthermore, because of the apparent absence of any nearest-neighbor superhyperfine interactions, we must conclude that the nearest-neighbors are not under significant strain and that they are approximately equivalent. This speculation necessarily makes the Poindexter model of the P_{bo} center less secure. We believe that the Poindexter P_{bo} model is actually a good candidate for P_{b1}, and the P_{bo} must also have six second-nearest neighbor silicon atoms to exhibit such nearly-axial behavior. We note that these prescriptions are in conflict with the picture of these dangling bond defects as surface states. Hence, we further speculate that they exist at either corners of steps in the interface. We are presently performing large-scale calculations on these new models.

REFERENCES

1. For a thorough though somewhat dated review of charge trapping in the MOS system, see E. H. Nicollian and J. R. Brews, MOS (Metal Oxide Semiconductor) Physics and Technology, John Wiley and Sons (New York, 1982) Chapter 11.
2. P. Winokur at The Natural Space Radiation and VLSI Technology Conference, Houston, Tx., 1986 (in press).
3. By intrinsic we mean those defects that involve only the elements silicon, oxygen and hydrogen. While an ideal MOS structure can be built of only oxygen and silicon, hydrogen is always present, and is so fundamentally important, that we include it in our study.
4. For reviews of intrinsic defects in a-SiO_2, see D. L. Griscom in: The Physics of SiO_2 and its interfaces, ed. S. T. Pantelides (Pergamon, New York, 1978) p 232;W. B. Fowler in: Structure and Bonding in Noncrystalline Solids, eds. G. E. Walrafen and A. G. Revesz (Plenum, New York, 1986) p157; A. H. Edwards in: Defects in Glasses, eds. F. L. Galeener, D. L. Griscom and M. J. Weber (MRS, Pittsburgh, 1986) p. 3.
5. For extended discussions of the dangling silicon orbital defect at the <111> Si/SiO_2 interface see A. H. Edwards, Phys. Rev. B Pt. II 36; A. H. Edwards in The Physics and Technology of a-SiO_2 (in press).
6. D. L. Griscom and D. B. Brown, this proceedings.
7. K. L. Brower, this proceedings.
8. A. H. Edwards and W. B. Fowler in: Structure and Bonding in Noncrystalline Solids, eds. G. E. Walrafen and A. G. Revesz (Plenum, 1986) p. 139; A. H. Edwards and G. Germann, Nucl. Inst. and Meth. in Phys. Res. (in press).
9. D. L Griscom, Phys. Rev. B 22, 4192 (1980).
10. J. Isoya, J. Weil, and L.Halliburton, J. Chem. Phys. 74, 5436 (1981).
11. J. K. Rudra, W. B. Fowler, and F. J. Feigl, Phys. Rev. Lett. 55 (1985) 2614.
12. M. G. Jani, R. B. Bossoli, and L. E. Halliburton, Phys. Rev. B 27 (1983) 2285.
13. R. H. Silsbee, J. Appl Phys. 32 (1961) 1459.
14. It is interesting to compare these results with the hyperfine data for the E_4' center (ref. 10). At low temperatures, the spin is localized predominantly on the long-bond side of the vacancy in contradistinction to the E_1' center. In this case Isoya *et al.* observed three weak hyperfine interactions. At higher temperatures, when the spin is delocalized onto both silicon atoms neighboring the vacancy, they observe five weak

hyperfine interactions. Again, Si3 in Fig. 2 is oriented so that there is strong cancellation of the defect wave-function at its nucleus.
15. J. K. Rudra and W. B. Fowler, Phys. Rev. B15 pt. II 35, 8223 (1987).
16. F. J. Feigl, W. B. Fowler, and K. L. Yip, Sol. State Commun. 14, 225 (1974).
17. K. L. Yip and W. B. Fowler, Phys. Rev. B 11 (1975) 2327.
18. A. H. Edwards and W. B. Fowler, J. Phys. Chem. Solids 46, 841 (1985).
19. The dependence on initial Si-Si separation is easy to explain using eq. (1). As this distance increases, the excitation energy decreases, while the force and coupling constants remain roughly unchanged.
20. O. F. Schirmer in The Physics of MOS Insulators, eds. G. Lucovsky, S. T. Pantelides and F. L. Galeener (Pergamon, New York, 1980) p. 102.
21. Y. Bar-Yam and S. T. Pantelides (unpublished).
22. D. C. Allan and M. P. Teter, Bull. Am. Phys. Soc. 33, 438 (1988).
23. D. L. Griscom, Phys. Rev B 22, 4192 (1980).
24. D. L. Griscom, E. J. Friebele, and G. H. Sigel Jr., Solid State Comm. 15, 479 (1974).
25. J. Vitko, J. Appl. Phys. 49 5530 (1978).
26. T-E Tsai and D. L. Griscom J. Non-Cryst. Solids 91, 170 (1987).
27. M. Stapelbroek, D. L. Griscom, E. J. Friebele, and G. H. Sigel, Jr., J. Non-Cryst. Solids, 32, 313 (1979).
28. E. J. Friebele, D. L. Griscom, M. Stapelbroek, and R. A. Weeks, Phys. Rev. Lett. 42, 1346 (1979).
29. D. L. Griscom and E. J. Friebele, Phys. Rev. B24, 4896 (1981).
30. E. H. Poindexter and P. J. Caplan, J. Appl. Phys. (in press).
31. W. Carlos, Appl. Phys. Lett. 49, 1767 (1986).
32. B. B. Triplett, T. Takahashi, and T. Sugano, Appl. Phys. Lett 50, 1663 (1987).
33. H. S. Witham and P. M. Lenahan, Appl. Phys. Lett. 51, 1007 (1987).
34. E. H. Poindexter, E. R.Ahlstrom, and P. J. Caplan in The Physics of SiO_2 and its Interfaces, ed. S. T. Pantelides (Pergamon, N. Y., 1978), p. 227.
35. K. L. Brower, Appl. Phys. Lett. 43, 1111 (1983).
36. E. H. Poindexter, G. J. Gerardi, M.-E. Rueckel, P. J. Caplan, N. M. Johnson, and D. K. Biegelsen, J. Appl. Phys. 56, 2844 (1984).
37. Lenahan and Dressendorfer, J. Appl. Phys., 55, 3495 (1984).
38. M. Cook and C. T. White, Phys. Rev. Lett. 59, 1741 (1987).
39. A. H. Edwards, Phys. Rev. B15 pt. II 36, 9638 (1987).
40. A. Redondo, W. Goddard, and T. McGill, J. Vac. Sci. Technol. 21, 649 (1982).
41. A. Redondo, W. Goddard, T. McGill, and T. Surratt, Solid State Comm. 20, 733 (1976).
42. Y. Bar-Yam and J. Joannopoulos, Phys. Rev. Lett. 56, 2203 (1986).
43. E. H. Poindexter, P. J. Caplan, B. E. Deal, and R. R. Razouk, J. Appl. Phys 52, 879 (1981).
44. G. J. Gerardi, E. H. Poindexter, P. H. Caplan, and N. M. Johnson, Appl. Phys. Lett. 49, 348 (1986).
45. K. L. Brower, Z. fur Physik. Chem. 151 S, 177 (1987).
46. J. S. Binkley, M. Frisch, K. Raghavachari, D. DeFrees, H. B. Schegel, R. Whitside, E. Fluder, R. Seeger, D. J. Fox, M. Head-Gordon, and S. Topiol, GAUSSIAN-82 Release C, Carnegie Mellon University.
47. P. J. Bishof, J. Am. Chem. Soc. 98 (1976) 6844.
48. R. C. Bingham, M. J. S. Dewar, and D. H. Lo, J. Am. Chem. Soc. 97, 1285 (1975).
49. As in ref. 39, the HOMO can be defined as the defect state because most (in this case 0.51 e)of the wave-function amplitude is on the defect atom.
50. A. Carrico, R. Elliott, and R. Barrio, Phys. Rev. B 34, 872 (1986).
51. S. T. Pantelides and M. Long in The Physics of SiO_2 and its Interfaces, ed. S. T. Pantelides (Pergamon, New York, 1978) p. 339.
52. C. T. White (private communication).
53. D. K. Biegelsen and M. Stutzmann, Phys. Rev. B15 33, 3006 1986.
54. M. Cook in Physics and Technology of a-SiO_2, eds. R. Devine and J. Arndt (in press).

IV. DEFECTS, IMPURITIES AND DAMAGE MECHANISMS IN Si-SiO_2 SYSTEMS

INTRODUCTION

E. H. Poindexter* and M. Schulz**

*U.S. Army ET&DL
Fort Monmouth, New Jersey 07703

**Universitat Erlangen Nornberg
Glukstr. 9, D8520 Erlangen, Germany

This chapter contains papers from symposium sessions and related special conference sessions in the area of defects, impurities, and damage mechanism in the Si/SiO_2 system and its interface. Bulk fused and crystalline SiO_2 are compared with thin film SiO_2 to develop an understanding of defect mechanisms (especially the important case of radiation damage); models are then transferred from the well-studied bulk to the more difficult thin film. On the well-characterized P_b centers, new information is presented on reaction kinetics and modeling of the microscopic configuration of this ESR-active moiety in (111), (110), and (100) planes. Formerly intractable electrical and optical properties of P_b centers are now yielding to sensitive new approaches. Defect transformations are found to occur on a range of time scales from seconds to several months.

The density of interface traps in state-of-the-art MOS field-effect devices is so low ($\leq 10^{10}cm^{-2}$) that the total number of traps in μm-sized devices is now countable. Individual traps have been studied and their respective electronic properties analyzed by techniques developed from the basic principles of transient spectroscopy. Analysis of the random telegraph noise signal in equilibrium in small devices and large capacitor structures, discussed here, yields information on this new type of slow trap, which causes 1/f noise in semiconductor electronics. A thorough study of thin tunneling oxides has led to determination of the neutral point of interface state distributions in the Cr-MOS system, and a basic quantum-mechanical study was performed to describe the generally-observed properties of the interface state distribution.

Significant progress in understanding detailed physical and chemical mechanisms is evident in the papers in this chapter, but further studies were judged to be very worthwhile to illuminate a number of remaining mysteries.

NATURE OF RADIATION-INDUCED POINT DEFECTS IN AMORPHOUS SiO_2 AND THEIR ROLE IN SiO_2-ON-Si STRUCTURES

D.L. Griscom, D.B. Brown and N.S. Saks

Naval Research Laboratory
Washington, DC 20375

ABSTRACT

Molecular-scale models for the post-irradiation buildup of interface states in SiO_2-on-Si structures are critically examined in light of (i) the natures of paramagnetic point defects in amorphous silicon dioxide known from electron spin resonance and (ii) recent electrical data[1,2] obtained for γ-irradiated MOSFETs. Evidence is cited for the production of atomic hydrogen in the oxide layer, powered by the recombination of excitons (bound e-h pairs) at the sites of OH groups. There is strong new evidence[2] that a small fraction (~1 %) of this H^0 reacts directly with the interface to form electrically-active interface states. However, the majority of the radiolytic H^0 may be converted to mobile protons by reaction with oxide trapped charges in the form of small polarons. Under such conditions, the field-aided drift of these protons to the interface is believed to be the rate limiting step in the long-term buildup of interface states.

I. INTRODUCTION

It is well known that metal-oxide-semiconductor (MOS) devices can be operationally degraded by exposure to ionizing radiations due to the formation of trapped charges in the oxide layer and/or electrically active interface states. In the case of thermal oxides on silicon, an accumulation of evidence, some very recent,[1,2] indicates that the primary mechanism for the generation of interface states involves the reaction with the interface of mobile species generated in the oxide.

The numbers of trapped oxide charges and interface states can be quantified electrically from the rigid shifts and "stretch-out", respectively, of the capacitance-voltage (CV) curves.[3] However, the precise identities of the microscopic species responsible for these effects must be determined or inferred by other means. By far the most powerful technique for characterizing the molecular-scale structures of such defect centers is electron spin resonance (ESR). The present paper begins with a condensed review of what is known from ESR about the structures of radiation-induced paramagnetic defect centers in bulk silica glass and thermally grown SiO_2 on silicon. In general, the same generic types of defects are seen in both forms of amorphous silicon dioxide (a-SiO_2). However, account must be taken of the fact that the quantitative radiation-induced defect yields in thin-film oxides can be orders of magnitude larger than in bulk silica glass. This is particularly true in the case of atomic hydrogen, a defect whose efficient production in relatively dry SiO_2 is probably due to the decay of mobile excitons at the sites of hydroxyl groups in the oxide.[4] A previously proposed neutral-hydrogen-diffusion model[5] for the post-irradiation buildup of interface states is then critically reconsidered and modified in light of a growing body of data pointing the importance of hole trapping in the oxide layer and the possible drifting of charged entities to the interface. One aspect of the original model is found to emerge with increased credibility, namely, the production of interface states by the reaction of radiolytic H^0 with some precursor structure (presumed to be an Si-H bond.)

II. ESR CHARACTERIZATIONS OF DEFECT CENTERS IN AMORPHOUS SiO_2

Several intrinsic radiation-induced point defects have been identified in a-SiO_2 by means of ESR techniques. These include the E′ center[6,7] (≡Si•), the peroxy radical (≡Si-O-O•),[8] and the nonbridging-oxygen hole center (≡Si-O•).[9] In the present notation, "≡" indicates three back bonds with oxygens in the glass network and "•" represents an unpaired electron. The requirement of an unpaired electron for ESR observation is not a major drawback, since any charge-trapping center must be paramgnetic either before or after trapping occurs. Due to the increased experimental sensitivity possible with large samples, most of the ESR studies which have resulted in the definitive structural characterizations have been performed on bulk fused silicas. However, the E′ center is detected in thermally-grown SiO_2 on silicon with adequate sensitivity for positive identification (e.g., ref. 10). Oxygen-associated hole centers (OHCs) are also observable in thermal oxides,[11, 12] although the small sample volumes make it difficult to distinguish among nonbridging-oxygen hole centers (NBOHCs), peroxy radicals, or other defect types (e.g., the putative small polaron) also involving holes trapped in oxygen 2p orbitals.

The principle defect at the SiO_2/Si interface, the P_b center, was also identified by ESR.[13, 14] The P_b center comprises an unpaired electron in a dangling orbital on the silicon surface; it will be represented here as Si_3≡Si• to distinguish it from the E′ center in bulk silica.

Three E′ variants have been spectroscopically distinguished in high-purity bulk a-SiO_2.[15] It seems certain that one of these variants comprises a hole (h^+) trapped at the site of a neutral oxygen vacancy:

$$\equiv Si\text{-}Si\equiv + h^+ \rightarrow \equiv Si\bullet \ {}^+Si\equiv. \qquad (1)$$

The fact that the unpaired electron is localized on just one of the two silicons at the site was confirmed by ESR observation of ^{29}Si hyperfine structure associated with only a single silicon nucleus.[7] Note in Eq. (1) that the trapped positive charge is lodged on the silicon which is not harboring the unpaired "spin". The simultaneous presence of both the spin and the charge has been supported by strong quantitative correlations between E′ centers counted by ESR and the radiation-induced trapped positive charge in thermally-grown SiO_2 on silicon.[10] However, under certain conditions the total trapped holes in MOS structures can significantly outnumber the E′ centers.[16]

In the temperature range ~150 - 350 C, E′ centers are thermally annealed by reaction with diffusing small molecules. For example, in very low-OH silicas the following diffusion-limited process takes place:[17]

$$\equiv Si\bullet \ {}^+Si\equiv + O_2 \rightarrow \equiv Si\text{-}O\text{-}O\bullet \ {}^+Si\equiv, \qquad (2)$$

where both the E′ on the left and the peroxy radical on the right of Eq.

(2) can be monitored by ESR.[9, 15, 17-19] An analogous process involving diffusion of water molecules has been proposed to occur in high-OH silicas.[20] In either case, the trapped positive charge is not affected by reactions involving neutral molecules.

There are perhaps several mechanisms for producing nonbridging-oxygen hole centers. However, most NBOHCs in high-OH silicas are likely to be the consequences of the radiolysis of hydroxyl groups:[21,22]

$$\equiv Si{-}OH \rightarrow \equiv Si{-}O\bullet + H\bullet. \quad (3)$$

Both the NBOHC and the atomic hydrogen on the right hand side of Eq. (3) have distinctive ESR signatures and hence can be quantitatively followed as a function of isochronal anneals. H^0 is stable long enough for observation only at temperatures below ~130 K (see, Fig. 1a). Strong evidence[21,22] has been given that above this temperature much of the atomic hydrogen dimerizes rather than participating in the back reaction of Eq. (3). Due to its larger size, H_2 is much less diffusive in a-SiO_2 than is H^0. Thus, the radiolytic molecular hydrogen formed in this process is relatively immobile until the temperature is raised above ~225 K, at which point the NBOHCs of Fig. 1a are consumed according to:

$$\equiv Si{-}O\bullet + H_2 \rightarrow \equiv Si{-}OH + H\bullet. \quad (4)$$

In fact, the reaction of Eq. (4) has been shown[21, 22] to be diffusion-limited in bulk a-SiO_2.

The atomic hydrogen which reappears on the right hand side of Eq. (4) is extremely reactive and short lived at elevated temperatures, thus eluding direct observation. However, its presence may be indirectly manifested in the generation of secondary reaction products. For example, this atomic hydrogen may react with physically dissolved CO molecules in fused silica to produce formyl radicals, •HCO. Note in Fig. 1a the "first stage" production of formyl radical due to the direct reaction of the initially observed atomic hydrogen with CO and the "second stage" process where the unseen H^0 is supplied by the "cracking" reaction of Eq. (4). (The formyl radical is believed to spontaneously dissociate above 300 K.)

III. MECHANISM OF ATOMIC HYDROGEN PRODUCTION IN SiO_2

As apparent from Fig. 1a, radiolytic atomic hydrogen plays a central role in defect production processes in bulk silica containing as little as 10^{20} OH/cm^3 (0.44 mol %). Since large concentrations of H^0 have also been observed by ESR in SiO_2-on-Si structures irradiated at cryogenic temperatures (Fig. 1b),[11] it is appropriate in the present context to attempt an understanding of the reasons for the highly efficient production of hydrogen atoms in relatively dry forms of a-SiO_2.

A recent paper[4] has investigated the γ radiolysis at 77 K of single crystal quartz containing 0.007 mol % OH and pure fused silica containing 0.4 mol % OH. Significantly, for doses below ~10^4 rads, the H^0 yield in the crystal was three times that in the glass, despite the unfavorable disparity in hydroxyl contents. This outcome was explained[4] by proposing that radiation creates mobile excitons (bound e-h pairs) which have long enough lifetimes in SiO_2 to "seek out" the few existing OH groups, where they trap and de-excite by fissioning the O-H bond. The mobility of excitons in the crystalline material is assumed to be much higher than in the glass, and the lifetime of excitons in quartz (~1 ms) is longer than that in glassy silica (~100μs).[23, 24] Both of these factors tend to account for the higher specific production efficiency of H^0 in quartz.

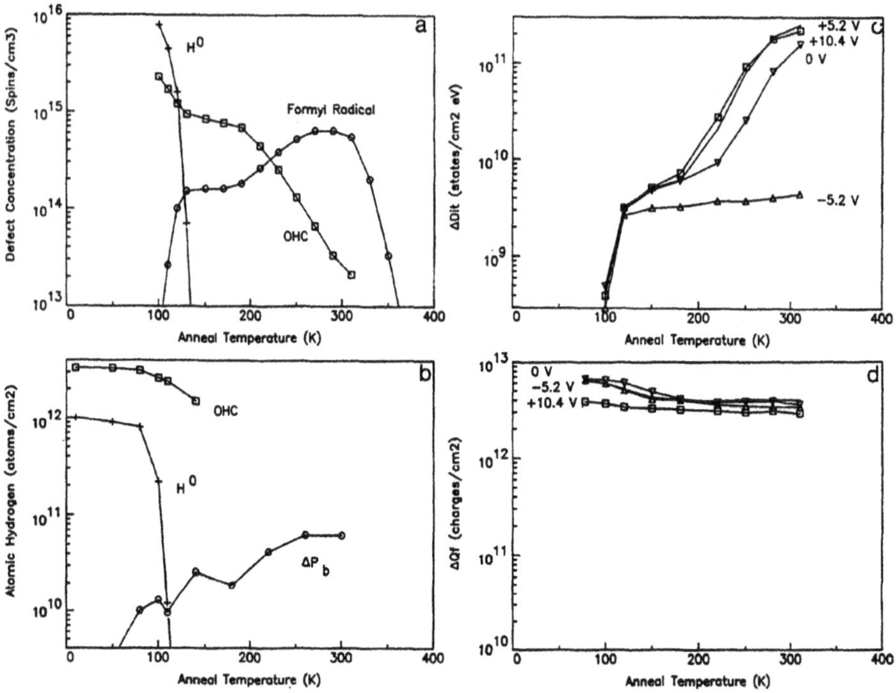

Fig. 1. Isochronal annealing of radiation-induced defect states in bulk silica glass and SiO_2-on-Si structures. (a) ESR data for bulk silica (0.44 mol % OH), x irradiated at 77 K.[21] (b) ESR data for 135-nm thermal oxide on Si, γ irradiated at 4 K with no applied bias.[11] (c) Change in interface-state density measured by charge pumping for p-channel MOSFETs with 26-nm dry-oxide gates, γ irradiated at 77 K under +5.2 V gate bias (note differing biases during anneals).[2] (d) Oxide trapped charge for the same devices as in (c).

The H^0 yield in quartz saturates at doses just above 10^4 rads,[4] while the yield in a-SiO_2 continues to grow linearly with dose up to at least 8 Mrads.[21] Apropos to a discussion in Section V., it is suggested here that the saturation phenomenon in quartz is due to a competition between the production of H^0 [Eq. (3)] and its consumption by reaction with newly created $[A\ell O_4]^0$ centers (also reported in ref. 4), thus forming diamagnetic $[A\ell O_4/H^+]^0$ centers.[25] (Aluminum impurities are nearly ubiquitous in crystal quartz but essentially absent in synthetic silica and MOS oxides).

IV. THE NEUTRAL-HYDROGEN-DIFFUSION MODEL REVISITED

It has been proposed[5] that if the "hot" H^0 appearing on the righthand side of Eq. (4) should be released near an SiO_2/Si interface, P_b centers may be created according to

$$Si_3{\equiv}Si\text{-}H + H\bullet \rightarrow Si_3{\equiv}Si\bullet + H_2. \qquad (5)$$

This suggestion was motivated by the parallelisms between the post-irradiation production of formyl radicals in fused silica (Fig. 1a)[21] and of P_b centers in SiO_2-on-Si structures (Fig. 1b).[11] The fact[26] that a positive gate bias is generally required to activate the buildup of interface states [whereas the diffusing species of Eqs. (3) through (5) are neutral] was rationalized by assuming that an electron must be provided by the substrate in order to drive the reaction of Eq. (5) to the right. However, preliminary theoretical calculations[27] indicate that the forward reaction of Eq. (5) is energetically favorable even in the absence of the extra electron. This eventuality might also have been anticipated from the fact that the strength of the H-H bond (4.5 eV) is 50 % greater than that of the Si-H bond (3.0 eV).[28]

In attempting to test the applicability of the model described above, the post-irradiation buildup of interface states in SiO_2-on-Si MOS structures has been described by a set of kinetic equations based on the forward reactions of Eqs. (4) and (5) and the back reaction of Eq. (3).[5,29] These equations yield to exact solutions in the limits (i) where the pre-irradiation $Si_3{\equiv}Si$-H interface bonds are only slightly changed in number or (ii) where they are totally converted to P_b centers. In the latter case, the (3rd-order-kinetic) solutions adequately describe previously published data[26] for the time dependence of the post-irradiation interface-state buildup at 290 K, using rate constants determined from the measured diffusivity of H_2 in a-SiO_2. However, a possible contradiction in this approach is that it implies a saturation in the number of interface states produced, whereas no evidence for such satuation appears in the literature.[30] Indeed, the data of ref. 26 which were fitted by the third-order-kinetic expression exhibit a final concentration of interface states ~$2.5 \times 10^{11}/cm^2$, whereas a simple model of an SiO_2 layer (β-tridymite) on a (111) silicon surface suggests there should be as many as 5×10^{14} dangling bonds which potentially might be converted to P_b centers.

There exist additional aspects of interface-state production which do not seem to be well explained by this model. Two examples will be mentioned here. First, several workers have advanced data indicating a role for trapped holes in the generation of interface states. For example, evidence has been presented[31-33] suggesting a one-to-one relationship between the number of interface states generated and the number of holes which moved to the Si-SiO_2 interface (although this observation must be balanced against other findings[34] which appear to show that holes can be driven to either interface with no significant effect on interface state generation). Second, indications have been given of interface state growth in very short times, e.g., <4 ms[35] or <100 μs.[1] These diverse findings clearly call for a critical review of the assumptions inherent in previous models for interface-state generation.

V. SOME FUNDAMENTAL DIFFERENCES BETWEEN THERMAL OXIDES BULK SILICA

An underlying premise of the neutral-hydrogen-diffusion model[5] was that radiation-induced processes in thermal oxides are likely to be the same as those observed in bulk silica. However, cautionary flags might have been raised by the significant quantitative differences in the radiation defect yields in the two forms of a-SiO_2 as apparent in Table 1 (refs. 2, 10, 11, 20, and 21). First, for similar conditions of irradiation, the concentrations of paramagnetic H^0 and E′ centers in 100 nm SiO_2 films exceed those in bulk silicas by approximately two orders of magnitude. (Since most of the E′ centers are evidently concentrated near the SiO_2-Si interface,[10, 12, 36] the local concentrations can be even higher than the uniformly averaged values tablulated here.) The high radiation yields of H^0 in thermal oxides are more understandable in light of very recent nuclear-reaction-profiling studies of MOS structures.[37] It was found that typical ultra dry thermal oxides contain about the same volume concentration of hydrogen, $\sim 10^{20}/cm^3$ (5×10^{14} H/cm^2 for a 50-nm oxide) as so-called "wet" silicas. However, sustantially larger concentrations $\sim 2\text{-}6\times10^{15}$ H/cm^2 were found at the metal/SiO_2 interface.[37] A very significant observation from Table 1 is that the H^0 yield in bulk "wet" silica exceeds the upper limit on the number of trapped holes (taken to be the sum of the NBOHCs and E′ centers) by a factor of ~2, whereas the oxide trapped charge in thin oxides can exceed the H^0 yield by a factor of 2 or more. Indeed, when thin thermal oxides are irradiated under bias, a major fraction of the e-h pairs created by the radiation may be converted to oxide trapped charges due to sweeping out of the electrons.[38, 39] By contrast, charge trapping in bulk silica is expected to be zero on the average, and the experimental upper limit on the trapping of carriers of both signs is seen in Table 1 to be negligible relative to what occurs in thin thermal oxides.

Table 1. Radiation-induced defect states in bulk silica glass and thermal oxides on silicon. X ray or γ ray dose approximately 0.75 Mrads. All measurements at room temperature except atomic hydrogen at 100 K. MOS structures irradiated under +2 MV/cm oxide field. Oxide thicknesses were 26 nm or ~120 nm(*). (After refs. 2, 10, 11, 20, and 21). Initial e-h pairs calculated from the dose assuming 16 eV/pair.

	Bulk Silica	Thermal Oxide on Silicon	
	Vol. Density	Vol. Density	Areal Density
Initial e-h Pairs	$6\times10^{18}/cm^3$	$6\times10^{18}/cm^3$	$2\times10^{13}/cm^2$
Trapped Holes	$<2\times10^{15}/cm^3$	$1\times10^{18}/cm^3$	$3\times10^{12}/cm^2$
Atomic Hydrogen	$4\times10^{15}/cm^3$	$4\times10^{17}/cm^3$	$1\times10^{12}/cm^2$*
E′ Centers	$1\times10^{15}/cm^3$	$2\times10^{17}/cm^3$*	$4\times10^{11}/cm^2$
Interface States	------	------	$3\times10^{11}/cm^2$

* Numbers for 120-nm oxide recalculated to 26 nm.

What then are the potential consequences to the neutral-hydrogen diffusion model of large-scale hole trapping in the oxide? The answer to this question will certainly depend on the nature(s) of the hole-trapping sites. If, for example, all trapped holes were E′ centers, the hydrogen reacting at these sites would become immobilized as Si-H bonds. But from Table 1 it can be seen that the numbers of holes trapped in thin thermal oxides can be almost an order of magnitude larger than the concentrations of E′ centers normally observed. In such cases, we suggest that the "excess" holes are trapped as small polarons.[40] Assuming the existence of such species in silica, their structure is almost certainly analogous to that of the $[A\ell O_4]^0$ center in quartz:[41]

$$\equiv Si{-}O{-}A\ell\equiv^{-} + h^{+} \rightarrow \equiv Si{-}O{-}A\ell\equiv , \qquad (6)$$

where the hole is trapped in a nonbonding 2p orbital of an oxygen bridging between a silicon and a substitutional aluminum. The aluminum binds the hole by virtue of its lower valence relative to silicon. Recent experiments have shown that holes do not self trap in impurity-free crystalline quartz above 4 K but can be stabilized up to 35 K near uncharged substitutional germaniums.[42] However, the randomness in the amorphous forms of SiO_2 should provide local potential minima capable of trapping holes even in the absence of impurities.[30] The OHCs observed by ESR in γ-irradiated SiO_2-on-Si structures, though unquantified, conceivably may be the postulated small polarons. In any event, the OHCs recorded in the experiment of ref. 11 evidently outnumbered any E′ centers (which were unobserved), and their annealing behavior (Fig. 1b) is seen to be similar to that of the oxide trapped charge in Fig. 1d.

The presence of large numbers of such self-trapped holes would profoundly affect processes based on neutral hydrogen diffusion, since reactions of the following type would be expected to swiftly convert all atomic hydrogens to protons:

$$\equiv Si{-}\dot{O}{-}Si\equiv + H\bullet \rightarrow \equiv Si{-}\overset{H^{+}}{O}{-}Si\equiv . \qquad (7)$$

This reaction is analogous to the conversion of $[A\ell]^0$ centers to $[A\ell/H^+]^0$ centers in quartz (see Section III.), except that the proton would not be strongly bound as it is in the Aℓ-center case. Subsequent chemistry would then be governed by the (electric field aided) transport of H^+, rather than diffusion of H^0 or H_2.

VI. SOME RECENT EXPERIMENTAL RESULTS[2]

Figure 1c presents the results of some recent studies[2] of the formation of interface states D_{it} during isochronal annealing (20 min) of MOSFETs (26 nm oxide thickness) exposed to ^{60}Co γ rays (dose 0.75 Mrads) at 80 K. The gate bias was held at +5.2 V during irradiation and was switched to -5.2, 0, +5.2, or +10.4 V during the anneals. D_{it} averaged over the central 1.0 eV of the Si band gap was determined by the charge pumping technique. Note the <u>bias-independent</u> "first stage"* increase in D_{it} on warming from 100 to 120 K. In light of the atomic hydrogen anneal curves of Figs. 1a and b (and the consequent production of formyl radical in Fig. 1a), this initial-warming effect is almost certainly due to a process involving reaction of atomic hydrogen with the interface. <u>We believe the reaction of Eq. (5) to be essentially confirmed by the initial annealing</u>

---

* Here, the term "first stage" refers to the first of two <u>temperature</u> stages evident in Figs. 1a and c, as opposed to the usage of ref. 30 signifying the rapid sweeping of holes to an interface under applied bias.

stage of Fig. 1c. However, the bias-independence of this low-temperature process effectively eliminates the need for an electron to be supplied by the substrate as proposed in ref. 5. Thus, while the growth in D_{it} recorded when the irradiated MOSFETs are warmed above 180 K (Fig. 1c) almost exactly parallels the growth in the formyl radical seen in bulk silica (Fig. 1a), different mechanisms must be at work in these two cases, since there is no longer any way to explain the influence of gate bias seen in Fig. 1c if we assume the diffusing species to be neutral hydrogen. Moreover, the activation energy determined for the growth in D_{it} in the range 180 - 300 K (~0.8 eV)[30,43] differs substantially from that of molecular hydrogen (~0.3 - 0.45 eV)[22] but agrees more closely with a value assigned to the diffusion of H^+ in a-SiO_2 (0.6 eV).[44]

Figure 1d displays the oxide trapped-hole concentrations ΔQ_f in the devices of Fig. 1c as estimated from the measured[2] threshold shifts ΔV_{th} by assuming the centroid to be fixed at the mid point of the film thickness. The temperature and bias dependences of ΔD_{it} and ΔQ_f are seen to be completely uncorrelated, and the hole concentrations exceed the final D_{it} values by more than an order of magnitude. Presumably the holes also outnumbered the radiolytic hydrogen atoms produced in this experiment (see Table 1). We suggest that the most important role of the holes is to convert atomic hydrogen to protons as per Eq. (7). We note that any molecular hydrogen which should be formed would be promptly "cracked" by reaction with the excess of polarons by analogy with Eq. (4). Thus, we explain the bias-dependent "second-stage" growth in D_{it} of Fig. 1c as due to the drift of protons to the interface (as suggested previously)[30] where they are reconverted to hydrogen atoms by electrons tunnelling from the silicon substrate. (The intermediate involvement of near-interface E′ centers in this process cannot be ruled out.) The released H^0 may then react with a $Si_3{\equiv}Si{-}H$ structure to form a P_b center according to Eq. (5) (similar to previous proposals).[45,46]

VII. INFLUENCE OF INITIAL BIAS

The results discussed in Section VI. were obtained for samples irradiated under a fixed initial oxide field of +2 MV/cm. Figure 2 presents some preliminary data showing the effect of varying the initial field (sustained during the irradiation) and then switching to a fixed final value of +2 MV/cm.[1] It appears that the data of Fig. 2 might be meaningfully decomposed into two components, one which is symmetrically peaked at 0 V, and one which increases between -2 to +2 V and remains roughly constant at higher positive biases. Presumably, the latter effect is due to the favorable influence of the rapid initial sweep of holes to the SiO_2/Si interface.[31-33] There are two possible causes for the decreases in ΔD_{it}/Dose at very large gate biases of either sign: First, fewer holes are trapped in the oxide due to their being totally swept out under high-field conditions,[47] and, second, smaller amounts of atomic hydrogen may be liberated in consequence of field ionization of the excitons which are evidently required to drive the radiolytic reaction of Eq. (3).

In recent time-resolved experiments,[1] the "short term" (<1 s) increases in D_{it} were almost totally suppressed when the short-pulse (1.5 µs) electron irradiations were carried out under negative gate bias. Since the short-term response (10 -30% of the final ΔD_{it}) appears to arise on a time scale of <10 µs,[1] and in view of its bias dependence, it was concluded to be due to rapid transport of holes to the interface.[1,48] On the other hand, the "long term" ΔD_{it} is reduced by only a factor of ~2 if the bias is held at a negative 2 MV/cm for the first 100 µs (or 10 ms) and then switched to a positive 2 MV/cm. However, this bias switching causes the time constant for the long term buildup to increase from ~50 s to ~200 s irrespective of switching time.[1] This "delay time" may be understood in terms of the proton-generating reaction [Eq. (7)] being spatially shifted farther from the interface when a negative initial bias is applied.

These considerations lead to the prediction that if irradiations were to be performed under conditions of precisely zero initial oxide field (so that neither holes nor electrons are swept out), the post-irradiation buildup of interface states in this special case might be determined by diffusion of neutral hydrogen (and hence become independent of final bias). Tests of this prediction are currently under way at this laboratory.

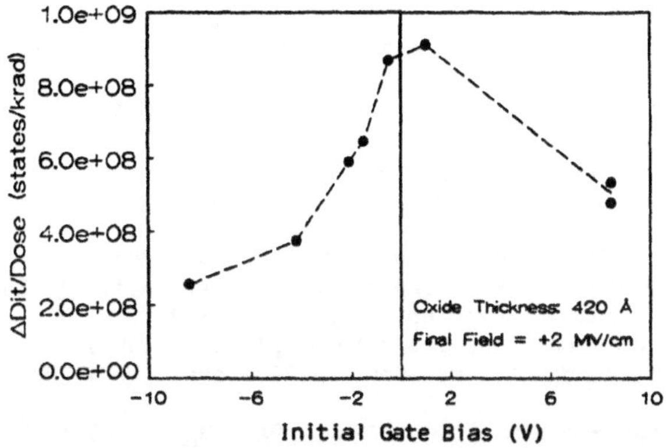

Fig. 2. Effect of the initial oxide field on the final number of interface states. Linac irradiations (35-50 krads) were carried out at room temperature under the indicated initial gate bias. The field was then switched at 10 ms to +2 MV/cm and the "final value" of ΔD_{it} was recorded after a lapse of $\sim 10^4$ s.

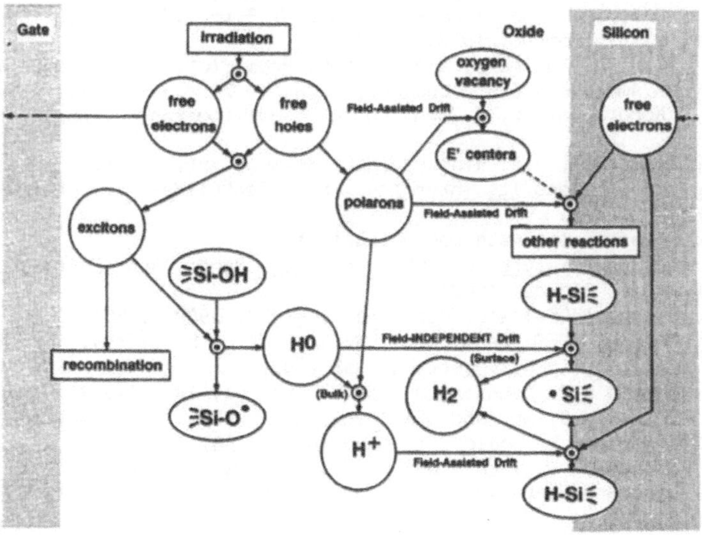

Fig. 3 Schematic summary of radiation damage processes in MOS oxides.

References

1. N.S. Saks, C.M. Dozier, and D.B. Brown, accepted for 1988 IEEE Nuclear and Space Radiation Effects Conf., Portland, OR, 11-15 July.
2. N.S. Saks and R.B. Klein, accepted for 1988 IEEE Nuclear and Space Radiation Effects Conf., Portland, OR, 11-15 July.
3. E.H. Nicollian and J.R. Brews, MOS (Metal Oxide Semiconductor Technology), (John Wiley, New York, 1982).
4. N. Azuma, T. Miyazaki, K. Fueki, I. Sakaguchi, and S.-I. Hirano, J. Am. Ceram. Soc. 69, 19 (1986).
5. D.L. Griscom, J. Appl. Phys. 58, 2524 (1985).
6. R.A. Weeks, J. Appl. Phys. 27, 1376 (1956).
7. D.L. Griscom, E.J. Friebele, and G.H. Sigel, Jr., Sol. State Commun. 15, 479 (1974).
8. E.J. Friebele, D.L. Griscom, M. Stapelbroek, and R.A. Weeks, Phys. Rev. Lett. 42, 1346 (1979).
9. M. Stapelbroek, D.L. Griscom, E.J. Friebele, and G.H. Sigel, Jr., J. Non-Cryst. Solids 32, 313 (1979).
10. P.M. Lenahan and P.V. Dressendorfer, J. Appl. Phys. 55, 3495 (1984).
11. K.L. Brower, P.M. Lenahan and P.V. Dressendorfer, Appl. Phys. Lett. 41, 251 (1982).
12. W.E. Carlos, Nucl. Inst. & Methods B1, 383 (1984).
13. E.H. Poindexter and P.J. Caplan, Prog. Surf. Sci. 14, 201-294 (1983).
14. K.L. Brower, Appl. Phys. Lett. 43, 1111 (1983).
15. D.L. Griscom, Nucl. Inst. & Methods B1, 481 (1984).
16. W.L. Warren and P.M. Lenahan, IEEE Trans. Nucl. Sci. NS-34, 1355 (1987).
17. A.H. Edwards and W.B. Fowler, Phys. Rev. B26, 6649 (1982).
18. R.A.B. Devine, J.J. Capponi, and J. Arndt, Phys. Rev. B 35, 770 (1987).
19. R.L. Pfeffer, in, The Physics and Technology of Amorphous SiO2, R.A.B. Devine, Ed. (Plenum, New York, 1987).
20. D.L. Griscom, in, Structure and Bonding in Noncrystalline Solids, A.G. Revesz and G.E. Walrafen, Eds. (Plenum, New York, 1986), p. 369.
21. D.L. Griscom, M. Stapelbroek, and E.J. Friebele, J. Chem. Phys. 78, 1638 (1983).
22. D.L. Griscom, J. Non-Cryst. Solids 68, 301 (1984).
23. D.L. Griscom, Proc. 33rd Freq. Control Symp. (Electronic Industries Asociation, Washington DC, 1979) p. 98.
24. N. Itoh, K. Tanimura, amd C. Itoh, Proc. Int. Conf. The Physics and Technology of Amorphous SiO_2, Les Arcs, France, 29 June-3 July, 1987.
25. M.A. Mondragon, C.Y. Chen, and L.E. Halliburton, J. Appl. Phys. (in press).
26. J.M. McGarrity, P.S. Winokur, H.E. Boesch, Jr., and F.B. McLean, in, The Physics of SiO2 and Its Interfaces, S.T. Pantelides, Ed. (Pergamon, New York, 1978), pp. 428.
27. A.H. Edwards, (private communication).
28. L. Pauling, Nature of the Chemical Bond, 3rd Edit. (Cornell University Press, Ithaca, 1960).
29. D.B. Brown, IEEE Trans. Nucl. Sci., NS-32, 3900 (1985).
30. F.B. McLean, IEEE Trans. Nucl. Sci. NS-27, 1651 (1980).
31. S.K. Lai, Appl. Phys. Lett. 39, 58 (1981).
32. G.J. Hu and W.C. Johnson, J. Appl. Phys. 54, 1441 (1983).
33. S.T. Chang, J.K. Wu, and S.A. Lyon, Appl. Phys. Lett. 48, 662 (1986).
34. P. Winokur and E. Boesch, (quoted in ref. 30).
35. H.E. Boesch, Jr., IEEE Trans. Nucl. Sci. NS-29, 1446 (1982).
36. C.L. Marquardt and G.H. Sigel, Jr., IEEE Trans. Nucl. Sci. NS-22, 2234 (1975).

37. A.D. Marwick and D.R. Young, J. Appl. Phys. 63, 2291 (1988).
38. J.M. McGarrity, IEEE Trans. Nucl. Sci. NS-27, 1739 (1980).
39. R.J. Powell and G.F. Derbenwick, IEEE Trans. Nucl. Sci. NS-18, 99 (1971).
40. R.C. Hughes and D. Emin, In The Physics of SiO_2 and Its Interfaces, S.T. Pantelides, Ed. (Pergamon, New York, 1978) p. 14.
41. J.A. Weil, Rad. Effects 26, 261 (1975).
42. W. Hayes and T.L. Jenkin, J. Phys. C 19, 6211 (1986).
43. N.S. Saks, C.M. Dozier, and D.B. Brown, 24th Nucl. Space Radiation Effects Conf., Snowmass, CO, July 28031, 1987.
44. S. Wolf and R.N. Tauber, Silicon Processing for the VLSI Era, Vol. 1: Process Technology, (Lattice Press, Sunset Beach, CA, 1986). press).
45. A.G. Revesz, IEEE Trans. Nucl. Sci. NS-24, 2102 (1977).
46. C.M. Svensson, In The Physics of SiO_2 and Its Interfaces, S.T. Pantelides, Ed. (Pergamon, New York, 1978), p. 328.
47. C.M. Dozier and D.B. Brown, IEEE Trans. Nucl. Sci. NS-27, 1694 (1980)
48. J.R. Schwank, P.S. Winokur, F.W. Sexton, D.M. Fleetwood, J.H. Perry, P.V. Dressendorfer, D.T. Sanders, and D.C. Turpin, IEEE Trans. Nucl. Sci. NS-33, 1178 (1986).

NOTE ADDED

Figure 3 provides a diagrammatic overview of the radiation damage processes in SiO_2-on-Si structures discussed in the body of the text. Here the circled entities are those presumed to be mobile, while the defect structures enclosed by elipses are those which are spatially fixed on the time scale of the experiment. The small circles containing asterisks are "reaction nodes" representing the reaction of the entities appearing at the origins of the incoming arrows to form the products shown at the points of the outgoing arrows. The decay of excitons (bound electon-hole pairs) is argued to power the fissioning of O-H bonds to produce atomic hydrogen. Only the H^0 which is formed nearest the interface can react directly with the interface; the remainder will be consumed in the bulk by dimerization (not illustrated) or by reaction with polarons to form protons. The indicated field-assisted drifts of charged entities are based on the application of a positive gate bias. The P_b center at the interface is represented by the notation "•Si≡"; its precursor is presumed to be a surface Si-H bond. As drawn, the diagram implies that protons driven to the interface are reconverted to hydrogen atoms by capturing an electron direcly from the substrate. However, the possible involvement of near-surface E′ centers and/or strained bonds in converting oxide charges to interface states must ultimately be considered. Most reactions terminate with the production of molecular hydrogen. However this radiolytic H_2 may be "cracked" by reaction with nonbridging-oxygen hole centers (≡Si-O•) or polarons to produce additional H^0 and/or protons.

CHEMICAL AND STRUCTURAL FEATURES OF INHERENT AND PROCESS-INDUCED DEFECTS IN OXIDIZED SILICON

Edward H. Poindexter, Philip J. Caplan, and Gary J. Gerardi

U.S. Army Electronics Technology and Devices Laboratory
Fort Monmouth, New Jersey 07703

ABSTRACT

Major inherent and process-induced point defects in oxidized silicon wafers have been examined by electron spin resonance (ESR). The P_b center is found to occur in new variations in native-oxidized wafers. Furnace thermochemical treatments and, notably, the resultant oxide water disposition have a strong effect on the initial occurrence of P_b centers (main source of interface traps) and on P_b generation by electric fields or rapid thermal stress. Brief or gentle postoxidation anneals can have very significant effects on durable passivation. The radiochemical creation and annealing of implant-induced oxide point effects---the E' (a charge trap), nonbridging oxygen hole center (NBOHC), and peroxy---are found to be generally similar, but not necessarily identical, to phenomena in well-studied bulk fused silica.

INTRODUCTION

The chemical and structural nature of electrically active point defects near the Si/SiO_2 interface is increasingly important as metal-oxide-silicon devices shrink to submicron dimensions. Atomic-scale or point defects are generated by modern radiation-based fabrication methods, by the increasing electrical stresses from nonscaling operational voltages and, in certain applications, by ambient radiation. There has been much recent progress in identification and characterization of the major inherent and process-related defects in the MOS system, primarily by electron spin resonance (ESR) in correlation with electrical measurements and molecular orbital defect modeling.[1]

The three sources of defects---processing, electrical stress, and ambient radiation---are closely interrelated. They have common features in their natures and in their combined or sequential effects on ultimate device reliability. This paper will present our own recent research findings on several types of Si/SiO_2 point defects. Their identity, structure, and thermochemical features will be presented and related to aspects of processing technology and device operation. These results will be incorporated in new interpretations, which include some recent work of other researchers, to lay a basis for improved models of the principal Si/SiO_2 defects. The discussion here will be focused on chemical and structural properties; electrical and optical properties of the major point defect---the P_b center---are presented elsewhere.[2]

EXPERIMENTAL APPROACH

Electron spin resonance has been the most generally successful source of defect identifications in oxidized silicon structures; it was used for the studies here. The special ESR techniques for MOS studies have been described elsewhere,[3] with additional details in other references cited later. Varian E-line Century X-band spectrometers with Princeton Applied Research signal averagers were used. Electrical measurements and ESR were made in dark room ambient at 23°C, except for α-irradiated samples, which were made at -196°C (77°K). Silicon wafer samples were prepared in industrial processing facilities when possible, together with special laboratory anneals and treatments when commercial approaches were unavailable or unsuitable. Although industrial IC facilities are generally capable of far higher quality, they are usually (and intentionally) much less versatile. Details of the various samples and treatments are given in pertinent sections to follow.

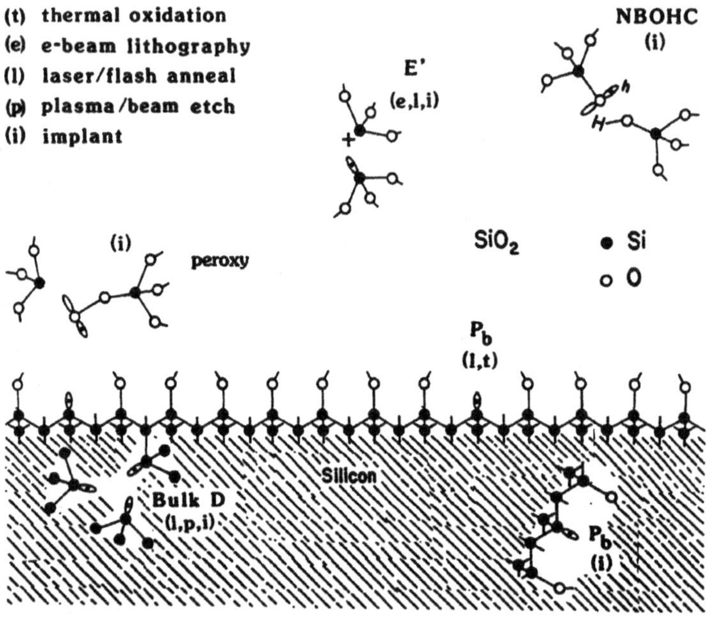

Fig. 1. Point defects observed by ESR in oxidized silicon, generated by or related to various wafer processing methods. The lower-case letters near each defect indicate the associated process technique.

RESULTS AND DISCUSSION

Structure of Si/SiO_2 Defects

The principal point defects in oxidized Si wafers observed so far by ESR are shown in Fig. 1. There are, apparently, other defects that are invisible to ESR, either because of low concentration or a usually nonparamagnetic charge condition; some of these have significant device electrical effects.[4]

The processing treatment that can generate each defect is noted by initial letter. It may be improper to say that P_b centers are generated by thermal oxidation, since they are an inherent consequence of mismatch between Si and SiO_2 lattices, and are developed during the growth of the oxide; this process is conceptually different from defect generation in an already oxidized wafer by the several radiative fabrication process. Actually, it is the interface P_b site that is inherent. The condition of the site, whether electrically or paramagnetically active or neutral, depends on specific details of oxidation and on subsequent brief or extended thermal treatments. All the other defects (including the buried P_b center) are definitely not inherent; they are present in damaged MOS systems only. Not shown is an oxide defect generated near the interface by x-rays;[5] it is a very recent finding, and little is known of its significance.

<u>The P_b Center</u>. The P_b center is the most thoroughly studied defect in the Si/SiO_2 system. It occurs in two different variations, depending on Si interface orientation (Fig. 2). On (111) silicon, it has one form: a Si atom bonded to three other silicon atoms in the Si surface at the interface, with nonbonded orbital normally containing one (paramagnetic) electron, and directed perpendicular to the interface into the SiO_2. It is designated $\cdot Si \equiv Si_3$. On (110) Si, the same center occurs in two of the four crystallographically possible orientations, along the nominal (111) bond directions aimed into the oxide. On (100) Si, the face of main IC device interest, there are two kinds of P_b center. The first, designated as P_{b0}, is essentially like the lone P_b variant occurring on (111), but is aimed along the two allowable (111) directions at the interface. The second, P_{b1}, has a different chemical structure. It was originally ascribed to $\cdot Si \equiv Si_2O$,[6] a partially oxidized Si atom with dangling orbital. However, this structure is not consistent with molecular orbital calculations of band-gap energy levels for dangling-orbital states.[7] At present, theory favors extended structures such as $Si_2{=}Si{:}\!\!-\!\!-Si \equiv Si_2O$. Observation of nuclear hyperfine structure in the ESR spectrum, done successfully for (111) Si,[8] may enable a rigorous modeling of the P_{b1} structure.

With the usual thermal oxidation, P_b centers occur at the main Si/SiO_2 interface only in the crystallographic orientations shown in Fig. 2. In

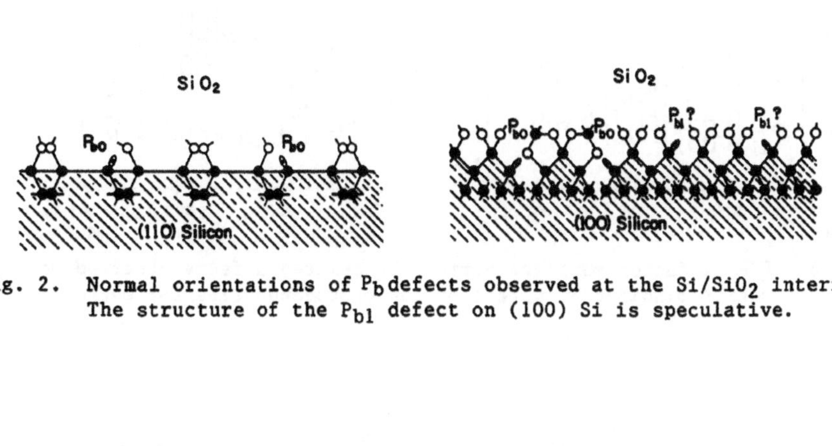

Fig. 2. Normal orientations of P_b defects observed at the Si/SiO_2 interface. The structure of the P_{b1} defect on (100) Si is speculative.

special cases, e.g., after very low-pressure oxidation[9] or for ESR observation of native oxides by optically enhanced spin-dependent recombination (SDR),[10] P_b centers with dangling orbitals along the other (111) directions have been seen. The g-factor plots for the so-called photoconductive ESR signal due to SDR are shown in Fig. 3. On (111) Si, P_b centers of the other three possible (111) directions are seen along with the usual perpendicular P_{b0}. On (110), the horizontal P_{b0} occurs together with the usual P_{b0}. However, (100) is quite different. Again, the usual P_{b0} occurs. (There are, of course, no other distinguishable orientations for P_{b0} on this face.) The P_{b1} center is absent; apparently, there is an entirely different center. The findings on (100) have been confirmed recently by direct SDR observation, i.e., not via ESR.[11] The meaning of these new orientations and defect center is not clear at this time.

Oxide Defects. The three damage centers most often observed in bulk SiO_2 are the E' ($\cdot Si{\equiv}O_3$ or $O_3{\equiv}Si{:}..{}^{+}Si{\equiv}O_3$), the nonbonding oxygen hole center, NBOHC ($O_3{\equiv}Si\text{-}O{:}...H\text{-}O\text{-}Si{\equiv}O_3$), and the peroxy radical ($O_3{\equiv}Si\text{-}O\text{-}O{:}...Si{\equiv}O_3$). The E' center and the two oxygen-related defects are easily produced in fused SiO_2; the former is a local oxygen-deficiency center; the latter two are local oxygen excess centers. The oxygen-related centers are not so common in crystalline SiO_2, since they depend on the initial presence of double oxygen linkages---a distinct misfit in quartz. Production of E' by displacement of an O atom in quartz is not a problem, of course. The identification of these defects is extensively discussed elsewhere.[12-14] Within the limitations of much poorer signal-to-noise ratio, generic types of all these oxide centers have also been observed in this-film SiO_2 on Si.

Silicon Defects. The only defect in pure bulk Si to be mentioned here is the so-called damage, or D signal, at g = 2.0055. Though studied widely,[15,16] its structure is not clear. It appears to be an ill-defined

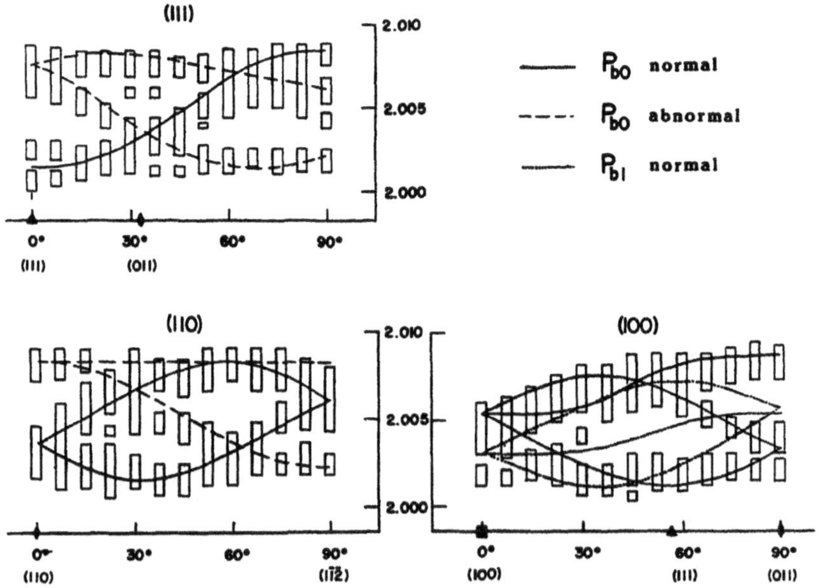

Fig. 3. ESR g-factor maps for various P_b-related defects observed on native-oxidized Si wafers by the photoconductive ESR method.

but stable array of several $\cdot Si \equiv Si_3$ centers that interact to eliminate the g-factor anisotropy, a feature otherwise characteristic of P_b centers and many other centers with $\cdot Si \equiv Si_3$ as a basic structure. The 2.0055 g-factor is approximately the weighted average for randomly oriented P_b-like centers. The other defect located in the Si bulk region is the oriented P_b center associated with oxide inclusions after O_2 implantation for silicon-on-insulator (SOI) technology.[17,18] Neither of these bulk Si centers will be examined here.

Electrical Properties of Defects

Of the defects shown, P_b, E', and D have been shown to be the atomic origin of electrical defects in the MOS system, such as interface states or traps,[19-22] radiation-induced oxide charge,[23] and bulk recombination centers,[24] respectively. The two oxygen-related hole centers are known to be important in defect reactions in damaged SiO_2, but electrical effects (if any) are unknown. It has been shown by ESR in gated structures that the P_b dangling orbital is the origin of about half of the interface traps in oxidized wafers undamaged by any additional radiation-like processes.[25] The P_b centers are also a very significant contributor to radiation- or injection-induced traps,[26] although they may not be the source of all such traps. The radiation-induced interface trap with density peak in the upper half band gap (the structure of which is not established) is invisible to normal ESR.[4]

Other important electrical defects which have not been ESR-detected, and are not necessarily detectable, are oxide fixed charge Q_f and ionic charge Q_m. Because E' centers are normally positively charged, they are a conceivable source of Q_f---the fixed charge near the interface after thermal oxidation. Regular ESR studies have not led to the detection of E' centers in concentration approaching Q_f, and E' signals observed by the very high sensitivity SDR method were later found to be due to implantation damage.[27] Finally, ionic charge will not be discussed here.

Thermal Oxidation

Thermal oxidation produces only one paramagnetic defect in quantities, namely, the P_b center. Since P_b centers are the major source of interface traps D_{it}, it is expected that P_b and D_{it} would show very similar thermochemical behavior. Initial D_{it} in oxidized wafers varies with oxidation temperature, according to the so-called Deal oxidation triangle.[28] The concentration of paramagnetically active P_b sites, i.e., P_b centers, varies the same way. If oxidation is in dry O_2, terminated by a fast pull (2-3 sec) in dry O_2 for cooling (O_2FP), the Si/SiO_2 interface is quenched at the lattice mismatch condition characteristic of the oxidizing temperature; P_b ranges from 2×10^{12} cm^{-2} at 800°C down to 2×10^{11} cm^{-2} at 1200°C. A slower cool allows a readjustment of interface mismatch to a value associated with a lower oxidation temperature. This increases P_b substantially in 1200°C oxides and slightly in 1000°C oxides; it decreases P_b slightly in 800°C oxides. The triangle relation apparently fails below 800°C; thus, the maximum thermal P_b for (111) Si is about 2.5×10^{12} cm^{-2}.

A nitrogen anneal immediately after oxidation, at the same temperature (5-15 min, followed by a 2-3 min slow cooling pull, N_2SP), allows adjustment of lattice strain and mismatch, free of the influence of continued oxide growth. This causes a great increase in P_b for 1200°C oxides. Added to oxidations at 800°C and 1000°C, the effect of N_2SP is less decisive, and a general statement is not possible. Numerous discrete postoxidation anneals were studied at 1000°C for varying times. Generally, P_b tends toward 1×10^{12} cm^{-2} for (111) Si, 0.4×10^{-12} cm^{-2} for (100) Si, regardless of previous oxidation procedures.[3]

The P_b centers are passivated by reaction with hydrogen. This may be achieved by thermal H_2 or forming gas, by atomic H plasma, or by heating of a metallized structure. In the latter case, aluminum reacts with surface OH groups to release atomic H; with poly-Si metallization, H_2 trapped in the Si overlayer is an apparent source. The poly-Si passivation is better in H_2 ambient; Al is not much affected by the ambient. The ultimate level of passivation achieved does not depend on the method, but the kinetics are different.[29] The resultant Si-H bond is rather weak ($\leq$20 kcal/mole); and it is easily broken by subsequent anneals in inert gases at 400°C or lower, by electric field stress,[30] by injection or radiation-initiated processes,[31] or by excessive annealing with atomic H.[29]

The concentration and disposition of H_2O in the oxide, especially near the interface, are prime factors in the reliability of the MOS structure against trap generation by hot carriers, high-field stress, or radiation. Oxidation in steam or postoxidation anneal in H_2 produces very low P_b, which may be regenerated by postoxidation thermal and other treatments. The effective desiccation of the interface in a steam-grown oxide, as indicated by regrowth of P_b, depends on the duration and temperature of N_2 anneal and the thickness of the oxide.[32] The interfaces of very thick oxides are harder to desiccate.

Ion Implantation

Ion implantation is a very damaging procedure. Every defect in Fig. 1 is produced in large quantity, except for interface P_b centers, where the effect is ambiguous.[3] Interface P_b sites have been depassivated by γ-radiation,[26] but active P_b are produced only in quantities not exceeding the thermal oxidation limit (2.5 X 10^{12} cm^{-2}); even this amount may require the application of an electric field across the interface. From several types of experiments reported on radiation-induced D_{it} and P_b, it seems that radiation of any kind, in doses below the amount necessary for virtual disintegration of the sample, may depassivate existing P_b sites, but does not generate new P_b sites. Perhaps direct breaking of existing Si-O bonds at the interface by radiation---especially implantation---is a gross process, generating a jumble of dangling orbitals without the required isolation and restricted orientation of distinguishable P_b centers.

In this present study, defects have been created by implantation at 23°C of Fe, P, B, BF_2, and He. Limited results for Fe, P, B, and BF_2 have been reported.[3,32] Here, we are concerned with annealing phenomena. In addition to the experiments on oxidized Si samples, a few concurrent tests were made with fused silica glass, in order to compare the thermal SiO_2; comparison of fused with thermal spectra allowed better identification of weaker and overlapping signals from thermal SiO_2.

A problem arises with the thermal films, in that the implant or other radiation can often easily penetrate the oxide and create high concentration of D centers in the Si; their ESR signals can overwhelm any oxide damage, especially the two oxygen-related hole centers. Implants of He ions (α-particles) were adjusted to avoid the problem; all the implanted ions (300 kV, 38° incidence) came to a stop within the 10,000 Å oxide. As a result, both oxygen-related defects could be examined. This set of experiments was planned to show if annealing processes observed in fused SiO_2 apply to thermal SiO_2; accurate temperatures were not required. The E' annealed to the 50% level at about 325°C. The NBOHC annealed at 200°C; concurrently, the peroxy signal grew in size. Finally, the peroxy itself was annealed at 525°C. In all cases, samples of fused silica implanted concurrently behaved like the thermal oxide films; thus, extensive published results on fused SiO_2 thermochemistry seem broadly applicable to the Si/SiO_2 system.

There are some unresolved issues, however. The high anneal temperature for E centers and the lower temperature for anneal of NBOHC and creation of peroxy suggest strongly that, in our case, the peroxy do not come from E' metamorphosis, but, rather, from NBOHC. Both schemes have been proposed;[33, 34] further study is needed to determine what favors one conversion or the other.

A second set of experiments was done to study details of E' annealing, ignoring the oxygen-related centers. Boron was implanted at 120 kV, 10^{13} cm^{-2}, into 4500 Å damp oxides. After extended aging (18 mos.) at 23°C, samples were annealed in N_2, air, O_2, steam, and forming gas (12% H_2 in N_2). The results are shown in Table 1.

To interpret these implants, the approach of Waite,[35] adapted by Griscom,[36] was used. For a diffusion-limited bimolecular annealing in solids, with concentration of damage centers [A] and annealing agent [B],

$$\frac{d[A]}{dt} = \frac{d[B]}{dt} = -4\pi r[A][B]\, D \left\{1 + \frac{r}{(\pi Dt)^{1/2}}\right\} \quad (1)$$

where r is effective damage center reaction capture radius, D is diffusivity, and t is time. In one common situation, the anneal agent is an atom ejected from the center during the damage event, or is a radiolytic product otherwise limited to the same concentration as A, i.e., $[A]_o:[B]_o = 1:1$. Then integration yields

$$\frac{[A]}{[A]_o} = \frac{1}{1 + 4\pi r\, Dt[A_o]} \quad (2)$$

We neglect the term involving $(\pi Dt)^{1/2}$, since it is small in all cases here; and signal-to-noise ratio on Si/SiO_2 films does not allow collection of anneal data over the wide range needed for subtleties. For a more general case, if excess annealing agent is available, $[B]_o > [A]_o$, we have

$$\frac{[A]}{[A]_o} = \frac{[A]_o - [B]_o}{[A]_o - [B]_o \exp\left\{-4\pi rDt([A]_o - [B]_o)\right\}} \quad (3)$$

or, in the limit of very large $[B]_o:[A]_o$ ratio

$$\frac{[A]}{[A]_o} = \exp\left(-4\pi r\, Dt\, [B]_o\right). \quad (4)$$

Diffusivity is assumed here to be of the simple form $D = D_o \exp(-E_a/kT)$, where E_a is an activation energy. In application of the formulae, we adopt $r = 5\times10^{-8}$ cm. In the case of 1:1 H_2O anneals, the fused SiO_2 experimental data and prediction of anneal temperature (T_a) for thermal SiO_2 were adapted directly from Griscom,[36] though more complex theory was used for wide-range fit to experiments in his case. For anneal with excess H_2O, permeation of H_2O was calculated to yield 6.6:1.[37] For 1:1 O_2, the data of Griscom[33] and Pfeffer[38] were used to derive D_o and E_a, and these values, in turn, used to predict T_a. For H_2, there are no directly parallel studies of E' annealing in fused SiO_2. Further, H_2 anneal of NBOHC was found to be exceedingly complicated.[39] For this situation, the values of D_o and E_a found in an intensive study of P_b passivation[40] are adopted for prediction of T_a; due to rapid permeation, $[H_2]_o$ was taken to be its saturated solubility in SiO_2.[41]

It is seen that there is excellent fit of theory and observation in the case of H_2O anneal over a range in $[B]_o:[A]_o$ ratio. The prediction for O_2 anneal in thermal SiO_2 is calculated similarly, although the behavior of O_2

Table 1. Comparison of E' anneal kinetics observed and derived from bulk fused SiO_2 with those predicted and observed for thin-film thermal SiO_2 on Si wafers. Fused SiO_2 γ-irradiated to 10^8 rad (Si); thermal SiO_2 boron-implanted at 120 kV to 10^{13} cm^{-2}. *NBOHC, not E'

Sample and Reference		Anneal Ambient	Anneal Agent	$[A]_o$ cm^{-3}	$[B]_o$:$[A]_o$	t sec	T_a C	D_o cm^2sec^{-1}	E_a eV
Observed for fused SiO_2									
Dry	38	air	O_2	5×10^{15}	1:1	600	200		
Dry	33	air	O_2	1.5×10^{16}	1:1	600	160		
Wet	36	air	H_2O	2×10^{17}	1:1	350	305		
Wet*	39	air	H_2	3×10^{14}	1:1	600	55		
O_2-starved	33	air	unk	$\sim 10^{17}$	unk	600	405		
Derived and <u>predicted</u> parameters									
			O_2	1.0×10^{18}	1.0:1.0	1200	50	1.5×10^{-5}	0.70
			H_2O	1.0×10^{18}	1.0:1.0	1200	215	1×10^{-6}	0.86
			H_2O	1.0×10^{18}	6.6:1.0	1200	170	1×10^{-6}	0.86
			H_2	1.0×10^{18}	1.0:1.0	1200	30	7×10^{-5}	0.58
Observed for thermal SiO_2 on Si									
Damp, 4500 A		N_2	H_2O	1×10^{18}	1:1	1200	220		
Damp, 4500 A		air	H_2O	1×10^{18}	1:1	1200	215		
Damp, 4500 A		O_2	H_2O	1×10^{18}	1:1	1200	210		
Damp, 4500 A		H_2O	H_2O	1×10^{18}		1200	160		
Damp, 4500 A		H_2	H_2	1×10^{18}	1:1	1200	65		
N_2-desiccated		N_2	unk	1×10^{18}	unk	1200	390		

in SiO_2 may well be quite different from a simple activated diffusion. Note that in a previous paper,[42] the anneal in N_2, air, and O_2 was suggested as being due to radiolytic O_2. This now seems unlikely; the observed T_a is too much higher than predicted. An unannealed steam oxide might *a priori* be expected to be host to the H_2O anneal process, rather than O_2. In contrast, a very similar steam oxide was desiccated in a dry N_2 anneal; and it would be expected that both H_2O and radiolytic O_2 would be drastically reduced. This thermal SiO_2 sample agrees very well with the analogous "starved" sample of fused silica. Finally, the H_2 anneal of E' is at the lowest temperature of all, and there is qualitative accord between theory and experiment. In this case, there is no known kinetic study of the anneal of E' centers in fused silica, and very complex theory has been applied to an anneal of NBOHC. In an overall comparison, however, thermal SiO_2 and fused silica seem in generally good concord, both qualitatively and quantitatively.

Electric Field Stress

Strong electric fields can generate defects in thermal SiO_2, or can act in concert with radiation for great enhancement of damage. We consider briefly here the creation of active P_b centers by negative bias stress at 23°C, with attention to the role of H_2O. Wafers were prepared with varying H_2O content in the oxidizing ambient, from <10 ppm in O_2 to 100% steam. Electric fields were generated by the corona method, which saturates at a field of 14 MV/cm for negative bias over thermal SiO_2. It was observed that P_b, if generated at all, reached a plateau at about 25 sec; and, in contrast to earlier findings, positive oxide charge was found to accumulate simultaneously.[30] Both very dry and very wet oxides were immune to P_b generation. In mid-range, at about 1000 ppm H_2O, P_b was activated to about $2 \times 10^{12}/cm^2$, i.e., The usual limit for thermal interfaces. The corona-induced depassivation was very sensitive to mild anneals following the oxidation. Although P_b might be thermally regenerated, it soon disappeared again, and the interface was thereafter very resistant to corona effects. It thus seems clear that H_2O is an essential part of the P_b activation process; but too much H_2O precludes the depassivation, perhaps by reducing the number of precursor P_b sites. Tentatively, it is proposed that adjacent H_2O molecules assist the depassivation by combining with protons which are easily dislodged from P_b sites upon capture of corona-attracted holes, thereby forming stable hydronium (H_3O^+) ions.[43] The energy of H^+ capture by H_2O is about 200 kcal/mole; and H_3O^+ has been detected in solids, besides being a long-established entity in water-soluble protonic acids.[44]

Rapid Thermal Annealing

Rapid thermal annealing (RTA) can reveal important transient chemical effects which elude furnace treatments.[32] Again, the interface water disposition is of special interest. In damp-oxidized samples, RTA at 600 to 800°C reduced P_b drastically from its initial value. However, RTA at 900°C and above brought the interface through this limited depassivation regime and restored P_b to its initial value. Hydrogen-passivated P_b was reactivated by RTA at 700°C, lower than for steam-passivated interfaces, and suggesting different mechanisms.

REFERENCES

1. E. H. Poindexter, Z. Phys. Chem., Neue Folge, 151, 165 (1987).
2. N. M. Johnson, in this proceedings.
3. E. H. Poindexter and P. J. Caplan, Prog. Surf. Sci., 14, 201 (1983).
4. E. F. da Silva, Y. Nishioka, and T.-P. Ma, Appl. Phys. Lett., 51, 270 (1987).
5. T. Takahashi, B. B. Triplett, K. Yokogawa, and T. Sugano, Appl. Phys. Lett., 50, 1663 (1987).
6. E. H. Poindexter, P. J. Caplan, B. E. Deal, and R. R. Razouk, J. Appl. Phys., 52, 979 (1981).
7. A. H. Edwards, in this proceedings.
8. K. L. Brower, Appl. Phys. Lett., 43, 1111 (1983).
9. A. Stesmans, Appl. Phys. Lett., 48, 972 (1986).
10. P. J. Caplan and E. H. Poindexter, Bull. Am. Phys. Soc., 32, 804 (1987).
11. R. L. Vranch, B. Henderson, and M. Pepper, Appl. Phys. Lett., 52, 1161 (1988).
12. R. A. Weeks, J. Appl. Phys., 27, 1376 (1956).
13. F. J. Feigl, W. B. Fowler, and K. L. Yip, Solid State Commun., 14, 225 (1974).
14. M. Stapelbroek, D. L. Griscom, E. J. Freibele, and G. H. Sigel, Jr., J. Non-Cryst. Solids, 32, 313 (1979).
15. M. F. Chung and D. Haneman, Surf. Sci., 19, 45 (1970).
16. S. T. Pantelides, Bull. Am. Phys. Soc., 32, 588 (1987).
17. W. E. Carlos, Appl. Phys. Lett., 50, 1450 (1987).
18. T. Makino and J. Takahashi, Appl. Phys. Lett., 59, 267 (1987).
19. E. H. Poindexter, P. J. Caplan, J. J. Finnegan, N. M. Johnson, D. K. Biegelsen, and M. D. Moyer, in "The Physics of MOS Insulators," G. Lucovsky, S. T. Pantelides, and F. L. Galeener, eds. (Pergamon, New York, 1980), p. 326.
20. C. Brunstrom and C. Svensson, Solid State Commun., 37, 399 (1981).
21. P. M. Lenahan and P. V. Dressendorfer, Appl. Phys. Lett., 41, 542 (1982).
22. N. M. Johnson, D. K. Biegelsen, M. D. Moyer, S. T. Chang, E. H. Poindexter, and P. J. Caplan, Appl. Phys. Lett., 43, 563 (1983).
23. P. M. Lenahan and P. V. Dressendorfer, J. Appl. Phys., 55, 3495 (1984).

24. R. J. Davis, R. Singh, S. J. Fonash, P. J. Caplan, and E. H. Poindexter, in "Thin Films and Interfaces," J. E. E. Baglin, D. R. Campbell, and W. K. Chu, eds. (North-Holland, New York, 1984), p. 604.
25. G. J. Gerardi, E. H. Poindexter, P. J. Caplan, and N. M. Johnson, Appl. Phys. Lett., 49, 348 (1986).
26. P. M. Lenahan, K. L. Brower, P. V. Dressendorfer, and W. C. Johnson, IEEE Trans. Nucl. Sci., NS-28, 4105 (1981).
27. B. Henderson, Appl. Phys. Lett., 44, 228 (1984).
28. B. E. Deal, M. Sklar, A. S. Grove, and E. H. Snow, J. Electrochem. Soc., 114, 266 (1967).
29. M. L. Reed and J. D. Plummer, Appl. Phys. Lett., 51, 514 (1987).
30. G. J. Gerardi, P. J. Caplan, E. H. Poindexter, and M. Harmatz, Bull. Am. Phys. Soc., 32, 826 (1987).
31. W. L. Warren and P. M. Lenahan, Appl. Phys. Lett., 49, 1296 (1986).
32. P. J. Caplan, E. H. Poindexter, P. K. Vasudev, and R. C. Henderson in "Science and Technology of Microfabrication," R. E. Howard, E. L. Hu, S. Namba, and S. W. Pang, eds. (Mater. Res. Soc., Pittsburgh, 1987), p. 241.
33. D. L. Griscom, Nucl. Instrum. Methods Phys. Res., B1, 481 (1984).
34. A. H. Edwards and W. B. Fowler, Phys. Rev. B, 26, 6649 (1982).
35. T. R. Waite, Phys. Rev., 107, 463 (1957).
36. D. L. Griscom, in "Structure and Bonding in Noncrystalline Solids," G. I. Walrafen and A. G. Revesz, eds. (Plenum, New York, 1986), p. 369.
37. A. J. Moulson and J. P. Roberts, Trans. Br. Ceram. Soc., 59, 388, (1960).
38. R. L. Pfeffer, in this proceedings.
39. D. L. Griscom, J. Non-Cryst. Solids, 68, 301 (1984).
40. B. J. Fishbein, J. T. Watt, and J. D. Plummer, J. Electrochem. Soc., 134, 674 (1987).
41. J. E. Shelby, J. Appl. Phys., 48, 3387 (1977).
42. E. H. Poindexter and P. J. Caplan, J. Vac. Sci. Technol., June 1988.
43. G. J. Gerardi, E. H. Poindexter, P. J. Caplan, and M. Harmatz, Bull. Am. Phys. Soc., 33, 444 (1988).
44. E. A. Moelwyn-Hughes, "Physical Chemistry," (Macmillan, New York, 1964), pp. 564,890.

CHEMICAL KINETICS OF HYDROGEN AND P_b CENTERS

K.L. Brower

Sandia National Laboratories
Albuquerque, New Mexico 87185-5800

ABSTRACT

Electron paramagnetic resonance (EPR) measurements indicate that P_b centers, which are a dangling-bond type of defect at the (111) Si-SiO_2 interface, can be passivated with H_2 at a rate which is proportional to the H_2 concentration in the thermal oxide and the interfacial density of P_b centers. This process is characterized by an activation energy, E_f, of 1.66 ± 0.06 eV with a second-order pre-exponential factor, $k^{(2)}{}_{of}$, of 1.94 (+2./-1.) x 10^{-6} cm^3/sec for temperatures between 230 and 260°C. The passivation process is demonstrated to be consistent with a chemical process in which H_2 molecules, during the course of their diffusional motion among the accessible interstices of the SiO_2 network and the reaction site at P_b centers, react directly with P_b centers. EPR measurements indicate that passivated P_b centers, denoted as HP_b, in effect dissociate in vacuum above 550°C.

I. INTRODUCTION

The P_b center is a defect that is associated with the (111) Si-SiO_2 interface. A model for the P_b center, that was first proposed by Poindexter's group [1] and is consistent with the preponderance of experimental and theoretical data [2,3,4,5,6,7,8], is illustrated in Fig. 1. According to this model, the tetrahedral bonding of the silicon atoms at the surface of the crystalline silicon is ordinarily accommodated by bonding to oxygen atoms of the oxide. However, for reasons which are not yet understood, approximately 0.5% of the (111) surface silicon atoms are not bonded to an oxygen atom of the oxide. Instead, an unpaired electron is localized on this defect silicon atom in a nonbonding sp^3-like hybrid orbital pointing in the [111] direction perpendicular to the interface. This paramagnetic interface defect is observable with electron paramagnetic resonance (EPR), which has been the main technique employed to deduce its atomic structure.

It is observed that the dangling-bond type of defects, which would include the P_b center [9], can be passivated with hydrogen [10]. In the case of the P_b center, Johnson, Biegelsen, and Moyer [11] demonstrated that P_b centers can be passivated with atomic deuterium. Poindexter et al. have shown that P_b centers are affected by annealing in various gas ambients including H_2 [2]. Nishi, who first observed the P_b center, also

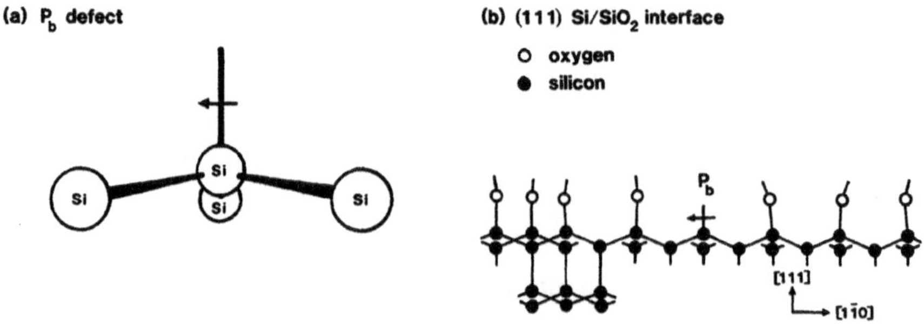

Fig. 1. (a) Ball-and-stick model of the P_b center. The arrow represents the unpaired electron localized in the dangling bond perpendicular to the interface. (b) Projection of the (111) Si-SiO_2 interface containing a P_b center looking along the [112] direction of the crystalline silicon. The oxide is represented only by the oxygen atoms bonding to the surface silicon atoms.

demonstrated its sensitivity to annealing in air [9]. However, specific chemical reactions and the values for the kinetic parameters that characterize the formation of passivated P_b centers and their dissociation have not previously been established.

The kinetic equations which describe the rate at which P_b centers are passivated by molecular hydrogen are presented in this paper. From EPR measurements of the P_b center, we are able to determine the values for the kinetic parameters (activation energy and the Arrhenius pre-exponential factor) that characterize the rate-limiting step in the passivation process (Sec. II). The chemical process by which P_b centers are passivated is identified by considering the rate equation and the associated kinetic values (Sec. III). Our conclusions are presented in Sec. IV.

II. KINETICS OF PASSIVATION

Silicon substrates with (111) surfaces were thermally oxidized in dry oxygen at 850°C for 5 hours to produce a thermal oxide on silicon that was between 500 and 600 Å thick. Samples prepared in this manner exhibited the P_b resonance as measured by EPR; this spectrum is shown in Fig. 2a. The EPR measurements were made with a K-band superheterodyne spectrometer at 273 K in the absorption mode. Further details concerning our experimental methods and sample preparation are presented elsewhere [12].

We observed that the intensity of the P_b resonance was unaffected if our thermally oxidized samples were subsequently annealed <u>in situ</u> in vacuum at temperatures up to the oxidation temperature; this spectrum is shown in Fig. 2b. These measurements indicate that P_b centers appear to be stable under thermal annealing for temperatures up to at least 850°C. Consequently the effects on the P_b center produced by annealing in hydrogen can be ascribed to chemical processes rather than thermal instability of the P_b center.

The P_b resonance associated with silicon samples which had been thermally oxidized as described above and subsequently annealed <u>in situ</u> in 99.999% pure H_2 at temperatures between approximately 220 and 500°C was reduced in intensity. The intensity of the P_b resonance was measured for several individual samples which had been annealed at the same temperature and H_2 pressure but for different times. A plot of the logarithm of the intensity of the P_b resonance versus time shows that the decay in the

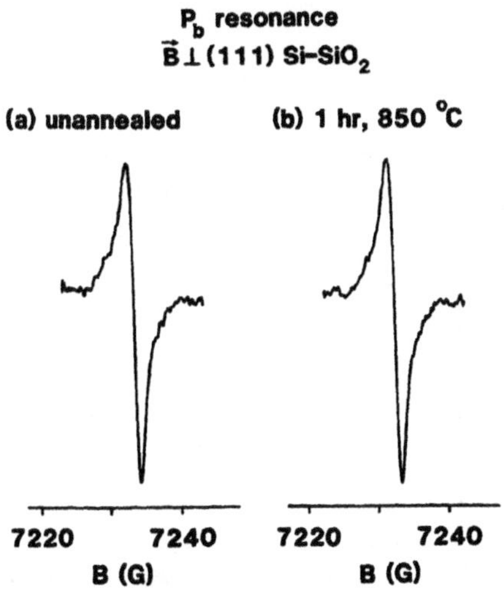

Fig. 2. (a) P_b resonance observed at K-band (20.26 GHz) after thermal oxidation of (111) silicon at 850°C for 5 hr. (b) P_b resonance from another sample oxidized at 850°C for 5 hr and, after cooling to room temperature, subsequently annealed <u>in situ</u> at 850°C for 1 hr in vacuum. Vertical scale is the same for both spectra.

intensity of the P_b resonance is exponential with the H_2 annealing time (Fig. 3). Thus, the rate at which P_b centers are passivated with H_2 can be expressed mathematically as

$$\frac{d[P_b]}{dt} = - k_f^{(1)} [P_b] \tag{1}$$

where $k^{(1)}{}_f$ is the first-order forward rate constant and $[P_b]$ is the interfacial density of P_b centers.

The functional dependence on the H_2 concentration in the oxide is embedded in $k^{(1)}{}_f$. In another series of measurements we determined $k^{(1)}{}_f$ for fixed temperature as a function of the H_2 pressure. These experimental results are plotted in Fig. 4 and indicate that the rate of passivation of P_b centers is linearly proportional to the H_2 pressure at fixed temperature. Thus, the rate at which P_b centers are passivated with H_2 is more completely expressed by the mathematical expression

$$\frac{d[P_b]}{dt} = - k_f^{(2)} [H_2][P_b] \tag{2}$$

where $k^{(2)}{}_f$ is the second-order forward rate constant and $[H_2]$ is the concentration of H_2 in the thermal oxide.

The value of $[H_2]$ corresponds to that for physical absorption of H_2 into vitreous SiO_2 which dominates for temperatures below approximately 600°C [13]. For low H_2 pressures the solubility of H_2 in vitreous silica is proportional to the gaseous H_2 pressure, p, at the surface and has been given by Shelby [14] as

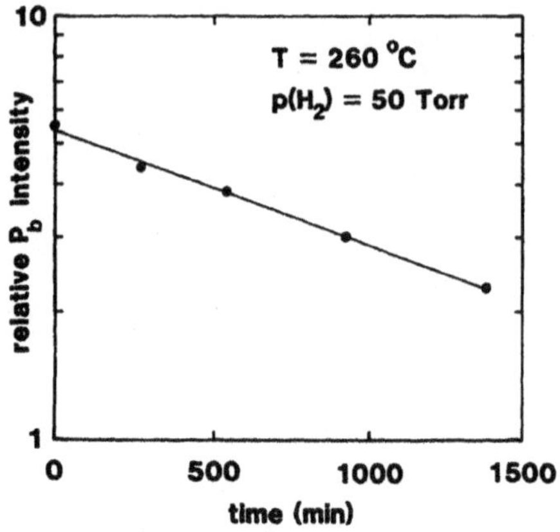

Fig. 3. Plot of the relative intensity of the P_b resonance as a function of time for post oxidation in situ annealing at 260°C and 50 Torr of H_2. One sample was prepared for each data point.

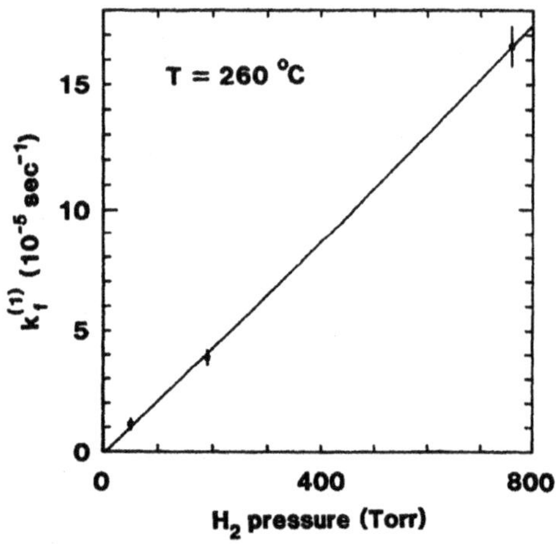

Fig. 4. Plot of the first-order rate constant in Eq. (1) versus H_2 pressure. The vertical extent of the data symbols indicates the standard deviation in $k^{(1)}{}_f$. A least squares fit of these data points to a straight lines shows the linearity in $k^{(1)}{}_f$ versus H_2 pressure.

$$[H_2] = K(T)pV^{-1} \tag{3}$$

where K(T) is specified by the expression [14,15]

$$K(T) = \left(\frac{h^2}{2\pi mkT}\right)^{3/2}\left(\frac{1}{kT}\right)\left[\frac{\exp(-h\nu/2kT)}{1-\exp(-h\nu/kT)}\right]^3 \exp(-\ \varepsilon/kT). \tag{4}$$

The parameter, V^{-1}, in Eq. (3) is the concentration of sites into which molecular hydrogen dissolves in vitreous silica and has a value of 1.27×10^{21} sites/cm^3 [14]. For a glass the value of V^{-1} depends on the size of the atomic species. The vibration frequency, ν, corresponds to 4.1×10^{12} Hz for H_2 in vitreous silica [14]. The binding energy, ε, for a molecule of hydrogen to a SiO_2 surface site is -0.105 eV [14] and corresponds to the heat of absorption at the surface. At 250°C and 760 Torr of H_2, the concentration of H_2 in vitreous silica according to Eqs. (3)-(4) is 4.16×10^{17} H_2/cm^3; at this temperature the H_2 concentration equilibrates within several milliseconds in a 500 Å thick oxide [12]. The solubility of H_2 is rather insensitive with regard to the type of vitreous silica [14]. We have therefore assumed that the behavior of H_2 in our thermal oxides is the same as that in vitreous silica and employed the calculated values for $[H_2]$ in the analysis of our experimental data.

Since the passivation process is believed to be thermally activated, $k^{(2)}{}_f$ is temperature dependent and is given by the Arrhenius expression where

$$k_f^{(2)} = k_{of}^{(2)} \exp(-\ E_f/kT) \tag{5}$$

In Eq. (5), E_f is the activation energy for the rate-limiting step in the passivation process, $k^{(2)}{}_{of}$ is the pre-exponential factor, and T is the absolute temperature. From our EPR measurements, we have determined values for $k^{(2)}{}_f$ at four different temperatures. These values for $k^{(2)}{}_f$ are plotted in Fig. 5 versus 1/T. A least squares fit of these data to Eq. (5) yield 1.66 ± 0.06 eV for the activation energy, E_f, and 1.94 (+2./-1.) x 10^{-6} cm^3/sec for the pre-exponential factor, $k^{(2)}{}_{of}$.

Thus, Eqs. (2) and (5) describe the kinetics of passivation of P_b centers by H_2, and the values we have obtained for E_f and $k^{(2)}{}_{of}$ characterize the rate at which P_b centers are passivated by H_2 for temperatures between 230 and 260°C.

III. CHEMICAL REACTIONS

There are a number of possibilities by which the passivation and dissociation processes can proceed chemically. Chemical reactions which might reasonably occur include the following:

$$P_b + H_2 \underset{b_1}{\overset{a_1}{\rightleftharpoons}} HP_b + H \quad (9)$$

$$HP_b \underset{b_2}{\overset{a_2}{\rightleftharpoons}} P_b + H \quad (10)$$

$$H_2 \underset{b_3}{\overset{a_3}{\rightleftharpoons}} H + H \quad (11)$$

$$T + H_2 \underset{b_4}{\overset{a_4}{\rightleftharpoons}} HT + H \quad (12)$$

and

$$HT \underset{b_5}{\overset{a_5}{\rightleftharpoons}} T + H \quad (13)$$

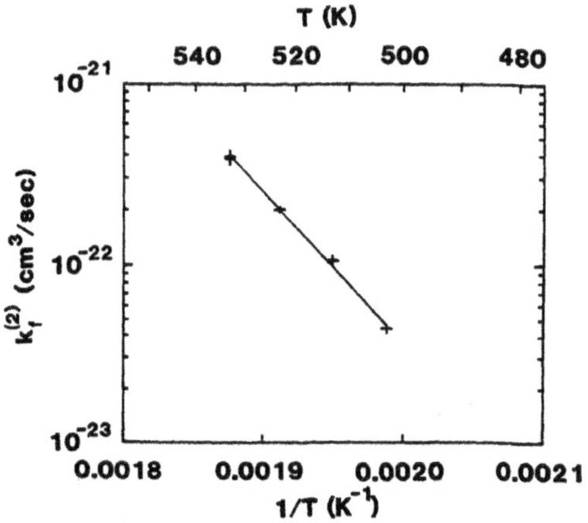

Fig. 5. Plot of the second-order rate constant in Eq. (2) versus the inverse absolute temperature. The extent of the data symbols indicates the standard deviation in $k^{(2)}{}_f$ and temperature. The straight line is a least squares fit of the Arrhenius equation to these data points.

In these chemical reactions HP_b represents the passivated P_b center, and T corresponds to a trap or misbonding within the thermal oxide. The parameters a_i and b_i correspond to the "forward" and "reverse" rate constants. Clearly, this set of chemical reactions [Eq. (9)-(13)] is not exhaustive, but is representative of the obvious possibilities. From the five chemical reactions listed above, it is not clear as to which chemical path the passivation process actually follows. We will consider three distinct chemical processes and formulate the corresponding rate equations. On the basis of a comparison between the rate equation derived for a particular process and that deduced from our experimental measurements, we will attempt to identify the dominant chemical process for the passivation processes.

Passivation process 1. In this process we imagine a two-step process. In the first step, atomic hydrogen is generated as a result of the dissociation of H_2 molecules [Eq. (11)]. This dissociation process might dominate either within the thermal oxide or at the Si-SiO_2 interface. The solubility of H_2 into crystalline silicon is negligible [16]. In the second step, some of the atomic hydrogen released in this process is captured by a P_b center resulting in its passivation [Eq. 10)]; the balance of the atomic hydrogen is assumed to recombine [Eq. (11)] in this simple process. In this process it is assumed that the HP_b centers do not dissociate ($a_2 = 0$). The rate equations for this two-step process are a pair of coupled first order non-linear differential equations,

$$\frac{d[P_b]}{dt} = - b_2 [H][P_b] \tag{14}$$

and

$$\frac{d[H]}{dt} = a_3[H_2] - b_2[P_b][H] - b_3[H][H] \tag{15}$$

which can be solved numerically for $[P_b]$ and [H] as a function of time. The H_2 concentration is, of course, known as a function of time. We have explored the solutions to these rate equations for various relative values of b_2, a_3, and b_3. In all of the solutions we explored, the decay in the number of P_b centers was non-exponential and the rate of decay in the number of P_b centers was approximately proportional to the square root of $[H_2]$. These features are significantly different from those we observe experimentally (Sec. II). Thus, the dominant mechanism for the passivation of P_b centers with H_2 does not proceed as a consequence of the dissociation of the H_2 molecule.

Passivation process 2. In this process H_2 molecules undergo partial chemical absorption at traps or misbondings, T, with the release of an atomic hydrogen [Eq. (12)]. This reaction supplies atomic hydrogen for reaction with P_b centers [Eq. (10)]. The balance of the atomic hydrogen is gettered by the traps or misbondings, T, in the SiO_2 [Eq. (13)]. We assume that in our temperature regime these three chemical reactions are irreversible ($a_2 = b_4 = a_5 = 0$) and that at time $t = 0$, $[H_2]$ goes from zero to a constant value. The rate equations for this process are

$$\frac{d[P_b]}{dt} = - b_2[H][P_b] \tag{16}$$

$$\frac{d[T]}{dt} = - a_4[H_2][T] - b_5[H][T] \tag{17}$$

and

$$\frac{d[H]}{dt} = a_4[H_2][T] - b_2[H][P_b] - b_5[H][T] \tag{18}$$

Solutions for $[P_b]$, [T], and [H] can be determined computationally using the method of finite differences. Although the decay in $[P_b]$ is typically non-exponential, the rate at which P_b centers decay is very nearly proportional to $[H_2]$. Analytical considerations indicate that the decay in $[P_b]$ will be very nearly exponential if the values for b_2, b_5, $[P_b(0)]$, and [T(0)] obey the constraint [12]

$$b_5 = \frac{b_2}{2} [1 - P_b(0)/T(0)] \tag{19}$$

Numerical solutions to Eqs. (16)-(18) with parameters satisfying Eq. (19) give solutions for which $[P_b]$ decays very nearly exponentially with time [12] and at a rate proportional to the H_2 concentration.

Although a mathematical solution consistent with our experimental observations is possible, the probability that this particular physical process occurs appears improbable because of the special relationship required [Eq. (19)] between two independent chemical reactions [Eqs. (10) and (13]. Furthermore, the relationship between b_2 and b_5 must be maintained as a function of temperature. Thus, this process by which P_b centers are passivated is deemed very improbable.

Passivation process 3. In this process the H_2 reacts directly with the P_b center according to Eq. (9). Since the average separation between P_b centers is 50 Å, the probability that the extra H atom is trapped at another P_b center is deemed negligible. This extra H atom can either readily escape from the oxide or become chemically bound within the oxide [17]. We assume in our temperature regime that this reaction [Eq. (9)] is irreversible ($b_1 = 0$); furthermore, the probability that the reverse reaction occurs is low due to the apparent scarcity of atomic hydrogen. The rate equation for this passivation process is very simply

$$\frac{d[P_b]}{dt} = - a_1 [H_2][P_b] \tag{20}$$

and agrees with Eq. (2). Thus, these considerations strongly suggest that the rate-limiting step in the passivation of the P_b center with H_2 is the chemical reaction in Eq. (9).

The measured value for $k^{(2)}_{of}$ is also reasonably consistent with passivation process 3. Let us assume that the the H_2 molecule can diffuse to P_b centers as easily as any other accessible sites in the SiO_2. Thus, the fraction of time the H_2 molecule spends on the average at a P_b reaction site is, to a first approximation, the same as it spends in any of the accessible interstices in the SiO_2. The passivation reaction [Eq. (9)] occurs only while the H_2 molecule is in the vacant site adjacent to the P_b center; this vacant site is referred to as the reaction site. Consequently the attempt frequency for the reaction, ν_s, is of the order of magnitude corresponding to the vibrational frequency of the H_2 molecule, 1.24×10^{14} Hz [18], multiplied by the fraction of time, f, the H_2 molecule spends next to the P_b center. Thus, $k^{(2)}_{of}$ in Eqs. (2) and (5) is approximated by the expression

$$k^{(2)}_{of} \sim \frac{\nu_s f}{[H_2]} \tag{21}$$

which reduces to a strictly geometric term

$$k^{(2)}_{of} \sim \frac{\nu_s}{V^{-1}} \tag{22}$$

where V^{-1} is the density of accessible sites in vitreous SiO_2 and corresponds to 1.27×10^{21} interstices/cm^3 [14]. This model predicts a $k^{(2)}_{of}$ on the order of 9.7×10^{-8} cm^3/sec. This is within approximately a factor of 20 for the measured value of $k^{(2)}_{of}$. The actual quantum process involving the chemical reaction in Eq. (9) is undoubtedly very complex.

Thus, it is an over-simplification to represent the rate of this quantum process by only the nearly free H_2 stretching frequency, ν_s. Thus, the order-of-magnitude agreement in $k^{(2)}{}_{of}$ which we have achieved between the measured and calculated values is not unreasonable.

IV. CONCLUSIONS

The P_b center is a specific interface defect associated with the (111) $Si\text{-}SiO_2$ interface that is observable with electron paramagnetic resonance (EPR). P_b centers can be passivated with H_2 above approximately 220°C, as evidenced by a decrease in the intensity of the P_b resonance. Passivated P_b centers, denoted as HP_b, appear to dissociate upon heating in vacuum above approximately 550°C as evidenced by a recovery in the intensity of the P_b resonance.

Electron paramagnetic resonance measurements of the intensity of the P_b resonance as a function of H_2 pressure, temperature, and time indicate that the rate at which P_b centers are passivated is linearly proportional to both the concentration of H_2 in the thermal oxide and the interfacial density of P_b centers [Eq. (2)]. The passivation reaction exhibits an Arrhenius temperature dependence. The passivation reaction of P_b centers with H_2 is characterized by an activation energy, E_f, of 1.66 ± 0.06 eV, and and a pre-exponential factor, $k^{(2)}{}_{of}$, of 1.94 (+2./-1.) x 10^{-6} cm^3/sec for temperatures between 230 and 260°C.

The kinetics of the passivation of P_b centers with H_2 are consistent with a chemical process in which H_2 reacts directly with P_b centers [Eq. (9)] during the course of its diffusional motion among the accessible interstices of the SiO_2 network and the reaction site at the P_b center for temperatures between 230 and 260°C. Two other chemical processes that involve the generation of atomic hydrogen, either by the dissociation of H_2 or the partial chemical absorption of H_2 by defects in the SiO_2 network, followed by reaction with P_b centers are shown to be either inconsistent or very improbable, respectively.

ACKNOWLEDGEMENTS

The extensive amount of sample preparation and technical support for this study was provided by R. Shrouf. Appreciation is expressed to C. I. H. Ashby, D. K. Brice, R. C. Hughes, S. M. Myers, and H. J. Stein for many helpful suggestions and ideas concerning this work. This work was supported by the U.S. Department of Energy under contract number DE-AC04-76DP00789.

REFERENCES

1. E. H. Poindexter, E. R. Ahlstrom, and P. J. Caplan, in The Physics of SiO_2 and its Interfaces, edited by S. T. Pantilides (Pergamon Press, New York, 1978), p. 227.
2. P. J. Caplan, E. H. Poindexter, B. E. Deal, and R. R. Razouk, J. Appl. Phys. 50, 5847 (1979).
3. K. L. Brower, Appl. Phys. Lett. 43, 1111 (1983); Zeitschrift für Physikalische Chemie Neue Folge 151, 177 (1987); in Materials Science Form. Defects in Semiconductors I, edited by H. J. Bardeleben (Trans Tech Publications Ltd., Switzerland, 1986), p. 181.
4. E. H. Poindexter, Zeitschrift für Physikalische Chemie Neue Folge 151, 165 (1987).
5. W. E. Carlos, Appl. Phys. Lett. 50, 1450 (1987).
6. A. Redondo, W. A. Goddard III, T. C. McGill, and G. T. Surratt, Solid State Communications 20, 733 (1976).

7. M. Cook and C. T. White, Phys. Rev. Lett. 59, 1741 (1987).
8. A. H. Edwards, Phys. Rev. B 36, 9638 (1987).
9. Y. Nishi, Japan. J. Appl. Phys. 10, 52 (1971).
10. S. J. Pearton, J. W, Corbett, and T. S. Shi, Appl. Phys. A 43, 153 (1987).
11. N. M. Johnson, D. K. Biegelsen, and M. D. Moyer, J. Vac. Sci. Technol. 19, 390 (1981).
12. K. L. Brower, submitted for publication.
13. J. F. Shackelford and J. S. Masaryk, J. Non-crystalline Solids 21, 55 (1976).
14. J. E. Shelby, J. Appl. Phys. 48, 3387 (1977).
15. J. F. Shackelford, P. L. Studt, and R. M. Fulrath, J. Appl. Phys. 43, 1619 (1972).
16. A. Van Wieringen and N. Warmoltz, Physica 22, 849 (1956).
17. K. L. Brower, P. M. Lenahan, and P. V. Dressendorfer, Appl. Phys. Lett. 41, 251 (1982).
18. J. Vitko, C. M. Hartwig, and P. L. Mattern, in The Physics of SiO_2 and its Interfaces, edited by S. T. Pantelides (Pergamon Press, New York, 1978), p. 215.

ELECTRONIC AND OPTICAL PROPERTIES OF SILICON DANGLING-BOND DEFECTS AT THE Si-SiO_2 INTERFACE

N. M. Johnson

Xerox Palo Alto Research Center
3333 Coyote Hill Road
Palo Alto, CA 94304

ABSTRACT

The Si-SiO_2 interface possesses residual electronic defects that are characteristic of the thermal oxidation of silicon. These defects contribute bands of interfacial electronic gap states and are paramagnetic. This paper reviews the current understanding of the electronic and optical properties of these technologically-relevant defects.

INTRODUCTION

Electrical measurements on metal-oxide-silicon (MOS) devices reveal both a continuous distribution of interface states that extends throughout the silicon forbidden energy band and a random spatial distribution of fixed charge situated in the oxide near the interface. In MOS devices fabricated for integrated-circuit applications, the interface-state distribution is generally found to be U-shaped with the density increasing monotonically with energy from midgap toward each band edge (e.g., Ref. 1). Such distributions are obtained by the combination of a high-temperature, post-oxidation anneal and a low-temperature, post-metallization anneal, with process optimization typically yielding midgap densities of less than 1×10^{10} eV^{-1} cm^{-2} on (100)-oriented silicon. Without these anneals the Si-SiO_2 interface possesses residual electronic defects that may be considered a consequence of or characteristic of the thermal oxidation of silicon. These defects contribute bands of interfacial electronic states and are paramagnetic. They can be ranked among the better understood deep-level defects in semiconductors and provide the only example where interfacial electronic gap states have been microscopically identified. This paper reviews the current understanding of these interfacial defects, with emphasis on their electronic and optical properties; findings from paramagnetic studies are summarized for completeness. A current review of the paramagnetic properties is presented by Brower [2]; chemical and structural aspects of interface defects are discussed by Poindexter and Caplan [3].

SAMPLE PREPARATION

Characteristic interface states were first observed [4,5] in MOS capacitors fabricated on (100)-oriented silicon wafers but have been most extensively studied on (111)-oriented silicon where the highest areal density of interfacial defects has been obtained. Thermal oxidation in dry O_2 and rapid cooling to room temperature in the same ambient contribute to the maximization of the density of these defects. Test-device fabrication must be completed without further perturbation of the interface. In particular, evaporation of the gate metal must be performed without exposing the wafer to ionizing radiation (e.g., soft X-rays in an electron-beam evaporator) since, in order to study the oxidation-related defects, a post-metallization anneal cannot be utilized to remove the radiation damage.

ELECTRONIC PROPERTIES

Both capacitance-voltage (CV) measurements and deep-level transient spectroscopy (DLTS) have been used to study characteristic interface defects [4-7]. Results from DLTS on as-oxidized (111) silicon are shown in Fig. 1 [6]. The interface-state distribution is dominated by two broad peaks that have equal peak densities within experimental uncertainty: one is centered approximately 0.3 eV above the silicon valence-band maximum (i.e., $E_V + 0.3$ eV) and the other at approximately 0.25 eV below the conduction-band minimum (i.e., E_C -0.25 eV). Each of these characteristic bands of interface states may be approximated by a Gaussian distribution centered at the peak energy with a standard deviation of ~0.1 eV. The cross sections for capturing an electron into the upper band or hole into the lower band are comparable and of the order of 10^{-14} cm^2 [8,9]. Peak densities between 1 and 2×10^{13} eV^{-1} cm^{-2} have been reported, and both peaks are removed by a post-metallization anneal at 450°C.

The oxidation conditions described in the last section also maximize the density of an interfacial paramagnetic defect as detected by electron spin resonance (ESR). The defect is historically termed the "P_b" center and displays a distinct resonance spectrum [10-13]. As-oxidized (111) Si typically possesses $\sim2\times10^{12}$ defects per cm^2. The principal parameters for the Zeeman interaction strongly indicate that the P_b defect is a trivalent silicon atom with a nonbonding or "dangling" orbital aligned along the single [111] direction, perpendicular to the interface. The hyperfine spectrum for oxidized (111) Si confirms that the chemical species on which 80% of the unpaired electron density is localized in the P_b center is a substitutional Si atom on the silicon side of the Si-SiO_2 interface and that the hybrid orbital at this Si site is 12% *s*-like and 88% *p*-like with the *p*-orbital pointing along the [111] direction [14-16]. Theoretical calculations of the hyperfine interactions for the trivalent silicon defect model of the P_b center are in excellent agreement with experiment [17,18]. However, an alternative model has been proposed in which the P_b center consists of a fivefold-coordinated Si atom, with the unpaired electron in a "floating-bond" state [19].

The characteristic interface states are an electrical manifestation of an oxidation-induced interfacial defect. Correlation between these interface states and P_b centers is demonstrated in Fig. 2 with isochronal anneals of oxidized Si (coated with Al) [6]. The results show that the spin density and the densities of the characteristic peaks have the same annealing kinetics. In addition, the areal density of interface states, obtained from integration of the energetic distribution for either band of characteristic gap states in Fig. 1, agrees with the P_b spin density to within experimental uncertainty. However, if P_b centers contribute localized states in the Si band gap, it must be possible to change the electron occupancy of the dangling orbital by electrical or optical techniques. This was indeed accomplished by use of the electric-field effect in the MOS structure to shift the intercept of the Fermi level at the Si-SiO_2 interface while monitoring the spin density by ESR [6,20-23]. The existence of gap states associated with P_b centers has also been demonstrated by spin-dependent recombination [24,25] and by optically-detected magnetic resonance [26], although these techniques do not establish the distribution of

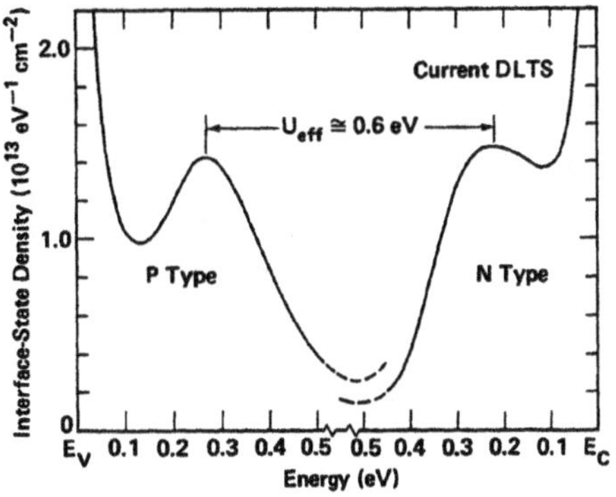

Fig. 1. Interface-state distribution measured by transient current spectroscopy on as-oxidized, (111)-oriented silicon. The distribution is a composite of measurements on identically processed *n*-type and *p*-type MOS capacitors.

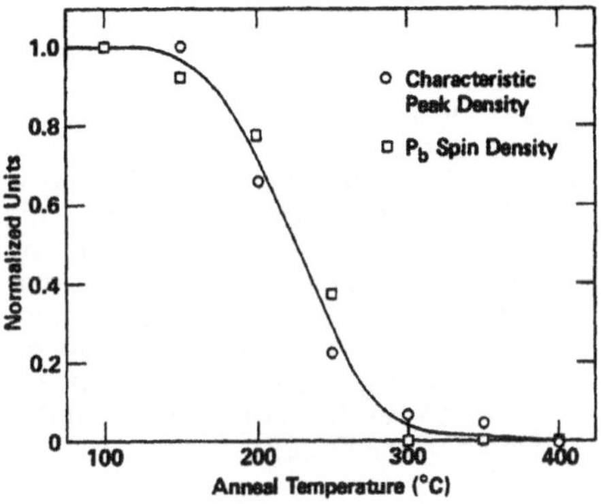

Fig. 2. Effect of isochronal anneals on the P_b spin density and the density of the $E_V + 0.3$ eV peak in the interface-state distribution on (111)-oriented, *n*-type silicon. All specimens were Al coated and annealed *in vacuo* for 15 min.

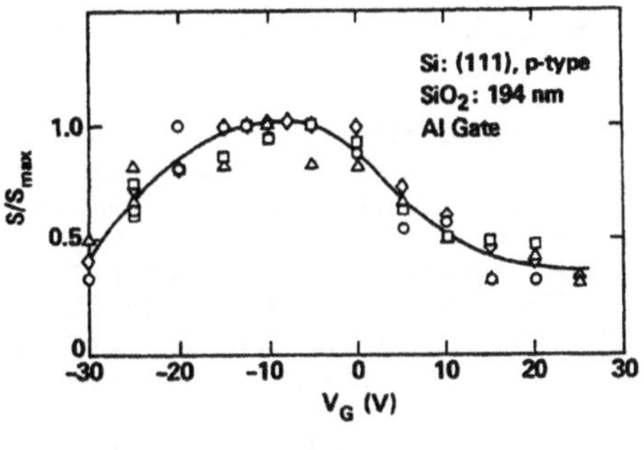

Fig. 3. Dependence of the P_b spin density S, normalized to the maximum density S_{max}, on the applied gate voltage V_G in a large-area MOS capacitor.

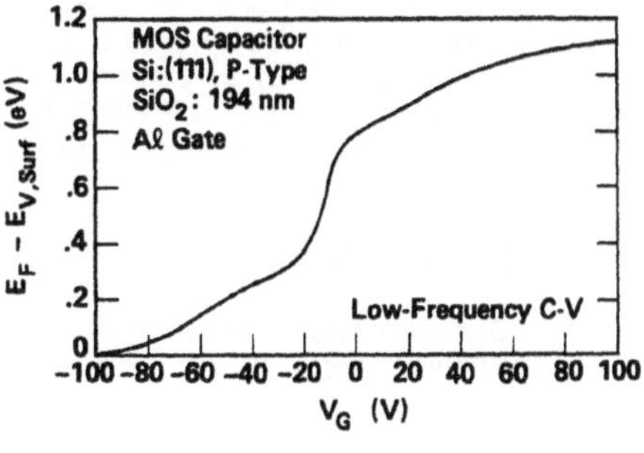

Fig. 4. Gate voltage dependence of the energy at which the Fermi level intersects the Si-SiO_2 interface, $E_F - E_{V,surf}$, for an as-oxidized MOS capacitor on (111)-oriented, *p*-type silicon.

the deep levels. In addition, theoretical calculations in general predict the existence of gap states for the interfacial trivalent silicon defect [18,27-33].

Figure 3 presents results for a large-area (e.g., 0.6 cm^2) MOS capacitor in which the interface-state occupancy was controlled with a gate voltage applied during the ESR measurement [6,7]. The P_b spin density decreases for either high negative or positive gate voltages which coincidentally correspond to strong accumulation and inversion, respectively, of the *p*-type MOS capacitor. Conventional CV analysis was used to relate the Fermi level E_F at the interface to the gate voltage (via the Si surface potential). A representative transfer function is shown in Fig. 4 [7]. Sweeping the Si band gap at the interface through the Fermi level reveals that the spin density is approximately constant as E_F shifts through the central region of the band gap but decreases as E_F approaches the band edges. This identifies an amphoteric defect with both electronic transitions in the band gap: when E_F intersects the interface near the Si valence band, the center is diamagnetic; for E_F near midgap, the center contains an unpaired electron and yields the P_b spin signal; and as E_F approaches the conduction band, the center acquires a second electron which pairs with the first to again produce a diamagnetic center. Gate-controlled ESR combined with CV analysis reveals that the electronic transition energies for the P_b center coincide with the characteristic peaks in the interface-state distribution [6,7]. Thus, both the annealing kinetics and the voltage dependence of the P_b center support the assignment of the $0 \leftrightarrow 1$ electron (or $2 \leftrightarrow 1$ hole) transition of the amphoteric center to the $E_V + 0.3$ eV peak and the $1 \leftrightarrow 2$ electron transition to the $E_C - 0.25$ eV peak; the singly occupied center is charge neutral. The characteristic peaks (Fig. 1) are separated by an effective correlation energy U_{eff} of approximately 0.6 eV. This is the difference between the energies required to add an electron to a singly occupied orbital and to an unoccupied center; it is the sum of two terms, a Coulomb repulsion and a compensating lattice relaxation energy. The measured U_{eff} is in good agreement with theoretical calculations for the trivalent silicon defect at the Si-SiO_2 interface [30,33].

Further support for the association of the characteristic peak at $E_V + 0.3$ eV with an electron transition of the P_b center was provided by spin-dependent DLTS [34]. However, this study reported a U_{eff} of ~0.1 eV; this discrepancy with the above cited work has not been resolved, although a theoretical interpretation has been suggested [18].

The band-gap energy distribution of P_b centers on (100) Si has been measured and compared with the interface-state density [35]. Two different P_b centers are observed on (100) Si [36,37]: P_{b0} which is essentially identical to the sole P_b center observed on (111) Si and P_{b1} which is as yet unidentified but clearly different in nature from P_{b0}. Results from electric-field controlled ESR and CV measurements are shown in Fig. 5 [35]. The electron transitions for P_{b0} are essentially the same as on (111) Si while P_{b1} has the $0 \leftrightarrow 1$ electron transition at $E_V + 0.45$ eV and the $1 \leftrightarrow 2$ transition at $E_V + 0.8$ eV. The P_b density is found to quantitatively correlate with the interface-state density from CV analysis: the nonbonding P_b orbitals accounted for ~50% of the characteristic gap states on the (100) Si wafers. The relative densities of P_{b0} and P_{b1} centers vary with wafer processing and device-operation stresses. Since their superimposed densities can be major contributions to the interface-state distribution, the shape of this distribution on (100) Si can vary widely, resembling the nearly symmetrical distribution of (111) Si for small P_{b1} densities, or becoming asymmetrical for large P_{b1} densities with a peak in the upper half of the gap.

OPTICAL PROPERTIES

Optical properties of interface states in the Si-SiO_2 structure have been extensively studied [38-42]. The general approach has been to use subband-gap

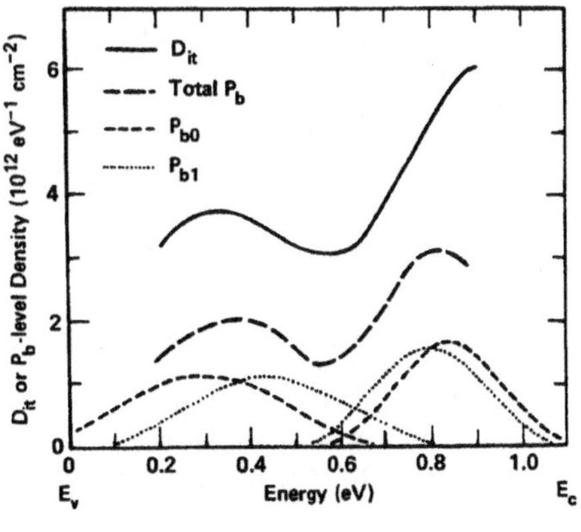

Fig. 5. Characteristic interface-state density D_{it} and bandgap energy-level densities of P_{b0} and P_{b1} spin centers on as-oxidized, (100)-oriented silicon.

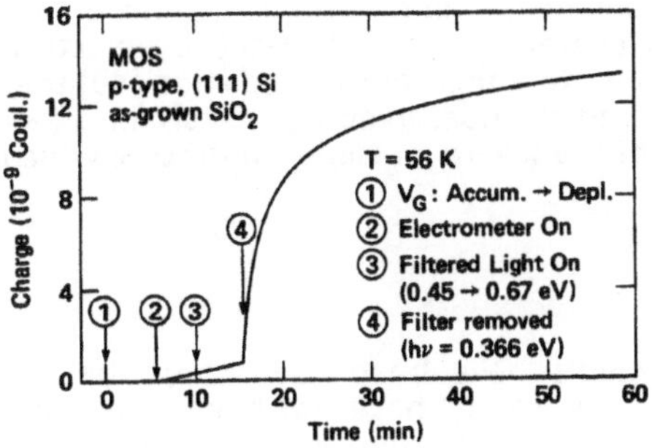

Fig. 6. Photodepopulation-induced charge collection for MOS capacitors with as-grown thermal oxides on (111)-oriented *p*-type silicon.

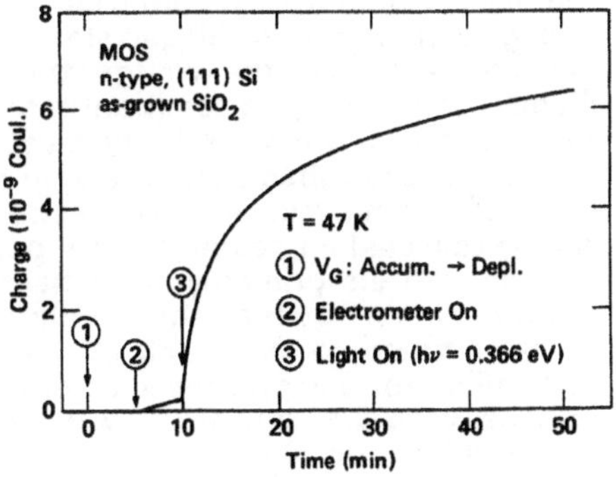

Fig. 7. Photodepopulation-induced charge collection for MOS capacitors with as-grown thermal oxides on (111)-oriented *n*-type silicon.

monochromatic light to optically-induce electronic transitions between the localized states of interface defects and the extended states of silicon. These transitions are detected by monitoring either photocurrent or photocapacitance transients in MOS devices. Data analysis yields interface-state distributions and photoionization cross sections. However, the above optical studies were conducted on fully-processed MOS devices which could not be expected to possess detectable concentrations of P_b centers. Only a few optical studies have so far been reported for P_b centers.

Optical Absorption Spectroscopies

Optical absorption by P_b centers on (111) Si has been measured by electroabsorption spectroscopy [43,44]. On MOS capacitors, electric-field modulation of the occupancy of the characteristic gap states was directly detected as a synchronous modulation of the absorption of subband-gap light. Both photon energy and surface-potential controlled trap occupancy were used to probe the interface-state distribution. The results on *p*-type Si yield a broad peak near $E_V + 0.3$ eV. In addition, the results provide an estimate of $\sim 2 \times 10^{-17}$ cm^2 (at 1 eV) for the effective optical cross section of the $0 \rightarrow 1$ electron transition of the interfacial defect, that is, the absorption is dominated by hole emission. This is consistent with an estimate from photothermal deflection spectroscopy that the cross section for electron emission is $<< 10^{-17}$ cm^2, which was inferred from the absence of detectable interfacial-defect absorption [45]. As an interesting contrast, absorption by the silicon dangling-bond defect in a-Si:H is significantly stronger than for the P_b center and is dominated by electron emission [44]. An unresolved issue pertains to a theoretical prediction [46] that optical transitions from the Si valence band to unoccupied P_b orbitals (i.e., $2 \rightarrow 1$ hole transition) should be the weakest of the several possible transitions, which disagrees with experiment.

Photoelectric Measurements

Transient photocurrent measurements have been used to probe the optical energy levels of the P_b center. The results have not been previously published, so a brief description of the experiment is necessary. The approach was to use a photon energy $h\nu$ that was less than one half of the Si bandgap so that only optical transitions between gap states and the immediately adjacent band of extended states were possible. The measurements were performed on both *p*-type and *n*-type MOS capacitors that possessed P_b spin densities of $\sim 10^{12}$ cm^{-2}, so that P_b was the dominant defect in the Si layer. The procedure involved first cooling a device to a low temperature T in order to freeze out thermal emission from the interface states. Then the capacitor was biased into strong accumulation to fill gap states with majority carriers. The measurement involved recording the charge transferred in the external circuit due to illumination through a semitransparent gate electrode. Results are shown in Figs. 6 and 7 for *p*-type and *n*-type capacitors, respectively. At point #1 in each figure the gate voltage was changed from accumulation to depletion. After a delay of several minutes the electrometer was connected (point #2). The charge collected immediately after point #2 is ascribed to thermal emission of majority carriers primarily from bandtail states at the interface which are shallower in energy than the dangling bond states. The light source was a He-Ne laser tuned to the 0.366 eV line To confirm that stray light was not responsible for any photoelectric effect, optical filters which transmitted light only between 0.45 and 0.67 eV were inserted between the sample and laser (point #3 in Fig. 6); the absence of any change at point #3 in Fig. 6 demonstrates that there was negligible stray light. A large photocurrent transient was induced when the 0.366 eV photons were directed onto the capacitors by removing the 0.45 eV cutoff filter; this is shown by point #4 in Fig. 6 and point #3 in Fig. 7. In both devices the amount of charge collected in the external circuit was comparable to the density of dangling bond defects. (Actually, the amount of charge released from interface states in an MOS structure is greater than the amount of charge collected in the external circuit. These amounts differed by a factor that was close to unity in the present measurements.) Thus, it appears that the $0 \rightarrow 1$ $(2 \rightarrow 1)$ electron transition of the P_b center can be induced by photoexcitation of a hole (electron) to the valence (conduction) band with $h\nu = 0.366$ eV; this interpretation assumes that there is no significant contribution to the collected charge from either a photothermal ionization process or photodepopulation of bandtail states. Such results should set limits on the differences between the thermal and optical energy scales for the P_b center, which relate to the amount of atomic relaxation experienced by the defect upon charge emission. This issue has been examined theoretically [30,47]. However, further work is required to settle this central issue in the microscopic identification of the P_b center [48].

CONCLUDING REMARKS

In addition to continuing interest in the properties and theory of P_b centers, there is currently strong technological motivation to understand the role of these defects in process-induced or operation-induced degradation of MOS devices [23,49-52]. This activity includes the study of hydrogen passivation of P_b centers [2,5,53]. Finally, the P_b center should begin to find use as a controllable, well-characterized interfacial defect to study other interface-related phenomena, such as low-dimensional transport in Si MOS devices or the detection of interface states by scanning tunneling microscopy

REFERENCES

1. E. H. Nicollian and J. R. Brews, "MOS (Metal Oxide Semiconductor) Physics and Technology," Wiley, New York (1982).
2. K. L. Brower, this proceedings.
3. E. H. Poindexter and P. J. Caplan, this proceedings.
4. N. M. Johnson, D. J. Bartelink, and M. Schulz, in: "The Physics of SiO_2 and Its Interfaces," S.T. Pantelides, ed., Pergamon, New York (1978).
5. N. M. Johnson, D. K. Biegelsen, and M. D. Moyer, in: "Physics of MOS Insulators," G. Lucovsky, S. T. Pantelides, and F. L. Galeener, eds., Pergamon, New York (1980).
6. N. M. Johnson, D. K. Biegelsen, M. D. Moyer, S. T. Chang, E. H. Poindexter, and P. J. Caplan, Appl. Phys. Lett. 43:563 (1983).
7. E. H. Poindexter, G. J. Gerardi, M.-E. Rueckel, P. J. Caplan, N. M. Johnson, and D. K. Biegelsen, J. Appl. Phys. 56:2844 (1984).
8. N. M. Johnson, Appl. Phys. Lett. 34:802 (1979).
9. N. M. Johnson, D. J. Bartelink, and J. P. McVittie, J. Vac. Sci. Technol. 16:1407 (1979).
10. Y. Nishi, Jpn. J. Appl. Phys. 10:52 (1971).
11. E. H. Poindexter, E. R. Ahlstrom, and P. J. Caplan, in: "The Physics of SiO_2 and Its Interfaces," S. T. Pantelides, ed., Pergamon, New York (1978).
12. P. J. Caplan, E. H. Poindexter, B. E. Deal, R. R. Razouk, J. Appl. Phys. 50:5847 (1979).
13. A. Stesmans, J. Braet, J. Witters, and R. F. Dekeersmaecker, Surf. Sci. 14:255 (1984).
14. K. L. Brower, Appl. Phys. Lett. 43:1111 (1983).
15. K. L. Brower, Phys. Rev. B 33:4471 (1986).
16. K. L. Brower and T. J. Headley, Phys. Rev. B 34:3610 (1986).
17. M. Cook and C. T. White, Phys. Rev. Lett. 59:1741 (1987).
18. A. H. Edwards, Phys. Rev. B, in press.
19. S. T. Pantelides, Phys. Rev. Lett. 57:2979 (1986).

20. E. H. Poindexter, P. J. Caplan, J. J. Finnegan, N. M. Johnson, D. K. Biegelsen, and M. D. Moyer, in: "Physics of MOS Insulators," G. Lucovsky, S. T. Pantelides, and F. L. Galeener, eds., Pergamon, New York (1980).
21. C. Brunstrom and C. Svensson, Solid State Commun. 37:399 (1981).
22. P. M. Lenahan and P. V. Dressendorfer, Appl. Phys. Lett. 41:542 (1982).
23. P. M. Lenahan and P. V. Dressendorfer, J. Appl. Phys. 55:3495 (1984).
24. B. Henderson, Appl. Phys. Lett. 44:228 (1984).
25. R. L. Vranch, B. Henderson, and M. Pepper, Appl. Phys. Lett. 52:1161 (1988).
26. K. L. Lee, L. C. Kimerling, B. G. Bagley, and W. E. Quinn, Solid State Commun. 57:615 (1986).
27. R. B. Laughlin, J. D. Joannopoulos, and D. J. Chadi, Phys. Rev. B 21:5733 (1980).
28. K. L. Ngai and C. T. White, J. Appl. Phys. 52:320 (1981).
29. T. Sakurai and T. Sugano, J. Appl. Phys. 54:2889 (1981).
30. A. H. Edwards, J. Elect. Mat. 14a:491 (1985).
31. A. S. Carrico, R. J. Elliott, and R. A. Barrio, Phys. Rev. B 34:872 (1986).
32. R. A. Barrio, R. J. Elliott, and A. S. Carrico, Phys. Rev. B 34:879 (1986).
33. W. B. Fowler and R. J. Elliott Phys. Rev. B 34:5525 (1986).
34. M. C. Chen and D. V. Lang, Phys. Rev. Lett. 51:427 (1983)
35. G. J. Gerardi, E. H. Poindexter, P. J. Caplan, and N. M. Johnson, Appl. Phys. Lett. 49:348 (1986).
36. E. H. Poindexter, P. J. Caplan, B. E. Deal, and R. R. Razouk, J. Appl. Phys. 52:879 (1981)
37. K. L. Brower, Z. Phys. Chem. Neue Folge (Germany) 151:177 (1987).
38. R. F. Pierret, Solid-Sate Electronics 19:577 (1976).
39. R. F. Pierret and B. B. Roesner, Solid-Sate Electronics 19:593 (1976).
40. D. W. Greve and W. E. Dahlke, Inst. Phys. Conf. Ser. 50:107 (1980).
41. D. W. Greve and W. E. Dahlke, Appl. Phys. Lett. 36:1002 (1980).
42. K. Blumenstock and M. Schulz, in: "Insulating Films on Semiconductors," M. Schulz and G. Pensl, eds., Springer Verlag, Berlin (1981).
43. N. M. Johnson, W. B. Jackson, and M. D. Moyer, Phys. Rev. B 31:1194 (1985).
44. W. B. Jackson and N. M. Johnson, Mat. Res. Soc. Symp. Proc. 46:545 (1985).
45. C. H. Seager, P. M. Lenahan, K. L. Brower, and R. E. Mikawa, Mat. Res. Soc. Symp. Proc. 46:539 (1985).
46. A. H. Edwards and W. B. Fowler, Mat. Res. Soc. Symp. Proc. 46:533 (1985).
47. J. Petit, M. Lannoo, and G. Allan, Solid State Commun. 60:861 (1986).
48. H. G. Grimmeiss, W. R. Buchwald, E. H. Poindexter, P. J. Caplan, M. Harmatz, G. J. Gerardi, D. J. Keeble, and N. M. Johnson, submitted for publication.
49. R. E. Mikawa and P. M. Lenahan, IEEE Trans. Nucl. Sci. NS-31:1573 (1984).
50. R. E. Mikawa and P. M. Lenahan, Appl. Phys. Lett. 46:550 (1985).
51. J. M. Sung and S. A. Lyon, Appl. Phys. Lett. 50:1152 (1987).
52. T. P. Ma, Y. Nishioka, and E. F. da Silva, Jr., this proceedings.
53. M. L. Reed and J. D. Plummer, Appl. Phys. Lett. 51:514 (1987).

TRANSIENT SPECTROSCOPY ON INDIVIDUAL INTERFACE TRAPS IN MOSFETS

A.Karwath and M.Schulz

Institut für Angewandte Physik
Universität Erlangen-Nürnberg
Glückstr.9
D-8520 Erlangen

ABSTRACT

Deep level transient spectroscopy (DLTS) has been applied to analyze defects in n- and p-channel MOSFETs having gate areas of the order of 1 • 1 μm^2. Transients due to electron or hole emission can be observed in these small devices by utilizing the internal amplification in the MOSFET. The number of interface traps is 10-20. The transient is quantized into steps. The properties of individual traps can be analyzed. Multiple charge carrier emission from a single trap is observed.

INTRODUCTION

Deep level transient spectroscopy (DLTS) is generally applied to analyze deep levels in bulk semiconductors /1-3/ and on SiO_2-Si interfaces /4,5/. The basic principle of DLTS is repetitive switching of the trap occupation from a completely filled state to a completely empty state. The emission transient of the trap charge is monitored as a capacitance transient.

We have applied the same basic DLTS measurement principle to MOS field effect transistors having gate areas of the order of 1 μm^2. The capacitances in such small devices are too low for a standard DLTS measurement. For the analysis of interface traps, we utilize the internal amplification of the MOSFET by measuring the conductance of the channel. Trapped charge in interface states affects the channel conductance of small MOSFETs. Periodic trapping and reemission causes the statistical switching and telegraph noise phenomenon in small MOSFETs /6-10/.

The DLTS measurement principle applied to MOSFETs is explained in Fig. 1 /11/. We bias the MOSFET into strong inversion by applying a gate voltage well above the threshold voltage of the transistor. All the interface traps are filled in this bias condition (Fig. 1a).

The gate bias is then reduced to a voltage close to the threshold for the channel formation. Interface states a few tenths of an electron volt below the band edge at room temperature then reemit the trapped charge carriers (Fig. 1b). This reemission is monitored as a transient in the channel conductance. The number of traps in this small area and the limited energy interval is low of the order 5-20. We therefore expect transients which are quantized into steps for the emission from only very few traps.

EXPERIMENTAL

We have studied n- and p-channel MOSFETs prepared by VLSI technology at the Siemens Research Laboratory in Munich. The channel length of the devices varied within L=0.5-1.5μm. The channel width was within W=0.5-10μm. The measurement was performed with battery-operated power supplies providing a constant current of 1-200nA to measure the channel conductance with a source-drain voltage of approx. 10mV. For this low voltage, the carrier density in the channel is not affected by the voltage drop. The threshold voltage typically is V_G=0.7-0.9V for the n-channel and V_G= -2.2V for the p-channel device. The source-drain voltage transients for the constant current are amplified and directly recorded on an xy-plotter with respect to time or stored by a transient recorder.

TRANSIENTS

Typical transients are shown in Figs. 2 and 3 for n- and p-channel devices, respectively. Strongly quantized transients are observed in both cases. Each step in the transient corresponds to an emission process for an individual trap. It is noted that the quantization steps vary

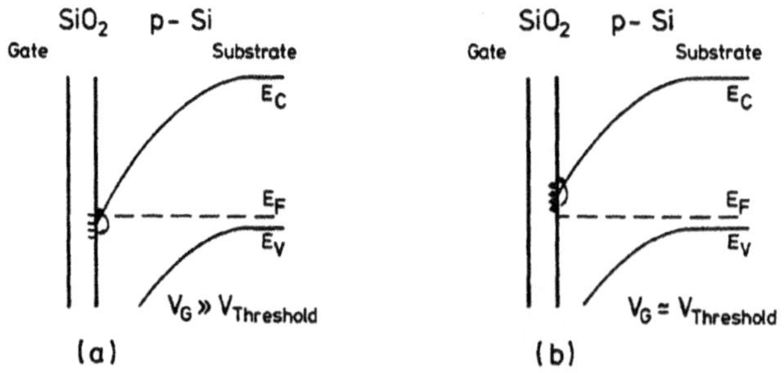

Fig.1. Schematic of the basic DLTS principle as applied to MOSFETs. a) Band structure for trap filling b) Band structure during reemission.

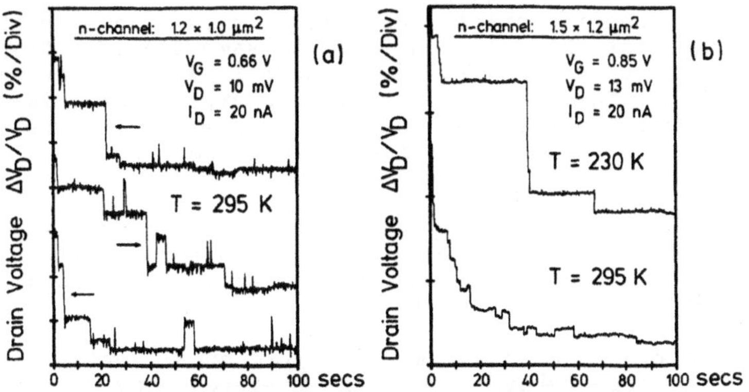

Fig.2. Quantized transients of electron reemission from individual traps in n-channel MOSFETs. The voltage drop across the channel is monitored vs. time for a constant current as indicated.
a) Sequential recordings of transients taken at the same conditions. The arrows mark the occurrence of the same step height.
b) Sequential recordings of transients taken at different temperatures as indicated.

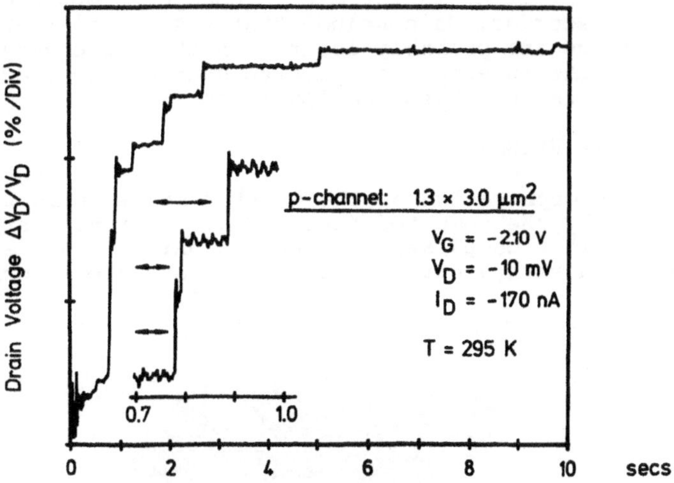

Fig.3. Quantized transient of hole reemission from individual traps in a p-channel MOSFET. The insert shows the "giant step" in an expanded time scale to demonstrate the resolution of sub-steps.

in magnitude from trap to trap. This effect is due to local variations of the internal amplification. It can be used to identify specific traps. The arrows in Fig. 2a mark the occurrence of the emission from the same specific trap in 3 different transients recorded sequentially with the same conditions. The statistics of the delay for the emission process has been studied for this specific trap. The distribution is random with an exponential (Poisson) distribution.

At room temperature, we occasionally observe a superposition of pulses due to trapping and reemission. Pure emission transients are observed at slightly reduced temperature 230K (Fig. 2b).

The transient observed for a p-channel device basically shows the same behavior with an inversed decay due to the opposite polarity. This example was chosen to demonstrate the occasional occurrence of "giant" steps like the one at 1 sec. in this device or the giant step at 35 sec. in Fig. 2b at 230K. These giant steps are composed of several (3 and 5, respectively) steps which are correlated. The traps observed therefore sometimes cluster with a correlated simultaneous emission of several holes or electrons, respectively. Multiple emissions sum up to the large step size. Individual steps could be fully resolved for the giant step in Fig. 3 and partially resolved (final 2 steps) in Fig. 2b.

It is noted that the emission time constants are extremely slow. Delay times well exceeding 100 sec. are observed. Fast times in the msec. range were observed by a transient recorder. We conclude that the traps under investigation are probably located in the oxide varying in depth from the interface to observe such a wide spectrum of time constants for these shallow traps.

QUANTUM STEP HEIGHT

The interpretation of the quantum steps is best visualized by the computed source-drain current distribution in Fig. 4. This computation assumes a uniform 2-dimensional channel area with a circular area in the center blocked for

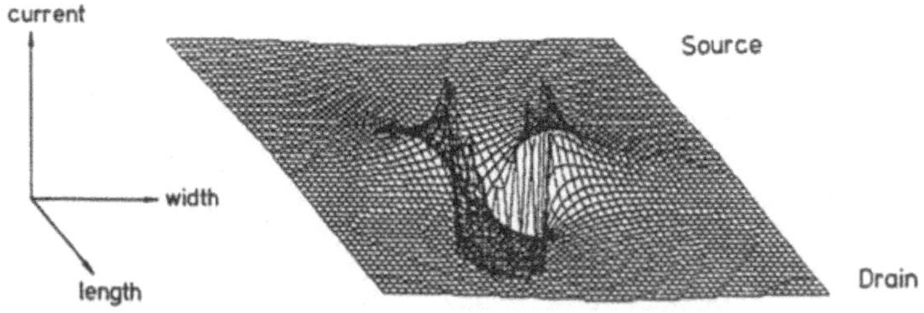

Fig.4. Computation of the source-drain current distribution around a circular blocked area in a 2-dimensional uniform MOSFET channel.

the current transport by a charge trapped in an interface state. The current from source to drain has to bypass this area. The current density on the side of the blocked area is therefore increased. The total current is slightly reduced. The reduction depends on the blocked area and the screening of this area by the mobile charge carriers in the channel.

We have used one specific trap center, i.e. a specific step in transients to monitor the effect of the channel carrier density on the magnitude of the step height (Fig. 5). The channel conductance is used as a measure of the mobile carrier density. The step height variation is roughly inversely proportional to the channel conductance G as indicated by the continuous line. This measurement is a direct observation of the screening effect in a 2-dimensional system using an individual trap as a local probe. Presently available theories are not applicable for the measurement range near the threshold of the transistor, however at high carrier densities and large conductivities, the inverse proportionality is well observed.

The statistical distribution of step heights observed is shown in Fig. 6. The counts are taken for 12 sequential transients so that a significant distribution is obtained. Three distinct peaks are observed rather than a smooth distribution. The analysis was performed at two different source-drain currents 20nA and 14nA as shown by the continuous and dashed lines, respectively. The general trend of a larger step height for a lower channel conductivity as shown in Fig. 5 is again observed by the shift of the peaks. It is noted, however, that the shift for the lowest peak marked "SD" is only marginal compared to the large shifts of the two other peaks marked "R1" and "R2". The small peak at step height 0.75% may be a double emission because it occurs

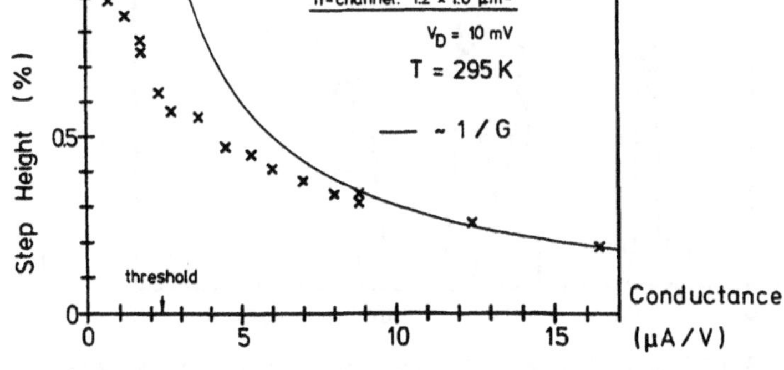

Fig.5. Step height of one specific trap as a function of the channel conductance which was varied by the gate bias voltage.

just at twice the value of the center peak. The statistics, however, is not as conclusive for multiple emissions as the resolution of correlated sub-steps visible in Fig. 2 and 3.

We interpret the unusual peak distribution on the basis of the model explained in Fig. 4 and the screening property of Fig. 5. The step height is expected to be a function of the location of the trap in the channel area depending on how much current is blocked by the charged center. A small step height is expected in areas with a large carrier density. A high carrier density occurs near the source and drain contact. In this rim area of the channel, the effect of the gate to increase the carrier density is reduced. We therefore conclude that the peak at the lowest step height represents trap centers in the channel rim near source and drain. The other 2 peaks can be attributed to interface states in the side rims, i.e. the bird's beak region of the device. In these regions, we expect a large density of traps due to the increased stress. This distribution of traps and of the quantization step heights is further supported by the observation that in short and wide channels the peak "SD" is dominant.

CONCLUSIONS

First results of DLTS measurements on µm-size MOSFETs are reported. Due to the low number of traps in these small devices the transients are quantized into steps. The steps vary in height. The step height is specific for individual traps so that they can be identified. Traps can be located in the channel by the shape of the statistical distribution.

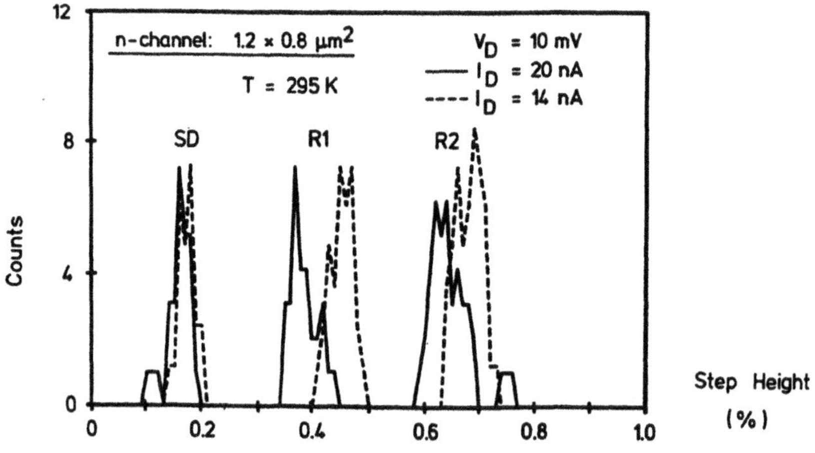

Fig.6. Statistical distribution of the measured step heights of all the traps emitting within 0.1 sec - 200 sec. The counts are averaged over 12 transients taken under the same conditions.

ACKNOWLEDGEMENTS

The authors are grateful to Dr.O.Jäntsch from Siemens AG, Munich who supplied the MOSFET samples studied in this experiment and who pointed out this interesting research topic to us. Partial support of the Stifterverband für die Deutsche Wissenschaft is gratefully acknowledged.

REFERENCES

1. Lang, C.V., J.appl.Phys. 45, 3023 (1974)
2. Miller, G.L., Lang, D.V., and Kimmerling, L.C., Ann.Rev.Mater.Sci. 7, 377 (1977)
3. Lefevre, H., and Schulz, M., IEEE Trans. ED-24, 973 (1977)
4. Schulz, M., and Johnson, N.M., Appl.Phys.Lett. 31, 622 (1977)
5. Schulz, M., and Klausmann E., Appl.Phys. 18, 169 (1979)
6. Uren, M.J., Day, D.J., and Kirton, M.J., Appl.Phys.Lett. 47(1985) 1195
7. Howard, R.E., Skocpol, W.J.,, Jackel, L.D., Mankiewich, A.M., Fetter, L.A., Tennant, D.M., Epworth, R., and Ralls, K.S., IEEE, ED-32 (1985) 1669
8. Kirton, M.J., Uren, M.J., Appl.Phys.Lett. 48, 1270 (1986)
9. Bollu, M., Koch, F., Madeneck, A., and Scholz, J., in: Insulating Films on Semiconductors, INFOS 87 (G.Declerck, R.DeKeersmaecker eds.) 1987 North-Holland Publ. p.142
10. Kirton, M.J., Uren, M.J., and Collins, S., ditto, p. 178
11. Karwath, A., Schulz, M., Appl.Phys.Lett. 52, 634 (1988)

OBSERVATION OF "1/f-NOISE STATES" IN CONDUCTANCE MEASUREMENTS ON MOS STRUCTURES

M.J. Uren, M.J. Kirton and S. Collins

Royal Signals and Radar Establishment
St. Andrew's Road, Great Malvern
Worcestershire WR14 3PS, United Kingdom

ABSTRACT

Defect states in the oxide, residing close to the $Si:SiO_2$ interface, are responsible for the generation of random telegraph signals in small-area MOSFETs and 1/f noise in large devices. By improving resolution and extending the frequency range of the conventional conductance technique, we demonstrate that these states are observable in measurements at room temperature and in device-quality oxides. A loss peak due to the normal interface states is seen, together with a plateau on the low-frequency side due to the trapping centres in the oxide.

INTRODUCTION

Capacitance-Voltage (C-V) and conductance (G-ω) measurements are the everyday experimental techniques used to assess the quality of the oxide:silicon interface in MOS structures[1,2]. The presence of interface states gives rise to the familiar phenomenena of stretch-out in the high frequency C-V curve and the loss peak in G-ω data. Conventional wisdom has it that at a given energy in the silicon band gap the interface states have only a *single*, well-defined cross-section. The value of this cross-section is ~ 10^{-16} - 10^{-15}cm^2 around mid gap, with some evidence to suggest that it falls off to ~ 10^{-19}cm^2 as the band edges are approached. The width of the conductance peak is much greater than one would anticipate from a distribution of interface states across the band gap with a single cross-section. Since these measurements are usually performed in depletion, the extra width is accounted for by the surface potential fluctuations produced by unscreened charges residing at the interface. In this paper, our purpose is to show that in device-quality oxides the defect states in the oxide – which reside close to the interface and are responsible for 1/f noise – give rise to an important contribution in G-ω data at low frequencies.

RANDOM TELEGRAPH SIGNALS, 1/f NOISE AND CONDUCTANCE MEASUREMENTS

In small MOSFETs, carrier trapping into localised defects in the oxide close to the $Si:SiO_2$ interface generates random telegraph signals (RTSs) in the drain current[3]. In large devices, the summation of many hundreds of these RTSs generates 1/f noise[4]. The observed time constants of the RTSs were in the range 10^{-3} to 10^2s, and the corresponding noise measurements (on large devices) showed a 1/f spectrum at all gate voltages in the measurement frequency range 1 to 10^4Hz. The cross-hatched area on figure [1] depicts the time and energy windows in which the RTSs were measured in n-channel devices; the straight line depicts the time constant of interface states measured with the conductance technique. To get a 1/f spectrum requires a uniform distribution of trap time constants on a log-time scale[5]. In figure [1], this corresponds to a uniform distribution along any vertical line within the cross-hatched area.

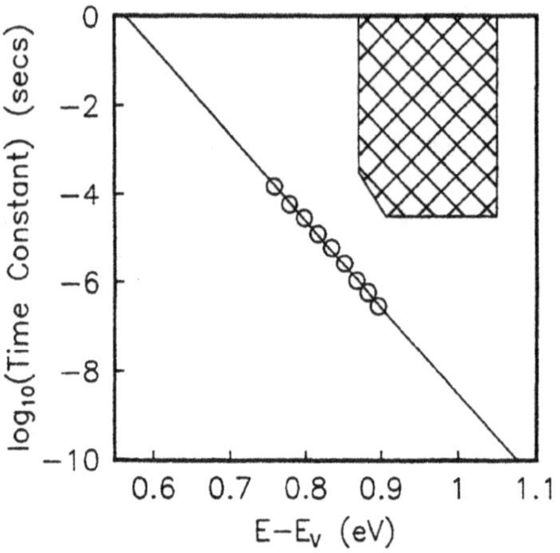

Fig. 1 Measured time constant versus surface Fermi level in the upper half of the band gap. Cross-hatched area shows the region typically accessed in noise measurements. (O) shows the interface state loss peak from G_P/ω measurements; the data corresponds to the capacitor shown in figure [2a].

Now since capture and emission events are recorded as a loss in G-ω measurements, the question that we addressed was the following: Are the defect states responsible for 1/f noise observable in the conductance technique? To our knowledge, Eaton and Sah[6], and Ngai and Liu[7] are the only previous workers to have posed and attempted to answer this question. Eaton and Sah found a marked asymmetry in their measured conductance spectrum, which they ascribed to the long time constants arising from carrier tunnelling from interface states into defect states in the oxide. They noted that this mechanism would also cause 1/f noise. Unfortunately, their samples were sufficiently non-standard that their work has been totally overlooked. They utilised 8nm-thick oxides with interface-state densities ~ 10^{13}cm^{-2}eV^{-1}, and the experiments were carried out at low temperatures. Since in the early 80's the accepted experimental evidence pointed to only a single, symmetrical broad peak in G-ω data and the origins of 1/f noise in MOSFETs were still uncertain, Ngai and Liu concluded that 1/f noise could not be caused by carrier trapping and must, therefore, be due to some form of dielectric loss mechanism.

An alternative way of formulating the problem is to ask whether the time constants (cross sections) of the 1/f-states shown in figure [1] extend to shorter times

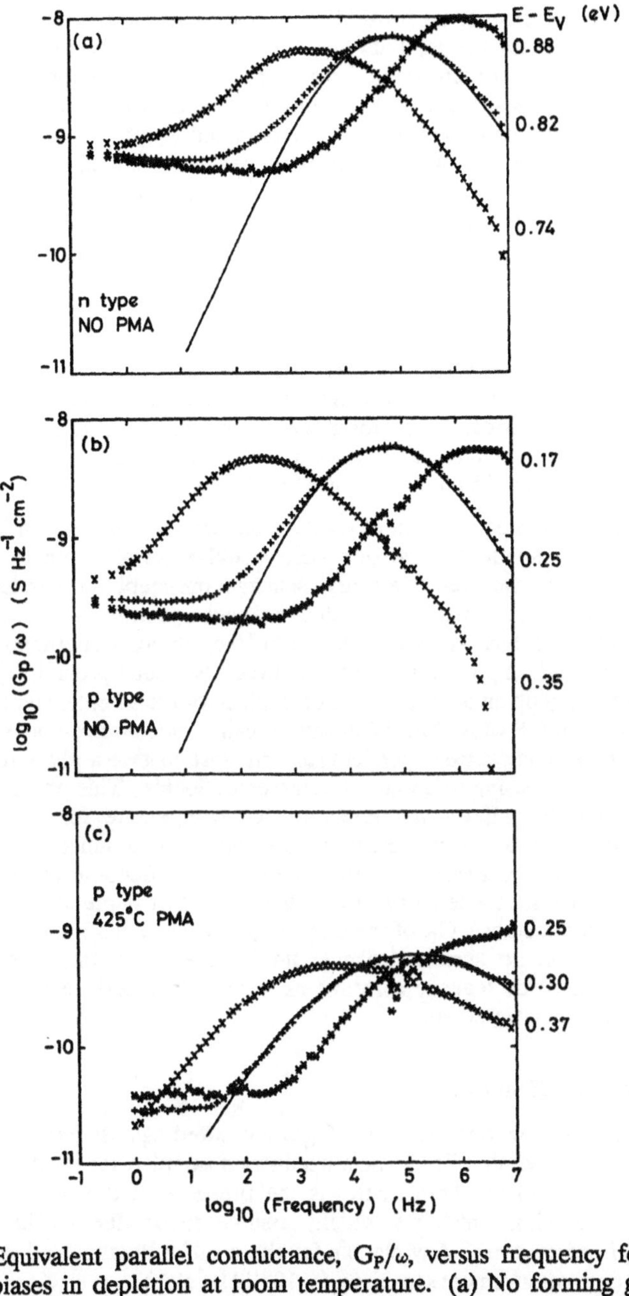

Fig. 2 Equivalent parallel conductance, G_P/ω, versus frequency for 3 gate biases in depletion at room temperature. (a) No forming gas PMA, n-substrate. For the fitted curve, interface state density $D_{it} = 1.6\times10^{11}cm^{-2}eV^{-1}$, potential fluctuations $\sigma_s = 1.5kT$, cross-section $\sigma = 5\times10^{-17}cm^2$. For the low-frequency plateau at 1Hz, $N_t = 5.6\times10^{17}cm^{-3}eV^{-1}$. (b) p-substrate, no PMA. $D_{it} = 1.3\times10^{11}cm^{-2}eV^{-1}$, $\sigma_s = 1.6kT$, $\sigma = 1.8\times10^{-17}cm^2$, $N_t = 2.4\times10^{17}cm^{-3}eV^{-1}$. (c) p-substrate, 425°C PMA. $D_{it} = 2.1\times10^{10}cm^{-2}eV^{-1}$, $\sigma_s = 2.8kT$, $\sigma = 6\times10^{-16}cm^2$, $N_t = 2.2\times10^{16}cm^{-3}eV^{-1}$.

to meet the single time constant measured for interface states in G-ω measurements. If there were no states in the the gap between the cross-hatched area and the straight line shown in figure [1], then there would be an easily measurable upper limit to the frequency at which a 1/f spectrum could be observed; such a limit has never been recorded. On the other hand, conventional understanding of G-ω measurements suggests that there is only a single, well-defined peak corresponding to a single, well-defined time constant (cross section). In order to resolve this apparent paradox, we have carried out G-ω measurements down to 0.25Hz such that this time-window gap is investigated for standard MOS structures.

EXPERIMENTAL

Measurements were carried out on conventional aluminium-gate MOS capacitors, of area $6.97\times10^{-3}cm^2$, fabricated on 0.3Ωcm n-type and 2.8Ωcm p-type [100] substrates. The dry gate oxides were grown for 3 hours at 900°C to a thickness of 35nm, followed by a 30min post-oxidation anneal in nitrogen. Some wafers received a post-metallisation anneal (PMA) in forming gas for 30mins at 425°C.

Above 50kHz, the measurement of conductance was entirely conventional and employed an HP4192 impedence analyser. Below 50kHz, an electrometer op-amp (Analog Devices AD515) was used as a charge sensitive pre-amplifier with an air-gap capacitor as the feedback capacitor. A dual-channel spectrum analyser system (HP3565S) provided the bias voltage with a 10mV excitation superimposed, and measured the phase and amplitude difference between two such pre-amp circuits: a reference circuit employing an air-gap capacitor in place of the sample, and the other circuit containing the MOS capacitor. With careful calibration and compensation for frequency and sample capacitance dependent gain, the system gave a phase resolution of 1mdegree at 1Hz increasing to about 10mdegrees at 50kHz. This phase accuracy was essential to see the effects we observed. To compensate for any residual inductances and the substrate series resistance, the measured conductance, G_m, and capacitance, C_m, were recorded at all measurement frequencies in strong accumulation and then subtracted from the values found in depletion to give the equivalent parallel conductance, G_P, of the trap. A guard ring and a nitrogen ambient ensured that there was no surface leakage; no low-frequency loss was seen in accumulation and the oxide leakage current was < 5fA. The surface potential was measured using a quasi-static sweep.

RESULTS AND DISCUSSION

Figures [2a,b,c] show our results for G_P/ω measured against frequency for p- and n-substrate samples without PMA and a p-substrate sample with PMA. The new, striking feature present in all three figures is the presence of the distinct plateau region at low frequencies, which we initially assume to be due to the uniform distribution of 1/f defect states. Any plateau for the n-substrate annealed samples is around or below our measurement sensitivity (~ $10^{-11}SHz^{-1}cm^{-2}$).

The peak in the three traces can be fitted to the standard G_P/ω formula as shown. Interface-state densities, D_{it}, for the un-annealed samples was typically ~ $10^{11}cm^{-2}eV^{-1}$ and for the annealed samples ~ $2\times10^{10}cm^{-2}eV^{-1}$. The variance of the surface potential fluctuations was between 1.5 and 3kT.

The G_P/ω spectrum arising out of a uniform distribution of time constants can be conveniently derived using the formalism in the appendix to Nicollian and Goetzberger[8]. A tunnelling model in which the defects are uniformly distributed in energy and distance into the oxide gives the appropriate distribution. This model then predicts a flat spectrum – as we have observed – and the value of G_P/ω is related to the trap density per unit volume, N_t, via the relation

$$N_t = \frac{4\kappa(G_P/\omega)}{q\pi} , \qquad [1]$$

where κ is the decay constant and is defined to be

$$\kappa = \frac{(8m^*\phi)^{1/2}}{(h/2\pi)} \qquad [2]$$

The barrier energy, ϕ, was taken to be 3.5eV and the effective mass was set to unity. Using equations [1] and [2] and the measured value of G_P/ω in the plateau region at low frequencies, we found $N_t \sim 2\times10^{16}cm^{-3}eV^{-1}$ for the annealed sample and 2×10^{17} - $5\times10^{17}cm^{-3}eV^{-1}$ for the un-annealed samples.

Our working hypothesis at this point is that the defect states responsible for the plateaux in G_P/ω shown in figures [2a,b,c] are also responsible for the measured 1/f spectrum in MOSFETs. To complete the argument, ideally one requires MOSFETs and MOS capacitors fabricated at the same time in the same process in order that the magnitude of the noise can be correlated with the magnitude of the plateau in G_P/ω. It must be borne in mind that in order to access the upper half of the silicon bandgap n-channel (p-substrate) MOSFETs are required to compare with n-substrate capacitors. Unfortunately, we do not have such a set of samples. We must, therefore, compare our values of N_t from G_P/ω with those that are measured from the noise in typical modern devices. Reimbold[9] has produced a simple trapping model of 1/f noise which allows N_t to be estimated from the magnitude of the power spectrum in both strong and weak inversion (which would correspond to accumulation and depletion for the capacitors). For the n-channel devices used in his study, he found values for N_t from 2×10^{15} - $2\times10^{16}cm^{-3}eV^{-1}$. Similarly, our noise measurements on n-channel devices (supplied to us by GEC Hirst Research Centre, Wembley, UK) gave $N_t \sim 10^{16}cm^{-3}eV^{-1}$ in weak inversion. Our G_P/ω measurements have a resolution which corresponds at best to about $5\times10^{15}cm^{-3}eV^{-1}$; so it is not surprising that we have not been able to see conclusively such a plateau for the n-substrate samples with PMA, although the plateau is quite clear when there is no anneal. For the valence-band edge case, we have no comparative MOS noise figures, since most PMOS devices are partly buried channel due to the threshold implant. But it is indicative that the p-substrate capacitors with PMA have comparable values of N_t to those seen in n-channel MOSFETs. Such comparisons are of limited value due to the different gate material (aluminium versus poly-silicon) and oxide growth conditions, but do serve to show that the values of N_t measured from G_P/ω are quite consistent with those derived from noise.

CONCLUSIONS

In this paper, we have demonstrated that the conventional view of the interface with only a *single* capture cross-section at a given energy is only part of the story. In practice a very wide spread of cross-sections is observed, with an upper limit of ~ $10^{-16}cm^2$ for states residing right at the interface. All of these states communicate with the inversion layer; and the capture and emission events are observed in both noise and conductance measurements. The agreed nomenclature for interface trapped charge[10] thus includes those states residing at the interface – which are observed in

C-V and G-ω measurements and which have hitherto been regarded as the *only* defect states present – and the oxide states responsible for low-frequency loss in G-ω measurements and 1/f noise. It is also quite possible that the fixed oxide charge has its origins in the long-time tail of the 1/f states. Overall, by extending the dynamic frequency range of G-ω meaurements down to frequencies ~ 0.25Hz, we have opened up the exciting possibility of being able to monitor the effects of different oxide treatments, hot-electron injection etc. on the oxide states and the interface states at the same time.

REFERENCES

1. E.H. Nicollian and J.R. Brews, MOS Physics and Technology, Wiley, New York, (1982).
2. M. Schulz, Interface States at the Si-SiO_2 Interface, Surf. Sci. 132:422 (1983).
3. K.S. Ralls, W.J. Skocpol, L.D. Jackel, R.E. Howard, L.A. Fetter, R.W. Epworth and D.M. Tennant, Discrete Resistance Switching in Submicrometer Silicon Inversion Layers: Individual Interface States and Low-Frequency (1/f?) Noise, Phys. Rev. Lett. 52:228 (1984).
4. M.J. Uren, D.J. Day and M.J. Kirton, 1/f and Random Telegraph Noise in Silicon MOSFETs, Appl. Phys. Lett. 47:1195 (1985).
5. M.J. Buckingham, Noise in Electronic Devices and Systems, Ellis Horwood, Chichester, (1983).
6. D.H. Eaton and C.T. Sah, Frequency Response of Si-SiO_2 Interface States on Thin Oxide MOS Capacitors, Phys. Stat. Sol. a12:95 (1972).
7. K.L. Ngai and S.T. Liu, Studies of Interface States of Silicon MOS Devices by Dynamic Conductance and Noise Mesaurements and Effects of Bias-Temperature Stress, Thin Solid Films 93:321 (1982).
8. E.H. Nicollian and A. Goetzberger, The Si-SiO_2 Interface – Electrical Properties as Determined by the Metal-Insulator-Silicon Conductance Technique, Bell. Syst. J. 46:1055 (1967).
9. G. Reimbold, Modified 1/f Trapping Noise Theory and Experiments in MOS Transistors Biased from Weak to Strong Inversion – Influence of Interface States, IEEE Trans. Electron. Devices ED-31:1190 (1984).
10. B. Deal, Standardised Terminology for Oxide Charges Associated with Thermally Oxidised Silicon, IEEE Trans. Electron. Devices ED-27:606 (1980).

THE GATE-VOLTAGE DEPENDENCE OF TRAPPING INTO INDIVIDUAL $Si:SiO_2$ INTERFACE STATES

M.J. Kirton, M.J. Uren and S. Collins

Royal Signals and Radar Establishment
St. Andrew's Road, Great Malvern
Worcestershire WR14 3PS, United Kingdom

ABSTRACT

Random telegraph signals are observed in small-area silicon MOSFETs due to the fluctuations in occupancy of single defects residing in the oxide close to the oxide:silicon interface. We review the general properties of these signals and show that carrier capture proceeds via a multi-phonon process. Anomalies in the gate-voltage dependence of the capture time and estimates of the trap distance into the oxide are considered. We demonstrate that the former has its origin in a gate-voltage dependent cross-section, and the latter in a gate-voltage dependent trap entropy of ionisation for the device operating around threshold.

INTRODUCTION

Localised trapping centres at the oxide:silicon interface have been under intensive investigation for many years. Their presence can be clearly observed in the standard electrical characterisation techniques of capacitance-voltage (C-V) and conductance (G-ω) measurements. Recently, it has become possible to investigate in detail individual interface defects through the study of small-area silicon MOSFETs[1-4]. By this means, it has been shown that defects in the oxide – residing close to the interface – are responsible for the 1/f noise observed in large-area devices[1,2]. Moreover, the random telegraph signals (RTSs) observed in small devices provide unique insight into the nature of these defects[3,4]. The relationship between the states responsible for RTSs (and 1/f noise) and those commonly observed in C-V and G-ω measurements will be considered in a companion paper[5]. Here, we wish to present new results from our further investigations of RTSs.

RESULTS AND DISCUSSION

General Observations

A standard RTS generated through the fluctuating occupancy of a single oxide defect is shown in the inset to figure [1]. Whilst a small but significant percentage of RTSs exhibit anomalous behaviour[4], the majority are of this form. The fractional occupancy, f, of a defect is governed by Fermi-Dirac statistics. If we write the mean capture and emission times as $\overline{\tau_C}$ and $\overline{\tau_E}$, then we have

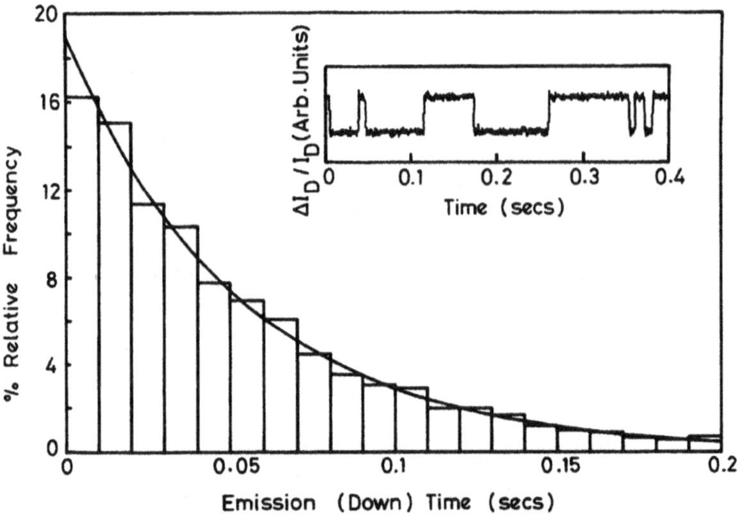

Fig. 1 Distribution of 4425 emission times for device H6 at 95K and V_G = 1.15V, showing that the time is distributed exponentially. $\bar{\tau}_E$ = 0.0528s, standard deviation = 0.0505s. The inset shows a portion of the I-t characteristic; the down time corresponds to emission.

$$f = \frac{1}{1 + g.\exp[(E_T - E_F) / kT]} = \frac{\bar{\tau}_E}{\bar{\tau}_C + \bar{\tau}_E} . \qquad [1]$$

and

$$\bar{\tau}_C/\bar{\tau}_E = g.\exp[(E_T - E_F) / kT] , \qquad [2]$$

where: E_T is the trap energy level; E_F is the Fermi energy; and g represents the trap degeneracy factor. As the gate bias changes, the separation of E_T and E_F alters with a corresponding change in the mark-space ratio, $\bar{\tau}_C/\bar{\tau}_E$. This allows one to identify the times in the up state with electron capture and the times in the down state with electron emission. As depicted in figure [1], the capture times – and, although not shown, also the emission times – are exponentially distributed as one would expect for a process governed by the two attempt rates $1/\bar{\tau}_C$ and $1/\bar{\tau}_E$. The slight departure from the theoretical curve at very small times is merely an artifact of the limited resolution of the experimental sampling rate.

An important and readily observable property of the capture process is that it is thermally activated[1]. Kirton and Uren[3] showed that if the capture time and cross-section, σ, are related by

$$\bar{\tau}_C = (\sigma \langle v \rangle n)^{-1} \qquad [3]$$

and

$$\sigma = \sigma_0 \exp[-\Delta E_B / kT] , \qquad [4]$$

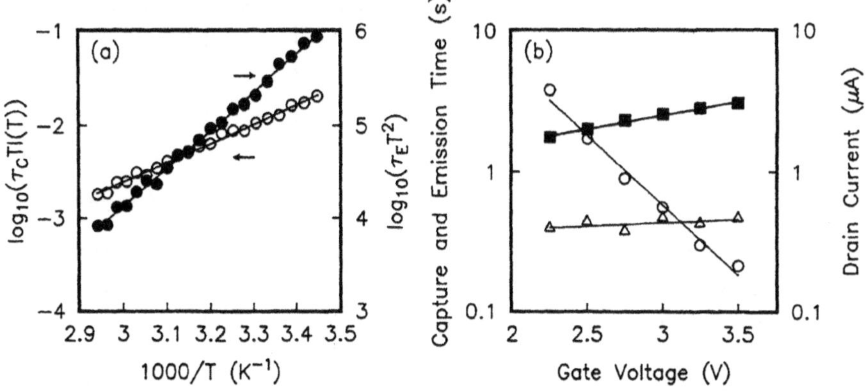

Fig. 2 (a) Temperature dependence of capture time (O) and emission time (●) for device H21. V_G = 3V, V_D = 50mV. (b) Gate-voltage dependence of the capture (O) and emission (▵) times for device H21 at T = 310K, V_D = 50mV. (■) indicates the drain current.

then the following two equations describe the temperature dependence of $\overline{\tau_C}$ and $\overline{\tau_E}$:

$$I(T)T\overline{\tau_C} = \frac{\exp[\Delta E_B / kT]}{\sigma_0 \alpha} \; ; \qquad [5]$$

$$T^2\overline{\tau_E} = \frac{\exp(\Delta S/k)}{\sigma_0 \beta} \times \exp[(\Delta E_B + \Delta H_{CT}) / kT] \; . \qquad [6]$$

In the above equations, n is the mean inversion-layer number density (per unit volume); <v> is the RMS thermal velocity; σ_0 is the cross-section pre-factor and is related to the tunnelling to the defect; ΔE_B is the activation energy for capture; ΔS is the trap entropy of ionisation, into which the degeneracy factor has been incorporated; and ΔH_{CT} is the trap binding enthalpy and is related to the trap binding energy, ΔE_{CT}, by $\Delta E_{CT} = \Delta H_{CT} - T\Delta S$. α and β are constants incorporating such parameters as drain voltage, gate length and width etc.[3]. The model also includes a T dependence to the inversion-layer thickness[6] not included in reference [3]. Figure [2a] shows plots of the logarithm of $I(T)T\overline{\tau_C}$ and $T^2\overline{\tau_E}$ versus 1/T. From the intercepts and gradients of these plots, one can evaluate σ_0, ΔS, ΔE_{CT} and ΔE_B. This model corresponds to the multi-phonon model familiar from studies on bulk defects and accounts for many of the observed properties of the RTSs[3,4].

Anomalous Gate-Voltage Dependence of the Capture Time

We would now like to turn our attention to the behaviour of the RTSs as gate bias is varied. The devices used were from the same wafer as those used previously[2-5], and had oxide thickness of 40nm and electrical dimensions of 0.5×0.75μm. On the basis of equation [3], it would be expected that the change in capture time on incrementing gate voltage would come about from changes in n, the inversion-layer number density (per unit volume). This assumes an invariant cross-section. If we define μ to be the electron mobility, q the electronic charge, V_D the applied drain voltage, w and l to be the gate width and length, and t to be the mean inversion-layer thickness, then we can write

$$I = nq\mu V_D t(w/l) \; . \qquad [7]$$

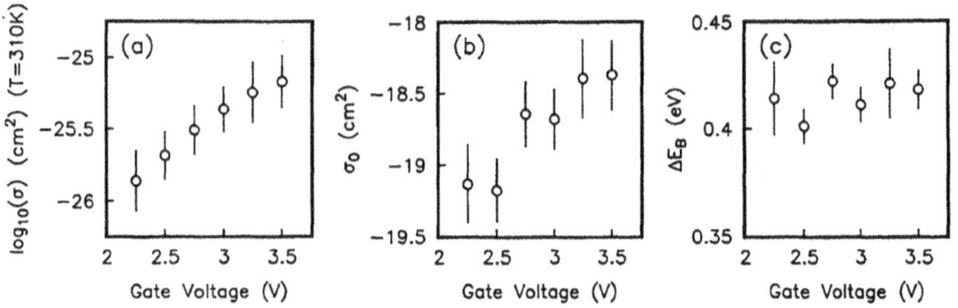

Fig. 3 Gate-voltage dependence of cross-section (σ), cross-section pre-factor (σ_0) and capture activation energy (ΔE_B) for device H21.

Thus as the gate bias increases, we expect n to be roughly proportional to I. Figure [2b] shows for one particular RTS that as the gate voltage is changed from 2.25 to 3.5 volts the current increases by a factor of 1.8, whilst there is a corresponding decrease in $\overline{\tau_C}$ by a factor of 15.7. Thus the capture time is decreasing much more rapidly than predicted by equations [3] and [7]. In fact this is quite a general phenomenon and is by no means restricted to the RTS of figure [2b]. This particular RTS was singled out for detailed investigation because it showed a more marked effect than most. In order to account for the data of figure [2b] on the basis of equations [3] and [7], it is required that the product μt decrease by a factor of 9 over the gate-voltage range 2.25 to 3.5 volts. However, the inversion-layer thickness changes by no more than a factor of two[7] and the mobility is found to change by 8%. Thus variations in μ and t alone are unable to account for the data presented in figure [2b].

Further investigations were carried out to measure σ, σ_0 and ΔE_B as a function of V_G. The results are shown in figures [3a-c]. In producing these figures, we utilised the formalism of equations [5] and [6] with a classically derived, V_G-dependent inversion-layer thickness. The mobility was taken to be a function of temperature only. Figure [3a] shows that the very strong dependence of $\overline{\tau_C}$ on V_G is accounted for by a monotonically increasing capture cross-section. Inspecting figures [3b,c], we see that this increase in σ is brought about by an increasing σ_0, with ΔE_B effectively constant throughout the range of V_G.

In a bulk solid, where the free carriers uniformly bathe a defect, the definition $\overline{\tau_C} = (\sigma\langle v\rangle n)^{-1}$ is perfectly sensible. In the case of a defect in the oxide of an MOS structure there are the following two problems: first, the inversion-layer charge is displaced from the defect site; and second, as the gate voltage changes so does the electric field in the inversion layer. As the electric field strength increases the charge-density peak moves closer to the interface, increasing wave function overlap and enhancing image-charge lowering of the barrier. In addition, the changing oxide field strength also lowers the barrier. The net result is an increase in value of σ_0. To quantify this explanation at room temperature will be a very difficult task, but it may be possible from the results of low temperature studies where only one sub-band is filled and for which model calculations of the electronic states are more tractable.

Anomalous Estimates of the Trap Depth

Clearly, it is important to know the spatial location of the defects responsible for RTSs and 1/f noise. Ralls et. al.[1] showed that by differentiating equation [2] with respect to gate voltage and evaluating the change in $E_T - E_F$ and the surface potential, ϕ_S, for a given change in V_G, the distance, d, of the trap into the oxide could be easily estimated. We show below that exactly the same result can be obtained from the

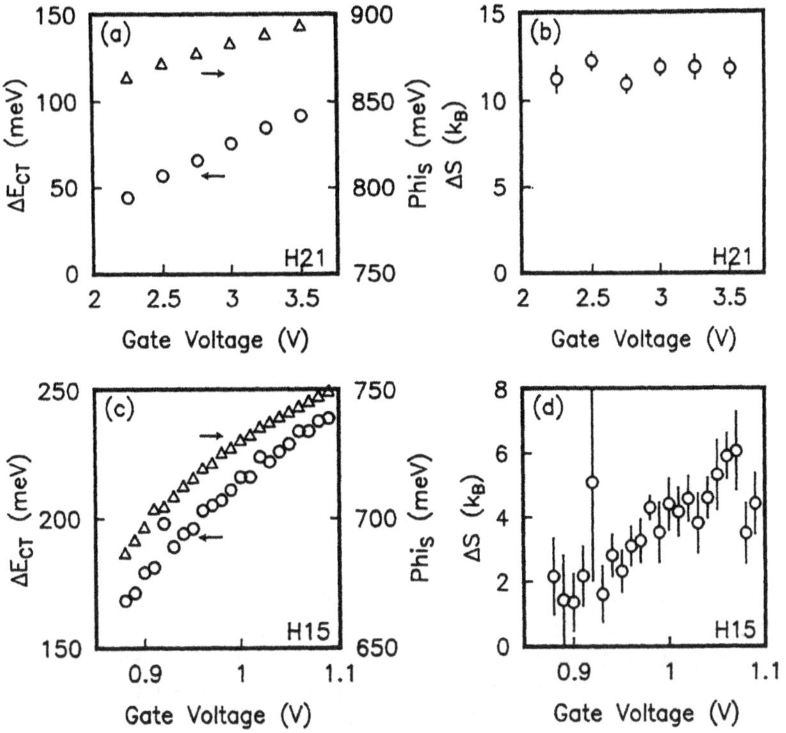

Fig. 4 (a), (c) show the gate-voltage dependence of trap energy level (ΔE_{CT}) and surface potential (Phi$_S$). (b), (d) show the gate-voltage dependence of trap entropy (ΔS). (a), (b) are for device H21 in strong inversion and (c), (d) for device H15 around threshold. V_T = 0.85V.

variation of ΔE_{CT} with gate voltage. However, the general point that we now wish to make is that whilst Ralls et. al's prescription produces perfectly sensible values for d (e.g up to 2nm) when the device is operating in strong inversion, anomalously large values (e.g up to 20nm) are obtained when the same analysis is applied to traps active around threshold.

To understand how this comes about consider figures [4a,c] which show the variations in ΔE_{CT} and ϕ_S for two defects operating in strong inversion and around threshold, respectively. From figure [4a], we see that for a ΔV_G of 1250meV ϕ_S changes by 30meV and ΔE_{CT} by 48meV. Now if the trap were located directly at the interface, and assuming the trap energy level moved rigidly with changing electrostatic potential, then no change in ΔE_{CT} would be recorded. The increase in ΔE_{CT} is due to the trap being located in the oxide such that it experiences the changing oxide field strength. With an oxide thickness, t_{ox}, of 40nm, $d = 48.t_{ox}/(1250 - 30) = 1.6$nm – a perfectly reasonable value. Applying the same analysis to the data for the trap of figure [4c], we find $d = 71.t_{ox}/(210 - 63) = 19.3$nm – a totally unreasonable value. The conclusion to be drawn is that in the threshold case the trap binding energy is changing as a function of gate voltage over and above what one would expect from simple electrostatics alone.

The trap binding energy, ΔE_{CT}, is calculated from the relation $\Delta E_{CT} = \Delta H_{CT} - T\Delta S$: ΔH_{CT} is the trap binding enthalpy and is obtained from the difference in the gradients of the emission and capture data; ΔS is the change in trap entropy on removing an electron from the defect and placing it in the inversion layer. Figures [4b,d] show that for the defect active in strong inversion ΔS is constant over a gate voltage range of 1.25V, whilst for the threshold trap ΔS changes by a factor of ~ 2.5 for a ΔV_G of 0.21V. In addition, the threshold trap's rate of change of ΔH_{CT} is inconsistent with electrostatics. Since ΔH and ΔS are recovered from temperature-dependent data obtained at fixed gate voltage their values are little affected by surface potential fluctuations.

The key to understanding this data probably lies in a consideration of the changes which take place in the local environment of the trap as the gate voltage is incremented. In strong inversion, the electron number density (per unit area) is high – typically $10^{12}cm^{-2}$, and the trapped charge is fully screened by the inversion-layer charge. However, around threshold the image charge of the trapped electron is shared between the gate, the channel and the depletion region, and the number density in the channel is very sensitive to changes in gate bias. Thus the location of the image charge is changing rapidly and so constitutes a rapidly changing local environment for the defect. This gives rise to modifications in the bonding and dynamical properties of the trap which are reflected in V_G-dependent enthalpy and entropy terms, respectively. A further test of this model would be to investigate the behaviour of d for traps operating in very weak inversion. Under such operating conditions, the trapped electron is screened by the gate and depletion-layer edge. When V_G changes, the inversion-layer charge is still too low to modify the screening field, and it would be expected that ΔH and ΔS would show no anomalous behaviour. Unfortunately, we have been unable to investigate in detail RTSs in this regime.

Gate-Voltage Dependence of the Emission Time

As shown in figure [2b], $\bar{\tau}_E$ rises only very slowly as the gate bias is increased. In general, $\bar{\tau}_E$ is found to show all variations from none at all to a sensitivity equal and opposite to that exhibited by $\bar{\tau}_C$. Using equations [2] and [3] and the fact that $n = N_C\exp[-(E_C - E_F) / kT]$, where N_C is the effective density of states in the conduction band, one finds

$$\frac{\bar{\tau}_E(V_{G2})}{\bar{\tau}_E(V_{G1})} = \frac{\sigma(V_{G1})}{\sigma(V_{G2})} \times \exp[(\Delta E_{CT}(V_{G2}) - \Delta E_{CT}(V_{G1}) / kT] \quad . \qquad [8]$$

As we have seen already, $\sigma(V_{G2}) > \sigma(V_{G1})$ and $\Delta E_{CT}(V_{G2}) > \Delta E_{CT}(V_{G1})$ for $V_{G2} > V_{G1}$. The behaviour of $\bar{\tau}_E$ depends on how these opposing effects cancel out. One would anticipate that the exponential term would generally dominate, corresponding to all of our observations up to this point. However if the σ term were to dominate, then $\bar{\tau}_E$ would decrease as V_G increased.

CONCLUSIONS

Telegraph-noise measurements on individual traps provide immense detail about the trapping process at the expense of information about the ensemble properties. In this paper, we have demonstrated using two example traps that the trap environment, and hence the trapping times, are strongly affected by the gate voltage. This manifests itself as a gate-voltage dependent capture cross-section and a gate-voltage dependent trap entropy around threshold.

REFERENCES

1. K.S. Ralls, W.J. Skocpol, L.D. Jackel, R.E. Howard, L.A. Fetter, R.W. Epworth and D.M. Tennent, Discrete Resistance Switching in Submicrometer Silicon Inversion Layers: Individual Interface States and Low-Frequency (1/f?) Noise, Phys. Rev. Lett. 52:228 (1984).
2. M.J. Uren, D.J. Day and M.J Kirton, 1/f and Random Telegraph Noise in Silicon MOSFETs, Appl. Phys. Lett. 47:1195 (1985).
3. M.J. Kirton and M.J. Uren, Capture and Emission Kinetics of Individual Si:SiO_2 Interface States, Appl. Phys. Lett. 48:1270 (1986).
4. M.J. Uren, M.J. Kirton and S. Collins, Anomalous Telegraph Noise in Small Area Silicon MOSFETS, Phys. Rev. B. (in press).
5. M.J. Uren, M.J. Kirton and S. Collins, Observation of "1/f-Noise States" in Conductance Measurements on MOS Structures, (this volume).
6. S.M. Sze, Physics of Semiconductor Devices, Wiley, New York, (1981).
7. T. Ando, A.B. Fowler and F. Stern, Electronic Properties of Two-Dimensional Systems, Rev. Mod. Phys. 54:437.

STEP HEIGHTS OF SWITCHING EFFECTS OF SINGLE INTERFACE TRAPS IN MOSFETS

O. Jäntsch and R. Kircher

Siemens AG, ZFE
Corporate Research and Development Laboratories
D 8000 Munich 83, Germany

ABSTRACT

MOSFETs with small gate areas show single electron switching by interface traps. The step height of the resistance change is calculated partly analytically and partly numerically with the device simulator GALENE II. It depends on the location of the interface traps, and decreases towards the edge of the bird's beak. Near threshold voltage the step height is influenced not only by the gate voltage directly but also by the pinch off effect of the channel under a trapped electron.

A MOSFET with a small gate area exhibits random telegraph signals (1,2,3). If such a MOSFET is biased in strong inversion for a long time and then switched back to a gate voltage of about 10 mV above threshold voltage the emission of single electrons from discrete interface traps can be observed (4). The step height of the resistance change may be different for different interface traps and depends on gate voltage (5).

Fixed Drain Current

For an interface trap localized in the middle of a rectangular gate and a continous channel under the whole gate area, an analytical expression for the step height can be derived at low gate voltages above threshold voltage.

Near source and drain of a MOSFET the drain current I_D includes a drift current and a diffusion current. But in the region of the inversion layer itself the drain current is a pure drift current:

$$I_D = \int_0^{x_c} e\, W\, \mu_n\, n \frac{dV}{dy}\, dx \,, \qquad [1]$$

where e is the elementary charge, x_c the thickness of the inversion layer, μ_n the mobility and n the electron density per cubic centimeter in the inversion layer.

The area under the gate is not homogenious. Near threshold voltage the inversion layer may disappear to some extent. Therefore even for rectangular gate areas the width W of the inversion layer is a function of y, where x,y,z are the coordinates perpendicular to the interface, from source to drain, and parallel to source and drain. Therefore in Eq.[1] W is an average value.

The distance between the metallurgical pn junctions of source and drain shall be the channel length L. The doping concentration of source and drain is much higher than the doping concentration in the channel region. Therefore the diffusion current near source and drain leads to a decrease in the length of the inversion layer. The difference between L and the effective length of the inversion layer L_{eff} is dependent on gate voltage. Within L_{eff} the transverse electric field is constant (6). The effective length of the inversion layer L_{eff} can be determined by the method of Hu et al.(7).

$$L_{eff}(V_G) = L - \Delta L(V_G). \qquad [2]$$

This dependence on gate voltage is important near threshold voltage and especially for LDD transistors.

The drain voltage V_D is assumed to be about 10 mV and small compared to the difference of gate voltage V_G minus threshold voltage V_{th}.

$$V_D \ll V_G - V_{th}. \qquad [3]$$

The diffusion regions near source and drain lead to a voltage drop of V_D. Therefore the effective drain voltage V_{Deff} applied to the inversion layer is smaller than the drain voltage between source and drain V_D. This effective drain voltage depends on gate voltage. It can be determined also by the method of Hu et al.(7).

Based on Eq.[3] it is assumed that the the field strength dV/dy is approximately constant in y and z direction. Near threshold it may depend on y and z, but analogously to W we take the average value of the field strength. The effective mobility is defined as usual. Then Eq.[1] leads to

$$I_D = e\, W\, \mu_{eff} \frac{V_{Deff}}{L_{eff}} \int_0^{x_c} n\, dx. \qquad [4]$$

The total dimensionless number of electrons in the inversion layer shall be N, which is obtained from

$$N = W\, L_{eff} \int_0^{x_c} n\, dx. \qquad [5]$$

Using Eq.[4] and [5] the effective drain voltage is

$$V_{Deff} = \frac{L_{eff}^2\, I_D}{e\, \mu_{eff}\, N} \qquad [6]$$

or

$$\frac{dV_{Deff}}{dN} = -\frac{L_{eff}^2\, I_D}{e\, \mu_{eff}\, N^2}, \qquad [7]$$

and

$$\Delta V_D = -\frac{L_{eff}^2\, I_D}{e\, \mu_{eff}\, N^2}\, \Delta N. \qquad [8]$$

Dividing Eq.[8] by [6] yields

$$\frac{\Delta V_D}{V_{Deff}} = -\frac{\Delta N}{N}. \qquad [9]$$

Using Eq.[4] and [5] it follows from Eq.[9]

$$\Delta V_D = -\frac{e\, \mu_{eff}}{I_D}\, \frac{V_{Deff}^2}{L_{eff}^2}\, \Delta N. \qquad [10]$$

If one trapped electron is emitted into the inversion layer and if this electron is completely added to the electrons in the inversion layer then $\Delta N = 1$. This is true, if the inversion layer is already formed under the negatively charged interface trap, and if the depletion zone under this inversion layer is already completed. It is not true, if the inversion layer is pinched off under the negatively charged interface trap, which occures near threshold voltage. In this case the charge of the emitted electron must first change the electrostatic potential in the depletion zone under the inversion layer. Then only part of the charge $\Delta N < 1$ is added to the charge of the electrons in the inversion layer.

The case $\Delta N < 1$ occures also, if the interface trap is situated not in the middle of the gate area, but is located at the edge near the bird's beak. Then the depletion zone under the bird's beak in the neighborhood of the interface trap must first be formed completely before that part of the charge $\Delta N < 1$ of the emitted electron is added to the charge in the inversion layer.

Fixed Drain Voltage

Eqs.[9] and [10] are valid if the drain current is kept constant and switching occurs only in drain voltage. However, if the drain voltage is kept constant, Eq.[6] leads to

$$I_D = \frac{e\, \mu_{eff}\, V_{Deff}}{L_{eff}^2}\, N \qquad [11]$$

and to

$$\Delta I_D = \frac{e\, \mu_{eff}\, V_{Deff}}{L_{eff}^2}\, \Delta N. \qquad [12]$$

Numerical Calculations

The two dimensional device simulator GALENE II (8) for arbitrary nonplanar structures has been used to investigate the influence of single electron emission from interface traps on the electron distribution at the surface region. For the simulation a cross section through an n-channel MOSFET perpendicular to the current flow in the channel has been chosen. The structure is shown in Fig.1. The channel width is

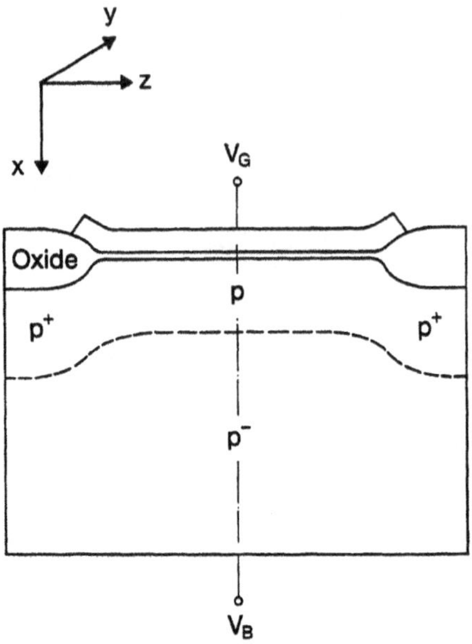

Fig.1. Schematic geometry of the cross section through a MOSFET used in the simulation. Note that from symmetry reasons, only one half of the device is used in the simulations

about 1 micron and the gate oxide thickness is 20 nm. The doping profiles are taken from SUPREM 3 simulations together with data from spreading resistance measurements. The threshold voltage of V_{th} = 0.35V has been determined from the electron density at the surface of the channel region. The charge of a single trapped electron has been assumed to be distributed over an area of less than $5x10^{-3}$ square microns, which corresponds to a charge density of $2x10^{10}$ cm^{-2}. The spatial distribution has been kept constant over a small polygon course along the interface. Because of symmetry of the problem with respect to the middle of the channel we restricted the simulated structure to one half of the cross section shown in Fig. 1. The grid has been refined near the charge distribution to eliminate the influence of the discretization on the effect. Because no current flow is taken into account, it is sufficient to solve Poisson's equation to obtain the electrostatic potential distribution and the carrier densities in the device. From the carrier densities the number of electrons and holes has been calculated by a post-processing integration for both conditions, with and without trapped electrons. This has been done for different locations of the trap in the interface to investigate the dependence relative to the bird's beak edge.

If the electrons in the inversion layer are distributed homogeneously along the channel and the one electron switching interface trap is situated in the middle of the gate area the change of the number of electrons by emission is $\Delta N = 1$. But the average value ΔN is smaller than one if the interface trap is situated near the edge of the gate at the bird's beak. The value of ΔN also strongly depends on the applied gate voltage. For voltages of $V_G - V_{th} \geq 0.5$ V it is $\Delta N = 1$. For this bias condition the gate field is 90% of its maximum value and the channel region is in strong inversion. For gate voltages $V_G - V_{th} < 0.5$ V the channel may be pinched off under the negatively charged trap, and ΔN will be less than 1. In this case the emitted electron must first form the depletion zone completely under the trap before an inversion layer can be formed. Fig. 2 shows the change of that part

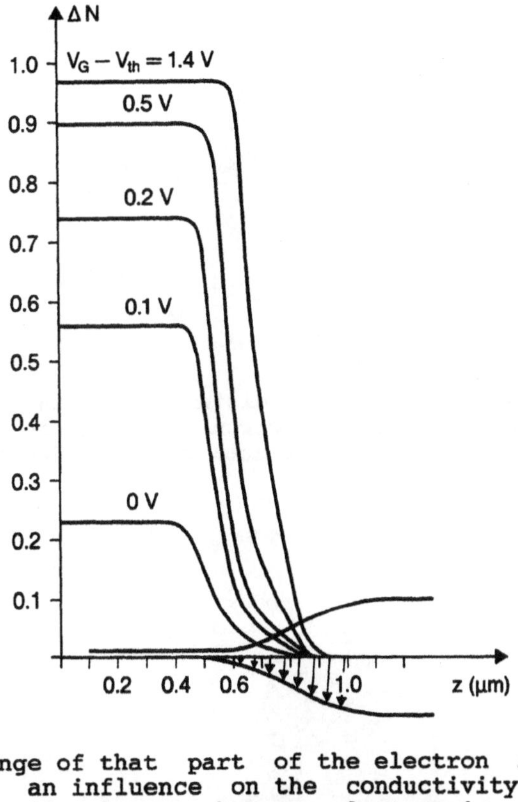

Fig.2. Change of that part of the electron ΔN which has an influence on the conductivity in the inversion layer when one electron is detrapped from an interface state as a function of the distance from the middle of the gate area to the bird's beak. The parameter is the bias $V_G - V_{th}$ at the gate contact

of the electron ΔN which has as an influence on the conductivity of the inversion layer as a function of the distance from the middle of the gate area to the bird's beak edge for several gate voltages. Here, the number of trapped electrons is 1 or 0 and ΔN is the difference in the number of electrons in the inversion layer calculated with and without charged trap. To complete the depletion layer under the inversion layer or near the bird's beak, the remaining fraction, 1 - ΔN, is necessary. The distance is measured from the middle of the gate area to the bird's beak edge. The simulations show a constant effect near the middle of the gate area and a decreasing influence towards the edge of the bird's beak. The decrease for larger distances reflects the effect of the bird's beak. As mentioned above, the height of the plateau near the middle of the gate area depends on the applied gate voltage. As long as the channel is pinched off

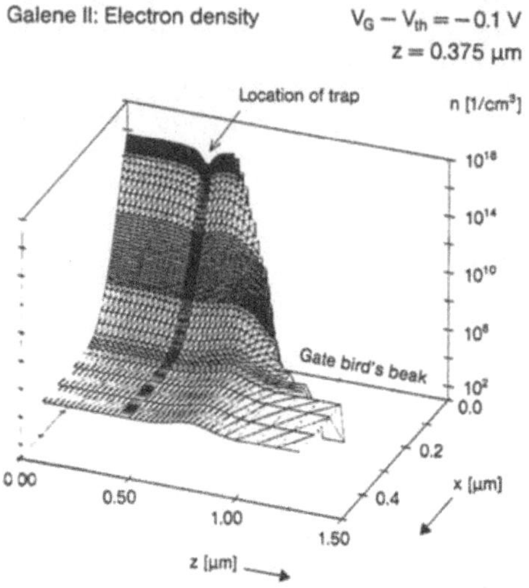

Fig.3. **Electron density distribution in the channel region for an electron trapped in the planar region of the gate area for $V_G - V_{th} = +0.1$ V**

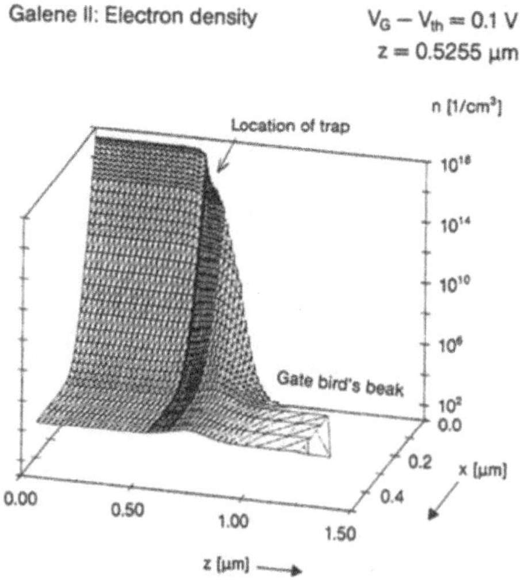

Fig.4. Electron density distribution in the channel region for an electron trapped near the bird's beak for $V_G - V_{th} = +0.1$ V

under the trapped electron an additional change of ΔN must be taken into account. To illustrate the influence of the charged trap on the electron distribution in the silicon region, the electron density has been plotted for several characteristic gate voltages at different locations of the trap. In Fig. 3 the electron distribution in the channel region is shown for an electron trapped in the planar region of the gate area for $V_G - V_{th} = 0.1$ V. One can see that the electron density along the surface is reduced around the location of the charged trap. By moving the trap towards the bird's beak, the bump in the density distribution becomes smaller, as shown in Fig. 4 for the same bias condition but at a distance of 0.5255 microns from the middle of the channel. If the bias $V_G - V_{th}$ becomes negative, the influence of the charged trap on the electron density is reduced by more than one magnitude, as shown in Fig 5. The maximum carrier density for this bias condition is about $5x10^{15}$ cm^{-3}, in contrast to the maximum of $1x10^{17}$ cm^{-3} for positive bias in Fig.3.

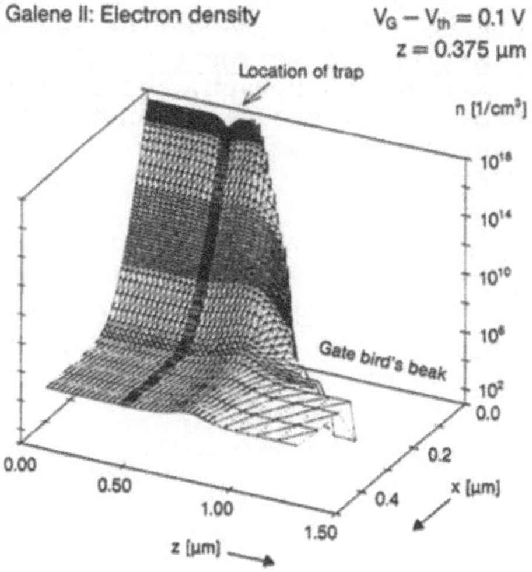

Fig.5. Electron density distribution as in Fig.3, but for negative $V_G - V_{th} = -0.1$ V

Conclusions

The comparision of the simulation results with experiments enables us to quantify the switching effects of single interface traps in MOSFETs depending on the location of the trap and the applied gate voltage. The interface traps can be localized under the gate area by measuring the step height. The equations are in good agreement with experimental data.

Acknowledgment

The authors would like to thank W.Bergner, A.Karwath, B.Moore, W.Neumüller, and M.Schulz for valuable discussions.

REFERENCES

1. K.S.Ralls, W.J.Skocpol, L.D.Jackel, R.E.Howard, L.A.Fetter, R.W.Epworth, and D.M.Tennant, Discrete resistance switching in submicrometer silicon inversion layers: Individual interface traps and low-frequency (1/f?) noise, Phys.Rev.Lett. 52:228 (1984).
2. M.J.Uren, D.J.Day, and M.J.Kirton, 1/f and random telegraph noise in silicon metal-oxide-semeconductor field-effect transistors, Appl.Phys.Lett. 47:1195 (1985).
3. M.Bollu and F.Koch, Quantum transport effect in small Si-MOS devices, in: "Proc.18th Int.Conf.Phys. Semiconductors," Stockholm, 1986.
4. A.Karwath and M.Schulz, Deep level transient spectroscopy on single isolated interface traps in field effect transistors, Appl.Phys.Lett. in press.
5. A.Karwath and M.Schulz, private communication.
6. C.G.Sodini, T.W.Ekstedt, and J.L.Moll, Charge accumulation and mobility in thin dielectric MOS transistors, Solid-State Electron. 25:833 (1982).
7. G.J.Hu, C.Chang, and Y.T.Chia, Gate-voltage-dependent effective channel length and series resistance of LDD MOSFET's, IEEE Trans.Electron Dev. ED-34:2469 (1987).
8. W.L.Engl, R.Kircher, K.H.Bach, and J.Götzlich, Simulations of 3D-devices with GALENE II, in: 1987 VLSI Proceess/Device modeling workshop, Tokyo.

LOW TEMPERATURE CONDUCTANCE OF Si-MOS DEVICES AFTER HOT CARRIER DEGRADATION

M. Bollu and F. Koch

Physik-Department
Technische Universität München
D-8046 Garching, Federal Republic of Germany

ABSTRACT

We consider the temperature dependence of hot carrier induced degradation effects in MOS devices. It is shown that the degradation is caused by interface states near drain with an energy dependent density strongly peaked at the conduction band edge. The degradation behaviour with increasing stress time is investigated up to a defect density of $6 \times 10^{12}/cm^2$. Below 10 K, the conductance gate voltage characteristic of a degraded device shows evidence for quantum effects in the form of a resonant tunneling process which causes a peak-to-valley conductivity change by many orders of magnitude.

1. INTRODUCTION

The operation of MOSFETs under conditions of high channel electric field is known to result in device degradation. The shift of threshold voltage and a reduced transconductance /1-5/ are a severe constraint for long term stability of modern VLSI circuits. There is general agreement that this degradation is due to some defects at, or close to the $Si-SiO_2$ interface caused by hot carriers in the high field region. These defects extend over a narrow strip of $\sim$ 0.1 µm in width just before the drain /4,5/.

In an earlier work /6/ we investigated whether these defects act as interface states which can exchange electrons or holes with the channel, or as trapped oxide charge. We analysed the subthreshold conductance of n-channel FETs at low drain bias (V_D = 1 mV) in the temperature range 40 - 300 K with the aid of a 2D device simulator. The "thermal scanning" of the Fermi level within the Si-band gap enables us to be sensitive to states even close to the conduction band edge. Figure 1a shows the cross-sectional view of the spatial configuration underlying our calculations. The defects are assumed to be Gaussian distributed with an extension of 0.1 µm near drain. In our case, $V_D < kT/e$, the current density can be determined by using Fick's law. From the calculated potential distribution (Fig. 1b), we obtained the channel current, and hence the conductance, by integration of the current density.

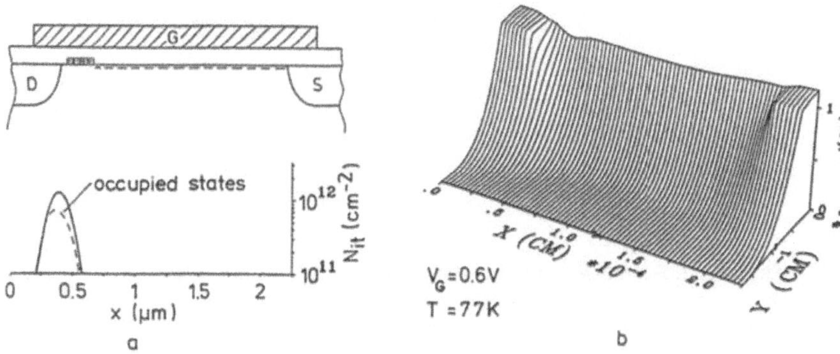

Fig. 1. a) Degraded device and the spatial distribution of defects underlying our calculations
b) Calculated 2D potential in a device assuming the configuration shown in Fig. 1a with an energetic distribution of interface states as given in Fig. 3

In the subthreshold regime the current is very sensitive to charges at the interface. Fig. 2 is taken from /7/ and shows the measured subthreshold conductance before (full lines) and after stress (dashed lines) at 300 K and 65 K. The post-stress conductance at 65 K is considerably more degraded than at 300 K. But the 65 K characteristic after stress is not simply shifted to more positive gate voltages which would more or less favour fixed oxide charges, but appears to be strongly bend down starting from a certain conductance level. We found that this behaviour can only be simulated (Δ in Fig. 2) by assuming acceptor-like interface states in the damaged strip with an energetic distribution as shown in Fig. 3. The steep increase of the density of states toward the band edge reflects the strange behaviour of the post-stress characteristic at 65 K. The total density,

$$N_{it} = \int_{E_m}^{E_c} D_{it}(E)\, dE,$$

for this case is $9.2 \times 10^{11}/cm^2$.

In Sec. 3 we investigate the hot carrier induced defects for various stress conditions as a function of stress time. At low temperature the damaged strip is such a bad conductor that the remainder of the channel can be regarded as a copper-wire connection to the external circuit. Resonant tunneling /8-10/ is going to be important. This will be discussed in Sec. 4.

2. EXPERIMENT

We used conventional Si n-channel FETs with poly-solicon gate. The gate oxide thickness is 42 nm, the channel width is 10 μm, and the gate length is 2.25 μm. Devices with arsenic source-drain implantations as well as with double implanted drains (DID) were used. DID-devices have suffered an additional phosphoric drain implantation with a dose of $2 \times 10^{14}/cm^2$. The effective channel lengths are 1.8 μm and 1.3 μm for the As-drain- and DID-devices, respectively. All test structures are passivated with 50 nm oxide covered by a 1 μm thick Si-nitride layer.

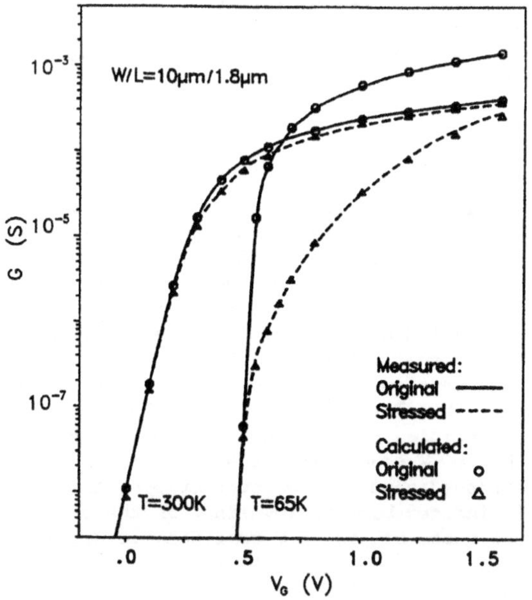

Fig. 2. Measured and calculated subthreshold characteristic of a device stressed at V_G = 3 V, V_D = 8 V, and T = 300 K for 30 min. For the calculation (Δ) we used an energetic distribution of acceptor-like interface states as given in Fig. 3.

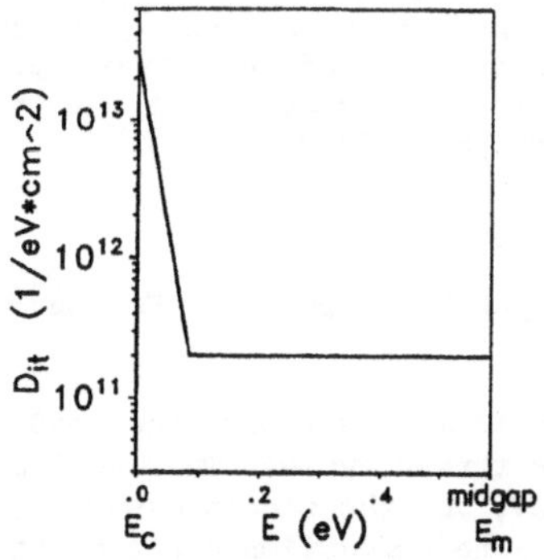

Fig. 3. Energetic distribution of interface states at the maximum of the Gaussian spatial distribution used for the calculation of the post-stress conductance in Fig. 2.

Source-drain electrical stress was carried out at 300 K and 77 K for times between 10 sec and 1000 h. Bias conditions during stress were chosen as V_G/V_D = 3V/8V and V_G/V_D = 8V/8V. Before and after each stress cycle the conductance has been measured with V_D = 1 mV at temperatures between 4.2 K and 300 K.

3. RESULTS

We used the method described in the introduction to investigate stress induced defects after different stress cycles. First we found that independent of the stress conditions (V_D = 8 V, V_G = 3V/8V, T = 300 K, T = 77 K after 300 K anneal) and stress time the defects are acceptor-like states. Even though for strong degradation a small amount ($\sim$ 10% of N_{it}) of oxide charge cannot be excluded. The reason is the following: For generated defect densities above 2 x $10^{12}/cm^2$ the low temperature ($\lesssim$ 100 K) conductance does not recover to the original value at the lowest experimentally accessible conductance level of 2 x 10^{-9} S. The resulting residual shift in V_G can, in principle, account for electrons trapped in the oxide. Although for the 300 K conductance there remains no residual shift between original and stressed conductance, some trapped oxide charge could be masked by drain induced barrier lowering. In the case of 77 K stress without 300 K anneal there can exist a considerable amount of negative oxide charge.

Up to a total density N_{it} of 2 x $10^{12}/cm^2$ the energetic distribution $D_{it}(E)$ can be determined with the aid of our 2D-simulation program. We found that the shape of the distribution is nearly independent of the way how the device has been stressed. There always appears the strong peaking $\sim$ 100 meV just before the band edge, as shown in Fig. 3. The distribution is only shifted in the log-scale corresponding to the total density N_{it}.

In the following we will concentrate on the total density N_{it} of hot carrier induced states and how it behaves with increasing stress time. At 4.2 K conduction starts when the Fermi level crosses the conduction band edge (apart from the groundstate energy of the lowest 2D subband). Then all states within the Si band gap are occupied by electrons, and hence negatively charged. One expects that the threshold voltage V_T for the onset of conduction is shifted by

$$\Delta V_T = eN_{it}/C_{ox} \tag{1}$$

C_{ox} being the oxide capacitance per cm^2. For our devices ΔV_T = 2V x $N_{it}/(10^{12}\ cm^{-2})$.

In Fig. 4 we show in a linear plot the conductance as measured at 4.2 K after successive stress cycles. This device has been stressed at 300 K with V_D = V_G = 8V, but the sequence shown in Fig. 4 is typical for all stress conditions. The conduction onset of the stressed devices is smeared out, so it is not quite clear how to fix the threshold voltage. For a better understanding, one has to realize that only the degraded, 0.1 µm wide strip near drain has suffered a threshold shift and acts as a blocking transistor in series with the undamaged remainder of the channel. This ultra-short-channel transistor is not able to hinder current flow totally before threshold is achieved. There are 1-dimensional leakage paths due to resonant tunneling, as will be discussed in the next section. Furthermore, localization due to the random potential at the interface can shift the onset of conduction to little higher gate voltages.

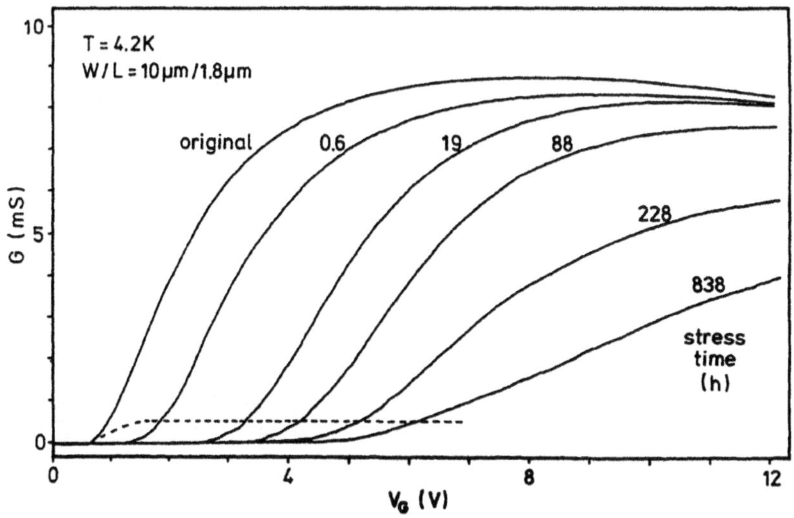

Fig. 4. 4.2 K conductance after successive stress cycles at $V_D = V_G = 8V$, $T = 300$ K. The dashed line marks the onset of metallic type of conduction.

We assume that the threshold which is related to N_{it} occurs at the transition between metallic and non-metallic type of conduction. At this care must be taken not to rise the temperature so much that a considerable amount of interface states will release their electrons and hence reduce the potential. We found metallic-type conduction whenever the conductance exceeds 0.5 mS for our 10 µm wide devices (dashed line in Fig. 4) independent of the degree of degradation. Although this level seems quite high in the usual understanding of a threshold voltage, 0.5 mS corresponds only to 1/5 of the minimum metallic conductivity scaled by 0.1 µm/10 µm. Despite some uncertainties in determining V_T, we expect ΔV_T to be proportional to N_{it} and that (1) holds approximately.

Figure 5 shows the threshold voltage shift ΔV_T at 4.2 K vs. stress time as obtained for DID- and arsenic drain-devices for various stress conditions. After the threshold has been shifted to ∿ 12 V ($N_{it} = 6 \times 10^{12}/cm^2$), the samples easily get leaky.

The data for 77 K stress has been measured immediately after stress. 300 K anneal of 77 K stressed devices results in a ∿ 30% relaxation of the threshold voltage shift and in a reduced degradation rate, $d(\Delta V_T)/dt$, during the following stress cycle performed under the same conditions. Figure 5 reveales that for smaller ΔV_T the degradation rate is higher. From this point of view the annealing behaviour is surprising. May be that defects, once annealed, are resistent to further stress.

Apparent from Fig. 5, the degradation process at 77 K is much faster than at room temperature, which is ascribed to the higher electron mobility at 77 K. The somewhat higher 77 K degradation rate at $V_G/V_D = 8V/8V$ compared to the rate at $V_G/V_D = 3V/8V$ has also been found by /11/.

For small stress times t, ΔV_T increases as t^n, with $0.4 < n < 0.8$, in agreement with results at 300 K /12/. The most interesting feature, however, appearing from Fig. 5, is that the degradation process is slowing down after approximately 8×10^{11} defects/cm^2 ($\hat{=} \Delta V_T = 1.6$ V) have been created. This behaviour is independent of the particular type of stress

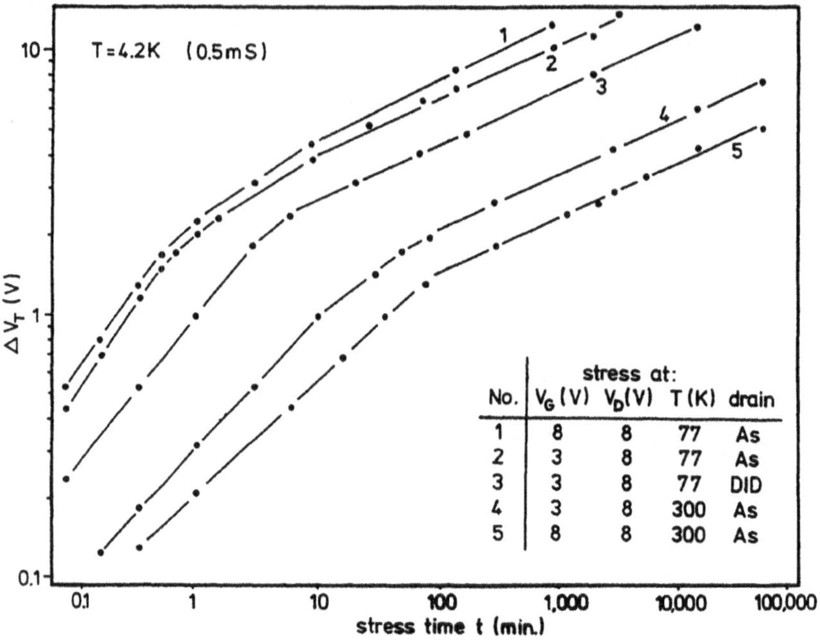

Fig. 5: Threshold voltage shift as measured at 4.2 K vs. stress time for various stress conditions. At 4.2 K, ΔV_T = 12 V corresponds to a threshold voltage shift of 1.5 V at 300 K.

or drain technology. We conclude that this change in sensitivity to stress at N_{it} = 8 x $10^{11}/cm^2$ reflects some property of the interface. May be there is some type of weak link with a density of 8 x $10^{11}/cm^2$ which is more sensitive to hot carrier stress than the remaining bonds at the interface or the microscopic degradation process changes at this critical density of defects.

Till now we have discussed generated interface states in the band gap. In principle, conductance measurements at 4.2 K should provide information about states within the conduction band of the silicon. But there appear some limitations for a quantitative analysis. The energetic density of defect states has to be nearly as large as the density of states of the 2D electron system (= 1.7 x 10^{14} eV^{-1} cm^{-2} for a (100)-Si surface) to be detectable. In addition, inhomogeneities and localisation effects in the damaged part of the channel complicate a theoretical description of current flow. Flicker noise, although it is strongly enhanced after stress /13,14/, provides no quantitative tool for determining the density of stress induced interface states.

4. 4.2 K SUBTHRESHOLD CONDUCTANCE - RESONANT TUNNELING

At low temperature the discrete nature of the interface charges becomes important. For a typical defect density of $10^{12}/cm^2$ the mean distance between trapped electrons is ∿ 10 nm, which has to be compared with the width of the damaged strip (∿ 100 nm).

The random distribution of charged states at the interface gives rise to potential fluctuations. As a result carriers are localized if only a small gate voltage is applied. Charge transport in this regime can be due

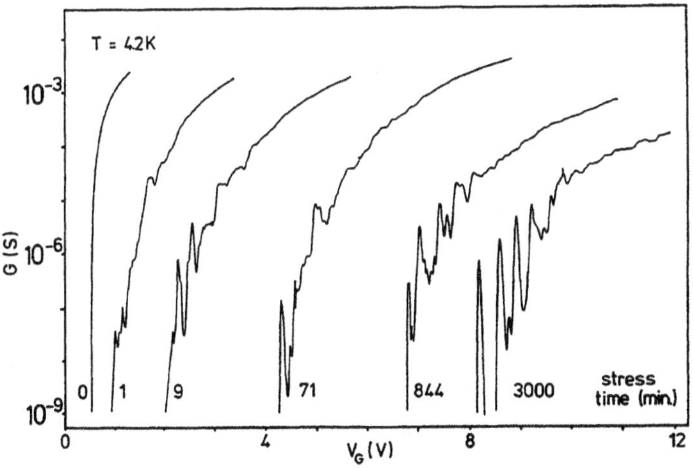

Fig. 6. 4.2 K conductance vs. V_G of a device stressed at V_G = 3 V, V_D = 8 V, T = 77 K. The device has not been annealed between subsequent stress cycles.

to thermal activation to the mobility edge /15/, hopping conduction /16/, or resonant tunneling /10/.

Figure 6 shows the 4.2 K subthreshold conductance as obtained after successive stress cycles. We note the distinct, randomly peaked structures that evolve with increasing stress. The peak-to-valley ratio of the strongest peaks is as much as 3 orders in magnitude.

It is the T-dependence that identifies the conductance peaks as the result of a resonant tunneling process, originally suggested by Azbel /8/ and modified by Stone and Lee /9/. Resonant tunneling can be understood in the following way. The damaged part of the channel where the carriers are localized acts as a potential barrier. Tunneling through the entire barrier (path 2 in Fig. 7) is very unlikely and is usually masked by hopping transport. However, if the Fermi energy coincides with the energy of some localized state in the barrier (path 1 in Fig.7) the transmission

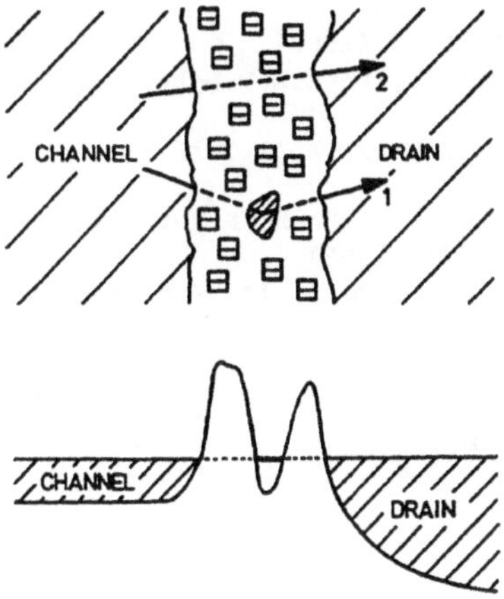

Fig. 7.
Tunneling paths across the damaged strip of a degraded device. Resonant tunneling via single localized states (path 1) is expected to dominate.

probability is strongly enhanced. Sweeping the gate voltage shifts the energy of the localized states with respect to the Fermi energy and gives rise to strong fluctuations in the conductance.

Resonant tunneling has already been observed by Fowler et al. /10/ who investigated 0.5 μm long unstressed MOSFETs in the 100 mK range. In our deliberately fabricated 0.1 μm long devices we observe resonant tunneling up to 20 K. The price we have to pay for our stress procedure are strong instabilities even after 200° C annealing. Peaks can shift or disappear just during measurements.

5. CONCLUSIONS

The temperature dependence of the subthreshold conductance after hot electron stress provides a sensitive tool for determining the energetic distribution of stress induced interface states. Independent of stress conditions the density of defect states strongly increases towards the conduction band edge of the silicon. After a defect density of $\sim 8 \times 10^{11}/cm^2$ has been created the degradation rate slows down in a characteristic manner. Because this behaviour does not depend on the type of stress and drain technology, it may indicate a particular property of the interface. In the low temperature subthreshold conductance of degraded FETs there appear peaked structures which we have attributed to resonant tunneling via localized states.

ACKNOWLEDGEMENTS

We thank A. Asenov for supplying the numerical data for Figs. 1-3 and W.Weber for stimulating discussions. Test structures and financial support have been obtained from the Siemens AG.

REFERENCES

1/ R.B.Fair, R.C.Sun, IEEE Trans. Electr. Devices ED-28(1), 83 (1981)
2/ S.L.von Bruns, R.L.Anderson, IEEE Trans. Electr. Devices ED-34(1), 75 (1987)
3/ W.Weber, C.Werner, G.Dorda, IEEE Electr.Dev.Lett. EDL-5, 518 (1984)
4/ A.Schwerin, W.Hänsch, W.Weber, IEEE Trans. Electr. Devices ED-34(12), 2493 (1987)
5/ T.Poorter, P.Zoestbergen, in IEDM Tech. Dig. 100 (1984)
6/ A.Asenov, M.Bollu, F.Koch, J.Scholz, Appl.Surf.Sci. 30, 319 (1987)
7/ M.Bollu, A.Asenov, F.Koch, to be published
8/ M.Ya. Azbel, Solid State Comm. 45, 527 (1983)
9/ A.D.Stone, P.A.Lee, Phys.Rev.Lett. 54, 1196 (1985)
10/ A.B.Fowler, G.L.Timp, J.J.Wainer, R.A.Webb, Phys.Rev.Lett. 57, 138 (1986)
11/ P.Heremans, Y.-C.Sun, G.Groeseneken, H.E.Maes, Appl.Surf.Sci. 30, 313 (1987)
12/ See for example C.Hu, S.C.Tam, F.Hsu, P.Ko, T.Chan, K.Terrill, IEEE Trans. Electr. Devices ED-32(2), 375 (1985)
13/ M.Bollu, A.J.Madenach, F.Koch, J.Scholz, H.Stoll, in: Proceedings of 9th Intern. Conference on Noise in Physical Systems, ed. C.M. van Vliet, p. 217, World Scientific, Singapore (1987)
14/ M.Bollu, F.Koch, A.Madenach, J.Scholz, Appl.Surf.Sci. 30, 142 (1987)
15/ T.Ando, A.B.Fowler, F.Stern, Rev.Mod.Phys. 54(2), 437 (1982)
16/ A.B.Fowler, A.Hartstein, R.A.Webb, Phys.Rev.Lett. 48, 196 (1982)

THE NEUTRAL LEVEL OF Si–SO$_2$ INTERFACE STATES

Sanjay Jain

AT&T Bell Laboratories
Allentown, PA 18103

From measurements of the temperature dependence of barrier height in $Cr - SiO_2 - n\text{–}Si$ structures, we have determined the oxide charge, the charge and density of $Si - SiO_2$ interface states, and the position of their neutral level. A negative oxide charge is observed which is attributed to acceptor-type states in the oxide populated by tunneling electrons. The extracted interface state density agrees with other techniques and the high temperature data support trap-assisted tunneling as the dominant recombination mechanism of holes in weak inversion. The neutral level is found 0.1 ± 0.07 eV above midgap of silicon. This is consistent with a distribution of donor states above and acceptor states below the midgap level.

1. INTRODUCTION

Little is known about the neutral level and absolute charge of interface states although they play important roles in determining the barrier height of Schottky diodes and threshold voltage of MOS field-effect transistors [1]. The existence of the neutral level is well-supported by theoretical calculations [2,3] of surface states spectra most of which predict bands of acceptor and donor states separated by a neutral level. In the metal-semiconductor system, Bardeen [4] has suggested that pinning of the fermi level to the neutral level may account, at high interface states densities, for the anomalous insensitivity of the Schottky barrier height to the metal-semiconductor work function difference. Applying Bardeen's model to a variety of metal-semiconductor contacts, Cowley and Sze [5], Turner and Rhoderick [6], and Ghosh and Chowdhary [7] have been able to estimate the neutral level and interface state density for silicon surfaces prepared by wet etching techniques. However, it is difficult to extract the neutral level with any precision from these data since work function and barrier height exhibit significant variation from one laboratory to another [8]. In metal-insulator-semiconductor structures, charged states exist both at the insulator-semiconductor interface and within the insulator bulk and it is necessary to separate the two charges to determine the interface states neutral level. Although a variety of techniques [9] can be applied to measure the energy spectrum of interface states, none can separate the absolute bulk and interface charges. Nevertheless, it is known [10] that both acceptor and donor type states exist at the $Si - SiO_2$ interface and, therefore, it is expected that a neutral level exists somewhere between the two types of states.

In this paper we show that the interface states charge and the oxide charge can be separated if two or more MOS devices with equal oxide charge but unequal interface state density are available. We study the temperature dependence of the capacitance and barrier height of $Cr - SiO_2 - n\text{–}Si$ tunnel diodes and vary oxide thickness to change interface state density and temperature to sweep the trap quasi-fermi level across the upper half of the silicon bandgap. From these data we determine the trap density in each device, the oxide charge, and, more importantly, the neutral level of $Si - SiO_2$ interface states. First we discuss interface trap statistics and MOS capacitance in section 2, then experimental details in section 3, and finally results in section 4.

After completion of this work [11], the author discovered that a similar method had been described earlier by Knoll et. al. [12,13] where the interface state density was varied by stress-induced damage in MOS structures with thick silicon dioxide films. Their results are discussed in section 4.

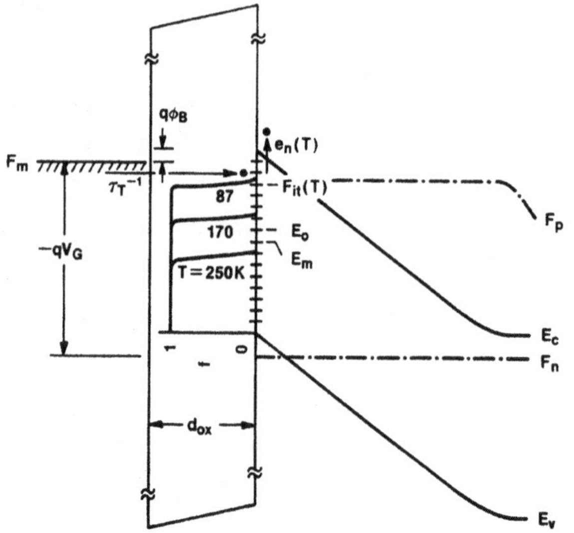

Figure 1. Energy band diagram of reverse biased $Cr - SiO_2 - n\text{–}Si$ tunnel diode with oxide thickness d_{ox}, showing the temperature dependent steady-state occupancy function f, fermi level F_{it} and the neutral level E_0 of interface states.

2. THEORY

Figure 1 shows the energy band diagram of a $Cr - SiO_2 - n\text{–}Si$ tunnel diode cooled to low temperature to freeze thermal emission of electrons and holes from $Si - SiO_2$ interface states and reverse biased to remove majority electrons from the $Si - SiO_2$ interface. At low temperature and oxide thickness less than 4nm, holes are recombined by tunneling faster than they are generated by thermal processes; as a result the surface density of holes is negligibly small. Since the difference $\phi_m - \chi_s = -0.06$ V [14] between the work function of chromium and the electron affinity of silicon is small, the barrier height ϕ_B is also small and interface states face a filled band in the metal.

The steady-state occupancy is derived by solving the Shockley-Read-Hall equation which has been modified to include tunneling transitions [15]. Inserting hole capture rate $\bar{p} \approx 0$ and $f_m \approx 1$ in eqs. (2,3) of [15] we obtain the steady-state occupancy,

$$f = 1/(1 + e_n\tau_T) \qquad E_m < E < F_m\,, \tag{1}$$

of interface states between midgap E_m and the metal fermi level F_m, in terms of the electron emission rate $e_n = \sigma_n v_{th} N_c e^{(E-E_c)/kT}$ into the silicon conduction band, and the electron tunneling rate $\tau_T = \tau_0 e^{2\eta_{ox}(d_{ox}+s)}$ into the metal conduction band. The tunneling distance $d_{ox}+s$ fluctuates with oxide thickness and interface state position, creating an approximately normal distribution $D(s)$ [16,17] of states in tunneling distance in addition to the well known energy distribution $D_{it}(E)$. The thermal capture crosssection σ_n [9] and the tunneling attempt-to-escape frequency τ_0^{-1} [17] are independent of energy near midgap but decrease rapidly near the conduction band edge; the product $\sigma_n\tau_T$ remains constant everywhere [18]. Thermal-emission controlled shallow levels with $e_n\tau_T > 1$ are empty while tunnneling controlled deep levels with $e_n\tau_T < 1$ are filled. Since the transition rates e_n and τ_T vary continuously over several orders of magnitude the occupancy can be approximated by

$$f = \begin{cases} 0 \\ 1 \end{cases} \qquad \text{for } E \; \substack{> \\ <} \; F_{it} - 2kT\eta_{ox}s\,, \tag{2}$$

where the quasi fermi level

$$F_{it} = E_c - kT\ln(N_c v_{th}\sigma_n\hat{\tau}_T) \tag{3}$$

of interface states is the energy at which thermal-emission and tunneling rates are equal, i.e. $e_n(F_{it})\hat{\tau}_T \equiv 1$, and $\hat{\tau}_T \equiv \tau_0 e^{2\eta_{ox}d_{ox}}$ is the tunneling time of states at the average tunneling distance d_{ox}. With rising temperature, tunneling controlled traps are discharged by thermal emission of electrons into the conduction band and the quasi-fermi level F_{it} falls, as shown in fig. 1.

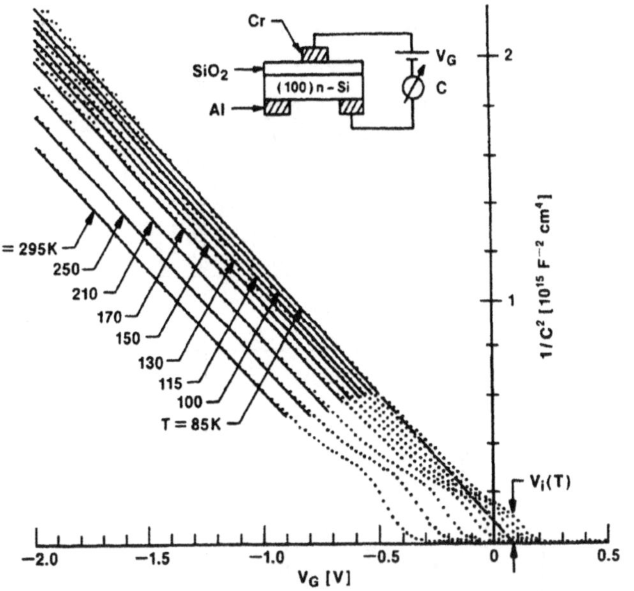

Figure 2. Measured $1/C^2 - V_G$ characteristics of MOS tunnel diode no. IB12 (see insert) as function of temperature T between 80 K and 300 K. The straight lines are least-square fits to the data with slope related to the dopant density N_D^+ and intercept voltage $V_i(T)$ decreasing as the interface state quasi-fermi level drops with rising temperature.

Integration of the trap occupancy f over trap energy and tunneling distance yields the interface states charge

$$Q_{it}/q \equiv \int_{-d_{ox}}^{\infty} dsD(s) \left[\int_{-\infty}^{E_0} dED_{it}(E) - \int_{-\infty}^{\infty} dED_{it}(E) f(E,s) \right] \tag{4}$$

which vanishes identically when all interface states upto the neutral level E_0 are filled with electrons. Substitution of eq. (2) for $f(E,s)$ reduces eq. (4) to

$$Q_{it}/q = \int_{F_{it}}^{E_0} dED_{it}(E) - \int_0^{\infty} dsD(s) \int_0^{2kT\eta_{ox}s} du [D_{it}(F_{it}+u) - D_{it}(F_{it}-u)] \,, \tag{5}$$

where the second term becomes negligible when D_{it} varies slowly on the scale of kT in energy. When D_{it} is approximately energy independent, as in our devices [17], the interface states charge becomes

$$Q_{it}/q = D_{it}(E_0 - F_{it}) \,. \tag{6}$$

According to eq. (6), the interface states charge is directly proportional to the fermi level F_{it} of states at the average tunneling distance d_{ox} and is independent of the magnitude of tunneling distance fluctuations. The measureable quantity is the total interface charge

$$Q_i/q \equiv Q'_{ox}/q + Q_{it}/q = Q'_{ox}/q + D_{it}(E_0 - F_{it}) \,, \tag{7}$$

which also depends on the unknown first moment Q'_{ox} of the oxide charge with respect to the $Si - SiO_2$ interface. While D_{it} is directly obtained as the slope, it is not possible to separate Q'_{ox} and E_0 from the $Q_i(F_{it})$ line of a single device. If, however, two or more devices with identical Q'_{ox} and E_0 but different D_{it} are available then, according to eq. (7), their $Q_i(F_{it})$ lines will intersect at the unique point defined by $F_{it} = E_0$ and $Q_i = Q'_{ox}$. For precision, this graphical method for determining E_0 could be replaced by either a multiple regression or the following two-step least-square fitting method. First, from the $Q_i(F_{it})$ line of each device we extract the slope D_{it} and the extrapolated charge Q_{ic} at $F_{it} = E_c$. Second, since

$$Q_{ic}/q = Q'_{ox}/q - D_{it}(E_c - E_0) \,, \tag{8}$$

from the line $Q_{ic}(D_{it})$ we extract the slope $-(E_c - E_0)$ and intercept Q'_{ox}.

TABLE 1. Device data [17]

Sample No.	T_{ox} [C]	d_{ox} [nm]	τ_T [sec]
IE12	875	3.97	$1.6\ 10^2$
ID12	875	3.64	$2.5\ 10^1$
IC12	875	3.25	1.0
IB12	875	3.11	$4.0\ 10^{-2}$
HA22	700	2.94	$7.1\ 10^{-3}$
HE12	700	2.51	$7.9\ 10^{-5}$

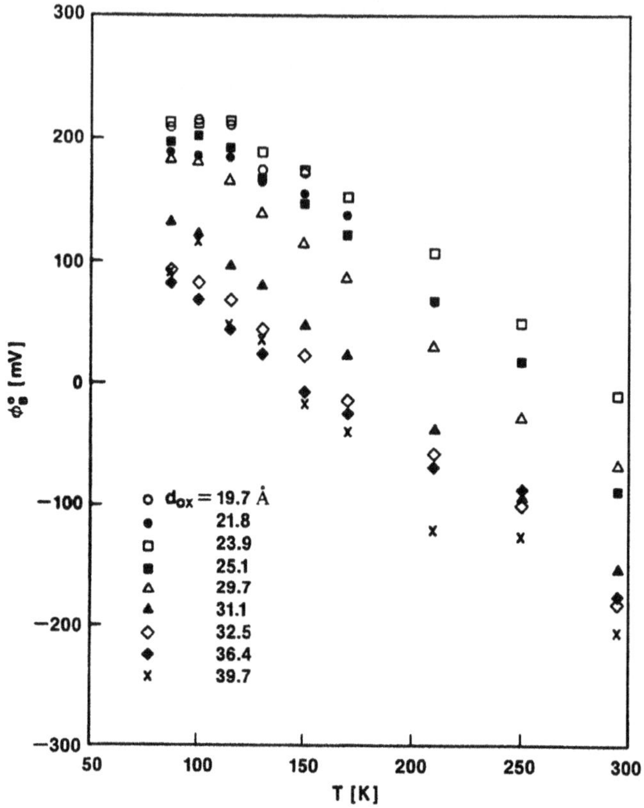

Figure 3. Measured barrier height ϕ_B^0 decreases linearly as function of temperature T due to thermal discharge of states which, at low temperature, are charged by filled tunneling.

The interface charge is determined from the small-signal high-frequency capacitance of the MOS tunnel diode. The interface state capacitance is completely negligible in our devices since the interface state fermi level F_{it} lies below F_m, the metal fermi level, and at the measuring frequency $\omega = 2\pi\times1$ MHz, $\omega\tau_T > 40$ for all our devices. This is different from the Schottky barrier diode where usually $F_{it} = F_m$ and $\omega\tau_T < 1$. From depletion layer theory the familiar linear $1/C^2 - V_G$ characteristic [8]

$$1/C^2 = 2(V_i - V_G)/\varepsilon_s q N_D^+ \,, \tag{9}$$

and the relation,

$$\phi_B^0 \equiv \phi_m - \chi_s - Q_i d_{ox}/\varepsilon_{ox} = V_i + \phi_n + kT/q - \varepsilon_s q N_D^+ d_{ox}^2 / 2\varepsilon_{ox}^2 \,, \tag{10}$$

between the flatband barrier height ϕ_B^0, the interface charge Q_i and the intercept voltage V_i are easily derived.

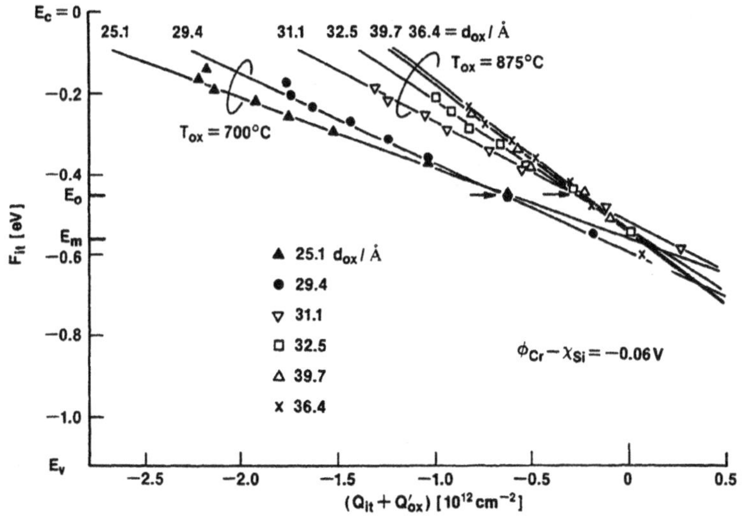

Figure 4. Interface charge $Q_i(T)$ as function of the interface state quasi-fermi level $F_{it}(T)$ for devices with oxides grown at $T_{ox} = 700\,^\circ$C and 875 °C. For each T_{ox}, all $Q_i(F_{it})$ lines intersect at a unique point whose coordinates give the oxide charge Q'_{ox} and the position of the neutral level E_0.

3. EXPERIMENT

We used the same set of samples described in [16,17]; important details of sample preparation are repeated here for completeness. (100) orientation float-zone single crystal silicon wafers were used to fabricate two sets of Cr – SiO_2 – n–Si capacitors with dry oxidation temperature $T_{ox} = 700\,^\circ$C and 875 °C respectively, and various oxidation times. After oxidation the oxide thickness d_{ox} was measured with a Rudolf Auto-EL^{II} ellipsometer and the sample was placed in an oil-free ion-pumped vaccuum system for evaporation of 99.999% pure chromium nuggets which were precleaned in dilute acid. The evaporation was done in several short cycles while maintaining the substrate temperature below 75 °C to prevent formation of "extrinsic" metal-impurity related interface states [19] by diffusion of metal through the oxide. The packaged devices were extensively characterized by our dark- and photo-capacitance transient spectroscopies and the sample parameters listed in table 1 of [16,17] were determined; we reproduce here the tunneling time $\hat{\tau}_T$, measured by photo-capacitance spectroscopy, in table 1, and the density of states D_{it} in fig. 5.

The high frequency capacitance-voltage characteristics were measured with a Boonton 72BD meter at nine temperatures between 80 K and 300 K. At each temperature, before the voltage was swept, the device was initially held in deep-depletion with gate voltage $V_G = -2$ V for a period longer than the tunneling time $\hat{\tau}_T$ (cf. table 1) to establish the steady-state interface state occupancy [17]. To minimize possible long-term variations in oxidation conditions, only samples fabricated on the same day were used in this study. Samples with oxide thickness outside the 2 to 4 nm range could not be used because of measurement temperature and time limitations.

4. RESULTS AND DISCUSSION

Figure 2 shows the measured $1/C^2 - V_G$ characteristics of device no. IB12 as function of temperature T. The small deviations from linearity observed at large reverse bias are believed to be due to unbalanced stray capacitance and dopant density fluctuations which are known to occur in float-zone silicon [20]. Straight lines were fitted within a fixed capacitance range to minimize these errors and the extracted dopant density $N_D^+(T)$ and intercept $V_i(T)$ inserted into eq. (15) to obtain the barrier height ϕ_B^0 shown in fig. 3 as function of temperature.

With decreasing temperature, the barrier height increases as the interface state quasi-fermi level F_{it} (cf. eq. (3)) rises and the interface charge shifts in the negative direction. With decreasing oxide thickness, the increase in the barrier height due to rising F_{it} is enhanced by an increase in the state density D_{it} (cf. fig. 5). For the thinnest oxides, ϕ_B^0 reaches a plateau near liquid nitrogen temperature because F_{it} becomes pinned as in a Schottky barrier, when it reaches the metal fermi level, and eq. (3) no longer holds. For the

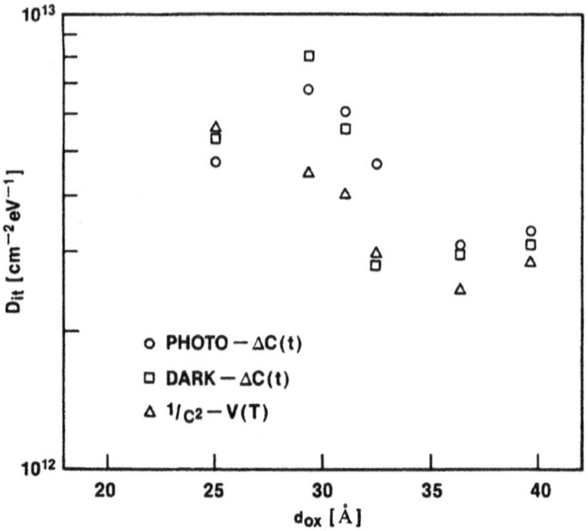

Figure 5. Density of interface states at midgap obtained by various methods.

thickest oxides, ϕ_B^0 continues to fall at a constant rate even near room temperature where F_{it} is below midgap E_m and the hole capture rate is nonnegligible because of inversion layer build-up. This confirms the inversion layer theory of Dahlke and Shimer [15] (their eq. (26) reduces to our eq. (3)), according to which eq. (3) for the quasi-fermi level remains valid at high temperature when hole capture processes dominate trap kinetics. An important implication is that in weak-inversion trap-assisted tunneling, and not band-to-band tunneling which was neglected in the theory [15], is the dominant hole recombination process; this result was also obtained by Dahlke and Jain [16] from transient spectroscopy.

Figure 4 displays the interface charge $Q_i(T)$, calculated from $\phi_B^0(T)$ using $\phi_m - \chi_s = -0.06$ V, as function of fermi level $F_{it}(T)$ computed from eq. (3) using $\sigma_n = 7\times10^{-14}$ cm^{-2} [17] and $\hat{\tau}_T$ data in table 1. For all samples linear $Q_i(F_{it})$ curves are observed whose slope, as shown in fig. 5, is in satisfactory agreement with the results of dark- and photo-capacitance spectroscopies. The weak energy dependence of D_{it} was also found in dark-capacitance spectroscopy [17] on the same samples. The common intersection point observed for the four devices with $T_{ox} = 875\,^\circ$C, occurs, as explained in section 2, at $F_{it} = E_0$ where $Q_{it} = 0$ and $Q_i = Q'_{ox}$. A negative oxide charge is observed whose density increases from -2.6×10^{11} cm^{-2} at $T_{ox} = 875\,^\circ$C to about -6.1×10^{11} cm^{-2} at $T_{ox} = 700\,^\circ$C. This charge is attributed to acceptor-type states in the oxide populated by electron tunneling from the metal, a mechanism which is absent in thick oxide MOS structures where a positive charge is usually observed [21].

Figure 6 shows the charge Q_{ic}, obtained by extrapolation to $F_{it} = E_c$, as function of interface state density D_{it}. From the slope of the $Q_{ic}(D_{it})$ lines we obtain $E_0 = E_c - 0.455 \pm 0.04$ eV at 875 °C and $E_c - 0.444$ eV at 700 °C. Systematic errors due to uncertainty in $\phi_m - \chi_s$, σ_n and temperature dependence of $\phi_m - \chi_s$ and $E_c - E_0$ are estimated to be of order 0.05 eV. Including the 2σ statistical error of 0.04 eV, we obtain $E_0 = E_c - 0.46 \pm 0.07$ eV for the position of the neutral level. This value is in good agreement with the data of Knoll et. al. [12,13] on thick oxide SiO_2 – (111)Si capacitors. Injecting electrons into the oxide to generate interface states, they observed a common intersection point in the high-frequency capacitance-voltage curves; from the capacitance at the intersection they found that the neutral level is exactly at midgap.

5. CONCLUSION

From the temperature dependence of barrier height in MOS tunnel diodes, we have determined the oxide charge, the density of $Si - SiO_2$ interface states, and the position of their neutral level. The observation of net negative oxide charge is attributed to acceptor-type states in the oxide which are filled by electron tunneling from the metal. The extracted interface state density agrees with the results of capacitance spectroscopies. The barrier height trend in the weak inversion regime implies trap-assisted tunneling as the dominant mechanism for recombination of holes. Considering both statistical and systematic errors, the neutral level is found 0.1 ± 0.07 eV above the midgap of silicon. This is consistent with a distribution of donor states above and acceptor states below the midgap level.

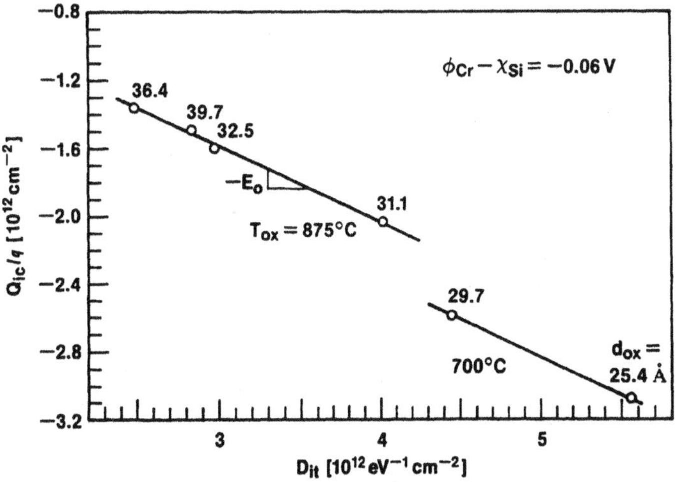

Figure 6. Extrapolated charge Q_{ic} as function of density D_{it} of interface states. The slope of the line $Q_{ic}(D_{it})$ is equal to the neutral level $E_0 - E_c$ and its intercept is equal to the oxide charge Q'_{ox}.

Acknowledgements: Helpful discussions with W. E. Dahlke are gratefully acknowledged.

REFERENCES

[1] S. M. Sze, *Physics of Semiconductor Devices* (Wiley-Interscience, New York, (1981).
[2] S. G. Davison and J. D. Levine, in *Solid Sate Physics* vol. 25, edited by H. Ehrenreich, F. Seitz, and D. Turnbull (Academic, New York, 1970).
[3] T. Sugano, in *Insulating Films on Semiconductors - Proc. of INFOS 83*, edited by J. F. Verweij and D. R. Wolters (Elsevier Science, Amsterdam, 1983).
[4] J. Bardeen, *Phys. Rev.* **B 71**, 717 (1947).
[5] A. M. Cowley and S. M. Sze, *J. Appl. Phys.* **36**, 3212 (1965).
[6] M. J. Turner and E. H. Rhoderick, *Solid-St. Electron.* **16**, 513 (1973).
[7] K. Ghosh and N. K. D. Chowdhary, *Int. J. Electron.* **54**, 615 (1983).
[8] E. H. Rhoderick, *Metal-Semiconductor Contacts* (Clarendon Press, Oxford, 1978).
[9] A. Goetzberger, E. Klausmann and M. Schulz, *CRC Crit. Rev. Solid St. Sci.* **6**, 1.
[10] E. H. Poindexter, P. J. Caplan, N. M. Johnson, D. K. Biegelson, M. D. Moyer and S. T. Chang, in *Insulating Films on Semiconductors - Proc. of INFOS 83*, edited by J. F. Verweij and D. R. Wolters (Elsevier Science, Amsterdam, 1983).
[11] S. Jain, Electrochem Soc. Mtg. Extended Abstract No. 250, Atlanta, May 1988.
[12] M. Knoll, D. Braünig and W. R. Fahrner, in *Insulating Films on Semiconductors - Proc. of INFOS 83*, edited by J. F. Verweij and D. R. Wolters (Elsevier Science, Amsterdam, 1983).
[13] M. Knoll, D. Braünig and W. R. Fahrner, *IEEE Trans. Nuc. Sci.* NS-29, 1471 (1982).
[14] S. Kar, *Solid-St. Electron.* **18**, 169 (1975).
[15] W. E. Dahlke and J. A. Shimer, *Solid-St. Electron.* **26**, 1129 (1983).
[16] W. E. Dahlke and S. Jain, *J. Appl. Phys.* **59**, 1264 (1986).
[17] S. Jain and W. E. Dahlke, *Solid-St. Electron.* **29**, 597 (1986).
[18] W. E. Dahlke and S. Jain, (this book).
[19] S. Kar and W. E. Dahlke, *Solid-St. Electron.* **15**, 221 (1972).
[20] S. K. Ghandhi, *Theory and Practice of Microelectronics* (Wiley, New York, 1968).
[21] B. E. Deal, *J. Electrochem. Soc.* **114**, 266 (1967).

MOS TUNNELING RATE AND INTERFACE STATE CAPTURE CROSS-SECTION

Walter E. Dahlke
Lehigh University
Bethlehem, PA 18015

Sanjay Jain
AT&T Bell Laboratories
Allentown, PA 18103

ABSTRACT

Considering electron tunneling as a recombination process, we calculate capture cross-section, σ_n, and transition rate, $\tau_T^{-1} = \tau_o^{-1} \exp(-2\eta_o d)$, for electron tunneling from an interface trap through the oxide ($d \approx 30$ Å) to the metal by employing Bardeen's matrix element on a simple one-dimensional model of the MOS structure. Calculated magnitudes and dependences on trap energy of dynamic parameters, τ_o^{-1}, σ_n, η_o agree reasonably well with measured data. Three-dimensional effects such as band tails are discussed qualitatively.

Introduction

Despite thirty years of intensive research on the Si-SiO_2 interface, important dynamical parameters of the interface traps remain unexplained or at best inadequately understood. They include the thermal capture cross-sections $\sigma_{n,p}$ and, in thin metal-oxide-semiconductor (MOS) structures, the tunneling transition rate,

$$\tau_T^{-1} = \tau_o^{-1} \exp(-2\eta_o d), \qquad (1)$$

with the tunneling attempt-to-escape frequency τ_o^{-1}, the tunneling attenuation $2\eta_o$ in the oxide of $d = 20$ Å to 40 Å thickness. The parameters $\sigma_{n,p}$ and τ_T^{-1} control the carrier exchange of the interface traps with the silicon conduction and valence bands and the metal conduction band, respectively.

Goetzberger et al. [1] and Klausman et al. [2] have experimentally determined the magnitude and the energy dependence of the cross-section $\sigma_n(E)$ using the conductance and the CC-DLTS techniques. They found the cross-section to be practically constant near silicon midgap and decreasing sharply toward the silicon conduction band edge. An orientation and processing dependence of $\sigma_n(E)$ was also observed. Several models were suggested to explain this effect. Tunneling to interface states spread out into SiO_2 [3] can be excluded since the measured conduction signal varies symmetrically with frequency. A model by Goetzberger et. al [4] of electrons bound to positive centers in SiO_2 near the interface seems also unlikely, since it should result in an increase of the cross-section toward the band edge and not a decrease. Schulz [5] suggested tunneling of electrons from the Si-SiO_2 interface to a discrete defect in SiO_2 less than 8 Å away from the interface. The capture probability of the tilted trap distribution, shown in Fig. 1 as dotted line, decreases toward the conduction band edge since the tunneling depth measured from the interface increases. This tilted distribution, however, is incompatible with our recently measured tunneling rate τ_T^{-1} from the metal into the interface obtained from dark- and photo-capacitance transient spectroscopy [6]. We found that both the tunneling attempt-to-escape frequency $\tau_o^{-1}(E)$ and the thermal electron capture cross-section $\sigma_n(E)$ exhibit a sharp decline toward the Si-SiO_2 interface.

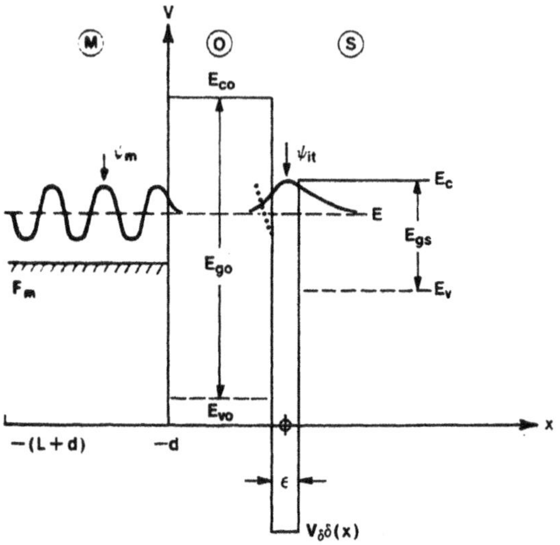

Figure 1. Energy band diagram of metal-oxide-semiconductor (MOS) structure. Interface states are modeled by the delta-function potential well $V_\delta \delta(x)$ located at the Si-SiO_2 interface x=0. ψ_{it} and ψ_m are the electron wavefunctions of the interface trap and the metal conduction band state at energy E.

In this paper we present a qualitative explanation of those observations based on a simple one-dimensional quantum mechanical model of the Si-SiO_2 interface. Independent of the detailed trap structure, the band structure of silicon causes decaying trap wave functions near midgap, which approach propagating waves at the silicon band edges. This in turn results in the observed rapid decrease of the cross-sections $\sigma_{n,p}$ and the attempt-to-escape frequency τ_o^{-1} near the silicon band edges.

Interface Model

Figure 1 shows schematically the conventional energy band diagram of an MOS structure where a delta-function potential energy well $V_\delta \delta(x)$ represents an interface state at the Si-SiO_2 interface. An electron of total energy E trapped in the well at x=0 is characterized by a wavefunction decaying slowly in the silicon forbidden band and faster in the oxide. Considering elastic tunneling from right to left into empty states of the metal conduction band, the electron wave function propagates in the metal and decays in the oxide. The transition probability of the electron is given by Fermi's golden rule,

$$\tau_T^{-1} = \frac{2\pi}{\hbar} \rho_m |M|^2 , \qquad (2)$$

with Planck's constant $2\pi\hbar$ and the density of states in the metal,

$$\rho_m = \frac{L}{\pi} \frac{m_m}{\hbar^2 k_m} . \qquad (3)$$

The metal has the thickness L, the effective mass m_m, and the wave number k_m. Bardeen's matrix element [7] is

$$M = \int \psi_{it}^* V_\delta \psi_m \delta(x) dx = \psi_{it}^*(o) V_\delta \psi_m(o) \qquad (4)$$

The trap wavefunction [8],

$$\psi_{it} = \left(\frac{2}{\eta_o^{-1} + \eta_s^{-1}}\right)^{1/2} x \begin{Bmatrix} \exp(-\eta_s x) & & x>0 \\ & \text{for} & \\ \exp(\eta_o x) & & x<0 \end{Bmatrix} , \qquad (5)$$

decays into the oxide with the decay coefficient η_o and into the silicon with the coefficient η_s. The magnitudes depend according to a Franz two-hand model on the corresponding bandgaps and effective masses, as described in Appendix 1. We note $\eta_o \geq 3\eta_s$, because of the different bandgaps in oxide and silicon, and further that the trap function $\psi_{it} \to 0$ at both the silicon band edges.

The metal wave function

$$\psi_m(x) = \frac{1}{L^{1/2}} \left(\frac{2k_m}{\eta_o + k_m}\right)^{1/2} e^{-\eta_o (x+L)} , \qquad (6)$$

decays in the oxide with the coefficient η_o.

Finally, the strength of the delta-function,

$$V_\delta = -(\eta_s + \eta_o) \frac{\hbar^2}{2m_{it}} , \qquad (7)$$

is obtained by integrating the Schroedinger equation for the trap from $-\epsilon/2$ to $\epsilon/2$ and $\epsilon \to 0$.

Tunneling Transition Rate and Capture Cross Section

Substitution of eqs. (3) to (7) in eq. (2) results in the final expression eq. (1) for the tunneling transition rate. The prefactor tunneling attempt-to-escape frequency,

$$\tau_o^{-1} = \frac{\hbar k_m}{m_m^*} \cdot \left(\frac{m_m}{m_{Si}}\right)^2 \cdot \frac{1+\eta_s/\eta_o}{1+\eta_s^2/\eta_o^2} \cdot 2\eta_s \quad , \tag{8}$$

is related to the normalized amplitudes of the wave functions eqs. (5),(6) and the densities of states in the metal, especially the Fermi velocity of electrons in the metal, $v_F = \hbar k_m/m_m^*$, and the ratio m_m^*/m_{it}^* of the effective masses in metal and trap. The energy dependence of τ_o^{-1} comes primarily from the attenuation coefficient η_s in the silicon band gap, which disappears at the silicon valence and conduction band edges, where the interface state wave function becomes a propagating wave and the de Broglie wavelength infinity. The other terms in eq. (8) are practically constant in the silicon bandgap.

Considering tunneling of metal electrons through the oxide into interface states as a recombination process [8], we obtain the recombination or tunneling rate,

$$\tau_o^{-1} = v_F \, \sigma \, n \quad , \tag{9}$$

with the tunneling capture cross-section σ and the electron density in the metal n_m. With τ_o^{-1} from eq. (8), we find from eq. (9) the tunneling cross-section,

$$\sigma = \left(\frac{m_m}{m_{it}}\right)^2 \frac{1+\eta_s/\eta_o}{1+\eta_s^2/\eta_o^2} \; \frac{2}{n_m} \, \eta_s \quad . \tag{10}$$

It disappears at the silicon band edges exhibiting the same energy dependence as the attempt-to-escape frequency eq. (8).

Comparison of Theory and Measurements

Substituting in eq. (8), (10) the numerical data shown in Table 1, we obtain the following theoretical results near silicon midgap: $\tau_o^{-1} = 8\text{x}10^{16}$ Hz, $\sigma_n = 4.6\text{x}10^{-15}$ cm^2 for trapped electrons, and $\tau_o^{-1} = 2\text{x}10^{16}$ Hz, $\sigma_p = 1.2\text{x}10^{-15}$ cm^{-2} for trapped holes. Calculated and measured data as function of trap energy are plotted in Fig. 2. We note the accuracy of measured data is about one order of magnitude, while processing may change data by several orders [9]. We recognize that the data for the measured σ_n in curve (1) and those for τ_o^{-1} in curve (2) are nearly proportional to each other, as predicted by the theoretical eq. (9). We further see that the magnitudes of the measured curves (1) and (2) differ from the calculated curve (3) for σ_n and τ_o^{-1} by about one order of magnitude, which is within error limits of calculated and measured data. Not plotted are data of τ_o^{-1} and σ_p for holes which would lie a factor of 4 below curve (3). Our model predicts similar decreases of the hole tunneling rate and the hole capture cross-section toward

Table 1

Data used to calculate τ_o^{-1} and σ_n in eqs. (8) and (10)

	E_c (eV)	m_n^* (m_o)	m_p^* (m_o)	η $(Å^{-1})$
Cr		2.3		
Si	1.1	0.26	0.26	0.18
SiO_2	9.0	0.49	0.49	0.52
Trap		0.5	1.0	

$n_{Cr} = 1.7 \times 10^{23} cm^{-3}$, $k_{Cr} = 1.7$ Å^{-1}

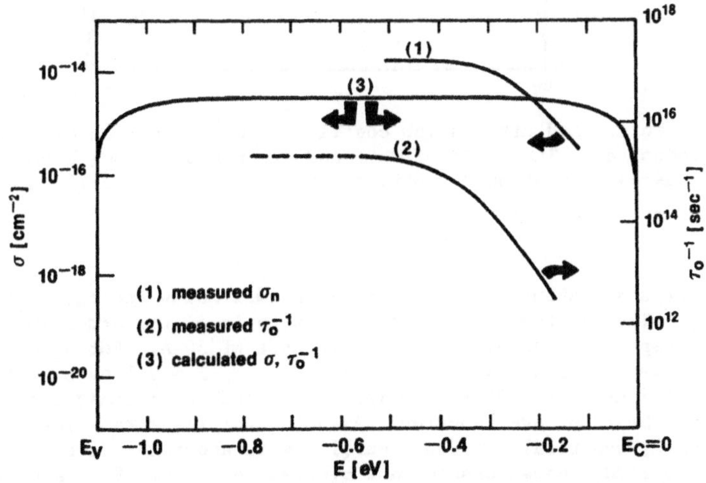

Fig. 2. Interface state capture cross-section σ and tunneling attempt-to-escape frequency τ_o^{-1} as function of energy in the forbidden gap of silicon. Measured data for the electron capture cross-section σ_n [1] are shown by curve (1) and for the tunneling rate τ_o^{-1} by curve (2), [6]. Curve (3) shows σ and τ_o^{-1} calculated from eqs. (10) and (8).

the valence band edge, but reliable measurements are not yet available. Comparing the energy dependence of measured and calculated curves (1), (2), (3), we observe that the decline of the measured curves (1) and (2) occurs at lower energy and is more gradual than the calculated curve (3).

This qualitative difference is probably caused by the simplistic one-dimensional model in Fig. 1, which we used since realistic three-dimensional trap wave functions were not available. We suspect the discrepancy is a consequence of silicon band tails at the interface, which have been pushed into the gap by the abrupt discontinuity in the periodic potential at the interface. A qualitative consideration confirms this. For example, states about 0.1 eV below the conduction band edge in Fig. 1, whose

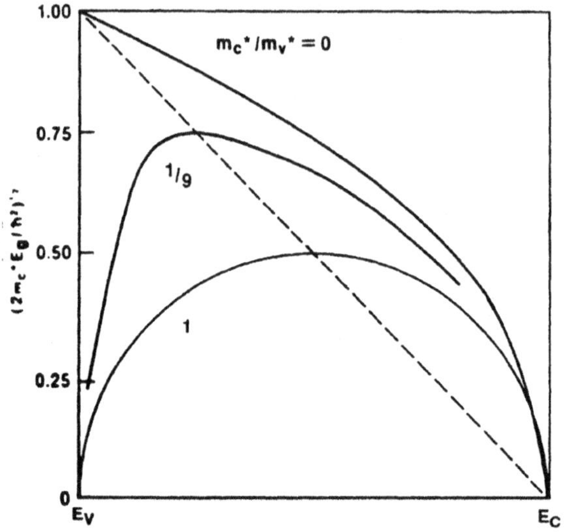

Figure 3. Normalized attenuation coefficient η as function of energy in the forbidden gap. The ratio of the conduction and valence band effective masses is assumed to be about unity for both silicon and silicon dioxide.

measured data are about two orders of magnitude below the calculated curve (3). The related measured trap density is of the order 10^{+13} $eV^{-1}cm^{-2}$, which corresponds to an average trap distance of 30 Å. For $\eta_s = 0.18$ Å^{-1} at midgap, we read from Table 1 and Fig. 3 at E_V-E = 0.1 eV an attenuation coefficient $\eta_s = 0.06$ Å^{-1}, which corresponds to a decay length of η_s^{-1} ≈ 15 Å for the trap wave function. Assuming for the three-dimensional trap a nearly spherical wavefunction [10] with about the same decay length of 15 Å, we find a considerable wave overlap for traps of 30 Å distance. This overlap according to Fig. 3 increases rapidly toward the conduction band edge; it causes a melting of the wave functions into each other and the propagating waves in the conduction band. The effect is a smooth decrease of the band gap and an increase of the effective density of states N_c of the conduction band. Therefore, the band-to-band tunneling current of metal electrons into the conduction band gap grows, while the trap-to-band tunneling current decreases. This current is proportional to τ_o^{-1} and σ, both of which drop at lower energy than for the linear model Fig. 1.

Conclusions

The observed similarity of the energy dependence of the thermal cross-section σ_n and the attempt-to-escape frequency τ_o^{-1} of an interface trap can be quantitatively explained by considering tunneling into traps as a recombination process, eq. (9). Correct estimates of σ_n, τ_o^{-1} near silicon midgap can be obtained from the model Fig. 1 and eqs. (10) and (8). The model predicts a very rapid decrease of τ_o^{-1} and σ_n toward the conduction band edges, where the de Broglie wavelength becomes infinite. The observed more gradual decrease of τ_o^{-1} and σ_n can be explained by a three-dimensional trap model at the interface. Band tails of the allowed silicon bands extend into the forbidden gap, which cause a smooth gradual decrease of the silicon bandgap at the interface and an earlier drop of τ_o^{-1} and σ toward the band edge than predicted by the linear model Fig. 1.

Appendix 1

The Franz [11] two-band model of the energy-momentum relation in the forbidden bandgap,

$$\frac{1}{\eta^2} = \frac{1}{\eta_c^2} + \frac{1}{\eta_v^2} = \frac{\hbar^2}{2}\left[\frac{1}{m_o^*(E_c-E)} + \frac{1}{m_v^*(E-E_v)}\right]^{-1} , \qquad \text{(A1)}$$

represents an interpolation between two known values at the conduction and valence band edges, which emerge from Eq. (A1) for $E \to E_c$ or $E \to E_v$. With

$$(E_c-E)/E_G = Z \text{ and } \beta = m_c^*/m_v^*$$

eq. (A1) assumes the form

$$\eta = \left[\frac{E_G\, 2m_c}{\hbar^2}\, \frac{Z(1-Z)}{1-Z(1-\beta)}\right]^{1/2} , \qquad \text{(A2)}$$

which obtains the maximum value

$$\eta_{max} = \left[\frac{E_G\, 2m_c}{\hbar^2}\right] Z_{max} \qquad \text{(A3)}$$

for

$$Z_{max} = \frac{1}{1+\sqrt{\beta}} . \qquad \text{(A4)}$$

The normalized attenuation coefficient, $\{Z(1-Z)/[1-Z(1-\beta)]\}^{1/2}$, is plotted in Fig. 3 as function of trap energy E for the parameter values $\beta = 0$, 1/9, 1.

References

[1] A. Goetzberger, E. Klausman and M. Schulz, CRC Crit. Rev. Solid St. Sci. 6, 1, (1976).

[2] E. Klausman, Insulating Films on Semiconductors 1979 (Inst. of Phys. Conf. Ser. 50, p. 97).

[3] M. Preier, Appl. Phys. Lett. 10, 362 (1967).
[4] A. Goetzberger, V. Heine and E. H. Nicollian, Appl. Phys. Lett. 12, 95 (1968).
[5] M. Schulz, Insulating Films on Semiconductors 1979 (Inst. of Phys. Conf. Ser. 50, p. 87).
[6] S. Jain and W. E. Dahlke, Solid-St. Electron 29, 597 (1986).
[7] J. Bardeen, Phys. Rev. Lett. 6, 57 (1961).
[8] L. B. Freeman and W. E. Dahlke, Solid-St. Electron. 13, 1483 (1970).
[9] E. H. Nicollian and J. R. Brews, MOS Physics and Technology, Wiley & Sons (1982).
[10] I. Lundström and C. Svensson, J. Appl. Phys. 43, 5045 (1972).
[11] W. Franz, Handb. d. Phys., S. Flügge, ed., Vol. 17, Springer, Berlin (1956).

DEFECT TRANSFORMATION PROCESS AT SiO_2/Si INTERFACES

T.P. Ma, Yasushiro Nishioka, and E.F. da Silva, Jr.

Center for Microelectronic Materials and Structures, and
Department of Electrical Engineering, Yale University
New Haven, Connecticut 06520-2157

ABSTRACT

The interface traps created by ionizing radiation or hot-electron injection in MOS capacitors have been found to undergo significant changes with time over an extended period (several months). Immediately after radiation or hot-electron damage, a interface trap peak above midgap ($\sim E_v + 0.75eV$) invariably appears. This peak, along with its background, would continuously change with time after damage, and one sailent feature of this change is the gradual conversion of this peak to a second peak located below midgap ($\sim E_v + 0.35eV$). The rate of this interfacial defect transformation process is a function of the device structure, its processing parameters, the details of the radiation or hot-electron treatment, the gate bias polarity, and the process is thermally activated. These and other pertinent results will be discussed.

1. INTRODUCTION

In recent publications, we reported on the long-term time dependent evolution of the interface traps in MOS capacitors after ionizing radiation or hot electron damage [1,2]. There we noted the observation that, in many cases, the post-radiation characteristic interface trap peak located above midgap would continuously decrease with time after irradiation or hot-electron damage, and accompanying that was the growth of a second peak located below midgap. We also noted that the formation and growth of the second peak appeared to be at the expense of the first peak, suggesting that some kind of defect transformation process might be taking place at the interface. Since then much more extensive work has been done, and we are convinced that not only does it indeed occur, but that the interfacial defect transformation process appears to be a fundamental process occuring generally in all the samples that we studied. Some of our recent results are presented and discussed here.

2. EXPERIMENTAL

The MOS capacitor samples used in this study covered a wide range of gate electrode materials (e.g., Al, Poly-Si, Mo, $TiSi_2$), various oxidation conditions (e.g., dry O_2, steam, with/without Cl, various oxidation temperatures), and processed at several different laboratories. Although due to the space limitation, the results to be presented here are for some specific sets of samples, qualitatively similar results have been observed in all the samples investigated.

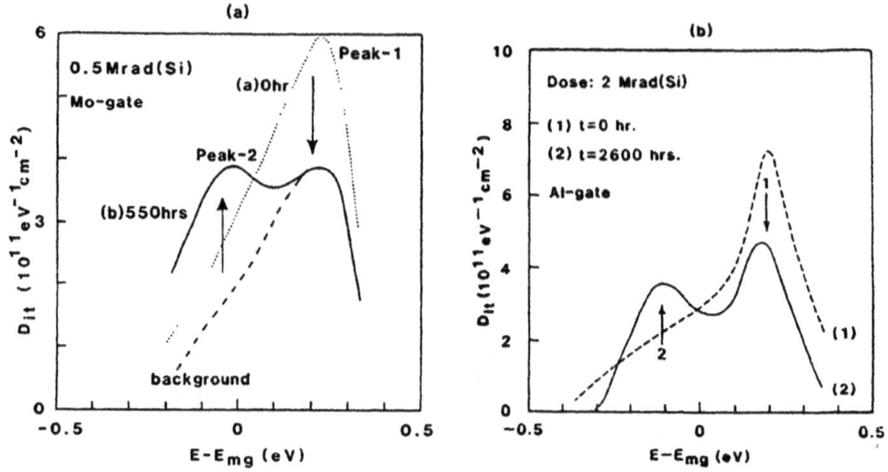

Fig.1 Interface trap distributions measured immediately after irradiation (dotted curves) and at later times (solid curves) for two different MOS capacitors: (a) Al-gate sample, with a gate oxide containing trace amount of fluorine; (b) Mo-gate sample.

The irradiation source used was an X-ray beam generated from a W target bombarded by 40 KeV electrons with a typical dose rate of 30 Krads(Si)/min. A total dose of up to 6 Mrads(Si) was used for some of the samples.

For the study of hot electron effects, a HP4145 parameter analyzer was used to inject Fowler-Nordheim tunneling current at a constant current level of 6×10^{-6} to 5×10^{-5} A/cm^2 for 200 seconds. Electrons were injected from the silicon substrate toward the gate (with a positive gate voltage).

The interface trap distributions were analyzed by measuring the high frequency and quasi-static C-V curves, with a ramp rate of 0.01 V/sec.

After X-ray irradiation or hot electron injection, the samples were stored at either room temperature, 50 °C, or 75 °C, and C-V measurements were made at several time intervals to monitor the time dependent evolution of the generated interface traps.

3. RESULTS

As shown in Figs. 1(a) and 1(b), a characteristic interface trap peak above midgap (dotted curves in both figures) appears right after irradiation or hot electron injection for two samples with different gate electrodes. This has been generally observed in all of the samples that we studied.

When measured some time later after irradiation (or injection), in some samples we observed an increase with time of this peak (designated peak-1 hereafter) for some extended period, followed by a decreasing trend; in some other samples peak-1 would continuously decrease with time right from the beginning. The detailed time dependence is a function of the processing history and damaging conditions, and is not the focus of this paper. What should be noted here is that, regardless whether peak-1 is rising or decreasing during the initial phase of this time dependent evolution, whenever peak-1 starts to decrease we observe the emergence and subsequent growth of a new peak below midgap (designated peak-2 hereafter), and the growth of this new peak is correlated with the reduction of peak-1.

The solid curves in both Figs.1(a) and 1(b) clearly illustrate the latent appearance of peak-2 in each of the two samples. Upon analyzing the data taken at much finer intervals, we found the growth of peak-2 was invariably at the expense of peak-1, and it looked as though a defect corresponding to peak-1 was gradually converting into another defect corresponding to peak-2.

It should be noted that peak-2 is riding on a background corresponding to the tail of the peak-1 distribution. Since this background is changing with time at the same rate as that of peak-1, we must subtract out this background effect to reveal more accurately the net change in peak-2. Using this procedure, we found that the amount of reduction in peak-1 corresponds almost exactly to the amount of growth in peak-2 for both samples exhibited in Figs.1(a) and 1(b), providing additional support to the defect transformation idea.

More convincing evidence for the defect transformation model is presented below. Figures 2(a) and 2(b) show the time dependent changes of peak-1 and peak-2, normalized with respect to the peak-1 value right after irradiation, for the two samples. If one follows the pair of peak-1 and peak-2 curves for either sample, one will notice that there is a strong correlation between the growth of peak-2 and the reduction of peak-1. In particular, note that the pair of curves for the Mo-gate sample (Fig. 2(a)) undergoes several up-and-down segments over the period of the experiment, and each up-going (down-going) segment in the peak-1 curve is correlated to a down-going (up-going) segment in the peak-2 curve. Moreover, the absolute changes in peak-2 are equal to those in peak-1 within experimental errors. These results suggest that the transformation process could go either way once peak-2 has achieved a sufficient size, although the peak-1 to peak-2 transition seems to be the dominant process in most cases.

Although we only showed here some results for X-ray irradiated samples, qualitatively similar results have been observed for hot-electron damaged samples as well.

After analyzing the data obtained on numerous samples produced in various laboratories, we found that, in addition to the defect transformation process, some samples exhibit a latent generation process (peak-1 and its background increases with time) after damage, while some other samples exhibit room temperature annealing behavior (the reduction in peak-1 exceeds the increase in peak-2), depending on the processing parameters and damaging conditions. In all cases, however, the transformation process is always present, leading us to believe that it is an universal phenomenon for samples damaged by radiation or hot electrons.

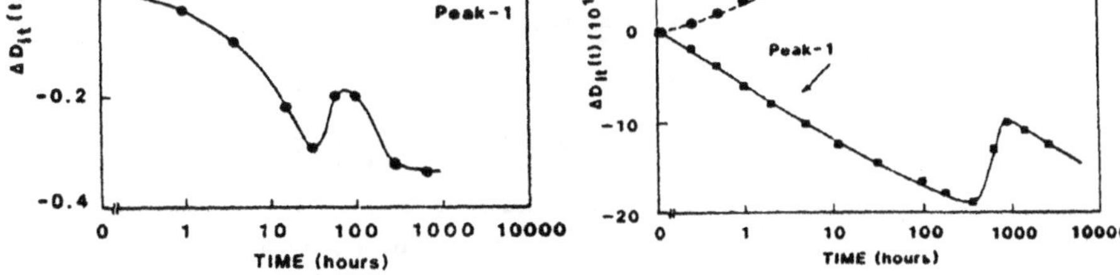

Fig.2 Time dependent evolution of peak-1 and peak-2 for two different MOS capacitors after irradiation, showing strong correlation between the two peaks: (a) Al-gate sample; (b) Mo-gate sample.

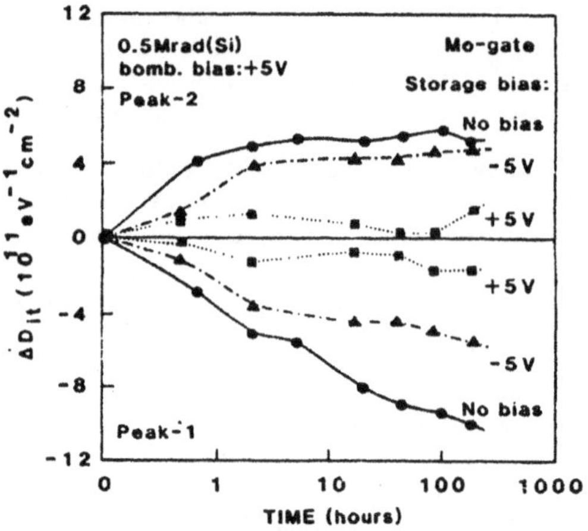

Fig.3 Time dependent evolution of peak-1 (lower portion of graph) and peak-2 (upper portion) for a Mo-gate MOS capacitor after irradiation and stored at room temperature under three different gate biases.

The latent generation and the annealing processes have been found to depend strongly on the initial damage level (the magnitude of peak-1 right after irradiation or injection): the annealing process becomes important when the initial damage level is high, while the generation process dominates when the damage level is low.

The rate of transformation has been found to depend on the presence of a gate bias during storage between measurements. As shown in Fig.3, the application of a gate bias with either polarity tends to suppress the transformation process, although positive bias gives rise to a stronger retardation effect.

In addition, raising the temperature accelerates the defect transformation process, as shown in Fig.4. In the temperature range studies so far (25-100°C), the thermal activation energy appears to be rather low ($< 0.1eV$) compared with the values typically reported for the thermal annealing process for these interface traps.

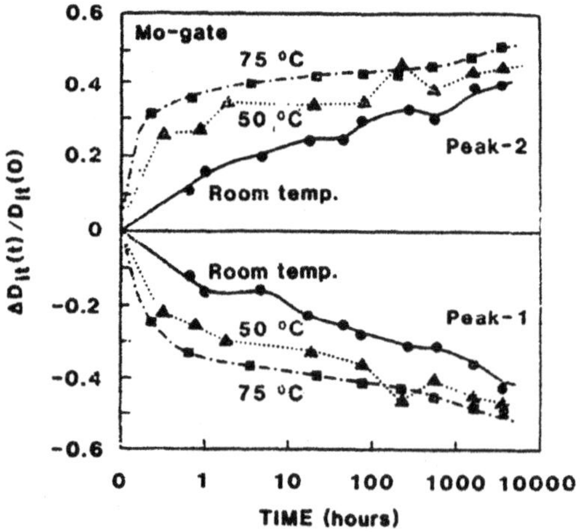

Fig.4 Time dependent evolution of peak-1 (lower portion of graph) and peak-2 (upper portion) for a Mo-gate MOS capacitor after irradiation and stored at three different temperatures.

One interesting feature we have observed is that, if the defect transformation is the only process taking place at the interface (i.e., there is no annealing or latent generation), the interface trap density located at midgap does not change with time even though other portions of the interface traps undergo drastic changes.

Two examples are shown below for an Al-gate sample stored at room temperature after hot-electron injection (Fig. 5(a)), and a Mo-gate sample stored at 75 °C after X-ray irradiation (Fig. 5(b)), respectively. It is apparent that in both cases the midgap interface trap density hardly changes throughout the entire defect transformation process. These examples are typical for devices in which only defect transformation process occurs.

As mentioned previously, in addition to the defect transformation process, some samples exhibit annealing or latent generation behavior under certain conditions. In the case where the annealing process dominates, the midgap interface trap density has been found to decrease with time; while in the case where the latent generation dominates, we observe an increase of the midgap density with time. Thus the time dependence of the midgap density provides a convenient indicator for us to tell these three processes apart without having to analyze the interface trap spectrum throughout the bandgap.

The next experiment provides strong evidence that peak-2 is not generated directly by ionizing radiation, even after repeated X-ray irradiation, and a necessary condition for its formation is the pre-existence of peak-1.

In this experiment, a set of Mo-gate MOS capacitors were first irradiated to a total dose of 2 Mrad(Si), followed by measurements of interface traps. These samples were then stored at 75 °C while their interface traps monitored. Afterwards, the same experimental sequence was repeated two more times on the same samples, and each time a X-ray dose of 2 Mrad(Si) was used.

As shown in Fig. 6(a), only peak-1 appears immediately after the first X-ray irradiation (labeled *0 hr*), which decreases continuously while peak-2 begins to appear more and more prominantly with time (see curves labeled *0.5 hr* and *23 hrs*). Note that for this particular sample, in addition to the defect transformation process, thermal annealing is appreciable at 75 °C, as the midgap density decreases significantly with time.

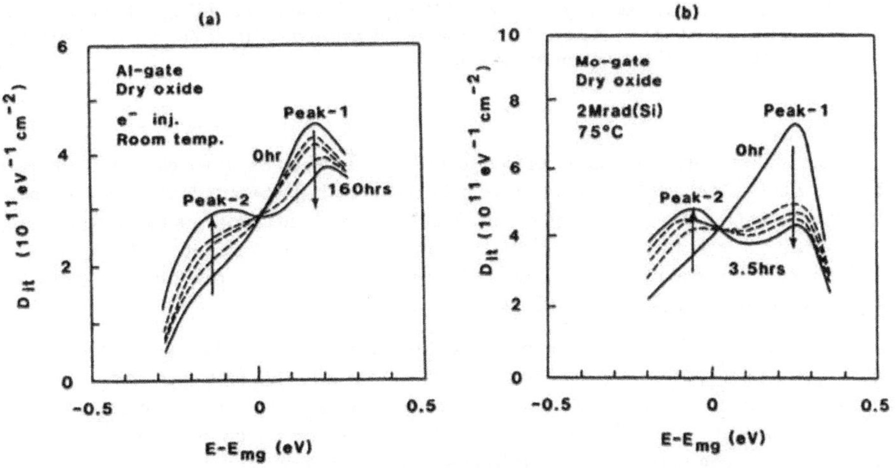

Fig.5 Time dependent changes of interface trap distribution for (a) Al-gate sample after hot-electron injection at a current density of 6×10^{-6} A/cm^2 for 200 sec and stored at room temperature; (b)Mo-gate sample after X-ray irradiation and stored at 75 °C. The arrows indicate the directions of change in D_{it} with time.

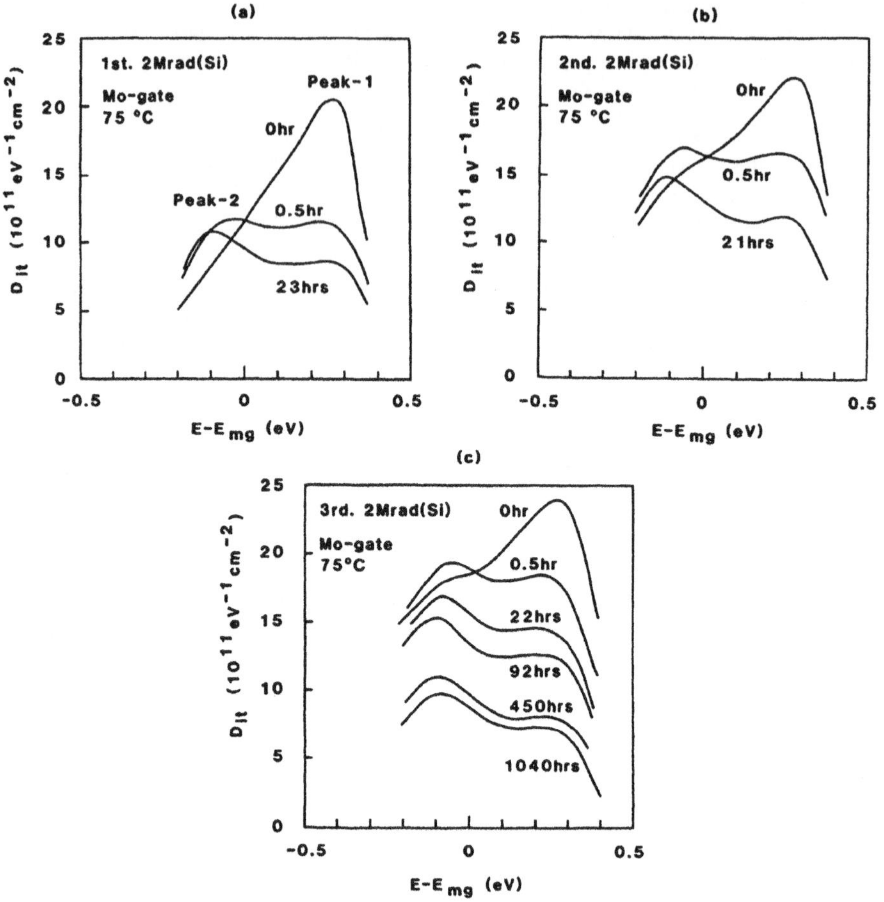

Fig.6 Interface trap distribution immediately after X-ray irradiation (curve marked *0 hr*) and some later times after irradiation: (a) after first 2 Mrad(Si) irradiation and storage; (b) after second 2 Mrad(Si) irradiation and storage; (c) after third 2 Mrad(Si) irradiation and storage. The sample was stored at 75 °C between measurements.

After the first X-ray irradiation and storage at 75°C for 23 hours, at which time the interface trap distribution looked like the curve labeled *23 hrs* in Fig.6(a), the sample was again irradiated to a dose level of 2 Mrad(Si). As shown in Fig.6(b), peak-1, along with its background, jumps back to a high level immediately after the second X-ray irradiation. In contrast, using the backbround subtraction procedure described earlier, we found that peak-2 did not increase at all as a result of the 2nd X-ray irradiation.

After the second irradiation, the post-irradiation time dependent behavior of the interface traps has been found to be qualitatively similar to that after the first irradiation: continued reduction of peak-1 with a corresponding increase of peak-2 (see curves labeled *0.5 hr* and *21 hrs* in Fig.6(b)). The peak height of peak-2 after the second storage period, however, has become higher than that after the first irradiation-stroage cycle.

The sample was then irradiated for the third time to a dose level of 2 Mrad(Si). This time the sample was stored and measured periodically for as long as 1000 hrs to investigate its longer-term time dependent behavior. As shown in Fig.6(c), the time dependent behavior is again qualitatively similar to those in Figs.6(a) and (b) up to 22 hrs of storage time. Thereafter, both peaks along with the background began to decrease at approximately the same rate, causing a downward parallel shift of the entire energy distribution (see curves below the one labeled *22 hrs* in Fig.6 (c)), suggesting that thermal annealing becomes the dominant process.

The effects of repeated irradiation-storage cycles are more clearly exhibited in Fig.7, where the changes in the peak-1 height (lower curve) and the peak-2 heights (with background effect subtracted, upper curve) are plotted as a function of time after irradiation. The storage temperature was again 75°C in all cases. It can be seen that, within each irradiation-storage cycle, there is a strong correlation between the growth of peak-2 and the reduction of peak-1 during the first hour; thereafter, peak-2 approaches saturation while peak-1 continues to decrease. Evidently peak-2 also decreases some later times after the apparent saturation level is reached (see peak-2 curve corresponding to the 3rd irradiation-storage cycle).

The important point to note here is that peak-2 does not grow immediately after each X-ray irradiation, despite the fact that each time peak-1 shows a large increase. Instead, peak-2 only grows when peak-2 decreases, suggesting that the reduction of peak-1 is a necessary condition for the growth of peak-2. This is consistent with the defect transformation model.

It should also be noted that, immediately after the 2nd and the 3rd irradiation, peak-2 actually showed a slight decrease, suggesting the possibility of a reverse transformation triggered by the irradiation. Evidence for this reverse transformation phenomenon also exists in some other samples (see Figs.2(a) and 2(b)), but its cause is not yet clear.

The saturation of peak-2 (which occurs after ~ 1 hour of storage at 75 °C for these samples) appears to indicate that only a large portion, but not all, of peak-1 eventually converts to peak-2. The other fraction of peak-1 does not seem to undergo the defect transformation process described above (its reduction does not contribute to the growth of peak-2). After each repeated irradiation, however, more of the convertible peak-1 defect is again generated, and the peak-2 growth process proceeds accordingly.

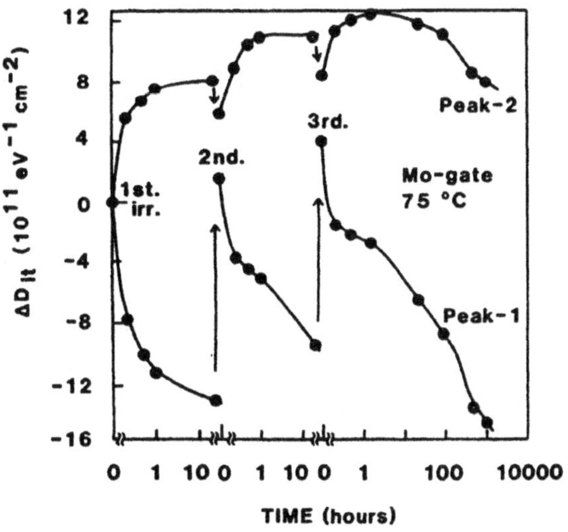

Fig.7 Time dependent change of interface strap peak-1 above midgap (lower curve) and peak-2 below midgap (upper curve). The three times at which irradiation was performed are marked. The sample was stored at 75 °C between measurements.

4. DISCUSSION AND CONCLUSION

The microscopic origin of the interfacial defect transformation process is not yet clear. Whether the transformation involves interactions of two neighboring sites, or simply a configuration change of the same defect site remains to be determined. Because of the very long time constant involved, we do not believe that the primary effect is due to the usual electronic processes (such as the simple charging and discharging of the interfacial defects); rather, some kind of chemical reactions or bonding alterations must be taking place. The fact that the transformation process proceeds with or without a gate bias, and that the presence of a gate bias with either polarity tends to retard this process, suggests that species of all three charge states (neutral, positive, and negative) may be involved. The low activation energy ($< 0.1eV$) involved in the transformation process may provide an important clue for its eventual understanding.

The fact that the midgap density does not change with time in samples where only defect transformation process takes place (i.e., no annealing or latent generation) may arise from two possibilities. One possibility is that the energy width of the decreasing peak-1 extends beyond midgap and into the lower half of the bandgap, while the width of the increasing peak-2 also extends beyond midgap and into the upper half of the bandgap. In this case, to obtain a midgap density that does not change with time, the rate of decrease of the peak-1 distribution at midgap must be exactly equal to the rate of increase of the peak-2 distribution there, so that the net change at midgap is zero. Another possibility is that the transformation process resembles a seasaw motion with the midgap as the pivot, such that one side goes up while the other side goes down, with the pivot point staying motionless.

While we have not yet been able to determine which of the two possibilities represents the real situation, the fact that the midgap density stays unchanged lends further support to the interfacial defect transformation model. From a practical point of view, our findings suggest that the time dependent behavior of the midgap density can be used to determine whether the dominant process in a given sample is defect transformation, annealing, or latent generation, which is much more convenient than having to analyze the energy distribution of the interface traps both above and below midgap.

ACKNOWLEDGMENT

This work was supported by the SRC under grant no. 87-SJ-079, and NRL under grant no. N00014-87-K-2028. Y. Nishioka was on leave from and supported by the Hitachi Central Research Laboratory. E. F. da Silva, Jr. was at Yale under the auspices of a doctoral fellowship grant from the Brazilian Government Agency CAPES.

REFERENCES

1. E.F. da Silva, Jr., Y. Nishioka, and T.P. Ma, Appl. Phys. Lett., **51**, 270 (1987)
2. Y. Nishioka, E.F. da Silva, Jr., and T.P. Ma, IEEE Electr. Dev. Lett., **8**, No.12, Dec. (1987)
3. Y. Nishioka, E.F. da Silva, Jr., and T.P. Ma, IEEE Trans. Nucl. Sci., **NS-34**, 1166 (1987)
4. E.F. da Silva, Y. Nishioka, and T.P. Ma, IEEE Trans. Nucl. Sci., **NS-34**, 1190 (1987)

V. EFFECTS OF PREOXIDATION SUBSTRATE QUALITY, SURFACE TREATMENTS, AND OXIDATION INDUCED POINT DEFECT GENERATION

INTRODUCTION

E. A. Irene

Chemistry Dept., CB#3290, Venable Hall
The University of North Carolina, Chapel Hill, NC 27599-3290

Present and past studies have shown that the precise method of cleaning of a Si surface can often modulate both the Si oxidation kinetics and electrical properties of the oxide and the Si-SiO_2 interface. Indeed, in a paper presented above, we have shown the results of in-situ studies in the aqueous Si cleaning environment. It was shown that "cleaning" is not "cleaning". Oftentimes impurities and/or surface roughness results. These issues were fully demonstrated in the subject session via a variety of experimental studies.

Preoxidation effects on silicon oxidation were discussed by Ruzyllo; the philosophy of different approaches to preoxidation were discussed. Hahn et al explored the process steps that give rise to surface roughness. Also, a variety of techniques were compared. Kao et al studied the specific Si cleaning procedures that are in industrial use. During discussions after the Kao et al paper, Al, a particularly abundant impurity in H_2O_2 was suggested by several symposium participants as the Si oxidation retarder discussed by Kao et al. The Ohmi et al paper presented possibly the cleanest Si oxidation procedure ever performed. They reported superior thin oxides and a higher Si-SiO_2 barrier height as well as nearly parabolic oxidation behavior for ultra dry O_2 oxidation. In the Duranko et al paper the comparison of anhydrous Hf treatments were compared with more conventional treatments. While C levels were reduced there were no significant improvements in electrical properties however. The Gambino et al paper examined the electrical implications of rough surfaces which were roughened due to RIE treatments. The results could be qualitatively explained based on field enhancement effects. The paper by Hasenack et al examined dielectric strength and interface state density on SiO_2 grown on Cz and Fz Si. High field breakdown was degraded with the highest oxygen content wafers. Long preoxidation anneals in Ar improved the situation. The Chenko paper discussed deleterious effects of long time DI water rinses on device oxide quality.

In addition to the effects of silicon surface cleaning, the oxidation process itself can have a major effect on properties microns below the Si surface. These effects are driven by the generation and annihilation of point defects by the growing oxide interface. The invited papers by Fair and Griffin and Plummer review these phenomena and in the final paper of this chapter Dunham presents implications of these effects for oxidation where more than one dimension is important.

PREOXIDATION SURFACE TREATMENTS IN THERMAL OXIDATION OF SILICON

Jerzy Ruzyllo

Center for Electronic Materials and Devices
Department of Electrical Engineering
The Pennsylvania State University
University Park, PA 16802

The role of preoxidation surface treatments in thermal oxidation of silicon is discussed. The nature of various treatments, and the effect they may have on the growth of an oxide, its structure and composition as well as electrical properties is evaluated. A systematic approach to the implementation of preoxidation treatments carried out outside the process chamber as well as in situ is presented for the case of the thermal growth of gate oxides. The possibilities of modifying the condition of the silicon surface through the preoxidation treatments are discussed.

INTRODUCTION

Over the years, the process of thermal oxidation of silicon was intensively studied primarily from the point of view of the effect of various parameters of oxidation on the growth kinetics and properties of silicon dioxide. Also, the impact of postoxidation annealings on the characteristics of $Si-SiO_2$ structure became well recognized. In contrast, considerably less attention was paid to the possible role of preoxidation treatments in the process of thermal growth of high quality oxides on silicon. The reason for this was a relatively low interest in the very thin oxide regime for which for a long time there was no application in silicon device technology. However, this is exactly the oxide thickness regime in which variations in the condition of silicon surface before oxidation are expected to have the most visible effect on the outcome of the thermal oxidation process. Currently, the interest in these type of interactions is significantly revived primarily due to the MOS device design requirement to reduce the thickness of the gate oxides to the 100Å level. Also, the emergance of new processing methods, which may leave the silicon surface contaminated and structurally disturbed, stems our interest in the development of treatments that may reestablish the quality of the silicon surface before oxidation.

For the purpose of discussion in this paper, the following three observations of general nature should be made at this point:

(1) The process of fabrication of high quality very thin silicon dioxide has to be viewed as a part a complex processing scheme rather than an individual, isolated step. Therefore, the possible effect on the oxide

growth and properties of operations performed on the silicon wafer, for instance etching, prior to thermal oxidation should not be overlooked.

(2) The thermal oxidation process has built-in surface modifying features due to the thermal energy accompanying the process, as well as conversion of the surface region of silicon into silicon dioxide. It should be noted, however, that the very same features of thermal oxidation may be responsible for the transformation of the originally neutral features of the silicon surface region into the active defects in the $Si-SiO_2$ structure.

(3) The fact should be recognized that the oxide fabrication procedure is actually a three step process consisting of (i) a set of preoxidation treatments, (ii) thermal oxidation during which interaction between silicon and oxidizing species is taking place, and (iii) the postoxidation treatments which can actually be narrowed down to various annealings. Each of these steps plays an important role in determining the final outcome of the entire operation.

The purpose of this paper is to consider various aspects of the preoxidation treatments in thermal oxidation of silicon. In general, the effect of preoxidation treatments is in modifying the chemical composition and crystallographic structure of silicon surface and subsurface region. As a result of these modifications, improved control over the growth of very thin oxide can be achieved along with reproducible chemical composition and structure of the oxide and the $Si-SiO_2$ interface. This paper at first briefly outlines possible silicon surface disturbances which have to be addressed during the preoxidation treatments. Then, various approaches to the implementation of preoxidation treatments are discussed. The remaining parts of this work are devoted to the review of experimental results and critical evaluation of techniques of silicon surface conditioning before oxidation.

CONDITION OF SILICON SURFACE BEFORE OXIDATION

Due to the generic features of silicon wafer manufacturing procedures, as well as due to the nature of treatments that the wafers are exposed to in the course of device manufacturing, the silicon surface subjected to thermal oxidation is never either structurally perfect or chemically clean. In other words, the situation shown in Fig. 1a in which the geometry of silicon bonds at the surface is undisturbed, and the surface is free of alien elements is highly unrealistic. First of all, it can be assumed that the crystal lattice at the surface is disrupted as compared to the bulk of the substrate (Fig. 1b). The degree of lattice deformation at the surface, or the surface roughness, varies substantially depending on the previously applied treatments including chemo-mechanical polishing in the case of the fresh wafer (1) or etching used to expose portion of the surface to oxidation during processing.

Due to the very high affinity of silicon for oxygen, the silicon surface in a normal ambient in terms of pressure, temperature and composition is always oxidized to form an oxide 5Å to 15Å thick (Fig.1c). The chemical content of the native, spontaneously grown oxide is difficult to determine as it varies significantly depending on the treatments given to the surface, ambient conditions, and methods of wafer handling. In general, it is nonuniform across the depth of the film, and in average does not reach the SiO_2 stoichiometry (2). Besides Si-O bonds and unoxidized silicon, the native oxide also includes Si-H and Si-OH groups (3,4).

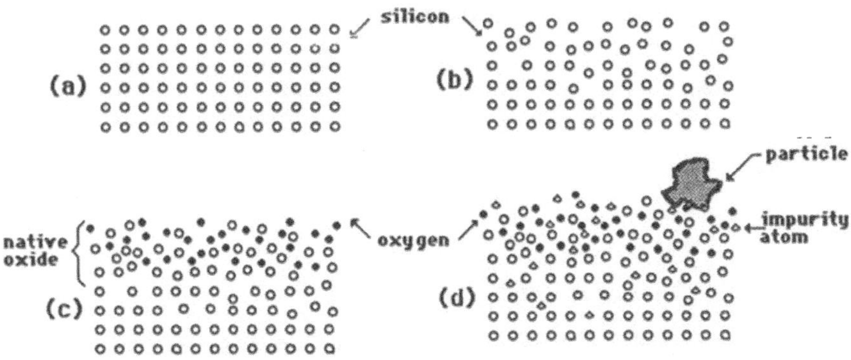

Fig. 1 (a) Simplified representation of an undisturbed silicon surface; (b) (c) and (d) possible disturbances of silicon surface before oxidation.

To complicate the situation even further, the vulnerability of the silicon surface to either physically or chemically adsorbed contaminants (Fig. 1d) accounts for one more destabilizing factor with potentially profound impact on the subsequently grown oxide. Again, the nature and content of contaminants is to a great extent controlled by the process conditions and wafer handling prior to oxidation. The common contaminants of concern are organic compounds, such as hydrocarbon $Si-CH_x$ groups (3), transition metals and alkali ions. Also, other elements, for instance gold and sulphur (2), can be found at the silicon surface. Due to the nature of the treatments preceding thermal oxidation, which often involve an HF dip resulting in a hydraphobic surface, the role of fluorine in the chemistry of silicon surface during oxidation should be also considered (see reference (5) in this volume).

All the above contaminants are enclosed within the native oxide and near surface region of silicon interacting in most cases adversely with growing thermal oxide. Beside these atomic scale contaminants, the particles of various sizes and composition are encountered even in the cleanest oxide fabrication environment (Fig. 1d). The micron-size particle, once allowed to contact the surface, is very difficult to remove since the total force of its adhesion exceeds gravitational force by factors greater than 10^6 (6). Clearly, even the smallest particle will eventually lead to oxide failure, usually observed by low-field breakdown (7). Therefore, protection of the silicon surface before oxidation against particles is a necessity which is difficult to overestimate.

The role of the features of the silicon surface subsequently subjected to thermal oxidation can not be generalized as it will depend on the response of each of them to the heat treatment and progressing surface penetration by the growing oxide. Typically, each can have a measurable effect on either the oxide growth kinetics or properties of an oxide and $Si-SiO_2$ interface, or both. For instance, the native oxide as such retards subsequent oxidation kinetics (8) while contaminants may adversely affect interface traps, minority carrier lifetime, and oxide dielectric strength. This last parameter is also known to deteriorate as a result of surface roughness (9). To complicate the situation even further, there is a possibility of structural imperfections interacting with the surface chemical contaminants to create electrically active defects during the subsequent oxidation.

As is seen from this discussion, an effort should be made to assure stability and reproducibility of the condition of silicon surface before oxidation. This is exactly an area in which the preoxidation surface treatments are coming into play. To make the best use of them, all aspects of their functioning have to be well understood. Further discussion in this paper is aimed at the consideration of the generalized approach to such treatments as well as an evaluation of the usefulness of some of them in the fabrication of high quality thermal oxides on silicon.

GENERAL OUTLINE OF PREOXIDATION TREATMENTS

For the purpose of this discussion, the preoxidation treatments are being defined as all the treatments to which the wafer is subjected after the completion of the last processing step preceding oxidation, in the device fabrication sequence this is usually a resist strip, and up to the onset of thermal oxidation inside the oxidation chamber.

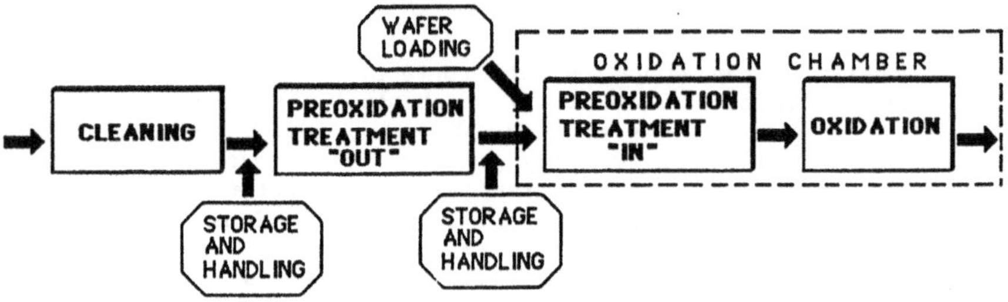

Fig. 2 Sequence of preoxidation treatments considered in this paper.

Figure 2 schematically shows consecutive steps which can possibility be performed on the wafer prior to thermal oxidation. They can be categorized in two different ways. First, they can be subdivided into intentional and unintentional treatments. While the concept of the former is clear, the latter are primarily related to the storage and handling of the wafer between operations as well as wafer loading into the oxidation chamber which includes setting the wafer in final position, and bringing it up to thermal equilibrium (Fig. 2). In this discussion we shall concentrate primarily on the intentional treatments, keeping in mind, however, the hazards to which the wafer is exposed during storage, handling and loading operations. These steps may account for process variables difficult to control and reproduce. Introducing particles, or even, in the extreme case, mechanical damage, or exposing the surface to the ambient of uncontrolled composition, may overshadow the effect of intentional preoxidation treatments.

The second criterion for distinguishing various types of preoxidation treatments is a location in which they are carried out. In these terms, they fall into two categories of operations performed outside of the reaction chamber (wafer "cleaning" and "additional preoxidation treatments out" in Fig. 2), and in situ, inside the reactor immediately before wafer exposure to the oxidizing agent at elevated temperature ("additional preoxidation treatments in" in Fig. 2). An important difference between these two cases is that the former will by definition be followed by storage, handling and loading steps. These are absent in the case of in situ treatments which, thus, usually determine the final condition of the silicon surface before oxidation.

Among the preoxidation treatments, clearly wafer cleaning plays the most important role in determining the outcome of thermal oxidation. Typically, cleaning consists of a series of steps aimed at the removal from the silicon surface of organic contaminants, alkali ions, and transition metals. A characteristic feature of chemical cleaning in the liquid phase is that it leaves the silicon surface covered with spontaneously grown chemical oxide. Initially 5Å to 12Å thick (10), this oxide can grow up to 20Å during several hours of wafer exposure to air at room temperature (8) creating a situation similar to that shown in Fig. 1c and d. Composition of the chemical oxide varies depending on the composition of the cleaning/etching solutions, and the sequence in which they are used. Although very thin, this oxide can affect growth reproducibility as well as electrical stability of SiO_2 thermally grown on the chemically cleaned silicon surface. Therefore, rather then directly subjecting the chemically cleaned wafers to thermal oxidation, an attempt should be made to stabilize the condition of the silicon surface by means of additional preoxidation treatments.

Any additional preoxidation treatments subsequent to cleaning (Fig. 2) should serve either one of the following purposes which are at the same time graphically presented in Fig. 3.

(1) To grow an ultrathin protective oxide in a controllable way instead of continued uncontrolled oxidation in the ambient air. Some kind of additional cleaning action in the gas phase should accompany such a process.

(2) To allow oxide growth in ambient air and then to remove it. Oxide removal should be immediately followed, without wafer exposure to ambient air, by oxidation in a clean oxidizing agent.

Clearly, the first approach relates to preoxidation treatments performed on the wafer outside to the oxidation chamber (Fig.2). In fact, these treatments should be closely coupled with cleaning (Fig. 3) to better protect the surface during subsequent wafer storage. In contrast, the second approach corresponds to in situ preoxidation treatments (Fig.2), and represents an attempt to compensate for the unfavorable byproduct of cleaning in the form of chemical oxide and for surface degradation during storage and handling between cleaning and oxidation.

Any preoxidation treatment must meet certain general requirements. First, it should not have an adverse effect on the already existing features of the device. For that reason it should be carried out at a thermal budget lower than that a thermal oxidation itself. Also, immediately before oxidation the surface should not be exposed to high energy particles or radiation or to an ambient causing nonuniform

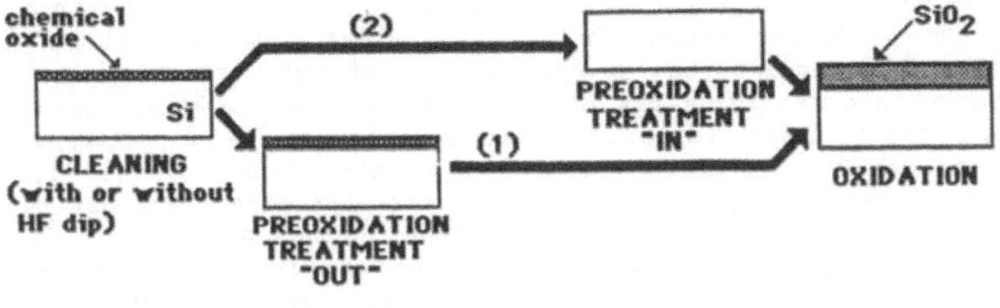

Fig. 3 Two ways of implementation of preoxidation treatments.

etching of the silicon surface. Moreover, the possibility of adding particles to the surface during the preoxidation treatments should be kept to an absolute minimum.

The ultimate solution to most of the current and future problems associated with fabrication of high quality silicon dioxide will be an integration of wafer cleaning, preoxidation treatments, if needed, and thermal oxidation within the same process chamber (Fig. 4). In this way the unintentional treatments (Fig. 2) can be eliminated to improve oxide reliability as well as growth reproducibility. This integrated approach implies the need to carry out wafer cleaning in the gas phase which is a challenging research task by itself (11).

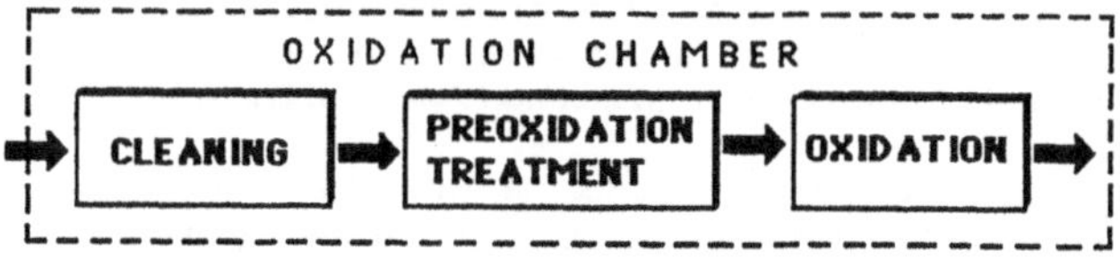

Fig. 4 Scheme of the integrated process of oxide fabrication.

SUMMARY AND DISCUSSION OF EXPERIMENTAL RESULTS

With a background of the general aspects of preoxidation treatments we are going now to review some experimental data related to the role of surface condition in the thermal oxidation of silicon. In this review we shall follow the outline presented in Fig. 2.

Surface Cleaning

Surface cleaning is achieved by immersing the wafer in liquid chemicals or by spraying it with these chemicals. The chemicals react selectively with surface contaminants to cause either their dissolution in the solvent or their conversion into a soluble compound that can be dissolved by the liquid ambient. The correlation between the preoxidation cleaning scheme and the oxide growth kinetics was noticed relatively early (12, 13), but only recently more detailed studies of the observed effects were undertaken (14, 15) with the most important findings summarized in Ref. (16).

One of the key issues in these studies appears to be the use of an HF dip at the end of the cleaning sequence in which chemical oxide is removed and a hydrophobic surface is created. Although there is not a clear picture of the effect of HF dip on the growth of very thin oxides (10), the trend for oxides thicker than about 250Å is well established (13-15). The results show an acceleration of oxide kinetics is the case of the surface receiving the HF dip last (Fig. 5). While these results concern conventional furnace oxidation featured by a relatively slow oxide growth rate, similar results have also been obtained for rapid thermal oxidation (Fig. 6) The reasons for this effect are not yet fully understood, and the possible explanations include intrinsic film stress of the oxide as well as the extension of the initial oxidation regime to a greater thickness than anticipated (14). In addition, the parallel oxidation model has been used to account for the observed differences (15). Consideration should also be given to the fact that the HF treated, and hence, hydrophobic silicon surface is highly reactive and

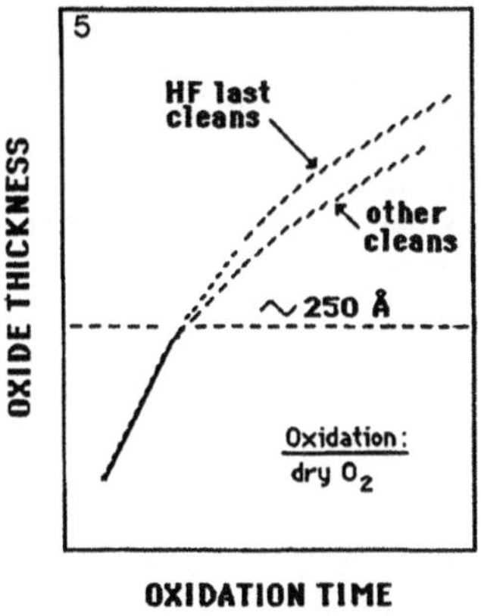

Fig. 5 Effect of the HF dip at the end of the cleaning sequence on the oxide growth during conventional furnace oxidation.

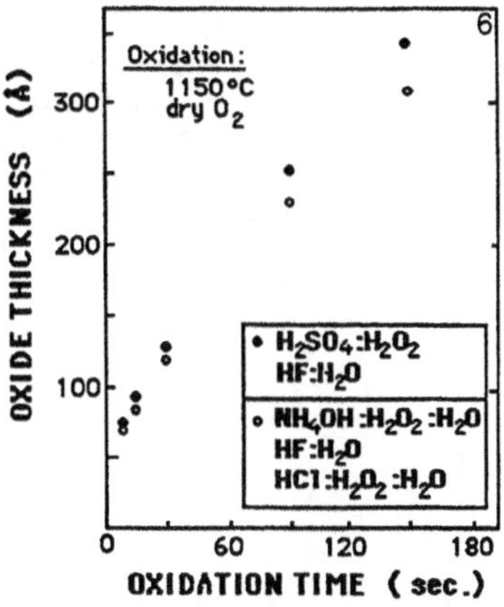

Fig. 6 Effect of wafer cleaning in rapid thermal oxidation of silicon.

contains a substantially higher amount of carbon than its hydrophilic counterpart (3). Upon dehydrogenization of Si-CH_x groups at elevated temperature, carbon remains at the surface and can possibly interact with the growing oxide.

In the light of these observations, the wet HF step at the end of the cleaning procedure does not seem to help in improving chemical stability of the silicon surface. On the other hand, if left unremoved by the HF dip, the chemical oxide resulting from cleaning may also introduce variables adversely affecting subsequent oxidation. These problems appear to be difficult to resolve within the preoxidation procedures carried out in the liquid agents. Thus, their resolution should be left to the additional to wet cleaning gas phase treatments performed prior to oxidation (Fig. 2). Otherwise, the possibly insufficient stability and reproducibility of the condition of the silicon surface will act as an uncontrolled factor obstructing precise modeling of thermal oxidation (17). Alternatively, the anhydrous HF rather than hydrous HF etching following cleaning should be considered or an attempt should be made to replace altogether wet cleaning with cleaning carried out entirely in the gas phase. While the feasibility of this former approach has already been proven (5), the implementation of the latter was until now only partially successful (18).

To conclude the brief discussion on preoxidation wet cleaning it is appropriate to point out that this treatment cannot be viewed as being entirely "inert", and as such should be applied with care. Clearly, cleaning is needed to remove surface contaminants, but on the other hand, because of the particulate contamination during wet cleaning, the overexposure of the surface to the liquid agents is not advisable. We were able to confirm experimentally this notion by studying dielectric strengths of the oxides grown on the surface subjected to multiply cleans. As a reference, the fresh wafer subjected only to the deionized water rinse, but otherwise uncleaned, was used. Other wafers were exposed to RCA cleans (20), up to four times in the row, followed each time by the spin drying. Then, all wafers were simultaneously oxidized and MOS capacitors were formed. The results of this test, shown in Fig. 7, clearly indicate that while the one cleaning step is critical to obtain high quality oxide, any additional cleaning gradually deteriorates oxide integrity. Eventually it deteriorates to the point where oxides in some test devices an completely shorted (black bars in Fig. 7). Obviously, this result can be less dramatic depending on the quality of the process environment, but nevertheless, the general trend is rather clearly demonstrated by this experiment.

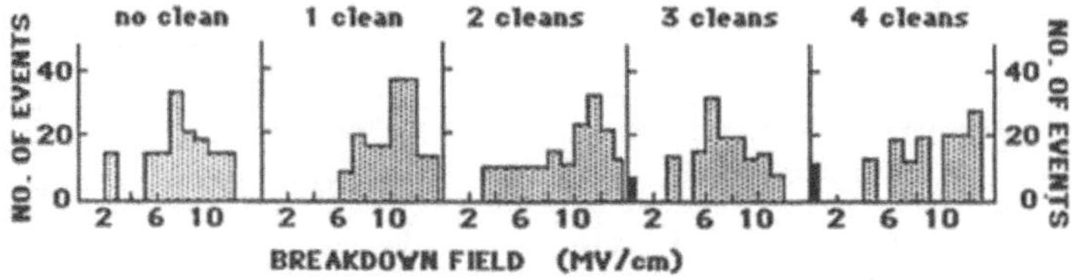

Fig. 7 Effect of number of consecutive preoxidation cleans of the breakdown events distribution in thermal oxides.

Preoxidation Treatments "Out"

The role of preoxidation treatments "out", i.e. carried out outside of the oxidation chamber is to improve stability and reproducibility of the surface condition after the wet chemical cleaning/etching process. The first group of methods falling into this category aims at the removal of the residual chemical oxide by either ion sputtering, reactive ion etching, high temperature H_2 annealing or high vacuum - high temperature annealing. These operations, however, if not carried out in situ immediately before thermal oxidation, are of limited value due to the uncontrolled regrowth of native oxide of poorly controlled composition in the ambient air.

The second group of procedures in this category, which are specifically designed to immediately follow wafer cleaning with HF dip last (approach 1 in Fig. 3), is exemplified by the UV/O_2 treatment. The exposure of chemically cleaned Si wafers to UV light in an oxidizing ambient results simultaneously in the growth of an ultrathin protective oxide, and removal of residual organic contaminants which, as mentioned earlier, tend to accumulate on the HF treated surface. The possible role of UV/O_2 exposure prior to thermal oxidation was studied (20), and some of its advantages, without apparent disadvantages, were determined. Improved oxide breakdown statistics were observed in the case of preoxidation exposure of chemically cleaned surfaces to UV light in the presence of wet O_2 (Fig. 8).

Preoxidation Treatments "In"

Once loaded into the oxidation chamber, the Si wafers can be subjected to in situ treatments that can enhance quality of the subsequently grown oxide. As mentioned earlier, they can be particularly effective as the high degree of control over the gaseous ambient composition between the given treatment and thermal oxidation can be achieved limiting in this way uncontrolled reactions at the silicon surface. The nature of such procedures, depicted as "approach (2)" in Fig. 3, will depend on the design of the oxidation reactor. For

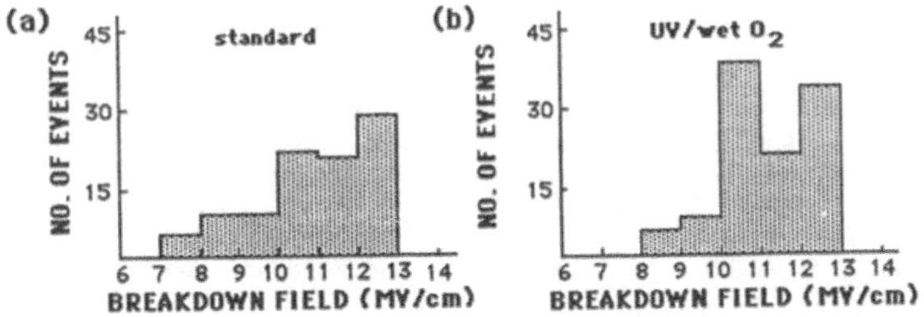

Fig. 8 Oxide breakdown distributions showing effect of preoxidation UV/O_2 treatment.

instance, in reactors with vacuum capabilities, the reduction of pressure at elevated temperature results in the removal of residual oxide through the formation of volatile SiO (21). Alternatively, providing a high enough temperature is used, etching of residual oxide in hydrogen can also be accomplished. The addition of HCl to the preoxidation annealing ambient in turn, results in the reduction of transition metal contamination. In the more complex reactors, the surface can be exposed to an ion beam prior to thermal oxidation for cleaning purposes. Whatever treatment is used, however, it should be immediately followed by oxidation to take the best advantage of in situ processing.

Besides the intentional treatments, substantial differences in the outcome of oxidation can be caused by conditions under which wafers reach thermal equilibrium at the temperature of oxidation. Here not only the rate of temperature ramp is important, but also the ambient may have an impact on the subsequent oxide growth process and oxide parameters. In this regard, advantages of using argon rather than nitrogen were experimentally proven (22).

SUMMARY

The goal of this paper was to systematically review the role of surface preoxidation treatments in the process of thermal growth of high quality silicon dioxide on silicon. In all cases the impact of these treatments on the oxide growth kinetics and the electrical characteristics must be carefully studied in order to improve the results of thermal oxidation. Also, the need to integrate preoxidation treatments and the oxidation process into the same reaction chamber was emphasized.

It appears that in the future, our approach to preoxidation procedures will have to be differentiated depending whether wet chemical cleaning or dry cleaning will be used during surface preparation for oxidation. The different nature of these two methods will cause different issues to be of concern. In the former, the chemical state of the surface will remain of main concern. In the latter, beside the ability of dry environment to remove contaminants from the surface, the surface roughness is expected to become the increasingly important issue.

ACKNOWLEDGMENTS

The author would like to thank Dr. Richard Tressler for critical comments and Sylvia Hossain for her assistance in carrying out some experiments reported in this paper.

REFERENCES

1. P.O. Hahn, M. Grundner, and A. Schnegg, First Workshop on Process-Related Defects in Semicond. Insul. Systems, Raleigh, 1987; also P.O. Hahn, M. Grundner, A. Schnegg, and H. Jacob, This Volume.
2. B.E. Deal and D.B. Kao, Proc. Workshop on Tungsten and other Refactory Metals for VLSI Applications, Palo Alto, 1986.
3. M. Grundner and H. Jacob, Appl. Phys. A 39, 73 (1986).
4. J. Ruzyllo, I. Shiota, N. Miyamoto, and J. Nishizawa, J. Electrochem. Soc., 123, 26 (1976).
5. G.T. Duranko, D.J. Syverson, L.A. Zazzera, J. Ruzyllo, and D.C. Frystak, This Volume.
6. R.A. Bowling, J. Electrochem. Soc., 123, 2208 (1985)
7. J.R. Monkowski and R.T. Zahour, Proc. IEEE Reliability Physics Symp., 1982, p. 244.
8. S.I. Raider, R. Flitsch, and M.J. Palmer, J. Electrochem. Soc., 122, 413 (1975).
9. C.M. Osburn, J. Electrochem. Soc., 121, 809 (1974); also J.P. Gambino, T.N. Nguyen, B. Cunningham, and J.F. Shepard, This Volume.
10. J. Ruzyllo, J. Electrochem. Soc., 134, 1869 (1987).
11. J. Ruzyllo, Microcontamination, 6 (3), 39 (1988).
12. F.N. Schwettman, K.L. Chiang, and W.A. Brown, Electrochem. Soc. Meeting, May 1978, Ext. Abstr. 78-1, 688 (1978).
13. F.J. Grunthaner and J. Maserjian, IEEE Trans. Nuc. Sci., NS-24, 2108 (1977).
14. G. Gould and E.A. Irene, J. Electrochem. Soc., 134, 1033 (1987).
15. J.M. deLarios, C.R. Helms, D.B. Kao, and B.E. Deal, in "Process Simulators for Silicon VLSI and High Speed GaAs Devices," Stanford Electronic Laboratory, Stanford, 1987.
16. D.B. Kao, B.E. Deal, J.M. deLarios, and C.R. Helms, This Volume.
17. H.Z. Massoud, J.D. Plummer, and E.A. Irene, J. Electrochem. Soc., 132, 2685 (1985).
18. J. Ruzyllo, A.M. Hoff, D.C. Frystak, and S.D. Hossain, submitted to J. Electrochem. Soc.
19. W. Kern, Semicond. Internatl., 7 (4), 94 (1984).
20. J. Ruzyllo, G.T. Duranko, and A.M. Hoff, J. Electrochem. Soc., 134, 2052 (1987).
21. F.W. Smith and G. Ghindini, J. Electrochem. Soc., 129, 1400 (1982).
22. J. Ruzyllo, J. Electrochem. Soc., 133, 1678 (1986).

THE Si-SiO_2 INTERFACE ROUGHNESS:

CAUSES AND EFFECTS

P.O. Hahn, M. Grundner, A. Schnegg, and H. Jacob

Wacker Chemitronic GmbH, Research Center
D-8263 Burghausen, FRG

ABSTRACT

The structural properties of silicon surfaces and interfaces have been studied from a macroscopic to an atomic scale by low energy electron diffraction (LEED), scanning tunneling microscopy (STM), and optical methods (lightscattering, magic mirror). The chemical state of silicon surfaces prepared by different cleaning procedures which influences the roughening was analyzed using x-ray photoelectron (XPS) and high resolution electron energy loss spectroscopy (HREELS) measurements.
It is shown that thermal oxidation of silicon, in the course of device production, leads to a pronounced roughness at the Si-SiO_2 interface. This interface roughness depends on the virgin silicon surface morphology and its chemical state, on bulk properties, and on the parameters of oxidation.
The influence of the interfacial structure in MOS inversion layers on electronic properties like mobility, interface states, fixed oxide charges, and on the dielectric breakdown behaviour will be demonstrated.

I. INTRODUCTION

In view of the importance of MOS structures in solid state electronics, thermal oxidation of silicon is one of the most important processing steps. Increasing integration densities in future devices request a perfect oxide grown on a perfect single crystal. The state of the art is documented in the proceedings of several topical conferences (1-3) and in review papers (4, 5).

Since 1960, when the first industrial field effect transistor appeared, the advances in MOS devices reflect the oxide quality improvement achieved by better material, equipment, and processing methods. This success was rendered possible by a tremendous research effort. The Si-SiO_2 interface belongs to the systems investigated best in solid state physics. The continuous scaling down of device dimensions and oxide layers, however, has renewed research activities and has lead to new characterization and production methods.
The use of very thin oxides in ultra large scale integrated (ULSI) circuits makes the interface between silicon and oxide layer more and more decisive for the electrical breakdown behaviour. Therefore, all factors influencing this interface have to be considered carefully.
The surface roughness and crystal distortion of the virgin wafer depends on the polishing process. Its chemical state, the saturation of the dangling

bonds with adsorbed atoms, is determined by the final cleaning step. Both influence the interface roughening during oxidation (fig. 1). This and the consequences on electronic properties will be the topic of this paper.

II. THE SMOOTHNESS OF POLISHED SILICON SURFACES

Besides perfect crystallinity a silicon substrate appropriate for ULSI must have a smooth surface at all correlation lengths, from the "macroscopic" (200 mm) down to the "atomic" range (0.1 nm). For each measurement range suitable techniques have to be applied. Fig. 2 gives a survey on relating methods in the ranges of centimeters down to nanometers.

Macroscopic Waviness: Magic Mirror

This very simple method using the optical reflection of parallel light on the wafer surface from a tungsten lamp provides fast information on the occurrence of long range undulations on the surface. The term "Magic Mirror" was coined for it because the wafer is used as a mirror, and images of defects appear like magic.
Details of the method are given in ref. (6). Surface defects like scratches, dimples, marks from prior treatment, "orange peel", and contours due to different attack of the polish, in general all deviations from the ideal flat surface can be seen immediately on the image reflected on a screen. Fig. 3a is the image of a perfect polished surface, fig. 3b shows defects like polishing marks and a slight contour structure. Such a surface would not be qualified for the manufacture of ULSI devices.

Microroughness: Lightscattering

We developed the method of elastic lightscattering to resolve the lateral surface structure down to the submicron range (0.16 μm). The basic principle and details are given in ref. (7). A laser beam hits the surface at a definite angle. The reflected beam is annihilated by a light trap, the diffuse scattered light is detected by a photomultiplier. Using two different la-

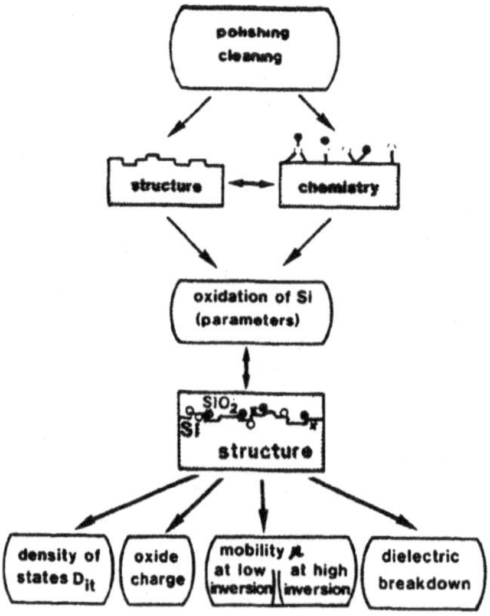

Fig. 1. Dependence of interface roughness and electrical properties on surface structure and chemical state

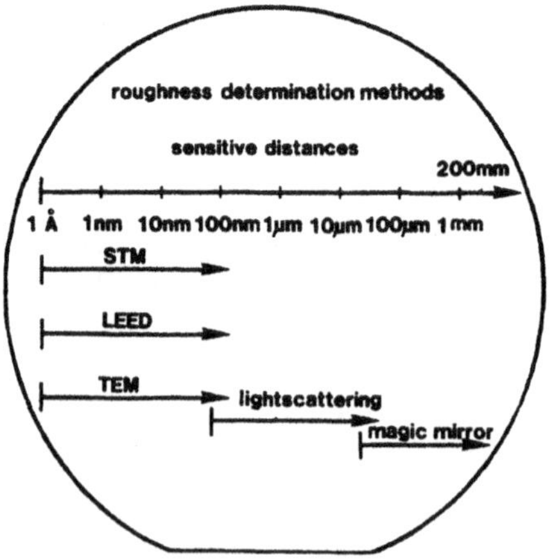

Fig. 2. Sensitive distances and related methods for surface roughness determination

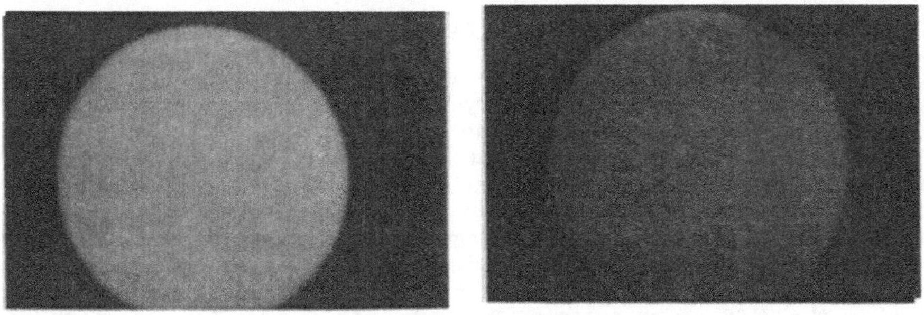

a b

Fig. 3. Magic Mirror images of a perfectly polished wafer (a) and with polishing marks (b)

sers with a penetration depth in silicon of 2 µm (HeNe laser, λ = 632 nm) and 0.01 µm (HeCd laser, λ =325 nm), respectively, superficial and subsurface defects, as well as periodicities of surface structures, can be analyzed.

Angle-resolved measurements provide informations about roughness components corresponding to different scattering vectors. The highest lateral resolution results at grazing incidence of the lightbeam, if the detector is placed in backscattering position. Using the HeCd laser structures are detectable down to a length of 0.16 µm.

Fig. 4 shows angle-resolved measurements at wafers from different polishing stages. The detector was moved stepwise from 20° to 160° (corresponding to correlation lengths of 0.35 µm to 14 µm; 170° is the position of specular reflected beam).

This figure demonstrates how the diffuse scattered intensity for all scattering angles is reduced in the course of polishing.

For calibration we compared our angle resolved measurements with Total Integrated Scattered (TIS) intensity measurements, a new ASTM-standard developed by Mattsson et al (8) to determine the root mean square (RMS) roughness.

In fig. 5 RMS values obtained by the TIS method are plotted versus the square root of the scattered intensity measured by our method. A linear relationship exists for the correlation length of 0.9 µm (shown in the figure) as well as for shorter ones. The RMS-roughness of a perfectly polished Si-surface is below 0.1 nm.

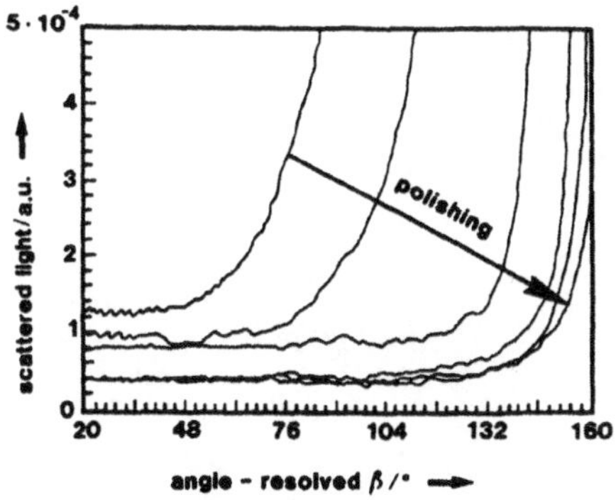

Fig. 4. Angle resolved measurements at wafer from different polishing stages (angle of incidence: 80°)

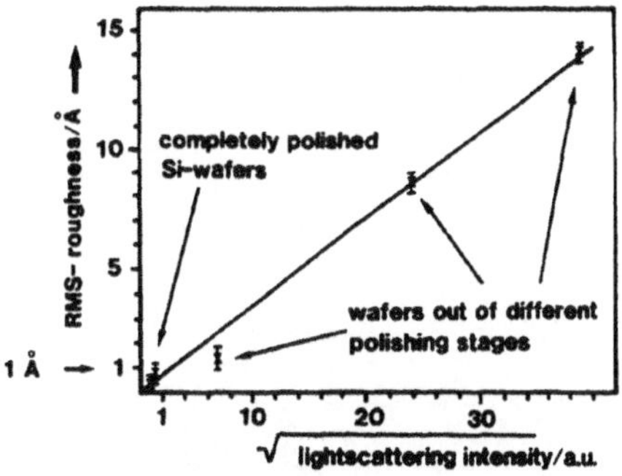

Fig. 5. RMS roughness determined by Total Integrated Scattered Intensity (TIS) vs square root of the scattered light in our system (8)

Integral Atomic Roughness: Low Energy Electron Diffraction (LEED)

With the low energy used in LEED (50 - 150 eV) the electrons are reflected at the topmost surface layers of a monocrystal, giving a pattern of sharp spots on the LEED screen at any energy in the case of an atomically flat, undisturbed surface. If steps are present the spot shape changes between "sharp" (in-phase scattering) and "characteristic" (out-of-phase) profiles. Profiling those spots precisely yields information on the distribution of terraces with respect to orientation, width, and regularity. Furthermore, point defects, domains and superstructures, planar defects like facets and bulk defects like mosaic structures and lattice strain can be identified by characteristic features in the diffraction pattern. The prospects and limitations of the method are reviewed and discussed in ref. (9).
We used this method to control the progress of surface flattening in etching and polishing. At any stage of etching and polishing we observed bright LEED-patterns exhibiting the (1x1) - structure (10).
The occurrence of diffraction spots with etched surfaces demonstrates that chemical etching generally produces undisturbed surfaces. The atomic step density, however, depends strongly on the etchant. Whereas acidic etched Si (111) surfaces show very high step densities, alkaline etched Si (111) surfaces are atomically very flat, disregarding macroscopic steps, which are not detectable by LEED. Fig. 6 demonstrates that polishing smoothens those surfaces, further increasing the length of terraces without steps. Perfectly polished samples exhibit less than one atomic step in a row of 30 atoms, which is the instrumental resolution. On purpose misoriented wafers show regular terraces. The regularity increases with polishing. The number of atoms per terrace correlates to the angle of misorientation, demonstrating the high quality of the polished wafer surface.

Local Atomic Roughness: Scanning Tunneling Microscopy (STM)

The LEED measurement integrates over an area of about 1 mm^2 and is therefore insensitive to a single defect. The method to choose in this case is STM, it provides the highest possible local resolution.
Fig. 7 shows the STM image of a perfectly polished wafer (p-type, (100)) (11) immediately after an HF-dip, which appears smooth on an atomic scale. The average of measurements on different spots yields an RMS-roughness value of 0.12 nm (atomic correlation lengths).
This wafer was selected by the lightscattering technique which gave an RMS-value of about 0.05 nm using TIS calibration (correlation length in the µm range).

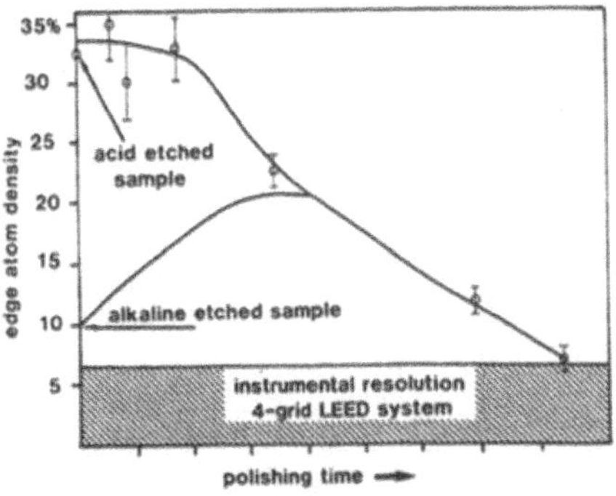

Fig. 6. Reduction of atomic step density by polishing (10)

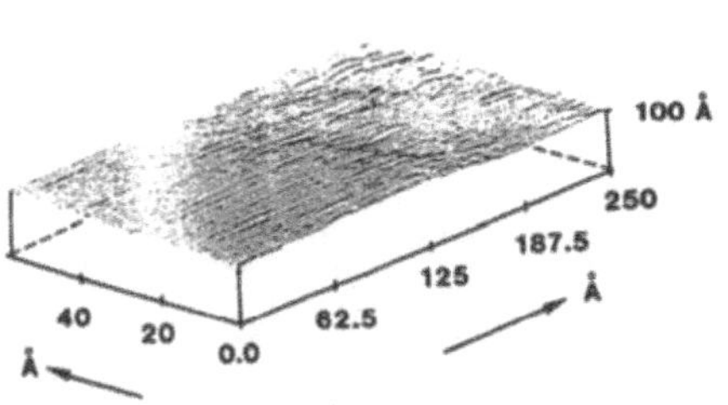

Fig. 7. STM image of a perfectly polished Si (100) surface showing 1.2 Å RMS roughness (11)

III. INFLUENCES ON ROUGHENING: THE CHEMICAL STATE OF THE SURFACE

Depending on the cleaning procedure the surface of a silicon wafer is either hydrophilic or hydrophobic, which can be measured by the contact angle of a water droplet on the surface. The hydrofluoric acid dip yields a hydrophobic surface, and a procedure such as the RCA cleaning gives a hydrophilic surface.

ESCA data demonstrate the chemical differences of these surfaces (12). Hydrophobic surfaces show a small amount of oxygen ($1 \times 10^{14}/cm^2$) preferably bonded to carbon. The fluorine coverage is surprisingly small ($1 \times 10^{14}/cm^2$ after 1 min in 40% HF). Less concentrated HF yields even lower coverages with fluorine (13). Longer exposure times do not raise the level of fluorine over $1.5 \times 10^{14}/cm^2$. This indicates that only 10% of the available bonds at (100) surfaces ($1.36 \times 10^{15}/cm^2$) can be saturated with fluorine.

In contrary hydrophilic surfaces give a large oxygen peak. The corresponding Si-peak shifts about 3.5-3.6eV, indicating a thin surface oxide. The oxide thickness, depending on the cleaning condition, was deduced from the ratio Si oxidized/Si elemental to be about 0.6 nm. Further and more detailed information was obtained by HREELS-measurements. Typical spectra are depicted in fig. 8. The oxide on the hydrophilic surface manifests itself in a strong asymmetric Si-O-Si vibration at 1150 cm^{-1}. This frequency is reduced by about 40 cm^{-1} compared to thermal oxides of similar thicknesses (12). Besides this shift the spectrum is equal to that of vitreous SiO_2. The magnitude and direction of the shift can be interpreted as a change in the angle of the Si-O-Si bridges (12), corresponding to a mean bridging angle of around 130 ° in contrary to 144° of bulk vitreous silica.

Beyond 3000 cm^{-1} we observe losses due to Si-OH groups. The OH band splits into singular groups at 3680 cm^{-1}, and associated groups around 3450 cm^{-1}. The wet chemical oxide thus contains structural units of the type $Si\text{-}O_x(OH)_y$ with x+y=4 in contrast to the network forming $Si\text{-}O_4$ tetrahedra of "dry" oxides. The hydrophilic properties are caused by the OH groups forming hydrogen bonds with water. Therefore, after storage in humid air the surface will be covered by a layer of hydrogen bonded water. Certainly these surfaces exhibit more Si-OH groups than those kept in UHV (approximately one OH group per tetrahedra), where Si-O-Si bridges are formed by condensation of associated groups.

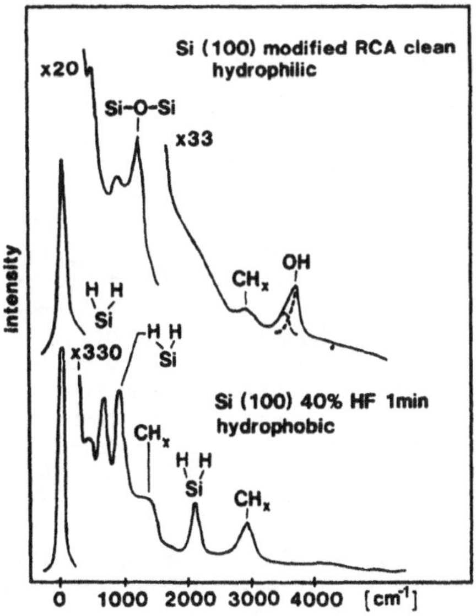

Fig. 8. Typical HREELS spectra of hydrophilic and hydrophobic surfaces

The same happens on aging in normal dry air. This condensation, however, is not homogenious across the wafer. The oxide layer on aged hydrophilic wafers dissolves lateral inhomogeniously after heating to about 800 °C in UHV (14). The HREELS spectrum of an hydrophobic (HF dip) Si (100) p wafer is depicted in the lower part of fig. 8. The hydride coverage shown is prevalent on this type of surface. The main lines are the Si-H_2(or Si-H) stretching vibration at 2100 cm^{-1}, a Si-H_2 scissor vibration at 900 cm^{-1}, and a Si-H_2(or Si-H) bonding vibration at 640 cm^{-1}. Note that no Si-O-Si excitation exists in the region 1050-1150 cm^{-1}.
ESCA measurements show that the coverage with elements besides H is less than 0.1 monolayer, consequently the hydrophobic property of the surface can be ascribed to the hydride. Hydrophobicity is retained if the wafer is rinsed with water. ESCA-measurements (13) show that fluorine on the surface is instantly displaced, the occurrence of OH-groups is proven by HREELS-measurements. These OH groups are the nuclei for further attack of water, which leads to a partial coverage with hydrous oxide within hours. Hydrophilic spots are formed within a hydrophobic matrix changing the local chemical reactivity. On storage in air the hydride groups are mostly retained but associated OH groups can be observed additionally (12). The Si-H groups are rather stable. They disappear on heating in high vacuum at temperatures around 750 °C simultaneously with the native oxide. At high temperatures and under oxidizing conditions the hydrogen may react with oxygen to form water, which enhances the oxidation rate by the cleavage of Si-Si bonds:

$$\text{Si-Si} + H_2O \qquad \text{SiOH} + \text{SiH}$$

We assume that either H is released again by thermal decomposition or that Si-H is oxidized to Si-OH. Then condensation leads to Si-O-Si and H_2O which again attacks Si-Si bonds.

Thus, there are mechanisms possible changing the local rate of oxidation, depending on the local variation of chemical species on the surface. These changes are only related to different cleaning procedures, storage, and annealing prior to oxidation. They will result in a roughness increase of the Si-SiO_2 interface after thermal oxidation.

IV. ROUGHNESS OF THE Si-SiO_2 INTERFACE

Atomic Roughness: LEED

The atomic roughness of the interface of oxidized silicon wafers was determined by the LEED method after removal of the oxide by HF (15). The results are summarized here:

1. Low oxidation rates (thick oxides, dry atmosphere) yield low roughness values, whereas the roughness is increased by high oxidation rates (thin oxides and wet atmosphere).
2. Fig. (9) demonstrates the influence of the post oxidation annealing temperature (nitrogen)on roughness.
3. There is a marked dependence on the pretreatment of the wafer before oxidation. A hydrophilic surface, after storage covered with inhomogeniously distributed Si-OH groups, shows tremendous roughening on an atomic scale after thermal oxidation compared to freshly prepared hydrophobic surfaces. Annealing in an UHV to the 7 x 7-structure followed by an HF-dip prior to oxidation leads to atomically smooth interfaces (15).

The addition of chlorine during oxidation causes slight roughening with moderate step atom densities (15), which agrees with the microroughness determined by lightscattering (see next chapter).

Fig. 9. Step atom density vs post oxidation annealing temperature (nitrogen) (15)

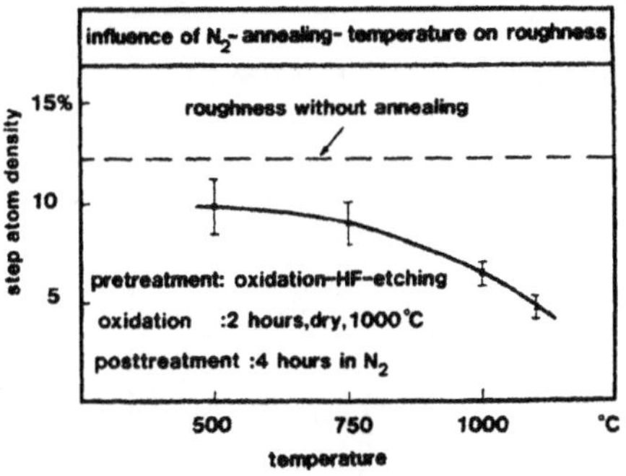

Microroughness: Lightscattering

Fig. 10 presents a relation between oxidation temperature, oxide thickness and microroughness (7).
Six wafers from each type of material were used to quantify the initial smoothness. All samples were subsequently RCA-cleaned. Immediately after cleaning the samples were oxidized in a dry atmosphere (4% Cl_2) at 800°C or 1000°C simultaneously. After oxidation the oxide was removed, the surface roughness determined, the cleaning procedure repeated, and the oxidation performed once again. This procedure was repeated several times.
The results demonstrate:

1. With increasing oxide thickness the roughness at the Si-SiO_2 interface increases substantially.
2. The roughness depends on the oxidation temperature. Oxidation at 1000°C leads to a pronounced roughness increase compared to the oxidation at 800°C.
3. Floating zone silicon shows a significantly smaller increase in roughness compared to Czochralski grown crystals.

The influence of other oxidation parameters is shown in fig. 11.
Dry oxidation roughens the surface substantially, whereas wet oxidation at the same temperature to the same thickness causes less roughening as determined by lightscattering. This is opposite to the behaviour on the atomic scale mentioned before. In accordance with the LEED-measurements are the results of chemical pretreatment prior to oxidation: A hydrophobic surface (Si-H groups due to an HF-dip) leads to a smoother interface compared to a hydrophilic surface (Si-OH groups due to HCl/H_2O_2 treatment) (7).
Fig. 12 shows the influence of pretreatment in a striking way. Before oxidizing to 700 Å (dry, 1000°C) a hydrophilic wafer was dipped halfway into deionized water. This dipped part of the wafer shows less roughness after oxidation and oxide removal.

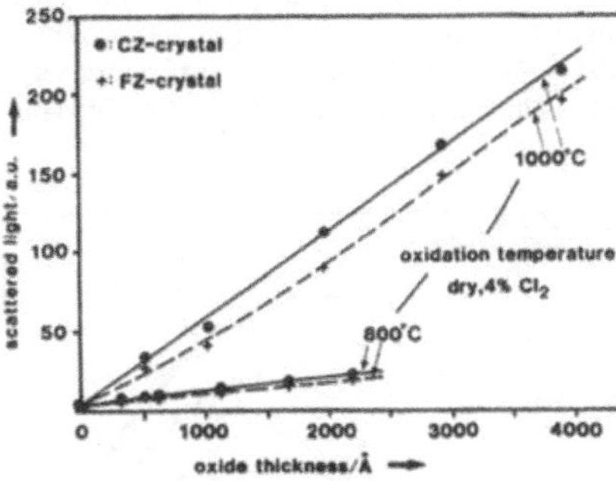

Fig. 10. Microroughness as a function of oxide thickness, oxidation temperature, and crystal growth condition (7)

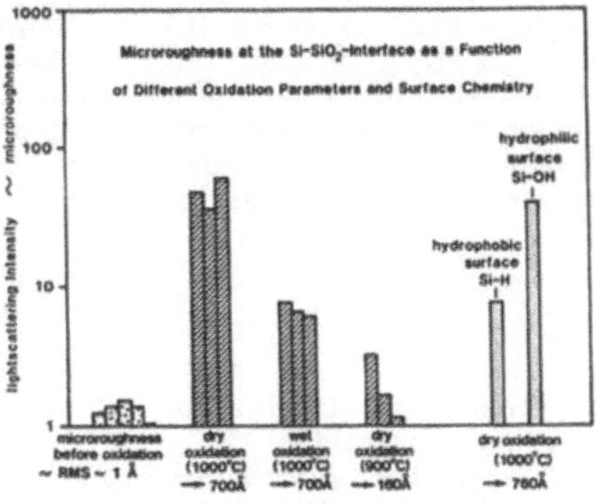

Fig. 11. Influence of different oxidation parameters on interface roughness (7)

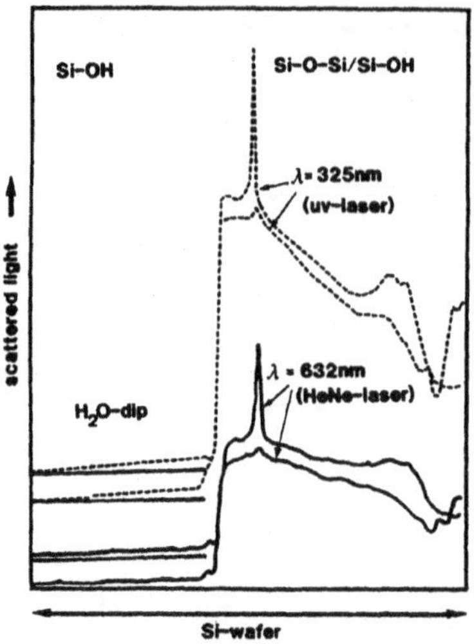

Fig. 12. Differences in interface roughness on a wafer dipped halfway into water prior to oxidation (7)

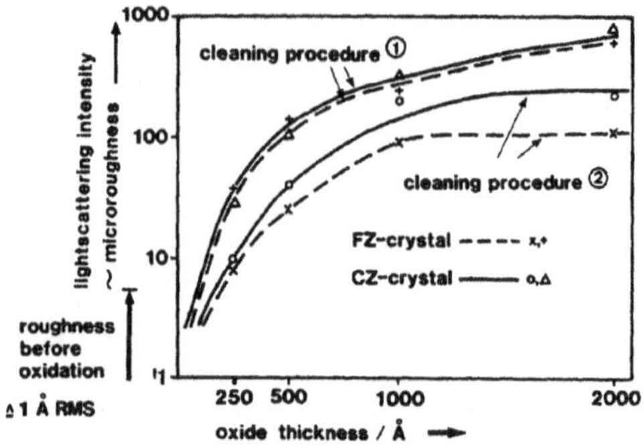

Fig. 13. Interface roughness vs oxide thickness, cleaning procedure, and growth condition 1: cleaning with NH_3/H_2O_2
2: additional treatment in HCL/H_2O_2 (7)

The figure gives two scans at a distance of 1.5 mm performed with both types of lasers, which shows the same structure, as well as the sharp peak caused by a scattering particle.

Fig. 13 presents interface roughness versus oxide thickness, at different cleaning procedures and crystal growth conditions (FZ grown and CZ grown, p (100), 15 Ohmcm). The initial smoothness of the polished wafers was in the 0.1 nm range.

We obtained the following results:

1. Again the interface roughness increases continously with oxide thickness for all samples investigated.
2. Roughness increase depends on cleaning procedures.
3. Using a cleaning method which influences roughness only slightly we observe less roughening on FZ grown compared to CZ grown material.

V. CORRELATED ELECTRONIC PROPERTIES

Due to its impact on electronic properties interface roughness is not only a mar. Fig. 14 shows the influence of the atomic roughness on the mobility in MOS-inversion layers. The mobility was deduced from Hall-measurements in liquid helium at special devices enabling roughness measurements on the same portion of the wafer (16). At low inversion we observe an increase of the mobility due to coulomb scattering, and at high inversion a decrease due to roughness scattering.

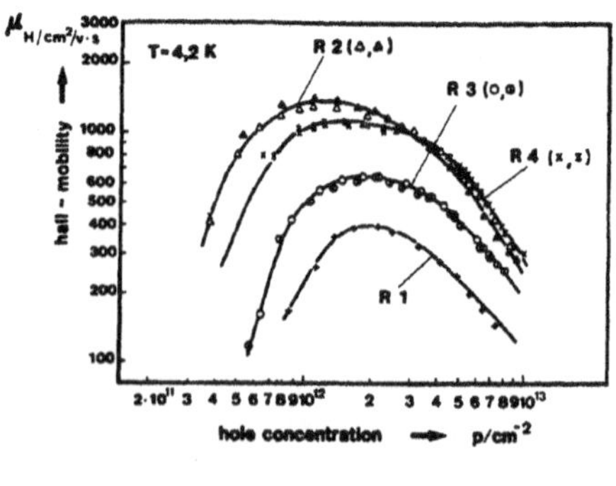

Fig. 14. Hall mobility vs hole concentration
Samples have the same oxide thickness but different atomic step densities:
R1=17%; R2=8%; R3=21%
R4=19%) (16)

After annealing in nitrogen, e.g, we measured twice the mobility and half the step density at sample R2 (8%; 280 $cm^2/v \cdot s$) compared to sample R1 (17%; 140 $cm^2/v \cdot s$) without annealing (atomic step density and mobility at high inversion in brackets: both samples were oxidized in a dry atmosphere simultaneously). Lightscattering measurements on these samples corroborated these results in terms of microroughness. Surprisingly the roughness reduces the mobility at all carrier concentrations. This suggests that charged centers have to be associated with steps or kinks at step edges preferably.
In spite of the highest atomic step density, however, we observe at R3 (21%; 230 $cm^2/v \cdot s$) a higher mobility than at R1 (17%; 140 $cm^2/v \cdot s$). In ref. (16) we explained this behaviour with hydrogen introduced by the wet oxidation of sample R3. Supplementary lightscattering data now show less roughening on the microscopic scale which also could give an explanation.
Additional N_2-annealing of the wet oxidized sample (R3) leads to sample R4. Now we observe a smooth interface for R4 (10% step atom density) and an increase in mobility for R4 (330 $cm^2/v \cdot s$), too.
The results demonstrate, that the product of atomic roughness and mobility at high inversion is constant approximately. In addition CV-measurements on the same devices show that both the interface state density and the fixed oxide charge depend on the atomic roughness (17) in a similar way.
Dielectric breakdown is another significant parameter to characterize thin gate oxides. Fig. 15 demonstrates the influence of the microroughness (lightscattering) on the dielectric breakdown. Samples from an intermediate polishing step with a rough but damage free (as indicated by LEED (18)) surface of about 8 Å RMS (TIS-calibration) show very early intrinsic breakdown (1-2 MV/cm) values. After final polishing (RMS < 1 Å) intrinsic breakdown fields are measured at about 10 - 12 MV/cm. FZ-material leads to a slightly higher intrinsic breakdown field strength compared to identically processed CZ-grown wafers.
The influence of roughness induced by the oxidation itself is demonstrated in fig. 16. All wafers used in fig. (13) received different degrees of microroughness by oxidation to various oxide thicknesses. After the roughness measurement those wafers were oxidized to 300 Å again for breakdown evaluation. Once again, with increasing roughness we observe a decrease of the intrinsic breakdown field strength.
These experiments demonstrate that the determination of the oxide thickness by optical methods depends on the interface roughness, too. The oxide thickness values used in fig. 16 were calibrated with TEM-measurements (19).

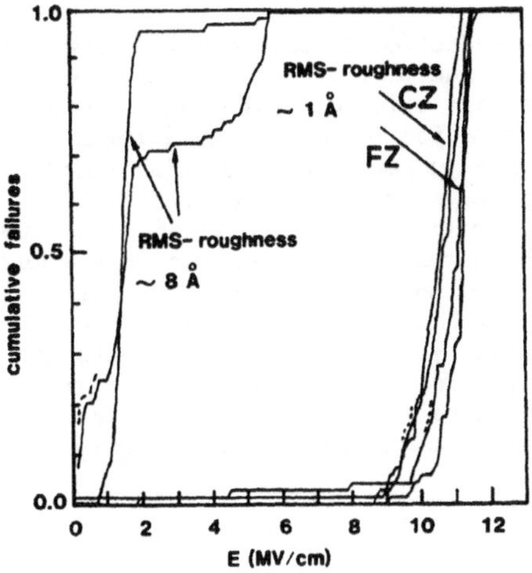

Fig. 15: Influence of interface roughness on the cumulative failures of MOS capacitors

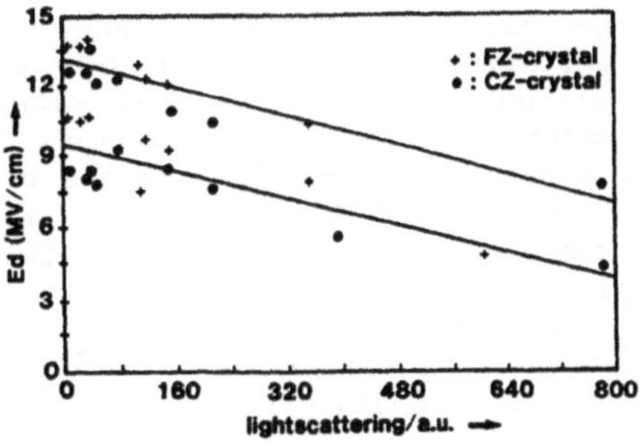

Fig. 16: Breakdown field strength vs interface roughness. The two curves correspond to different methods for breakdown evaluation (7)

VI DISCUSSION

Chemo-mechanical polishing produces excellent surfaces with respect to atomic roughness (LEED, STM, TEM) and microroughness (lightscattering). All atoms maintain their (1 x 1) surface structure at the topmost surface layer. From High-Resolution-LEED-data we can deduce that on average a distance of more than 100 Å is step free and the vertical distribution of steps in a stretch of 1000 Å is confined to a few atomic layers only (10). The surface atoms occupy regular lattice sites with a measurable accuracy of less than 0.3 % (10). However, the STM-image (fig. 7 and (11)) exhibits some residual roughness. Considering the STM-measurements the surface appears rougher than expected from the LEED data. This difference can be explained by statistically adsorbed atoms which may pretend a roughness in the STM-image and do not respond to LEED.

Feenstra (20) reports a roughness of about 1.8 Å in agreement with our best STM-value of 1.2 Å. Lightscattering measurements confirm the smoothing effect of polishing in the submicron range, but angle resolved measurements are still able to show distinct differences for special correlation lengths.

Oxidation leads to a pronounced roughening at the Si-SiO_2 interface. The following model may explain the results (15): Oxidation is a random process of diffusion and reaction of oxygen at the interface, so we expect a steady roughening with growing oxide thickness. On the other hand in thermal equilibrium a smooth, step-free surface is energetically favoured, as is the free surface of silicon, where, on annealing in UHV, step-free surfaces are produced. The balance between roughening and smoothing determines the final roughness.

Therefore, both atomic and microroughness depend on oxide thickness, oxidation parameters, and bulk properties. However, wet oxidation shows a paradoxical behaviour, as we observe a higher atomic roughness (15) and a lower microroughness (7) compared to dry oxidation.

In comparison the oxidation produces less roughening on the atomic scale than in the submicron range. Up to now the submicron range has been neglected due to a lack of appropriate techniques.

In addition, adsorbed atoms and molecules given by the cleaning procedure influence the chemical reactivity and consequently interface roughness.

There is an indication that even "contaminations" in the bulk like oxygen, carbon, nitrogen, and inhomogeniously distributed dopants (striations), may influence the interface roughening. Interface roughness, therefore, is a very sensitive indicator for all parameters influencing the oxidation.

The presented roughness data are in contradiction to the general belief, that the Si-SiO_2 interface appears very sharp and abrupt. Many TEM-data and some STM-measurements (20) have established this opinion in the past. We believe the resolution of these methods is simply too high and the information is confined to such a small distance making a judgement on the total wafer surface impossible. A tremendous number of images would be needed.

Furthermore electrical properties like interface states, fixed oxide charges and carrier mobility depend on the atomic roughness. Our data confirm calculations by Schrieffer (21), Stern and Howard (22), and Cheng (23) who predicted these results theoretically. Additional TEM-measurements by Goodnick (24) confirm the correlation to mobility. Breakdown measurements show a systematic dependence on the interface microroughness. Our results agree with TEM-data obtained by Carim (25) and Ravindra (26), too. The degradation of the oxide quality with increasing oxide thickness was reported in many papers (27).

VII CONCLUSION

The results demonstrate that the interface roughness has a decisive influence on the electronic properties of MOS-structures.

The interface roughness results from the interplay of many parameters as demonstrated. More detailed investigations have to be carried out. In general, oxidation increases the roughness. Therefore, polished wafers for ULSI circuits have to be perfect in terms of surface structure. Surface cleaning methods and the subsequent processing steps have to observe minimal roughening. The lightscattering technique is an easy and reliable tool to control and improve processing steps on line for better device performance.

Acknowledgement

The authors thank M. Henzler for very helpful discussions regarding the LEED and lightscattering measurements.

References

1. Springer Series in Electrophysics 7, edited by M. Schulz and G. Pensl (Springer, Berlin, 1981)
2. S.T. Pantelides (ed.): "The Physics and Chemistry of SiO_2 and Its Interfaces", (Pergamon, New York 1980)
3. Proceedings of the International Conferences on Insulating Films on Semiconductors INFOS (1987 Leuven, 1985 Toulouse, 1983 Eindhoven) North-Holland, Amsterdam
4. T. Ando, A.B. Fowler, and F. Stern, Rev. Mod. Phys., 54, 437 (1982)
5. A. Qurmazd, in "SiO_2 and Its Interfaces" ed. by G. Lucovsky and S. Pantelides (Mater. Res. Soc. Proc. to be published, Boston 1987)
6. K. Kugimiya, J. Electrochem. Soc., 130, 2123 (1983)
7. P.O. Hahn, I. Lampert, and A. Schnegg, in "SiO_2 and Its Interfaces" ed. by G. Lucovsky and S.T. Pantelides (Mater. Res. Soc. Proc. to be published, Boston 1987)
8. L. Mattsson, in "Thin Film Technologies", SPIE Proceedings, 652, Innsbruck (1986); (ASTM standard: F 1048)
9. M. Henzler, in "Advances in Solid State Physics", 27, 185, ed. by P. Grosse, Vieweg, Braunschweig (1987)
10. P.O. Hahn, in "Thin Films - Interfaces and Phenomena ed. by R.J. Nemanich and P.S. Ho (Mater. Res. Soc. Proc., 54, 645 (1986)
11. E. Hartmann, F. Koch, P.O. Hahn, and J. Behm, Journ. of Appl. Phys., to be published
12. M. Grundner and H. Jacob, Appl. Phys. A, 39, 73 (1986)
13. M. Grundner and R. Schulz, Proc. of the Topical Conf. "Deposition and Growth: Limits for Microelectronics" ed. by G.W. Rubloff, (AVS Meeting, Anaheim, 1987)
14. M. Grundner, J. Vac. Sci. Technol. A, 5, 2011 (1987)
15. P.O. Hahn and M. Henzler, J. Appl. Phys., 52, 4122 (1981) and J. Vac. Sci. Technol. A, 2, 574 (1984)
16. P.O. Hahn and M. Henzler, J. Appl. Phys., 54, 6492 (1983)
17. P.O. Hahn, S. Yokoyama, and M. Henzler, Surf. Sci., 142, 545 (1984)
18. G. Pietsch and M. Henzler to be published
19. F. Le Goues private communication
20. R.M. Feenstra and G.S. Oehrlein, Appl. Phys. Lett., 47, 97 (1985)
21. J.R. Schrieffer, Phys. Rev., 97, 641 (1955)
22. F. Stern and W.E. Howard, Phys. Rev., 163, 816 (1967)
23. Y.C. Cheng, Proc. of the 2nd International Conf. on Solid Surfaces, Kyoto, Jpn. J. Appl. Phys., 2, 363 (1974)
24. S.M. Goodnick, R.G. Gann, J.R. Sites, D.K. Ferry, C.W. Wilmsen, D. Fathy, and O.L. Krivanek, J. Vac. Sci. Technol. B, 1, 803 (1983)
25. A.H. Carim and R. Sinclair, Proc. Electrochem. Soc. 86-4, 458, (1986)
26. N.M. Ravindra, J. Narayan, D. Fathy, J.K. Srivastava, and E.A. Irene, J. Mater. Res., 2, 216, (1987)
27. M.S. Liang and J.Y. Choi, Appl. Phys. Lett., 50, 2, (1987)

DEFECTS AND IMPURITIES IN SiO_2 INTERFACE

FOR OXIDES PREPARED USING SUPERCLEAN METHODS

Tadahiro Ohmi, Mizuho Morita, and Takeo Hattori*

Department of Electronics, Tohoku University
Sendai 980, Japan

*Department of Electrical and Electronic Engineering
Musashi Institute of Technology
1-28-1 Tamazutsumi, Setagaya-ku, Tokyo 158, Japan

ABSTRACT

A superclean oxidation technology characterized by hydrogen and nitrogen contamination-free oxidation has been developed. It was experimentally demonstrated that an Si-SiO_2 interface with a high energy barrier height of 3.5 eV for electron emission from Si into SiO_2 could be realized. X-ray photoelectron spectroscopy analysis revealed that the valence-band discontinuity of the Si-SiO_2 interface is 4.5 eV which is smaller than that of the conventional oxide. These results suggest that the direct tunneling and Fowler-Nordheim tunneling current densities through the oxide can be controlled by interface doping for requirements of each target device oxide such as gate oxide, DRAM storage capacitor oxide, E^2PROM oxide, etc.

INTRODUCTION

Control of the electrical properties of the Si-SiO_2 interface in the MOS system is one of the major factors for the fabrication of stable and high performance silicon devices and integrated circuits. In the advanced ULSI where the thickness of gate oxide and DRAM storage capacitor oxide will become thinner, the energy barrier height for electron at the Si-SiO_2 is very essential in addition to the traps located at or very near the Si-SiO_2 interface. Because the energy barrier height which is one of parameters controlling the direct tunneling current densities through the oxide decides the minimum oxide thickness that may lead to the device miniaturization limit. The electrical properties at the Si-SiO_2 interface have been generally accepted to be strongly dominated by defects and impurities in the Si-SiO_2 interface. However, no one still has realized the defects-free and impurities-free Si-SiO_2 interface, and the effect of defects and impurities in the Si-SiO_2 interface on the energy barrier height for electron at the interface has not been well understood. In order to approach the nature of the Si-SiO_2 interface and realize the high performance ULSI devices, we have developed the superclean oxidation technology. In this paper, we discribe the superclean oxidation technology characterized by hydrogen and nitrogen contamination-free oxidation and the energy barrier height for electron at the Si-SiO_2 interface for oxides prepared using the superclean method.

EXPERIMENTAL

We have designed and set up the superclean oxidation furnace[1] as illustrated in Fig. 1. The oxidation system is comprised with the source gas sylinder, the gas supply system including the gas purifier, the quartz reactor, the lamp heater, the pumping system and the loadlock system. In order to decrease the densities of nitrogen-related traps in the SiO_2 and in the Si-SiO_2 interface, the high purity oxygen gas which was produced by the electrolysis of ultrapure deionized water was employed as an oxidant. The content of nitrogen in the oxygen gas supplied to the reactor was below 2 ppm, which is below 1/10 of that in the commercial oxygen gas produced by the conventional air separation technique. The gas supply system had been also developed based on the concept of the ultra clean technologies.[2] The materials used in this system, which are in contact with the gases, are the stainless steel (SUS316L) from which the degassing rate is very low. The inner surfaces of the stainless steel in contact with the gases were polished by the electro-polishing[3] or ultra clean treatment,[4] the roughness (Rmax) of which is below 0.5 μm. The amount of the moisture adsorbing on the inner surface of this tube is smaller than that of the conventional blight anneal tube, because the area of the polished surface is small. So this gas supply system can supply the ultra pure gases to the reactor by the short time purging of the purified gas even after the gas flow stopping.[5] The substitution of a gas to other kind of gas in this system is fast due to emplying the valves in which the dead space for the gas stream is little.[6] Particles having effective diameters greater than 0.1 μm could not be detected in the gas of this system by the particle counter. The external leak of this system was less than $2x10^{-11}$ atm cc/sec. The oxygen gas was purified by a purifier in the gas supply system in order to decrease the densities of moisture-related traps in the SiO_2 and in the Si-SiO_2 interface. The content of moisture in the oxygen gas supplied to the reactor was below 0.2 ppm (-85°C in dew point). The cold-wall-type quartz tube reactor was employed in order to avoid the contamination by the diffusion of impurities from and through the quartz tube. The halogen

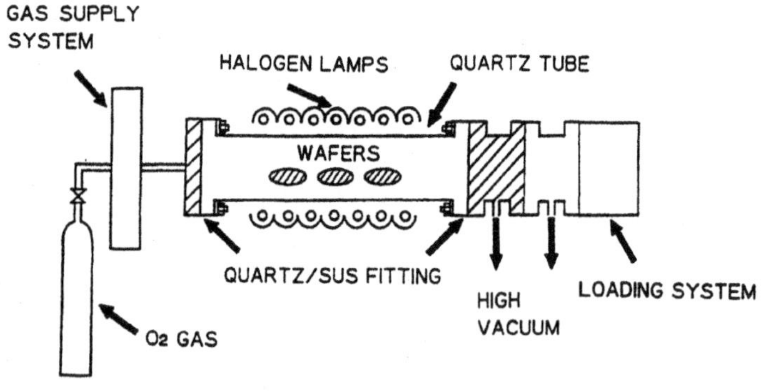

Fig. 1 Superclean oxidation furnance. The external leak of reactor is below $2x10^{-11}$ atm cc/sec and extremly tight.

lamps were utilized as a heater, from which the divergence of impurities and/or particles is little. The fitting between the reactor and the stainless steel flange which we have been designed is extremely tight and so the external leak of the reactor was also below $2x10^{-11}$ atm cc/sec. The pumping system consists of the turbo molecular pump and the stainless steel (SUS316L) tubes with the electrochemically buffered inner surfaces, which connects between the reactor and the turbo molecular pump. Before oxidation, the reactor was pumped down to below $1x10^{-9}$ Torr by the high vacuum system. The setting of the wafer into the reactor was performed by using the automatic loadlock system with the magnetic floating arm without the mechanical rubbing, which enables to the particle-free loading, so as to prevent the contamination of the reaction chamber by the impurities in air during the transportation of the wafer.

P-doped (3.6-5.4 Ω cm) Si (100) wafers having a diameter of 33 mm were used as substrates. Prior to the oxidation, the substrates were chemically cleaned and etched. The oxidation was carried out at a temperature of 800°C in an atomospheric pressure. The ultra pure Ar gas in which the contents of moisture and other impurities were less than 10 ppb[5] was used as a inert gas for the purging of the reactor before and after the oxidation. Doped poly-Si/SiO_2/Si (MOS) devices were fabricated for the evaluations of the electric characteristics of the SiO_2 and the Si-SiO_2 interface. The oxide thickness was measured by ellipsometry and checked from the accumulation capacitance of the MOS devices. The band lineup at the Si-SiO_2 interface was determined from the quantitative analysis of highly resolved valence band spectra. The detailed chemical structures of SiO_2 and its interface, which was not detected previously because of low resolution,[7,8] were also determined from highly resolved Si 2p photoelectron spectra.[7,8] The electric properties of the SiO_2 and the Si-SiO_2 interface were evaluated through the measurements of the high frequency capacitance-voltage (C-V), the quasi static C-V, the ramp current-voltage (I-V), and the dc I-V characteristics of the MOS devices.

RESULTS AND DISCUSSION

Figure 2 shows the Fowler-Nordheim plots of the currents through the oxide from Si for the superclean oxide having the thickness of 14.3 nm and for the conventional oxide, where the conventional oxide was grown using the commercial electric furnace as a reference. The current at the average electric field in the range over about 7 MV/cm is found to be dominated by the Fowler-Nordheim tunneling. The energy barrier height of the superclean oxide at the Si-SiO_2 interface for the electron emission from Si into SiO_2, which was derived from the gradient of the Fowler-Nordheim plots postulating that the relative effective mass is 0.42,[9] is 3.5 eV which is higher than 3.2 eV of the conventional oxide. This result implies that the superclean oxide has an advantage for its application to the gate oxide and DRAM storage capacitor oxide in the ULSI devices compared with the conventional oxide. In Fig. 2, the data of the superclean samples coincide with the ideal line over six orders of the current densities. This indicates that the superclean Si-SiO_2 interface has an ideal potential configuration.

The valence-band discontinuity is also very sensitive for the interface doping.[10] X-ray photoemission spectra taken in the region of the valence-band edges reveal that the valence-band discontinuity of the superclean samples at the Si-SiO_2 interface is 4.5 eV, where this value is smaller than the conventional samples value of 4.7 eV, as seen from Fig. 3. Variations of the valence-band discontinuties due to the interface doping produced by the annealing of the superclean oxide are

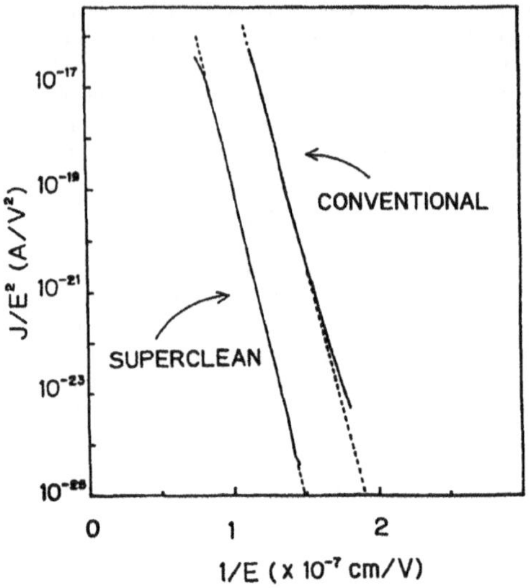

Fig. 2 Fowler-Nordheim plots of the current through the oxides.

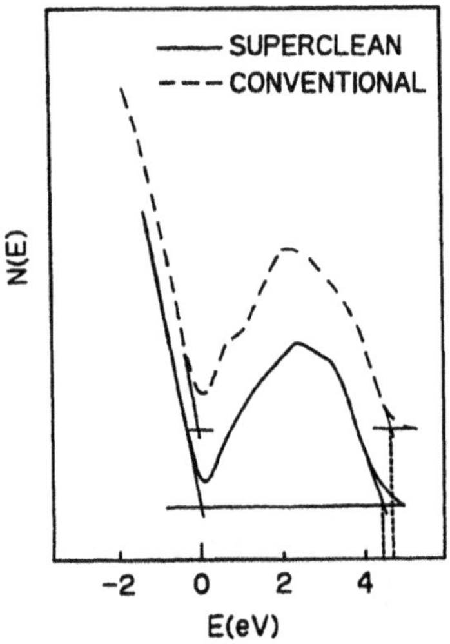

Fig. 3 X-ray photoemission spectra taken in the region of the valence-band edges.

Table 1 Valence-band discontinuities at the $Si-SiO_2$ interface.

Oxidation	Annealing	Band discontinuities (eV)
Superclean	- - -	4.5
	H_2/Ar, 450°C, 30min	4.7
	N_2, 800°C, 60min	5.1
Conventional	- - -	4.7

summarized in Table 1, where the oxide layer thicknesses are approximately 2 nm. It is worthwhile to note in Table 1 that the valence-band discontinuities increase due to the interface doping of hydrogen and nitrogen. This result implies that the conventional oxide inevitably includes hydrogen or nitrogen impurities at the $Si-SiO_2$ interface due to the poor process environment. The change in the energy barrier height at the $Si-SiO_2$ interface for an electron emission from Si to SiO_2 can be correlated with the change in the valence-band discontinuity.

Typical interface trap densities as a function of the energy for the superclean $Si-SiO_2$ interface are shown in Fig. 4, where the densities were measured by the quasi-static C-V technique. The interface trap densities at midgaps are of order of 10^9 $cm^{-2}ev^{-1}$. The dielectric

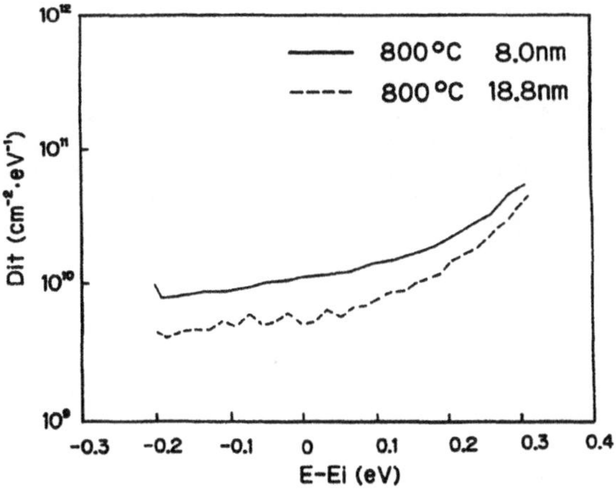

Fig. 4 Interface trap densities as a function of energy for the superclean samples.

breakdown strength of the superclean oxide films is above 10 MV/cm, where the breakdown strength is defined as an average electric field at which a current level exceeds 1 μA.

From the results that Fowler-Nordheim plots of the current through the superclean oxides fit with the ideal line in the region over six orders of the current densities and that the interface trap densities at the Si-SiO_2 interface are very low, it is derived that no charges related to the interface trap contribute to the formation of the potential configuration at the Si-SiO_2 interfaces and so the superclean Si-SiO_2 interface has an ideal potential configuration. One interpretation of the high barrier height is that the SiO_2 band gap energy becomes wide and/or the locatized state near the SiO_2 conduction band edge disappear, because the order of SiO_2 structure near the Si-SiO_2 interface prepared by the moisture-free oxidation may be close to the crystal.[9] By the conventional oxidation which act as one of the interface doping, the energy barrier height at the Si-SiO_2 interface decreases with the increase in the valence band discontinuity as can be seen from Figs. 2 and 3. These results lead to another interpretation that the band gap energy of SiO_2 near the Si-SiO_2 interface is not varied by the interface doping treatment. Further investigation is, however, necessary to clarify the effect of the interface doping on the energy band structure at the Si-SiO_2 interface.

CONCLUSION

We have demonstrated that an Si-SiO_2 interface with a high energy barrier height of 3.5 eV for electron emission from Si to SiO_2 is prepared using the superclean oxidation technology characterized by hydrogen and nitrogen contamination-free oxidation. The valence-band discontinuities of the Si-SiO_2 interface were found to be changed by the annealing treatments as the interface doping. It may be deduced that the change of the barrier height for electron emission at the interface correspond to that of the valence-band discontinuities. These suggest that the direct tunneling and Fowler-Nordheim tunneling current densities through the oxide can be controlled by the interface doping, suiting to requirements of each target device oxide such as gate oxide, DRAM storage capacitor oxide, E^2PROM oxide, etc.

ACKNOWLEDGMENTS

The authors wish to thank Assoc. Prof. J. Murota and Assoc. Prof. T. Shibata for useful discussions, T. Hatayama, E. Morikawa, K. Suma, K. Tanaka and E. Hasegawa for their help in the experiments. This study was performed in the Super Clean Room in Laboratory for Microelectronics, Research Institute of Electrical Communication, Tohoku University.

REFERENCES

1. T. Ohmi, Y. Kanno, T. Hatayama, E. Morikawa, K. Suma, and M. Morita, Proceedings of the 4th ULSI Ultra Clean Technology Symposium, Tokyo 1987, p.203.
2. T. Ohmi, N. Mikoshiba, and K. Tsubouchi, Proceedings of the First International Symposium on Ultra Large Scale Integration Science and Technology, Pennsylvania 1987 (to be published).
3. H. Maehata, Proceedings of the 1st ULSI Ultra Clean Technology Symposium, Tokyo 1985, p.232.

4. Y. Kanno, Proceedings of the 1st ULSI Ultra Clean Technology Symposium, Tokyo 1985, p.14.
5. T. Ohmi, J. Murota, Y. Kanno, Y. Mitsui, K. Sugiyama, K. Kawasaki, and H. Kawano, Proceedings of the First International Symposium on Ultra Large Scale Integration Science and Technology, Pennsylvania 1987 (to be published).
6. M. Toda, Y. Kanno, and T. Ohmi, Proceedings of the 4th ULSI Ultra Clean Technology Symposium, Tokyo 1987, p.100.
7. T. Suzuki, M. Muto, M. Hara, K. Yamabe and T. Hattori, Jpn. J. Appl. Phys. 25, 544(1986)
8. T. Hattori, M. Morita and T. Ohmi, to be presented at Intern. Conf. Solid State Devices and Materials, Tokyo, 1988
9. M. Lenzlinger and E. H. Show, J. Appl. Phys. 40, 278(1969).
10. P. Perfetti, C. Quaresima, C. Coluzza, C. Fortunato, and G. Margaritondo, Phys. Rev. Lett. 57, 2065(1986).
11. F. Rochet, S. Rigo, M. Froment, C. d'Anterroches, C. Maillot, H. Roulet, and G. Dufour, Advances in Physics 35, 237(1986).

THE EFFECT OF CHEMICAL CLEANING ON THE KINETICS OF THERMAL OXIDATION

D. B. Kao
and
B. E. Deal

Fairchild Research Center
National Semiconductor Corp.
Santa Clara, CA 95052

J. M. de Larios
and
C. R. Helms

Stanford Electronics Labs
Department of Elec. Engr.
Stanford University
Stanford, CA 94305

ABSTRACT

A detailed study concerning the effect of chemical cleaning on the kinetics of silicon thermal oxidation is presented. This investigation included a wide range of temperature, thickness, and chemicals for both (100) and (111) silicon wafers. The results indicate that treatment by ammonium peroxide causes substantial retardation of dry oxide growth as compared with acidic solutions such as HCl, H_2SO_4, HNO_3, and HF. The effect of chemical cleaning varies with thickness, temperature, doping, ambient, and crystal orientation. Analysis based on the Deal-Grove model shows that the principal effect of chemical cleaning is on the linear-rate constant and the initial rapid oxidation. A series of two-step oxidation experiments was also conducted with different thermal and chemical treatments between steps. The results demonstrate that the oxidation is affected by the surface film produced on the silicon by chemical cleaning and that this thin film remains on the outer surface of the growing oxide throughout the oxidation process.

INTRODUCTION

Preprocess wet chemical cleaning is used extensively in integrated circuit processing to reduce contamination and improve yield and reliability. The surface films produced by the various chemical treatments can also affect device performance. For example, it has been shown (1,2) that the chemical oxide produced by chemical cleaning is critical to the properties of poly-emitter bipolar devices and contact structures using silicidation of tungsten. It has also been reported (3-7) that the kinetics of thermal oxidation can be affected by chemical cleaning. Continued scaling of integrated circuits demands thinner oxides and requires stringent control and better understanding of not only growth kinetics, but also the $Si\text{-}SiO_2$ interface.

In this study, an extensive set of experiments was conducted to examine the effect of chemical cleaning on the kinetics of thermal oxidation. A wide range of temperature, thickness, and chemicals were included for both dry and wet oxidations, and for (100) and (111) silicon orientations. The kinetic data were analyzed according to the Deal-Grove model (8). An explanation is proposed based on the extracted kinetic parameters and the results of chemical effects on two-step oxidations.

EXPERIMENTAL PROCEDURES

Silicon wafers used in this investigation were SEH 100 mm p-type (100) and (111) wafers, with resistivity in the range of 11 - 18 ohm-cm. The chemical cleaning procedures examined in this study, as listed in Table I, consisted of RCA clean (9), reverse RCA (RRCA), sulfuric peroxide, and MPR, each with an optional final HF-dip. Samples as received from the manufacturer were also included as a reference and designated as "No clean". The wafers were chemically cleaned, rinsed in a dump rinser or an overflow bath, and then spin-dried. The gentle overflow bath was used only after the final HF-dip, because the reactive silicon surface produced by the HF is vulnerable to contamination in the dump rinser. The spin rinser/drier was chosen over manual blow-dry for thorough and consistent drying. To minimize contamination, a getter wafer remained in the HF bath at all times. In addition, test wafers were sandwiched by baffle wafers in the teflon cassette throughout the chemical cleaning, rinsing, and drying processes.

Thermal oxidations were carried out at temperatures of 800°, 1000° and 1200°C, with durations chosen to produce oxide thicknesses of 6 nm to 1 um. The samples were slowly pushed into the oxidation furnace in argon, oxidized in dry O_2 or pyrogenic steam, annealed in argon for 15 minutes and slowly pulled out in argon ambient. In order to minimize the effect of thermal gradient in the oxidation furnace, three groups of samples were oxidized in each run. Oxide thickness was determined by ellipsometry, averaged over the three samples with three to five data points from each. A series of two-step oxidations incorporating different thermal and chemical treatments were also carried out. Their procedures and results will be described later.

RESULTS FROM KINETICS EXPERIMENTS

Effects of Nine Different Chemical Treatments at 1000°C

Significant differences were noticed in oxide growth on p-type (100) wafers in dry O_2 at 1000°C following the nine different cleaning procedures listed in Table I. Samples cleaned with the RRCA procedure exhibited the slowest growth rate. All the other cleans resulted in substantially thicker oxides that were much closer in thickness. In that group, RCA and RCA+HF appeared to result in the thinnest oxides, and RRCA+HF the thickest, although some irregularities were observed. Little effect of chemical cleaning was apparent for thin oxides less than 175 A. A closer examination, however, revealed a slight reversal in thickness: the samples cleaned by the reverse-RCA procedure grew thicker oxides than the samples cleaned by RRCA+HF. Beyond this thin region, the percentage difference in oxide thickness increases with thicker oxide, reaches a maximum around 1000 A, and then decreases with still thicker oxides up to 1 micrometer.

Table I. Chemical cleaning procedures used in this study

RCA	RRCA (Reverse RCA)	Sulfuric Peroxide	MPR	No Clean
$NH_4OH:H_2O_2:H_2O$	$HCl:H_2O_2:H_2O$	$H_2SO_4:H_2O_2$	$H_2SO_4:H_2O_2$	as received
$HCl:H_2O_2:H_2O$	$HF: H_2O$		$HCl:HNO_3$	
	$NH_4OH:H_2O_2:H_2O$			
optional HF	optional HF	optional HF	optional HF	

The thinnest oxide results from the RRCA clean in which ammonium peroxide is used in the last step. Thicker oxides prevail, however, if in the cleaning procedure the ammonium peroxide is followed by HCl (as in the RCA procedure), H_2SO_4 (verified in a separate experiment), or HF. This observation leads to the conclusion that the last chemical used in the preoxidation clean dominates the kinetics of silicon oxidation.

<u>Chemical Effects on Oxidation at Different Temperatures</u>

The overall effect of chemical cleaning on oxidation kinetics is illustrated by oxide growth data of samples cleaned by RRCA and RRCA+HF. They are plotted in log-log format in Fig. 1 for both (100) and (111) orientations at three temperatures (800°, 1000°, and 1200°C) over the thickness range of 60 A to 1 micrometer. Important results are summarized as follows.

(a) <u>Temperature dependence</u> -- As can be seen in Fig. 1, no difference in thickness is apparent at 1200°C. But, for thicknesses around 3000 - 4000 A, a 15% reduction by ammonium peroxide exists at 1000 C, and a 20% reduction occurs at 800°C. The maximum retardation, however, is approximately 23% at both temperatures. It takes place when the thickness is about 900 A at 1000°C, and near 1250 A at 800°C.

(b) <u>Thickness dependence</u> -- At 1000°C, no reduction in thickness exists in oxides thinner than 175 A. The retardation by ammonium peroxide becomes more severe with thicker oxides, reaches a maximum of about 23% at 1000 A, and decreases with still thicker oxides. At one micrometer, the different is only 5-6 %. This diverging-then-converging trend also exists at 800°C, but it occurs in thicker oxides. Little difference, or even a slight reversal, in thickness is observed for oxides thinner than 175 A at 1000°C and 200 A at 800° C. In this thin region, a small upward bending in the kinetics curves is believed to be caused by oxide formation during wafer loading.

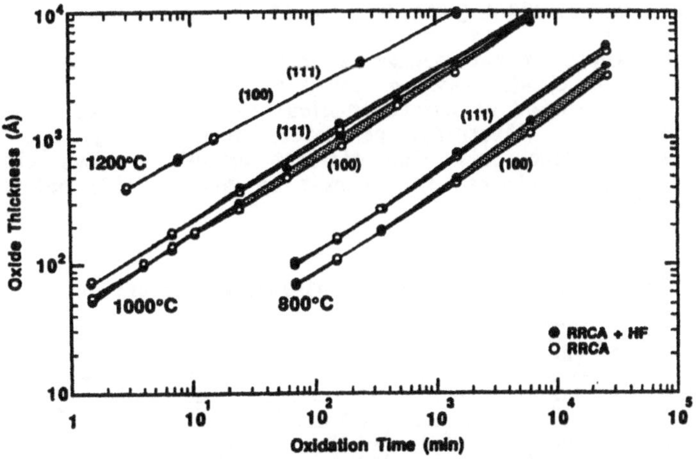

Fig. 1 Effect of chemical cleaning on oxidation kinetics. Dry oxide thickness versus oxidation time for (100) and (111) silicon at 800°, 1000°, and 1200°C after RRCA and RRCA+HF cleans.

(c) Orientation dependence -- The percentage retardation by ammonium peroxide is smaller on the (111) than (100) surface. Figure 1 also confirms the well-established orientation dependence that oxides grow thicker on (111) than (100) surfaces. On both surfaces the effects of chemical cleaning follow the same temperature and thickness dependence. However, the maximum retardation is less than 15% for the (111) surface and about 23% for (100).

ANALYSIS OF KINETIC DATA

The analysis of the kinetic data is based on the linear-parabolic model by Deal and Grove (7), in which the oxide growth is described by the following expression,

$$x_o^2 + Ax_o = B\ (t+\tau)$$

where x_o is the oxide thickness at time t, and A, B, and τ are constants. Kinetic parameters for thicker dry oxides were extracted and listed in Table II. It can be seen that the main effect of chemical cleaning is on the linear-rate constant B/A and which is associated with the initial rapid growth of dry oxides. The parameters for (111) surface at 800°C are not included in Table II because of insufficient data points for accurate parameter extraction. The effects of chemical cleans, however, still exist and are consistent with the description in the previous section. Results from experiments described in the next sections will shed more light on how chemical treatment affects these kinetic parameters.

DOPING EFFECTS, WET OXIDES, NATIVE OXIDES, AND ELECTRICAL PROPERTIES

Results from experiments using doped wafers indicate that (a) lightly doped n- and p-type wafers exhibit similar effects of chemical cleaning, and (b) heavy $POCl_3$ doping increases the oxidation rates, but reduces the amount of retardation by ammonium peroxide.

Preliminary data of steam oxidation show little or no effect from chemical cleaning in thick oxides at 800° and 1000°C. Unexpected large reversals were noticed among thin oxides at 800°C: for example, 196 A (RRCA) versus 176 A (RRCA+HF), and 126 A versus 99 A.

The thickness of native/chemical oxides after chemical cleans and their subsequent growth were monitored by ellipsometry. RRCA-cleaned silicon surface showed no unique chemical oxide thickness or growth at room temperature. Electrical parameters such as V_{FB}, N_F, and D_{it} measured by CV techniques also did not exhibit obvious differences caused by chemical cleans, in contrast to results (10) obtained from rapid thermal oxidation.

Table II. Linear-parabolic parameters for dry oxides

Temperature	Orient	clean	B (A²/min)	B/A (A/min)	τ(min)	Xi(A)
800°C	100	RRCA	1120	0.17	1325	220
	100	RRCA+HF	1100	0.31	271	84
1000°C	100	RRCA	16900	5.3	43	230
	100	RRCA+HF	16200	11.1	13	140
	111	RRCA	17600	10.2	20	200
	111	RRCA+HF	17400	21	4	80
1200°	100	both	74100	235	0.9	210
	111	both	74350	256	1.0	260

EXPERIMENTS TO ISOLATE THE ORIGIN OF RETARDED OXIDATION

The kinetic data and analyses in the previous sections indicated that ammonium peroxide in the reverse-RCA cleaning procedure is the cause for the retarded oxidation. It was also noticed that this retarding action is removed if, following the reverse-RCA clean, the wafer is treated in another chemical such as HCl, H_2SO_4, or HF. To further clarify the properties of the surface film produced by ammonium peroxide, a series of experiments was conducted in which the surface film was subject to high temperature anneal, oxidation, or chemical treatments between two oxidation steps.

High-Temperature Anneal in Argon

After four different chemical cleans, (100) and (111) silicon wafers were annealed in argon for different times at 1000° or 1200°C before being oxidized at 1000°C for 160 minutes in dry O_2. The oxide thicknesses for the different cleaning and annealing conditions are listed in Table III. Retarded oxidation occurred in both (100) and (111) samples, following the standard preoxidation anneal in argon for 10 minutes. Anneal in argon at 1000°C for 130 minutes partially removed the retarding properties of the RRCA clean, but caused no obvious changes in samples cleaned by the MPR, RCA, or RRCA+HF procedures. The retarding action of RRCA on (111) samples was completely erased by argon anneal at 1200°C, but only partially on (100) samples. Visible pitting and color changes were observed on samples subjected to the 1200°C anneal. Perhaps this treatment was too drastic, and anneal at 1100°C should be investigated. This experiment does demonstrate, however, that the RRCA clean produces a unique chemical film on the silicon surface that retards thermal oxidation. This film can be removed by high temperature preoxidation anneal.

Two-Step Oxidation at 1000° and 1200°C

A two-step oxidation was conducted to determine whether the retarded oxidation rate is observed at 1000°C after a short initial oxidation at 1200°C. RRCA and RRCA+HF cleaned samples were first oxidized at 1200°C in dry O_2 for 16.5 minutes to grow about 100 nm of oxides. They were then oxidized at 1000°C for 320 minutes. The oxide thicknesses after each oxidation are listed in Table IV.

It can be seen that the first oxidation at 1200°C produced slightly thicker oxides for (111) than (100) wafers, but virtually identical thickness for the two cleans. The (111) oxides were substantially thicker, however, after the second oxidation at 1000°C,

Table III. Oxide thickness in A after argon anneal and oxidation

Orientation	clean	Ar anneal time and temperature		
		10 min 1000°C	130 min 1000°C	Ramp to 1200°C
100	MPR	1040	1044	998
	RCA	1043	1048	1003
	RRCA+HF	1094	1074	1001
	RRCA	845	943	915
111	MPR	1277	1254	1213
	RCA	1304	1280	1234
	RRCA+HF	1324	1300	1237
	RRCA	1114	1211	1224

Table IV. Oxide thickness in A after the two oxidations

orientation	clean	first oxidation 1200°C	second oxidation 1000°C
100	RRCA+HF	1004	2238
	RRCA	1005	2234
111	RRCA+HF	1013	2422
	RRCA	1013	2417

indicating that crystal orientation affected oxide growth. On the other hand, the effect of cleaning was erased, and samples cleaned by RRCA and RRCA+HF produced oxides of essentially the same thickness.

Two-Step Oxidations with Etch Back

As mentioned previously, the retarding action of the film produced by ammonium peroxide could take place either at the Si-SiO2 interface or at the outer surface of the oxide. Our earlier work (11) indicated that removing the outer layer of the oxide eliminates the retarding properties in the subsequent oxidation. In the following experiment, the properties of the outer surface are further investigated. After a first oxidation in dry O_2 at 1000°C for 120 minutes, the oxide on the wafer was etched into a wedge by continuously feeding diluted HF into a beaker in which the wafers were held vertically. Figure 2 shows the profiles of oxide thickness after the etch back and after a second oxidation. It can be seen that etching back as little as 3 nm of the oxide removes the retarding properties of ammonium peroxide. This result clears identifies a rate-controlling outer layer, which is further confirmed by the next experiment.

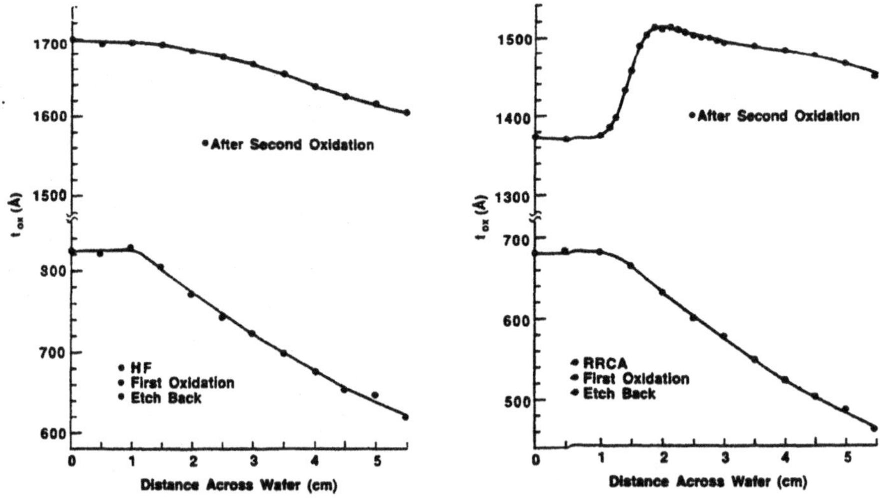

Fig. 2 Two-step oxidation with etch back. The initial oxide across the wafer was etched into a wedge of decreasing thickness. The profile of the second oxide indicates that the outer surface of the oxide controls the oxide growth.

Effect of Chemical Clean between Two Sequential Oxidation Steps

Two groups of samples, one cleaned in RRCA and the other with an additional HF dip, were oxidized in dry O_2 for 120 minutes to thicknesses of 700 and 850 A, respectively. They were cleaned in different chemicals before being loaded in the furnace for a second oxidation at 1000°C for 220 minutes. The oxide thicknesses are listed in Table V. Several observations can be made.

(a) Chemical treatment of the outer surface of the initial oxide affects subsequent oxidation.
(b) Treatment by ammonium peroxide after the first oxidation retards the second oxidation, regardless of the initial clean.
(c) The retarding action of ammonium peroxide no longer occurs, if HCl is used after the ammonium peroxide as in the RCA clean.
(d) After the first oxidation, the retarding property of ammonium peroxide in the original clean can not be removed by HCl. (c) and (d) suggest that the surface film produced by the initial ammonium peroxide was hardened during the first thermal oxidation.
(e) Treatment in boiling water removes this hardened surface film. As a result, subsequent oxidations are not retarded.

PROPOSED MODEL

Analysis of kinetic data determined that treatment by ammonium peroxide reduces the linear-rate constant in the oxidation process. The annealing and two-step oxidation experiments further established that the rate-dominating action exists in a surface film that remains on the outer layer of the oxide during oxidation. An understanding of the chemical cleaning effects can now be obtained based on a two-layer model. A thin outer layer, with its own parameters for oxidant transport, limits the overall oxidation rate. For thicker oxides, presence of this film of fixed thickness reduces the linear-rate constant (12), as can been seen from the following equation.

$$\frac{dx_o}{dt} = \frac{C^*N}{1/k + 1/h + x_1/D_1 + x_o/D}$$

where D_1 is the oxidant diffusivity in the surface film of fixed thickness x_1. The effects of the surface film on thin oxide growth are under further investigation.

Table V. Oxide thickness in A after two oxidation steps with a chemical clean between the steps

Initial clean	First Oxidation	Second clean	Second oxidation
RRCA	706	None	1425
RRCA+HF	850	None	1785
RRCA	701	RCA2	1378
RRCA+HF	862	RCA2	1826
RRCA	703	RCA1+RCA2	1400
RRCA+HF	853	RCA1+RCA2	1776
RRCA	703	RCA1	1339
RRCA+HF	854	RCA1	1585
RRCA	730	Boiling DI	1572
RRCA+HF	872	Boiling DI	1713

SUMMARY

Kinetic data of cleaning effects and results from two-step oxidations are presented. It is shown that the effect on oxidation by chemical cleaning is mainly on the linear-rate constant. A likely mechanism is that ammonium peroxide produces a unique thin surface film which stays on the outer surface of the oxide throughout the oxidation process and retards the oxidation. This surface film can be removed by acids such as HF, HCl, or H_2SO_4. It can also be partially removed by high temperature Ar anneal or boiling water. In two-step oxidations, chemical treatment of the first oxide can change subsequent oxidation rates. Further investigation is currently underway to identify the physical and chemical properties of this surface film.

ACKNOWLEDGEMENT

The authors wish to thank D. Waterbury for her valuable assistance in the experiments. The work at Stanford was supported in part by grants from Fairchild Semiconductor Corporation (now part of National Semiconductor Corporation) and by SRC contract No. 86-11-101.

REFERENCES

1. B. E. Deal and D. B. Kao, in Tungsten and Other Refractory Metals for VLSI Applilcations II, Material Research Society, p. 27, (1987)
2. M. Wong, N. Kobayashi, R. Browning, D. Paine, and K. C. Saraswat, J. Electrochem. Soc., 134, 2339, (1987).
3. F. J. Grunthaner and J. Maserjian, IEEE Trans. Nucl. Sci., NS-24, 2108 (1977).
4. F. N. Schwettman, K. L. Chiang, and W. A. Brown, Paper No. 276 in Spring Meeting of The Electrochemical Society, p. 688 Extended Abstracts 78-1, Seattle, WA (1978).
5. G. Gould and E. A. Irene, J. Electrochem. Soc., 134, 1031, (1987).
6. J. M. deLarios, C. R. Helms, D. B. Kao, and B. E. Deal, Appl. Sur. Sci., 30, 17 (1987).
7. J. Ruzyllo, G. t. Duranko, and A. M. Hoff, J. Electrochem. Soc. 134, 2052 (1987).
8. B. E. Deal and A. S. Grove, J. Appl. Phys., 36, 3770 (1965).
9. W. Kern and D. A. Puotinen, RCA Review, 31, 187 (1970).
10. M. Moslehi, S. Shatas, K. C. Saraswat, and J. D. Meindl, IEEE Trans, Electron Devices, ED-34, No. 6, 1407 (1987)
11. J. M. deLarios, C. R. Helms, D. B. Kao, and B. E. Deal, Appl. Phys. Lett., submitted for publication.
12. L. M. Landsberger and W. A. Tiller, Appl. Phys. Lett., 49, 143 (1986).

STUDIES OF ANHYDROUS HF PREOXIDATION
TREATMENT OF SILICON SURFACES

G.T. Duranko*, D.J. Syverson and L.A. Zazzera
FSI International, Chaska, MN 55318

J. Ruzyllo and D.C. Frystak
CEMD, Dept. of Electrical Engineering
Penn State University
University Park, PA 16802

ABSTRACT

The effect of application of anhydrous HF gas to chemically cleaned silicon surfaces prior to thermal oxidation was examined. A comparison to a dilute hydrofluoric acid dip-rinse shows that the anhydrous HF pre-oxidation treatment leaves the silicon surface fluorinated but with slightly less carbon contamination and lower particulate contamination. SIMS and ESCA measurements have shown that in the course of subsequent thermal oxidation, fluorine undergoes redistribution in the oxide. Performance of gate oxides in terms of oxide breakdown statistics and interface charge density were studied using MOS capacitors.

INTRODUCTION

Recently, in the area of thin thermal oxide fabrication, more attention has been given to chemical cleaning and its effects on the silicon surface (1-3), oxidation kinetics (4-7), and gate oxide electrical properties (8-10). Trends in MOS device design are requiring closer inspection of the effects of various chemical and gas surface modifications prior to gate oxidation on device characteristics. One issue of importance in this regard is a concern about the so-called chemical oxide, usually about 10Å thick, which grows and remains on the silicon surface subjected to treatments in various chemistries.

Leaving a chemical oxide on the bare silicon surface prior to high temperature operations protects the surface from contamination due to the saturation of the dangling silicon bonds. However, this oxide, of undetermined composition and varying uniformity may effect the characteristics of the subsequently grown gate oxide. Therefore, its removal from the surface prior to thermal oxidation is thought to improve the integrity of the gate oxide.

*Present address: AT&T Microelectronics, Allentown, PA 18103

Traditionally, removal of this chemical oxide is accomplished through etching in a dilute hydrofluoric acid solution. Unfortunately, if close attention is not paid to proper filtering, rinsing, and drying techniques, the "HF last" step followed by a rinse-dry process can leave the silicon surface rich in carbon and high in particulates (11). With reduced gate oxide temperature growth, carbon will be more difficult to desorb from the surface, resulting in carbon contamination of the gate oxide. In addition, remaining particulate can lead to yield and reliability failures.

With the advent of anhydrous HF etching, the advantages of "HF last" processing are retained along with overcoming the disadvantages of hydrous HF etching at the same time. A mixture of anhydrous HF gas, water vapor, and nitrogen is capable of reducing chemical oxides to comparable levels as a hydrofluoric acid dip. In addition, the process parameters can be selected such that the silicon surface becomes saturated with fluorine, inhibiting native oxide regrowth and extending cue times. Lower carbon levels have been demonstrated as compared to hydrofluoric acid immersion processing (12) which will lead to reduced carbon interface effects. Since the process is totally dry, there is no need for rinsing and drying and, hence, there is no increased particulate generation associated with this operation.

Introducing fluorine into the furnace as NF_3 or dipping a wafer into hydrofluoric acid prior to oxidation has been shown to increase the oxidation rate (13) and improve certain electrical characteristics (14), respectively.

A characteristic of anhydrous HF etching is that it leaves the surface fluorinated. In order to justify this treatment as a viable preoxidation treatment, the effects of fluorine must be investigated on the subsequently grown gate oxide. The purpose of this experiment is to evaluate the effectiveness of using anhydrous HF as a preoxidation treatment.

EXPERIMENTAL

In this experiment, p-type, 2-5 ohm-cm, (100) silicon wafers were used. All wafers received a sacrificial oxidation of 400Å at 1000°C in a dry oxidation ambient. After the sacrificial oxide was removed, the wafers were divided into three different processing groups. The treatments are as follows: (A) H_2SO_4-H_2O_2, HF:H_2O (1:100), NH_4OH-H_2O_2-H_2O, and HCl-H_2O-H_2O sequence (B clean), (B) same as (A) with anhydrous hydrogen fluoride (AHF) gas treatment added as the final step, and (C) same as (A) with HF:H_2O (1:100) immersion with a cascade rinse, rinser/dryer step as the final treatment (HF R/D). Anhydrous HF treatments were carried out in a commercial single wafer etcher. During the 15 second etch, the gas flows were set at 7.5 lpm N_2, 1 lpm of vapor and 200cc of HF gas. Back pressure in the chamber was set at 25 IWC (inches of water column). Wafers from each group were then simultaneously loaded into the furnace and oxidized at 900°C to a thickness of approximately 275Å. Wafers were given a 5 min. push-in and 5 min. pull-out cycle with a 10 min. preoxidation anneal in a nitrogen ambient.

X-ray Photoelectron Spectroscopy (XPS) and Secondary Ion Mass Spectroscopy (SIMS) were used to determine atomic concentrations after cleans and after thermal oxidation. Compositional depth profiles were performed with both analytical techniques. XPS measurements were performed using a Perkin-Elmer Model 5000 XPS spectrometer. Angular dependent XPS was used to determine the relative concentrations of O, Si, F, and C prior to oxidation and through the as-grown oxide via argon sputtering. The profile depth scales were calibrated by dividing the ellipsometer oxide thickness by the time required to reach the $Si-SiO_2$ interface. The time required to reach the interface is defined by a 10% decrease in the oxide signal. SIMS measurements were performed with a Perkin-Elmer Model 6300 SIMS Spectrometer, equipped with a duoplasmatron Cs^+ ion source to enhance negative secondary ion yield. The depth profile scales were calibrated using a method similar to the XPS calibration.

Vacuum evaporated aluminum gate contacts, 1mm in diameter, were used to evaluate MOS capacitors. Post-metallization anneal at 450°C in forming gas was performed after metallization. Oxide and interface properties were extracted from capacitance-voltage measurements. Interface state density was determined using the quasi-static method combined with the high frequency C-V measurements (15). Oxide breakdown field measurements were carried out using an appropriately calibrated memory scope to monitor the current and voltage at breakdown. The voltage ramp rate was 1V/S.

RESULTS AND DISCUSSION

For the sake of clarity, the discussion of obtained results is broken into sections which address specific issues of interest in this study.

Oxide Growth

It has been reported in the literature that preoxidation cleaning has an effect on the oxidation kinetics (4-7). More specifically, wafers that are immersed in hydrofluoric acid and rinsed in DI water following cleaning tend to grow thicker oxides than wafers receiving a base or an acid treatment as the final step of a clean. One might expect variations in cleaning sequence to affect the linear oxidation regime where oxidation is controlled by the reaction of the silicon with the oxygen and not the parabolic regime in which oxidation is controlled by the diffusion of the oxidant through the oxide. Surprisingly, it has been shown that chemical preoxidation treatment effects extend into the parabolic regime (4-7). In this experiment, the similar studies are expanded to include AHF as the final treatment prior to oxidation.

Figure 1 shows thickness vs. time data for oxidations performed at 1000°C in dry oxygen. Each oxidation includes wafers from all preoxidation treatments used in this experiment. As expected, the wafers that received an HF immersion last grew the thickest oxide for all oxidation times. The greatest percent change between the HF dip and the B clean and the AHF and the B clean occurs at about 1800Å and 550Å respectively. Interestingly, a small difference exists between the AHF and the B clean. It is reasonable to assume that the differences noticed between the HF immersion and the AHF samples clearly indicate a different silicon surface condition resulting from these two treatments. Although both surfaces were hydrophobic prior to insertion into the furnace, the nature of the interaction between the fluorine, carbon, oxygen, and silicon bonds may be different in each case. This difference then affects both oxidation regimes. The linear oxidation regime may be affected by the different surface chemistry and the parabolic oxidation regime by altering the oxide structure or inhibiting diffusion through the outermost part of the oxide (representing the surface structure prior to oxidation). Surface analysis techniques were implemented to enhance our understanding of the effects accounting for the observed differences.

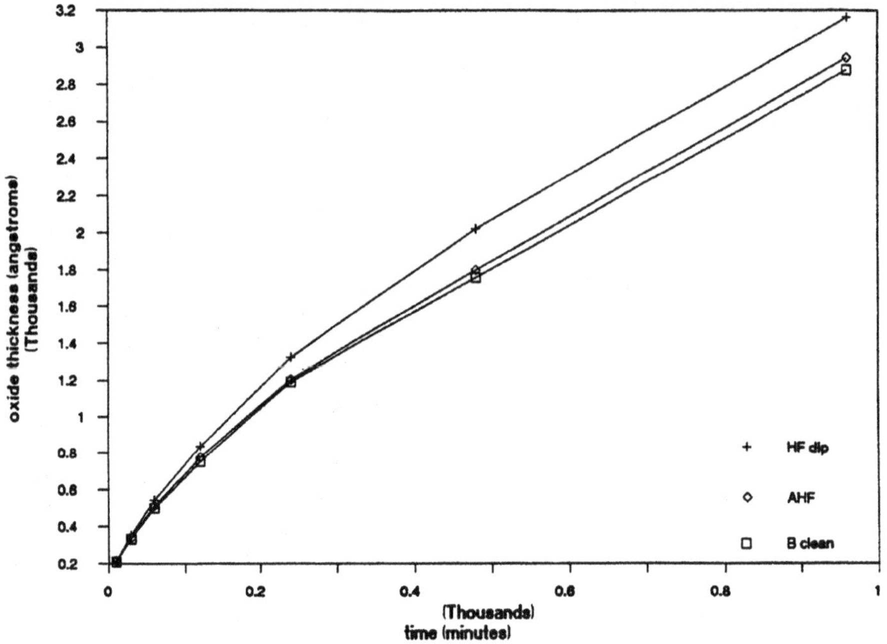

Fig. 1. Plot of oxide thickness vs. time for 3 preoxidation treatments.

Surface Analysis and Oxide Composition Characterization

XPS and SIMS were used to examine the effects of the preoxidation treatments used in this experiment. Fluorine, carbon, oxygen, and silicon were monitored after cleaning, heat treatments in pure argon, and finally oxidation.

XPS was used specifically to study silicon wafers exposed to different preoxidation treatments (12) and also wafers subjected to temperature treatments in an oxygen-free ambient after cleaning. Results shown in Fig. 2 indicate that the AHF cleaning procedure leaves approximately twenty-five times more fluorine and less carbon on the silicon surface than the HF immersion rinse-dry (HF R/D) process. An interesting result is noticed when examining the fluorine concentration after the B clean. There is approximately ten times more fluorine on the surface of the B cleaned samples compared to the HF R/D treated wafers in spite of the fact that in the B clean sequence, an HF step precedes both an ammonium hydroxide-hydrogen peroxide step and a hydrochloric acid-hydrogen peroxide step. The tenacious nature of fluorine on the silicon surface results in its incorporation into the chemical oxide that is grown during the subsequent ammonium-peroxide step.

The B cleaned and anhydrous HF treated wafers were then subjected to argon annealing at elevated temperatures and examined with XPS. Argon rather than nitrogen was used to avoid problems related to Si-N phase formation during nitrogen annealing. In addition, special attention was given to the elimination of oxygen in the argon. Fluorine and carbon concentrations at the surface were monitored after annealing at 392°C, 500°C, 750°C and 1000°C. XPS surveys indicate that the fluorine and carbon concentrations are reduced as the anneal temperature is increased. Relative carbon concentrations were reduced to 0.93% and 0.98% for the B clean and AHF

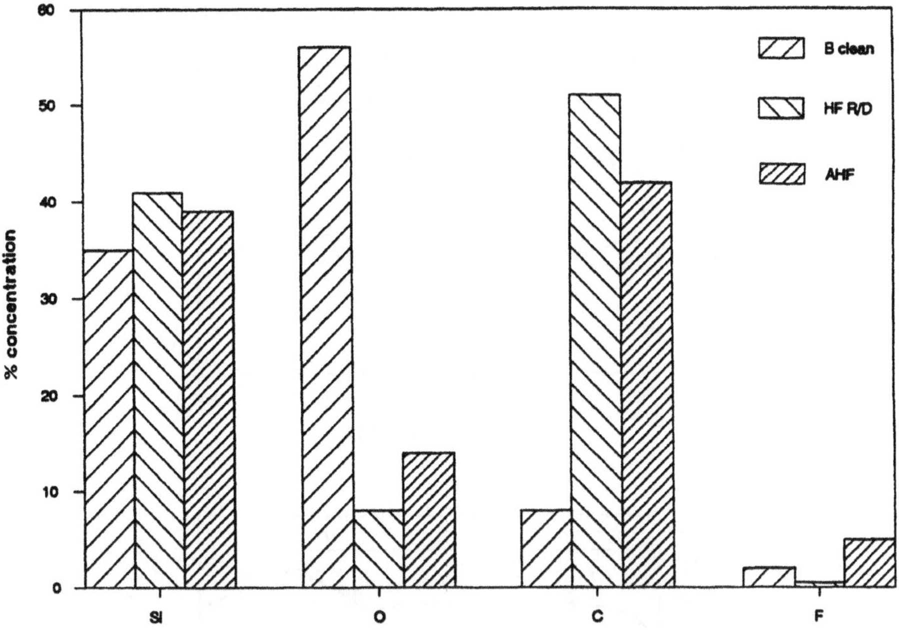

Fig. 2. Relative percent concentrations for Si, O, C, and F after various pretreatments as determined by XPS.

respectively after annealing at 1000°C. Relative fluorine concentrations, in turn, were reduced to 0.06% and 0.08% for the B clean and the AHF respectively. It is clear that the amount of fluorine that resides on the silicon surface for both the B clean and the AHF treatment prior to turning on the oxygen in the furnace to initiate oxidation is present in very small quantities.

To evaluate the composition of the thermal oxides grown on the surfaces prepared in the way presented above, SIMS and XPS surveys and profiles were taken.

Both SIMS and XPS surveys show that the B cleaned wafers have the most fluorine at the oxide surface. The fact that the fluorine may be incorporated into the chemical oxide, where it remained even after the heat treatment, might explain why it was found at the oxide surface as the chemical oxide could be pushed to the top of the thermal oxide during thermal oxidation.

This result is also confirmed by SIMS profiles which show, in addition, that the relative intensities of fluorine through the bulk of the different oxides do not reflect the surface fluorine concentrations (Fig. 3). No significant variations in the fluorine content in the bulk of the oxide was detected for the different pretreatments. At the Si-SiO_2 interface, the HF immersion pretreatment appears to result in less fluorine when compared to the B clean and the AHF treatments. In the vicinity of the interface, SIMS quantitation is made difficult by matrix effects (16), hence this study cannot be conclusive with regard to the absolute amounts of fluorine at the interface.

On the other hand, XPS profile results have shown that no fluorine is present in the oxide as the fluorine was below the detection limit of this technique.

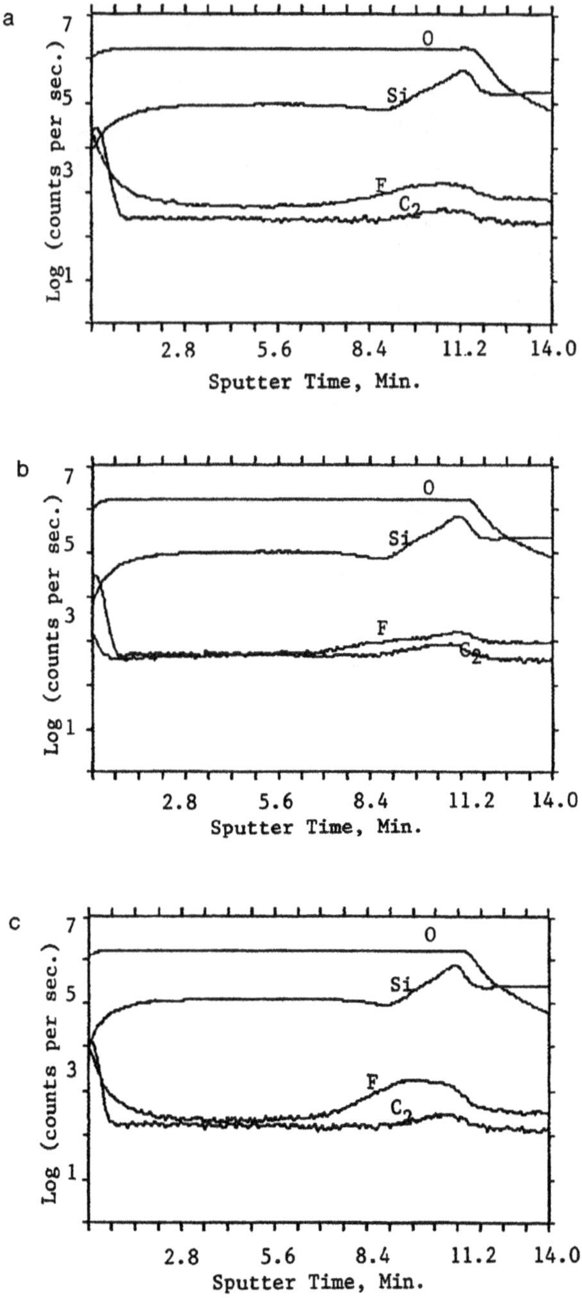

Fig. 3. Negative SIMS profiles of a) B clean, b) HF R/D, and c) AHF pretreated thermal oxides.

Electrical Characterization

It was experimentally demonstrated (17) that wafers immersed in a dilute HF solution and loaded directly into the furnace without a deionized water rinse-dry step showed improvements in breakdown distribution, slightly reduced fixed oxide charge, and significant reductions in sodium concentration in a sodium-rich environment. In addition, improvements in radiation hardness have also been reported as a result of this treatment (14). It is apparent that fluorination of the surface before oxidation can result in improved quality of the subsequently grown thermal oxide.

Since anhydrous HF treatments will fluorinate the silicon surface to a greater degree than an HF dip without a final rinse (12), electrical characteristics of thermal oxides grown on this surface were examined.

Improvements in capacitors grown with an immersion HF dip without a final rinse may also be due to lower particulate levels in these samples since rinsing and drying after an "HF last" process is very critical with regards to particulate contamination (18). Evidence exists that particulate generation after HF processing does lead to a reduction in breakdown (10). Hosaka et al. shows a significant increase in mid-field breakdowns and some increase in low-field breakdown. Breakdown could be occurring due to a weakening of the oxide at the place where the particle is incorporated into the oxide.

It was noticed in this work that there were no significant differences in interface state density and fixed oxide charge amongst the B clean, HF R/D, and AHF pretreatments. The anhydrous HF pretreatment yielded slightly higher average dielectric strengths when compared to the B clean and the HF R/D pretreatments. More mid-field breakdowns occurred in the HF R/D step which correlates well with other available data (10). Since the B clean and the AHF pretreated samples resulted in much lower particulate contamination (particle counts on the three different pretreatments have shown that the HF R/D had much more particulate than the B clean or the AHF pretreatment), the mid-field breakdowns could be attributed to particulate contamination, although no further testing was done to show this.

SUMMARY

The feasibility of incorporating anhydrous HF as the final step of a preoxidation clean has been initially investigated. Surface analysis has shown that the application of the anhydrous HF treatment to cleaned silicon surfaces results in more fluorine and less carbon when compared to the HF R/D. The B clean, surprisingly, also showed more fluorine than an HF R/D. The AHF and B clean also have less particulate on the silicon surface when compared to the HF R/D. Upon high temperature annealing, the relative fluorine concentrations in the B clean and the AHF treatments were significantly reduced to only 0.06% and 0.08%, respectively. SIMS and XPS surveys of the thermal oxides show a significant amount of fluorine on the surface which has been attributed to the incorporation of fluorine into the B clean chemical oxide which is pushed to the surface during oxidation. There may be more fluorine at the Si-SiO_2 interface in the AHF treated samples but matrix effects precludes determination of this exactly. Electrical measurements indicate that there is very little effect on the interface state density and the fixed oxide change with variation in the pretreatment. Dielectric strengths were slightly improved for the AHF treatment when compared to the B clean and the HF R/D treatment. This was attributed to particulate contamination. Oxide growth on AHF treated samples differs only slightly from the B cleaned samples. Differences in growth rates suggest a different silicon surface condition prior to oxidation which has been established using XPS and SIMS surveys.

REFERENCES

1. G. Gould and E. A. Irene, "An In-Situ Study of Aqueous HF Treatment of Silicon by Contact Angle Measurement and Ellipsometry", Office of Naval Research, Technical Report No. 17, June (1987).
2. W. Kern, "Purifying Si and SiO_2 Surfaces with Hydrogen Peroxide", Semiconductor International, 94, April (1984).
3. I. Lampert, "Influence of the Cleaning Method on the Chemical Behavior of Hydrophilic Silicon Surfaces", Electrochem. Soc. Abstracts 87-1, 381 (1987).
4. G. Gould and E. A. Irene, "The Influence of Silicon Surface Cleaning Procedures on Silicon Oxidation", J. Electrochem. Soc., 134, 1031 (1987).
5. J. M. deLarios, C. R. Helms, D. B. Kao, and B. E. Deal, Effect of Silicon Cleaning Procedures on Oxidation Kinetics and Surface Chemistry, in: "Process Simulators for Silicon VLSI and High Speed GaAs Devices", Stanford Electronics Laboratory, Stanford, CA (1987).
6. F. J. Grunthaner and J. Maserjian, "Experimental Observations of the Chemistry of the SiO_2/Si Interface", IEEE Trans. Nucl. Sci., NS-24, 2108 (1977).
7. F. N. Schwettmann, K. L. Chang, and W. A. Brown, "Variation of Silicon Dioxide Growth Rate with Pre-Oxidation Clean", Electrochem. Soc. Abstracts, 78-1, 689 (1978).
8. J. Ruzyllo, G. T. Duranko, and A. M. Hoff, "Preoxidation UV Treatment of Silicon Wafers", J. Electrochem. Soc., 134, 2052 (1987).
9. I-W. Wu, J. C. Mikkelsen, Jr., and M. Koyanagi, "Processing Effects on the Breakdown Characteristics of Thin Gate Oxides in CMOS Technology", Electrochem Soc. Extended Abstracts Vol. 87-2, 868 (1987).
10. T. Hosaka, Y. Kato, Y. Kojima, Y. Hattori, S. Endo, M. Kamiya, and Y. Hayashi, "Ultra Thin Reliable CVD SiO_2 Film for VLSI's", Electrochem. Soc. Extended Abstracts Vol. 86-2, 620 (1986).
11. D. Burkman, W. R. Schmidt, C. A. Peterson, and B. F. Phillips, "Critical Wafer Cleaning", Technical Report 217, FSI International, Inc., Chaska MN.
12. L. A. Zazzera, J. F. Moulder, and J. S. Hammond, "XPS and SIMS study of Anhydrous HF and UV/Ozone Modified Silicon Surfaces", to be presented at 173rd Electrochemical Society Meeting, Atlanta, GA, Abstract #188 (1988).
13. M. Morita, S. Aritome, M. Tsukude, and M. Hirose, "A New SiO_2 Growth by Fluorine-Enhanced Thermal Oxidation", IEDM, 144 (1984).
14. E. F. daSilva, Jr., and T. P. Ma, "Strong Effects of Pre-oxidation Fluorine Treatment on the Radiation Response of MOS Capacitors", SRC paper #C87062, Feb. 1987.
15. R. Castagne and V. Vapaille, "Description of the SiO_2-Si Interface Properties by Means of Very Low Frequency MOS Capacitance Measurements", Surf. Sci., 28, 157 (1971).
16. K. Wittmaack, "The Use of Secondary Ion Mass Spectroscopy for Studies of Oxygen Absorption and Oxidation" Surface Science, 68, 118 (1977).
17. Y. Wang and R. C. Barker, paper presented at the 1985 IEEE-SISC, December 1985.
18. D. J. Syverson and G. T. Duranko, "Contamination Aspects of Anhydrous HF Gas Processing", to be presented at Institute of Environmental Sciences Meeting, May 1988.

THE INFLUENCE OF THE Si-SUBSTRATE CHARACTERISTICS ON THE QUALITY OF POLY-Si AND Al GATED MOS OXIDES

C. Hasenack, M. Heyns, R. Falster* and R. De Keersmaecker

Interuniversity Microelectronics Center (IMEC vzw)
Kapeldreef 75, B-3030 Leuven - Belgium

*Monsanto Electronic Materials Company
Featherstone Road, Wolverton Mill South
Milton Keynes, Buckinghamshire MK12 5TB - United Kingdom

ABSTRACT

The dielectric strength and the Si/SiO_2 interface state density were determined for thermally grown silicon dioxide layers grown on various Si-substrates (FZ, CZ with varying oxygen content and EPI material). A clear dependence of the high-field breakdown characteristics on the starting oxygen content of the Si-substrate was established. The effects of cleaning, pre-oxidation heat treatment and gate electrode material were also investigated.

INTRODUCTION

Oxide degradation and breakdown phenomena of SiO_2 layers during high-field stressing have been the subject of several studies (see e.g. refs.1-7). It has already been established that low and medium-field breakdowns are related to gross oxide defects (1-3). Since these defects can effectively be suppressed by taking the appropriate precautions (8), increased attention is being paid to the high-field breakdowns which are commonly attributed to intrinsic properties of the oxide (9). However, as device dimensions become smaller, other factors than bulk properties of the oxide - such as the Si/SiO_2 interface roughness - may increasingly affect the high-field ("intrinsic") breakdowns (8,10). Since this can lead to serious reliability problems, the effect of any initial wafer characteristic as well as pre-oxidation treatments which can affect the final oxide characteristics should be carefully studied in order to determine the necessary conditions to guarantee an optimum oxide quality. Therefore, the main goal of the present study is to reveal the effect of the Si-substrate and of specific pre-oxidation treatments on the high-field breakdown characteristics of thermally grown SiO_2.

EXPERIMENTAL

The Si substrates used in this work were all 4-inch, p-type wafers with (100) orientation, and a resistivity of 5-10 ohm-cm. The experiments included float-zone (FZ), Czochralski (CZ) and epitaxially grown (EPI) silicon. The CZ substrates were divided in three groups depending on the interstitial oxygen content: <27 ppma (CZ-27), 30±2 ppma (CZ-30) and CZ-35±2 ppma (CZ-35). The interstitial oxygen concentrations are given according to the ASTM-F121-76 calibration, as determined by Fourier Transform

Infrared Spectrometry. The interstitial oxygen content of the p^+ substrates in the EPI wafers was 30 ppma.

Before oxidation, the wafers underwent various combinations of cleaning treatments and some wafers also underwent a pre-oxidation heat treatment (POHT) after cleaning. Cleaning treatments included the conventional RCA-clean (11) with or without a subsequent 10 sec 2% HF-dip (called RCA+HF and RCA, respectively). The POHT was performed in Ar at 1025°C for 10 hours in a double-walled quartz-tube furnace and was intended to produce an oxygen-depleted surface region.

Oxidation was performed in the same furnace where part of the batch underwent the POHT. For that purpose, the wafers that did not undergo the POHT were simply added to the boat already carrying the pre-oxidation heat treated samples so that all wafers were oxidized simultaneously. Oxidation was performed for 75 min at 900°C in dry O_2 and immediately after oxidation the samples received an in-situ anneal in Ar at 900°C, during 10 min. Warming-up and cooling-down was also performed in flowing Ar.

After the oxide growth, Al and poly-Si gate capacitors with various areas were prepared using standard wet lithography. The poly-Si layers were doped with $POCl_3$ at 800°C. Annealing and drive-in was performed at 900°C during 30 min. The Al-gated structures underwent a sintering process in forming gas at 420°C for 20 min.

As determined from capacitance-voltage (C-V) measurements, the oxide thickness of the non-POHT and the POHT samples measured 19.5 and 21.9 nm respectively, thus indicating that either the POHT caused a slight oxidation of the wafers or that the oxidation rate of the POHT wafers was enhanced. The same difference in oxide thickness was also observed by means of ellipsometric measurements assuming a fixed oxide refractive index of 1.46.

The dielectric strength of the oxides was assessed by means of breakdown measurements, constant-current stressing and constant-voltage wear-out measurements. The breakdown measurements were performed on 240 capacitors for each wafer. The Si/SiO_2 interface states density was extracted from the combined high-low frequency C-V measurements (12). The RCA+HF-cleaned wafers were repeated to check the reproducibility of our results.

RESULTS

Dependence of the breakdown characteristics on starting oxygen content

Figure 1 shows the mean breakdown field as a function of the substrate type for non-POHT RCA+HF cleaned wafers. It is evident that a drastic reduction of the mean breakdown field occurs for the CZ-35 substrates. This behaviour was observed independently of the gate material and of the pre-oxidation treatments which the wafers underwent. From constant-voltage wear-out and constant-current stressing measurements it follows that for the same CZ-35 samples, a reduction is observed of both the life-time of the oxides and of the charge Q_{bd} that can be injected into the oxide before breakdown, respectively.

The current-voltage (I-V) curves measured for capacitors prepared on the highest oxygen-content substrates that did not undergo the POHT were also drastically different from those measured for capacitors prepared on the other substrates. A shoulder was observed starting around 5 MV/cm (vertical arrow in Fig.2), which was never observed on I-V curves measured for capacitors prepared on the other substrates. Its appearence at 5 MV/cm for the Al-gated capacitors indicates that the influence of the starting oxygen content is stronger than can be expected from the mean breakdown field values alone.

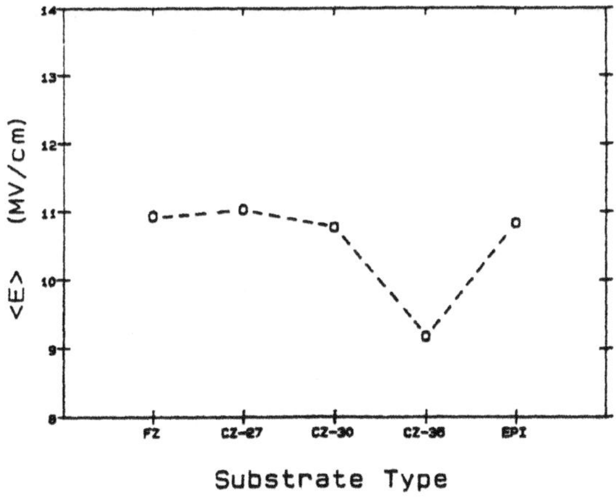

FIGURE 1 - Mean breakdown fields as a function of the substrate type. Cleaning method: RCA+HF. No POHT. Gate area=$1.075x10^{-3}$ cm^2; gate material: Al.

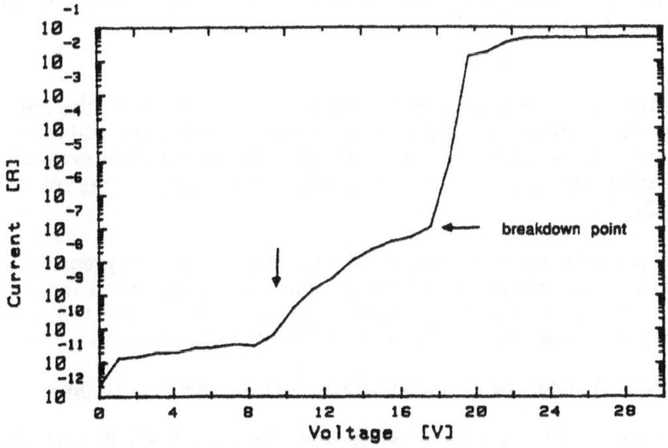

FIGURE 2 - Typical I-V curve of capacitors prepared on non-POHT CZ-35 substrates. Gate area=$1.075x10^{-3}$ cm^2; gate material: Al.

Dependence of the breakdown characteristics on the POHT

For all substrates, the POHT promoted an increase of the mean breakdown fields of the oxides as shown in Fig.3. However, its effect on the breakdown distributions was different for the CZ-35 substrates in conparison to any of the other substrates. This is demonstrated in Fig.4: the singly-peaked breakdown distribution corresponding to the non-POHT CZ-35 substrate is found to be shifted entirely by 1 MV/cm towards higher field values for CZ-35 wafers that underwent the POHT (Fig.4a). For the other substrates, the main effect of the POHT is the reduction of the small peak in the failure distribution at the left of the main distribution peak, as exemplified by the curve for the EPI-substrate in

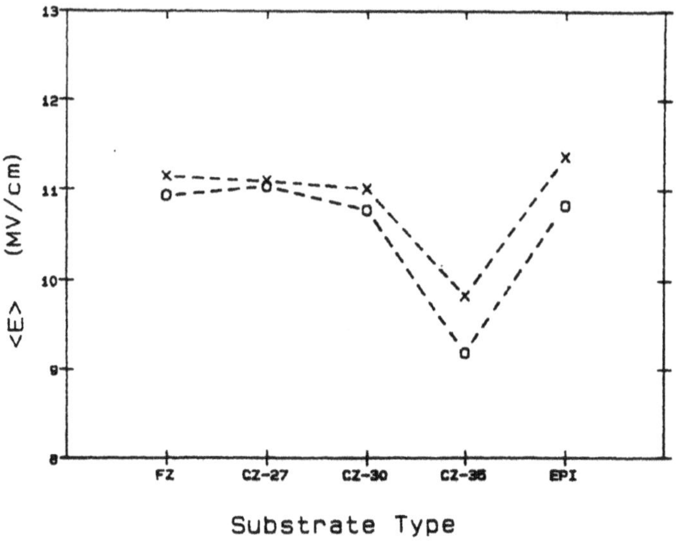

FIGURE 3 - Mean breakdown fields as a function of the substrate type showing the effect of the POHT: with (x) and without (o) the POHT. Cleaning method: RCA+HF. Gate area=1.075×10^{-3} cm^2; gate material: Al.

Fig.4b. It is due to the reduction of these failures and/or due to the shift of the main breakdown peak towards higher fields that the mean breakdown fields were always seen to increase for the POHT case. It is, however, worth to mention that even after this POHT, the mean breakdown fields still exhibit the substrate oxygen content dependence as shown in Fig.3.

Also the shoulder observed on the I-V curves of capacitors prepared on non-POHT CZ-35 substrates was reduced and often completely suppressed for substrates that underwent the POHT, yielding in the latter case I-V curves similar to those obtained for capacitors prepared on any of the FZ, CZ-27, CZ-30 or EPI substrates.

Dependence of the breakdown characteristics on the cleaning procedure

The influence of the cleaning procedure on the breakdown characteristics becomes clear from Fig.5 where the breakdown histograms are compared for RCA+HF and RCA cleaned, non-POHT samples, prepared on a CZ-27 substrate. It is seen that the use of the HF-dip results in a shift of the whole breakdown distribution towards lower fields, with a consequent lowering of the mean breakdown field. In addition, as generally observed, the breakdown distribution of the RCA+HF cleaned wafer was slightly broader than that of the RCA cleaned wafer.

Dependence of the Si/SiO_2 interface states density on the pre-oxidation treatments

In contrast to the oxygen dependence observed for the breakdown results, no dependence of the Si/SiO_2 interface states density (neither of the midgap value nor of the distribution) of the Al-gated capacitors on the starting oxygen content was found. However, a systematic dependence on the pre-oxidation treatments was observed as shown in Fig.6. Because the interface states densities were very low and almost at the limit of the measurement accuracy (12) and because the values were widely spread for the RCA+HF+POHT wafers, this behaviour should be seen as a trend only. The midgap Si/SiO_2 interface states density of the poly-Si gated capacitors remained always below 10^{10} $cm^{-2} \cdot eV^{-1}$ and no correlation could be established with either the starting oxygen content of the substrates or the pre-oxidation treatments.

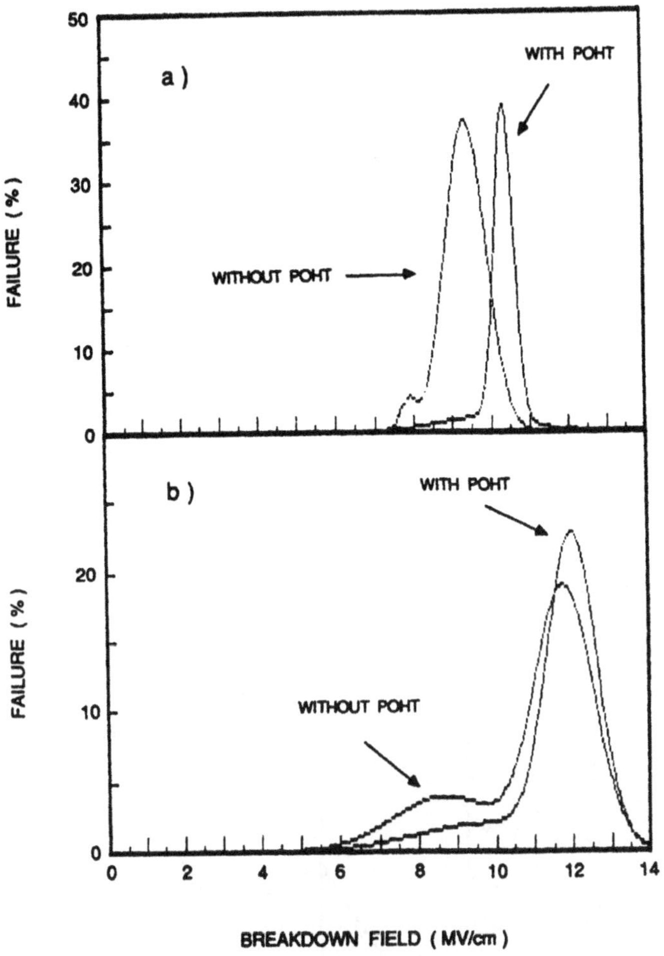

FIGURE 4 - Fitted breakdown distributions of RCA+HF cleaned (a) CZ-35 and (b) EPI substrates, with and without POHT. Gate area=1.075×10^{-3} cm^2; gate material: Al.

DISCUSSION

Reduction of the dielectric strength as well as enhanced electron injection into the oxide (as seen for the CZ-35 sample in Fig.2), has already been correlated with the presence of impurity-decorated stacking faults and dislocation loops at the Si/SiO_2 interface (13-16). Referring to Fig.4, however, it must be stressed that breakdown occurred at higher fields (~9 MV/cm, with no secondary breakdown distribution at lower fields) than those where these defects have been reported to induce breakdown (~5 MV/cm) (13). Thus, these defect structures are not likely to be the cause of the low mean breakdown fields and of the characteristic I-V curves observed for the capacitors prepared on CZ-35 substrates.

Although the exact nature of the defects giving rise to the low mean breakdown fields and the enhanced injection rate starting from approximately 5 MV/cm is still under investigation, the fact that these phenomena were only observed for the CZ-35 substrates is indicative of the presence of weak spots at the Si/SiO_2 interface, the origin of which is likely to be the very high oxygen content of these substrates. In fact, weak spots have been

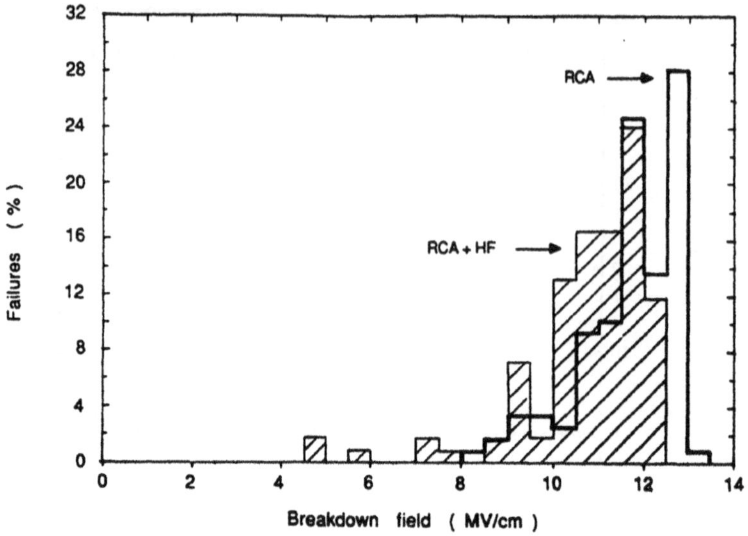

FIGURE 5 - Breakdown distribution of RCA+HF and RCA cleaned CZ-27 substrate, without POHT. Gate area=1.075×10^{-3} cm^2; gate material: Al.

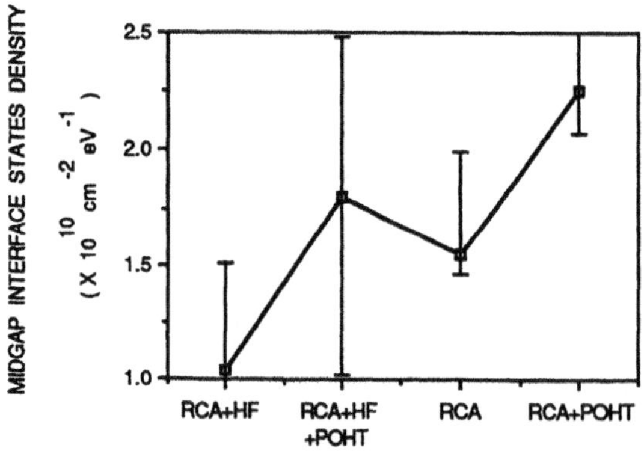

FIGURE 6 - Si/SiO_2 interface states density at midgap as a function of the pre-oxidation procedures. Gate area=4.414×10^{-3} cm^2; gate material: Al. Error bars reproduce the spread over all wafers of the interface states density values as determined from the analysis of the low and high-frequency C-V curves taken from 3-5 capacitors on each wafer.

shown to be caused for example by oxygen precipitates which, during the oxide growth, are included in the oxide (17-19). During an irreversible breakdown measurement, the local injection rate is enhanced at these spots (20). Due to the large trapping rate and consequent charge build-up, a screening effect occurs, lowering the injection rate and thus giving rise to the observed shoulder-like shape of the I-V curve.

For all wafers, the POHT was observed to improve the dielectric strength of the grown oxides. This may be attributed both to the POHT-induced lowering of the defect density in the surface region of the Si-substrate and to the thicker oxide present on the POHT wafers. In the latter case, if small precipitates of a fixed size and density are embedded into the oxide, the ratio between their diameter (assuming spherical shapes) and the oxide thickness would be lowest for the POHT samples and highest for the non-POHT ones. Together with the assumption that the oxide above a precipitate determines the breakdown event, it follows that the measured dielectric strength is deteriorated for the non-POHT samples (18,19).

Concerning the effect of the cleaning procedures, our work showed an HF-dip, as a final cleaning treatment, to be detrimental from a dielectric strength point of view. This can be explained based on the fact that an HF-dip leaves the silicon surface highly reactive by removing the protective cleaning-induced or native oxide, so that the plating of the exposed hydrophobic surface with contaminants is favored (21,22). The latter have been shown to worsen the breakdown characteristics of SiO_2 (13-16), as confirmed in our present work.

CONCLUSIONS

It has been shown in this work that the starting oxygen content of silicon wafers as well as the pre-oxidation treatments - heating and cleaning procedures - can dramatically influence the high-field breakdown characteristics of SiO_2 layers thermally grown on such substrates. At present, the main findings of this ongoing study may be summarized as follows:

- A drastic degradation of the high-field breakdown characteristics was observed for the highest oxygen content substrates (35 ppma) in comparison to the lower oxygen content ones ($\leq$ 30 ppma);
- For all substrates, a pre-oxidation heat treatment in Ar for 10 hours at 1025°C improved the dielectric strength of the grown oxides. However, it did not remove the dependence of the dielectric strength on the starting oxygen content of the substrates;
- A reduction of the mean breakdown fields was observed when the wafers underwent an HF-dip as a final cleaning step before oxidation.

ACKNOWLEDGEMENTS

The authors would like to express their thanks to P. Bal for his help with the measurement systems, J. Wils for assistance with the measurements and Guido Groeseneken for helpful discussions. C. Hasenack is also indebted to CNPq (Conselho Nacional de Desenvolvimento Científico e Tecnológico - Brazil) for his fellowship.

REFERENCES

1. N. Klein, IEEE Trans. Electron Devices ED-13, 788 (1969).
2. N. J. Chou and J. M. Eldridge, J. Electrochem. Soc. 117, 1287 (1970).
3. C. M. Osburn and D. W. Ormond, J. Electrochem. Soc. 119, 591 (1972).
4. D. R. Wolters and A. T. A. Zegers-van Duijnhoven, J. Vac. Sci. Technol. A5, 1563 (1987).
5. M. M. Heyns and R. F. DeKeersmaecker, "The role of electron and hole traps in the degradation and breakdown of thermally grown SiO_2 layers", paper presented at the MRS-meeting, Boston, December 1987.
6. S. Holland, I. C. Chen, J. Lee, Y. Fong, K. K. Young and C. Hu in "Silicon nitride and silicon dioxide thin insulating films", Eds. V. J. Kapoor and K. T. Hankins (The Electrochem. Soc., Pennington, NJ, 1987), p.361.
7. Z. A. Weinberg and T. N. Nguyen, J. Appl. Phys. 61, 1947 (1987).
8. C. M. W. Hillen, R. F. De Keersmaecker, M. M. Heyns, S. K. Haywood and I. S. Daraktchiev in "Insulating Films on Semiconductors", Eds. J. F. Verweij and D. R. Wolters (North-Holland, Amsterdam, 1983), p. 274.
9. D. N. Klein, Thin Solid Films 100, 335 (1983).
10. P. O. Hahn, I. Lampert, and A. Schnegg, Material Research Society meeting, Dec. 1987.

11. W. Kern and D. A. Puotinen, RCA Rev. 31, 187 (1970).
12. E. H. Nicollian, and J. R. Brews, MOS (Metal Oxide Semiconductor) Physics and Technology (Wiley, New York, 1982).
13. P. S. D. Lin, R. B. Marcus, and T. T. Sheng, J. Electrochem. Soc. 130, 1878 (1983).
14. P. S. D. Lin, and H. J. Leamy, Appl. Phys. Lett. 42, 717 (1983).
15. K. Honda, A. Ohsawa, and N. Toyokura, Appl. Phys. Lett. 45, 270 (1984).
16. K. Honda, A. Ohsawa, and N. Toyokura, Appl. Phys. Lett. 46, 582 (1985).
17. K. Yamabe, K. Taniguchi, and Y. Matsushita, Proceedings of the Symposium on Defects in Silicon, vol. 83-9, 629 (1983).
18. H. Abe, F. Kiyosumi, K. Yoshioka, and M. Ino, Proceedings of the International Electron Devices Meeting, 372 (1985).
19. H. Abe, and F. Kiyosumi in "Semiconductor Silicon 1986", Eds. H. R. Huff and T. Abe (The Electrochem. Soc., Pennington, NJ, 1986), p. 1011.
20. P. Solomon, J. Appl. Phys. 47, 2098 (1976).
21. K. D. Beyer and T. H. Yeh, IBM Techn. Discl. Bull. 20, 1741 (1977).
22. I. Lampert, Extended Abstracts Electrochem. Soc. p. 381, May 1987.

DIELECTRIC BREAKDOWN OF SiO_2 GROWN ON ROUGH Si SURFACES

J.P. Gambino, T.N. Nguyen*, B. Cunningham, and J.F. Shepard

IBM East Fishkill
Hopewell Jct., NY 12533

* IBM Research Division, T.J. Watson Research Center
Yorktown Hts., NY 10598

ABSTRACT

SiO_2 grown on CCl_2F_2 reactive ion etched Si has very low breakdown fields, due to reactive ion etching (RIE)-induced surface roughness. Oxidation of the rough Si surface results in a rough Si-SiO_2 interface and a rough SiO_2 surface. The low breakdown fields are due to field enhancement from bumps at the injecting electrode. The sharpness of the bumps, and hence the field enhancement, is greatest at the Si-SiO_2 interface. The interface state density, the minority carrier generation lifetime, and the oxide flatband voltage, mobile charge, and hole trapping, are comparable for metal-oxide-semiconductor (MOS) capacitors fabricated on CCl_2F_2-etched Si and on control Si (no RIE).

INTRODUCTION

SiO_2 grown on rough Si has low dielectric breakdown fields due to field enhancement at the injecting interface (1,2,3). For programmable nonvolatile memory applications (i.e., electrically erasable programmable read-only memory (EEPROM)), a gate oxide with field enhancement at the injecting interface is beneficial because the programming voltage of the device is decreased (2). However, for other gate oxide applications, field enhancement is detrimental because of the lower breakdown voltage.

Roughness can be introduced into a Si substrate by wet etching (2) or reactive ion etching (RIE) (3,4,5,6). Olcer et al. (2) and Fong et al. (3) performed detailed electrical measurements for SiO_2 grown on rough Si produced by wet etching and RIE, respectively. Both found greater field enhancement for electron injection at the Si-SiO_2 interface than at the gate-SiO_2 interface, but the reason for the polarity dependence was unclear.

In this study, the origin of polarity-dependent field enhancement for SiO_2 grown on rough Si is examined. SiO_2 grown on a rough Si surface, produced by CCl_2F_2 RIE, exhibits low breakdown fields and polarity-dependent field enhancement, in agreement with the earlier studies (2,3). Cross-sectional transmission electron microscopy (TEM) reveals that bumps at the Si-SiO_2 interface are sharper (i.e., have a smaller radius of curvature) than those at the gate-SiO_2 interface, resulting in the polarity dependent field enhancement.

EXPERIMENT

Etching was performed in an asymmetric diode reactor made of stainless steel with a Si cathode (99.9999% pure) and a perforated Al anode (97% pure). The Al anode consists of an open-ended can, with the opening facing the cathode and having the same diameter as the cathode. During etching, Si wafers were on the rf powered (13.56 MHz) Si electrode, while the Al anode was grounded.

After an RCA clean, Si wafers (phosphorous-doped, <100>, 0.8-2.0 ohm-cm, 82 mm diameter) were exposed to CCl_2F_2 RIE for 5 min with a Si etch rate of ~47 nm/min. The power density was 0.25 W/cm^2, the self-bias voltage 60 V, the flow rate 200 sccm, and the pressure 70 mtorr. Half of each wafer was masked from the etch by 250 nm of SiO_2 as a control.

After RIE, the mask oxide was stripped and the wafers were cleaned. The cleaning (7) consisted of (i) an O_2 ash, (ii) an H_2SO_4/HNO_3 clean, and (iii) an RCA clean. Steps (i) and (ii) were followed by an HF dip and a deionized water rinse, while step (iii) was followed by a deionized water rinse only. The cleaning treatment removes contaminants on the surface as well as ~1.5 nm of Si.

Metal-oxide-semiconductor (MOS) capacitors were fabricated on both the masked and the CCl_2F_2-etched halves of each wafer ("control oxides" and "RIE oxides", respectively). Gate oxides, 40 nm thick, were grown at 900°C in a dry O_2 ambient (2% HCl). Al dots (0.018 cm^2) were evaporated by resistance heating on the frontside, and blanket Al was deposited on the backside of each wafer, followed by a 400°C N_2/H_2 (90%/10%) anneal for 30 min.

Prior to gate oxide growth, Si surfaces both with and without RIE were inspected for surface roughness by scanning electron microscopy (SEM). For a quantitative measure of surface roughness, cross-sectional transmission electron microscopy (TEM) was performed on both post-RIE and post-oxidation samples. For TEM of oxide samples, Ti was deposited on the SiO_2 to provide high contrast at the SiO_2 surface. The SiO_2 thickness and index of refraction were also measured by ellipsometry.

A variety of electrical measurements were used to characterize the MOS capacitors. Dielectric breakdown was measured with a current-ramp method (8). Current was ramped (in an exponential staircase) from $1x10^{-11}$ to $1x10^{-1}$ A while monitoring the voltage, with breakdown defined as a sudden drop (of at least 0.05 V) in the measured voltage. The current and voltage at which the sudden drop occurs are defined as the breakdown

Fig. 1. Cross-section TEM micrograph of SiO_2 grown on region of wafer masked from RIE (control oxide).

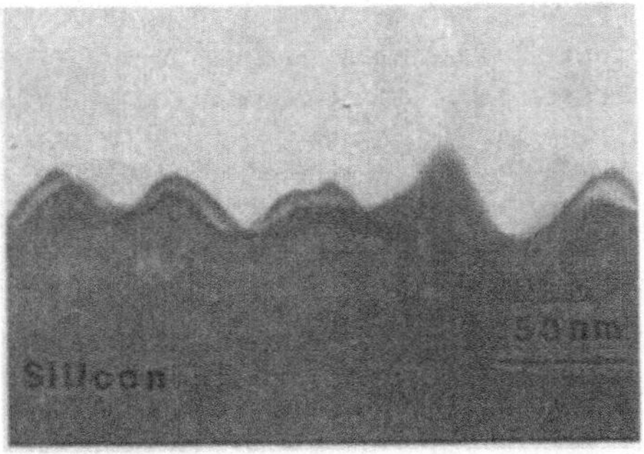

Fig. 2. Cross-section TEM micrograph of CCl_2F_2-etched Si.

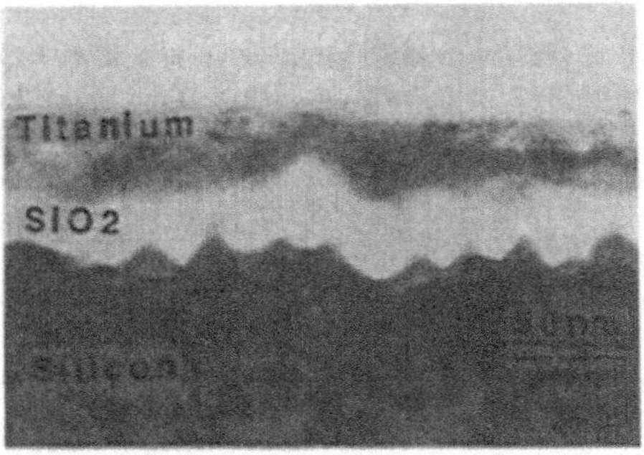

Fig. 3. Cross-section TEM micrograph of SiO_2 grown on CCl_2F_2-etched Si (RIE oxide).

current and breakdown voltage. Current-voltage data were taken using a linear voltage ramp. Hole trapping in the SiO_2 was determined by injecting holes into the oxide (avalanche injection) and measuring the flatband shift (9). The Si-SiO_2 interface state density was measured by comparing high-frequency and quasi-static capacitance-voltage characteristics (10). The surface recombination velocity and bulk carrier generation lifetime were obtained from capacitance transient measurements and Zerbst's analysis (10). The mobile ionic charge concentration was measured using a charge-voltage technique (11). The measurement is similar to the triangular voltage sweep method (10), except that step voltages are used rather than a triangular voltage ramp, and the ionic current peak is integrated automatically with a digital coulombmeter.

RESULTS

SEM and TEM observations reveal that for the control side of the wafer, both the original Si surface and the Si-SiO_2 interface after gate oxide growth are smooth (Figure 1). The few bumps that are observed are typically less than 1 nm in height. In contrast, the surface is extremely rough after the CCl_2F_2 etch (Figure 2), with bumps 17 to 31 nm in height, spaced about 50 nm apart. Oxidation smooths the surface to some degree but the Si-SiO_2 interface is still extremely rough (Figure 3), with bumps 12 to 29 nm in height. Lattice fringe imaging indicates that the bumps are crystalline Si having the same orientation as the substrate. The SiO_2 surface is approximately conformal to the Si-SiO_2 interface, and is therefore also rough.

For oxidized samples, the average SiO_2 thickness measured by TEM is ~40 nm for control oxides and ~30 nm for RIE oxides. (The thickness for RIE oxides was measured from "valleys" in the Si substrate, rather than from "peaks".) The SiO_2 thicknesses measured by ellipsometry are 50 nm and 57 nm for control and RIE oxides (Table I), respectively, and are significantly higher than the thicknesses measured by TEM.

The accumulation capacitance is about 1.11 times higher for SiO_2 on CCl_2F_2-etched Si than for controls (Table I). The interface state density, flatband voltage, and mobile charge were low for all capacitors. The oxide flatband voltage is essentially identical for both RIE and control oxides, while the mobile charge and the density of interface states are slightly higher for RIE oxides. The minority carrier generation lifetime is high and about the same for the Si underneath the RIE and the control oxides (Table I).

The breakdown voltages for RIE oxides, compared to control oxides, are much lower and much more sensitive to gate voltage polarity (Figure 4). For RIE oxides, positive gate bias (electron injection from substrate) results in much lower breakdown voltages than negative gate bias (electron injection from gate). Interestingly, the corresponding breakdown currents (tunneling current just before breakdown) are one to three orders of magnitude larger for RIE oxides than for control oxides (Figure 5). The voltage needed to induce 10^{-12} A tunneling current through the oxide, is lower in RIE oxides than control oxides by a factor of 2.4 and 5.0 (Figure 6), for electron injection from the Al gate and

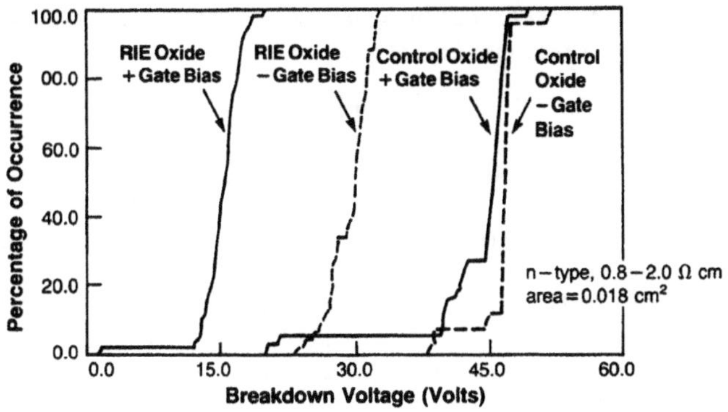

Fig. 4. Breakdown voltage for control and RIE oxides measured at positive and negative gate bias.

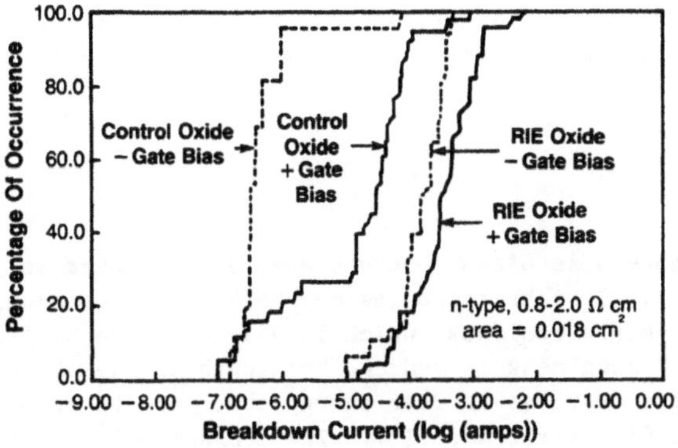

Fig. 5. Breakdown current for control and RIE oxides measured at positive and negative gate bias.

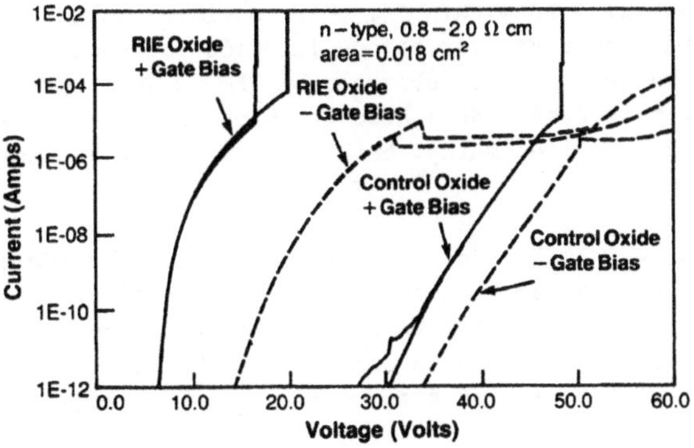

Fig. 6. Oxide leakage current for control and RIE oxides measured at positive and negative gate bias.

Table I. Optical and electrical properties of SiO_2 grown on control Si and CCl_2F_2 reactive ion etched Si.

	Control Oxide	RIE Oxide
Ellipsometry Thickness (nm)	50	57
Index of Refraction	1.49	1.54
Accumulation Capacitance (pF)	1194	1329
Flatband Voltage (V)	-0.222	-0.223
Flatband Voltage Shift (V)	-3.8	-3.0
Mobile Charge (cm^{-2})	$1x10^{10}$	$2x10^{10}$
Midgap Dens. of St. ($cm^{-2}eV^{-1}$)	$1x10^{10}$	$2x10^{10}$
Generation Lifetime (10^{-6} s)	530	690
Surface Recomb. Vel. (cm/s)	0.036	0.036

from the Si substrate, respectively. For RIE oxides compared to control oxides, the hole trapping characteristics are similar, though the magnitude of the flatband shift after avalanche injection is smaller (Table I).

DISCUSSION

Surface roughness is often observed when Si is etched in gases containing Cl (3,5,12). The roughness may be due to the nonuniform deposition of polymeric residues, which locally mask the Si from etching (12). The roughness is reduced by growth of the 40 nm oxide, possibly due to the decrease in free energy that results from the decrease in area of the $Si-SiO_2$ interface (13). The mechanism by which the roughness is reduced during oxidation is unclear.

The discrepancy in control oxide thicknesses measured by TEM and ellipsometry is substantial (~20%), and is not understood. The TEM measurements were calibrated using lattice fringe images of the Si substrate, and therefore are expected to be very accurate. The optical measurements were made with three different ellipsometers, which all gave the same thickness value to within 5%. One possible explanation for the discrepancy is that there are variations in oxide thickness across the wafer, and that the small region observed by TEM is not representative of the average oxide thickness. Note however that the variation in oxide thickness measured by ellipsometry, both between wafers and across wafers, was less than 5%.

The discrepancy in RIE oxide thicknesses measured by TEM and ellipsometry is even greater than that for control oxides. However, both TEM and ellipsometry are expected to be less accurate due to surface roughness. The RIE oxide thickness measured by TEM is probably a lower limit, because of overlapping of the bumps in the TEM image (Figure 3).

The high generation lifetimes and low interface state densities for capacitors fabricated with RIE oxides indicate that there is little if any metallic contamination from the RIE reactor (4,14). The high accumulation capacitance for RIE oxides probably results form a greater effective surface area due to roughness (2). The difference in the avalanche injection-induced flatband voltage shift between RIE and control oxides may be due to the difference in accumulation capacitance.

The reduced breakdown voltages for RIE oxides cannot be due to bulk defects in the SiO_2, because the breakdown currents for RIE oxides are higher than those for control oxides. In addition, the flatband voltage, hole trapping, and mobile charge densities are low for RIE oxides (and comparable to control oxides), indicating that the density of bulk SiO_2 defects is low. The higher tunneling current at a given applied voltage for RIE oxides, implies higher electric fields at the injecting interface (i.e., field enhancement). The data indicates that the low breakdown voltages for RIE oxides are due to enhanced electron injection at the rough, injecting electrode. The higher breakdown currents for RIE oxides compared to control oxides, are probably due to thin regions in the RIE oxides. (Breakdown current increases with decreasing oxide thickness.)

The amount of field enhancement depends on the dimensions of the bump. For a fixed bump height, the field enhancement increases as the radius of curvature at the tip of the bump decreases (15). The smallest radius of curvature is ~2.5 nm for bumps at the Si-SiO_2 interface, and ~7.0 nm for bumps at the gate-SiO_2 interface. Hence, the field enhancement is expected to be greatest for injection from the Si-SiO_2 interface, consistent with the current-voltage characteristics. For control oxides, the lower breakdown voltage and higher leakage current for positive gate bias compared to negative gate bias are due to a higher barrier height for Al on SiO_2 (3.2 eV) compared to Si on SiO_2 (3.1 eV) (16).

CONCLUSION

SiO_2 grown on CCl_2F_2 reactive ion etched Si has very low breakdown fields, due to RIE-induced surface roughness. Oxidation of the rough Si surface results in a rough Si-SiO_2 interface and a rough SiO_2 surface. The low breakdown fields are due to field enhancement from bumps at the injecting electrode. The sharpness of the bumps, and hence the field enhancement, is greatest at the Si-SiO_2 interface. The interface state density, the minority carrier generation lifetime, and the oxide flatband voltage, mobile charge, and hole trapping, are comparable for metal-oxide-semiconductor (MOS) capacitors fabricated on CCl_2F_2-etched Si and on control Si (no RIE).

ACKNOWLEDGEMENT

The authors thank Ed Vishnesky, Carol Galli, and Bob Hicks for assistance in processing the wafers, and Steve Cohen for performing the avalanche injection measurements.

REFERENCES

1. D.J. DiMaria and D.R. Kerr, Appl. Phys. Lett., 27, 505 (1975).
2. M. Olcer, H.J. Buhlmann, and M. Ilegens, J. Electrochem. Soc., 133, 621 (1986).
3. Y. Fong, A.T. Wu, R. Moazzami, P.K. Ko, and C. Hu, Recent News Paper No. 26.8, IEEE Int. Electron Dev. Meet., Wash., D.C., December 1987.
4. L.M. Ephrath and R.S. Bennett, J. Electrochem. Soc., 129, 1822 (1982).
5. G.C. Schwartz and P.M. Schaible, J. Vac. Sci. Tech., 16, 410 (1979).
6. C. Koburger, F.R. White, L. Nesbit, and S.D. Emmanuel, IEEE Trans. Elec. Dev., ED-33, 766 (1986).
7. T. Hosoya, Y. Ozaki, and K. Hirata, J. Electrochem. Soc., 132, 2436 (1985).
8. T.N. Nguyen, Recent News Paper No. 701, Electrochemical Society Spring Meeting, Boston, Mass., May 1986.
9. D.R. Young, E.A. Irene, D.J. DiMaria, R.F. DeKeersmaecker, and H.Z. Massoud, J. Appl. Phys., 50, 6366 (1979).
10. E.H. Nicollian and J.R. Brews, "MOS Physics and Technology", John Wiley & Sons, N.Y. (1982).
11. D.R. Kerr, unpublished.
12. B. Gorowitz and R.J. Saia, in "Plasma Processing for VLSI", p. 297, N.G. Einspruch and D.M. Brown, ed., Academic Press, Inc., N.Y. (1984).
13. A.H. Carim and R. Sinclair, J. Electrochem. Soc., 134, 741 (1987).
14. S.W. Pang, D.D. Rathman, D.J. Silversmith, R.W. Mountain, and P.D. DeGraff, J. Appl. Phys., 54, 3272 (1983).
15. T.J. Lewis, J. Appl. Phys., 26, 1405 (1955).
16. D.J. DiMaria, in "The Physics of SiO_2 and its Interfaces", p. 160, S.T. Pantelidas, ed., Pergamon Press, N.Y. (1978).

EFFECTS OF DEIONIZED WATER RINSES ON GATE OXIDE LEAKAGE CURRENTS

Mark A. Chonko

Motorola, Inc.
3501 Ed Bluestein Blvd.
Austin, TX 78721

INTRODUCTION

Deionized (DI) water is used extensively in the manufacture of integrated circuits. It is particularly important in clean-ups prior to gate oxidations where it often follows an HF dip. DI water rinses can reduce the cleaning efficiencies of subsequent chemical treatments[1] and reduce the oxidation rate of silicon at room temperature.[2] Significant impact on the I-V characteristics of silicon dioxide by deionized water rinses are reported here. Leakage currents through a gate oxide increase with rinse time. Electron injection from a poly-silicon electrode increase much faster than from the substrate electrode. Currents can increase by many orders of magnitude. These increases are interpreted as arising from roughened oxide interfaces caused by a contamination layer.

EXPERIMENTAL

MOS capacitors were fabricated on 100 mm, 14-22 ohm-cm, p-type silicon. A dry HCl oxidation process was used to grow 650 angstroms of gate oxide. Phosphorous diffused poly-silicon was patterned to form gate electrodes. Prior to oxidation, the wafers were given a short dip in 50:1 water:HF, followed by various rinses in DI water. The HF bath was filtered and recirculated. The DI rinses were done in an overflow style tank. The quality of the DI water was very good as measured by conventional standards.

Electrical measurements were performed on an HP 4145 parametric analyzer and LOMAC automated test system. All measurements with positive gate bias were performed under illumination.

RESULTS

Figure 1 shows the current-voltage characteristics resulting from a 30 minute hot (70 C) DI rinse as compared to no pre-gate clean at all. Leakage currents with the poly-silicon biased negatively (Fig 1a) increased by several orders of magnitude for the rinsed wafers. The positive gate bias case (Fig 1b) shows little difference between the two treatments.

Figure 2 shows the electric field strength required to maintain a fixed current density decreasing with increasing rinse times in room temperature DI water. The gate bias is negative. Figure 3 illustrates the difference

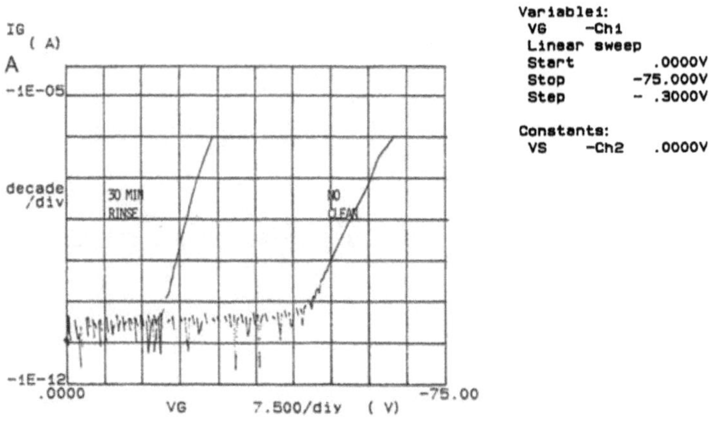

FIGURE 1A

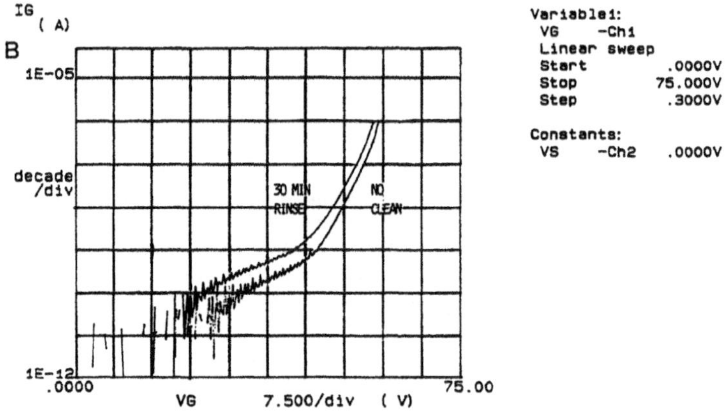

FIGURE 1B

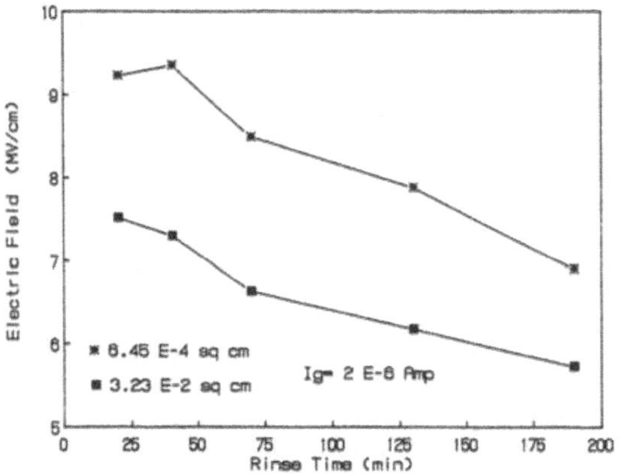

FIGURE 2

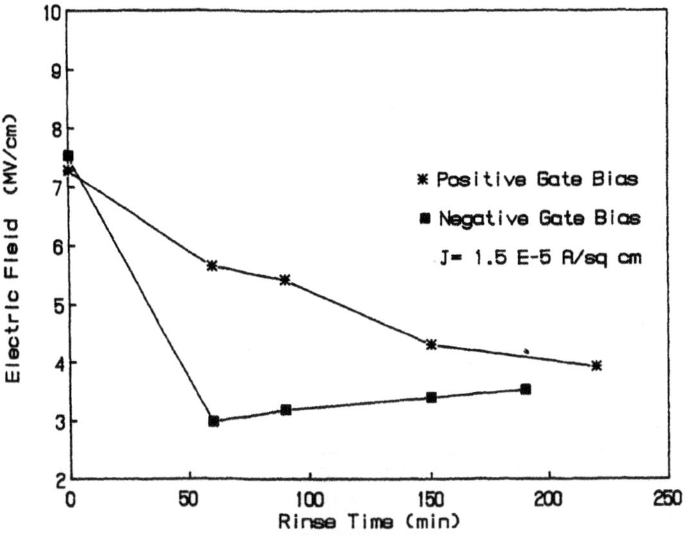

FIGURE 3

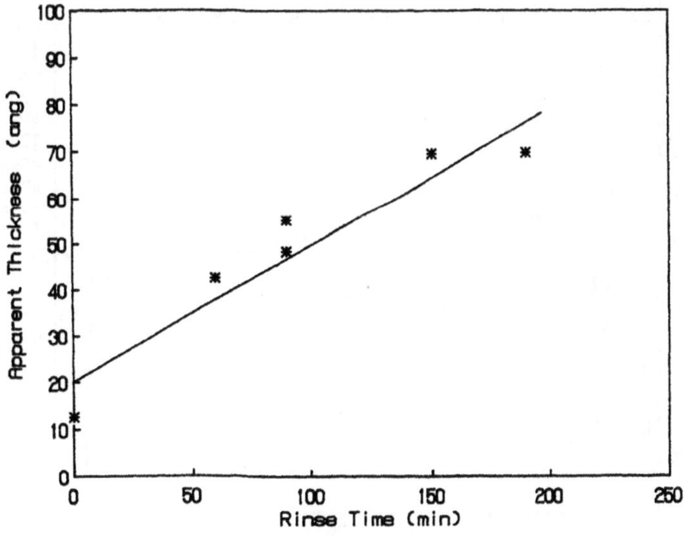

FIGURE 4

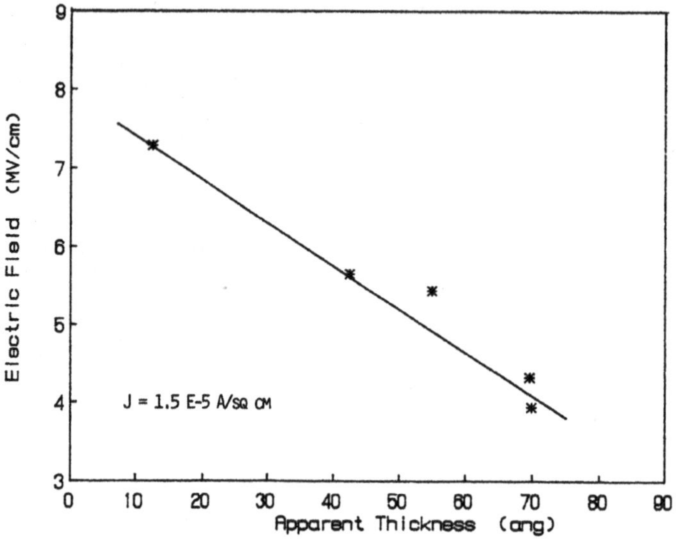

FIGURE 5

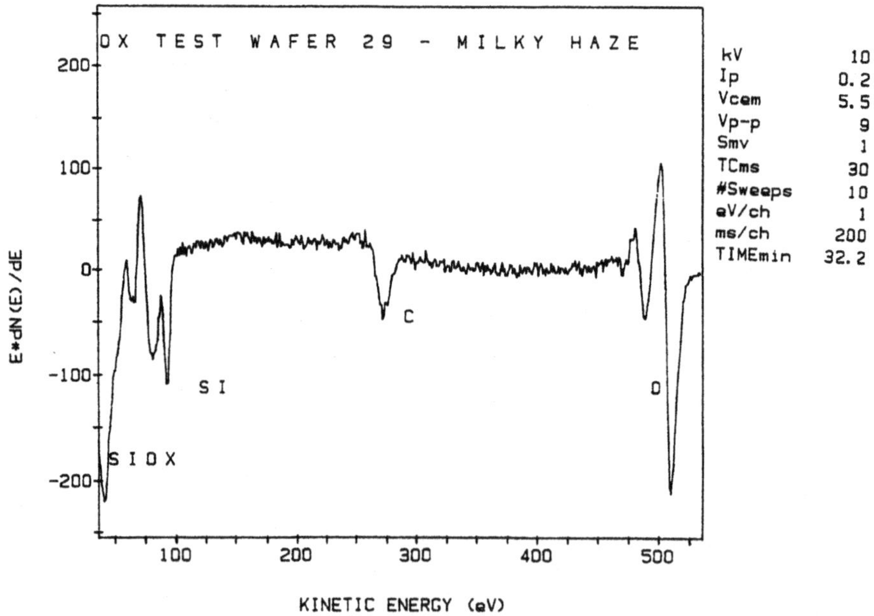

FIGURE 6

between positive and negative gate biases. The leakage currents increase at different rates, eventually saturating for both polarities.

DISCUSSION

Increased leakage currents have been observed for oxides grown on poly-silicon and have been attributed to surface roughness at the 1st poly-oxide interface.[3] The I-V characteristics in Figure 1 can also be explained in terms of surface roughness at the poly-oxide interface. The roughness originates from a non-uniform contamination layer which is deposited on the surface of the silicon during the DI water rinse prior to oxidation. This layer is incorporated into the intial oxide and is transported to the top surface as oxide growth proceeds.[2] If the contamination layer is thin enough, substrate oxidation can proceed normally. As the thickness of the layer increases with longer rinse times, oxygen diffusion to the substrate is retarded. This means that the substrate-oxide interface will no longer be smooth. Its roughness will depend on the thickness and uniformity of contamination, as well as the diffusivity of oxygen through it. This can explain the differences in leakage currents seen between positive and negative gate biases as a function of rinse time.

Figure 4 shows the apparent thickness of a layer versus rinse time in hot DI water. The thickness was measured with a Rudolph AUTO EL III ellipsometer. The refractive index was fixed at 1.465. Figure 5 compares the field required to maintain a fixed current density as a function of apparent thickness of a contamination layer. Results for positive gate bias are shown. Auger analysis of a rinsed surface shows only oxygen, silicon and carbon. This is similar to the ESCA results reported in Ref. 2.

SUMMARY

Deionized water rinses have been observed to have a significant effect on gate oxide leakage currents. The results are interpreted in terms of a non-uniform contamination layer on the surface of the silicon prior to oxidation which produces a rough poly-oxide interface for short rinse times and a rough substrate-oxide interface for longer rinse times.

ACKNOWLEDGEMENTS

Valuable discussions with Arnie Lesk and Doug Keitz are gratefully acknowledged.

REFERENCES

1. K.D.Beyer and R.H. Kastl, J. Electrochem. Soc., 129, 1027 (1982)

2. Licciardello et al, App. Phys. Lett., 48, 41 (1986)

3. R.M. Anderson and D.R. Kerr, J. App. Phys., 48, 4834 (1977)

OXIDATION-INDUCED DEFECTS AND EFFECTS IN SILICON DURING LOW THERMAL-BUDGET PROCESSING

Richard B. Fair

Microelectronics Center of North Carolina
Research Triangle Park, North Carolina 27709

Department of Electrical Engineering
Duke University, Durham, North Carolina 27706

ABSTRACT

Low thermal-budget processing of *Si* can produce point-defect concentrations that depend on time, temperature and ambient in ways that are unlike observations from high temperature/long time anneals. Following ion implantation, anneals in oxidizing ambients may produce point-defect supersaturations that are not significant compared to levels produced by dissolving point-defect clusters or extended dislocations. A reverse oxidation-enhanced diffusion (OED) effect is observed for 750°C anneals of *B* an *As*-implanted *Si*. Other sources of point-defects in low thermal-budget processes include dissolving precipitates and metal silicide reactions which can assist in eliminating process-induced defects.

INTRODUCTION

Oxidation of the *Si* surface leads to the generation of excess self-interstitials which saturate the *Si* substrate and cause the growth of extrinsic defects such as oxidation-induced stacking faults (OSF). OSF growth times are typically hours at temperatures greater than 1100°C.

In advanced, scaled technologies, the need for high temperature, long time anneals is diminishing. For submicron silicon processing, low temperatures or short-time processes are required to keep doping profiles shallow. For example, a typical junction depth for a $2\mu m$ CMOS process is $0.5\mu m$. Assuming a diffusion coefficient of $10^{-12} cm^2/sec$, the process designer is allowed 2500 sec of processing time. By comparison, a one-half micron process with a $0.15\mu m$ junction can only tolerate 225 sec![1] The question then becomes - what is the impact of oxidation and point-defect supersaturation in low thermal budget processes?

To address this question it is first necessary to understand the nature of point defect kinetics for such processes. Several studies will be reviewed. It will be shown that point defect supersaturations produced by short oxidation cycles increase with time in the linear growth region. When oxidation is combined with the annealing of implanted junctions, the dissolution of ion implant damage produces point-defect transients that may dominate over oxidation-produced point-defect supersaturations. It will be shown that such effects can produce a reverse oxidation-enhanced diffusion effect for *B* and *As*. Other sources of point defects such as precipitate dissolution or metal silicide reactions can dominate the growth/shrinkage of OSF's and implantation damage. Thus, the low thermal-budget regime can produce point-defect concentrations and extended defects that depend on time, temperature and ambient in a way that is unlike high temperature, long-time processes.

OXIDATION-INDUCED STACKING FAULT DYNAMICS

Short-Time Oxidations and Point Defects

Recently, Hu[2] performed an analytical study of the kinetics of the supersaturation of self-interstitials in short-time/low-temperature oxidation of Si. He proposed that in contrast to long-time/high-temperature oxidation where the self-interstitial supersaturation decreases with time as t^{-n} (n=0.2–0.3), short-time oxidation produces self-interstitials that increase in concentration with time. When linear-parabolic oxide growth begins, the increase in self-interstitial concentrations reaches a maximum.

Experimental measurements of stacking fault growth in the parabolic oxidation regime show that the concentration of excess self-interstitials, C_I, decreases with decreasing oxidation rate according to the expression[3,4]

$$C_I(t)=K\left[\frac{dx_o}{dt}\right]^{2n}. \tag{1}$$

However, this is only an empirical expression fitted to a limited range of experimental data. Thus, it is unwarranted to extrapolate Eq. (1) to the short-time or the linear oxidation regime.[2] But, there is no reliable data available on the growth of oxidation stacking faults under these conditions because of experimental difficulties. Available data cover the range T > 1050°C and t > h hour. As a result, the kinetics of self-interstitial supersaturation are unclear.

Hu[2] has calculated the self-interstitial supersaturation for the linear-parabolic region. The transition time constant, t_o, from the linear to the parabolic growth regimes is given as

$$t_o=\frac{k_p}{4k_L^2} \tag{2}$$

where k_P and k_L are the parabolic and the linear rate constants. Ignoring bulk recombination of generated self-interstitials, the time dependence of the surface concentration of excess interstitials is shown in Fig. 1 for various values of t_o.[2] The increase in interstitial concentration is evident in the linear growth region.

The next step is to relate C_I to the rate of growth of a stacking fault, dr/dt:

$$\frac{dr}{dt}=\pi a_o^2 D_I(C_I-C_I^1) \tag{3}$$

where r is the radius of the stacking fault, a_o is the capture radius of the bounding dislocation of the fault and C_I^1 is the interstitial concentration in equilibrium with the stacking fault. For small faults, where r<1000Å, Hu[2] has pointed out that C_I becomes infinite as r becomes zero. This means that if no nucleation centers exist, it requires an infinite supersaturation of interstitials to start the fault growth! Since so little is known about the nature of nucleation centers and they are required for initiating faults, it is not possible currently to analyze stacking fault growth in short-time/low-temperature oxidation.

Experimentally, Rogers[5] recently studied stacking fault growth during 1100°C, 30 sec rapid thermal oxidation (RTO). No growth of pre-existing OSF's could be detected during RTO.

Time Dependence of Surface Concentration of Self-Interstitials for Linear Parabolic Oxidation (after Hu)

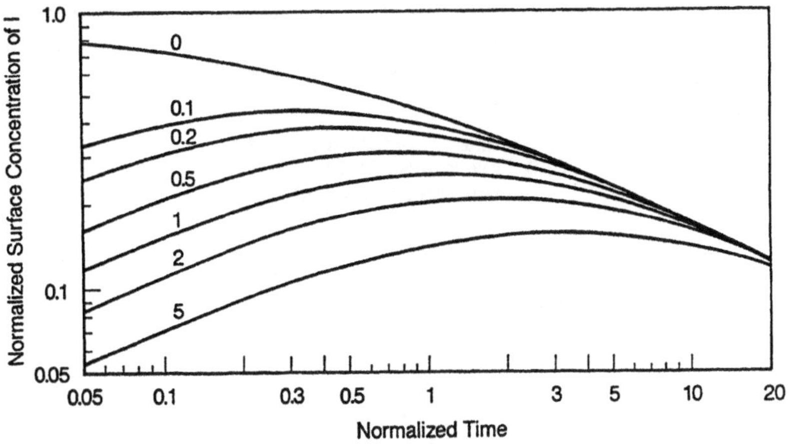

Figure 1

OSF Shrinkage for Short-Times

Other observations regarding short-time or low-temperature oxidation-induced defects include the work of Nishi and Antoniadis.[6] They observed rapid initial shrinkage of OSF's in *Si* at temperatures down to 950°C when anneals were performed in N_2. The mechanism is believed to be due to surface-annihilation of self-interstitials emitted from the dislocation that bounds the fault. An example of accelerated shrinkage is shown in Fig. 2 for N_2 anneals of OSF's at 950°C and 1000°C.[6] The activation energy for rapid fault shrinkage was found to be 2.3eV compared to 4.3eV for long-time, slow fault shrinkage. Nishi and Antoniadis proposed that there are two components of self-interstitial flux: one towards the surface, F_f, and one towards the silicon bulk, F_B. If $F_f > F_B$ then fault shrinkage is surface-annihilation limited. Otherwise, the diffusion of interstitials to the bulk dominates. The cause of surface-controlled shrinkage is due to the elongated semi-elliptical shape of an OSF with the major axis along the *Si* surface, as shown in Fig. 3. If the portion of the elliptical dislocation near the surface shrinks rapidly because of its proximity to the surface (self-interstitial sink), then the OSF dislocation line becomes rounder toward the surface. After that the observed length of the OSF is determined by F_B because the etch used to delineate the OSF always reveals the maximum dimension. Therefore, for short-time anneals, the apparent rapid dissolution of OSF's has little impact on the deep lying portion of the OSF bounding dislocation. These defects still have the potential of penetrating junctions and creating potential device yield problems.

EFFECT OF OXIDATION ON IMPLANT DAMAGE

Ion implantation processes introduce damage in the *Si* substrate. Annealing times, temperatures and ambients are crucial to removing the damage so as to allow high quality junctions to be produced. These process variables are also important in minimizing dopant diffusion for shallow junction formation. In low-thermal budget processes, point-defect generation from dissolving point-defect clusters or extended-defect structures may dominate over point-defect production at an oxidizing surface.

As an example of the dominance of implantation damage annealing over oxidation enhanced diffusion (OED), consider the B profiles in Fig. 4. BF_2^+ was implanted at $2x10^{14}cm^{-2}$, 40 keV into <100> Si covered with 150Å of SiO_2. Annealing was performed at 750°C, 1 hr in either N_2 or O_2. It is obvious that substantial diffusion has occurred, well beyond that predicted by normal thermally-assisted diffusion processes (30Å diffusion length compared to 800Å measured). It has been shown that the period of enhanced

OSF Shrinkage Under Surface and Bulk Fluxes of Self-Interstitials (Nishi and Antoniadis)

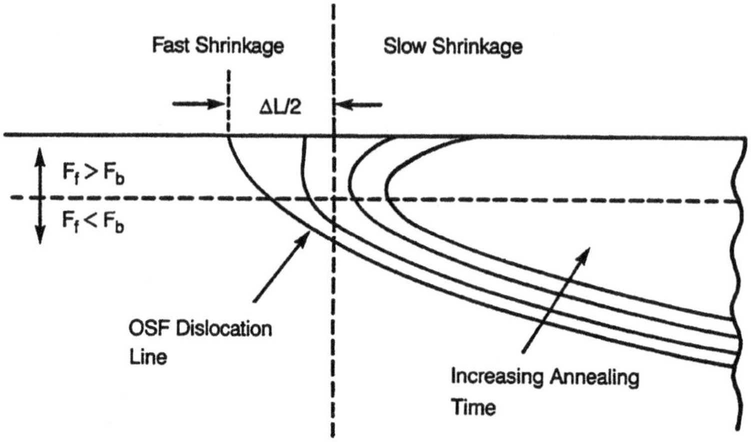

Figure 2

OSF Length vs. Annealing Time (Nishi and Antoniadis)

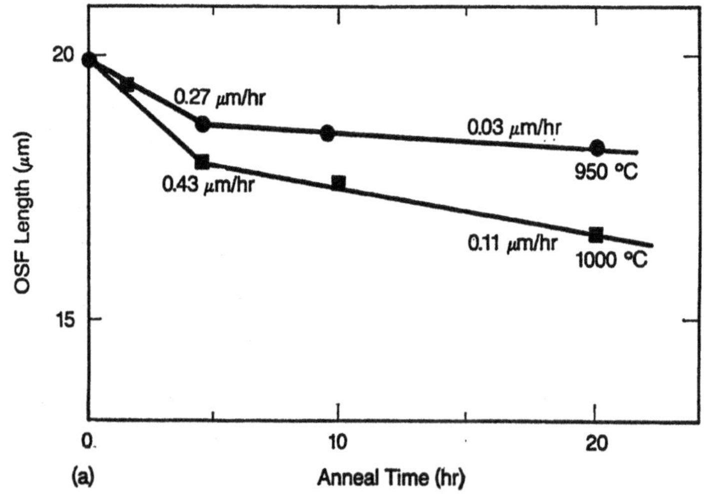

Figure 3

Reverse Oxidation-Enhanced Diffusion at 750 °C for Boron

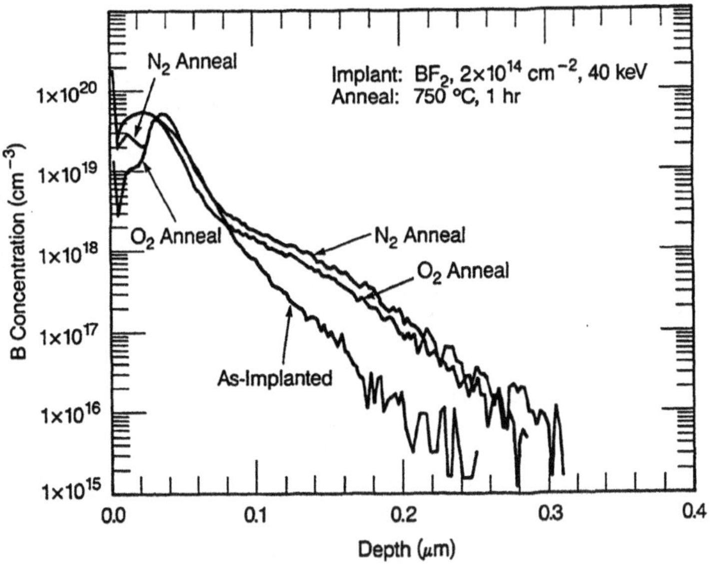

Figure 4

diffusion at 750°C is 30 minutes, and is associated with a self-interstitial supersaturation from the dissolution of end-of-range self-interstitial clusters and small dislocations.[7,8] Cross-section TEM showed that considerable extended dislocations at 400Å depth remained in the sample after 30 minutes, marking the location of the original amorphous-crystalline interface. Thus, the shrinkage of these extended defects had little effect on the enhanced B diffusion at 750°C.

It can be seen in Fig. 4 that more diffusion occurred during N_2 annealing than during O_2 annealing - a reverse OED effect. For implants of $7x10^{14}$ and $2x10^{15}cm^{-2}$ no difference was detected between the B profiles obtained by annealing in O_2 and N_2. This result is consistent with the concentration dependence of OED previously reported.[9]

The reverse OED effect is also seen during low temperature annealing of As implants. Figure 5 shows the results for 4 hr, 750°C anneals of As implanted at $7x10^{14}cm^{-2}$, 40 keV. It should be pointed out that during the dry O_2 anneal, 130Å of SiO_2 were grown, consuming 60Å of Si. This accounts for one-half of the observed difference between the O_2 and N_2 annealed profiles. A 10% uncertainty in the depth of the SIMS profiles could mean that no effect can be detected at all! However, the reverse OED effect was consistently observed on all 8 samples measured.

Oxidation has been shown to slow the removal of end-of-range extended defects created by high dose implants that form amorphous surface layers.[10] Dissolution rates of extended dislocations depend on temperature, annealing ambient and proximity to the Si surface. Surface proximity directly affects the flux of self-interstitials from the defects to the surface. In a neutral ambient the self-interstitial flux is towards the surface (from the dissolving defect). In an oxidizing ambient the self-interstitial concentration at the surface is raised, and the flux changes direction or is simply reduced. This effect is illustrated by the

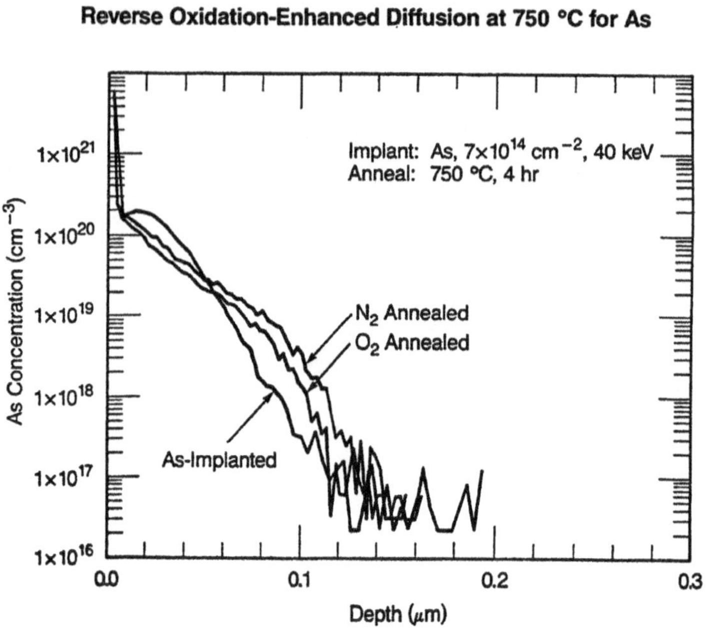

Figure 5

cross-section micrographs shown in Figs. 6a and 6b.[11] A two-dimensional, masked structure was implanted with Ge^+ at 150 keV and $9x10^{14}cm^{-2}$ and annealed at 450°C, 30 min. followed by RTA at 1100°C for 10 sec in either Ar or O_2 ambients.[12] For the Ar anneal, dislocation loops under the mask edge were dissolved during RTA. When RTA was performed in O_2, the end-of-range loops were still present under the mask edge. The O_2 anneal changed the point-defect boundary condition at the Si surface as is illustrated in Fig. 7. Thus, the reverse OED effect can be explained on the basis of a reduction in the net self-interstitial concentration beyond the depth of the implant damage location (400-600Å in this study) as the result of reduced damage dissolution in the presence of an oxidizing Si surface.

OTHER POINT DEFECT SOURCES FOR DISLOCATION DISSOLUTION

The growth or shrinkage of OSF's or implantation-induced defects can be significantly influenced by the generation of vacancies or self-interstitials from these additional sources and/or surface boundary conditions:

- oxygen or dopant precipitate growth/shrinkage[5, 12, 13]
- high concentration P diffusion[14, 15]
- silicon nitride or silicon dioxide surface covering[13] or a free surface[16]
- Silicide formation for contacts to junctions.[17, 18]

Two of these effects are particularly important for low thermal budget processing - the dissolution of precipitates and silicidation. Both of these phenomena have been shown to assist in the rapid removal of end-of-range dislocations (type II defects) during RTA.

When an amorphous layer is formed by ion implantation, end-of-range defects are formed at the amorphous-crystalline interface. These defects are normally very stable.

XTEM illustrating influence of ambient on mask edge defect density following 450 °C, 30 min + 1100 °C, 10 s anneal: (a) argon and (b) oxygen ambients

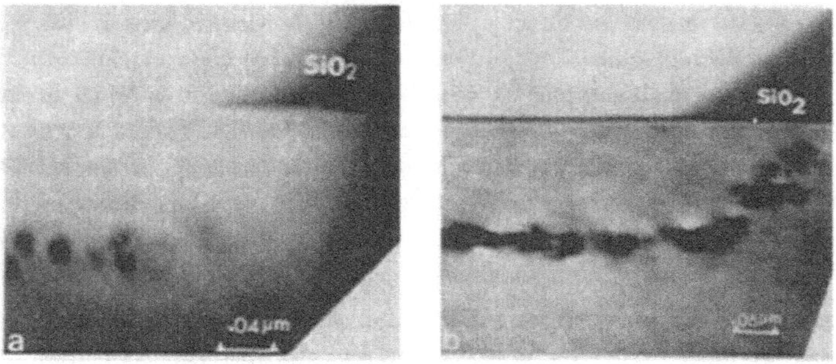

Figure 6

Surface Effect on End-of-Range Damage Annealing

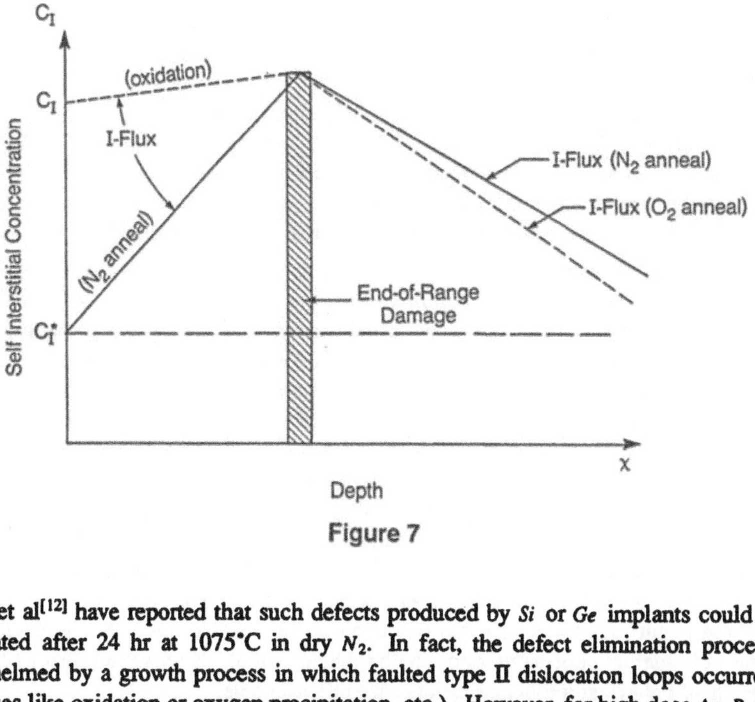

Figure 7

Jones, et al[12] have reported that such defects produced by *Si* or *Ge* implants could not be eliminated after 24 hr at 1075°C in dry N_2. In fact, the defect elimination process was overwhelmed by a growth process in which faulted type II dislocation loops occurred (via processes like oxidation or oxygen precipitation, etc.). However, for high dose *As*, *P* and *Ga* implants, exceeding the solid solubility limit of the implanted species resulted in rapid dissolution of the type II defects during RTA (typically 1100°C for 2 sec).[12, 19] This is believed to be due to the release of vacancies or generation of an interstitial sink during precipitate dissolution.[12] The defect shrinkage process has a 5 eV activation energy which correlates with the activation energy for *Si* self-diffusion. The removal of projected-range defects (type V) caused by the dopant concentration exceeding solid solubility was not enhanced by precipitate dissolution if the defects evolved into relatively stable half-loop dislocations.[12] Judging from the rate of shrinkage of the type II defects, it is expected that similar results would be obtained even if the *Si* surface were being oxidized during RTA.

If the extended defects are located beyond the high-concentration diffusion layer, point defect generation can enhance their growth. For the case of high concentration *P* implantation, Tsai, et al [14] have shown that stacking faults grow during *P* diffusion at the same time that enhanced *As* buried layer diffusion occurs. Thus, the *P* diffusion produces a supersaturation of self-interstitials that flood the bulk of the *Si* substrate. Experiments show that even with a Si_3N_4 cap covering the *Si* surface, fault growth and enhanced buried layer diffusion occur.[20] Therefore, the *P* diffusion itself produces the excess self-interstitials beyond what might be produced by oxidation.

The injection of vacancies during silicidation of refractory metals on the *Si* surface has recently been reported.[17, 18] Wen, et al [19] have reported that $TiSi_2$ formation results in the elimination of end-of-range (type II) damage during RTA. Hu[18 reported that $TaSi_2$ formation on single crystal *Si* at 950°C causes an enhancement in the diffusivity of *Sb* buried marker layers. Since *Sb* is believed to diffuse primarily via a vacancy mechanism, Hu concluded that silicidation produces a vacancy, supersaturation. This hypothesis was also proposed by Wen, et al [17] and is diagrammed in Fig. 8.

Recently Fahey and Dutton[21] studied buried *P* and *B* marker diffusions during the oxidation of WSi_2 on silicon. The oxidation of WSi_2 causes an efflux of *Si* atoms from the *Si* substrate to the surface of the silicide where oxidation occurs. This situation is assumed to be analogous to the efflux of *Si* atoms during $TaSi_2$ and $TiSi_2$ formation. Fahey and Dutton verified that *Si* atoms were consumed from the *Si* substrate by the advancement of the WSi_2 layer past the original WSi_2/Si interface during oxidation. However, no enhanced or retarded *B* or *P* diffusion was detected during 1100°C oxidation of the WSi_2. They concluded that no appreciable amount of vacancies were generated during the process. In addition they pointed out that vacancy/self-interstitial supersaturations or undersaturations produced by the *Si* efflux process depend largely on the surface properties of the silicide-*Si* interface.

The *Si* efflux during direct nitridation of *Si* occurs in the presence of an interface that apparently allows a significant build up of vacancies or a significant undersaturation of self-interstitials. This has been verified by observations of corresponding changes in the diffusivity of dopants in the *Si*.[22] For the case of oxidation of WSi_2, the large *Si* efflux was either not sufficient to cause a vacancy supersaturation, or a corresponding self-interstitial supersaturation might also have been present. It is also not clear what role oxygen and oxidation has. Wen, et al [23] have detected the complete elimination of end-of-range defects during two step silicidation when the annealing ambient was *Ar*. Sheet resistance of the $TiSi_2$ was 2 ohms/sq. However, if N_2 was the annealing ambient during silicidation, a *Ti*-rich *TiON* layer formed on the surface. For the same annealing steps, the sheet resistance was 4.5 ohms/sq, the silicide layer was thinner by 30%, and the end-of-range defects were not totally eliminated. The dependence of $TiSi_2$ sheet resistance on O_2 concentration is shown in Fig. 9.[24] These results suggest that some oxidation of the metal or silicide may decrease the resulting point-defect supersaturations in the underlying *Si*. This would also explain the WSi_2 results of Fahey, although no proof is currently available.

CONCLUSIONS

In low thermal budget processing, point-defect generation can be quite different from generation during traditional high temperature processing. Oxidation in the linear region will produce self-interstitial concentrations which increase with time. However, little growth of existing oxidation-induced stacking faults should occur and little can be said about stacking fault nucleation.

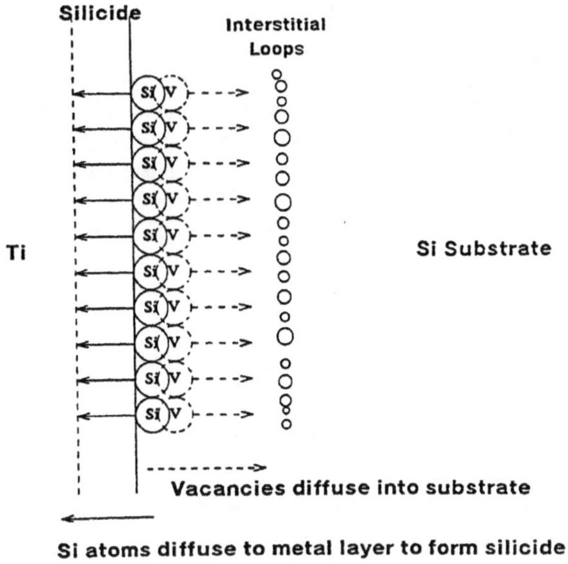

Figure 8

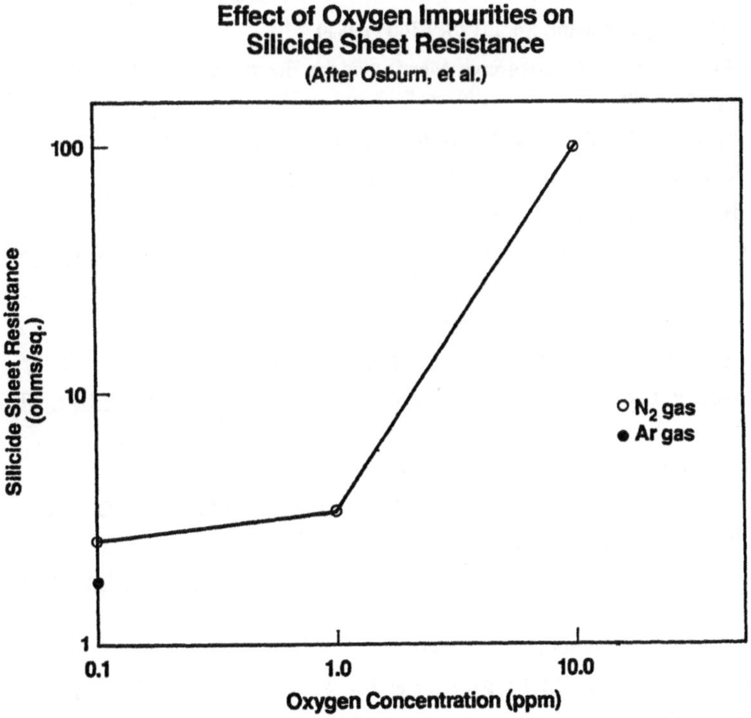

Figure 9

An enhanced OSF surface shrinkage mechanism has been reported for snort-time anneals in inert ambients. However, the OSF dislocation line in the *Si* bulk is unaffected.

Other sources of point defects can dominate self-interstitial production during oxidation. Ion implantation damage annealing can produce large point-defect supersaturations, and even produce a reverse OED effect where oxidation reduces the damage annealing rate and, therefore, enhanced dopant diffusion. Extended defect shrinkage or growth can also be dramatically influenced by point defects generated by precipitate growth or dissolution, *Si* surface layers and metal silicidation reactions. In the latter case, oxidation processes in the silicides may affect the supersaturation levels of vacancies in the underlying *Si*.

REFERENCES

1. C. M. Osburn and A. Reisman, J. of Supercomputing, 1, 149 (1987).
2. S. M. Hu, J. Appl. Phys., 57, 4527 (1985).
3. K. Taniguchi, K. Kurosawa, and M. Kashiwagi, J. Electrochem. Soc., 127, 2243 (1980).
4. A. M. Lin, R. W. Dutton, D. A. Antoniadis, and W. A. Tiller, J. Electrochem. Soc., 128, 1121 (1981).
5. W. B. Rogers, unpublished
6. K. Nishi and D. A. Antoniadis, Appl. Phys. Lett., 46, 516 (1985).
7. Y. Kim, R. B. Fair, H. Z. Massoud, unpublished.
8. M. Servidori, R. Angelucci, F. Cembali and S. Solmi, J. Appl. Phys., 61, 1834 (1987).
9. R. B. Fair, J. Electrochem. Soc., 128, 1360 (1981).
10. A. C. Ajmera and G. A. Rozgonyi, Meeting of the Electrochemical Soc., Boston, Abs. #239, Spring 1986.
11. A. C. Ajmera and G. A. Rozgonyi, Appl. Phys. Lett., 49, 1269 (1986).
12. K. S. Jones, S. Prussin and E. R. Weber, J. Appl. Phys., 62, 4114 (1987).
13. S. M. Hu, J. Appl. Phys., 51, 3666 (1980).
14. J. C. C. Tsai, D. G. Schimmel, R. B. Fair, and W. Maszara, J. Electrochem. Soc., 134, 1508 (1987).
15. K. Nishi and D. A. Antoniadis, J. Appl. Phys., 59, 1117 (1986).
16. J. Narayan and K. Jagannadham, J. Appl. Phys., 62, 1694 (1987).
17. D. S. Wen, P. L. Smith, C. M. Osburn and G. A. Rozgonyi, Appl. Phys. Lett., 51, 1182 (1987).
18. S. M. Hu, *Appl. Phys. Lett.*, 51, 308 (1987).
19. T. E. Seidel, D. J. Lischner, C. S. Pai, R. V. Knoell, D. M. Maher, and D. C. Jacobson, Nucl. Instrum. Methods Phys. Res. B., 7/8, 251 (1985).
20. J. C. C. Tsai, D. G. Schimmel, R. E. Ahrens and R. B. Fair, J. Electrochem. Soc., 134, 2348 (1987).
21. P. Fahey and R. W. Dutton, unpublished.
22. P. Fahey, G. Barbuscia, M. M. Moslehi and R. W. Dutton, Appl. Phys. Lett., 46, 784 (1985).
23. D. S. Wen, P. Smith, C. M. Osburn and G. A. Rozgonyi, unpublished.
24. C. Osburn, H. Berger, R. Donovan and G. Jones, unpublished.

IMPLICATIONS OF OXIDATION MODELS ON THE POINT DEFECT BEHAVIOR IN THE SILICON SUBSTRATE

P. B. Griffin and J. D. Plummer
Integrated Circuits Laboratory
Stanford University, Stanford, CA 94305

Abstract

There is a strong coupling between oxidation and point defect processes in the bulk of the silicon. Some insight into the oxidation process itself can be gained by examining the interstitial generation at the Si/SiO_2 interface. Oxidation models which span the growth kinetics of the oxide films and also the interstitial kinetics in the silicon substrate deserve to be taken seriously. We review some models which attempt this, and determine some essential common features of any plausible model.

Introduction

The link between oxidation and point defect generation was first established by Hu [1]. Silicon interstitials generated at the oxidizing interface flow into the substrate and cause anomalous dopant diffusion and stacking fault growth. These effects are clearly important for reasons relating to device fabrication. Mechanisms for the point defect generation and experimental data on the point defect kinetics can provide an atomic level insight into the details of the oxidation reaction itself. An ideal oxidation model would successfully account for both the oxide growth and point defect kinetics in a quantitative manner.

The insights obtained from the point defect behavior might provide a means to distinguish between the many different models for silicon oxidation. Most of these models were proposed in order to explain the fast initial growth kinetics which is not treated by the original Deal-Grove model for oxidation [2]. Models with different physical origins (space-charge, oxide structure, oxide stress, oxidant solubility, surface layers, for review see [3]) often predict similar oxide growth kinetics. Blanc [4] has shown that some of the models for oxide growth are in fact mathematically identical. In this paper, we discuss point defect mechanisms and examine the correlation between interstitial generation and the atomic processes at the $Si - SiO_2$ interface.

Oxidation Enhanced Diffusion

A schematic of the effects that occur during the oxidation of silicon is shown in Figure 1. The dopants boron and phosphorus exhibit enhanced diffusivity (OED) and stacking faults grow in size under the oxidizing regions. The growth of stacking faults is taken as direct evidence that oxidation injects interstitials into the bulk of the silicon. It implies that boron and phosphorus have substantial interstitial components of diffusion [5] which allows these dopants to be used as markers for the interstitial concentration in the substrate. Many investigators have examined the time and temperature dependence of OED and related it to the oxidation rate as follows

$$OED \propto (dx/dt)^n \tag{1}$$

where n is found to be sub-linear. There has been a tendency in the literature to assume that $n = 0.5$, in part because simple models can fit this time dependence. Experimental results covering a wide range of oxidation times at two temperatures are shown in Figure 2. Whether there is any one power dependence of OED on oxidation rate is unlikely. Rather, the apparent power dependence is a reflection of a combination of non-linear processes occurring at the $Si - SiO_2$ interface. Models which provide physical explanations for this sublinear power dependence can provide an atomic level insight into the oxidation interface reaction. There are several plausible models in the literature, invoking such different phenomena as viscous relaxation of strain, stress from free volume requirements and atomistic chemical reactions at the interface. It is possible to rule out some of the proposed models by experimental methods. The remaining models have some fundamental common properties which are required of any comprehensive oxidation model.

A General Flux Model for OED

In order to investigate the common properties of models for interstitial generation, it is necessary to take a higher level view of the general process. A descriptive model for the point defect injection process is shown in Figure 3. The letters represent the fluxes of the point defects near the $Si - SiO_2$ interface. For example, $\mathcal{G}$ is the time dependent flux due to the generation of interstitials at an active source. The $\mathcal{R}$ flux at the active surface corresponds to some reabsorption of the injected species. $\mathcal{R}'$ is the surface recombination flux at an inert interface. $\mathcal{D}$ is the diffusion flux of interstitials into the bulk of the silicon. This flux model is phenomenological and general in nature. Once the individual flux terms are known, the behavior of the point defects in the bulk can be predicted easily. In particular, if the time dependence of each flux component is known, the time dependence of ΔC_i (the interstitial excess during injection) can be determined. For instance, the ratio $\mathcal{G}/\mathcal{R}$ is a dimensionless quantity and if ΔC_i is involved as a term in either the numerator or the denominator of this ratio, then the time dependence of ΔC_i must match the net time dependence of the other terms in the expression for dimensional consistency. This allows the time dependence of ΔC_i to be obtained for any given flux balance of interest. Exactly which flux balance will dominate at a particular time is determined by the magnitude of the various fluxes.

This simple procedure allows one to interpret the fundamental assumptions of the different proposed models for interstitial generation in terms of a more general higher-level model. It gives a unified basis for comparing the different models. The approximate behavior of the more detailed physical model can be predicted without having to tediously solve the full differential equations.

In the next section, we see that models with very different starting assumptions about the generation, diffusion and recombination fluxes can have the same overall time dependence for OED. However, it is possible to experimentally confirm certain features of the general flux model which indicate which flux terms are dominant. This provides a set of conditions that any specific model must meet, and which might provide a link to certain models for the oxide growth kinetics.

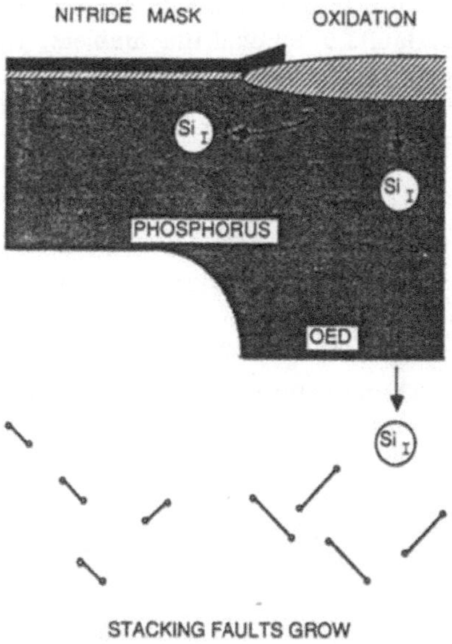

Figure 1. Effect of interstitials generated at an oxidizing interface on phosphorus diffusion and stacking fault growth.

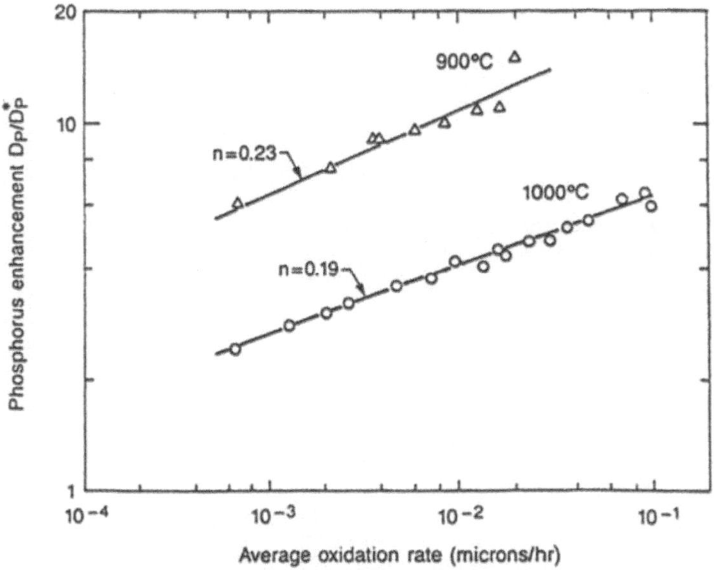

Figure 2. Experimental OED data from Dunham and Plummer (1986).

Review of Proposed Models for OED

To provide a simple example of the application of the flux model, we examine the early paper by Hu [1] which provided a unified description of the OED of dopants and the growth of stacking faults in an oxidizing ambient. Hu proposed that the oxidation reaction was slightly incomplete and that the fraction of unoxidized atoms became interstitials and saturated the lattice. The generation rate of interstitials depended on the oxidation rate so

$$\mathcal{G} = A\frac{dx}{dt} \tag{2}$$

and the recombination of interstitials occurred in proportion to their excess concentration;

$$\mathcal{R} = K_s(\Delta C_I) \tag{3}$$

In addition, Hu assumed that the diffusion of the interstitials was fast, leading to a short diffusion relaxation time allowing the diffusion term to be ignored. Taking the ratio of the fluxes at the interface gives

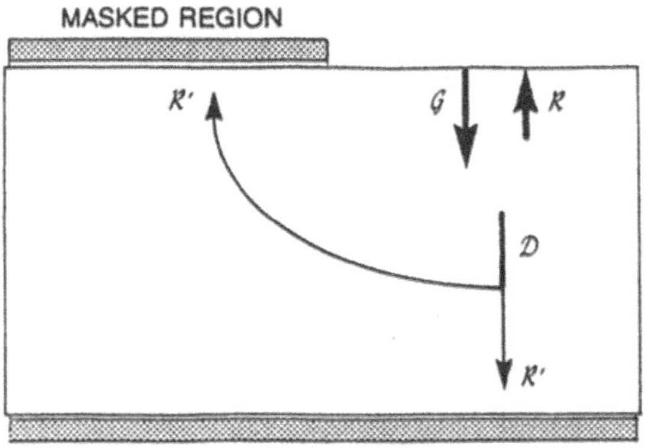

Figure 3. Diagram of the important point defect fluxes in silicon during oxidation (or nitridation) of the silicon surface.

$$\mathcal{G}/\mathcal{R} = A\frac{dx}{dt}/K_s(\Delta C_I) \tag{4}$$

so that

$$\Delta C_I \propto \frac{dx}{dt} \tag{5}$$

The basic time dependence of the OED in this case is linearly proportional to the oxidation rate. At the time the experimental evidence was not strong enough to suggest a sub-linear oxidation rate dependence and Hu returned to the fundamental ideas in this model and made suitable modifications when the experimental data were more conclusive.

We now consider three specific models which have been proposed to explain OED. These models were chosen because they provide detailed explanations of the sublinear power law kinetics of OED based on different physical assumptions. Hu's later model [7],[8] made the intuitively appealing assumption that the generation of interstitials was directly correlated with the oxidation rate and that the surface loss was proportional to the excess interstitials. Lin *et al.* [9] assumed that the generation rate was due to free volume requirements at the interface and a surface loss term for many of the defects was assumed to occur via further oxidation reactions in a non-stoichiometric boundary layer. Dunham and Plummer [10] assumed that molecular dissociation at the interface provided a supply of O_2 and O which had different effects on the point defect kinetics. Since we are dealing with interstitial concentrations near an oxidizing interface, the only flux components of interest are $\mathcal{G}$, $\mathcal{R}$ and $\mathcal{D}$. A detailed derivation of the time dependence of each flux component is provided in [6]. Here we state only the pertinent results for each of the models. Table I shows how ΔC_i varies with the oxidation rate for the pertinent flux balance ratios.

We will now consider the problem of distinguishing between the different models which have been proposed. One way to do this is to push the atomic level mechanisms to regions where they are testable. For example, Dunham's point defect model predicts that the reaction with molecular oxygen at high temperatures introduces a $\sqrt{dx/dt}$ dependence of ΔC_i on the oxidation rate. Reactions with atomic oxygen produced by the dissociation reaction give rise to a constant interstitial supersaturation. Thus the model predicts how the slope should change with temperature.

According to the model of Ghez and van der Meulen [11] for the partial pressure dependence of oxide growth kinetics, both atomic and molecular oxygen contribute strongly to the oxidation reaction at 1000 °C. This is in qualitative agreement with the OED power dependence of 0.19 on oxidation rate in Figure 2. The model can be tested by measuring the time dependence of OED at lower and higher temperatures. A correlation between the predicted dependence of the O and O_2 concentrations in two different models would be strong evidence that the basic physical mechanism is correct.

Examining Lin's model shows that the interstitial supersaturation is directly correlated with the stoichiometry of the boundary layer through the reaction

$$xO + Si \rightleftharpoons SiO_x \tag{6}$$

and

$$\Delta C_i \propto (dx/dt)^{1-\frac{x}{2}} \tag{7}$$

Since the experimental data on ΔC_i is reasonable well modeled by a single value of the power dependence n, it implies that the stoichiometry of the boundary layer is not changing with time. If variations in the power dependence of ΔC_i were found as a function of temperature, it would imply corresponding variations in the boundary layer stoichiometry. Physical measurements on the composition of any boundary layer might indicate whether or not this is a likely model.

Hu's model is different than the others in that the diffusion flux of interstitials into the silicon is accounted for, so that all three flux components are important. This model can be tested by thinning the wafer so the influence of the backside is no longer negligible, or by using special two-dimensional test structures [12]. In both cases, the OED under an oxidizing interface remains unaltered, as shown schematically in Figure 4. Simulations of this 2-D structure are shown in Figure 5, and indicate that to model the experimental results, most of the generated interstitials must flow back towards the growing oxide. This means that , to first order, the diffusion flux of interstitials into the silicon is negligible compared to the interface

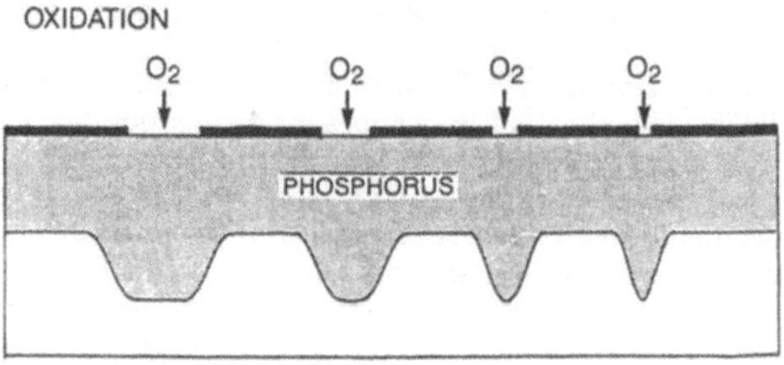

Figure 4. Schematic diagram of the OED of phosphorus as the oxidizing window shrinks. No fall-off in OED is observed [13].

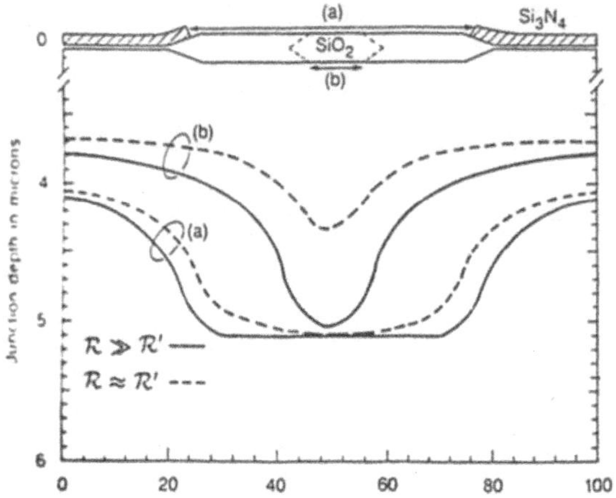

Figure 5. SUPREM IV simulations of OED under wide and narrow oxidizing windows. Only the case where most of the generated interstitials are reabsorbed at the same interface match the experimental results $\mathcal{R} >> \mathcal{R}'$.

TABLE I. Summary of the time dependence of ΔC_i.

AUTHOR	OXIDATION REGIME	PRIMARY FLUX BALANCE	TIME DEPENDENCE OF ΔC_i
Dunham *et al.*[10]	High Temp.	$\mathcal{G}/\mathcal{R}_1$	$t^{-0.25}$
	Low Temp.	$\mathcal{G}/\mathcal{R}_2$	Constant
Lin *et al.*[9]	Linear	$\mathcal{G}/\mathcal{R}$	Constant
	Parabolic	$\mathcal{G}/\mathcal{R}$	$t^{-0.2}$
Hu [7,8]	Linear $t < t_0$	$\mathcal{G}/\mathcal{D}$	$t^{+0.5}$
	Linear $t > t_0$	$\mathcal{G}/\mathcal{R}$	Constant
	Parabolic $t < t_0$	$\mathcal{G}/\mathcal{D}$	Constant
	Parabolic $t > t_0$	$\mathcal{G}/\mathcal{R}$	$t^{-0.5}$

reaction process occurring for the point defects. Since Hu's model relys on a delicate balance between the $\mathcal{G}$ and $\mathcal{D}$ flux components, it can be ruled out. Models whereby large numbers of interstitials are generated during oxidation and flow back into the growing oxide to be further oxidized provide qualitative and quantitative agreement with the experimental results for OED.

With regard to the generation rate of interstitials, Leroy [13] has pointed out that macroscopic tensile stresses at the $Si - SiO_2$ interface cannot be responsible, because of the linear dependence of the interface stress on thickness of the wafer. The relief of localized compressive stress during the oxidation of kinks and jogs on the silicon surface was suggested as a mechanism for the interstitial generation.

Necessary Features of an OED Model

This brings us to a necessary feature of any specific model for the point defect kinetics. In terms of the general flux model, the principle flux balance must be between the generation of interstitials and a surface loss at that same interface. This also leads to a problem in distinguishing between the different models, since it is only the behavior of the ratio $\mathcal{G} \ / \ \mathcal{R}$ that is important in determining the time dependence of OED. Thus a given $\mathcal{G} \ / \ \mathcal{R}$ ratio can predict the experimental time dependence of ΔC_i for many different values of the individual flux components.

A general model which allows for these variations in the oxidation rate dependence has been implemented in the SUPREM-IV [14] simulator as follows

$$G = \Theta \left(\frac{V_{int}}{B/A} \right)^{G_{pow}} V_{mol} V_{int} \tag{8}$$

$$R = K_s \left(K_{rat} \left(\frac{V_{int}}{B/A} \right)^{K_{pow}} + 1 \right) \Delta C_i \tag{9}$$

where V_{int} is the interface velocity, B/A is the linear rate coefficient (the maximum velocity) from [2], V_{mol} is the molar volume, Θ is a fraction of the injected atoms, K_s is the surface recombination velocity at an inert interface and K_{rat} is a multiplying factor that gives a higher value of surface recombination velocity under a moving interface. The exponents G_{pow} and K_{pow} determine the power dependence of the generation and recombination fluxes on the oxidation rate.

Different time dependencies for $\mathcal{G}$ and $\mathcal{R}$ can predict the same overall time dependence for ΔC_i. The choice for the time dependence of the individual $\mathcal{G}$ and $\mathcal{R}$ fluxes can be narrowed down somewhat by the following arguments. It seems that an $\mathcal{R}$(t) increasing with time is not physically correct. It is true that the recombination at the active surface is much greater than at an inert surface ($\mathcal{R} \gg \mathcal{R}'$). But it seems that in the limit of long oxidation times, the oxidizing interface is changing very slowly so that $\mathcal{R} \approx \mathcal{R}'$ and $\mathcal{G} \approx 0$. This implies that $\mathcal{R}$ is a decreasing function of time. To model the experimental results which indicate that

$$\Delta C_i \propto A \, (dx/dt)^{0.2 \rightarrow 0.5} \tag{10}$$

requires that either $\mathcal{G}$(t) or $\mathcal{R}$(t) or both are non-linear functions of the oxidation rate.

From this discussion, it is clear that there is a problem distinguishing the "correct" time dependence for $\mathcal{G}$ and $\mathcal{R}$. Only the time dependence of the ratio is experimentally accessible. What is certain is that there are large point defect fluxes at an oxidizing interface and that most of the generated point defects are reabsorbed at the growing interface. The physical mechanism for the power law dependence of

ΔC_i on oxidation rate is not clear, but several possibilities have been proposed in the literature. Efforts should continue to be made to link both oxidation and point defect models in a consistent framework at an atomic level.

References

[1] Hu, S, M., *J. Appl. Phys.*, **45**, 1567, (1974).

[2] Deal, B. E., and A. S. Grove, *J. Appl. Phys.*, **36**, 3770, (1965).

[3] Massoud, H. Z., J. D. Plummer, and E. A. Irene, *J. Electrochem. Soc.*, **132**, 2693, (1985).

[4] Blanc, J., *J. Electrochem. Soc.*, 1981, (1986).

[5] Fahey, P., G. Barbuscia, M. Moslehi,, and R. W. Dutton, *Appl. Phys. Lett.*, **46**, 784, (1985).

[6] Fahey, P., P. B. Griffin and J. D. Plummer, to be published.

[7] Hu, S. M., *Appl. Phys. Lett.*, **43**, 449, (1983).

[8] Hu, S. M., *J. Appl. Phys.*, **57**, 1069, (1985).

[9] Lin, A., R.W. Dutton, D.A. Antoniadis and W.A. Tiller, *J. Electrochem. Soc.*, **128**, 1124, (1983).

[10] Dunham, S. T., and J.D. Plummer, *J. Appl. Phys.*, **59**, 2541, (1986).

[11] Ghez, R., and Y. J. van der Meulen, *J. Electrochem. Soc.*, **119**, 1100, (1972).

[12] Ahn, S. T., P. B. Griffin, J. D. Shott, J. D. Plummer, and W. A. Tiller, J. Appl. Phys., **62**, 4745, (1987).

[13] Leroy, B., *Phil. Mag.* **B**, **55**, 159, (1987).

[14] SUPREM IV Computer Code was developed in the Integrated Circuits Laboratory, Stanford University.

INTERSTITIAL FLUXES DURING SILICON OXIDATION

Scott T. Dunham

Electrical, Computer and Systems Engineering Department
Boston University
Boston, MA 02215

ABSTRACT

In extending IC fabrication models from 1D to 2D, it is essential to understand the behavior of point defects. Recent experimental work on enhanced diffusion during the thermal oxidation of silicon demonstrates that altering generation/recombination processes in the silicon substrate does not affect interstitial supersaturation during oxidation. In contrast, experimental observations of enhanced diffusion during the thermal nitridation of SiO_2, which results in the growth of a thin interface oxide, indicate that changes within the oxide have a strong influence on interstitial kinetics. These observations suggest that diffusion into the oxide, rather than diffusion into the silicon or surface regrowth, is the dominant sink for interstitials generated by the oxidizing interface. Further, the sublinear dependence of OED and OISF growth on oxidation rate is shown to be a direct result of assuming a steady-state balance between interstitial generation at the interface and diffusion into the oxide. The model also explains the retarded diffusion of phosphorus observed for very long oxidation times.

INTRODUCTION

Accurate modeling of point defect interactions is essential for developing thermal fabrication processes for submicron devices in which 2D effects are important. Since silicon oxidation enhances the diffusion of most dopant impurities (OED) [1-6] and causes the growth of interstitial-type stacking faults (OISF) [7-9], it has been concluded that the oxidizing interface injects interstitials into the silicon [1,10-13]. There remains, however, much controversy over the physical mechanisms which determine the amount of interstitial injection into the silicon under various processing conditions.

The oxidizing interface is thought to cause interstitial generation because of the large difference in molecular volume between silicon and SiO_2. There are three possible paths for those interstitials to follow: diffusion into the bulk, diffusion into the oxide, or regrowth at the interface. Each of these possibilities has been proposed in the effort to explain the sublinear dependence of interstitial supersaturation on oxidation rate [13-16]. The supersaturation of interstitials near the oxidizing interface depends on a balance between the generation rate and losses due to those three mechanisms. By observing how changes in the oxide or in the silicon affect the interstitial concentration, the dominant sink for interstitials generated at the oxidizing interface can be determined.

If diffusion into the silicon dominates, then alteration of generation or recombination processes for interstitials in the silicon or at its interfaces should alter the interstitial supersaturation, while changes within the oxide should have little effect. Similarly, if the dominant sink for interstitials created by the interface reaction is diffusion into the oxide, then the interstitial supersaturation should be insensitive to changes in the silicon, but would vary with changes in the oxide. Finally, if regrowth at the surface dominates, then the supersaturation should be independent of changes in both the silicon and the oxide, depending only on changes in the properties of the interface.

In this paper, we examine a number of experimental observations and their implications in order to arrive at a clearer picture of silicon interstitial kinetics during silicon oxidation. Based on these observations, we conclude that most of the interstitials generated by the oxidizing silicon interface flow back into the oxide where they react with the incoming oxidant species, completing the oxidation reaction. This conclusion is further supported by a simple analysis which shows that the sublinear dependence of interstitial supersaturation on oxidation rate is a direct result of assuming a steady-state balance between interstitial generation at the interface and diffusion into the oxide. This model also accounts naturally for the retarded diffusion of phosphorus for very long oxidations [17].

INTERSTITIALS AND FREE VOLUME DURING OXIDATION

Due to the difference in molecular volume between Si and SiO_2, there is a need for free volume at the Si-SiO_2 interface in order for the larger SiO_2 network to fit in the place of the silicon lattice [18,19,20]. The average distance between silicon atoms in the SiO_2 is approximately 1.3 times larger than in the silicon lattice. Expansion perpendicular to the interface can occur readily; however, at the interface, the oxide is constrained by its connection with the rigid silicon network. Without a supply of free volume to relieve the stress in the plane of the interface, the resulting excess free energy would overcome the normal driving forces and strongly retard the oxidation process. The free volume constraint at the interface can be met by a net flow of silicon away from the interface in the form of silicon interstitials. Viscous flow of the oxide has been suggested as an alternative release mechanism by several authors [17,21]; however, at the interface, viscous flow would require breaking bonds between the substrate and the oxide.

The interstitials generated by the oxidation process can diffuse either into the silicon or into the oxide. Since the interstitials diffusing into the oxide will react with the oxidant species there, the concentration gradient into the oxide would be expected to be much larger than that into the silicon, suggesting that most of the interstitials will diffuse back into the oxide. In addition, the SiO_2 has a much more open and flexible structure than the silicon, which would make it more able to accomodate interstitial species and cause interstitial species to segregate to the oxide.

It is also possible to show that the observed magnitude of OED effects [16,22] and OISF growth [15] imply that only a small fraction of the oxidized silicon atoms flow into the substrate, far too few to account for the free volume constraint at the interface.

Although a sizable percentage of the the atoms in the oxidized silicon may diffuse into the oxide as interstitials, the actual concentration of interstitials in the oxide may still be very small since the interstitials are continually reacting with the incoming oxygen. This additional oxide growth due to the oxidation of silicon interstitials also requires free volume, but, since the reaction takes place away from the interface and the reactants are not bonded to the silicon lattice, the expansion can be relaxed by means of viscous flow within the oxide. In contrast, regrowth at the interface, would appear to be inconsistent with a net flow of material away from the interface as is required by the need for free volume at the interface.

EXPERIMENTAL OBSERVATIONS

In this section, we will examine some experimental observations in order to reach some conclusions about the behavior of interstitials generated by the oxidation process. The observations of enhanced diffusion during the oxidation of silicon membranes [23] and two-dimensional structures [24] demonstrate that interstitial supersaturation near the oxidizing interface is independent of changes in generation/recombination processes in the silicon. In contrast, the large effect that the thermal nitridation of SiO_2 [25,26,27] has on diffusion in the silicon indicates that interstitial supersaturation in the silicon depends critically on processes occurring in the oxide. All of these experimental observations point to the oxide as the dominant sink for the interstitials created at oxidizing silicon interfaces.

Oxidation Enhanced Diffusion in Silicon Membranes

Ahn *et al.* [23] recently measured OED of phosphorus in silicon membranes (20 to 500μm thick) where front and backside conditions were varied. The backsides of high resitivity, p-type, float-zone ⟨100⟩ silicon wafers were etched back to form thin membranes. Both the front and back surfaces were patterned to produce alternating stripes of bare silicon and a thin oxide/nitride sandwich. The wafers were then oxidized at 1100°C in dry O_2. The bare silicon regions oxidized, generating interstitials at the interface, while the nitride layer acted as an oxidation mask and the Si-SiO_2 interface beneath provided recombination sites. This structure resulted in four combinations of front and backside conditions.

The diffusion of an implanted phosphorus profile, located near the front side of the wafer, was measured by angle lapping and staining to find junction depths. Ahn *et al.* found that the phosphorus diffusion near the oxidizing surface was independent of whether the backside interface caused generation (oxidizing interface) or recombination (nonoxidizing) of interstitials, even for membranes as thin as 20μm. In contrast, phosphorus diffusion near a nonoxidizing interface was strongly influenced by backside reactions (as has previously been reported by Mizuo and Higuchi [28]). Thus, interstitial supersaturation near the interface during oxidation is not influenced by adding an additional source or sink of interstitials, indicating that the interstitial **concentration**, rather than the flux of interstitials into the silicon, is controlled by the oxidation process.

Two-Dimensional Oxidation Enhanced Diffusion

A recent investigation by Griffin and Plummer [24] of OED in two-dimensional structures results in similar conclusions. Starting with 1-3 Ω-cm n-type, float-zone ⟨100⟩ silicon wafers implanted with boron, they formed alternating stripes of a thin oxide/nitride sandwich and bare silicon. After oxidation in dry O_2 at 900-1100°C, they measured the boron diffusion as a function of lateral position by angle-lapping and junction staining. They found that the enhanced diffusion of boron remained constant under the entire oxidizing region. Even at the edge, where the oxidizing and nonoxidizing surfaces met, the diffusion was not influenced by the surface recombination at adjoining areas. Once again, the oxidizing interface determines the concentration of interstitials in the silicon near the interface rather than the flux into silicon.

From these observations, it is clear that diffusion into the silicon is not the dominant sink for interstitials generated at the interface during oxidation. The two remaining possibilities for the dominant interstitial sink are the regrowth at the interface and diffusion into the oxide. However, in order to model the data with interface regrowth dominant, it is necessary to assume that the recombination velocity at the oxidizing interface is orders of magnitude larger than at a nonoxidizing interface [23,24].

Enhanced Diffusion during Thermal Nitridation of SiO_2

If interstitials diffuse back into the oxide, it should be possible to alter interstitial supersaturation at a fixed oxidation rate and temperature by changing the properties of the oxide as a sink for interstitials. This is difficult under normal oxidation conditions since the diffusion length for interstitials into the oxide would be expected to be very short. However, the thermal nitridation of SiO_2 provides the desired circumstances.

During thermal nitridation, the nitrogen species react quickly to form a nitrogen-rich layer at the surface and at the silicon interface. A bulk replacement reaction in the bulk acts much more slowly, displacing oxygen, some of which diffuses to the silicon interface, growing a thin oxygen-rich layer and pushing the nitrogen-rich layer away from the interface [29,30]. As for normal oxidation, interstitials are generated at the interface, causing enhanced diffusion and stacking fault growth [26,27]. Despite the very slow oxidation rate (only $\sim$ 30 Å after 2 hours at 1100°C [29]), the diffusivity enhancement is very large ($\sim$ 4.8 for phosphorus at 1100°C for 1 hour [27]), more than twice the enhancements observed for normal oxidation even when the oxidation rates were 100 times as fast [2]. This can be accounted for by observing that the thin nitride layer is an effective barrier to silicon diffusion [31]. Since the nitride diffusion barrier is so close to the interface ($<$ 30 Å), there is very little oxide volume in which the interstitials can be annihilated by reaction with oxidizing species and therefore, as observed, the interstitial concentration builds up to much higher supersaturations. This evidence further supports our assertion that most of the interstitials generated at the interface diffuse back into the oxide, rather than diffusing into the silicon or regrowing at the interface.

MODELING INTERSTITIAL SUPERSATURATION DURING OXIDATION

Currently, the most common assumption made in modeling point defect kinetics is that the concentration of interstitials at the oxidizing interface is determined by the balance of two large fluxes: generation due to oxidation and interface regrowth. As noted above, these models require that the recombination velocity for interstitials at the oxidizing interface is much larger than at a nonoxidizing interface. In addition, such models fail to account for the sublinear dependence of interstitial supersaturation on oxidation rate. They deal with such observations by assuming some nonlinear generation rate or recombination rate. Consideration of the steady-state balance between interstitial generation at the interface and interstitial diffusion into the oxide, however, leads directly and simply to the observed dependence of interstitial supersaturation on oxidation rate.

We will consider here only a single oxidant species (a more detailed analysis for multiple species was included in previous work [16]). Since the need for free volume is directly proportional to the interface reaction rate, the rate of interstitial generation at the interface would also be expected to depend linearly on the interface reaction rate,

$$G_I = \alpha k_i (C_{O_2}(0) - C^*_{O_2}), \tag{1}$$

where α is the fraction of interface reactions which result in interstitial generation, k_i is the interface oxidation reaction rate, $C_{O_2}(0)$ is the concentration of the oxidizing species near the oxidizing interface, and $C^*_{O_2}$ is the concentration of oxidizing species in equilibrium with the oxidizing interface. Under normal oxidizing conditions, $C_{O_2}(0) \gg C^*_{O_2}$. Under some conditions $C_{O_2}(0)$ is less than $C^*_{O_2}$. For example, annealing at high temperature in a vacuum results in decomposition of the oxide [32]. We have indicated the oxidizing species to be molecular oxygen, which is thought to be the dominant diffusing species for oxidation in dry O_2, but the analysis is valid for any single oxidizing species (e.g., H_2O for wet oxidation).

As the interstitials diffuse into the oxide, they will react with the incoming oxygen species to complete the oxidation process. The diffusion length of interstitials into the oxide is usually short enough that the concentration of the oxidizing species is nearly constant [16]. Thus, in steady-state, the continuity equation for interstitials in the oxide is

$$D_I^{SiO_2} \frac{d^2 C_I}{dx^2} = k(C_{O_2}(0) C_I - C^*_{O_2} C^*_I), \tag{2}$$

where $D_I^{SiO_2}$ is the diffusion constant for interstitials in the oxide, C_I is the concentration of interstitials and k is the reaction rate between interstitials and the oxidizing species in the oxide. Equation (2) has a solution of the form,

$$C_I(x) = C_0 \exp\left(-\frac{x}{L_I^{SiO_2}}\right) + \left(\frac{C^*_{O_2} C^*_I}{C_{O_2}(0)}\right), \tag{3}$$

where x is the distance away from the oxidizing interface and the diffusion length for interstitials in the oxide is

$$L_I^{SiO_2} = \left(\frac{D_I^{SiO_2}}{k C_{O_2}(0)}\right)^{1/2}. \tag{4}$$

In steady-state, the generation rate equals the flux away from the interface. Therefore, using Equations (1), (3) and (4),

$$\alpha k_i (C_{O_2}(0) - C^*_{O_2}) = \frac{D_I^{SiO_2} C_0}{L_I^{SiO_2}} = \left(D_I^{SiO_2} k C_{O_2}(0)\right)^{1/2} C_0 \tag{5}$$

and the supersaturation of interstitials in the oxide at the interface is given by

$$C_I(0) = C_0 + \left(\frac{C^*_{O_2} C^*_I}{C_{O_2}(0)}\right) = \frac{\alpha k_i}{(D_I^{SiO_2} k)^{1/2}} \left(\frac{C_{O_2}(0) - C^*_{O_2}}{C_{O_2}(0)^{1/2}}\right) + \left(\frac{C^*_{O_2} C^*_I}{C_{O_2}(0)}\right). \tag{6}$$

The oxidation rate is proportional to the interface reaction rate,

$$\frac{1}{\Omega_{SiO_2}} \frac{dx_o}{dt} = (1 + \alpha) k_i (C_{O_2}(0) - C^*_{O_2}), \tag{7}$$

where Ω_{SiO_2} is the molecular volume of SiO_2 and x_o is the oxide thickness. The $(1+\alpha)$ factor accounts for the interstitials which diffuse into the oxide and oxidize, contributing to the film growth. In terms of the oxidation rate, therefore, the interstitial supersaturation in the silicon is

$$\frac{C_I^{Si}(0)}{C_I^{Si^*}} = \frac{K_1 \dfrac{dx_o}{dt}}{\left(\dfrac{dx_o}{dt} + K_2\right)^{1/2}} + \frac{K_2}{\dfrac{dx_o}{dt} + K_2}, \tag{8}$$

where

$$K_1 = \frac{m_I \alpha}{C_I^{Si^*}} \left(\frac{k_i}{(1+\alpha) D_I^{SiO_2} k \Omega_{SiO_2}} \right)^{1/2} \tag{9}$$

and

$$K_2 = (1+\alpha) k_i \Omega_{SiO_2} C_{O_2}^*. \tag{10}$$

m_I is the segregation coefficient for interstitials between the silicon and the oxide.

Under normal oxidation conditions, $C_{O_2}(0) \gg C_{O_2}^*$ $(dx_o/dt \gg K_2)$ and Equation (8) reduces to

$$\frac{C_I^{Si}(0)}{C_I^{Si^*}} = K_1 \left(\frac{dx_o}{dt} \right)^{1/2} \tag{11}$$

Thus, simply assuming that the dominant sink for interstitials generated at the interface is diffusion into the oxide predicts the observed sublinear dependence of interstitial supersaturation on oxidation rate, without requiring any assumptions of nonlinearity. The square root power dependence is consistent with the 0.4 to 0.5 power dependence commonly found when fitting enhanced diffusion [1,3,5] and stacking fault growth data [7,8,33,34].

In contrast with the common assumption that the oxidizing interface determines the flux of interstitials into the silicon substrate, the proper boundary condition for the oxidizing Si-SiO_2 interface under most experimental conditions is that reactions in the oxide fix the interstitial concentration in the silicon near the interface. Only when modeling very short times (perhaps important for rapid thermal processing) or very slow oxidation rates (very low partial pressure of oxidant) is diffusion into the silicon a significant flux. In such cases, as was done in our analysis of the nitridation of SiO_2 [31], the time required to "fill up" the substrate determines the duration of the transient period. For oxidation in dry O_2 at 1000°, the transient period can be estimated to be about 1 second [22], short enough to be ignored for existing processes. This transient period would increase at higher temperatures due to the increase in the equilibrium interstitial concentration, but the increase would be offset somewhat by increased oxidation rates and smaller interstitial supersaturation ratios.

We may note that in nonoxidizing ambients $(dx_o/dt = 0,\ C_{O_2}(0) = C_{O_2}^*)$, the interstitial concentration reverts to its equilibrium value (Equation (8) reduces to $C_I^{Si}(0) = C_I^{Si^*}$). However, for very slow oxidation, the interstitial concentration can actually be reduced by an oxidizing interface resulting in retarded rather than enhanced diffusion for phosphorus and boron. This would occur when the annihilation of interstitials by oxygen in the oxide is stronger than the interface generation. Whether significant retarded diffusion ever occurs for a given temperature and orientation depends on the relative values of reaction and diffusion coefficients.

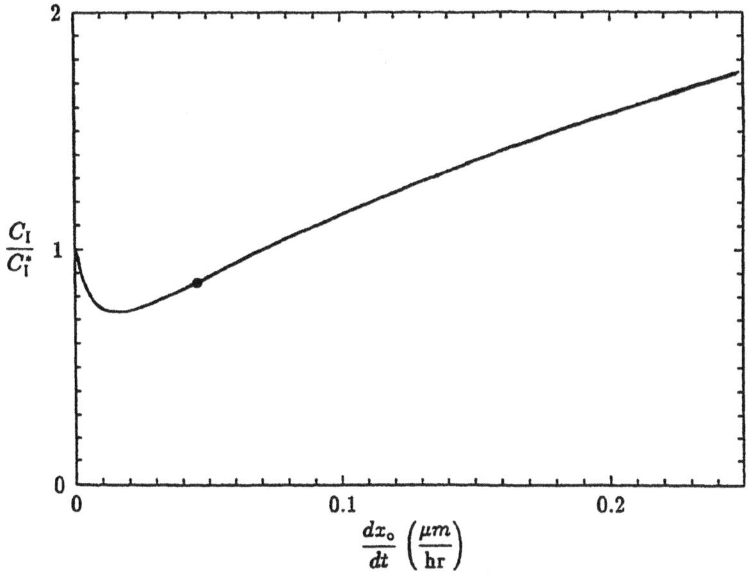

Figure 1: Interstitial supersaturation versus oxidation rate (Equation(8)). Comparison to data from Francis and Dobson, 1160°C.

Retarded diffusion of phosphorus has been observed for very long oxidations at high temperatures (above 1150°C) [17]. Figure 1 plots Equation (8) using a value of K_1 (3.5) extrapolated to 1160°C from lower temperature data [35]. The value of K_2 (0.01) was chosen to fit the observation of retarded diffusion of phosphorus in ⟨111⟩ silicon at 1160°C made by Francis and Dobson [17]. In determining the location of the data point, phosphorus was assumed to diffuse nearly totally by interstitial mechanisms and it was assumed that average oxidation rate and interstitial concentration could be used. The phosphorus ORD data is obviously very limited, but it appears that the model developed here can account for the observed effect.

CONCLUSIONS

Considered together, recent experimental investigations of enhanced diffusion during silicon oxidation and SiO_2 nitridation indicate that the primary sink for interstitials generated by the oxidation process is the growing oxide film. Further, the observed sublinear dependence of OED and OISF growth on oxidation rate follows simply from the assumption of a steady-state balance between interstitial generation at a rate proportional to the oxidation rate and diffusion of interstitials into the oxide. The model also accounts for the retardation of phosphorus diffusion which has been observed for long oxidation times at high temperatures.

A practical consequence of these results is that, in order to correctly model point defect kinetics, fabrication process simulators should utilize boundary conditions at oxidizing interfaces which result in interstitial supersaturation depending only on processes in the oxide and at the interface. Those processes determine the concentration of interstitials at the interface, not the flux into the substrate, and interstitial supersaturation near an oxidizing interface is independent of processes within the silicon.

References

[1] A. M. Lin, D. A. Antoniadis and R. W. Dutton, *J. Electrochem. Soc.* **128**, 1131 (1981).

[2] K. Taniguchi, K. Kurosawa and M. Kashiwagi, *J. Electrochem. Soc.* **127**, 2243 (1980).

[3] D. A. Antoniadis and I. Moskowitz, *J. Appl. Phys.* **53**, 6786 (1982).

[4] Y. Ishikawa, Y. Sakina, H. Tanaka, S. Matsumoto and T. Numi, *J. Electrochem. Soc.* **129**, 644 (1982).

[5] D. A. Antoniadis, A. G. Gonzales and R. W. Dutton, *J. Electrochem. Soc.* **125**, 813 (1978).

[6] D. A. Antoniadis, A. M. Lin and R. W. Dutton, *Appl. Phys. Let.* **33**, 1030 (1978).

[7] S. P. Murarka, *Physical Review B* **16**, 2849 (1977).

[8] S. M. Hu, *Appl. Phys. Let.* **27**, 165 (1975).

[9] H. Shirarki, *Jap. J. Appl. Phys.* **15**, 1 (1976).

[10] R. B. Fair, *J. Appl. Phys.* **51**, 5828 (1980).

[11] T. Y. Tan, U. Gösele and F. F. Morehead, *Appl. Phys.* **31**, 97 (1983).

[12] D. A. Antoniadis, *J. Electrochem. Soc.* **129**, 1093 (1982).

[13] S. M. Hu, *J. Appl. Phys.* **45**, 1567 (1974).

[14] S. M. Hu, *Appl. Phys. Let.* **43**, 449 (1983).

[15] A. M. Lin, R. W. Dutton, D. A. Antoniadis and W. A. Tiller, *J. Electrochem. Soc.* **128**, 1121 (1981).

[16] S. T. Dunham and J. D. Plummer, *J. Appl. Phys.* **57**, 2541 (1986).

[17] R. Francis and P. S. Dobson, *J. Appl. Phys.* **50**, 280 (1979).

[18] W. A. Tiller, *J. Electrochem. Soc.* **127**, 619 (1980).

[19] W. A. Tiller, *J. Electrochem. Soc.* **127**, 625 (1980).

[20] W. A. Tiller, *J. Electrochem. Soc.* **128**, 689 (1981).

[21] T. Y. Tan and U. Gösele, *Appl. Phys. Let.* **39**, 86 (1981).

[22] S. T. Dunham, to appear in *J. Electrochem Soc.*

[23] S. T. Ahn J. D. Shott and W. A. Tiller, *Proceedings of the Electrochemical Society Meeting*, San Diego, CA, October 1986.

[24] P. B. Griffin and J. D. Plummer, *Proceedings of the Electrochemical Society Meeting*, San Diego, CA, October 1986.

[25] P. Fahey, R. W. Dutton and M. Moslehi, *Appl. Phys. Let.* **43**, 683 (1983).

[26] S. Mizuo, T. Kusaka, A. Shintani, M. Nanba and H. Higuchi, *J. Appl. Phys.* **54**, 3860 (1983).

[27] P. Fahey, G. Barbuscia, M. Moslehi and R. W. Dutton, *Appl. Phys. Let.* **46**, 784 (1985).

[28] S. Mizuo and H. Higuchi, *J. Electrochem. Soc.* **129**, 2292 (1982).

[29] C. J. Han, M. M. Moslehi, C. R. Helms and K. C. Saraswat, *Appl. Phys. Let.* **46**, 641 (1985).

[30] R. P. Vasquez, A. Madhukar, F. J. Grunthaner and M. L. Naiman, *Appl. Phys. Let.* **44**, 969 (1984).

[31] S. T. Dunham, *J. Appl. Phys.* **62**, 1195 (1987).

[32] R. Tromp, G. W. Rubloff, P. Balk, F. K. LeGoues and E. J. van Loenen, *Phys. Rev. Lett.* **55**, 2332 (1985).

[33] D. A. Antoniadis, *J. Electrochem. Soc.* **129**, 1093 (1982).

[34] T. Y. Tan and U. Gösele, *Appl. Phys. Let.* **40**, 616 (1982).

[35] S. T. Dunham and J. D. Plummer, *J. Appl. Phys.* **57**, 2551 (1986).

VI. ELECTRON AND HOLE TRANSPORT AND TUNNELING IN SiO_2

INTRODUCTION

Paul V. Dressendorfer

Sandia Laboratories, P. O. Box 5800
Albuquerque, NM 87185

Tunneling, transport, and trapping properties of carriers in SiO_2 are of significance not only from a scientific point of view but also from a practical perspective. Study of these properties lend insight into the chemical and structural nature of defects in the SiO_2/Si system, and of fundamental interactions between carriers and the atomic "lattice". Nonvolatile memory integrated circuits depend upon the injection or tunneling of carriers into SiO_2 and their subsequent transport and trapping. With the reduction in feature sizes of advanced integrated circuits, hot carrier injection into the gate dielectric has become a significant concern for long term reliability. Processes used in the manufacture of integrated circuits, such as ion implantation and sputtering, create traps in the oxide which can adversely affect its properties. For circuits which must operate in radiation environments, the transport and trapping characteristics determine how long and how well those circuits will perform.

The papers of this chapter touch on a number of aspects of these problems. Three papers provide overviews. "Charge Trapping in SiO_2" by D. R. Young describes experimental results on both bulk and interface traps, and provides information on some of the process dependencies of these traps for thermally grown oxides. "Historical Perspective on Tunneling in SiO_2" by J. Maserjian demonstrates that tunneling into SiO_2 can produce quantum interference effects, perhaps somewhat surprisingly given the amorphous structure of SiO_2. This result is interpreted to imply that the SiO_2 near the Si/SiO_2 interface is more ordered than that in the bulk. "Hot Electron Transport in Silicon Dioxide" by D. J. DiMaria and M. V. Vischetti summarizes results showing that stabilization of a hot electron distribution in SiO_2 occurs because of electron interactions with polar and non-polar phonons, thereby preventing velocity runaway. Ballistic transport of electrons is also observed in hot electron minimum energy of 2.2±0.2 eV.

The four other papers in the session treat specific aspects of these topics in detail. "Ion Implantation and Ionizing Radiation damage produced by energy dissipation in atomic processes is significantly more efficient at creating oxygen vacancy centers than that produced by ionizing processes, and that high temperatures (>800°C) are required to irreversibly anneal these centers. A different stress/anneal sequence is examined in "Interface Improvement by Hot Carrier Degradation - Forming Gas Annealing Cycle" by Q. Y. Ye, A. Zrenner, and F. Koch. In this case a high rf field stress at 4.2K is followed by a forming gas anneal at 400°C, and produces better interface characteristics than those of a virgin sample. The other two papers of the session examine characteristics of oxides prepared by alternative techniques. "Interface and Bulk Trapping Centers in Low Temperature Ion-Beam Sputtered Silicon Dioxide Films" by M. E. Zvanut and F. J. Feigl illustrates that the trap characteristics of these films are similar to those of thermally grown films which have been degraded by ionizing radiation. "Effects of X Irradiation and High Field Electron Injection on the Properties of Rapid Thermal Oxides" by W. K. Schubert and C. H. Seager provides evidence that these oxides show a similar dependence of damage sensitivity on post oxidation anneal conditions to that of normal thermal oxides. However, injection of electrons into trapped holes does not result in the creation of interface traps, in contrast to results on standard thermally grown oxides.

CHARGE TRAPPING IN SiO_2

D. R. Young

Sherman Fairchild Center
Lehigh University
Bethlehem, PA 18015

Avalanche injection techniques are extensively used to inject electrons and holes from silicon into the thin SiO_2 layer of the type used in MOS transistors. This makes it possible to evaluate the electron and hole trapping kinetics in this material using only simple MOS devices. The simplicity of these devices makes it relatively easy to study the effect of variations in the processing technology used. The results show that processing dependencies are different for "bulk" traps and traps located at the Si-SiO_2 interface. It is interesting to note that both the hole and electron traps located at the interface have the same processing dependence and are very sensitive to treatments in oxygen suggesting that these traps are related to an oxygen deficiency in the SiO_2 near the interface.

Introduction

Charge trapping (hole and electron) in the thin SiO_2 layers used in contemporary silicon semiconductor technology can be a source of instabilities in these devices. The traps involved can be classified as being in the bulk of the SiO_2 or in the vicinity of the Si-SiO_2 interface. The bulk trap concentration depends on the process used to make the devices in a different way than the interface traps. It has also been observed that traps are generated at the interface by passing electron current through the SiO_2. The bulk traps to be discussed will be electron traps and the interface traps can be either electron or bulk traps. It is interesting to note that the process dependence for the interface traps is identical for the electron and the hole traps.

The technique will not be described in detail in this paper but consists of passing current through the SiO_2 using the method first used by Nicollian and Berglund [1] that injects the charge carriers into the SiO_2 as a result of an avalanche process in the semiconductor. This technique has been improved by changing the wave form used from the sine wave used by Nicollian and Berglund to a saw tooth wave form as suggested by Nagai et al. [2]. Experience has shown that the wave form is not critical for electron trapping measurements but is for hole trapping and the sawtooth waveform is optimum [3].

The current is passed for a set period of time and this process is interrupted and a flat band capacitance measurement is made to measure the trapped charge. In a typical run the charge measurement is made 200-500 times. This entire process has been highly automated including data acquisition. In addition, programs have been written to analyze the results to obtain the trap densities and cross-sections. The same apparatus is used for electron and hole trapping with the exception that p type silicon wafers are used for electron trapping and n type silicon for hole trapping. In the latter case, the connection to the sample is reversed to effectively reverse the polarity of the voltages applied. The means to extract the trap cross section and density results from the C-V measurements has been described earlier and will not be repeated here [4].

The traps are labeled in terms of the cross-section measured as a result of passing current through the insulator. It is realized that there are many possible errors in this method but this has proved a useful means of cataloging the traps. The cross-sections quoted are probably more accurate for the electron traps than for the hole traps. Although useful methods have been developed to separate traps with close but different cross-sections it should be emphasized that in some cases the results might be a composite of several traps that could not be resolved. The hole trapping rate at the Si-SiO_2 interface is relatively high and it is likely that this masks the effects of hole traps in the bulk of the SiO_2 and therefore bulk hole traps have not been extensively studied.

The samples are MOS capacitors with aluminum gate electrodes and are made using generally accepted techniques with the exception that the resistivity of the silicon is .1 to .2 ohm-cm as suggested by Nicollian [5]. Changes in the usual procedure are made as the process dependence studies are carried out.

The trap location has been determined by thickness dependence studies and by the photo I-V technique developed by DiMaria [6].

Bulk Electron Traps

The first work was done by Nicollian et al. [7], who showed that the electron trapping is controlled by the "water" content of the SiO_2 the exact form of the "water" was not specified but it was thought to be in the form of Si-OH. This was also indicated by subsequent work by Gdula [8] and also by Ushirokawa et al. [9]. Nicollian et al. [7] have shown that the capture of an electron on a "water" related center results in the release of atomic hydrogen.

More recent work by Lai et al. [10] has shown that very low trapping rates can be obtained if very carefully controlled processing procedures are used to reduce the "water" content and by long post oxidation annealing treatments in N_2 at the oxidation temperature. The precautions that need to be made to obtain this low trapping rate consist of using a double wall oxidation furnace to avoid the contamination by water diffusing through the quartz tube, the pre-burning and cold trapping of the gases, and cooling down the sample in a N_2 ambient in the oxidation apparatus to avoid the pick up of water while the sample is warm. Using these techniques the effective trapped charge has been reduced to $8x10^{10}/cm^2$ as compared to $1x10^{12}$ observed previously. An electron fluence of 1.10^{18} electrons/cm^2 is used for this work (see Figure 1). Subsequent work by Weinberg et al. [11] using a rapid thermal annealing furnace have shown that these long nitrogen anneals are not required. In this case a time of 10 sec. at 600°C was sufficient to remove the "water" from the sample. It is suggested that in a conventional furnace during the annealing you are drying out the furnace walls as well as the sample which does not occur with a rapid thermal annealing furnace with cold walls. Another technique to obtain very low trapping rates is to use silicon gate electrodes based on the work by Young et al. [4]. The reason for this improvement may be due to the fact that this process is carried out at a relatively high temperature and as a result the "water" may be removed from the sample.

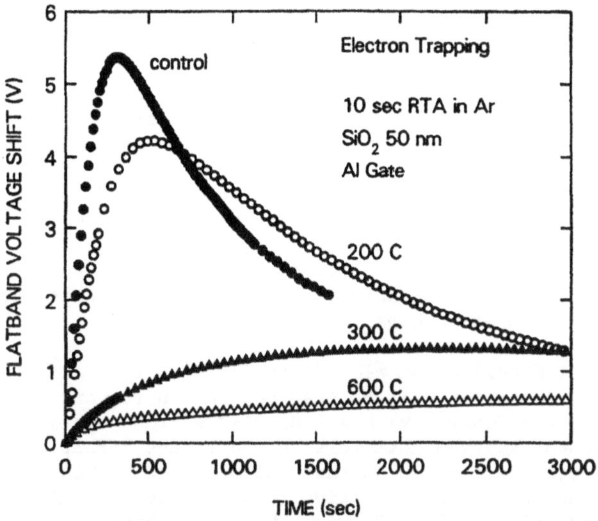

Fig. 1. Electron trapping in water-diffused oxides after rapid thermal annealing (RTA) in Ar for 20 sec. at various temperatures. The electron trapping is measured monitoring the flatband voltage shift as a function of the electron avalanche injection time. The injection current was $3x10^{-7}$ Å. All samples had an Al gate area of $5.19x10^{-3}$ cm^2, an oxide thickness of 50 nm, and were injected at room temperature. (Reproduced with permission from the Journal of Applied Physics.)

One complication that arises in this work is the observation first reported by Gdula [8] that the charge build up does not occur monotonically but a turnaround occurs (see for example the results of Young et al. [4]). This has been shown to result from the formation, as a result of the passage of electron current, of donor states near the interface which contribute a positive charge to partially compensate the negative charge due to the trapped electrons. The magnitude of this effect depends on processing in a similar way to the electron trapping rate. It has also been observed that these states are not charged if the measurement is made at 120°C and this means is used to avoid this complication in the measurements. These donor states can be charged and discharged at will as a result of the application of a moderate electric field. The generation of these states has been studied by Feigl et al [12] and it was observed that this was closely connected with the trapping process but was significantly enhanced when "water" was diffused into the sample. Lai and Young [13] have studied these states in more detail including a comparison with fast interface state generation which is also observed. Sah et al. [14] have studied the kinetics of generation and annealing of both the donor states reported above and also the fast donor like state at .25eV above midgap at the interface. In a subsequent publication Sah et al. [15] have reported that the trapping rates and the "donor" generation rates all increase as a result of keV electron irradiation. They have also observed the enhancement of the generation of these states as a result of water exposure. Weinberg et al. [16] have observed that these states can also be generated if the samples are exposed to non-penetrating vacuum ultraviolet but in this case with a negative voltage applied to the gate electrode which is opposite to the usual case with positive voltage applied indicating that the diffusing species is neutral that causes this effect.

The movement of hydrogen to the interface as a result of the passage of an electron current has been observed by Gale et al. [17] using SIMS techniques which substantiates the model based on atomic hydrogen as the source of these states. Ushirokawa et al. [9] observed that the passage of electron current through the oxide decreased the effective doping in the silicon. This effect has been extensively studied by Sah et al. [18] and they have shown that this is due to hydrogen released by the trapping process entering the silicon and reducing the effective doping concentration. Their work shows that this is a result of complexes formed between the hydrogen and the boron dopant atoms.

The work of Nicollian et al. [7] indicated a trap cross-section of $1x10^{-17}$ cm^2 for the water related trap but the subsequent work of Ushirokawa et al. [9] and Young [19] has indicated a cross-section of $2x10^{-18}$ cm^2. It has been observed that this is due to the effect of a post metallization heat treatment at 300°C that was not used by Nicollian. Ultra sensitive attenuated total reflection IR absorption techniques have been applied by Hartstein et al. [20] with the observation that the effect of this heat treatment is to change the OH radicals to loosely bound H_2O. It seems reasonable that the trap cross-section would be reduced as a result.

Implantation of various species into SiO_2 leads to additional bulk traps that depend on the species being implanted. In this work the samples are all annealed at 1000°C after the implantation to anneal out the implantation damage. Arsenic has been extensively studied by De Keersmaecker et al. [21] and by Alexandrovia et al. [22]. Arsenic provides traps with cross-sections of 10^{-15} cm^2 and $2\text{-}3x10^{-16}$ cm^2. It was observed that optical excitation would remove the trapped charge with an absorption threshold of 3.3 eV. The earlier reference suggested that the arsenic was substitutional for oxygen while more recent work by Alexandrovia suggests that it is substitutional for silicon. Other implanted species that have been studied are tungsten by Kahng et al. [23] and by Young et al. [24], aluminum by Young et al. [25] and by DiMaria et al. [26], and phosphorus by DiMaria et al. [27].

Trapping centers can also be introduced into SiO_2 by exposure to radiation. The dominant effect of the exposure to radiation is the buildup of positive charge at the $Si\text{-}SiO_2$ interface [28,29]. The radiation may be from any source that has an energy component beyond the 9 eV band gap of SiO_2. This positive charge is due to holes trapped at this interface. The trapping probability of holes is around .1 as contrasted with electrons which have a trapping probability of less than 10^{-5} depending on the processes used. In addition, neutral traps are formed that require a higher temperature to remove by annealing treatments than the trapped holes (600°C vs. 100-200°C). These traps were first observed by Aitken and Young [30]. Subsequent work by Schmitz and Young [31] has shown that there are two types of neutral traps involved. One type has a cross-section of $1.7x10^{-16}$ and is located in the bulk of the SiO_2 and the other trap has a cross-section of $1.6x10^{-15}$ and is located at the Si-SiO_2 interface.

One major area of concern is the inadvertent exposure to radiation as a result of the processes used to make useful devices. One example of this is reactive ion etching which has been studied in this context by Ephrath and DiMaria [32]. Their conclusions were that traps were introduced in an unprotected oxide exposed to a CF_4 plasma used to clean the oxide layers that can be removed by annealing at 600°C. This damage did not occur if the oxide was protected by an aluminum or n^+ polysilicon gate electrode. This damage did not occur in a plasma etching system which indicated that during etching the damaged material is removed by the etching process. The effect of low energy x-ray exposure and 10 keV electron exposure has been studied by Reisman et al. [33] and they have observed both neutral bulk trap generation and positive charge buildup due to trapped holes. These results lead to concerns with respect to the use of electron beam photolithographic techniques, electron beam evaporators, and x-ray photolithography as process tools and suggest the need to be aware of the possibility of damage to the insulator if these tools are used.

Interface Traps

There are three traps that are associated with the Si-SiO_2 interface. These traps are discussed here as separate traps as a result of the way they are observed but it seems likely that they are actually different manifestations of the same trap.

1. Shallow electron trap

This electron trap can only be filled at temperatures below room temperature according to the observations by Ning [34] and by Young et al. [4]. Using thermal stimulated re-emission measurements, Ning observed a trap depth of 300 meV below the conduction band edge. Another interesting result was that a small fraction of the trapped electrons were not thermally re-emitted at room temperature. He measured a capture cross-section of $2x20^{-15}$ cms for these traps.

2. Small Cross-Section Trap

A trap with a cross-section of $5x10^{-19}$ has also been observed at the interface as reported by Young [19]. It has been shown by Aslam et al. [35] that a fraction of the electrons trapped into the shallow traps described above transfer into this trap and as a result these traps seem to be intimately related. This agrees with the conclusions reported earlier by Ning. A bimodal trap model has been proposed to describe the results. In this case the electrons are first trapped in the shallow trap and then transferred into the deeper trap, probably by a tunneling process.

3. The Hole Interface Trap

Holes trapped at the interface can also act as colombic positively charged traps as studied by Aitken and Young [36]. However, more recent work by Lai [37] suggests that the electrons do not annihilate the trapped holes but instead compensate them with the formation of new interface states.

The process dependence of these three traps is identical. The earlier observation based on measurements of the trapping behavior at 77°K was that prolonged treatments in a N_2 ambient at 1000°C increased the trapping rate for this trap. The same treatment decreased the trapping rate for the bulk trap [4]. Subsequent studies have indicated that the other two traps respond in the same way.

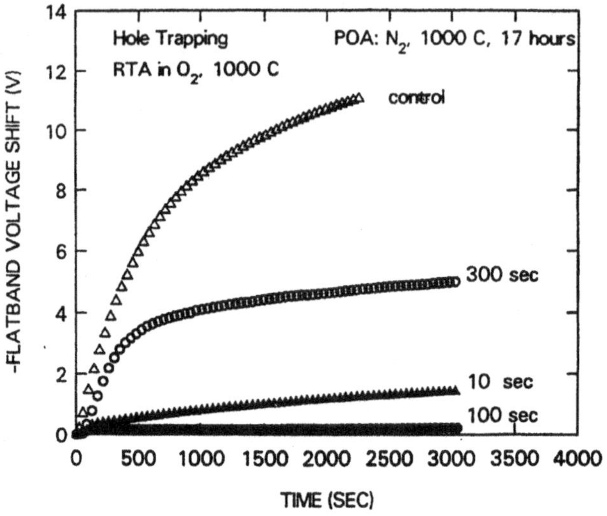

Fig. 2. Negative flatband voltage shifts (indicating positive oxide charge) measured as a function of hole avalanche injection time, after RTA in O_2 at 1000°C, for various times. These oxides had a long-post-oxidation anneal (POA) in N_2 before the oxygen treatment. The injection current was $1x10^{-11}$ A and the gate area was $5.186x10^{-3}$ cm^2. (Reproduced with permission from the Journal of Applied Physics.)

Balk et al. [38] observed that short time annealing in a O_2 ambient at 1000°C would reverse the effects of the long time N_2 anneal. This was suggested as a result of the belief that hole traps were excess Si in the SiO_2. Their work indicated that this can be accomplished in 3 minutes. These studies have been extended by Weinberg et al. [11] and they have found optimum times of 60-100 sec. Another interesting result of Weinberg's work is that whereas the N_2 treatment by itself increased the trapping rate, the combination of this long treatment followed by the very short O_2 gave the best results as far as minimizing the trapping. If the treatment is too long the results approach those obtained with a sample withdrawn from the oxidation furnace without treatment as would be expected. The hole trapping rates for samples prepared with the combined N_2 and O_2 treatments are the lowest the authors have observed, as shown in Figure 2, and the optimum time is around 100 sec. The trapping probability for the optimum case was .4% as compared with 1-10% for the usual case.

A model is proposed to explain these results which is based on excess Si as being the source of these traps. It is suggested that the oxidation process leaves behind excess Si in the SiO_2 and as the oxidation proceeds this excess Si is oxidized but more is formed in the vicinity of the new interface. In addition, it is well known that nitrogen at the interface will inhibit to some extent the oxidation process and reduce the SiO_2 growth rate. This would decrease the rate of forming new excess Si regions but at the same time make it possible to oxidize the excess Si already present which would explain the advantage of the combined treatment. Measurements of the nitrogen profile have been made using SIMS techniques [11] indicating the buildup in nitrogen concentration at the interface as a result of the prolonged N_2 treatment. The term excess Si is used very loosely in this context. The excess Si may actually be trivalent silicon as discussed by Svensson [39].

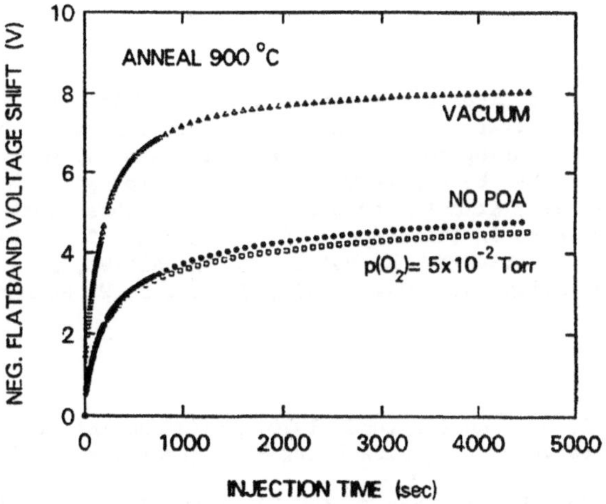

Fig. 3. Comparison of hole trapping as a function of avalanche injection time for samples without POA (post oxidation annealing), with POA at 900°C in ultrahigh vacuum, and with POA in the presence of $5x10^{-2}$ Torr O_2 in the vacuum chamber. (Reproduced with permission from the Journal of Applied Physics.)

The delicate role of oxygen annealing is shown also by the work of Hofmann et al. [40]. They compared the hole trapping rate after heating in an ultrahigh vacuum with that obtained by heating in an O_2 ambient. The ultrahigh vacuum treatment resulted in a large increase in the trapping rate even at temperatures at low at 900°C, but an ambient pressure of $5x10^{-2}$ torr of oxygen was sufficient to avoid the deleterious effect. These results are shown in Figure 3. This again demonstrates the crucial role of O_2 in controlling the trapping at the interface.

Conclusions

The traps can be classified as bulk traps and traps located at the Si-SiO_2 interface. Electron trapping in the bulk for the "standard" SiO_2 is controlled by the concentration of the "water" related species which may be in the form of OH radicals or "loosely bound H_2O." This trapping can be minimized by careful procedures to make "water free" SiO_2. Bulk traps can be also introduced by radiation exposure or by ion implantation. Interface traps are both electron and hole traps. The electron traps are shallow traps that can only be filled at temperatures below room temperature and also very small cross-section traps. Charge is also transferred from the shallow traps to the small cross-section traps. In addition, there are hole traps located at the interface. The interface trap concentration (either electron or hole) can be significantly reduced by delicate treatments in an O_2 ambient which suggests that they are due to partially oxidized Si located near this interface. The most effective treatment for reducing the concentration of these traps consists of a relatively long treatment in N_2 followed by a very short O_2 anneal, both carried out at 1000°C.

Acknowledgments

The author is particularly indebted to J. A. Calise who worked closely with him for many years, designing and constructing apparatus, making special samples, measuring their characteristics and for many helpful suggestions. There has been a very close association with the following who have provided many stimulating discussions and whose work has been frequently referred to in this paper: D. J. DiMaria, Z. A. Weinberg, S. K. Lai, R. F. De Keersmaecker, K. Hofmann, J. M. Aitken, F. J. Feigl, R. Gale, P. Balk, M. Aslam, S. Alexandrovia, C. T. Sah, W. Schmitz.

References

1. Nicollian, E. H., and Berglund, 1970, Avalanche Injection of Electrons into Insulating SiO_2 Using MOS Structures, J. Appl. Phys., 41:3052.
2. Nagai, K., Hayashi, Y., and Tariu, Y., 1969, Jap. J. Appl. Phys. Lett., 15:174 (1969).
3. Aitken, J., and Young, D. R., 1977, Avalanche injection of holes into SiO_2, IEEE Trans. N-24., 2128.
4. Young, D. R., Irene, E. A., DiMaria, D. J., De Keersmaecker, R. F., and Massoud, H. Z., 1979, Electron trapping in SiO_2 at 295 and 77°K, J. Appl. Phys., 6366.
5. Nicollian, E. H., and Brews, J. R., "MOS (Metal Oxide Semiconductor) Physics and Technology," John Wiley and Sons, New York (1981), p. 505.
6. DiMaria, D. J., Young, D. R., De Keersmaecker, R. F., Hunter, W. R., and Serrano, C. M., 1978, Centroid location of implanted ions in the SiO_2 layer of MOS structures using the photo I-V technique, J. Appl. Phys., 49:5441.
7. Nicollian, E. H., Berglund, C. N., Schmidt, P. F., Andrews, J. M., Electrochemical charging of thermal SiO_2 films by injected electron currents, 1971, J. Appl. Phys., 426:5654.
8. Gdula, R. A., The effects of processing on hot electron trapping in SiO_2, 1976, J. Electrochem. Soc., 123:42.
9. Ushirokawa, A., Suzuke, E., and Harashima, M., Avalanche injection effects in MIS structures and realization of n-channel enhancement type MOS FETS, 1973, Jap. J. Appl. Phys., 12:398.

10. Lai, S. K., Young, D. R., Calise, J. A., and Feigl, F. J., Reduction of electron trapping in silicon dioxide by high-temperature nitrogen anneal,
11. Weinberg, Z. A., Young, D. R., Calise, J. A., Cohen, S. A., DeLuca, J. C., and Deline, V. R., Reduction of electron and hole trapping in SiO_2 by rapid thermal annealing,
12. Feigl., F. J., Young, D. R., DiMaria, D. J., Lai, S., Calise, J., The effects of water on oxide and interface trapped charge generation in thermal SiO_2 films, 1981, J. Appl. Phys., 52:5665.
13. Lai, S. K., and Young, D. R., Effects of avalanche injection of electrons into silicon dioxide--generation of fast and slow interface states, 1981, J. Appl. Phys., 52:6231.
14. Sah, C-T., Sun, J. Y-C., Tzou, J. J-T., Generation-annealing kinetics of the interface donor states at 0.25 eV above the midgap and the turn-around phenomena on oxidized silicon during avalanche electron injection, 1983, J. Appl. Phys., 54.
15. Sah, C-T, Sun, J. Y-C., Tzou, J-T., Effects of keV electron irradiation on the avalanche-electron generation rates of three donors on oxidized silicon, 1983, J. Appl. Phys., 54:4378.
16. Weinberg, Z. A., Young, D. R., DiMaria, D. J., and Rubloff, G. W., Exciton or hydrogen diffusion in SiO_2, 1979, J. Appl. Phys., 50:5757.
17. Gale, R., Feigl, F. J., Magee, C. W., Young, D. R., 1983, Hydrogen migration under avalanche injection of electrons in Si metal-oxide-semiconductor capacitors, J. Appl. Phys., 54:6938.
18. Sah, C-T, Sun J. Y-C., Tzou, J. J-T., 1983, Deactivation of the boron acceptor in silicon by hydrogen, Appl. Phys. Lett., 43:204.
19. Young, D. R., 1981, Characterization of electron traps in SiO_2 as influence by processing parameters, J. Appl. Phys., 52:4090.
20. Hartstein, A., and Young, D. R., 1981, Identification of electron traps in thermal silicon dioxide films, Appl. Phys. Lett., 38:631.
21. De Keersmaecker, R. F., and DiMaria, D. J., 1980, Electron trapping and detrapping characteristics of arsenic-implanted SiO_2 layers, J. Appl. Phys., 51:1085.
22. Alexandrovia, S., and Young, D. R., 1983, Microscopic location of electron traps induced by arsenic implantation in silicon dioxide, J. Appl. Phys., 54:174.
23. Kahng, D., Sundbury, W. J., Boulin, D. M., and Ligenza, J. R., 1974, Interfacial dopants for dual-dielectric charge-storage cells, Bell Syst. Tech. J., 1723.
24. Young, D. R., DiMaria, D. J., and Bojarczuk, N. A., 1977, Electron-trapping characteristics of W in SiO_2, J. Appl. Phys., 48:3425.
25. Young, D. R., DiMaria, D. J., Hunter, W. R., and Serrano, C. M., 1978, Characterization of electron traps in aluminum-implanted SiO_2, J. of Res. and Dev., 22:285.
26. DiMaria, D. J., Young, D. R., Hunter, W. R., and Serrano, C. M., 1978, Location of trapped charge in aluminum-implanted SiO_2, 22:289.
27. DiMaria, D. J., Young, D. R., De Keersmaecker, R. F., Hunter, W. R., and Serrano, C. M., The Electrochemical Society Fall Meeting, Atlanta, Georgia, 1977, Abstract No. 212 (unpublished).
28. Snow, E. H., Grove, A. S., and Fitzgerald, D. J., 1967, Effects of ionizing radiation on oxidized silicon surfaces and planar devices, Proc. IEEE, 55:1168.
29. Mitchell, J. P. Wilson, D. K., 1967, Surface effects of radiation on semiconductor devices, Bell System Tech. j., XLVI:1.
30. Aitken, J. M., Young, D. R., 1976, Electron trapping by radiation-induced charge in MOS devices, J. Appl. Phys., 47:1196.
31. Schmitz, W., Young, D. R., Radiation induced electron traps in silicon dioxide, 1983, J. Appl. Phys., 54:6443.
32. Ephrath, L. M., DiMaria, D. J., 1981, Review of RIE induced radiation damage in silicon dioxide, 1981, Solid State Technology.

33. Reisman, A, Merz, C. J., Maldonado, J. R., Molzen, W. W., Jr., 1984, Low energy x-ray and electron damage to IGFET gate insulators, J. Electrochem. Soc., 131:1404.
34. Ning, T. H., 1978, Thermal reemission of trapped electrons in SiO_2, J. Appl. Phys., 49:5997.
35. Aslam, M., 1987, Electron self-trapping in SiO_2, J. Appl. Phys., 62:159.
36. Aitken, J. M., and Young, D. R., Electron trapping by radiation-induced charge in MOS devices, J. Appl. Phys., 47:1196.
37. Lai, S. K., 1983, Interface trap generation in silicon dioxide when electrons are captured by trapped holes, J. Appl. Phys., 54:2540.
38. Balk, P., Aslam, M., and Young, D. R., 1984, High temperature annealing behavior of electron traps in thermal SiO_2, Solid State Electronics, 27:709.
39. Svensson, Christer M., in The Physics of SiO_2 and Its Interfaces, ed. Sokrates T. Pantelides, Pergamon Press, New York (1798), p. 329.
40. Hoffman, K., Young, D. R., Rubloff, G. W., 1987, Hole trapping in SiO_2 films annealed in low-pressure oxygen atmosphere, J. Appl. Phys. 62:925.

HISTORICAL PERSPECTIVE ON TUNNELING IN SiO_2

J. Maserjian*

Jet Propulsion Laboratory
California Institute of Technology
Pasadena, CA 91109

ABSTRACT

This paper examines tunneling in SiO_2 in light of work since 1974. It considers both its near-ideal behavior and departures from the ideal. The trapezoidal tunneling barrier is a good first order model provided it has not been degraded by excessive tunnel injection. However, the tunneling dispersion relation $\kappa(E)$ is not constant throughout the oxide film, but different near each interface. For a specific barrier region, good agreement is obtained using a two valley , two effective mass $\kappa(E)$ dependence. Quantum interference effects are clearly observed at the Si-SiO_2 interface, but not at the metal-SiO_2 interface.

INTRODUCTION

The study of tunneling in solids began with work in the thirties on metal-semiconductor contacts (1). It wasn't until the fifties that appropriate theoretical formulations for tunneling in thin insulating films were first published (2,3), followed in the sixties by a flurry of experimental work. The results at that time were disappointing for the most part because of material limitations and problems such as pin holes, non-reproducibility, and a general lack of agreement with theoretical predictions (4). This experience contributed to a not uncommon belief that only tunneling in single crystalline materials could be quantitatively treated (5,6).

The first widely accepted results on tunneling in SiO_2 came in the late sixties from work of Lenzlinger and Snow (7,8), who showed Fowler-Nordheim (FN) tunneling into relatively thick SiO_2 films (≈1000Å) used at that time in MOS (metal oxide semiconductor) devices. During that period, early work on tunneling through very thin oxides (<50Å) began with the study of Si-SiO_2 interface states (9-11) and of the effect of minority carriers at the interface (12,13).

In 1974 my coworkers and I showed that thin oxide (20-40Å) MOS structures exhibited a highly reproducible behavior that could be analyzed in terms of a simple trapezoidal energy barrier and a tunneling dispersion relation derived from the data (14-16). Furthermore, the FN tunnel current was shown to exhibit an oscillatory behavior around the classical dependence which we interpreted as resonant reflections of the electron wave at the Si-SiO_2 interface (16-18). We showed this result to be quantitatively consistent with an ideal trapezoidal barrier and an abrupt interface. These were the first observations of quantum interference effects in solids -- now being widely investigated in heterojunction structures grown by molecular beam epitaxy. The implications of these results contradicted the common belief at the time that tunneling through amorphous insulators was ill-defined, and that the Si-SiO_2 interface was smeared out due to diffusion during thermal oxidation. Moreover, our results suggested that the expected barrier rounding due to image forces was absent.

*Presently on assignment at University of California, Santa Barbara, California

We proposed that image forces are probably negligible in such structures, reasoning that the electron wave function is not sufficiently localized to sense a significant image effect. These findings and interpretations were controversial and not widely accepted at that time.

Subsequent work over the years has verified the main conclusions of the above. Today extremely thin SiO_2 films are being incorporated in VLSI technology (19) with their tunneling properties setting a lower limit on thin gate oxides (≈60Å) for scaled-down submicron transistors. It also plays a key role in EAROM (electrically alterable read only memory) technology which uses tunnel injection into buried gate electrodes (20). Tunnel injection into SiO_2 has also become a very useful technique in recent years for investigating oxide and interface trap generation at high fields (21-25), and complements information obtained from the more widely used technique of avalanche injection (26-29).

Now that tunneling in SiO_2 is more generally accepted as a well-defined process, there may be a tendency to go too far to the extreme of oversimplifying the MOS tunneling model. The purpose of this paper is to re-examine MOS tunneling in light of work since 1974. In particular, I will first review some general considerations for MOS tunneling, and then describe results (previously unpublished) which hopefully will resolve confusion over the SiO_2 barrier effective mass.

MOS TUNNELING CONSIDERATIONS

The simple trapezoidal barrier shown in Fig. 1 is a good first-order model for FN tunneling, even for extremely thin oxide (e.g.,30Å) MOS structures (16), provided excessive current injection has not degraded the SiO_2 by trap generation and charge accumulation (21,22). Unfortunately, much of the reported work has not taken this limitation into account, or have resorted to steady state conditions in which changes induced in the oxide are allowed to saturate. Fig. 1 indicates the x-component of the wave vector k for each of the three regions (i.e., k= k_m, k_{ox}, k_s) to emphasize that the tunneling properties are uniquely determined by how k varies with x and E. We will come back to this key point shortly. The trapezoidal barrier model can also be considered free of any barrier rounding effect from image forces. This has now been conclusively shown to be negligible by an elegant photon-assisted tunneling experiment of Weinberg and Hartstein (30). Subsequently, they explained this result using an approximate quantum mechanical treatment that takes into account the effect of the delocalized electron wave function on its image force (31).

Fig. 2 schematically shows the key structural features of the MOS barrier. Resonant FN tunneling measurements (16-18,21) and extensive studies using various interface analysis techniques (32,33) have established the Si-SiO_2 interface to be approximately a one monolayer SiO_x chemical transition. This transition is approximated in Fig. 1 as a linearly graded barrier region of width Δx. Departures from this ideal interface occurs in the form of atomic steps at the interface (roughening) that depends on crystal orientation and oxidation process (33). The SiO_2 near the Si interface is seen to be highly strained into a denser structure by its attempt (during thermal growth) to accommodate the Si bonds at the substrate. This strain,extending some 20 to 50Å from the interface,

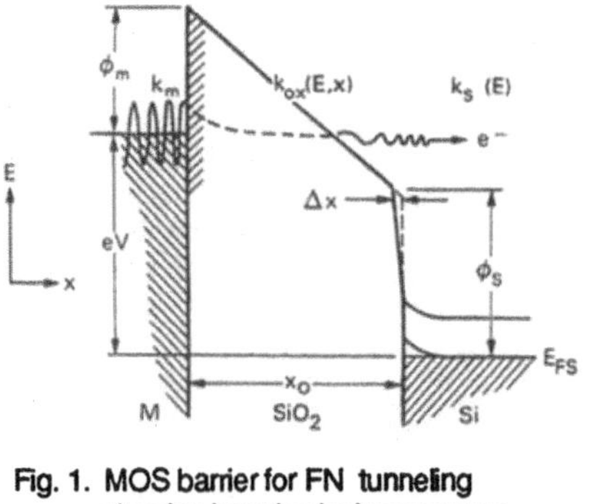

Fig. 1. MOS barrier for FN tunneling showing key physical parameters.

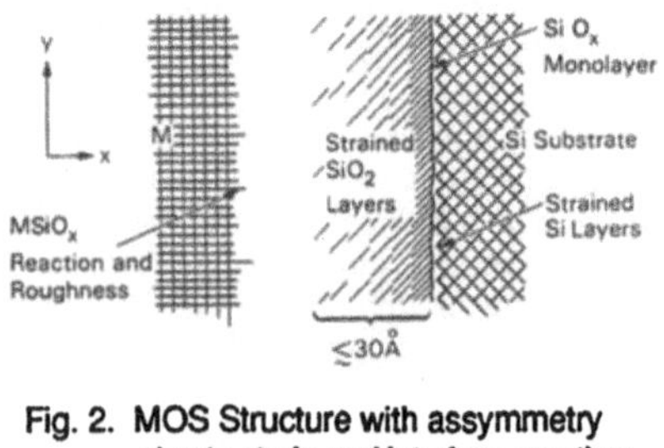

Fig. 2. MOS Structure with assymmetry due to strain and interface reaction.

causes distinct changes in the physical properties of the SiO_2 as seen, for example, by changes in the band gap (18), index of refraction, chemical reactivity, electron diffraction, and photoemission spectra (33). Hydrogen also plays a major role at the Si-SiO_2 interface by tying up dangling Si bonds that cannot be accommodated by the strained SiO_2. Hydrogen densities at the interface in excess of 10^{14} cm^{-2} have been estimated from indirect evidence (33). The metal-SiO_2 interface, on the other hand, is not subject to the same constraints from thermal growth and might be viewed as a bulk-like amorphous oxide terminated abruptly by the metal. However, it can be modified by chemical reaction, especially during high energy deposition (e.g., metal sputtering or vapor deposition of poly Si) or during post metal anneal (PMA). The latter effect has been shown in the case the Al-SiO_2 interface (34,35). Such reactions can change both the barrier height and the tunneling dispersion relation in the oxide near the metal interface. The PMA can also significantly effect the barrier heights due to reaction of hydrogen at the interfaces (33,36). Such effects confuse comparisons of tunneling results obtained from different processes.

The above structural considerations certainly complicate our ability to treat MOS tunneling with any quantitative legitimacy. However, if we consider tunneling from each interface separately and specify the MOS fabrication process, the problem becomes quite manageable. Let us start by examining the basic tunneling process. In the case of tunneling from the metal, we have

$$J = 2e/(2\pi)^3 \iiint_{\text{all } \mathbf{k}} (hk/2\pi m)\ D(k)\ d\mathbf{k}^3, \qquad (1)$$

where J is the tunnel current density, e is the electron charge, h is Plank's constant, m is the electron mass, k is the x-component of the wave vector **k** in the metal, and D(k) is the tunneling transmission coefficient for k. The integral is over all **k** for allowed tunneling transitions between occupied and empty states. The prefactor $2/(2\pi)^3$ is the electron density in phase space (with spin degeneracy) for an electron-gas-like metal. The factor $(hk/2\pi m)$ inside the integrand is just the x-component of electron velocity impinging on the barrier. Using a constant free electron mass for the metal, and conserving energy and transverse **k**, one can immediately integrate over all values of transverse **k** in the metal and obtain for low temperatures (i.e., a Fermi Step-function) the relation

$$J = (4\pi em/h^3) \int_0^{eV} D(E')E'dE', \qquad (2)$$

where E' is the energy difference measured downward from the metal Fermi level (energy due to the transverse components of **k**). D(E') is usually formulated by the WKBJ approximation (in its simplest form); that is

$$D \approx \exp\{-2 \int_0^t \kappa(x)\, dx\}, \qquad (3)$$

where $\kappa(x)$ is the value of imaginary k_{ox} at the energy E(x) in the oxide energy gap corresponding to position x, and t is the turning point where the electron emerges into the oxide conduction band. Unfortunately, this approximation does not retain quantum interference effects at the interfaces. However, the interference effect can only be seen in sufficiently thin oxides (<70Å) where electrons tunneling into the oxide conduction band can reach the opposite interface without losing their wave coherence by scattering (18). Therefore, Eq.3 is quite adequate for thicker oxides if we know the correct relation for $\kappa(x)$. If one assumes a trapezoidal barrier, then for an *arbitrary* $\kappa(E)$ dependence one obtains the familiar expression for FN tunneling

$$J_o = AF^2 \exp\{-C/F\}, \qquad (4)$$

where F is the electric field, and A and C are constants dependent on the model used (8,16,37,38). In the case of tunneling from the Si interface, the electrons are quantized in the narrow accumulation layer of the Si at FN fields. This leads to a physically simpler monoenergic tunneling process. The result in the WKBJ approximation is essentially identical, except for a modified expression for the prefactor A which gives almost the same numerical value (38).

We wish to evaluate D more precisely than by Eq.3 to retain the quantum interference effects and to introduce appropriate barrier modifications. In general, this is done by solving Schroedinger's equation with the appropriate boundary conditions to obtain the electron wave function. Let us make the usual assumption that the important properties are retained by the plane wave part of the wave functions. Then we may write Schroedinger's equation in the form

Table1. Parameters for MOS tunneling. Typical values for FN tunneling from metal to Si.

Parameter	Values	Source
$\varnothing_m$	Al: 3.15 ± 0.10eV Cr: 3.45 ± 0.10eV	Photoemission, C-V
$\varnothing_{so}$	4.05 ± 0.10eV	Photoemission, C-V
$\alpha \Rightarrow \Delta\varnothing_s(F)$	$\alpha = 2.4 \pm 0.1$eV(Å/V)$^{2/3}$	Theory, C-V
x_o	30 -100 $\pm$ 0.5Å	Theory, C-V, Ellipsometer
Δx	2.5 ± 0.5Å	Resonant FN Tunneling, Interface Analysis
k_m	1.5 ± 0.5Å^{-1}	Electron gas model
k_s	0.8 ± 0.2Å^{-1}	Band Theory, Resonant FN Tunneling
m_1 $m_2 \Longrightarrow \kappa, k_{ox}$ E_2	$m_1/m = 0.85 \pm 0.10$ $m_2/m = 0.27 \pm 0.05$ $E_2 = 0.75 \pm 0.25$eV	Band Theory, J (V, x_o), Classical FN Tunneling, Resonant FN Tunneling

$$\psi'' + k^2(x)\,\psi = 0\,, \tag{5}$$

and it is clear that our solution depends entirely on how we model k(x) throughout the MOS structure. The wave functions at the boundaries may be written (for each k)

$$\psi_m = a\exp\{ik_m x\} + b\exp\{-ik_m x\}\,, \qquad \text{in the metal,}$$

and

$$\psi_s = c\exp\{ik_s x\} + d\exp\{-ik_s x\}\,, \qquad \text{in the Si,}$$

with a, b, c, and d constant coefficients. As will be discussed later, we also treat k_m and k_s as constants equal to their values at the Fermi level. Then we have for tunneling from the metal, $D=(k_s/k_m)\,|c/a|^2$, or from the Si, $D=(k_m/k_s)\,|b/d|^2$, where the ratios of coefficients are obtained by solving for the matching wave function in the oxide. This solution can be obtained numerically on a microcomputer most conveniently (21) by dividing the barrier into narrow layers of constant k and propagating the wave function through the barrier by successively matching boundary conditions. The solution for arbitrary barriers can be readily obtained in this way.

The physical parameters that define k(x) throughout the MOS barrier are listed in Table 1, together with their typical values and sources of information, for the case of FN tunneling from the metal to the Si interface. The indicated barrier heights for Al and Si have been measured by internal photoemission by several investigators (38); however, they are subject to the uncertainties of the measurement and sample processing differences. In the case where we wish to compare tunneling results with different metal contacts but with otherwise identical processing, it is quite direct to compare C-V curves to determine the barrier height difference $\Delta\varnothing_m$. Fig.3 shows this result for Al and Cr (i.e., Au/Cr) contacts, where no PMA was used to avoid interface reaction. The higher densities of interface states (with no PMA) introduces some uncertainty in $\Delta\varnothing_m$ due to differences of slopes, but we still detect a clear increase in the barrier for Cr of at least 0.30eV over that of Al. This is in contrast to the suggestion that it should be less than that of Al (38).

Continuing down the list in Table 1, the oxide thickness x_o and the shift of Si barrier height with field $\Delta\varnothing_s(F)$ can be accurately determined from measurements taken in degenerate accumulation

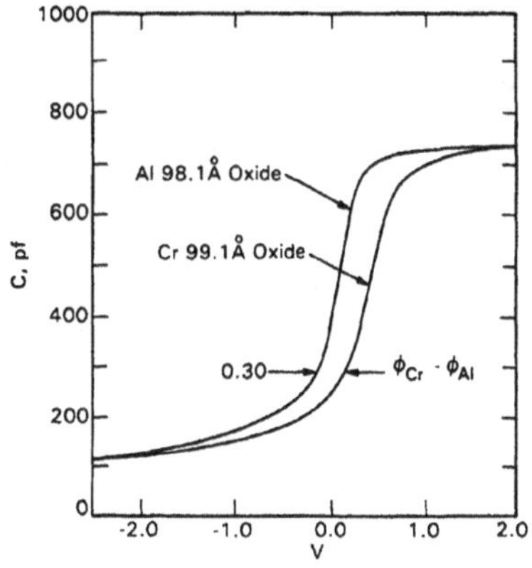

Fig. 3. Al - Cr barrier height difference obtained from C-V data.

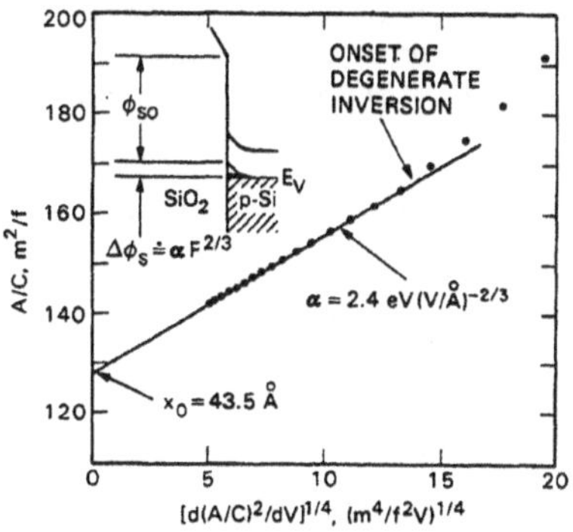

Fig. 4. Oxide thickness x_o and band bending $\Delta Ø_s$ from C-V data.

(14,35) as illustrated in Fig. 4. The form of the dependence is a consequence of quantum confinement of states at the Si interface (14). The oxide thickness x_o is obtained from the intercept x_o/ε_{ox} (ε_{ox} is the oxide permittivity) and agrees within a fraction of an ångstrom with ellipsometer measurements taken before metallization, and when no PMA is used (35). The shift of the Si barrier height $\Delta Ø_s$ ($= Ø_s - Ø_{so}$) due to band bending is given by

$$\Delta Ø_s = \alpha F^{2/3} , \tag{6}$$

where α is determined from the slope s ($=d(A/C)/d[d(A/C)^2/dV]^{1/4}$) and the relation

$$\alpha = (3\varepsilon_{ox} s^2 /2)^{2/3} , \tag{7}$$

and the oxide field is obtained iteratively from

$$F = (V + Ø_m - Ø_{so} - \alpha F^{2/3}) / x_o . \tag{8}$$

The transition layer width Δx listed in Table 1 is accurately determined by resonant FN tunneling measurements (21). Fig. 5 illustrates the observed oscillatory dependence of FN tunneling around the classical dependence J_o with constant slope C (see Eq.4). In Fig. 6, the classical current density J_o is divided out so that we see only the oscillatory factor which depends entirely on quantum interference effects at the interface. The theoretical fit is most sensitive to the value of Δx, especially with the other parameters constrained to a small range of uncertainty (21).

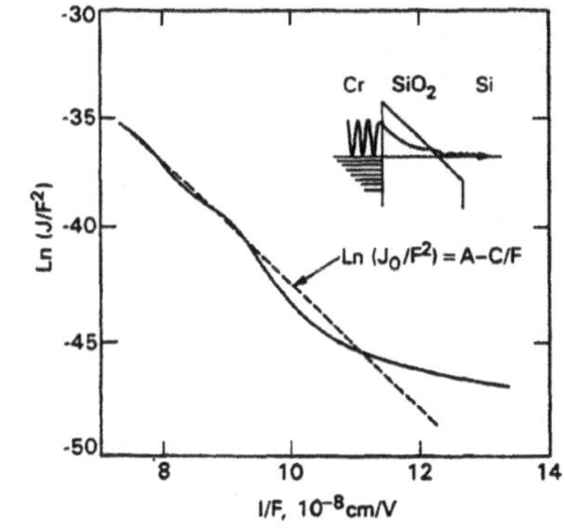

Fig. 5. FN plot showing oscillations about classical dependence J_o.

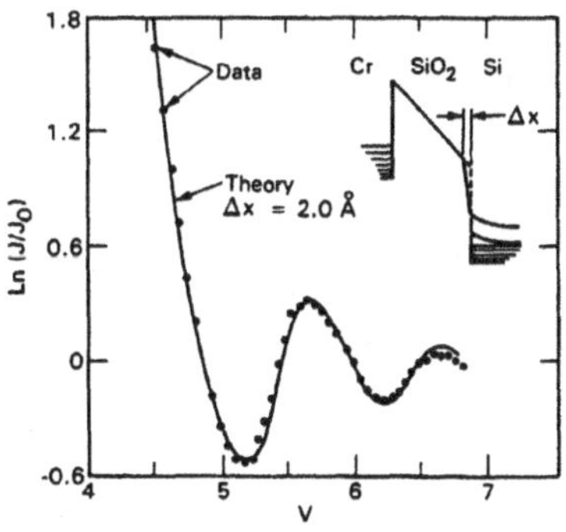

Fig. 6. Oscillatory factor for FN tunneling. Theoretical fit for 2Å transition layer.

We use the free electron gas model to calculate a constant k_m at the metal Fermi level. This model is usually sufficient since it retains the essential behavior of the electron flux in the metal impinging upon the tunneling barrier, and k_m has only a weak effect on the result. The value of k_s in Si is less clear. Band structure calculations of Si at energies well above the bottom of the conduction band ($\approx$2eV) are less reliable. The $k_s(E)$ dependence is also subject to a large applied field and the influence of other interfacial disturbances such as strain, impurities (H), and atomic steps. In the same spirit that we use a constant k_m in the metal, we postulate a constant k_s at the Si side of the interface, choosing a value (e.g.,$0.8\pm0.2Å^{-1}$) consistent with some average (for the energy range) from band calculations (39). We have tested the sensitivity of this constant k_s assumption by comparing with various models of $k_s(E)$, and have found little effect on the result. Therefore, we conclude that a constant k_s is a sufficiently good approximation.

The $k_{ox}(E)$ and $\kappa(E)$ dependences are commonly assumed to be parabolic with a constant effective mass m_{ox}; that is

$$k_{ox} \approx (2\pi/h)\,[2m_{ox}(E-E_c)]^{1/2}, \qquad E > E_c \tag{9}$$
$$\kappa \approx (2\pi/h)\,[2m_{ox}(E_c-E)]^{1/2}, \qquad E < E_c$$

where E_c is the energy at the bottom of the oxide conduction band. Since the form of the FN dependence (see Eq. 4) is independent of the $\kappa(E)$ dispersion relation, one needs other information to test the validity of this or any assumed dependence. In our early work (14,16), we had examined the thickness dependence $J(x_o)$ and found that it was inconsistent with a parabolic dependence, but more consistent with a Franz type of dependence given by

$$\kappa = (2\pi/h)\,[2m_{ox}(E_c-E)(1-(E_c-E)/E_g)]^{1/2}, \tag{10}$$

where E_g is the energy gap. Franz had originally proposed this dependence to account for the influence of both the conduction and valence bands throughout the band gap (40). This approximation is much more likely to apply to narrow gap semiconductors with similar electron and hole effective masses than to SiO_2. Therefore one should interpret our early results as an empirical "Franz-like" fit that showed a departure from a parabolic dependence, rather a strict Franz dependence. In fact, we had shortly thereafter suggested that a model with two conduction band valleys could better explain the observed results (18). Indeed band structure calculations of α-quartz and other allotropic forms of SiO_2 such as tridymite reveal such low lying conduction band valleys (41) that are quite consistent with this suggestion. Fig. 7 shows the relation of the k(E) dependence to the "band structure" of SiO_2. We use the term band structure loosely for amorphous SiO_2, with the understanding that we are really averaging over variations of long range structure, while preserving short range order. The band structure diagram illustrates the characteristic two low-lying conduction band valleys of SiO_2: the lowest valley (minimum at the Γ-point) having a smaller E(k) curvature, and the the higher lying valley (minimum at the A-point) having a larger curvature. These conduction bands are felt by electrons as two effective masses as shown to the left in Fig. 7. Note that the E

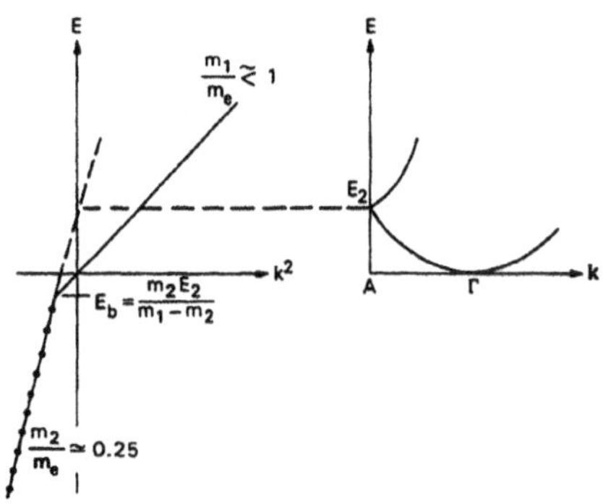

Fig. 7. E-k dependence of two-valley conduction band structure of SiO_2.

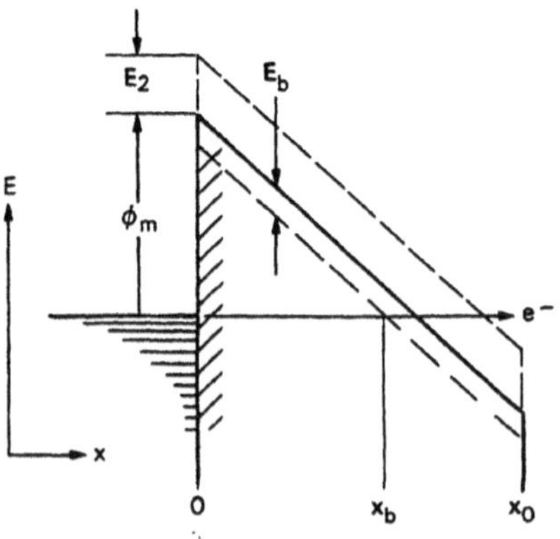

Fig. 8. Two-valley barrier showing branch energy E_b and branch point x_b.

versus k^2 plots are straight lines for constant E(k) curvature (proportional to the inverse of a constant effective mass), and that κ^2 equals $-k^2$. The earlier results (14,16) obtained for $\kappa(E)$ were replotted with k_{ox} ($\equiv$ k) in the two-valley form shown in Fig.7, giving a surprisingly good fit with reasonable values of effective masses (m_1,m_2) and energy E_2 compared with band calculations (18).

This two-valley dependence can be analyzed quite straightforwardly as follows. The WKBJ approximation is accurate for describing the attenuation of the wave function in the barrier before the electron emerges into the conduction band. Kane has shown (43) that the tunneling exponent (of Eq.3) for two bands becomes

$$\int_0^t \kappa \, dx = \int_0^{x_b} \kappa_2 \, dx + \int_{x_b}^t \kappa_1 \, dx \tag{11}$$

where κ_1 and κ_2 corresponds to parabolic bands with effective masses m_1 and m_2, respectively, and x_b corresponds to the point where the electron crosses the branch energy E_b. Fig.8 shows this relation more clearly, with the higher lying band edge and the branch energy indicated by the dashed lines. Beyond the tunneling distance t, the electrons emerge into the lower valley band with higher mass m_1. *It is this larger mass that is observed by the quantum interference effect of resonant FN tunneling* (16,18,21).

Using the parabolic, constant mass form for each valley, κ_1 and κ_2, one can readily evaluate Eq.11 and obtain for the FN tunneling constant (of Eq. 3)

$$C = (8\pi \sqrt{2m_2}/3h) \; [(\emptyset_m + E_2)^{3/2} - \sqrt{m_1/(m_1 - m_2)} \, (E_2)^{3/2}]. \tag{12}$$

An interesting property of the two part integral of Eq.11 is that for thin oxides (<70Å) the second integral of κ_1 is entirely within the strained (dense) SiO_2 region, while the first integral of κ_2 averages over the remaining oxide. Any chemically induced change of the barrier region adjacent to the metal (indicated by shading in Fig. 8) would have some influence on the κ_2 integral and thus the effective value of m_2.

In arriving at our two-valley $\kappa(E)$ dependence, we have used thus far : band theory, our early work on thickness dependence $J(x_o)$, and resonant FN tunneling results. We also have required that the dependence yield the same value of the slope C for classical FN tunneling. In the next section we further test the validity of the two-valley $\kappa(E)$ dependence, using different injecting contacts, and precise fitting of theory to the J(V) dependence.

EXPERIMENTAL RESULTS

The results presented below (Figs.9 to 13) were obtained by the same procedures as described in Ref.21. In addition, Al evaporations were carried out from radiatively heated high purity C crucibles (which achieved low Na contamination levels $<3 \times 10^{10}$cm^2). Other key procedures were as follows: The MOS structures were fabricated from dry oxides grown at 800°C on (100) Si with fast withdrawal into N_2 and no anneal or PMA. Arrays of MOS devices of 2.1×10^{-3}cm^2 area were obtained by evaporating through a precision shadow mask. The J(V) measurements were made at 77°K with voltages applied at sufficiently short times to avoid oxide degradation.

Fig.9 shows J (A/cm^2) vs.V results for Cr contacts, comparing theoretical fits with data. In Fig.10, the result for the 48.1Å oxide is replotted with the classical FN dependence J_o divided out to show the oscillatory factor (as in Fig. 6). In this case we also extend this plot below the FN voltage range by using the two-valley $\kappa(E)$ dependence to calculate $J_o(V)$ at the lower voltages. The cusp occurs at the onset of FN tunneling and is introduced by the classical response (J_o) to the sharp barrier corner of our linearly graded Δx approximation. The theoretical fit is seen to agree almost exactly with the data over this range. At still lower voltages (not shown) the data drops below the theoretical curve. This departure was originally attributed to the effect of tunneling into the indirect conduction band of Si (16), where conservation of transverse-k restricts the tunneling current. Since then, we have attempted to model this effect for (100) Si and found it too weak to account for the difference. In the case of (111) and (110) orientation, the effect should be very large, yet only small differences from the (100) orientation have been observed (16,38). It appears that the transverse-k conservation rule is not applicable in this case (38), probably because the interface is not precisely lattice matched and thus the transverse-k must undergo considerable scattering as it encounters departures from

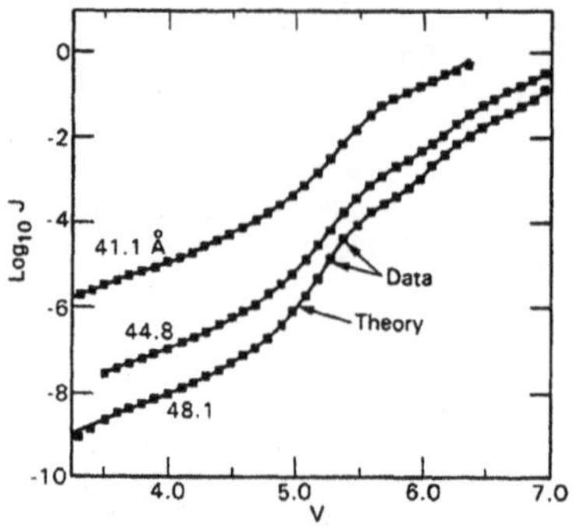

Fig. 9. Tunneling J(V) dependence for Cr contact and three oxide thicknesses.

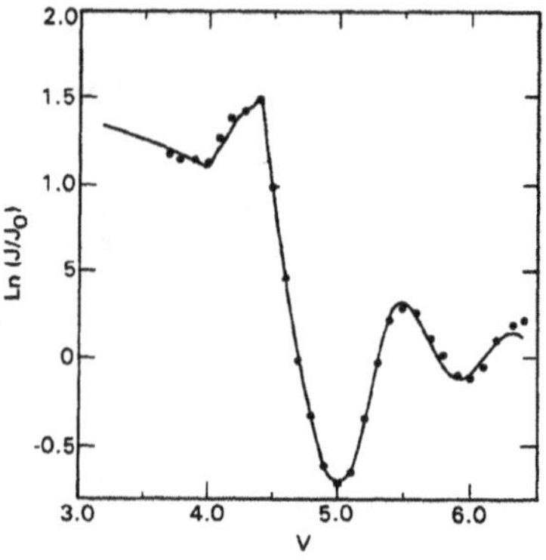

Fig.10. Oscillatory factor and theoretical fit for 48.1Å oxide sample of Fig.9.

periodicity. A likely explanation for the deviation at low voltages is the increasing contribution of the strained interfacial oxide as the tunneling path traverses more of this barrier region at lower voltages. Figs.11 and12 show similar results for Al contacts and a thicker oxide plotted in the same way.

In Table 2, the parameters are listed for the theoretical fits shown in Figs. 9-12. The oxide thickness x_o and band bending coefficient α were independently obtained from C-V measurements as described above. The value of k_m (not shown) was set equal to 1.5$Å^{-1}$. The remaining parameters were fitted within the limits shown. A computer routine (Simplex) for fitting multiple parameters within specified ranges was used to arrive at a least square fit to the data. This procedure is limited by the ability of the program to find the deepest least square valley for the best fit, so that the parameter set may not be entirely unique and some further adjustment may be possible. With this qualification, the parameters are seen to be reasonably consistent among themselves and within the limits imposed. The overall fit is surprisingly good considering the complex J(V) structure and the simple tunneling model assumed. The differences between $Ø_{Cr}$ and $Ø_{Al}$ is quite consistent with C-V data shown in Fig.3. The value of $Ø_{so} \approx 4.0$ eV agrees with the value previously reported for oxides thinner than 65Å (18), which was found to decrease at the rate of about 10 meV/Å from 4.2 eV at 65Å. The values of m_1 agree with early work (16), except for one Cr sample (0.72). The larger m_2 and lower E_2 values for Al compared with Cr contacts was found consistently (in other samples as well), and may be due either to the effect of some reaction at the Al-SiO_2 interface or to Al-induced strain effects in the oxide (43). The values of Δx for Al was also found significantly larger than for Cr, suggesting that Al influences the structure of the Si-SiO_2 interface after oxide growth even with no PMA. This rather puzzling finding may also be related to Al-induced strain effects.

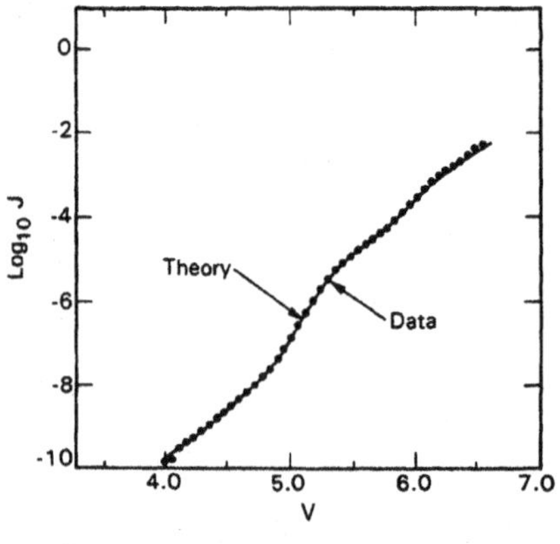

Fig.11. Tunneling J(V) dependence for Al contacts and 50.4Å oxide.

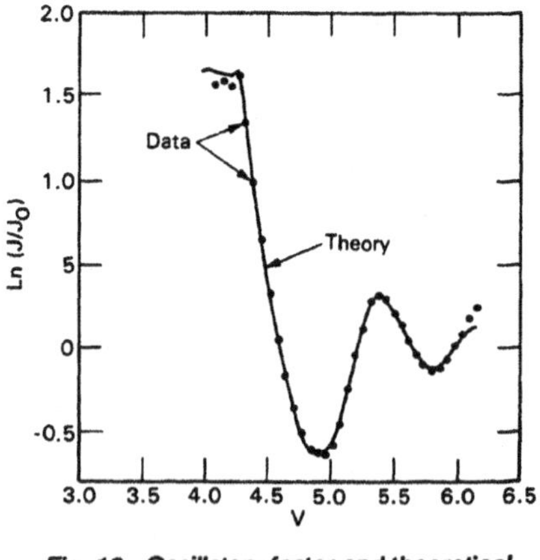

Fig. 12. Oscillatory factor and theoretical fit for results of Fig. 11.

Table 2. Parameters for theoretical J(V) dependence of Figs. 9-12.

	Metal :	Cr	Cr	Cr	Al
From C-V measurements:	x_0, Å :	41.4	44.8	48.1	50.4
	α , eV(Å/V)$^{2/3}$:	2.4	2.4	2.4	2.4
Fitted with no constraints:	Δx, Å :	1.93	1.87	1.92	2.92
Fitted within range:	$Ø_{Cr}$ [3.0-3.8eV] :	3.55	3.51	3.34	-
	$Ø_{Al}$ [3.0-3.4eV] :	-	-	-	3.17
	$Ø_{so}$ [3.9- 4.3eV] :	3.99	4.03	4.06	3.95
	k_s [0.5-1.2Å^{-1}] :	0.90	0.75	0.82	0.94
	m_1/m [0.5 -1.0] :	0.72	0.87	0.85	0.80
	m_2/m [0.1 - 0.5] :	0.225	0.225	0.226	0.32
	E_2 [0.0 -1.5eV] :	0.75	1.00	0.99	0.50
Calculated from above fit:	E_b , eV :	0.34	0.35	0.36	0.33
	C , V/Å :	2.53	2.73	2.55	2.42
	m_{ox}/m :	0.32	0.37	0.38	0.40

Calculated values of parameters are also listed at the bottom of Table 2. The branch energy E_b, $=m_2E_2/(m_1-m_2)$, appears very well behaved and might be preferred over E_2 as the starting parameter. The constant C calculated from Eq. 12 is also the slope obtained when fitting to the classical FN plot (as in Fig. 5), and therefore can be compared directly with other published results such as reviewed in Ref.38. The values of m_{ox}/m based on the single effective mass model are also calculated for the purpose of comparison with other work.

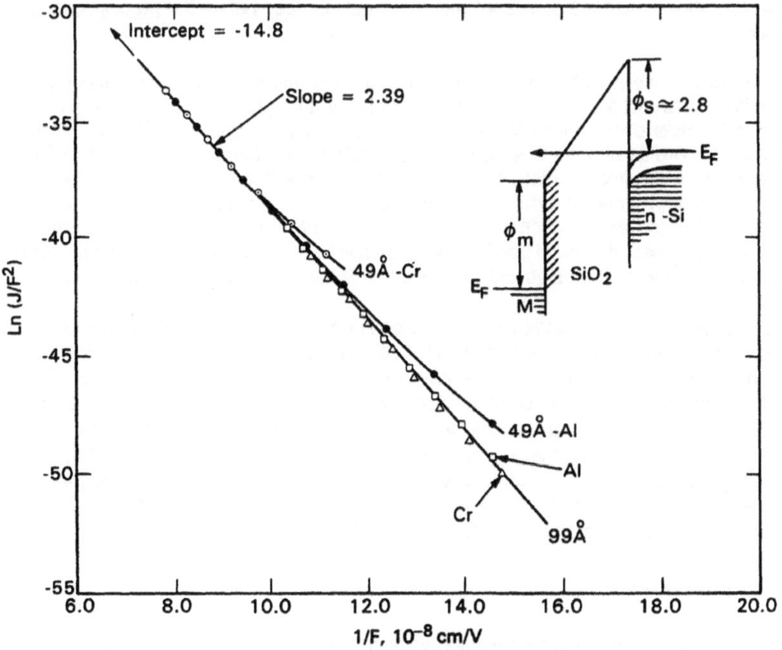

Fig. 13. FN injection from the Si interface for Cr and Al contacts and different oxide thicknesses.

In Fig. 13, results are shown for FN tunneling from the Si interface with Al and Cr metal contacts, and for different oxide thicknesses. First of all, we observe remarkable consistency for the FN slopes and positions of the different curves. (The departures from the straight line in the 49Å oxide samples occur below FN voltages.) This slope agrees almost exactly with the value of 2.385 V/Å obtained by Weinberg (38). He obtains m_{ox}/m=0.50, using the single effective mass model with his calculated correction for $Ø_s$ (≈2.9eV); we obtain m_{ox}/m=0.56 with our measured correction for $Ø_s$ (≈2.8eV) .

Secondly, we observe no hint of resonant FN oscillations as in Fig.5, even though we include thin oxide (49Å) samples. In fact, *we have never observed oscillations when injecting from the Si interface*, although we have examined many devices and process runs (with Al and Cr contacts). This result suggests that the more bulk-like amorphous SiO_2 near the metal interface introduces much greater scattering in the conduction band that destroys the electron wave coherence. Otherwise, since it has been shown that the metal-SiO_2 interface is quite abrupt when no PMA is used (35), the electron wave interference effect should have been seen.

Recall that the form of the classical FN relation (see Eq. 4) is independent of the κ(E) dependence. If we use the two-valley κ(E) dependence and calculate the FN slope C from Eq. 12 using $Ø_s$=2.8eV and values of m_1, m_2, and E_2 from Table 2 for tunneling from the metal, we obtain C≈2.0 V/Å. This is considerably lower than the value of 2.9 V/Å obtained in Fig. 13, suggesting that the barrier in the strained SiO_2 near the Si interface is a much more effective barrier for tunneling, with a correspondingly larger value of m_2. This interpretation is also consistent with the smaller tunneling currents observed from the metal interface at voltages below the FN region. Even though we cannot use resonant tunneling from the Si interface to test the κ(E) dependence in the strained SiO_2 region, the two-valley dependence must also apply here to account for the larger effective mass m_1 seen for $k_{ox}(E)$ in the conduction band near the Si interface, using resonant FN tunneling from the metal.

CONCLUSIONS

In retrospect, tunneling in SiO_2 is still seen as a well-defined process to first order. It can be quantitatively analyzed with precision, if we limit our observations to a specified MOS process and to tunneling from one interface. However, we need to examine carefully the effect of the different SiO_2 structure near the Si interface compared with the metal interface before the process can be more fully described.

The fact that we are able to observe and analyze quantum interference effects at the Si-SiO_2 interface is rather remarkable in view of the supposedly amorphous structure of the thermal SiO_2. This and many other observations (33) lead us to characterize the SiO_2 near the Si-SiO_2 interface as almost pseudomorphic with the Si substrate and therefore, also highly strained and denser than bulk SiO_2. It must be this more ordered structure that is responsible for the long scattering distances required for electron wave interference to be observed. In contrast, this is not the case at the metal-SiO_2 interface and consequently, the quantum interference effect is not observed there. Likewise, the SiO_2 near the metal interface provides a weaker tunneling barrier than the SiO_2 near the Si interface, when the barrier heights at the metal and Si interfaces are taken into account.

Our results cannot be made consistent with the single effective mass model (38) for SiO_2 at either interface. The two-valley, two-mass κ(E) dependence has been shown to give a precise fit to the resonant FN tunneling results for the case of tunneling fom the metal interface. However, the same κ(E) dependence does not correctly describe tunneling from the Si interface. It appears that a larger value of m_2 (and probably a different value of E_2) is required to account for the difference. Therefore, the strained SiO_2 near the Si interface needs to be described separately by a different two-valley κ(E) dependence over some effective thickness. Likewise, the κ(E) dependence for SiO_2 at the metal side needs to be reevaluated from resonant tunneling data with this addition. Clearly, further work is required to complete this description of tunneling in SiO_2.

ACKNOWLEDGMENTS

The author wishes to express his appreciation to N. Zamani for his major contributions to this work, to D. B Lawson for careful sample fabrication, to F. J. Grunthaner for many stimulating discusions, and to the many other collaborators and colleagues who have contributed over the years. This research was performed at the Jet Propulsion Laboratory, California Institute of Technology, as part of the Center for Space Microelectronics Technology, and sponsored by the National Aeronautics Administration (NASA) and recently by the Strategic Defense Initiative Organization/ Innovative Science and Technology Office.

REFERENCES

1. A.H. Wilson, Proc. Roy. Soc. (London), 136A, 487 (1932)
2. R. Holm, J. Appl. Phys., 22, 569, (1951).
3. E.L. Murphy and R.H. Good, Jr., Phys. Rev., 102, 1464 (1956).
4. I. Giaever, Ch. 3 in: *Tunneling Phenomena in Solids*, E. Bursten and S. Lundqvist, ed., (Plenun Press, New York, 1969).
5. C.B. Duke, *Tunneling in Solids*, (Academic Press, New York, 1969).
6. S.L. Kurtin, T.C. McGill, and C.A. Mead, Phys. Rev. B., 3, 3368 (1971).
7. E.H. Snow, Solid-St. Comm., 5, 815 (1967).
8. M. Lenzlinger and E.H. Snow, J. Appl. Phys., 40, 278 (1969).
9. W.E. Dahlke, Appl. Phys. Lett., 10, 261 (1967).
10. H.C. Card and E.H. Rhoderick, Solid-St. Electron., 15, 993 (1972).
11. T.P. Ma and R.C. Barker, J. Appl. Phys., 45, 317 (1974).
12. R.A. Clarke and J. Shewchun, Solid-St. Electron., 14, 957 (1971).
13. S. Kar, Appl. Phys. Lett., 25, 587 (1974).
14. J. Maserjian and G. Petersson, Appl. Phys Lett., 25, 50 (1974).
15. J. Maserjian, G. Petersson, and C. Svensson, Solid-St. Electron., 17, 335 (1974).
16. J. Maserjian, J. Vac. Sci. Technol., 11, 996 (1974).
17. G. Petersson, C.M. Svensson, and J. Maserjian, Solid-St. Electron., 18, 449 (1975).
18. G. Lewicki and J. Maserjian, J. Appl. Phys., 46, 3032 (1975).
19. E.H. Nicollian and J.R. Brews, *MOS Physics and Technology* (Wiley, New York, (1982).
20. D.J. Di Maria, K.M. De Meyer, C.M. Serrano, and D.W. Dong, J. Appl. Phys., 52, 4 (1981).
21. J. Maserjian and N. Zamani, J. Appl. Phys., 53, 559 (1982).
22. J. Maserjian and N. Zamani, J. Vac. Sci. Technol., 20, 743 (1982).
23. A. Badiki, B. Eitan, I. Cohen, and J. Shappir, Appl. Phys. Lett., 40, 396 (1982).
24. Y. Nisson-Cohen, J.Shappir, and D.Frohmon-Benthkowsky, Solid-St.Electron., 28, 717 (1985).
25. J.M. Lung and S.A. Lyon, Appl. Phys. Lett., 50, 1152 (1987).
26. E.H. Nicollian, G.H.Berglund, P.F.Schmidt, and J. M. Andrews, J. Appl. Phys., 42, 5654 (1971).
27. M. Bakowski, R. Cockrum, N. Zamani, J. Maserjian, and C.R.Viswanathan, IEEE Trans. Nucl. Sci., NS-25,1233 (1978).
28. S.K. Lai and D.K. Young, J. Appl. Phys., 52, 6231 (1981).
29. M.V. Fischetti, R. Gastaldi, F. Maggioni, and A. Modelli, J. Appl. Phys., 53, 3136 (1982).
30. Z.A. Weinberg and A. Hartstein, Solid-St. Comm., 20, 179 (1976).
31. A. Hartstein and Z.A. Weinberg, Phys. Rev. B., 20, 1335 (1979).
32. F.J. Grunthaner and J. Maserjian, in: *Physics of SiO_2 and its Interfaces*, S.T.Pantalidis, ed., (Pergamon Press, New York, 1978); p. 60.
33. For a comprehensive review see: F.J. Grunthaner and P.J. Grunthaner, Materials Science Reports, 1, pp.65-160 (1986).
34. M.H. Hecht, F.J. Grunthaner, and J. Maserjian, Mat. Res. Soc. Symp. Proc., 25, 317 (1984).
35. M.H. Hecht, R.P.Vasquez, F.J.Grunthaner, N.Zamani, and J.Maserjian, J. Appl. Phys., 57, 5256 (1985).
36. Z.A. Weinberg and A. Hartstein, J. Appl. Phys., 54, 2517 (1983).
37. M. Av-Ron, M. Shatzkes, T.H. DiStefano, and R.A. Gdula, J. Appl. Phys., 52, 2897 (1981).
38. Z.A. Weinberg, J. Appl. Phys., 53, 5052 (1982).
39. M.L. Cohen and T.K. Bergstresser, Phys. Rev., 141, 789 (1966).
40. W. Franz, in: *Handbuch der Physik*, S. Flugge, ed., (Springetr, Berlin, 1956); p.155.
41. W.B. Fowler, in: *Physics of SiO_2 and its Interfaces*, S.T.Pantalidis, ed., (Pergamon Press, New York, 1978)
42. E.O. Kane and E.I. Blount, Ch.6 in *Tunneling Phenomena in Solids*, E. Burstein and S. Lundqvist,ed., (Plenum Press, New York, 1969).
43. V.Z. Zekeriya and T.P. Ma, IEEE Trans. Nucl. Sci., NS-31, 1261 (1984); Appl. Phys. Lett., 47, 54 (1985).

HOT ELECTRON TRANSPORT IN SILICON DIOXIDE

D.J. DiMaria and M.V. Fischetti

IBM Thomas J. Watson Reasearch Center
P.O. Box 218
Yorktown Heights, N.Y. 10598

ABSTRACT

Hot electron transport in silicon dioxide is examined with emphasis on current experimental and theoretical results. For oxide layers thicker than 100 Å, steady-state transport has been shown to control the carrier flow at all fields studied. The transition from a nearly thermal electron distribution at electric fields less than approximately 1.5 MV/cm to significantly hot distributions with average energies between 2 and 6 eV at higher fields of up to 16 MV/cm is discussed. The significance of non-polar phonon scattering in controlling the dispersive transport at higher electric fields, thereby preventing run-away and avalanche breakdown, is reviewed. For oxide thicknesses $\lesssim$ 100 Å, a transition from the steady-state to the ballistic transport regime occurs with the observation of quantum size effects and single phonon scattering events, as predicted theoretically. Also, both interface and bulk trap generation in SiO_2 are shown to be caused by hot electrons with energies $\gtrsim$ 2.2 eV.

INTRODUCTION

Transport studies in wide bandgap insulators have recently generated much interest for both scientific and technological reasons. Experimentally, work has been performed on materials such as SiO_2, Si_3N_4, Si-rich SiO_2, and SiO_xN_y which are used in the electronics industry,[1-13] polymer films used in the high-power insulation industry,[14-18] and rare gas solids.[19] Theoretically, this experimental work has generated rethinking of the hot carrier transport in insulators, with most of the calculations to date performed for SiO_2 (references 8-12). However, the fundamental picture for electron transport that has evolved appears to apply universally to all these different wide-band gap dielectrics. The discussion here will be limited to SiO_2 on which most of our work has been performed. Silicon dioxide is currently the best understood wide-band-gap insulator, and a necessary material component of all silicon-based device technologies used in the electronics and computer industries.

Our understanding of high-field steady-state electron transport in SiO_2, generated mostly by the work at IBM over the past 6 years, is quite different from what had been believed previously.[2] The motion of electrons in the oxide layer is completely controlled by their interaction with the lattice and not by electronic excitation or their interaction with defects and trapping states in the forbidden gap.[8-10] Both polar (inelastic) and non-polar (quasi-elastic) collisions occur. Energy gained from the applied electric field is lost mostly in large amounts to the polar modes (in particular, LO-phonon emission at 0.153 eV). The non-polar phonons seem to have a role mostly as a catalyst in stabilizing the electron energy at high fields. They cause large angle scattering and therefore long path lengths characteristic of dispersive transport. Electrons have near-thermal energies at electric fields of $\lesssim$ 1.5 MV/cm. For fields greater than this value, the distribution heats up to a steady-state condition where the average energy can be as large as 6 eV at the highest fields attained before destructive breakdown occurs. The scattering of the hot carriers by the non-polar modes is essential for the stabilization of the distribution above this 1.5 MV/cm threshold. Prior to the IBM work, the importance of the non-polar modes at high fields had been neglected.[20-27]

This resulted in the incorrect idea that most electrons in SiO_2 were near thermal energies until breakdown. Then the electrons could accelerate to energies greater than that of the band-gap ($\approx$ 9.5 eV) and cause impact ionization. The holes produced by the ionization process were thought to cause a current runaway condition and the destruction of film.[24-27] This older model for high-field transport has now been conclusively eliminated for SiO_2 films $\lesssim$ 500 Å in thickness.

In other insulators such as Si_3N_4 which have much smaller low-field mobilities as compared to SiO_2 (which is $\approx$ 30 cm^2/V-sec), transport is dominated by defects or traps.[13,28] With these materials, most electrons move only small distances before they are captured into shallow trapping sites. Subsequently, these carriers can be re-emitted either thermally, by tunneling, and/or a combination of these two processes. However, a small fraction of the electrons can behave in a manner similar to the hot carrier distributions observed in SiO_2.[13] Hole transport in SiO_2, which is very dispersive and has a low mobility, is also expected to be similar to that observed for trap-modulated electron conduction.[29]

In this review, we will be concerned mainly with the work at IBM which was used to directly study the energy distributions of hot electrons in SiO_2 films. Theoretical Monte-Carlo simulations will be compared to these results. After briefly reviewing the measurement techniques and our original work on steady-state transport in thicker oxide films, we will discuss more recent results on ballistic transport in thin SiO_2 layers and trap creation by hot carriers.

EXPERIMENTAL TECHNIQUES

Three different experimental techniques were used during the course of the our studies. These techniques in chronological order of use are electroluminescence,[1,2,30] carrier separation,[1,2,31,32] and vacuum emission.[2,4,33] The latter two, which were used to obtain the data presented here, will be reviewed. A discussion of the electroluminescence technique can be found in the literature.[30] Carrier separation measures only the average energy of the hot electron distribution while vacuum emission measures the energy distribution itself. Each technique has its limitations, but their combination yields a quantitatively consistent picture for transport in SiO_2.

Carrier separation measures the average energy of the electron distribution by counting the number of electron-hole pairs produced in single-crystal Si by hot electrons injected from the SiO_2. For these experiments, a metal-oxide-semiconductor field-effect-transistor (MOSFET) with a p-channel is used under negative gate voltage bias to separate the electron/hole pairs produced in the Si before they recombine.[2,31] The hole current is measured from the p^+ source/drain contacts to the channel while electron current is measured from the substrate contact. The number of pairs produced is converted to the average energy of the incoming hot electron distribution using the theory of Alig et al..[34]

Vacuum emission directly measures the hot electron distribution ejected into vacuum through the thin metal gate of an MOS capacitor. An Einzel lens configuration is used to focus and energy analyze these hot electrons.[2,4,33] Data are obtained by counting the total number of hot electrons emitted into vacuum as a function of the analyzer potential for different oxide fields. The derivative of these data with respect to the analyzer voltage gives the energy distribution. Unless cesiation is used, the first 0.9 eV of information about the distribution is lost due to the small positive electron affinity of SiO_2. After extensive studies with different metal gate compositions and thicknesses, we have concluded that most of the emitted electrons we detect escape through thin metallic regions near the edges of the polycrystalline grains. These few electrons ($\approx$ 1 in 10^5-10^6 of those collected by the metal gate) are believed to be unattenuated by scattering in the metal as will be demonstrated later in some of our ballistic studies.

For many of the steady-state experiments on thick oxide films, Si-rich SiO_2 injectors were used.[35] The use of Si-rich SiO_2 injectors for cathodic field enhancement and breakdown reduction made many of our early experiments possible. Since the energy relaxation length for electrons in SiO_2 is $\approx$ 30 Å, the cathodic injection mode is not important in determining the electron energy in thick oxides.[2] The oxide field near the anode interface determines the average energy of the electron distribution. For cases where structures with and without injectors were compared, identical results were obtained. For all the ballistic transport studies on films $\leq$ 100 Å in thickness, injectors were not used because of the difficultly in correcting for the small voltage drop across this layer. Also in many experiments on thicker oxides, the electric field range was extended by using the additive internal field associated with electron charge trapping on energetically deep sites in the oxide film.

STEADY-STATE TRANSPORT

If the SiO_2 layer thickness is large compared to the characteristic scattering length of the hot carrier distribution, the carriers would be expected to be in a steady-state condition for all electric fields, except possibly near breakdown. For oxide films with thickness in the range from 100 to 5,000 Å, the work done at IBM has shown this to be the case.[1-5,8-10] The average energy characteristic for steady-state transport in 1500 Å thick SiO_2 layers determined using both the carrier separation and vacuum emission techniques is shown in Fig. 1. As can be seen from this data, runaway from the polar LO-phonon modes (the dominant mode having an energy of 0.153 eV) occurs at a very low field (about 1.5 MV/cm) as initially predicted by Fitting and Frieman.[3,5,20] However, the runaway of most of the electrons does not proceed to cause impact ionization even up to fields as high as 16 MV/cm. From 4 to 16 MV/cm, the average energy of the distribution increases very slowly from about 3 to 6 eV with respect to the bottom of the oxide conduction band, as shown in Fig. 1. The scattering processes responsible for stabilizing the distribution at the higher fields are the non-polar collisions (acoustic phonons). From the slope of these data in this regime, an energy relaxation length of ≈ 30 Å can be obtained. This quantity corresponds to the distance over which an electron of the distribution must lose an amount of energy equivalent to its average energy value so as to remain in a steady-state condition.

These data from the IBM group totally contradicted the old model of transport in which LO phonon scattering keep electrons near thermal energies until runaway occurs at destructive breakdown.[21-27] Using Ridley's earlier suggestion that acoustic or non-polar phonon scattering should also be considered,[36] Fischetti extended the Monte-Carlo calculation of Fitting and Frieman to include it.[8-10] His initial results are shown in Fig. 1 (solid line) and compared to the data. Later, other groups (particularly that of D.K.Ferry) performed similar Monte-Carlo simulations.[11,12] However, Ferry's group assumed that non-polar intervalley scattering to a second conduction band of SiO_2 was limiting the hot electrons to the observed steady-state condition. Our view is that there is no experimental or theoretical justification for the high value of the deformation potential and for the high density of states employed by Porod and Ferry. The conventional Monte Carlo approach of Fischetti gives the best agreement with data at low fields near the threshold.[8,9] However, more recent calculations using a Quantum Monte Carlo approach give better agreement with the measured energy distributions at high fields, particularly in predicting the high energy tail.[10] On thicker oxides (≳ 500 Å), a small number of hot carriers out in the high energy tail have energies exceeding the SiO_2 bandgap, but there is little evidence that they trigger impact ionization.[4]

The important physical aspect of the acoustic or non-polar phonon interaction is large angle scattering (momentum reversal) which gives the electrons much larger path lengths than the actual length of material traversed. LO phonon scattering still accounts for most of the energy loss. This is depicted in Fig. 2 using Monte-Carlo calculations for the trajectories of several electrons.[9] At low fields, "streaming-type" transport is observed where most of the electronic paths are in the direction of the applied field. At fields significantly above the threshold for electron heating, dispersive transport dominates due to the large angle scattering caused by the acoustic phonons. Although the scattering rate for the polar phonon modes is decreasing with increasing electron energy, the distribution is stabilized against velocity runaway by the rapid increase in the path length due to the increasing scattering rate of the non-polar phonon modes.

Hot-electron transport in amorphous SiO_2 films does not depend strongly on the mode of oxide growth (thermal and deposited) or material for the anode contact. On very thick oxide films (> 1500 Å) which have not been densified properly, deviations from the expected behavior have been observed.[3]

BALLISTIC TRANSPORT

As the SiO_2 thickness is further reduced, a non steady-state condition is reached where some of the electrons of the distribution are not hot enough to be scattered by acoustic or non-polar phonon modes.[2,6,7] This occurs for oxides under 100 Å in thickness. The characteristic length at which this occurs is a measure of the average distance an electron of the distribution must move through, starting from thermal energy, in order to reach a steady-state condition. This length which is ≈ 30 Å is called the "heat-up" distance.[2] Similarly if hot electrons are injected into an oxide at a low electric field condition (≲ 1.5 MV/cm), one would expect to measure a characteristic "cool-down" distance of similar dimensions. This distance has also been deduced to be ≈ 25-35 Å from internal photoemission measurements where photons are used to inject hot electrons.[37-39] For the determination of both of these lengths, the distribution is in a non-steady-state condition where LO phonon modes dominate the hot electron scattering.

Decreasing the oxide thickness even more results in a condition where distances are so short that some electrons will not collide with any phonon mode, either polar or non-polar. These electrons, which can gain kinetic energies equal to the applied potential they fall through, are called "ballistic" electrons.[6,7] When the oxide layer is thin enough that nearly all the electrons in the distribution traverse the SiO_2 layer ballistically, a characteristic length called the "mean free path" can be measured with values ranging from 7 to 15 Å depending on the oxide field.[6,7,40,41] The remainder of this section will be concerned only with transport in the ballistic or near-ballistic regime in very thin SiO_2 layers.

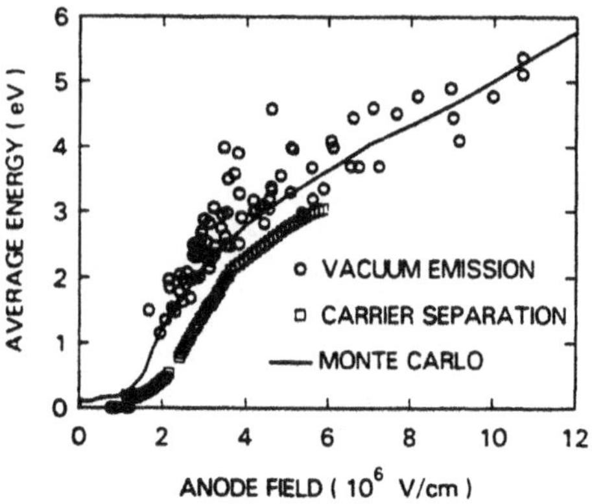

Fig. 1. Comparison of vacuum emission and carrier separation data to a Monte-Carlo simulation for steady-state electron transport in a thick SiO_2 layer. This figure shows that significant carrier heating begins at electric fields much lower than that necessary for destructive breakdown of the oxide

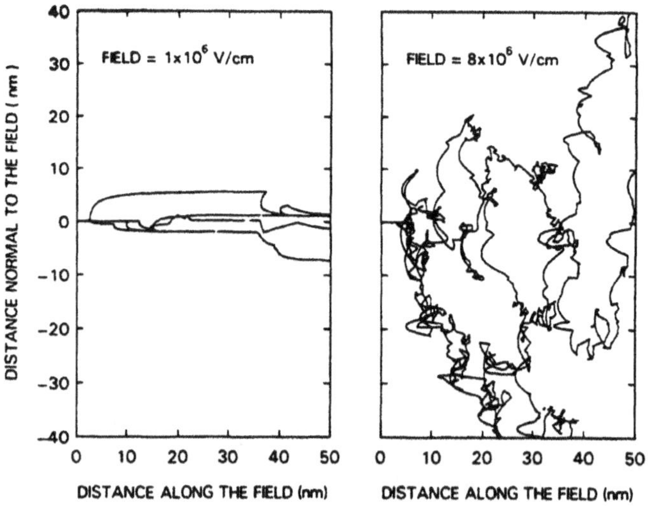

Fig. 2. Trajectories of four electrons in real space at two values of the electric field. The low-field (1 MV/cm) trajectories show that the transport is dominated by the forward polar scattering. At 8 MV/cm, the non-polar scattering at large angles randomizes the direction of motion and effectively increases the actual path length of the electrons.

In 1975, Lewicki and Maserjian reported the first evidence for ballistic transport in SiO_2.[40] This was based on observations of oscillations in current-voltage (I-V) data for very thin layers (30-75 Å) of SiO_2 incorporated into MOS capacitors. They argued that these oscillations were the result of quantum reflection and interference of ballistic electron wavefunctions at an abrupt anode interface. Over the past 2 years, we have extended their earlier work and confirmed the original interpretation by directly measuring ballistic transport in SiO_2 using the vacuum emission technique.[6,7]

Figures 3 and 4 show recent vacuum emission data and Monte-Carlo simulations for energy distributions of the hot carriers emerging from a very thin oxide layer (56 Å in thickness). In Fig. 3, the gate voltage dependence of the distribution at 83 K is shown while in Fig. 4 the temperature dependence at fixed voltage is studied. All data clearly show a peak at an energy corresponding to the ballistic condition (which is approximately the gate voltage minus the 3.1 eV energy barrier for the substrate/SiO_2 interface). Also, additional peaks shifted from the ballistic peak by energies of 0.063 and 0.153 eV are observed. These are what would be expected for single phonon scattering by the two dominant LO-phonon modes. Comparison to a Monte-Carlo simulation confirms this interpretation, as demonstrated in this figure. Peaks at energies less than the ballistic peak correspond to single phonon *emission* while those at larger energies, but with greatly reduced amplitude, correspond to the less probable absorption of single LO phonons by the hot electrons. An example of this latter type of phenomenon is demonstrated by the shoulder observed in both the data and simulation in Fig. 4 which is shifted up in energy from the ballistic peak by 0.153 eV, particularly at 295 K.

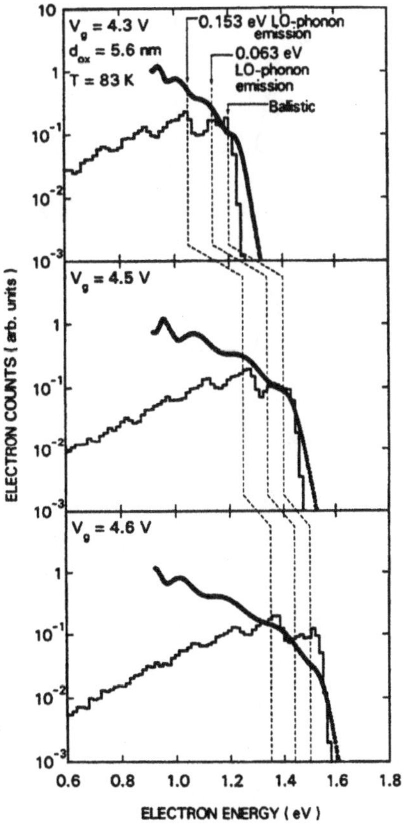

Fig. 3. Comparison of vacuum emission data with Monte-Carlo simulations for near-ballistic transport in a very thin oxide at 83 K. The ballistic peak and phonon-induced side bands are shown to track the applied gate voltage.

Other trends consistent with the Monte-Carlo simulations are observed in data like those in Figs. 3 and 4. The structure in these results attributed to the ballistic peak and phonon-induced sidebands shifts by the change in applied voltage and becomes less pronounced with increasing temperature, as expected. With thinner oxides or lower voltages, the amplitude of the ballistic peak increases with respect to that of the sidebands while the opposite trend is observed with thicker films or higher voltages. Modulation of the ballistic peak due to quantum reflection at the SiO_2/gate interface has also been observed as predicted in our simulations and seen in the I-V data. A gradual disappearance of the structure and broadening in the energy distribution, consistent with that expected for the ballistic to steady-state transition, is seen as the oxide thickness approaches 100 Å.

The only clear discrepancy between the data and theory in Figs. 3 and 4 occurs at low temperature and electron energy. The calculations used in the Monte-Carlo simulation solve exactly for the 3-dimensional electron energy distribution at the Si/SiO_2 interface and its tunneling probability into the SiO_2 layer. Although corrections for the quantum reflections in the oxide layer are taken into account, we have made no attempt to include contributions to the electron source function from lower-lying, 2-dimensional quantum levels in the Si accumulation layer. Qualitatively, the low energy structure in the hot electron distribution at low temperature behaves in a manner consistent with the contributions expected from these quantum states. Although difficult, we will attempt to add these effects to future simulations.

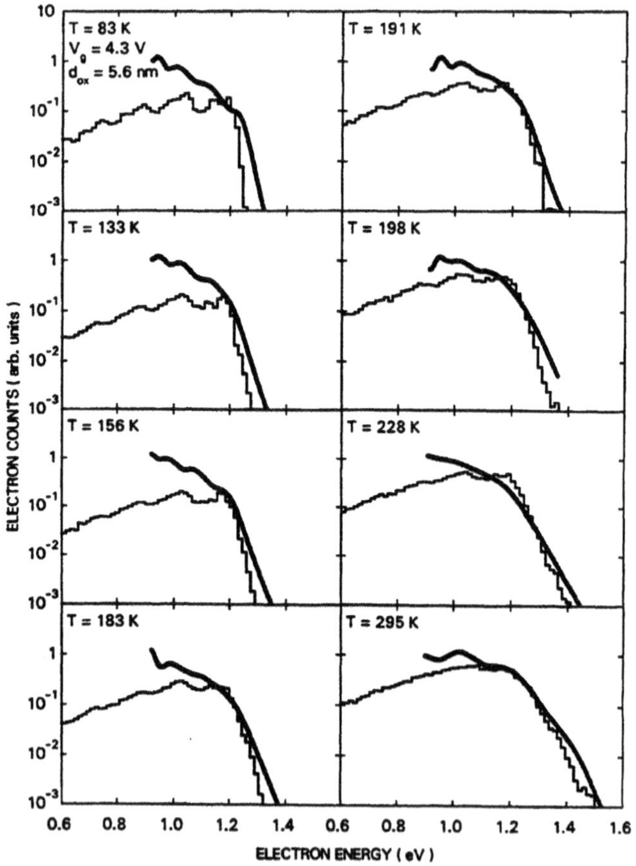

Fig. 4. Comparison of vacuum emission data with Monte-Carlo simulations for near-ballistic transport in a very thin oxide at variable temperatures under fixed gate voltage. These data show the temperature broadening of the ballistic peak, phonon-induced sidebands, and Fermi source function at the Si/SiO_2 interface.

To further confirm the existence of ballistic transport in very thin SiO_2 layers, we used the carrier separation technique. This allows us to study even thinner oxides and lower applied voltages than used in the vacuum-emission experiments, since we do not have to overcome the electron affinity of the SiO_2. However, only a measure of the average electron energy is obtained and not the full distribution. From these data, the ballistic condition appears to hold for most electrons in the distribution for the first 0.8 eV of energy gained from the applied field. In addition to studying the transition from the ballistic to the steady-state regimes, we are also able to study ballistic electron behavior in the transition region from direct tunneling (gate into Si substrate) to Fowler-Nordheim tunneling (gate into SiO_2 conduction band). Again, undulations in the data are believed to be due to quantum reflection effects at the contact/oxide interfaces as previously described.[6,7]

TRAP GENERATION

Since the early 1980's, many groups have claimed to observe trap generation at high electric fields in SiO_2.[42–52] Many of these results were controversial. In some cases, background trapping in the as-fabricated oxides was not separated out from trapping in the generated sites. In other cases, interface charging (where the charge state can be either positive or negative depending on the Si/SiO_2-interface Fermi level position) was not clearly separated from electron trapping in the oxide bulk. Since many of these researchers looked at only a limited range of high electric fields, the effects due to the generated traps were not distinct. Although some of these researchers discussed the possibility of trap creation by hot electrons, heating characteristics like that shown in Fig. 1 were not available until 1986.

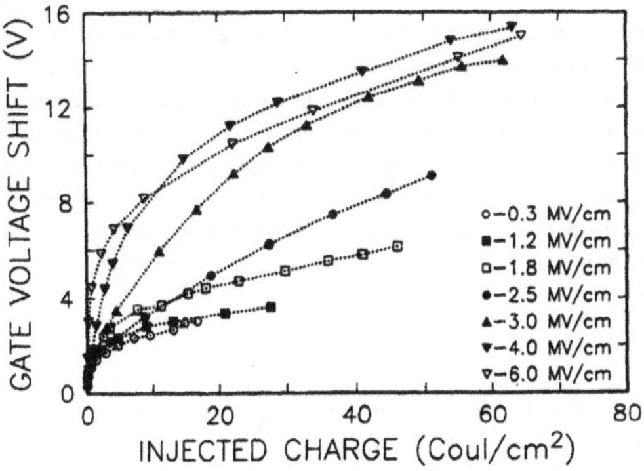

Fig. 5. Electron trapping characteristics for bulk SiO_2 layers with a thickness of 670 Å for several different average electric fields. These data show a sudden increase in trapped electron build-up for fields above the heating threshold of ≈ 1.5 MV/cm.

Recently, the IBM group has demonstrated a direct correlation of trap generation (both interface and bulk) with hot electrons in SiO_2 (references 51 and 52). This is shown for bulk electron trapping in Fig. 5 where the charge build-up is studied from voltage shifts of I-V characteristics. At the threshold field for heating (≈ 1.5 MV/cm), there is a sudden increase in electron trapping, particularly for injected electron fluence ≳ 1 coul/cm². Positive interfacial charges and interface states can be separated from bulk electron charging by using both capacitance-voltage (C-V) and I-V characteristics. Their behavior as a function of field is similar to that in Fig. 5. From data like that shown in this figure, the number of generated traps as a function of electric field can be shown to follow a trend similar to our heating results in Fig. 1.

The data in Fig. 5 on thick oxides does not unequivocally determine whether the electric field or the hot electron energy is directly responsible for trap creation. By additionally studying charge build-up in very thin oxide layers, this separation can be made. As previously discussed in the section on ballistic transport, electrons injected into thin oxides at high electric fields still require a distance of ≈ 30 Å to reach a steady-state condition. By varying the oxide thickness, the average energy of a near-ballistic distribution impinging on the anode/SiO_2 interface can be changed from 0 to 4 eV almost independently of the field. This is shown in Fig. 6 where the trap generation rate as a function of electron energy is plotted for several oxide thicknesses and both voltage polarities. A minimum electron energy of ≈ 2.2±0.2 eV for trap creation can be deduced from the data. These results also imply that the bulk of the SiO_2 layer is not necessary for the generation mechanism. Consistent with our results, other recent studies of hydrogen evolution during electron injection of SiO_2 layers[53] suggest that a possible cause of trap creation in SiO_2 is the cracking of hydrogen bound to impurity sites near the anode interface. The mobile hydrogen can then diffuse into the oxide layer to produce the sites causing the observed charging phenomenon. Similar arguments have been made for trap creation in polymer films exposed to hot electrons.[17]

CONCLUSIONS

This review has summarized both the experimental and theoretical work performed on electron heating during the past 6 years, mostly at IBM. Electrons injected into SiO_2 can gain significant energy from the applied field for ≳ 1.5 MV/cm but do not undergo velocity runaway. Stabilization of the hot electron distribution is attained by interaction with the surrounding oxide lattice via polar and non-polar phonons. There is no evidence at any oxide thickness that impact ionization in the SiO_2 layer can occur for an "average" electron of the distribution, even out to fields of 16 MV/cm. From vacuum emission studies, a small fraction of the hot electrons can be found out in the high energy tail of the distribution. Some of these very hot electrons are predicted by quantum Monte-Carlo simulations where the scattering rate can decrease significantly with increasing energy.

On very thin oxides, ballistic transport has been observed. SiO_2 is the only material other than GaAs (references 54-56) where a direct observation of this important phenomena has been reported. Phonon induced side-bands in the hot electron distributions caused by single LO phonon scattering have been seen for the first time in any material. This latter observation was made possible by minimizing electron-electron interactions with control of the number of electrons introduced into the oxide conduction band during these experiments. Also, the vacuum emission experiments have allowed us to gain information on the electrons at the substrate-Si/SiO_2 interface that can subsequently tunnel into these thin oxides under the appropriate applied voltage. The expected behavior of the Fermi function with temperature has been observed in these studies.

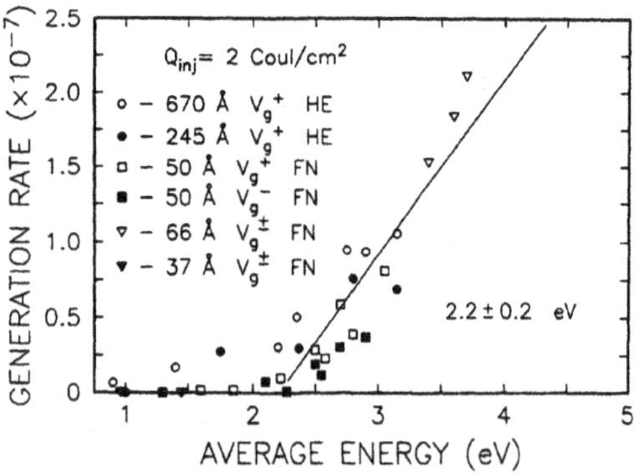

Fig. 6. Trap generation rate (number of created traps divided by number of injected electrons) in SiO_2 as a function of average electron energy after carrier injection of 2 coul/cm^2 using Fowler-Nordheim (FN) tunneling or optically-induced hot-electron (HE) injection. These data (which are nearly independent of voltage polarity, means of injection, and oxide thickness) show a minimum energy of 2.2±0.2 eV is necessary for site creation.

Trap creation in SiO_2 has been unequivocally related to the energy of the hot electrons. A minimum energy of 2.2±0.2 eV is necessary to generate both interface and bulk oxide traps. On very thin SiO_2 devices, this problem is minimized because injected electrons may not have a large enough distance in the oxide to heat-up to ≳ 2.2 eV before they enter the anode contact, depending on the gate voltage condition. In future MOSFET technologies where hot electron injection into the oxide from near the drain contact is a problem, lower power supply voltages or thinner SiO_2 layers will be necessary to circumvent trap generation.

Although all the oxides we studied were amorphous, we have still been able to observe ballistic electron transport, at least on the scale of the characteristic transport lengths (≈ 10-30 Å). This observation could be giving us information on the short range order of the SiO_2 films. Our results are consistent with the idea that amorphous SiO_2 may have ordered regions which are nearly crystalline on a scale of 10-30 Å. As suggested by Mott, ballistic transport and the large value of the low field mobility for electrons in SiO_2 are also consistent with a lack of disorder scattering caused by a negligible number of localized states near the conduction band edge.[57] This latter explanation would be expected for s-like wavefunctions (which are believed to mainly form the bottom of the SiO_2 conduction band) and small variations of Si-O and O-O interatomic distances even with the large variation in the Si-O-Si bond angle.

Acknowledgements: The authors wish to acknowledge the critical reading of this manuscript by A. R. Williams.

REFERENCES

1. T.N.Theis, D.J.DiMaria, J.R.Kirtley, and D.W.Dong, Phys. Rev. Lett. 52, 1445 (1984).
2. D.J.DiMaria, T.N.Theis, J.R.Kirtley, F.L.Pesavento, D.W.Dong, and S.D.Brorson, J. Appl. Phys. 57, 1214 (1985).
3. D.J.DiMaria, M.V.Fischetti, M.Arienzo, and E.Tierney, J. Appl. Phys. 60, 1719 (1986).
4. S.D.Brorson, D.J.DiMaria, M.V.Fischetti, F.L.Pesavento, P.M.Solomon, and D.W.Dong, J. Appl. Phys. 58, 1302 (1985).
5. D.J.DiMaria, M.V.Fischetti, E.Tierney, S.D.Brorson, Phys. Rev. Lett. 56, 1284 (1986).
6. D.J.DiMaria, M.V.Fischetti, J.Batey, L.Dori, E.Tierney, and J.Stasiak, Phys. Rev. Lett. 57, 3213 (1986).
7. M.V.Fischetti, D.J.DiMaria, L.Dori, J.Batey, E.Tierney, and J.Stasiak, Phys. Rev. B 35, 4404 (1987).
8. M.V.Fischetti, Phys. Rev. Lett. 53, 1755 (1984).
9. M.V.Fischetti, D.J.DiMaria, S.D.Brorson, T.N.Theis, and J.R.Kirtley, Phys. Rev. B 31, 8124 (1985).
10. M.V.Fischetti and D.J.DiMaria, Phys. Rev. Lett. 55, 2455 (1985).
11. W.Porod and D.K.Ferry, Phys. Rev. Lett. 54, 1189 (1985).
12. W.Porod and D.K.Ferry, 4th Intl. Conf. Hot Electrons, Innsbruck, Austria, July 1985, unpublished.
13. D.J.DiMaria and J.R.Abernathey, J. Appl. Phys. 60, 1727 (1986).
14. H.R.Zeller, P.Pfluger, and J.Bernasconi, IEEE Trans. Elect. Insul. EI-19, 200 (1984).
15. P.Pfluger, H.R.Zeller, and J.Bernasconi, Phys. Rev. Lett. 53, 94 (1984).
16. E.Cartier and P.Pfluger, Phys. Rev. B 34, 8822 (1986).
17. E.Cartier and P.Pfluger, IEEE Trans. Electr. Insul. EI-22, 123 (1987).
18. E.Cartier, P.Pfluger, J.J.Pireaux, and M.Rei Vilar, Appl. Phys. A 44, 43 (1987).
19. E.Cartier and P.Pfluger, to be published in Physica Scripta.
20. H.H.Fitting and J.U.Frieman, Phys. Status Solidi A 69, 349 (1982).
21. D.K.Ferry, Appl. Phys. Lett. 27, 689 (1975).
22. D.K.Ferry, J. Appl. Phys. 50, 1422 (1979).
23. D.K.Ferry, in *The Physics of SiO_2 and Its Interfaces*, edited by S.T.Pantelides (Pergamon, New York, 1978), pp. 29-34.
24. T.H.DiStefano and M.Shatzkes, Appl. Phys. Lett. 25, 685 (1974).
25. T.H.DiStefano and M.Shatzkes, J. Vac. Sci. Technol. 13, 50 (1976).
26. P.Solomon and N.Klein, Solid State Commun. 17, 1397 (1975).
27. P.Solomon, J. Vac. Sci. Technol. 14, 1122 (1977).
28. D.J.DiMaria and P.C.Arnett, IBM J. Res. Develop. 21, 227 (1977).
29. R.C.Hughes, Phys. Rev. B 15, 2012 (1977).
30. T.N.Theis, J.R.Kirtley, D.J.DiMaria, and D.W.Dong, Phys. Rev. Lett. 50, 750 (1983).

31. A.S.Ginovker, V.A.Gritsenko, and S.P.Sinitsa, Phys. Status Solidi A 26, 489 (1974).
32. Z.A.Weinberg, W.C.Johnson, and M.A.Lampert, Appl. Phys. Lett. 25, 42 (1974).
33. P.M.Solomon, in *The Physics of SiO_2 and Its Interfaces*, edited by S.T.Pantelides (Pergamon, New York, 1978), pp. 35-39.
34. R.C.Alig, S.Bloom, and C.W.Struck, Phys. Rev. B 22, 5565 (1980).
35. D.J.DiMaria and D.W.Dong, J. Appl. Phys. 51, 2722 (1980).
36. B.K.Ridley, J. Appl. Phys. 46, 998 (1975).
37. C.N.Berglund and R.J.Powell, J. Appl. Phys. 42, 573 (1971).
38. R.Poirier and J. Olivier, Appl. Phys. Lett. 21, 334 (1972).
39. P.M.Solomon and D.J.DiMaria, J. Appl. Phys. 52, 5867 (1981).
40. G.Lewicki and J.Maserjian, J. Appl. Phys. 46, 3032 (1975).
41. J.Maserjian and N.Zamani, J. Appl. Phys. 53, 559 (1982).
42. C.S.Jeng, T.R.Ranganath, C.H.Huang, H.S.Jones, and T.T.L.Chang, in *Int. Elec. Dev. Meeting Technical Digest* 1981 (IEEE, New York, 1983), pp. 388-391.
43. M.-S. Liang and C.Hu, in *Int. Elec. Dev. Meeting Technical Digest* 1981 (IEEE, New York, 1981), pp. 396-399.
44. A.Badihi, B.Eitan, I.Cohen, and J.Shappir, Appl. Phys. Lett. 40, 396 (1982).
45. M.-S. Liang, C.Chang, W.Yang, C.Hu, and R.W.Brodersen, in *Int. Elec. Dev. Meeting Technical Digest* 1983 (IEEE, New York, 1983), pp. 186-189.
46. M.M.Heyns, R.F.DeKeersmaecker, and M.W.Hillen, Appl. Phys. Lett. 44, 202 (1984).
47. M.M.Heyns and R.F.DeKeersmaecker, J. Appl. Phys. 58, 3936 (1985).
48. M.-S.Liang, J.Y.Choi, P.K.Ko, and C.Hu, in *Int. Elec. Dev. Meeting Technical Digest* 1984 (IEEE, New York, 1984), pp.152-156.
49. Y.Nissan-Cohen, J.Shappir, and D.Frohman-Bentchkowsky, J. Appl. Phys. 60, 2024 (1986).
50. S.Horiguchi, T.Kobayashi, and K.Saito, J. Appl. Phys. 58, 387 (1985).
51. D.J.DiMaria, Appl Phys. Lett. 51, 655 (1987).
52. D.J.DiMaria and J.Stasiak, unpublished.
53. R.Gale, F.J.Feigl, C.W.Magee, and D.R.Young, J. Appl. Phys. 54, 6938 (1983).
54. A.F.J.Levi, J.R.Hayes, P.M.Platzman, and W.Wiegmann, Phys. Rev. Lett. 55, 2071 (1985).
55. M.Heiblum, M.I.Nathan, D.C.Thomas, and C.M.Knoedler, Phys. Rev. Lett. 55, 2000 (1985).
56. M.Heiblum, I.M.Anderson, and C.M.Knoedler, Appl. Phys. Lett. 49, 207 (1986).
57. N.F.Mott and E.A.Davis, in *Electronic Processes in Non-Crystalline Materials*, 2nd edition (Clarendon Press, Oxford, 1979), pp. 512-516.

ION IMPLANTATION AND IONIZING RADIATION EFFECTS IN THERMAL OXIDES

Roderick A. B. Devine

Centre National d'Etudes des Télécommunications
BP 98, 38243 Meylan
France

ABSTRACT

Electron spin resonance data on oxygen vacancy defect (positive fixed oxide charge) creation in thin, thermal oxides produced by ion implantation is reviewed. It is found that energy dissipated by atomic displacement processes creates defects between 400 and 700 times more efficiently than that lost by ionizing processes. It is furthermore determined that the energy required in a displacement process to create an oxygen vacancy center is ∿ 28 eV. Isochronal and isothermal annealing data combined with subsequent irradiation with ionizing radiation demonstrate that although oxygen vacancy centers dissappear for anneal temperatures above 400°C, they are not irreversibly removed unless temperatures ≥ 800°C are used.

INTRODUCTION

The technique of ion implantation is widely used in the manufacture of silicon based integrated circuits. In particular, for high density integration (VLSI) applications, implantation is essential to produce well defined source and drain areas (auto-aligned gate technology) or, together with more sophisticated methods, to produce graded, lightly doped drains so as to reduce electric fields and the risk of hot electron injection into the gate oxide (degradation) at the gate/drain edge. Figure 1 shows a schematic cross section of a MOSFET indicating the various strategic points at which ion implantation is used. Potentially dangerous steps are

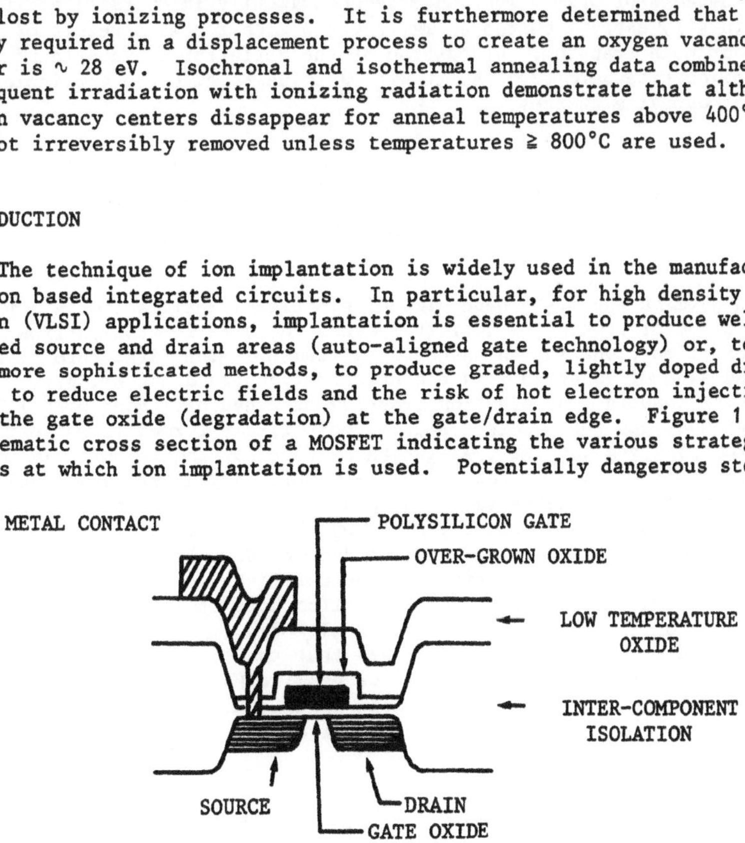

Figure 1. Schematic cross-section of a finished MOSFET

those in which implantation is made through an existing oxide layer which is not subsequently etched off. Examples of such implants are 10^{12} B^+/cm^2 at 30 keV (threshold voltage adjustment) through the grown gate oxide and $\sim 3 \times 10^{15}$ P^+/cm^2 at $\sim$ 170 keV (source and drain contact formation) at the edge of the gate through an oxide thickness $\sim$ 500 Å (thickness of the gate oxide plus the over-grown oxide layer). One is therefore interested in understanding the creation and annihilation kinetics of the damage produced in thin, amorphous SiO_2 by ions having the indicated energies, i.e. energies at which the primary mechanism of energy loss and defect creation is via nuclear stopping[1] (atomic displacement) by O and Si atoms.

The manufacture of the circuit shown in figure 1 requires not only different stages of implantation, but also metallization steps and, of course, lithographic steps. In the former case, the use of evaporators having electron beam heaters may result in the generation of X rays, a radiation to which the circuit will thus be subjected during the metallization process. More importantly, high resolution lithography is relying increasingly on electron beam and X ray (short wavelength) sources. Although primarily designed to expose organic photoresists, stray radiation could leak through to underlying oxides and result in damage/defect creation. Finally, it should be remembered that the finished microcircuit might be destined to operate in a hostile environment in which it may be naturally subjected to high energy radiation (satellite applications for example).

The defects of prime importance technologically in MOSFET's are those which naturally carry a charge or may develop a charge as a result of electron or hole trapping. Discussion of the different defects found in SiO_2 is given elsewhere [2,3] so it will suffice to cite two particular defects of opposite charge:

$$\equiv\text{Si-O-Si}\equiv \;\rightarrow\; \equiv\text{Si}^{+}\;{}^{\bullet}\text{Si}\equiv + \text{O} + e \;. \tag{1}$$

$$\equiv\text{Si-OH} \;\rightarrow\; \equiv\text{Si-O}^{-} + \text{H}^{\circ} \;. \tag{2}$$

The first term on the right hand side of equation 1 defines the oxygen vacancy (E_1') center which is positively charged, here it results from some interaction producing an oxygen displacement. The first term on the right hand side of equation 2 is the negatively charged non-bridging oxygen which results from electron trapping at a hydroxyl bond with liberation of hydrogen [4]. This defect is most likely formed in wet oxides where Si-OH linkages may exist as growth defects [5] and is unlikely to be relevant to thin gate oxides which are usually dry unless substantial amounts of hydrogen diffused into the oxide during post-metallization annealing is incorporated into the oxide by the opening of Si-O-Si bonds or attachment to existing dangling Si-O bonds. The most troublesome technological defect for dry oxides is the positively charged E_1' center (equation 1). This defect may also arise from hole trapping at an existing defect precursor (the B_2 center) $\equiv$Si$\nleftrightarrow$Si$\equiv$ although laser irradiation experiments on bulk, dry silica [6] indicate that the saturation density of E_1' defects so produced $\sim 10^{15}/cm^3$. If we examine this density in terms of a possible voltage shift in the threshold voltage (ΔV_{th}) of a transistor [7] produced by a uniform fixed oxide charge density, N_F ($=10^{15}/cm^3$) we have:

$$\Delta V_{th} = \Delta V_{FB} = -\frac{N_F\, d_{ox}^2\, q}{2\,\varepsilon_o \varepsilon_r} = -0.9\ \text{mV} \;. \tag{3}$$

where d_{ox} is the gate oxide thickness (taken in the example calculation as 200 Å) and q is the charge on the electron (absolute value). Such a small threshold shift is negligible. Clearly, defect densities $\sim 10^3$ larger are required before the threshold voltage shift becomes significant (>1V) Consequently, we are not concerned with the creation of E_1' defects by precursor transformation but by their direct creation throuh O displacement either by displacive or ionizing radiation.

A study of defects found in SiO_2 requires, ideally, a spectroscopic technique which enables identification of specific defects. For example, the E_1' center is positively charged but it is not the only possible positively charged, intrinsic defect [8]. However, the defect is also paramagnetic and has a unique electron spin resonance signature [3]. A spin resonance technique is thus ideally suited to the study of the creation and annihilation of E_1' defects. Unfortunately, spin resonance methods are limited to the detection of a total number of spins irrespective of the sample volume unlike electrical methods (e.g. C(V)) [7] which detect a specific number of defects per unit volume. In consequence, in dealing with oxides of a technological thickness ($\leq$ 1000 Å) one is limited to a sample volume at best $\sim 10^{-5}$ cm^3 in the microwave cavity. Assuming a minimum number of detectable E_1' defects $\sim 2 \times 10^{12}$ sets the minimum detectable density by resonance to be $\sim 4 \times 10^{16}/cm^3$, nearly two orders of magnitude larger than detectable electrically. In the following we shall review E_1' defect data which has been obtained on thin oxides despite the forementioned experimental difficulties. Prior to the studies reviewed very little data of this kind existed and certainly, no systematic studies were available. The data reviewed have been obtained primarily at room temperature using an X band frequency ($\sim$ 9.5 GHz) microwave spectrometer. Details of typical experimental conditions and precautions can be found in reference 9. Samples were in general wet or dry (<1000 Å) oxides grown on Si substrates (> 20 Ωcm) using standard technological processing and temperatures. Ion implantations were carried out using an HVE 500 implanter and γ irradiations , where referenced, were made using ^{60}Co γ's at a rate of 0.8 Mrad/hour.

DISPLACIVE RADIATION

Systematic studies of the growth of the numbers of E_1' defects in SiO_2 as a function of implanted ion, ion dose and ion energy have been carried out [9-11] for ions having atomic masses in the range 14 to 209. The results obtained for the E_1' defect density as a function of cumulated ion dose are shown in figure 2, note that the density per cm^2 of sample surface is quoted. Various parameters related to the implanted ions are given in table 1. The thickness of the thermal oxide into which the ions were implanted was always significantly larger than the damage range for atomic displacement so that in all cases the energy lost in displacement of ions was contained in the oxide. Using the measured saturation density of defects shown in figure 2 for all the ions studied together with the longitudinal straggling data for the ions quoted in table 1 we deduce the volumetric saturation densities shown in table 1. Although the ions had widely varying masses and longitudinal straggling depths varied by a factor of almost two, the density of defects created at saturation by atomic displacements (N) is always $0.7 \leq N \leq 1.7 \times 10^{19}/cm^3$. Further relevant physical information may be extracted if we use the initial slope results ($dN(E_1')/dD$) quoted in table 1 for the low dose ($D \to 0$) regime where individual damage cascades for the ions do not overlap. In figure 3 we plot $dN(E_1')/dD$ versus energy density into atomic displacement processes calculated [1] from the different ion energies and cascade volumes (table 1). From the slope of this curve one deduces that the total energy deposited

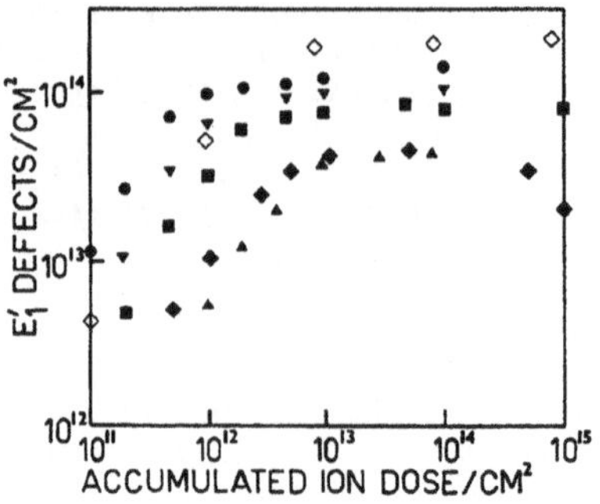

Figure 2. E'1 defect density per cm² for samples of thermal SiO_2 implanted with:

●	Bi^+	270 keV
▼	Sb^+	160 keV
■	As^+	100 keV
◇	Ar^+	125 keV
◆	O^+	30 keV
▲	N^+	30 keV

as a function of accumulated dose of ions.

into displacive processes per E'_1 created is ∿ 2100 eV. This figure, one notes, combines displacements of both Si and O ions. In figure 4 we plot the initial slopes, $dN(E'_1)/dD$ versus the calculated number of O displacements in the implanted ion cascade. The slope of the curve indicates that for each E'_1 defect observed, 37 O displacements have occured. Assuming the energy for displacement of an O atom is approximately half of that for a Si and that half as many Si atoms are displaced as O's then from the figure of 2100 eV per E'_1 created, we conclude that ∿ 28 eV is actually consumed in creating the remnant defect.

Table 1. A summary of various parameters relating to ions implanted into thermal SiO_2 studied in this work.

Ion	Energy (keV)	Longitudinal straggling (Å)	Saturation density x $10^{19}/cm^3$	Cascade† volume x $10^{-18} cm^3$	$dN(E'_1)/dD$ (per ion)
Bi	270	283	1.7	15.9	73
	240	257	nm	11.8	62
Sb	160	209	1.9	10.6	34
Ar	150	506	1.7	55.8	26
	100	342	1.7	21.8	18
As	100	219	1.4	6.3	16
Ne	43	285	1.1	5.6	7
N	30	294	0.7	2.2	8
O	30	255	0.7	2.5	7

† Corrected for transport effects following reference 12

The previous results indicate that of 37 O's displaced, hence vacancies created, only one ends up in the form of a "permanent" E'_1 defect - one is then tempted to ask what has happened to the other 36 vacancies/interstitial O's. Intuitively, one might argue for vacancy-interstitial rec-

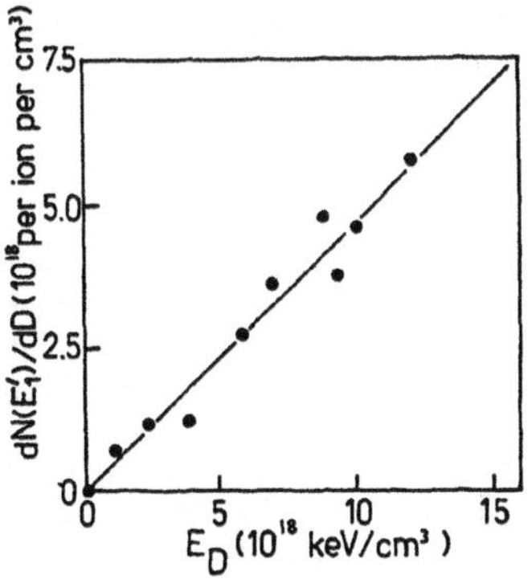

Figure 3. Plot of the initial slope of defect creation ($dN(E_1')/dD$) per ion per cm^3 as a function of energy density into atomic displacement in an individual cascade (E_D).

ombination due, perhaps, to the presence of a thermal spike [13] but one might also argue that oxygen interstitials dimerize (perhaps with dissolved O) following a displacement as described by equation 1, leaving B_2 defects:

$$2{\equiv}Si{-}O{-}Si{\equiv} \quad \rightarrow \quad 2{\equiv}Si{\nleftrightarrow}Si{\equiv} + O_2 \; . \tag{4}$$

Since the B_2 defects are hole traps which can transform to E_1' defects a second, ionizing radiation exposure ought to reveal them. To test this hypothesis, double irradiation experiments were performed [14] in which samples were first implanted with Ar^+ ions then subsequently implanted with 80 keV H^+ ions, the energy being chosen so that the main mechanism of energy loss in the oxide was by ionization and creation of electron-hole pairs. The results obtained indicate no substantial increase in the numbers of E_1' defects present before and after the second implantation. One is thus lead to conclude that either there are few B_2 defects resulting from the Ar implantation or that B_2 defects transformed into E_1' defects by hole trapping as E_1' defects transformed into B_2 defects by electron trapping. This process would have to have occured at equal rates. More

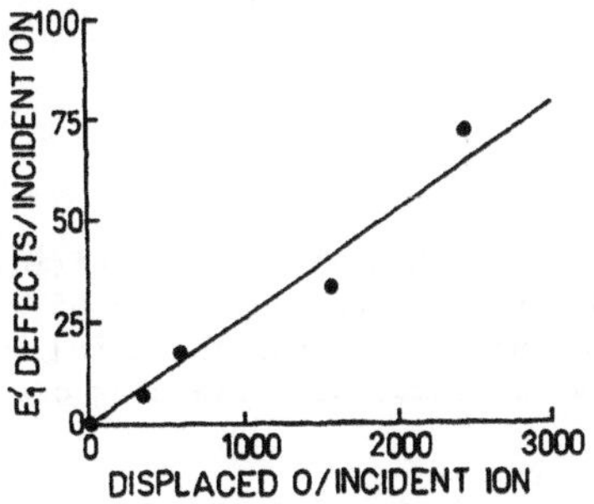

Figure 4. Plot of the measured initial slope for E_1' defect creation per incident ion ($dN(E_1')/dD$) as a function of calculated number of displaced O atoms per incident ion. Displacement is assumed due to loss of energy through atomic processes.

sophisticated measurements, using electron and hole injection methods are needed to clarify the situation exactly.

IONIZING RADIATION

The effects of ionizing radiation on amorphous SiO_2 films are important, as mentioned previously, because circuits may be subject to such radiation either during the process of manufacture or during their subsequent use. Various authors [15-18] have studied the effects of irradiation on amorphous SiO_2 and the dynamics of E_1' defects creation - the results are summarised in figure 5 where the volume density of defects created as a function of dose of ionizing radiation (in Mrad) is plotted. The low dose results were obtained (< 1000 Mrad) through studies on bulk silica irradiated with ^{60}Co γ rays [18] whilst the higher dose results were obtained using thin films irradiated with electrons [15] and protons [16,17]. For wet oxide, the increase in defect density with dose (< 10^5 Mrad) is essentially linear with dose whilst results for dry oxide are sublinear below $\sim$ 1000 Mrad but linear above (as best we can estimate).

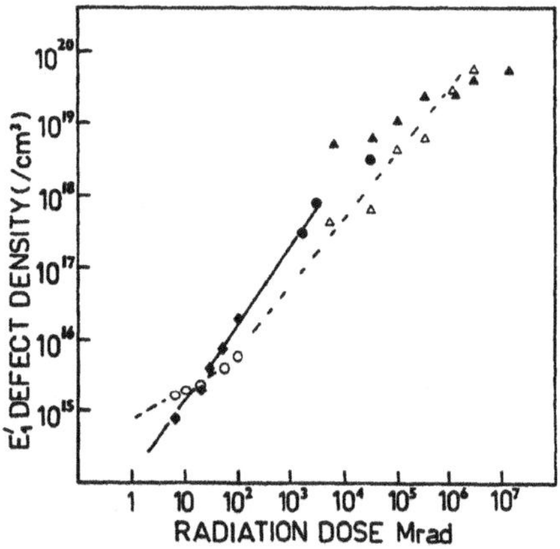

Figure 5. E_1' defect densities produced in SiO_2 by ionizing radiation, open symbols : dry oxide closed symbols : wet oxide. Open circles and closed diamonds from reference 18, open and closed triangles from reference 15 and closed circles from reference 16.

We believe the sublinearity at low dose (and enhanced creation efficiency as compared to the wet oxide case) is due to the transformation of B_2 defects formed in the growth process and due to the transformation of E_1' defects into peroxy radicals by trapping of diffusing, molecular oxygen [19] :

$$\equiv Si^+ \; {}^\bullet Si\equiv \; + \; O_2 \quad \rightarrow \quad \equiv Si^+ \; {}^\bullet O\text{-}O\text{-}Si\equiv \; . \qquad (5)$$

At high dose, the creation of E_1' defects tends to be linear with dose because all growth precursors have been transformed and because E_1'→peroxy radical transformation is restricted because of a lack of an adequate amount of dissolved oxygen.

Using the data presented in figure 5 we can estimate the ionizing energy dissipation necessary to create an E_1' defect. Taking the results for wet oxide and assuming the defect density grows linearily with dose we deduce a necessary energy deposition per defect of $\sim 9 \times 10^5$ eV. The high dose data for the dry oxide [15] yields a figure of 1.4×10^6 eV per defect. These results, when compared with the value of 2100 eV found for atomic displacement mechanisms indicate that ionizing radiation is between 430 and 670 times less efficient than displacive radiation in producing E_1' defects. It is interesting to compare this result with data obtained from macroscopic measurements of radiation induced compaction in thin films [20]. Here, for a fractional volume compaction of $\sim 3.4 \times 10^{-3}$ ($-\Delta V/V$) an energy deposition of 7×10^{18} keV/cm^3 is needed into atomic displacement or 7×10^{21} keV/cm^3 into ionizing processes. The ratio of the efficiencies is 1000 which is within a factor of two of the result obtained for E_1' defect creation. It is simple to demonstrate that given the numbers of E_1' defects and the volume changes, the whole volume compaction cannot be attributed to these defects alone but the result does emphasise that both macroscopic and microscopic damage mechanisms have a similar dependence on the nature of the energy deposition process inducing the damage.

ANNEALING

Various authors [21-23] have discussed the annealing of E_1' defects produced in wet and dry oxide by X or γ radiation. It is assumed that in these cases, annealing occurs via the trapping of diffusing O_2 molecules in dry oxide and OH molecules in wet [23]. The former process is described by equation 5 whilst the latter is described as :

$$\equiv Si^{\bullet} + H_2O \rightarrow \equiv Si{-}OH + H^{\circ} \ . \qquad (6)$$

Data [17,24] on the isochronal annealing of ion implanted- thin oxides is shown in figure 6. In general, little distinction is observed between wet and dry oxides. Isothermal annealing curves [17,24] demonstrate that the annealing process is non-exponential with time and has a rather sharp initial drop followed by a long tail. Various theories have been used [17, 23,24] to model the isothermal curves but no clear distinction between them (their applicability) can be made. This is certainly due to the fact that in amorphous SiO_2 the activation energy for molecular diffusion or thermal de-trapping has a distribution of values [25,26]. Such a distribution results in the fact that a set of parameters for almost any model one choses can be found enabling a fit to the experimental results.

If "annealing" as defined by equations 5 and 6 is applicable, one might expect it to be reversible in the presence of ionizing radiation. In figure 6 we show the results of experiments performed on samples implanted with 10^{14} Ar^+ ions/cm^2 at 75 keV or 125 keV then annealed at various temperatures and susequently re-implanted with different doses of 70 keV H^+ ions in numbers themselves insufficient to create significant numbers of E_1' defects [16]. One observes, indeed, a re-activation phenomena due to the ionizing radiation produced by the H^+ implant provided that the anneal temperature has not exceeded 700 → 800°C (for 10 minute anneals). If this temperature is exceeded, the E_1' defects cannot be recovered. Data obtained for isochronal annealing of peroxy radicals in dry, bulk silica [27] follows the annealing of the "re-activation" curve of the E_1' defect as one would expect, assuming equation 5 is valid. One thus concludes

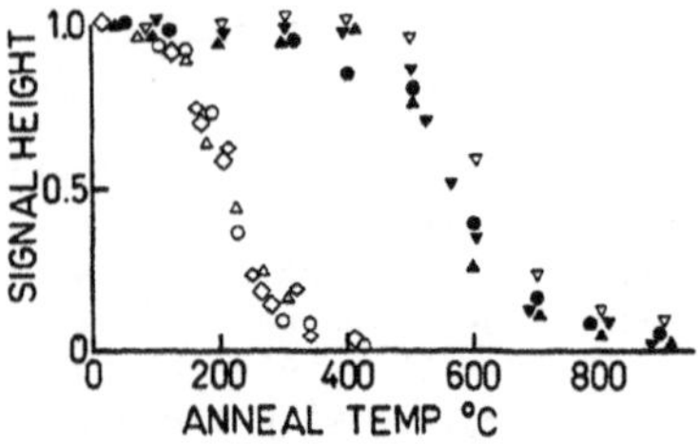

Figure 6. 10 minute isochronal anneal results for Ar implanted, thermal oxides:
○ 50 keV $10^{15}/cm^2$; ◇ 75 keV $10^{14}/cm^2$;
△ 125 keV $10^{12}/cm^2$; ◇ 150 keV $10^{12}/cm^2$.
In the following set, the first data coincide with the energy of the Ar implant and the second with the dose of 70 keV H^+ ions following the anneal at the indicated temperature.
▽ 125 keV $10^{14}/cm^2 \rightarrow 5 \times 10^{14}/cm^2$
▲ 125 keV $10^{14}/cm^2 \rightarrow 10^{14}/cm^2$
▼ 75 keV $10^{14}/cm^2 \rightarrow 10^{14}/cm^2$
● 75 keV $10^{14}/cm^2 \rightarrow 5 \times 10^{14}/cm^2$

that in the range 500°C ≤ T ≤ 800°C dissociation of E_1'/peroxy radical defects occurs and true structural annealing results. This sets a lower limit to the annealing conditions required technologically to eliminate E_1' defects and/or peroxy radicals.

We will not discuss optical bleaching of defects although some data on this exists. Suffice to say that it would seem most likely that E_1' centers trap electrons excited from the valence band of SiO_2 to form B_2 centers. Little detailed analysis of this phenomenon exists at present.

CONCLUSIONS

Despite the fact that the application of spectroscopic techniques such as electron spin resonance is difficult in thin films for sensitivity reasons, it has been possible to extract a finite amount of physical data.

It is clear that displacement damage produced by energy lost in atomic processes is much more efficient in creating oxygen vacancy centers than is energy lost in ionizing processes. For a given equivalent dose of radiation we estimate the ratio of efficiencies to be between 400 and 700 times dependent upon the OH content of the oxide. We are also able to estimate that it requires approximately 28 eV of energy into atomic processes to create an oxygen vacancy.

Isochronal annealing data suggest that E_1' defects are "annealed" by O_2 or OH trapping for temperatures ≥400°C - they can, however, be reactivated if the oxide is subsequently subjected to ionizing radiation. If annealing is carried out above 800°C, the defects are irreversibly annealed.

ACKNOWLEDGEMENTS

The author wishes to express his sincere thanks to Dr. Bob Pfeffer of the US Army, Fort Monmouth for his help and advice at different stages of the work.

REFERENCES

1. K. Bruce Winterbon "Ion Implantation Range and Energy Deposition Distributions" (Plenum, New York 1975) Volumes 1 and 2
2. K. L. Yip and W. Beall Fowler, Phys. Rev. B11 2327 (1975)
3. D. L. Griscom, this volume
4. E. H. Nicollian, C. N. Berglund, P. F. Schmidt and J. M. Andrews, J. Appl. Phys. 42 5654 (1971)
5. F. J. Feigl, D. R. Young, D. J. DiMaria, S. Lai and J. Calise, J. Appl. Phys. 52 5665 (1981)
6. J. H. Stathis and M. A. Kastner, Phys. Rev. B29 7079 (1983)
7. See for example: E. S. Yang, "Fundamentals of Semiconductor Devices" (McGraw Hill, New York 1978) Chapter 8
8. E. P. O'Reilly and J. Robertson, Phys. Rev. B27 3780 (1983)
9. R. A. B. Devine and A. Golanski, J. Appl. Phys. 54 3833 (1983)
10. R. A. B. Devine and A. Golanski, J. Appl. Phys. 55 2738 (1984)
11. A. Golanski, R. A. B. Devine and J. C. Oberlin, J. Appl. Phys. 56 1572 (1984)
12. D. A. Thompson and R. S. Walker, Rad. Eff. 36 91 (1978)
13. G. J. Dienes and G. H. Vineyard, "Radiation Effects in Solids" (Interscience, New York 1957)
14. R. A. B. Devine and C. Fiori, J. Appl. Phys. 57 5162 (1985)
15. R. Pfeffer, J. Appl. Phys. 57 5176 (1985)
16. R. A. B. Devine, Appl. Phys. Lett. 43 1056 (1983)
17. R. A. B. Devine, J. Appl. Phys. 56 983 (1984)
18. R. A. B. Devine, J. Non-Cryst. Solids (submitted)
19. A. H. Edwards and W. B. Fowler, Phys. Rev. B26 6649 (1982)
20. E. P. EerNisse and C. B. Norris, J. Appl. Phys. 45 5196 (1974)
21. D. L. Griscom, Nucl. Inst. Meth. Phys. Res. B1 481 (1984)
22. R. A. B. Devine, Phys. Rev. B35 9783 (1987)
23. D. L. Griscom in "Structure and Bonding in Non-Crystalline Solids" editors G. E. Walrafen and A. G. Revesz (Plenum, New York 1986) p 369
24. R. A. B. Devine, Nucl. Inst. Meth. Phys. Res. B1 378 (1984)
25. J. E. Shelby and S. C. Keeton, J. Appl. Phys. 45 1458 (1974)
26. R. A. B. Devine, J. Appl. Phys. 58 716 (1985)
27. R. A. B. Devine and C. Fiori, J. Appl. Phys. 58 3368 (1985)

INTERFACE AND BULK TRAPPING CENTERS IN LOW TEMPERATURE ION-BEAM SPUTTERED SILICON DIOXIDE FILMS

M.E. Zvanut[a] and F.J. Feigl

Sherman Fairchild Center, Solid State Studies
Lehigh University
Bethlehem, PA 18015

Current and capacitance voltage data obtained using a metal-sputtered oxide-semiconductor capacitor revealed trapping centers similar to those observed in irradiated thermal oxides. A positive charge density of approximately $5 \times 10^{12} cm^{-2}$ exists in the as-sputtered film and an equivalent amount of charge is present at the silicon/silicon dioxide interface. Analysis of current-induced trapping yields a $5 \times 10^{-15} cm^2$ electron capture cross section for a bulk trapping center.

INTRODUCTION

Thin films of sputter deposited silicon dioxide are of interest to both the optical and microelectronic industries. For instance, ion beam sputtered silicon dioxide has been employed as an optical coating material [1] and the use of sputter deposition has been investigated for the low temperature fabrication of microelectronic device grade oxides [2]. This increased interest in the use of sputter deposited oxides has motivated the investigation reported here concerning the electrical response of such films. Electrical techniques have been used to study the ion-beam sputter deposited SiO_2 layer of aluminum-sputtered oxide-native oxide-silicon capacitors.

The electrical characteristics of as-deposited RF sputtered SiO_2 films have been reported recently [2]. Those for ion-beam sputter deposited films are presented in the appendix. Below, we describe the charge trapping behavior of sputtered oxides subjected to constant current electron injection. The results are compared with those of similar investigations using thermal and deposited oxides. In particular, trapping centers detected through the electrical measurements are shown to be similar to those found in irradiated thermal oxide films.

EXPERIMENTAL DETAILS

The SiO_2 films were ion beam sputter deposited from an SiO_2 target onto polished (100) or (111) silicon substrates at room temperature. P-type and n-type substrates with resistivities of 0.2, 3, and 70 ohm-cm were used. The substrates were not acid cleaned prior to deposition; therefore, an 0.5-1.5 nm thick native oxide layer existed between the Si surface and the deposited SiO_2 film [3,4]. The thickness of the total

SiO_2 layer was measured on an ellipsometer to be (59 +/- 2) nm. An array of Metal Oxide Semiconductor (MOS) capacitors was fabricated on each substrate by sputtering onto the surface of the deposited SiO_2 $0.01cm^2$ circular Al electrodes.

The charge in the SiO_2 film was measured during constant current electron injection into the oxide. Electrons were injected by means of high field tunneling through the Si/SiO_2 interfacial barrier. To maintain a constant current during injection, a feedback circuit was employed to continually adjust the voltage applied to the MOS capacitor.

Upon passage of a current through the oxide, electrons are trapped in the SiO_2 film. The trapped charge concentration is proportional to a shift along the voltage axis of a capacitance-voltage or current-voltage characteristic. Several different techniques were used to measure the trapped charge density. First, the feedback voltage shift was continuously monitored throughout the current injection process. This voltage shift is equivalent to the voltage-axis shift of the positive bias current-voltage (IV) characteristic discussed by Solomon [5]. Second, a portion of the high frequency capacitance-voltage (CV) curve is automatically measured at preset time intervals during the charge injection process [6]. This process yields the flatband voltage shift. Third, the photocurrent-voltage (PIV) characteristic of the device was measured before and after charge injection to a specified concentration. The data obtained from the first two methods are reported in terms of a voltage shift with respect to an uninjected sample versus cumulative time of injection.

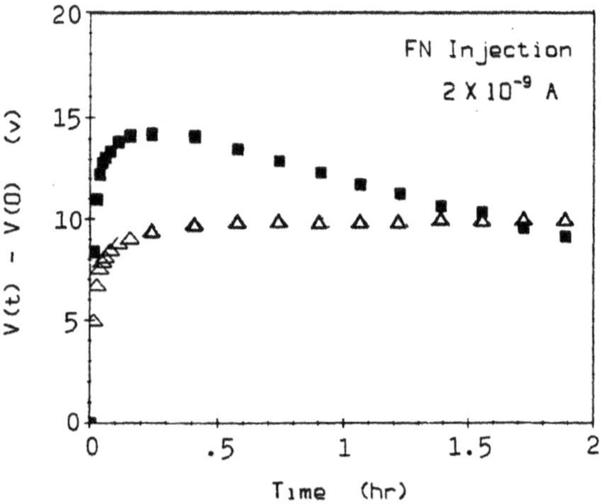

Fig. 1. Flatband (squares) and Feedback (triangles) voltage shifts measured during constant current Fowler-Nordheim (FN) electron injection.

Figure 1 illustrates the charging curves obtained by monitoring the feedback voltage (triangles) and the flatband voltage (squares).

ANALYSIS

Quantitatively, the centroid-weighted areal charge concentration N, was related to the voltage shift, ΔV, through the relation, $N = (-\varepsilon/ed)\Delta V$. ε is the dielectric constant for SiO_2, d is the oxide thickness and e is 1.6×10^{-19}C. The time dependence of the charge trapping in the oxide film during electron injection was analyzed in terms of a first order kinetic process [7]. Assuming that the areal density of charge captured at a particular trapping center depends only on the injected electron fluence, the following equation describes the increase of the charge density.

$$N(f) = \sum_n N_n(1-\exp(6_n f)) \qquad (1)$$

Here, 6_n is the electron capture cross section for trap n with saturated areal density, N. For the constant current injection experiments, the fluence, f, is the product of the current density and the time.

The injection data, consisting of various oxide charging curves, were analyzed in terms of the growth of charge at the Si/SiO_2 interface, N_{it}, and of charge within the oxide film, N_{ot}. The latter component, the oxide trapped charge N_{ot}, was separated into two components which were differentiated by their location with respect to the silicon surface. N_{oti} is the charge density located within 3 nm of the Si/SiO_2 interface and is referred to as near interface trapped charge. N_{otb}, bulk oxide charge, is the oxide charge density located between the Al/oxide

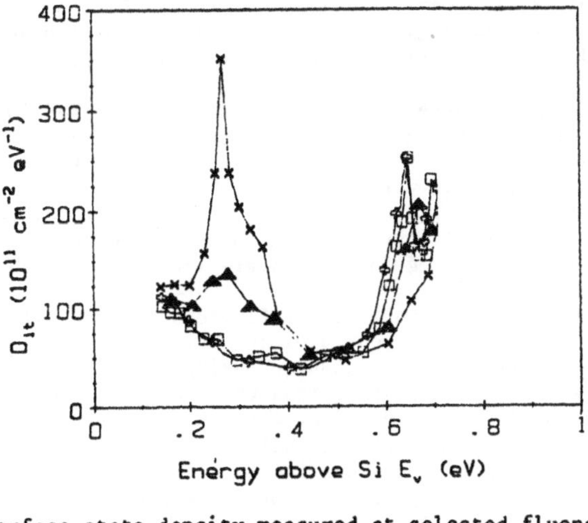

Fig. 2. Interface state density measured at selected fluences during current injection; plus signs, 0 C/cm^2; squares, 7×10^{-6} C/cm^2; triangles, 1.5×10^{-4} C/cm^2; multiplication signs, 14×10^{-4} C/cm^2.

interface and the 3 nm boundary of the near interface component. The N_{ot} components represent contoid-weighted amounts of charge, with the centroid origin at the Al/SiO_2 interface [7,8].

The several charge components are operational definitions determined by the sensitivity of the several experimental techniques used to measure oxide charge (PIV, IV, and CV). The PIV voltage shift was used to determine the bulk charge N_{otb}, as suggested by the analysis of Powell [9]. The difference between the PIV and IV voltage shifts was used to estimate the near-interface charge N_{oti}. This procedure was based on Solomon's analysis showing that IV shifts monitor all of the bulk charge and some of the near-interface charge (i.e., $N_{otb} + N_{oti}$) [5]. As was mentioned above, the feedback voltage shift (triangles in fig. 1) was used as the equivalent of the IV voltage shift. Finally, the CV shift (the flatband voltage shift in fig. 1) measures the sum of all charge components ($N_{otb} + N_{oti} + N_{it}$) and can therefore be used to determine N_{it} by comparison with the IV shift [7,8].

An upper bound on the interface trap charge N_{it} was determined indirectly from an analysis of the shape for a high frequency CV characteristic. This analysis provides the spectral density of interface traps, D_{it}, as a function of energy within the silicon band gap, E [7,8]. N_{it} may be determined from the integral of the interface state spectrum $D_{it}(E)$ if the absolute charge state of the interface component is known at each gate bias condition. In the present experiments, the PIV, IV, and CV voltage shifts were used to define the location of charge trapping centers present in the sputtered oxide films. The absolute charge ambiguity which confounds the separation of interface and near interface traps was unresolved.

EXPERIMENTAL RESULTS

The primary results concerning the trapping centers in the sputtered oxide films are the data in figures 1 and 2. Figure 1 depicts the charging curves and fig. 2 shows the variation of the interface state density, D_{it}, during the fluence regime shown in fig. 1.

The charging curve of figure 1 will be discussed in terms of three regions determined by the time or fluence of the injection. The first consists of times less than 0.01 hr. Analysis of this initial injection region shows that this portion of the charge trapping involves bias induced band-to-trap tunneling rather than a current induced band-to-band tunneling. Details of the analysis and the model developed to study this bias dependent tunneling phenomenon are presented elsewhere [10].

The second region consists of fluences between 0.1 and 1×10^{-4} C/cm^2 (times between 0.01 and 0.2 hr. in fig. 1). The flatband and feedback voltage shift measured during this fluence region are approximately equal indicating that the trapped charge is located in the bulk of the oxide film. PIV data were obtained before and after samples had attained a flatband voltage shift typical of this fluence region. In agreement with the flatband and feedback voltage data, the PIV results reveal an increase in N_{otb}. All voltage shifts indicate that the sign of the bulk oxide charge is negative.

In this fluence region a peak at 0.3 eV above the silicon valence band appears in the interface state density spectra (solid triangles in figure 2). Assuming that all the traps responsible for this peak have the same charge state, the integral of the interface state density yields an interface trap concentration approximately one-third that of the charge concentration detected by the flatband and feedback voltage shifts. It was stated above that the two voltage shifts are approximately equal. Thus, in the simplest case, the interface states are neutral under conditions required for measurement of the flatband voltage. For the samples used here, under such conditions the fermi level at the Si/SiO_2 interface is 0.2 eV above the silicon valence band edge. Thus, the interface states at 0.2 eV are empty at the flatband condition. From the above statements, it is concluded that the interface states are neutral when empty. That is, they are acceptor-like states.

The third region includes the data obtained at fluences greater than 1×10^{-4} C/cm^2 (for the data of fig. 1, times greater than 0.2 hr). This section of the charging curve is referred to as the turn-around region since the incremental flatband voltage shift changes sign. During this fluence regime, the incremental feedback voltage shift approaches zero and the 0.3 eV peak exhibited by the interface state density spectra grows (see crosses, fig. 2). (The decrease in an apparent peak at 0.65 eV represents a small concentration of interface charge and has not been included in this analysis.)

The flatband and feedback voltage data of region three indicate that positive charge is accumulating within 3 nm of the Si/SiO_2 interface. Since it was concluded above that the interface charge is neutral at flatband conditions, the positive charge responsible for the turn-around phenomenon is near interfacial charge. The increase in the interface state peak correlated with an increase in the post-turn-around flatband voltage shift.

In summary, the constant current injection data is interpreted in terms of three fluence regimes characterized by different spatial locations of the trapped charge: 1. At fluences less than 0.1×10^{-4} C/cm^2, band-to-trap tunneling results in electron trapping in the bulk of the SiO_2 film within 3 nm of the Si/SiO_2 interface. 2. At fluences between 0.1×10^{-4} C/cm^2 and 1×10^{-4} C/cm^2, current injection results in charge trapping in the bulk of the oxide and at the Si/SiO_2 interface. The interface states are acceptor-like states located 0.3 eV above the Si valence band edge. 3. At fluences greater that 1×10^{-4} C/cm^2, the generation of positive charge in the near interfacial region and traps at the abrupt interface dominate the trapping process.

Note that the turn-around portion of the charging curve and the associated D_{it} analysis were obtained on only a few samples. Oxide breakdown at the large injection fluence limited the data acquisition.

Regions 2 and 3 of the flatband voltage charging curves were shown to be independent of the current density over a current range covering at least 2 orders of magnitude. Thus, equation (1) was used to fit the data at times greater than 0.01 hr. Table 1 contains the results for the electron capture cross section, σ, and saturated areal density of trapped charge, N, for the dominant trapping centers of region 2. The oxide layers of samples 1 and 2 were fabricated separately under nominally the same conditions. The substrate of sample 1 was p-type (111) 0.2 ohm-cm Si and that for sample 2 was n-type (100) 70 ohm-cm Si. Capacitor to capacitor variation for the numbers reported in table 1 is approximately 20%. For the data of region 3 the effective capture cross section for positive charge is approximately 1×10^{-17} cm^2 and the areal charge density is 75×10^{11} cm^{-2}.

DISCUSSION

The interface and bulk trapping properties observed for the ion beam sputter deposited films may be compared with those reported for other thin SiO_2 films. The trapping behavior exhibited in region 1 of the charging curve is similar to that observed in oxides fabricated by means of photostimulated deposition [11]. Since samples fabricated by such methods necessarily have a native oxide at the silicon surface and since the sputtered oxide traps are located in the region of the native oxide/sputtered oxide interface, it may be concluded that this interface is the source of the trapping center.

Table 1. Kinetic Parameters

	σ_1 (10^{-15} cm^2)	N_1 (10^{11} cm^{-2})	σ_2 (10^{-14} cm^2)	N_2 (10^{11} cm^{-2})
Sample 1	6	38	4	23
Sample 2	4	57	4	15

The turn-around phenomenon exhibited by the data in region 3 has been observed previously in thermal oxides [12,13]. In such films, the species responsible for this phenomenon is thought to be located in the near interfacial region. Similarly, positive charge associated with the flatband voltage shift of fluence region 3 is near interfacial charge. Also note that in both the thermal [13] and sputtered deposited oxides, interface charge is related to, but not the same as, the charge responsible for the turn-around of the flatband voltage shifts.

Most significantly, the experimental results obtained here may be compared with those previously observed in irradiated thermal oxides. An interface state density on the order 1×10^{12} cm^{-2} eV^{-1} is typically found in thermally grown SiO_2 films which have been intentionally exposed to 25 keV electron beams or γ-irradiation [14,15]. The interface trap density spectra shown fig. 2 (plus signs) indicate that a similar value for D_{it} is obtained for as-deposited ion beam sputtered oxides. The bulk trapping centers for themal oxides subjected to x-irradiation exhibit trapping centers with capture cross sections between 1×10^{-13} and 1×10^{-15} cm^2 [16]. These cross sections cover a range whith includes those listed in table 1. It is apparent that the radiation generated during keV ion beam sputtering creates in the deposited SiO_2 film trapping centers similar to those found in intentionally irradiated thermal oxides.

A similar conclusion was deduced by Hickmott studying defect centers in RF sputter-deposited SiO_2 films. Using optical absorption and electron paramagnetic resonance spectroscopy he concluded that the defects observed in the sputtered films are similar to those found in irradiated bulk silica [17]. The work reported here extends the work of Hickmott to include the electrical response of deposited films. It would be of interest to determine whether the centers observed by Hickmott are responsible for those found using the electrical techniques.

In summary, we have studied the trap spatial location and trapping kinetics of ion beam sputtered SiO_2 thin films. The results of the electrical measurements during current injection indicate that the sputter deposited films have trapping centers similar to those observed in SiO_2 films fabricated by chemical deposition and in thermal oxides subjected to ionizing radiation.

---

a) Present Address: Dept. of Physics and Astronomy, Phillips Hall, University of North Carolina, Chapel Hill, NC 27599-3255.

ACKNOWLEDGEMENTS

We thank Dr. J. D. Zook and Mr. D. Smith of the Honeywell Physical Science Center for invaluable assistance. This research was supported in part by the U. S. Office of Naval Research Electronic and Solid State Science Program. Dr. Zvanut gratefully acknowledges support as a graduate fellow of the IBM corporation.

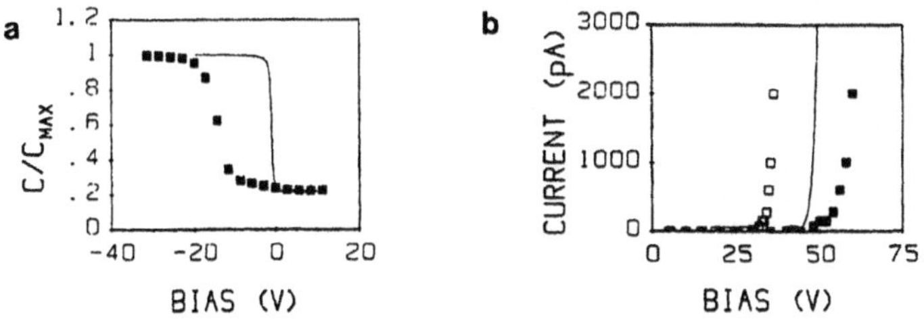

Fig. A1 Experimental (symbols) and ideal (solid) characteristics of an as-deposited ion beam sputtered oxide. (a) experimental capacitance normalized to the maximum capacitance versus gate bias and (b) IV characteristic.

APPENDIX

The experimental and ideal CV and step-IV characteristics for an ion-beam sputter deposited oxide before current injection [18] are presented in figures A1(a) and A1(b), respectively. The ideal curves were calculated based on the analysis presented by Nicollian and Solomon [8,5]. The CV data were obtained as described in reference 8. The IV data were obtained by measuring the current and flatband voltage approximately 30 seconds after the application of each bias step. The solid squares represent the current plotted against the applied voltage. The open squares represent the current versus the applied voltage less the flatband voltage shift with respect to the unbiased sample. The latter representation of the data accounts for the effect of charge trapping on the interface field. Comparison of the ideal and experimental characteristics indicates that the interface state density at midgap is approximately 3×10^{12} cm^{-2} eV^{-1} and that a positive charge density of 5×10^{12} cm^{-2} exists in the bulk of the oxide film.

REFERENCES

1. J. R. Sites, P. Gilstrap, and R. Rujkorakarn, Opt. Engr. 22, 447 (1983).
2. S. Suyama, A. Odamoto, and T. Serikawa, J. Appl. Phys. 62, 2360 (1987).
3. A. Ourmazd, D. W. Taylor, J. A. Rentschler, and J. Bevk, Phys. Rev. Let. 59, 213 (1987).
4. S. I. Raider, R. Flitsch, and M. J. Palmer, J. Electrochem. Soc. 122, 413 (1975).
5. P. Solomon, J. Appl. Phys. 48, 3843 (1977).
6. R. O. Gale, "Electron Currents in $Al/SiO_2/Si$ MOS Capacitors" (Ph.D. Dissertation, Lehigh University, 1984), unpublished.
7. F. J. Fiegl in VLSI Electronics Microstructure Science, N. Einspruch and G. Larrabee, eds. Vol. 6 (Academic, New York, 1983), p. 147.

8. E. H. Nicollian and J. R. Brews, MOS Physics and Technology (Wiley, New York, 1982).
9. R. J. Powell and C. N. Berglund, J. Appl. Phys. 42, 4390 (1971).
10. M. E. Zvanut, F. J. Fiegl, and J. D. Zook, J. Appl. Phys., submitted for publication.
11. U. Sharma, RCA Rev. 47, 551 (1986).
12. D. R. Young, in Insulating Films on Semiconductors, G. G. Roberts and M. J. Morant, eds. (Institute of Physics, London, 1980), p. 28.
13. M. V. Fischetti, J. Appl. Phys. 56, 575 (1984).
14. T. P. Ma, G. Scoggan, and R. Leone, Appl. Phys. Lett. 27, 61 (1975).
15. P. S. Winokur, J. M. McGarrity, and H. E. Boesch, Jr., IEEE Trans. Nucl. Sci. NS-23, 1580 (1976).
16. J. M. Aitken and Dr. R. Young, J. Appl. Phys. 47, 1196 (1976).
17. T. W. Hickmott, Appl. Phys. Lett. 15, 232 (1969).
18. M. E. Zvanut, "The Development of Spectroscopic Techniques to Study Defects in Thin Films Silicon Dioxide", (Ph.D. Dissertation, Lehigh University, 1988), unpublished.

INTERFACE IMPROVEMENT BY
HOT CARRIER DEGRADATION - FORMING GAS ANNEALING CYCLE

Q.Y. Ye, A. Zrenner, and F. Koch

Physik-Department
Technische Universität München
D-8046 Garching, FRG

ABSTRACT

We study Si-SiO_2 interfaces with surface channel carriers subjected to strong, lateral rf fields ($E_{rf} \approx 10^4$ V/cm; f≈3GHz) at 4.2K. Interface degradation is found to be caused by hot holes in a two step process. Positive charge appears in the oxide at 4.2K and is subsequently converted to interface traps by warming up to 300K. When the severely degraded sample is now subjected to a 400°C forming gas anneal cycle, substantial improvement over the original predegradation quality results. The effect has been examined for various types of oxides on n- and p-type samples.

I. INTRODUCTION

Carriers accelerated by the high electric fields in the surface channel of a micrometer-sized field effect transistor (FET) device are known to damage and degrade the Si-SiO_2 interface with time. The degradation appears in the form of charge injected into and trapped in the oxide. Interface traps, capable of exchanging charge with the channel, are created when the FET is operated under hot carrier conditions. In a device the source drain electric field is highly inhomogenious and the degradation is very nonuniform along the channel.

In a recent publication /1/ it has been shown that lateral electric field heating of the carriers can also be achieved using a pulsed, 30kW magnetron source coupled to the sample in a stripline geometry (Fig.1). In the microwave frequency field (f≈3GHz) the carriers are heated homogeniously in a macroscopically large (i.e. mm^2) area at the interface. This allows us to study various physical effects, such as cyclotron resonance, magnetotransport and the inversion layer capacitance. It permits us to examine both the electron and hole transport conductivity in the surface channel using the capacitive coupling scheme /2/.

II. EXPERIMENTAL NOTES

A detailed description of the sample and the coupling arrangement shown in Fig. 1 will be found in Ref. 1. It suffices here to recall that the sample is a n- or p-type MOS capacitor with a resistive gate layer of poly-crystalline Silicon over an area of $2x2mm^2$. The resistive gate is covered with two metallic pads leaving a central gap of 0.5mm width as a channel region. Care has been taken to cover the exposed oxide edges around the active area with a very thick dielectric in order to avoid charging and parasitic channel conductance.

A gate voltage $\pm V_g$ can be applied during the microwave stress and for the subsequent measurement. All surface channel conductivities are measured at 4.2K to assure complete carrier freeze out in the substrate. The measurements are done with a self-balancing rf bridge operating at 30MHz.

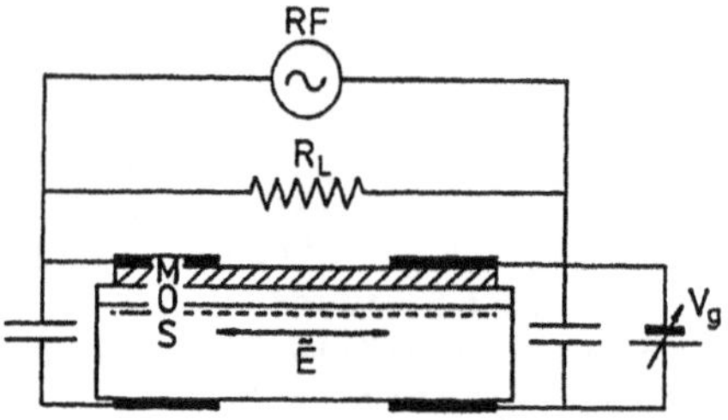

Fig.1. Experimental arrangement and sample geometry for both the microwave heating and surface conductivity measurements.(after Ref./1/)

Samples have been chosen from different wafers and oxides have been prepared for our purposes in the "line" of the Siemens Central Research Laboratories. The wafers for samples A, B, C and D were selected from different batches. Each of the wafers received a lightly doped poly-crystalline silicon gate layer and was annealed subsequently in Ar/H_2 at 900°C. In Table I we list the important parameters of both wafers and oxides according to the oxidation "protocol".

Table I. Specification of wafers and oxides

Sample	Substrate Resistivity type	Oxide Growth T	Oxide Type	Oxide Thickness	Annealing
A	n-type 8 Ω·cm	900°C	Dry	650A	900°C Ar/H_2
B	n-type 8 Ω·cm	1000°C	Wet	1300A	900°C Ar/H_2
C	p-type 10 Ω·cm	1000°C	Wet	1300A	900°C Ar/H_2
D	p-type 20 Ω·cm	950°C	Dry	550A	950°C Ar/H_2

The strength of the maximum rf field in the exposed channel region is estimated as a few times 10^4 V/cm. An external attenuator allows the field strength to be reduced to any small fraction of this value. The absolute value of the rf field strength is known to a factor of two only. The pulse duration time is 0.5μsec. A very low duty cycle (10^{-5}) has to be chosen in order to avoid lattice heating. The rf stress is done for a time in the order of minutes up to an hour with the sample held at 4.2K. The accumulative stress time is therefore typically 0.1 second.

A great deal of work remains to be done on the details of correlating a given electrical stress with the resulting degradation. It remains to study how the degradation depends on time, strength of the rf field and the applied gate voltage. For present purposes it suffices to state that at V_g=0V no measureable degradation occurs on a virgin sample. The surface density N_s is adjusted to $3x10^{12}cm^{-2}$ of surface channel electrons or holes for the electrical microwave stress. The effect for electrons is relative small. Holes in the surface layer, however, consistently give a large shift in the conductance characteristric. It is for this reason that we choose to confine the present work to the heating of holes in the surface layer.

The surface channel conductivity measurements are taken both before and after rf field stress at 4.2K to check the changes in oxide charges N_{ox}, interface traps and carrier mobility. After the microwave stress the samples are warmed up to room temperature in about 20 minutes and are held at room temperature for a period of time (hours to days) to allow the interface traps to develop. Afterwards the degraded samples remain permanently in the degraded condition if stored at room temperature or below. A subsequent 400°C forming gas (80%N_2, 20%H_2) anneal is carried out under flowing gas condition for one hour.

In this work we emphasize the stress-degradation-anneal behavior for various different oxides. Several different samples from each type have been investigated in order to assure that the observations are reproducible.

III. THE STRESS-ANNEAL CYCLE FOR VARIOUS Si-SiO_2 INTERFACES

Remarkably different response to the stress-anneal cycle has been found in different wafers and oxides. We start in Fig.2 with the consideration of a species from wafer A. Measurements of the conductivity characteristics for electrons and holes are carried out at 4.2K using the capacitive coupling technique/2/. As the separation of the electron and hole conductivity onset is a qualitative measure of the interface trap density, only a small range of V_g is shown in Fig.2. The conductivity thresholds as marked have been used to evaluate the effective electron mobility μ_{eff} in the upper part of Fig. 2. The four $\sigma(V_g)$ curves show a typical stress-degradation-anneal behavior. The original data (heavy solid line) shows that the virgin sample has a relative high interface trap density. The microwave stress was done on a surface hole channel of $N_s=3x10^{12}cm^{-2}$ at 4.2K for 15 minutes. The remeasurement of the conductivity at 4.2K (thin dashed line) with the sample still at He temperature reveals that $\sigma(V_g)$ curve has shifted to lower voltage indicating an increase of positive oxide charge. The slope of $\sigma(V_g)$ has effectively not changed

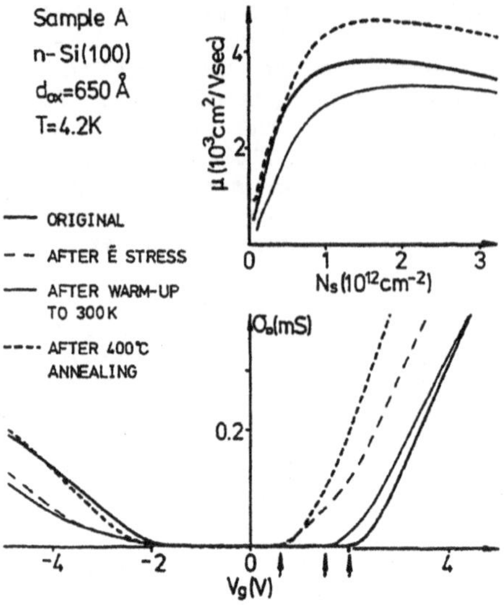

Fig.2. Electron and hole conductivity vs. V_g of sample A for the stress-anneal cycle. The upper curves give the elecrton mobilities vs. N_s.

except of some distortion at low N_s whose origin is not clear till now. The sample was then warmed up to 300K for a time of three days and recooled for measurement again (thin solid line). Both the electron and hole mobility appear to be reduced measurably. The onset of the hole conductivity has remained fixed, whereas the onset of electron conductivity has increased to a value near the original curve. The final step is the 400° C forming gas anneal cycle (heavy dashed line). The smallest separation between the onset of electron and hole conductivity is achieved after this final anneal step. It is nearly a factor of two less than the original! The sample has been improved significantly as reflected also in the electron mobility. We record a maximum of 4670cm²/V·sec, compared to the previously measured "original" of 3780cm²/V·sec. There can still be a small error in threshold marking, but there is no doubt that a higher transconductance has been achieved. This is a remarkable and unexpected behavior. A control sample prepared along with the stressed species but subjected only to the anneal showed no change.

The present work has been devoted to explore how different oxides react to the stress-anneal cycle. Fig. 3 gives the variation found by working through the same procedure with a species from the wafer B. There are small differences in detail. The final anneal results in an electron conductivity threshold that differs only little from that in the degraded state. Nevertheless, the interface state density has been reduced (note the shift of the hole conductivity) and the electron mobility is up from 3510cm²/V·sec to 4260cm²/V·sec. After the initial degration the sample is improved considerably over its original state.

That not all samples can be changed according to the above cycle is demonstrated in Fig 4. This particular wafer was found to have a peculiar structure in the original quasi-static C-V characteristic. Neither A nor B showed such a structure. This

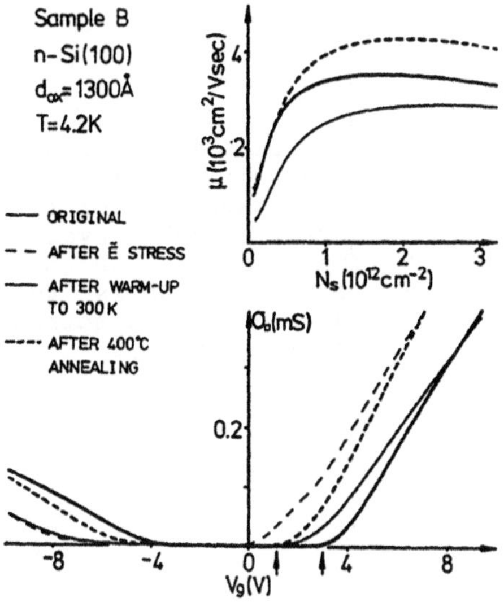

Fig.3. Electron and hole conductivity vs. V_g of sample B for the stress-anneal cycle. The upper curves give the elecrton mobilities vs. N_s .

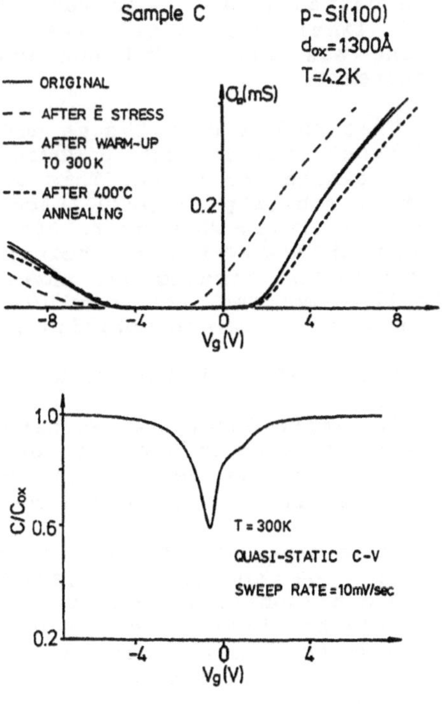

Fig.4. Electron and hole conductivity vs. V_g of sample C for the stress-anneal cycle. The lower curve is quasi-static C-V curve measured at room temperature.

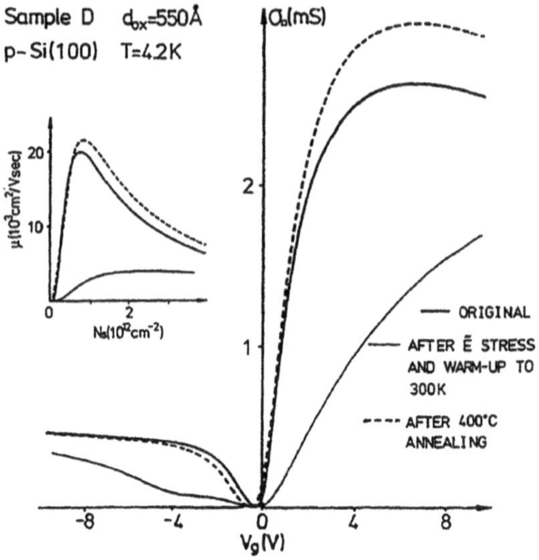

Fig. 5. Electron and hole conductivity vs. V_g of sample D for the stress-anneal cycle. The insert on the left side shows the corresponding electron mobility vs. N_s.

species from wafer C reacted to the microwave stress with a voltage-shift to lower V_g in exactly the same way as A and B (thin dashed line). Upon warming up to 300K the curve (thin solid line) reproduced the original characteristic. The forming gas anneal did not materially change the original conductance.

That such a result does not apply for all p-type samples is demonstrated in Fig. 5 where the data from a very unusual p-type sample are shown. This species has an original maximum electron mobility of μ=20,183cm^2/V·sec at 4.2K. After undergoing the stress-warming to 300K cycle this value was reduced to only 3900cm^2/V·sec as shown in the insert on the left side of Fig.5. The forming gas anneal has not only restored, but actually improved the conductance! There is a further 6% increase to a maximum mobility μ_{eff}≈21,370cm^2/V·sec.

IV. SPECULATIONS. A MODEL FOR THE STRESS-ANNEAL CYCLE

The improvement that we have demonstrated is an unexpected and very striking effect. Many of the details remain to be investigated, but some general characteristics do emerge from the data. Some basic trends can be understood in general terms.

It is clear that the most decisive change in the conductance after the low temperature stress is a shift to lower V_g, without a change of mobility. We can only conclude that hot holes have been injected into the oxide and that these are on the average at a distance larger than the Bohr radiaus $a_B^* \approx 20$A away from the interface /3/. The reduced mobility and the increased number of interface traps in the 300K warm-up step is then the result of a conversion of injected positive oxide charge into interface traps. This two step process has been described in the literature also by other authors using different methods/4/.

It is not surprising that exposure to hydrogen in the

forming gas anneal undoes some of the damage caused by the stress-warm-up to 300K sequence. What is, however, new and unexpected is that the sample ends up considerably better than what had been the original state. There is a significant interface quality improvement for even very good samples. This remains a question to which we can only provide a speculative answer. We suppose that the stress creates initially an undesirable species, which then is repassivated by hydrogen. As a guess, one may think of an interfacial OH^- bondy to Si in some manner. The disruption of the O and H by electrical stress opens up the possibility to properly tie up O with an interfacial Si. The additional hydrogen supplied by the anneal serves to passivate the remaining defects.

We believe that additional information on the process is contained in the line shape of the mobility vs. N_s curve. In particular, it appears that at high N_s the annealed version of $\mu_{eff}(N_s)$ lies above the original. This permits to a reduced surface roughness scattering in convertional thinking. This nominal surface roughness could be an electric dipole which has similar characteristics, for example short range scattering potential and no screening by the Ns carriers. Such a connection between chemical interface reaction and electronic scattering characterics might be analysed in future work to gain new insight into the real nature of surface roughness.

V. CONCLUSIONS

We have demonstrated a stress-degradation-anneal behavior on several different $Si-SiO_2$ interfaces. The high rf field stress is done laterally on a surface hole channel at 4.2K. Degradation occurs in a two step process: positive charges are trapped in oxide at low temperature and are subsequently converted into interface traps after warming up to room temperature. The final 400°C anneal not only restores the severely degraded sample but improves the mobility in the order of 30% and decreases the interface trap density.

VI. ACKNOWLEDGEMENTS

The work has been supported in part by the Siemens AG. We thank G.Dorda and C.Zeller for supplying many different Si wafers. We thank M.Bollu for helpful discussion. A. Zrenner thanks the Siemens AG for financial support via the Ernst von Siemens Stipendium.

REFERENCES

1.Ye Qiu-Yi, A. Zrenner, and F. Koch, Appl. Phys. Lett.52,561 (1987).
2.V.Dolgopolov, C. Mazure, A. Zrenner, and F. Koch, J. Appl. Phys.55,4280 (1984).
3.A. Gold and W. Götze, Phys. Rev. B33, 2495 (1986).
4.Genda Hu and Walter C. Johnson, Appl. Phys. Lett. 36(7) 590 (1980).

EFFECTS IF X IRRADIATION AND HIGH FIELD ELECTRON INJECTION ON THE PROPERTIES OF RAPID THERMAL OXIDES

W.K. Schubert and C.H. Seager

Sandia National Laboratories
Albuquerque, NM 87185

The effects of x-ray exposure and high field electron injection on the electrical properties of capacitors incorporating rapid thermal oxides show a considerable dependence on the particular post oxidation annealing (POA) conditions used during oxide processing. Most important seem to be the POA ambient and temperature. POA temperatures from 1150°C to 800°C have been investigated, with lower temperatures resulting in reduced midgap voltage shifts for a given x-ray dose or injected electron fluence. Low temperature high field injection experiments on the present samples have failed to show a direct conversion from trapped holes to interface states formed after photoinjection of electrons as has been reported for conventionally grown thicker oxides. Rather, the injected electrons seem to simply neutralize the trapped holes, returning the capacitance-voltage curve to near its original position and shape.

INTRODUCTION

SiO_2 gate dielectrics of a few hundred angstroms or less in thickness are already an integral part of metal oxide semiconductor field effect transistors (MOSFETs) and will become even more important as device dimensions continue to shrink below one micron. Rapid thermal oxidation (RTO) may offer a processing alternative for the growth of such thin oxides due to its inherently more precise time-temperature profile control (when compared to conventional furnace processing). It is recognized that the high field and ionizing radiation response of oxides is strongly dependent upon the processing conditions used. The high heating and cooling rates of rapid thermal processing may result in a different oxide and interface structure than conventional processing, and thus affect the stress response. Although considerable literature exists on the more routine properties of rapid thermal oxides, not much has been done to date to characterize in detail how the RTO conditions affect the radiation or high field response. In this paper, we report the results of a variety of experiments dealing with these responses in rapid thermal oxides from 20 to 30 nm thick. We look at the effects of oxidation temperature and POA temperature. Low temperature, high field injection experiments have been carried out to look at the relationship between hole trapping and interface state creation in these oxides.

A large body of literature has developed concerning the effects of ionizing radiation on SiO_2 in MOS devices (1). The major effects are the buildup of trapped positive charge, manifested by a parallel, negative shift of the capacitance-voltage (C-V) curve, and the generation of interface states, which cause a stretchout of the C-V curve. In high field electron injection (hereafter referred to as Fowler-Nordheim or FN injection) one observes much the same sort of damage (2-5). At much higher electron fluences one has the added possibility of electron trapping which occurs with much lower efficiency than hole trapping and thus requires the presence of much higher injected electron densities (6,7).

A complete review of the above literature cannot be done in a short paper, but it is known that regardless of whether FN injection or ionizing radiation is used to generate the damage, the degradation of the MOS capacitor properties can be minimized through careful processing at certain critical steps. For example, hole traps near the Si-SiO_2 interface have been associated with oxygen deficient defects created when the oxide is annealed in a reducing (i.e. low oxygen) atmosphere such as inert gas or vacuum (8,9). Inert gas POAs are commonly used to reduce processing-induced fixed oxide charge (10). Simple oxygen vacancy models for hole traps have been proposed (11-13), a hypothesis supported by electron spin resonance (ESR) results for the E' center (14,15). Since several models for interface state formation rely on the presence of H in the oxide (16-19), other examples of critical processing steps might include wet oxidations or forming gas anneals (10,20-24).

EXPERIMENTAL TECHNIQUE

The MOS capacitors used in this study were fabricated on <100>, p-type, 3 to 6 Ω-cm Si substrates. Substrates were 8 X 16 mm pieces cut from 100 mm wafers. Backside B implants were carried out (prior to oxidation) to help insure good ohmic contacts for electrical measurements. Standard RCA cleaning (25) followed by a brief, dilute HF dip was carried out immediately prior to insertion of the substrate into the RTO furnace. The rapid thermal processing was carried out in a Heatpulse 610 furnace built by A. G. Associates. The small samples were held in the chamber by a 100 mm support wafer which had been cleaned in the same manner as the substrates.

Samples were inserted into the furnace under a N_2 flow. After closing the furnace door, an O_2 purge was employed before heating commenced. Ramp rates were typically about 200°C/sec. Nominally dry oxidation conditions were used with oxidation temperatures ranging from 1000 to 1150°C. An <u>in situ</u> POA in N_2 (with an intervening cooldown to ~200°C) was employed immediately after sufficient purging. All gases were obtained directly from ultrahigh purity cylinders. POA times and temperatures ranged from one to four minutes and 800 to 1150°C respectively.

As quickly as possible after completion of the rapid thermal processing, Al was thermally evaporated through a 32 dot shadow mask to ~100 nm thickness to form the front contacts. For low temperature experiments where a semitransparent gate was needed, this Al thickness was reduced to about 15 nm. The gate area was 0.006 cm^2. Al was also used for the back contact. The final processing step was a 30 min., 450°C post metallization anneal in N_2.

Electrical characterizations of initial properties consisted primarily of quasistatic and high frequency (1 MHz) capacitance-voltage measurements to determine fixed oxide charge and interface state density, D_{it} (26,27). To test for a high quality oxide and metal-oxide interface, breakdown fields were measured on several capacitors from each run using the voltage ramp method with the capacitor in accumulation.

X-ray exposures were implemented using a 60 kV, Mo target machine (dose rate ~ 50 krad(SiO_2)/sec.). X-ray induced damage was tracked out to doses as high as 50 Mrads(SiO_2), with periodic interruptions to measure C-V shifts. A small positive oxide field of about 1.6 MV/cm was applied during the irradiation experiments to drive the holes toward the Si-SiO_2 interface. FN injection experiments were carried out in accumulation. A computer controlled set-up monitored the charge injection level and periodically interrupted the injection to determine V_{fb} and V_{mg} using the capacitance.

Low temperature injection experiments of the type previously reported by Chang, et al. (4), were carried out in a cryostat which allowed in situ C-V and I-V measurements. In these experiments, integrated interface state densities were measured using the Jenq technique (28,29) at 90 K. Briefly, in this technique the capacitor is swept from accumulation to deep depletion with electrons frozen into all interface states deeper than ~0.2 eV (at 90 K) below the conduction band edge. After reaching deep depletion, the sample is illuminated to produce an inversion layer and bring the interface to its most positive condition. Then the capacitor is swept back toward accumulation. The presence of interface states creates a ledge between the positive and negative going curves whose width is determined by the amount of charge in the interface states. After measuring the initial C-V curve at 90 K, FN injection of electrons from a semitransparent Al gate is carried out until ΔV_{mg} is -0.5 to -1.0 volts. At this point another C-V curve is measured. Then photoinjection of electrons from the Al gate is carried out using 325 nm photons obtained from an arc lamp and a small monochromator. A small negative gate voltage is applied during the photoinjection. Photoinjection is continued until the C-V curve stops shifting, and then a final C-V curve is measured.

RESULTS AND DISCUSSION

Oxides grown by our rapid thermal oxidation process generally proved to be of good quality as indicated by flatband voltage, interface state density (D_{it}), and breakdown field. Accumulation breakdown fields (arbitrarily defined as the field at which leakage currents of 1μA through the capacitor were obtained) were in the range of 9 to 11 MV/cm. With 1 min., 1150°C N_2, POAs, initial fixed oxide charge was reduced to less than 2×10^{11} cm^{-2} and midgap D_{it} values were less than 5×10^{10} $eV^{-1}cm^{-2}$.

As we have reported before (5,30), 1150°C POAs produced the minimum initial fixed oxide charge, but lower POA temperatures yielded reduced midgap shifts upon electron injection or x-ray exposure. Figure 1 shows the results of room temperature FN injection experiments for different POA temperatures. The oxides used were grown at 1050°C to thicknesses of 25 to 30 nm and all POAs were done in pure N_2 for four minutes at the indicated temperature. Within a run (defined here as samples cleaned together and processed sequentially within about 3 hours), POA temperatures below about 1000°C consistently resulted in smaller midgap voltage shifts when compared to samples annealed at higher temperatures. Samples subjected to the lower POA temperatures sometimes exhibited small positive midgap voltage shifts at low fluences, while this was never observed for POA temperatures ≥1050°C where hole trapping always dominates. This seems to indicate a balance between electron and hole trapping in these oxides which is dependent upon the POA temperature.

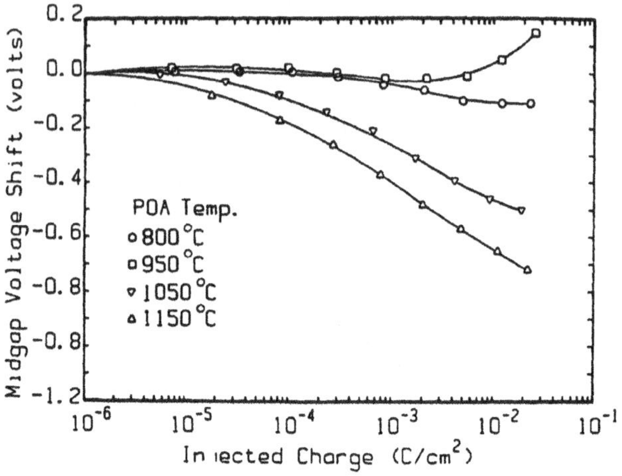

Figure 1. Midgap voltage shift due to FN injection of electrons in 25 to 30 nm thick SiO_2 films grown at 1050°C and post oxidation annealed at the indicated temperatures in N_2.

The above results indicate that the POA treatment strongly affects the final properties of our oxides. To demonstrate this further, oxidation temperatures were varied from 1000 to 1150°C with oxidation times appropriately adjusted to yield oxides from 25 to 30 nm thick. POAs were done at either 1150 or 950°C. The results are depicted in Fig. 2, where we have converted midgap shift to trapped hole density (N_h) assuming a sheet of charge near the interface. Fig. 2 shows the same general trends with respect to POA temperature as in Fig. 1, although the differences for the 1100 and 1150°C oxidation temperatures are insignificant. The POA temperature effects are especially strong at the lower oxidation temperatures. For example, at an injection level of 10^{-4} C/cm^2 for the 1000 or 1050°C oxides, one can reduce the hole trapping by about 10 times by using the lower POA temperature. It is interesting to note that in terms of hole trapping, the higher oxidation temperatures are better for the highest POA temperature, while the lower oxidation temperatures are better for the lowest POA temperature.

The data of Fig. 1 suggest that for 1050°C oxidation conditions, an optimal POA temperature exists in the range of 800 to 950°C. However the very small high fluence response indicated in Fig. 1 was difficult to consistently reproduce. The data in Fig. 2 corresponds to larger midgap shifts for the same injection levels than that shown in Fig. 1, even when identical processing protocols were used (1050°C oxidation and 950 or 1150°C POAs). In addition to their apparent increased electron trapping, the samples used in Fig. 1 also exhibited about 10 times greater interface state buildup during FN injection than those shown in Fig. 2. This run-to-run variability may indicate the importance of some other variable which we are not monitoring or controlling (such as H_2O partial pressure during oxidation and post oxidation annealing).

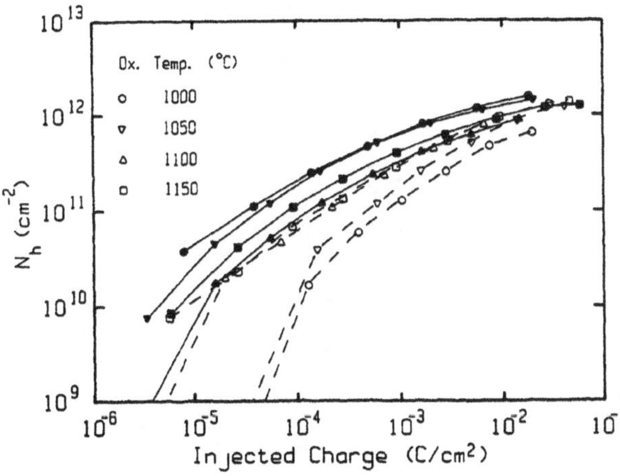

Figure 2. Trapped hole densities due to FN injection in oxides grown at indicated temperatures and post oxidation annealed either 950 (open symbols) or 1150°C (closed symbols).

Typically the shape of the interface state density curve in the gap after injection of about 10^{-2} C/cm^2 was a featureless U-shaped distibution rising rapidly toward the conduction and valence band edges. As mentioned above, when very small C-V shifts were observed even after high injection levels, we also usually observed much larger interface state buildup (integrated from midgap to flatband, levels of 3 to 5 x 10^{11}cm^{-2} as compared to 5 x 10^{10}cm^{-2}).

A similar dependence upon POA temperature is seen for the x-ray induced hole trapping for these oxides, however the trends with POA temperature are more clear cut, as shown in Fig. 3. The x-ray data make it clear that for 1050°C oxidation, the hole trapping rates continue to decrease with POA temperature down to at least 800°C. The hole trapping in the hardest of these samples is even lower than we reported in earlier work where a small amount of O_2 was added to the POA ambient to reduce the hole trapping rates (30). The shape of D_{it} after 50 Mrads (SiO_2) is similar to that observed in the FN case, but midgap levels were somewhat higher.

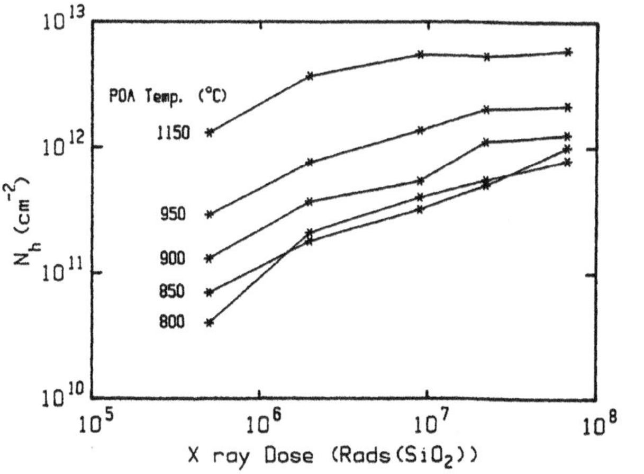

Figure 3. Trapped hole densities due to x-ray exposure for oxides grown at 1050°C and annealed at the indicated temperatures.

To investigate the connection between the trapped holes and the interface states formed during FN injection, low temperature injection experiments similar to those reported by Chang, et al. (4) were carried out. Typical results of such an experiment are indicated in Fig. 4. Note that the initial C-V curve (solid) shows no interface states (i.e. no ledge). After FN injection the C-V curve (dash-dot-dash) is shifted to more negative voltages indicating the presence of trapped holes as expected. However, after photoinjecting electrons until the C-V curve stops shifting (10^{12} to 10^{13} electrons/cm^2), the C-V curve (dashed) simply returns to near its initial position and shape with no ledge to indicate the formation of interface states. This is in contrast to the results of Chang, et al. (4) for thicker, conventionally grown oxides on n-type Si, where photoinjection immediately converted the trapped holes into interface states. In our oxides, the photoinjected electrons simply seem to neutralize the formerly trapped holes, returning the oxide to essentially its virgin condition.

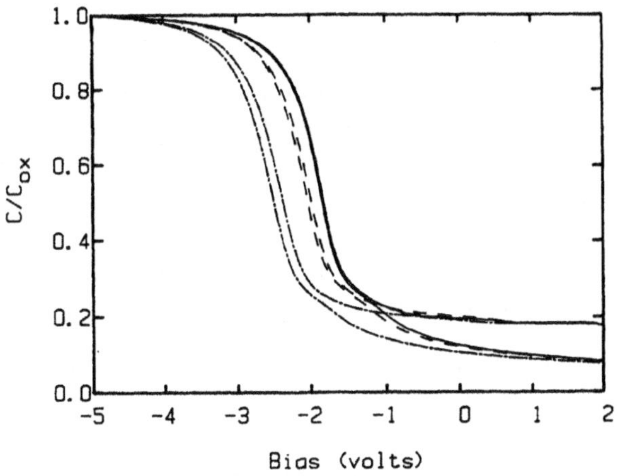

Figure 4. Low temperature (89 K) C-V curves for a sample before (solid) and after (dash-dot-dash) low temperature FN injection, and after photoinjection of electrons from the Al gate until the C-V curve stopped shifting (dashed). The photoinjected electrons simply annihilate the positive charge trapped during FN injection, with no evidence for interface states formation.

CONCLUSIONS

We have investigated the effects of oxidation and POA temperatures on the radiation and high field electron injection response of rapid thermally processed, thin SiO_2 films on p-type silicon. We have studied POA temperatures ranging from 800 to 1150°C, and oxidation temperatures from 1000 to 1150°C. The results show that particularly for oxidation temperatures ≤1050°C, the POA treatment largely controls the oxide properties. POA temperatures below 1000°C result in reduced hole trapping during irradiation or FN electron injection, especially at lower doses (< 3 Mrad (SiO_2)) and fluences (<2 x 10^{16} electrons/cm^2). Although the same general trends with respect to hole trapping appear in both FN injection and x-ray exposure data, we have not observed a consistent relationship between injection level, x-ray dose and trapped positive charge. In these oxides, photoinjection of electrons after low temperature FN injection simply annihilates the trapped positive charge, returning the oxide to near its virgin state. This is in contrast to results in the literature for thicker, conventionally grown oxides on n-type Si (4), where photoinjection immediately converted each trapped positive charge into one amphoteric defect. This may indicate that the stress induced defect is fundamentally different in the present rapid thermally grown oxides. Further investigation of these differences is continuing.

ACKNOWLEDGEMENTS

The authors thank J. Perry and W. Bayless for their invaluable technical help. This work was performed at Sandia National Laboratories, supported by the U.S. Department of Energy under contract DE-AC04-76DP00789.

REFERENCES

1. For a review of much of this literature see: "MOS (Metal-Oxide-Semiconductor) Physics and Technology," by E. H. Nicollian and J. R. Brews, (John Wiley and Sons, New York, 1982) pp. 549-577 or see the December issues of IEEE Trans. Nucl. Sci. from 1966 to the present.
2. H. E. Boesch and J. M. McGarrity, IEEE Trans. Nucl. Sci. NS-26:4814 (1979).
3. M. Knoll, D. Bräunig, and W. R. Fahrner, J. Appl. Phys. 53:6946 (1982).
4. S. T. Chang, J. K. Wu and S. A. Lyon, Appl. Phys. Lett. 48:662 (1986).
5. C. H. Seager and W. K. Schubert, to be published, J. Appl. Phys. (1988).
6. D. R. Young, E. A. Irene, D. J. DiMaria, and R. F. DeKeersmaecker, J. Appl. Phys. 50:6366 (1979).
7. S. Pang, S. A. Lyon, and W. C. Johnson, Appl. Phys. Lett. 40:709 (1982).
8. Z. A. Weinberg, D. R. Young, J. A. Calise, S. A. Cohen, J. C. DeLuca and V. R. Deline, Appl. Phys. Lett. 45:1204 (1984).
9. K. Hofmann, D. R. Young and G. W. Rubloff, J. Appl. Phys. 62:925 (1987).
10. R. R. Razouk and B. E. Deal, J. Electrochem. Soc. 126:1573 (1979).
11. B. E. Deal, J. Electrochem. Soc. 121:198c (1974).
12. W. C. Johnson, IEEE Trans. Nucl. Sci. NS-22:2144 (1975).
13. M. H. Woods and R. Williams, J. Appl. Phys. 47:1082 (1976).
14. P. M. Lenahan and P. V. Dressendorfer, Appl. Phys. Lett. 44:96 (1984).
15. H. S. Witham and P. M. Lenahan, Appl. Phys. Lett. 51:1007 (1987).
16. J. M. McGarrity, P. S. Winokur, H. E. Boesch, Jr., and F. B. McLean in "The Physics of SiO_2 and Its Interfaces," ed. S. T. Pantelides (Pergamon, New York, 1978) p. 428.
17. P. S. Winokur and H. E. Boesch, Jr., IEEE Trans. Nucl. Sci. NS-27:1647 (1980).
18. F. B. McLean, IEEE Trans. Nucl. Sci. NS-27:1651 (1980).
19. D. L. Griscom, J. Appl. Phys. 58:2524 (1985).
20. T. Takahashi, B. B. Triplett, K. Yokogawa and T. Sugano, Appl. Phys. Lett. 51:1334 (1987).
21. J. R. Schwank, D. M. Fleetwood, P. S. Winokur, P. V. Dressendorfer, D. C. Turpin and D. T. Sanders, IEEE Trans. Nucl. Sci., December (1987), to be published.
22. P. S. Winokur, E. B. Errett, D. M. Fleetwood, P. V. Dressendorfer and D. C. Turpin, IEEE Trans. Nucl. Sci. NS-32:3954 (1985).
23. J. R. Schwank, D. M. Fleetwood, P. S. Winokur, P. K. Hund and E. C. DasKalos, J. Radiation Effects; Research and Engineering, 5, to be published.
24. A. G. Revesz, IEEE Trans. Nucl. Sci. 24:2102 (1977).
25. W. Kern and D. A. Puotinen, RCA Rev. 31:187 (1970).
26. E. M. Terman, Solid State Electron. 5:295 (1962).
27. R. Castagne and A. Vapaille, Surf. Sci. 28:557 (1971).
28. C. S. Jenq, Ph.D. dissertation, Princeton University (unpublished).
29. G. Hu and W. Johnson, Appl. Phys. Lett. 36:590 (1981).
30. W. K. Schubert and C. H. Seager, to be published, Materials Research Society Symposia Proceedings, SiO_2 and Its Interfaces (1988).

INDEX

GPSR Compliance

The European Union's (EU) General Product Safety Regulation (GPSR) is a set of rules that requires consumer products to be safe and our obligations to ensure this.

If you have any concerns about our products, you can contact us on ProductSafety@springernature.com

In case Publisher is established outside the EU, the EU authorized representative is:

Springer Nature Customer Service Center GmbH
Europaplatz 3
69115 Heidelberg, Germany

Batch number: 10761599

Printed by Printforce, the Netherlands